MW01201262

# LIQUID DETERGENTS
## Second Edition

# SURFACTANT SCIENCE SERIES

# LIQUID DETERGENTS

## Second Edition

Edited by
Kuo-Yann Lai
*Colgate-Palmolive Company*
*Piscataway, New Jersey, U.S.A.*

Taylor & Francis
Taylor & Francis Group

Boca Raton   London   New York   Singapore

A CRC title, part of the Taylor & Francis imprint, a member of the
Taylor & Francis Group, the academic division of T&F Informa plc.

Published in 2006 by
CRC Press
Taylor & Francis Group
6000 Broken Sound Parkway NW, Suite 300
Boca Raton, FL 33487-2742

International Standard Book Number-10: 0-8247-5835-8 (Hardcover)
International Standard Book Number-13: 978-0-8247-5835-6 (Hardcover)
Library of Congress Card Number 2005044033

### Library of Congress Cataloging-in-Publication Data

Liquid detergents / edited by Kuo-Yann Lai.-- 2nd ed.
    p. cm. -- (Surfactant science series ; v. 129)
  Includes bibliographical references and index.
  ISBN 0-8247-5835-8 (acid-free paper)
  1. Detergents. I. Lai, Kuo-Yann, 1946- II. Series.

TP992.5.L56 2005
668'.14--dc22                                 2005044033

Taylor & Francis Group
is the Academic Division of T&F Informa plc.

Visit the Taylor & Francis Web site at
http://www.taylorandfrancis.com

and the CRC Press Web site at
http://www.crcpress.com

# About this book ...

This revised and expanded edition of *Liquid Detergents* covers all fundamental theories, practical applications, and manufacturing aspects of liquid detergents, from hand dishwashing liquids, liquid laundry detergents, to shampoos and conditioners. More than 30% of new material has been added, and this covers all the advances in liquid detergent products and technologies in the last decade.

Over 1800 relevant and up-to-date references are cited; these include books, book chapters, journal articles and patents for each product category. A wealth of information is presented in 300 helpful figures and tables.

Twenty-three international researchers from academia and industry have contributed their expertise to the book. This second edition of *Liquid Detergents* will continue to serve as a convenient, comprehensive and useful reference for researchers, and product development chemists and engineers, in the detergent field.

# Preface

Since its publication in 1996, the first edition of *Liquid Detergents* has been well received around the world by researchers in the detergent field. However, since its first publication there have been significant advances in this area. This second edition is intended to capture these advances and maintain the book as a useful, up-to-date reference.

Every chapter has been updated and expanded. This is true for both theoretical and application aspects. Over 30% of the information is new and updated. Chapter 2, "Hydrotropy," has been rewritten to incorporate a significant number of updated references. In Chapter 3, "Phase Equilibria," the discussion of emulsion has been expanded, and a section on nanoemulsions added. For Chapter 4, "Rheology of Liquid Detergents," in addition to general updating, data on the rheology of current commercial detergent raw materials and finished products are included, emphasizing those with particularly unique properties. Chapter 5, "Rheology Modifiers and Thickeners for Liquid Detergents," has been rewritten with expanded coverage of all the different rheology modifiers and thickeners for detergent applications. Chapter 6, "Nonaqueous Surfactant Systems," has been expanded to give a more comprehensive theoretical review of aggregation in nonaqueous solvents. Chapter 7, "Light-Duty Liquid Detergents," has been significantly rewritten and expanded; the new trends in recent years, including the success of antibacterial products and sensorial products, are a major focus of discussion. An extensive review of recent patent trends and a new discussion of "high-efficiency detergents," "color/fabric care" and "wrinkle reduction" have been added to Chapter 8, "Heavy-Duty Liquid Detergents." Chapter 9, "Liquid Automatic Dishwasher Detergents," has been updated to cover the evolution of products in recent years, and includes complete summaries of a large number of new patents granted since the mid-1990s. Chapter 10, "Shampoos and Conditioners," has been completely rewritten to align with the other application chapters; it has also been significantly expanded, with extensive summaries of patents for various new technologies and new products for shampoos and conditioners. Chapter 11, "Liquid Hand Soap and Body Wash," is a newly written chapter that covers not only liquid hand soaps, but also the exciting developments in shower gel/liquid body wash products in recent years. Chapter 12, "Fabric Softeners," has been rewritten and updated. New sections, especially on household cleaning wipes, have been added to Chapter 13, "Specialty Liquid Household Surface Cleaners." New discussions of continuous vs. batch process, aeration avoidance and microbial contamination have been added to Chapter 14, "Manufacture of Liquid Detergents."

It is hoped that, with these updates and additions, the second edition of *Liquid Detergents* will continue to serve as a useful and handy reference for researchers in the field.

# Acknowledgments

I would like to express my gratitude to Colgate-Palmolive Company for allowing me to undertake this project. I am especially grateful to Dr. Robert Pierce who introduced me to the world of detergents nearly 30 years ago. It was in his laboratory and through his teaching and many stimulating discussions, that I have learned much about detergents.

My sincere thanks go to all the contributors for taking time from their busy schedule and making personal sacrifices to contribute to this second edition.

I also would like to thank Anita Lekhwani, Senior Acquisition Editor, for her support, patience and encouragement. Thanks also to Jill Jurgensen and Tao Woolfe for all their help in coordinating the book for production.

Finally, I would like to thank my wife, Jane, and my children, Melody, Amy, and Peter, for their loving support throughout the project.

**Kuo-Yann Lai**
Colgate-Palmolive Company

# About the editor ...

**Kuo-Yann Lai**, **Ph.D.**, is Worldwide Director at Colgate-Palmolive Company's Global Technology, Piscataway, New Jersey. He has nearly three decades of industrial R&D experience in consumer products, spanning from basic research, product development, technology transfer, manufacturing operation, and purchasing, to product commercialization. The products that he has worked on include detergents, soaps, and dentifrices.

Most recently, he served as Technical Director for Colgate-Palmolive Company's Greater China operation. He is the author of *Material and Energy Balances* as well as being author or co-author of several scientific papers and book chapters, including a chapter on foam additives in *Foams: Theory, Measurements, and Applications* (Marcel Dekker, Inc.), and two chapters in the first edition and four chapters in this second edition of *Liquid Detergents*.

Dr. Lai holds seven U.S. and numerous international patents. He was the recipient of Colgate-Palmolive's "President's Award for Technical Excellence" in 1984, the Organization of Chinese Americans' "National Asian American Corporate Achievement Award" in 1992, and was the New Jersey Inventors' Hall of Fame "Inventor of the Year" in 1994.

Dr. Lai received his B.S. degree in Chemical Engineering from National Cheng Kung University (1969) in Taiwan, Republic of China, his M.S. degree in Physical Chemistry from the University of Texas at El Paso (1974), and a Ph.D. degree in Colloid and Surface Science from Clarkson College of Technology (1977), Potsdam, New York.

# Contributors

**Evangelia S. Arvanitidou**  Global Technology, Colgate-Palmolive Company, Piscataway, New Jersey

**Irena Blute**  YKI, Institute for Surface Chemistry, Stockholm, Sweden

**Guy Broze**  Advanced Technology Department, Colgate-Palmolive Research and Development, Inc., Milmort, Belgium

**Arno Cahn (deceased)**  Arno Cahn Consulting Services, Inc. Pearl River, New York

**André Crutzen**  Advanced Technology Department, Colgate-Palmolive Research and Development, Inc., Milmort, Belgium

**Nagaraj Dixit**  Research and Development, Global Technology, Colgate-Palmolive Company, Piscataway, New Jersey

**Stig E. Friberg**  Institute for Formulation Science, University of Southern Mississippi, Hattiesburg, Mississippi

**Joan Gambogi**  Global Technology, Colgate-Palmolive Company, Piscataway, New Jersey

**Philip A. Gorlin**  Research and Development, Global Technology, Colgate-Palmolive Company, Piscataway, New Jersey

**Subhash Harmalker**  Global Technology, Colgate-Palmolive Company, Piscataway, New Jersey

**Santhan Krishnan**  Research and Development, GOJO Industries, Inc., Akron, Ohio

**Kuo-Yann Lai**  Global Technology, Colgate-Palmolive Company, Piscataway, New Jersey

**Paul Reeve**  Rohm and Haas France S.A.S., Sophia Antipolis, France

**Charles Reich**  Advanced Technology/Hair Care, Global Technology, Colgate-Palmolive Company, Piscataway, New Jersey

**R.S. Rounds**  Fluid Dynamics, Inc., Flemington, New Jersey

**Amit Sachdev**  Research and Development, Global Technology, Colgate-Palmolive Company, Piscataway, New Jersey

**Jan Shulman**  Rohm and Haas Company, Research Laboratories, Spring House, Pennsylvania

**Marie Sjöberg**  Institute for Surface Chemistry, Stockholm, Sweden

**Jiashi J. Tarng**  Advanced Technology/Hair Care, Global Technology, Colgate-Palmolive Company, Piscataway, New Jersey

**Tom Tepe**  Rohm and Haas Company, Research Laboratories, Spring House, Pennsylvania

**Torbjörn Wärnheim**  ACO HUD AB, Upplands Väsby, Sweden

**Karen Wisniewski**  Research and Development, Global Technology, Colgate-Palmolive Company, Piscataway, New Jersey

**Len Zyzyck**  Research and Development, Global Technology, Colgate-Palmolive Company, Piscataway, New Jersey

# Contents

# 1

# Liquid Detergents: An Overview

†**ARNO CAHN**   Arno Cahn Consulting Services, Inc., Pearl River, New York

**KUO-YANN LAI**   Global Technology, Colgate-Palmolive Company, Piscataway, New Jersey

## I.  INTRODUCTION

Liquid detergents provide convenience in our daily life ranging from personal care of hand and body cleansing and hair cleaning and conditioning to home care in dishwashing and cleaning of various household surfaces to fabric care in laundering and fabric softening. Compared with powdered detergents, liquid detergents dissolve more rapidly, particularly in cold water, they generate less dust, and they are easier to dose. It is not surprising, therefore, that liquid forms of cleaning products have been gaining in popularity since their introduction in the late 1940s.

With the exception of fabric softeners and shampoos, the solid form of cleaning products preceded the liquid form. This is true of manual and automatic dishwashing, laundering, and general personal cleansing products. As a result, the technical history of liquid detergents is to a large extent one of emulating the performance features of the powder models.

---

†Dr. Cahn passed away on October 26, 2004. This overview chapter is based on his earlier work in the first edition with an update since that time. We would like to acknowledge Dr. Cahn for the enormous contributions that he made to the detergent industry over the last few decades.

1

All other factors — soiling, water hardness, and temperature — being equal, cleaning performance is a function of concentration and type of active ingredients that are delivered into the cleaning bath. Almost by definition, the liquid form involves a dilution of the active ingredients, that is, a given volume of a powdered detergent can generally deliver more active ingredients than an equal volume of a liquid detergent. The task of providing performance equality with powders is therefore not insignificant. It is made even more difficult when salts often pose problems of solubility and compatibility with any organic surfactants of the formulation. Finally, formulation problems are most severe when the active components are less stable in an aqueous environment than in a solid matrix.

These considerations apply principally to the heavy-duty liquids, the largest of the liquid detergent categories, but they also come into play with automatic dishwasher liquid detergents.

The situation is different for products designed for light duty, such as for hand dishwashing and softening fabrics. These liquids are generally superior in performance to their powder counterparts to the extent that these existed in the first place. This is also true of shampoo formulations, for which there is no common solid equivalent.

Since the mid-1990s there have been numerous new products launched around the world and there have been many advances in technology in this field. Liquid detergents have further gained popularity around the world replacing many traditional products in solid, powder, or other forms. Detergent manufacturers have introduced a large number of new products in every category. These products not only offer continuous improvement in cleaning performance but also incorporate more and more additional benefits. This is true for all products. Chapters 7 to 14 provide a detailed review of these new products and the advances in new technologies in every area.

This chapter is intended to give readers a historical overview of the various products as well as the new developments in the last decade (1995–2004).

## II.  LIGHT-DUTY LIQUID DETERGENTS

On a truly commercial scale, the age of liquid detergents can be said to have begun in the late 1940s when the first liquid detergent for manual dishwashing was introduced. This liquid consisted essentially of a nonionic surfactant, alkylphenol ethoxylate. It produced only a moderate amount of foam when in use.

This proved to be a serious detriment. To be successful, consumer product innovations must show a large measure of similarity to the conventional products they are intended to displace. In this case, copious foam was the essential performance attribute that needed to be as close as possible to that generated from powders and soap chips.

The requirement for copious foam levels has a technical basis and is more than a mere emotional reaction to a visual phenomenon. With soap-based products the appearance of a persistent foam signals that all hard water ions have been removed by precipitation as calcium and magnesium caboxylates and that excess soap is available to act as a surfactant.

The foaming requirements for light-duty liquids were met by the next series of product introductions in the early 1950s. These formulations were based on high-foaming anionic surfactants. They were capable of maintaining adequate levels of foam throughout the dishwashing process and possessed sufficient emulsifying power to handle any grease to produce "squeaky clean" dishware. This was accomplished by a mixture of anionic surfactants — alkylbenzenesulfonate, alcohol ether sulfate, and alcohol sulfate — sometimes in combination with nonionic surfactants. To maintain foam stability alkanolamides were incorporated. In some products alkanolamides were subsequently replaced by long-chain amine oxides.

The formulation of light-duty liquids overcame a second major technical hurdle inherent in the formulation of all liquid detergents: to maintain homogeneity in the presence of significant levels (about 30% or more) of moderately soluble organic surfactants. Coupling agents or hydrotropes (see Chapter 2) were introduced for this purpose, specifically short-chain alkylbenzenesulfonates, such as xylene-, cumene-, and toluenesulfonate, as well as ethanol.

Light-duty liquids have maintained a significant market volume to this day. This is in spite of the introduction and increasing popularity of automatic dishwashing machines and the detergents formulated for these machines. In fact, the use of both has increased greatly since their introduction in the late 1950s. This can be explained in part by the fact that some consumers use the light-duty liquids for washing delicate laundry items by hand in addition to continued use of them for washing small loads of dishes.

Over the years, minor additives have been incorporated into light-duty liquid formulations, principally to support marketing claims for special performance features. For a period in the 1960s, antimicrobials were incorporated into some products designed to prevent secondary infections of broken skin during dishwashing. After an absence of some 30 years antimicrobials are again appearing in light-duty liquids, and antimicrobial-containing formulations have become an important product segment. This is clearly a result of the increasing awareness of the possible presence of bacteria in foods, especially in chicken.

Improving the condition of skin as a result of exposure to light-duty liquid solutions proved to be technically very difficult. Exposure times are relatively short, about 20 minutes, three times a day in the best circumstances, and use concentrations are low, about 0.15%. The combination of low use levels and short exposure times makes it difficult to overcome the adverse effects of skin exposure to other influences, such as dry air in heated homes and strong household chemicals.

Generally speaking, light-duty liquid compositions are relatively nonirritating to skin. Mildness to skin could therefore be claimed for these products with reasonable justification. During the 1960s and 1970s the cosmetic image was further enhanced by making light-duty liquids more opaque, and imparting to them the ability to emulsify grease, combined with a persistent foam, has been the main objective of technical improvement.

In line with cleaning efficacy, solid particles have also been incorporated into some light-duty liquid formulations with the objective of increasing the effectiveness of the products in removing solid caked-on or baked-on soiling from articles.

Since the mid-1990s a great wave of evolution has taken place in the hand dishwashing liquid detergent market. The new products not only include "smarter" surfactants and surfactant mixtures, but also address multiple consumer needs offering multidimensional benefits. While consumers are in general quite satisfied with the primary cleaning function of dishwashing detergents, they have started looking for additional benefits beyond cleaning. New products introduced to the market incorporate various benefits including antibacterial and hand care properties and cleaning of tough-to-remove soiling.

A number of nontraditional ingredients have been introduced to light-duty liquid detergent formulations. These include some novel surfactants, antimicrobial agents, special polymers, and enzymes. Novel surfactants such as mid-chain branched ethoxy sulfates, ethylene diaminetriacetate, ethoxylated/propoxylated nonionic surfactants, Gemini surfactant, bridged polyhydroxy fatty acid amides, and the amphoteric surfactant sultaine are used for enhancement of cleaning or foaming performance.

The antimicrobial agent most commonly used in light-duty liquid detergents is triclosan (2,4,4′-trichloro-2′-hydroxydiphenyl ether). Other antimicrobial agents such as triclorocarban (TCC) and *para*-chloro-*meta*-xylenol (PCMX) are also used in some products, although to a much lesser extent.

Many polymers are used in light-duty liquid detergents to give various benefits. For example, polyoxyethylene diamine is used to increase grease cleaning, polyacrylate to aggregate and suspend particles, amino acid copolymer to tackle resistant soiling, polyethylene glycol to increase solubility, and ethylene oxide–propylene oxide copolymer to increase solubility, grease cleaning, or foam stability, or to improve mildness.

The other major development in light-duty liquid detergents since 2000 has been the introduction of experiential products, with different colors and fragrances that enhance a cleaning task. Colgate-Palmolive launched the Spring Sensations line in the U.S. market in the spring of 2000. New variants in colors and fragrances such as Orchard Fresh and Green Apple have been added to the line. Procter & Gamble followed with Joy Invigorating Splash and Tropical Calm and in the spring of 2001 with Dawn Fresh Escapes featuring Citrus Burst Apple Blossom and Wildflower Medley.

More recently, the aromatherapy benefit offered by personal care products has been extended to hand dishwashing products. Colgate-Palmolive launched Ultra Palmolive Anti-Stress Aromatherapy Dish Liquid with lavender and ylang-ylang extracts claiming "a whole new sensation in dishwashing."

## III. HEAVY-DUTY LIQUID DETERGENTS

Once light-duty liquid products had established an attractive market position, the development of heavy-duty liquids could not be far behind. As with light-duty liquids, the requirement of similarity to existing products also had to be met. In this case these products were powdered laundry detergents. The powdered laundry detergents of the 1950s were characterized by the presence of high levels of builder, specifically pentasodium tripolyphosphate (STPP), and relatively low levels, about 15%, of surfactants. In formulating a heavy-duty liquid, therefore, the major technical objective was to find ways of stably incorporating maximum levels of builder salts.

The first commercially important heavy-duty liquid was introduced into the U.S. market in 1958. The product incorporated tetrapotassium pyrophosphate, which is more soluble than STPP. Even so, in the presence of a surfactant system of sodium alkylbenzenesulfonate and a mixture of alkanolamides the formulation could tolerate only 15 to 20% of tetrapotassium pyrophosphate.

Incorporation of an antiredeposition agent, another ingredient present in laundry powders, proved to be another major technical hurdle. Antiredeposition agents, generally carbohydrate derivatives such as carboxymethylcellulose, had been introduced into laundry powders to prevent graying after a number of repeat wash cycles. In one product the patented solution to this problem consisted of balancing two antiredepostion agents of different specific gravity such that the tendency of one to rise in the finished product was counterbalanced by the tendency of the second to settle in the product [1].

Although the first major commercial heavy-duty liquid composition was formulated with a builder system, the concentrations of builders and surfactants it delivered into the washing solution were lower than those provided by conventional detergent powders. As a liquid, however, the product possessed a unique convenience in use, particularly for full-strength application to specific soiled areas of garments. Convenience was accompanied by effectiveness, because the concentration of individual ingredients in the neat form approached that of a nonaqueous system.

This is illustrated by the following consideration. Recommended washing product use directions lead to washing solutions with a concentration of about

0.15% of the total product. At a surfactant level of about 15% in the product, the final concentration of surfactant in the wash solution is about 0.0225%. The efficacy of surfactants in providing observable cleaning at such a low concentration attests to the power of the interfacial phenomena that underlie the action of surfactants.

By contrast, a heavy-duty liquid containing 20% surfactant, applied full strength, leads to a surfactant concentration of 20%, some three orders of magnitude larger than in the case discussed above. At these (almost nonaqueous) concentrations solution phenomena, such as those occurring in nonaqueous dry cleaning, are likely to be responsible for cleaning efficacy. The popularity of heavy-duty liquids for pretreating stains was thus based not only on convenience but also on real performance.

In the mid-1960s branched-chain surfactants were replaced by more biodegradable analogs in all laundry products. In heavy-duty liquids sodium alkylbenzenesulfonate, derived from an alkylbenzene with a tetrapropylene side chain, was replaced by its straight-chain analog, referred to as sodium linear alkylbenzesulfonate (LAS).

The conversion to more biodegradable surfactants was prompted by the appearance of foam in rivers. The appearance of excessive algal growth in stagnant lakes prompted a second environmental development that proved to be beneficial to the expansion in use of heavy-duty liquids: the reduction or elimination of the sodium tripolyphosphate builder in laundry detergents. Restrictions on the use of phosphate in laundry detergents were imposed by a number of states and smaller administrative agencies beginning in 1970. Because no totally equivalent phosphate substitute was immediately available, the performance of heavy-duty laundry powders was adversely affected. As the whole-wash performance differential between powders and liquids narrowed, the usage of heavy-duty liquids for the whole wash expanded, markedly so in areas where phosphate had been banned.

In the first nonphosphate version of a commercial product, phosphate was replaced by NTA (trisodium nitrilotriacetate), a powerful builder, comparable to condensed phosphate in its efficacy in sequestering calcium ions in the washing solution. Because of reports of adverse teratogenic effects in laboratory experiments, this builder was withdrawn from the market toward the end of 1971. It was replaced by sodium citrate, an environmentally more acceptable but inherently less powerful calcium sequestering agent. At the same time surfactant levels were increased by a factor of about three. What had happened in practice (if not in theory) was that higher levels of surfactants had been introduced to compensate for the loss in the builder contribution to washing efficacy provided previously by phosphate.

The 1970s saw the introduction of several heavy-duty liquids that carried this substitution to its limit, being totally free of builder and consisting solely of surfactants at levels ranging from 35 to 50%. These compositions were distinguished from light-duty liquids by the presence of laundry auxiliaries, such as fluorescent whiteners and antiredeposition agents. With the exception of a few products based on surfactants only, most heavy-duty liquids are formulated with a mixture of anionic and nonionic surfactants, with anionics predominating.

The steady expansion of the banning of phosphate across the U.S. accompanied by an increase in the convenience and efficacy of heavy-duty liquids led to an expansion in the use of this product category in the 1970s and 1980s. This expansion was fueled not only by the publicity that normally accompanies the introduction of new brands but also by some significant product improvements. The first of these to appear in the early 1980s was the incorporation of proteolytic and, later, amylolytic enzymes. In liquid detergents, with their relatively high amounts of water, proteolytic enzymes must be stabilized to prevent degradation during storage [2,3]. Enzymes make a significant and demonstrable contribution to washing efficacy, not only in the removal of enzyme-specific stains, such as grass and blood, by proteinases, but also in an increase in the level of general cleanliness. The latter effect is the result of the ability of a proteolytic enzyme to act upon proteinaceous components of the matrix that binds soils to fabrics.

Enzymes had been used in detergent powders in the U.S. and Europe as early as 1960. They were subsequently withdrawn in the U.S., but not in Europe, when the raw proteinase used at the time proved to have an adverse effect on the health of detergent plant workers. Improvements in the enzymes, specifically encapsulation, eliminated their dustiness and made it possible to use these materials in detergent plants without adverse health effects.

Since the 1990s enzyme mixtures have been commonly used in heavy-duty liquids. Most products contain a minimum of a protease for removal of proteinaceous soils and an amylase to facilitate starchy food-based soil removal. Some products contain lipases for degrading fatty or oily soils and cellulases to improve fabric appearance by cleaving the pills or fuzz formed on cotton and synthetic blends.

The second product innovation was the incorporation of a fabric-softening ingredient. Again, a powdered version of a "softergent" that had been on the market for some time served as the model product. In a powder the mutually antagonistic anionic surfactants and cationic softening ingredients could be kept apart so that they would not neutralize their individual benefits in the wash cycle. In a liquid this proved to be unattainable. As a result, the choice of surfactants in liquid softergents was restricted to nonionics.

Although the incorporation of enzymes and fabric softeners strengthened the market position of heavy-duty liquids, it did not solve the basic problem of limited general detergency performance in normal washing. As noted earlier, heavy-duty liquids came close to the performance of the first nonphosphate laundry powders. With time, however, the performance of nonphosphate laundry powders improved as new surfactant systems and new nonphosphate builders, notably zeolite in combination with polycarboxylate polymers, were introduced.

From the mid-1980s to the mid-1990s some major brands of heavy-duty liquids were converted from builder-free to builder-containing compositions. The first of these products employed a builder system consisting of sodium citrate in combination with potassium laurate [2]. Later, potassium laurate was replaced by a small-molecule ether polycarboxylate sequestrant, a mixture of sodium tartrate monosuccinate and sodium tartrate disuccinate [3]. In these builder-containing products the stabilization of enzymes is technically more difficult than in builder-free systems. A combination of low-molecular-weight fatty acids, low-molecular-weight alcohols, and very low levels of free calcium ions proved to be the solution to this problem.

In the U.S. heavy-duty liquids have grown at about 3% volume share of market a year in the last decade replacing powder laundry detergents that have dominated the market for years. By 1998 liquids had surpassed powders for the first time, and by 2001 liquid products accounted for 72% volume share of the U.S. laundry detergent market while powder laundry detergents declined to only 28% [4]. In Canada the heavy-duty liquid detergent volume share of the market grew from 15% in 1997 to 35% in 2001 [4]. In other parts of the world the volume share of heavy-duty liquid detergents grew at varying degrees.

There has been a significant technological development in heavy-duty liquid detergents in the last decade. Several thousand patents in this area were granted during this period. While many of these advances continue to focus on improvement in cleaning efficacy with conventional approaches using alternative surfactant systems, optical brighteners, or enzymes, there has been a greater emphasis on additives incorporated into the detergent formulation at low concentrations that deliver other significant, consumer-perceivable benefits. A strong emphasis in recent years has been on fabric and color care benefits, with the goal of preserving fabric appearance after multiple launderings. The market has also shifted toward consumer-friendly products that reduce fabric wrinkling and eliminate the need for ironing or reduce ironing time. Procter & Gamble developed a "Liquifiber" technology using a hydrophobically modified cellulosic to help reduce wrinkles in clothes. There has also been a continuous effort to find novel polymers that reduce dye transfer in the wash or rinse. Several patents on soil release technologies have been granted, with the focus being shifted from synthetics or blends to cotton garments. Novel enzymes are routinely finding new uses in liquid detergents, with efforts aimed at reducing allergenicity also being actively pursued. Polymers have been employed to modify the rheology of various liquid formulations for improving product aesthetics through suspension of visual cues. Incorporating encapsulated fragrances and additives into heavy-duty liquids for masking or eliminating malodors is another important development in recent years.

## IV.  LIQUID AUTOMATIC DISHWASHER DETERGENTS

Liquid automatic dishwasher detergents (LADDs) were first introduced to the U.S. and European markets in 1986. Prior to that, all dishwasher detergents were in powder form. LADDs have slowly gained popularity since their introduction. At the same time there has been an increase in the number of households with dishwashers, especially in the U.S. and Europe. By the early 2000s about half of U.S. households had dishwashers. LADDs account for about 40% of the dishwasher detergent market; 40% is accounted for by the powder form and 20% by the new unit-dosed form.

There has been an evolution in the technology of LADDs from clay hypochlorite bleach form to gel hypochlorite bleach form to gel enzyme nonbleach form.

The first LADDs were essentially powder compositions in a liquid form, in which functional components were suspended or dispersed in a structured liquid matrix. The liquid matrix consisted of water and the common structuring additives used were bipolar clays and a co-thickener comprising a metal salt of a fatty acid or hydroxy fatty acid. These liquid products, although minimizing some of the shortcomings of powders, suffered from two major disadvantages. First, the rheological properties of these products were such that the product needed to be shaken prior to dispensing. Second, the shelf life stability of these products did not meet consumer expectations. These problems were recognized by the manufacturers and aesthetically superior, non-shake, stable, and translucent products were introduced to the market in 1991 as "gels." All the liquid products marketed in the U.S. today are essentially in "gel" form using polymeric thickeners.

## V.  SHAMPOOS AND CONDITIONERS

Shampoos are liquid detergents designed to clean hair and scalp. They bear some resemblance to hand dishwashing liquids in that they are essentially builder-free surfactant solutions.

The history of shampoos is long, beginning well before the days of synthetic surfactants. The advent of synthetic surfactants greatly expanded the options for formulators and at the same time improved the aesthetics of the products.

Aesthetic properties, such as appearance (clear or pearlescent), viscosity, and fragrance, are perhaps more important in this product group than in any other product category discussed in this book. Development and maintenance of an adequate foam level is a performance property and also an aesthetic property in that it is noticed and evaluated by users.

Shampoos almost always contain additives with activity in areas other than cleaning and foaming, designed to provide specific performance attributes such as hair luster and manageability and elimination of dandruff.

The use concentration of shampoos is estimated as near 8%. This is an order of magnitude greater than that of laundry and dishwashing liquids. Mildness to skin and low irritation to eyes are therefore important requirements for shampoos.

Salts, generally sodium but also triethanolammonium, of long-chain alcohol sulfates and alcohol ether sulfates are the most widely used surfactants in shampoo formulations. Alkanolamides act as viscosity regulators and foam stabilizers.

The most general benefits associated with the use of conditioners are a reduction in static charge on hair and hence a greater ease of combing, that is, improved manageability. Cationic, quaternary surfactants and cationic polymers provide these benefits as a result of electrostatic adsorption on hair. Analogous to "soft-ergents," the mutual antagonism of the cationic conditioners and the anionic surfactants that provide the primary shampoo function of removing oily deposits on hair presents a problem in the development of conditioning shampoos. Some anionic surfactants, notably carboxylated nonionics, have been found to be more tolerant toward cationic surfactants than alcohol sulfates or alcohol ether sulfates.

Like all other liquid detergents, shampoos have evolved from basic cleaning products into products with multiple benefits. "Two-in-one" shampoos that combine cleaning and conditioning benefits in one product have gained increasing acceptance since their development in the late 1980s and have become the major product type on the market. Consumers like the convenience and the savings from this kind of product in contrast to using shampoo and conditioner separately. The primary conditioning agent used in most two-in-one shampoos is dimethicone. Other related silicones such as dimethiconol, amodimethicone, and dimethicone copolyol have also been used, either in a primary or secondary capacity. Because many of these materials are not soluble in water, it is necessary to incorporate these ingredients into the product with emulsifying agents or stabilizers. Therefore, two-in-one shampoos are typically oil-in-water emulsions. There have been significant technological advances in two-in-one shampoos focusing on improving cleaning or conditioning benefits and improved stability. There have been numerous patents relating to these kinds of products, especially since the 1990s.

Shampoos are also formulated with antidandruff agents. Water-insoluble anti-dandruff agents, such as zinc pyrithione (ZPT), selenium sulfide, climbazole, coal tar derivatives, and sulfur, have been used in many products for treating dandruff. In the last decade there have been many new developments in this kind of product providing improved antidandruff efficacy. Three-in-one shampoos are also available, which provide cleaning, conditioning, and antidandruff benefits in one product.

Shampoos for particular individual needs have been increasing in acceptance among consumers with specific cosmetic or health concerns. The demand for specialty products is driven by race, age, gender, image, personality, lifestyle, health, well-being, fashion, etc. New specialty shampoos that have been developed and are appearing on the market include those offering volume control, color protection, sun protection, revitalization or repair of damaged hairs or split-ends, frizz and flyaway reduction, and styling control.

The growing trend of using natural ingredients in personal care products in recent years also holds true for shampoos and conditioners. Keratins, vitamin E, essential oils, green tea, rosemary, grapefruit, grape seeds, saw palmetto, lotus, honey, chitosan, and ginseng are examples of some of the ingredients used in "natural" shampoos. Some of these shampoos only contain minute amounts of these ingredients for making "ingredient claims" with no real substantiated benefits.

## VI. LIQUID HAND SOAP AND BODY WASH

The initial development of liquid hand soap may be dated to as early as the 1940s. In the 1960s and 1970s liquid soaps started to appear as institutional and hospital health care hand washing products, some using simple liquid fatty acid coco soaps and some using blends of synthetic surfactants. In the late 1970s liquid soap was developed and launched on the mass market in the U.S. With the advantages liquid soaps offer over conventional bar soaps, they soon gained consumer acceptance and became increasingly popular.

Liquid soap can be stored and dispensed with the convenience characteristics of all liquids. Beyond these generic attractions, they possess an aesthetic advantage over conventional bar soaps in that during use, and particularly during occasional use, they are not subject to the visual and physical deterioration in appearance of bar soaps. Stored in an aqueous matrix (residual water from washing), soap bars tend to slough and crack to various degrees. The cracks, in turn, can collect dirt, which leads to a less than attractive appearance.

As liquid soap has gained popularity, its application has extended beyond washing hands to body cleansing and liquid body wash/shower gel products. These have become a growing product subcategory.

Since the mid-1990s liquid hand soap and body wash/shower gel products have experienced probably the biggest increase in use among all the liquid detergents. This is especially true for liquid body wash/shower gel products. While there has not been a dramatic change in the cleaning chemistry and formulation, this product category has expanded with ever-growing new consumer benefits. The growing usage of liquid soap and body wash products is not merely at the expense of traditional bar soaps but is an additional usage. Consumers started using these new products for benefits that they did not get or expect to get from traditional bar soaps.

New liquid hand soap products introduced to the market in the last decade continue to focus on superior cleaning plus antibacterial and skin moisturizing benefits. Triclosan is the universal choice of antimicrobial agent for these products.

With the advent of liquid body wash or shower gel, the rapid pace of innovation in the bath and shower market in the last decade has transformed traditional bathing and showering practice from the necessity of basic cleaning and hygiene to pampering and caring for the well-being of body and mind. The skin care benefits that were being delivered via products sold only in specialty stores and for indulgences such as spas are now coming onto the mass market. Relaxation of body and mind is being offered in the shower with the introduction of aromatherapy shower gels based on essential oils, traditionally known to soothe the nerves and relax the muscles. A desire for youthful appearance and willingness to pay for products that promise such a benefit are leading to the development and introduction of a multitude of antiaging shower products based on firming, exfoliation, etc. (see Chapter 11).

## VII.  FABRIC SOFTENERS

Fabric softeners or conditioners are designed to deliver softness to washed clothes and to impart a pleasant smell. They first made their appearance in the U.S. market in the 1950s. The softening effect is typically accomplished using cationic surfactants, "quats" (quaternary ammonium surfactants), which adsorb onto fabric surfaces. Di-hard tallow dimethylammonium chloride (DHTDMAC) has been the most commonly used softening ingredient for several decades. The positive charge on the nitrogen atom combined with the high molecular mass associated with the long alkyl chain ensure adsorption of the compound on the substrate and a soft feel of the conditioned fabric.

In contrast to most other liquid detergents, fabric softeners are not true solutions. The long-chain quaternary salts do not dissolve to form an isotropic solution.

Cotton is the primary target substrate for fabric softeners. With repeated washing the fine structure of cotton at the surface of a fabric becomes dendritic, that is, many fine spikes of cotton fibers are formed that protrude from the surface of the textile. Electrostatic repulsion holds these spikes in place, but in the presence of a cationic softening agent they are smoothed out. Synthetic fabrics, such as polyester and nylon, are not subject to this phenomenon. Much of the "softening" with these substrates is provided by the mechanical flexing action in the drier. However, the mechanical action of the drier causes a buildup of static electricity on synthetic fabrics, which can result in considerable sparking when garments made of synthetic fibers are withdrawn from the clothes drier. Fortunately, the agents that confer softening to cotton fibers also reduce the buildup of static charges on synthetics.

In a conventional fabric softener formulation the level of the quaternary surfactants is about 5%. Low concentrations of leveling agents can also be present. These materials, often nonionic surfactants, assist in the uniform deposition of the softening quats. In addition, a buffering system is used to ensure an acidic pH. Finally, a solvent, such as isopropanol, present at a level of about 10 to 15%, ensures a viscosity range suitable for easy dispensing from the bottle.

As additives to improve ease of ironing and to reduce the wrinkling tendencies of a treated textile, silicone derivatives, such as polydimethyl siloxanes, have been incorporated into liquid fabric softener compositions [5].

As alternative softening quaternaries, imidazolinium compounds have been introduced with a claim of superior rewet performance. This can be a useful performance feature because with continuing usage and buildup of cationics on the substrate, the water absorption of the substrate can be adversely affected. The use of anionic detergents in the main wash can mitigate this phenomenon because the anionic surfactant can combine with the cationic fabric softener to form a combination that is removed as part of the oil on the fabric.

Since the late 1970s concentrated fabric softener products have been marketed in the U.S. and Europe. The concentration of the softening cationic in these products is about three times as high as in conventional products.

As more and more attention was paid to the environmental impact of every product, the biodegradability profile of DHTDMAC was scrutinized. In the early 1990s, as the result of changes in European regulations, fabric softener manufacturers in Europe voluntarily replaced DHTDMAC with the more biodegradable esterquats. Since 1996 manufacturers in the rest of the world have also started to remove DHTDMAC from products and to replace it with esterquats. Replacing DHTDMAC with esterquats is not a simple one-to-one replacement in a formula. It requires full reformulation to maintain product aesthetics and performance. This is discussed in detail in Chapter 12.

Over the years consumers' expectation of and demand for this kind of product have been increasing. Like all other liquid detergents, more and more benefits have been added to fabric softener products. These added benefits include ease of ironing, wrinkle reduction, fiber care and protection, antibacterial properties, color protection, long-lasting freshness, deodorization, soil release, and dye transfer inhibition. There are significant differences in consumer needs and expectations from different parts of the world. In spite of all these developments, fragrance remains the most important attribute of the product on which consumers base their purchasing decision. Manufacturers offer products with various new fragrance variants as line extensions on a continuing basis.

## VIII. SPECIALTY LIQUID HOUSEHOLD SURFACE CLEANERS

Detergents for cleaning various household surfaces are considered specialty cleaners. These include all-purpose cleaners for floors and surfaces, and cleaners for bathrooms, kitchens, toilet bowls, and glass.

Early versions of specialty liquid cleaners were based on low levels of tetrapyrophosphate builder and surfactant, and additions such as alkanolamides and a sufficient amount of hydrotrope to keep the composition homogeneous. For sanitizing products, the additions included compounds with antimicrobial efficacy, such as pine oil or antimicrobial cationics. With the advent of phosphate bans, sodium citrate has emerged as the most common phosphate replacement in these products.

For increased efficacy in removing particulates adhering to substrates, some general-purpose cleaners incorporate a soft abrasive, such as calcium carbonate. The resulting products are milky suspensions with about 40 to 50% of suspended calcium carbonate [6]. Keeping these compositions homogeneous through extended storage is a technical challenge. One approach to solving this problem is to provide "structure" to the liquid medium. Surfactants present as a lamellar phase are capable of structuring liquids. U.S. patent 4,695,394 discloses a composition containing both soft abrasive and bleach.

Solvent cleaners are generally free of builder salts. The cleaning efficacy depends on solvent-type compounds, such as glycol ethers. Solvent cleaners are less effective on particulate soiling, such as mud on floors; however, they are effective against oily soiling, particularly on modern plastic surfaces.

Window cleaners constitute a specialty within the solvent cleaner category. Because any residue left on glass after drying leads to streaking or an otherwise undesirable appearance, these products are highly dilute aqueous solutions containing extremely low surfactant levels — most often nonionic surfactants — and a combination of glycol ethers and isopropyl alcohol as the solvent system.

Bathroom cleaners, sometimes referred to as tub-tile-and-sink cleaners, represent "subspecialty" liquids that must be effective against a combination of sebum soil deposited from skin detritus during bathing or showering and the hardness deposits deriving from hard water or from the interaction of hard water with soap, that is, calcium salts of fatty acids (soap scum). One subset in this group depends on acids for removing this combination of soiling. The acids contained in these products range from strong hydrochloric and phosphoric acids to moderately strong organic acids such as glycolic acid. Other products are formulated at a basic pH, incorporating calcium sequestrants, such as the sodium salt of ethylenediaminetetraacetic acid (EDTA), surfactants, and, in the case of products with disinfecting action, antimicrobial quaternaries.

Toilet bowl cleaners, like bathroom cleaners, are formulated to remove mineral deposits, principally iron salts that form an unsightly deposit at the water level. Again, acids ranging in strength from hydrochloric to citric are found in these products.

Like other liquid detergents, household surface cleaners have been produced in recent years with added benefits beyond their simple cleaning action. These added benefits include disinfection, surface shine, prevention of tenacious soil adhesion, and reduced fogging.

With so many different kinds of cleaning tasks in the home, consumers are look-ing for convenience, efficiency, and time savings from products. To satisfy these needs many products are moving toward more dilute, ready-to-use form. Sprays are popular forms to meet these needs. There is a large array of spray cleaners now available, including all-purpose cleaners and cleaners for bathrooms, kitchens, furniture, and glass.

The biggest change in household cleaners around the turn of the millennium, largely in the developed markets of Europe and North America, was the rise of wipes as a product form. These take the convenience factor even further, presenting the cleaner at its use concentration (like spray cleaners) but already impreg-nated in the cleaning implement. Wipes constitute yet another delivery system for liquid cleaners.

The use of wipes eliminates the need to rinse the surfaces on which they are used. Consumers expect wipes to give streak-free cleaning and quick drying of surfaces. The use of volatile solvents is an easy way to achieve effective cleaning with no residue, but the solvents contribute significantly to the odor of the product and can be limited by volatile organic compound considerations. Therefore, some developments are concerned with lower levels of solvent.

The area in which these types of wipe products have made the biggest impact is that of floor cleaning. The main advantage of these systems is that they represent an essentially "bucketless" floor cleaning method, which was first mentioned in the literature almost 10 years ago. There are wet and dry wipes. Both are used in conjunction with a resilient slightly spongy pad on the end of a long handle. In the wet system, wipes are supplied saturated with the cleaning solution. The wet wipe is secured to the bottom of the pad to clean the floor. In the dry system, dry nonwoven wipes are supplied separately from the cleaning solution, which is bottled. The dry nonwoven wipe is attached to the bottom of the pad at the end of the handle, and the cleaning solution is fixed in some way to the handle, either in a holder for the bottle or in a reservoir.

This type of system has led to one of the biggest changes in consumer cleaning habit and practice in the last decade. First, the system makes floor cleaning imme-diately available, cutting out the setup phase of getting out a bucket, cleaner, and mop and then making the solution. Second, it eliminates the need to clean the mop and bucket. Third, because minimal solution is used on the floor and the wipe is highly absorbent, the cleaned floor does not need rinsing. For many consumers this has completely changed the way they clean floors.

The formulations of the liquids impregnated in the wipes and the liquids sup-plied in bottles are similar. Typically, foam suppressors such as silicones are added to minimize foaming during the cleaning so as not to leave consumers with the impression that rinsing may be needed.

There have also been significant packaging innovations that have contributed to the new products in terms of convenience and aesthetics.

**TABLE 1.1** Major Raw Materials Used in Various Liquid Detergents

| Product | Surfactants | Foam stabilizers | Hydrotropes/ solvents | Builders/ sequestrants | Other additives |
|---|---|---|---|---|---|
| Light-duty liquids | Linear alkylbenzenesulfonate salts (LAS), alkyl ether sulfate salts (AEOS), betaines, alkylpolyglycoside (APG), paraffin sulfonate salts, alcohol ethoxylates, fatty acid glucoamides, alkyldimethylamine oxides | Fatty acid alkanolamides, alkyldimethylamine oxides | Sodium xylenesulfonate, sodium cumenesulfonate | EDTA, sodium citrate | Triclosan (antibacterial), enzymes (cleaning aid), lemon juice (cleaning aid), protein (skin care), abrasives (cleaning aid), polymers (skin care) |
| Heavy-duty liquids | Linear alkylbenzenesulfonate salts (LAS), alkyl ether sulfate salts (AEOS), alkyl sulfate salts, alcohol ethoxylates, N-methylglucamides | | Sodium xylenesulfonate, sodium cumenesulfonate | Sodium citrate, sodium tripolyphosphate | Enzymes (stain remover), borax (cleaning aid), sodium formate, calcium chloride (enzyme stabilizing system), hydrogen peroxide (bleach), soil release polymers (soil release), polyvinylpyrrolidone (dye transfer inhibition) |
| Liquid automatic dishwasher detergents | Alkyldiphenyl oxide disulfonate salts, hydroxy fatty acid salts | | | Pentasodium tripolyphosphate, tetrasodium pyrophosphate, sodium carbonate, sodium silicate, sodium citrate | Sodium hypochlorite (bleach), polyacrylate sodium salts (rheology modifier), carbopol (rheology modifier), enzymes (cleaning aid), monostearyl acid phosphate (suds depressant) |

| Product | Surfactants | Fatty acid derivatives | Solvents/hydrotropes | Builders/chelating agents | Other additives (function) |
|---|---|---|---|---|---|
| Shampoos and conditioners | Alkyl sulfate salts, alkyl ether sulfate salts (AEOS), betaines, alpha-olefinsulfonate salts (AOS), polysorbate 20, PEG-80 sorbitan laurate | Fatty acid alkanolamides, amine oxides | | Citric acid, EDTA, polyphosphates | Polyquaternium 7 (conditioner), polyquaternium-10 (conditioner), fatty alcohols (conditioning aid), silicones (conditioner), climbazole (antidandruff), zinc pyrithione (antidandruff), glycol monostearate (opacifier), aloe vera (luster promoter), jojoba (luster promoter) |
| Liquid hand soap and body wash | Alcohol sulfate salts, alcohol ether sulfate salts, alpha-olefinsulfonate salts (AOS), alkylbenzenesulfonte salts (LAS), sodium isethionate, fatty acid salts, alkylpolyglucoside, betaines | Fatty acid alkanolamides | | EDTA, sodium citrate | Triclosan (antibacterial), glycerin (moisturizer), essential oils (aromatherapy), glycol distearate (pearlescent agent), citric acid (pH adjuster), sodium chloride (viscosity adjuster), microparticles (exfoliant), dried fruit particles (exfoliant), vitamins (antioxidant) |
| Fabric softeners | Di-hard tallow dimethylammonium chloride (DHTDMAC), esterquats, imidazolinium salts, diamido quaternary ammonium salts | | Ethanol, isopropanol, polyethylene glycol | | Fatty alcohol (co-softener), fatty acid ester (co-softener), fatty amides (co-softener), amido amines (co-softener), polyethylene terephthalate (soil release agent), PVP-type polymers, (dye transfer inhibitor) |
| Specialty liquid household surface cleaners | Linear alkylbenzenesulfonate salts (LAS), alcohol sulfate salts, alkylsulfonate salts, alkyl ether sulfate salts (AEOS), alkylphenol ethoxylates, alcohol ethoxylates | | Glycol ether, ethanol, isopropanol, sodium xylenesulfonate, sodium cumenesulfonate | Sodium carbonate, sodium sesquicarbonate, sodium citrate, EDTA | Pine oil (disinfectant), orange oil (cleaning), benzalkonium cationics (antimicrobial), sodium hypochlorite (bleach), calcium carbonate (cleaning), acids/alkalis (cleaning) |

## IX.  MANUFACTURE AND RAW MATERIALS

In principle, the manufacture of unstructured liquid detergents in general is relatively simple, as it involves mainly good mixing of aqueous solutions. For light- and heavy-duty liquids, which contain sodium salts of surfactant acids, neutralization can be carried out *in situ*, that is, as a first step in the mixing process. The heat of neutralization must be dissipated before addition of more temperature-sensitive ingredients such as the fragrance. Heat must also be dissipated in the manufacture of products that require heat input to solubilize individual ingredients. In contrast, the manufacture of structured liquid detergents can be quite difficult because of the complexity of their rheological profiles. Both structured and unstructured liquids can be manufactured using either batch or continuous processes depending on the specific production and volume requirements. There can be significant manufacturing challenges, such as overfoaming, aeration of product, and long batch cycle times. Detailed discussions on all aspects of liquid detergent manufacture can be found in Chapter 14.

The raw materials used in the production of liquid detergents are discussed in some detail in Chapters 7 to 13. Table 1.1 provides a summary of the major raw materials used for various product categories. The similarities and differences between these products are evident.

## REFERENCES

1.  Reich, I. and Dallenbach, H., U.S. Patent 2,994,665 to Lever Brothers Company, 1963.
2.  Letton, J.C. and Yunker, M.J., U.S. Patent 4,318,818 to Procter & Gamble Company, 1982.
3.  Bush, R.D., Connor, D.S., Heinzman, S.W., and Mackey, L.N., U.S. Patent 4,663,071 to Procter & Gamble Company, 1987.
4.  Grime, J.K., in *5th World Conference on Detergents*, Cahn, A., Ed., AOCS Press, Champaign, IL, 2003, pp. 21–22.
5.  Dumbrell, R.J., Charles, J.P., Leclerg, I.M., de Bakker, R.M.A., Goffinet, P.C.E., Brown, B.A., Atkinson, R.E., and Hardy, F.E., British Patent 1,549,180 to Procter & Gamble Company, 1979.
6.  Clark, F.P., Johnson, R.C., and Topolewski, J., U.S. Patent 4,129,527 to Clorox Company, 1978.

# 2
# Hydrotropy

**STIG E. FRIBERG and IRENA BLUTE**   Institute for Formulation Science, University of Southern Mississippi, Hattiesburg, Mississippi and YKI, Institute for Surface Chemistry, Stockholm, Sweden

## I.  INTRODUCTION

Hydrotropes are an essential ingredient of cleaning and laundry products, serving to reduce excessive thickening of the former and to improve the dirt-removing action of the latter.

This chapter provides a short review of the early development in the knowledge of the function of these compounds. This is followed by a discussion of the fundamentals of their action and a section devoted to phenomena specific to their action in cleaners and detergents.

## II.  HISTORICAL REVIEW

Hydrotropes are molecules traditionally with a structure of a short hydrocarbon chain, often aromatic, combined with a polar group that in the early days of their development was ionic. Figure 2.1 shows a few typical structures of these kinds of molecules and in addition gives examples of more recent developments.

The historical development of the science of these molecules has been amply described [1–3] and the following treatment is, hence, condensed. The evolution of knowledge of these compounds and their action may be characterized as taking place in three distinct periods, the first of which was the introduction by Neuberg in 1916 [4,5]. Neuberg described the hydrotropes as compounds enhancing the solubility of organic compounds in water and investigated a large number of them.

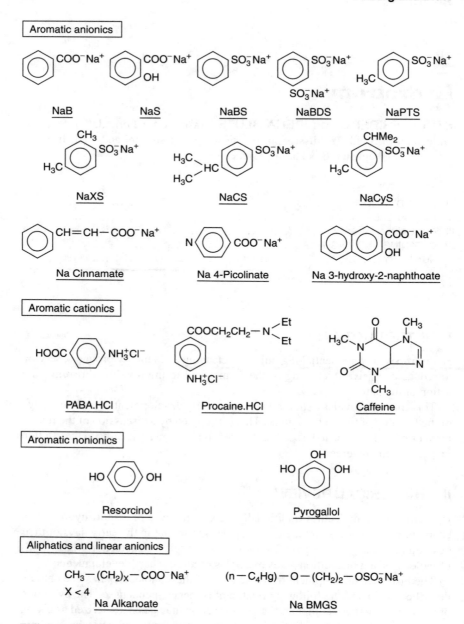

**FIG. 2.1** Structure of some hydrotropes.

The next period came 30 years later with efforts geared toward chemical engineering focusing on applied aspects. McKee [6] noted that hydrotrope solutions during dilution have a tendency to separate the dissolved compound leaving the hydrotrope in the aqueous solution to be used anew for extraction purposes. In addition, during this period the structure of hydrotrope solutions and the specific mechanism of the enhanced solubility were the aim of an initial discussion with Lumb [7] advocating the view that the enhanced solubility is due to solubilization: a well-known colloidal phenomenon in surfactant solutions. Licht and Wiener [8] supported the view of McKee in describing the enhanced solubility as a "salting-in" effect.

The third period came when Lawrence [9], Friberg and Rydhag [10], and Pearson and Smith [11] presented phase diagrams for hydrotrope solutions. The interpretation of the results from the determination of phase diagrams [10] introduced a new view of the hydrotrope solubilizing action. Instead of the earlier attempts to relate the increased solubility to the association of the hydrotrope molecules *per se*, the results showed that the superior solubilization in a hydrotrope solution compared to that in a surfactant solution (Figure 2.2) is in fact an outcome of the hydrotrope action on the collodial association structure of surfactants. The very large solubilization of a hydrophobic amphiphile, octanoic acid, in a hydrotrope solution is caused by the influence of the hydrotrope molecule on the packing conditions in colloidal association structures, especially a lamellar liquid crystal (Figure 2.3). In fact the hydrotrope molecule was seen as an entity that would not only be unsuited to form such a liquid crystal, but actually by its presence would prevent the formation. This result has had a bearing on the practical applications of hydrotropes, which will be briefly discussed in the following paragraphs.

Many studies of these applications have been reported reflecting the importance of these compounds in the commercial realm. Among the recent examples of new molecules with hydrotropic action may be mentioned vitamin C [12], useful for sunscreen formulations [13,14]. Other hydrotropes reported as new are diisopropylnaphthalene sulfonates [15], while the application of hydrotropes to solubilize pharmaceuticals continues to be extensive [16–24]. Long-chain amphihiles [25] and polymers [26] have been shown to exert hydrotropic action under certain conditions. Among the most recent developments should be mentioned the alkyl polyglucosides as hydrotropes [27]. They have been shown to be useful in strongly alkaline systems [28].

Investigations using hydrotrope solutions in reaction kinetics have varied from the direct analysis of the influence on the kinetics by the solubilization *per se* [29] of aromatic esters to more elaborate reaction systems. Microwave heating was early shown as an efficient way to enhance organic reactions [30,31]. However, the use of common organic solvents causes environmental problems in connection with microwave heating and aqueous solutions offer safe and convenient

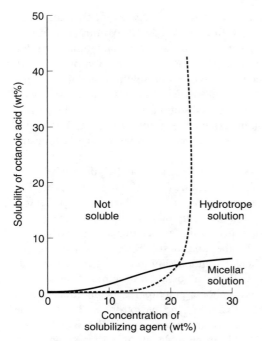

**FIG. 2.2** Solubilization of octanoic acid into an aqueous micellar solution of a surfactant, sodium octanoate (——), is limited, while the solubilization into a corresponding hydrotrope (sodium xylenesulfonate) solution (- - - -) is very large at high concentrations of the hydrotrope.

reaction media as demonstrated by the Hantzch dihydropyridine ester synthesis [32]. Hydrotrope solutions also offer a useful medium in the scale-up process [33]. In addition, hydrotrope solutions have been involved in reactions concerning solid particles. As examples may be mentioned the template-free synthesis of microtubules [34], important materials in nano-technology, and the more sophisticated role of hydrotropes to concurrently optimize the interfacial tension and the colloidal stabilization of rhodium particles in biphasic liquid–liquid alkene hydrogenation catalysis [35]. Finally reaction kinetics has been used as a means to follow the association of hydrotrope molecules in aqueous solutions [36].

The use of hydrotrope solutions for extraction was introduced by McKee [6] (Figure 2.4). More recent studies have been concerned with optimization [37] and with the separation of *o*- and *p*-chlorobenzoic acids [38–40]. The latter separation is excellent even for eutectic mixtures of the two compounds [39]. Results such as these may at first lend support to an earlier suggestion [41] of complex formation

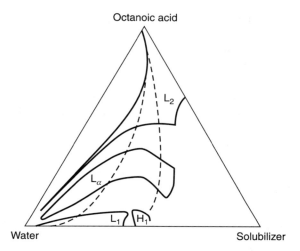

**FIG. 2.3** Solubilization of octanoic acid into a surfactant micellar solution ($L_1$) is limited because the addition of the acid leads to the formation of a lamellar liquid crystal ($L_\alpha$).

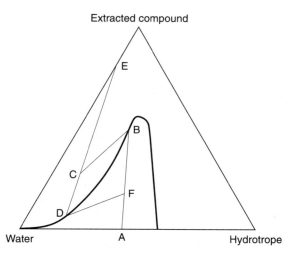

**FIG. 2.4** Extraction is initiated with the aqueous hydrotrope solution at A, saturation extraction at B. The total composition along BC when water is added. The extracted compound, saturated with water, is separated E, while the aqueous solution changes along BD. Evaporation of water gives DF and the process is repeated.

between the hydrotrope and the solubilizate. However, vapor pressure values of the solubilizate [42] do not support such an interpretation, but instead show the solubilizate to be located in a colloid association structure. Hydrotrope solutions have also been used to study the solubility and mass transfer coefficient for butyl acetate [43] and methyl salicylate [44]. In the production of semisynthetic antibiotics, intermediates may be difficult to separate. A good example is the production of 6-aminopenicillanic acid, the separation of which from phenoxyacetic acid is difficult with traditional systems. However, the use of aqueous solutions of sodium monoglycol sulfate has been shown to be efficient for the process [45].

Finally, hydrotrope solutions have been involved in natural product separation, e.g., in lipase purification and the evaluation of its thermal stability [46], as well as in the extraction of piperine from black pepper [47].

## III.  FUNDAMENTALS

The structure of hydrotrope molecules, as mentioned earlier, is characteristic (Figure 2.1). One finds a short, predominantly aromatic hydrophobic chain and in most cases an ionized polar group. With this structure in mind, it is not surprising that the association structures of the hydrotrope molecules in water have attracted some interest over the years, even if these may not be the decisive feature in the practical applications of these compounds.

It is of interest to note that the focus of research on the self-association of hydrotrope molecules was due, in part, to the early discussions about the fundamental nature of the solubility-enhancing capacity of hydrotrope molecules in aqueous solutions. These early attempts at clarification argued for the phenomena of colloidal solubilization versus molecular dispersion [6–8]. This dispute was resolved by the results from traditional surface chemistry analysis of interfacial tension, etc., which favored a colloidal association of molecules at high concentrations [48,49], and from vapor pressure measurements [42].

On a more detailed scale, osmotic vapor pressure measurements and light scattering determinations [50] gave results that were interpreted as arising from the formation of dimers and trimers at the initial association of nicotinamide in water while at higher concentrations an aggregation number of 4.37 was found. As expected, the trimerization constant was significantly greater, about two orders of magnitude, than the dimerization constant. It was tacitly assumed that the association takes place through stacking of the molecules, an expected conclusion considering the molecular structure of these compounds. However, this assumption was to some extent cast in doubt by Balasubramanian and coworkers [51], who determined the crystalline structure of sodium *p-tert*-butylbenzenesulfonate dihydrate, sodium cumenesulfonate semihydrate, sodium toluenesulfonate hemihydrate, and sodium 3,4-dimethylbenzenesulfonate. In none of these crystalline structures was a stacking of the molecules found and it was concluded that the notion of stacking of the molecules during association in aqueous solutions should not be assumed *a priori*. As for the conditions in a solubilized system at high concentrations, the determination of vapor pressure of the solubilizate phenethyl alcohol in sodium xylenesulfonate solutions [42] showed a constant vapor pressure at hydrotrope concentrations above the association concentration, indicating a colloid association without structure changes, once the association and solubilization take place. These results, once more, justify the emphasis on the influence of the hydrotrope molecule on colloidal association structures as a meaningful exercise.

The comparison with surfactant associations is a relevant theme and Srinivas and Balasubramanian [52] have evaluated this difference by observing the surface tension of and solubilization by a series of sodium alkylbenzenesulfonates. Varying the alkyl chain length gives a range of compounds with properties changing from those of a hydrotrope to those of a traditional surfactant. The results were interpreted to indicate that the transition is gradual.

In this context it is appropriate to caution against routine interpretation of such results. Mechanical analysis of the variation of surface tension versus the logarithm of the amphiphile concentration may be misleading, as exemplified by the results for a series of alcohols, which were interpreted as indicating micellar association [53]. The correct visualization in the form of a plot of surface tension versus the activity of the amphiphile [54] shows no indication of a sudden association (Figure 2.5). Hence, while a plot of the surface tension values from Srinivas and Balasubramanian [52] against the logarithm of the concentration certainly gives the knick-points characteristic of micellar association behavior (Figure 2.6), a correct interpretation must await information about the activity of the amphiphiles taking into consideration the very high concentrations of hydrotrope for association to take place.

Unfortunately, information about the activity of hydrotrope molecules in the concentration range of interest is not available. The only determination in existence, to our knowledge, is concerned with a more complex associated system [55]. This is in contrast to the case for traditional long-chain surfactants, which have been thoroughly investigated [56–59], the results of which justified the approach to use concentrations instead of activities in the common plot of surface tension to determine critical micellization concentrations. The closest to hydrotrope molecules should be bile salts, which have been investigated [60].

The main concern of the investigations discussed so far was the self-association of the hydrotrope molecules. Although such a subject constitutes an interesting area of research, it must be kept in mind that the hydrotrope molecule functions as a modifier of surfactant association structures in the majority of its applications. It is, hence, of interest to review available material on the alteration of surfactant association structures by addition of a hydrotrope.

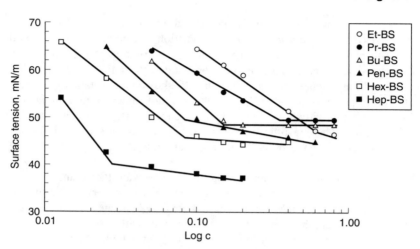

**FIG. 2.5**   Curve of surface tension vs. logarithm of mole fraction of amphiphiles at first indicates micellization. (Adapted from Srinivas, V. and Balasubramanian, D., *Langmuir*, 14, 6658–6661, 1998.)

The initial publications did not emphasize the specific action of the hydrotrope molecules in different applications. Instead they considered the structural modification of aqueous micelles by the addition of hydrotrope. Assessing the results from this point of view [61–64] the conclusion was that the reduction of electrostatic repulsion is the main cause of the modification of surfactant micelles from spherical to cylindrical shape after addition of a hydrotrope with opposite charge.

It should be noted that the effect is present at hydrotrope concentrations well below the self-association concentration. A more elaborate and sophisticated investigation of this phenomenon has recently been presented by Kaler and collaborators [65,66]. They analyzed the influence of added salt and added hydrotrope on a solution of worm-like micelles and were able to reveal the alteration of relevant length scales of the micellar system (contour length, entanglement length, mesh size, persistence length, and cross-sectional radius) by a combination of rheological, flow birefringence, and small-angle neutron scattering measurements, (Figure 2.7). The interpretation is similar to that of more simple systems [61–64], with the important difference that information now was obtained also for the number of branches, which increase due to the change in relative energy of branching versus that of the formation of end caps.

Although this research on ionic surfactant micelles is of high quality and fundamentally relevant, the research on nonionic micellar systems has a more direct bearing on the application of hydrotropes. One essential function of hydrotropic

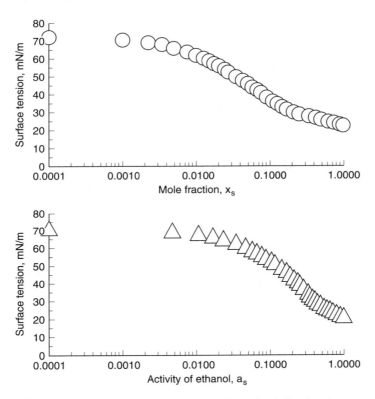

**FIG. 2.6**  Curve of surface tension vs. logarithm of mole fraction shows a pattern indicating micellization (top), while the curve vs. the logarithm of activity shows no such features (bottom).

molecules in practice is to clarify slightly turbid systems of high water content: an important problem as regards applications. This phenomenon has been described as a "coupling" or "linking" of organic and aqueous regions in a liquid vehicle. In this context a recent publication [67] is of interest, relating the action of hydrotropes to a general scheme of interactions with the oil and aqueous regions in emulsions and microemulsions [68,69]. There is no doubt that this manner of describing the phenomena is valuable, but it must be emphasized that a purely molecular modeling approach may significantly contribute to a better understanding of the clouding phenomenon. The approach by Shinoda and Arai [70] interpreting the cloud point behavior as a consequence of the curvature of the surfactant layer, defined the cloud point as an increase of the layer radius toward the hydrophobic region to a degree such that a normal micelle could not be formed because of packing considerations.

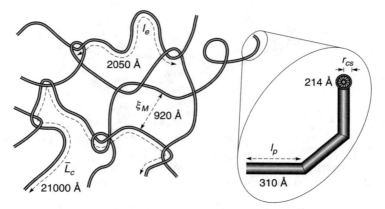

**FIG. 2.7**   Relevant length scales for the colloidal structure of worm-like micelles: contour length, $\overline{L}_c$, entanglement length, $l_e$, mesh size, $\xi_m$, persistence length, $l_p$ and cross-sectional radius, $r_{cs}$. Values shown are those measured for a solution with 1.5% total surfactant at a cetyltrimethylammonium tosylate (CTAT)/sodium dodecylbenzenesulfonate (SDBS) ratio of 97/3 with 0.l0% added sodium tosylate. (Reproduced from Schubert, B.A., Kaler, E.W., and Wagner, N.J., *Langmuir*, 19, 4079–4089, 2003. With permission.)

The recent investigation [71] of a nonionic system, hexaoxyethylene dodecyl ether and water, showed a hydrotrope molecule to be introduced into the micelle first at concentrations at which the hydrotrope self-associates.This increase of the minimum concentration at which the hydrotrope molecule enters the micelle from the values in ionic systems [61–66] is in all probability due to electrostatic effects. One essential result of the investigations into nonionic systems [71] is that the presence of the hydrotrope reduces the size of the micelle; i.e., the radius of the curvature toward the hydrophobic region is reduced and, hence, the cloud point is enhanced in accordance with the views of Shinoda and Arai [70]. Investigations of block copolymer systems [72–76] may now be interpreted in a similar manner and the coupling or linking action of a hydrotrope in a nonionic system is given a simple explanation in the form of a modified micellar structure.

The clarifying action of a hydrotrope in an aqueous system of ionic surfactants, which is generally described as a coupling action of the hydrotrope between the organic and aqueous regions, has recently been given a simple explanation [77,78]. It was shown that in a number of systems the cloudiness at high water content is due to the formation of a lamellar liquid crystalline phase and that the addition of a hydrotrope destabilizes the lamellar structure in accordance with the original interpretation of this phenomenon [10], (Figure 2.8).

These results have added to the understanding of the action of hydrotropes in the clouding in aqueous solutions, but have also had a significant influence on other

8

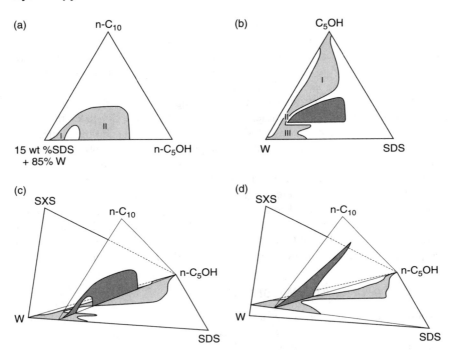

**FIG. 2.8** Microemulsion regions for the system containing (a) water (W), sodium dode-cylsulfate (SDS), pentanol (C₅OH), decane (n-C₁₀), and sodium xylenesulfonate (SXS). The composition of the aqueous solution of the surfactant and the hydrotrope is given in the left-hand corner. (b) The system containing water (W), sodium dodecylsulfate (SDS), and pentanol (C₅OH) shows an isotropic liquid solubility region with three kinds of amphiphilic association structures. I: inverse micellar region; II: bicontinuous micellar region; III: aque-ous micellar solution; LLC: lyotropic liquid crystal. (c, d) Combination of the diagrams in (a) and (b). (Reproduced from Friberg, S.E., Brancewicz, C., and Morrison, D.S., *Langmuir*, 10, 2945–2949, 1994. With permission.)

applications of hydrotropes. At first, the influence on the method of preparation of oil-in-water microemulsions should be mentioned. The early phase diagrams showing the oil-in-water microemulsion areas [79] could not be interpreted, but did indicate significant problems in practical formulation efforts, because the areas were narrow and there were serious difficulties establishing the areas of thermo-dynamically stable formulations. These problems were resolved once the influence of a hydrotrope was established [77]. Addition of hydrotrope in small amounts (of the order of 5% by weight) gives wide areas of microemulsion formulations, which is very useful for practical applications (Figure 2.8).

The early realization that the tremendous solubilization of octanoic acid into an aqueous solution of sodium xylenesulfonate [10], as distinct from the modest solubilization into a surfactant solution, is in fact, due to the inability of the hydrotrope molecule to form liquid crystals with the hydrophobic straight-chain amphipile, led to the later insight that hydrotropes would be useful to prepare vesicles at high concentrations in a simple one-step process [80,81]. In this context it is useful to refer to the early studies of bile salts and their relation to lecithin liquid crystals [82,83] and the resulting vesicles. Although lecithin and other double-tail surfactants were popular compounds to prepare vesicles, it is obvious that the traditional oxyethylene adducts with sufficiently short hydrophobic chains to make them "insoluble" in water also should be useful to prepare vesicle solutions [84]. Actually such surfactants were the medium in the determinations of the vesicle formation kinetics [85,86]. The realization of these phenomena led to a renewed interest in the phase diagrams of systems with hydrotropes and liquid crystal-forming surfactants [87] and to an interesting study of the change in the rheological properties of a lamellar liquid crystal due to the addition of a hydrotrope [88].

## IV. CLEANING AND WASHING

Cleaning and washing processes are mainly concerned with the removal of "oily dirt," depending to a high degree on the complex phase equilibria encountered in the surfactant–water–oily dirt system [89].

In addition to the progress in the area of traditional hydrotropes [90,91] one finds two treatments [92,93] on the action of a nontraditional hydrotrope structure in cleaning and laundry systems. Instead of the short and bulky molecule (Figure 2.8), this compound [94] is a dicarboxylic acid of considerable chain length (Figure 2.9).

The fundamental action of this hydrotrope in a liquid cleaner has been investigated. In such an application, the hydrotrope functions in the formulation concentrate by preventing gelation. In addition, under the dilute conditions in the washing process, the hydrotrope facilitates the removal of oily dirt from the fabric. In the following discussion these two functions are related to the phase equilibria of water–amphiphile systems.

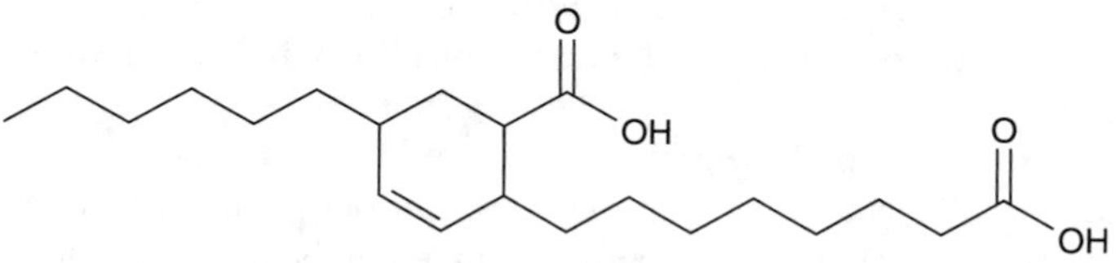

**FIG. 2.9** Dicarboxylic acid hydrotrope with an elongated structure.

The formula for the dicarboxylic acid (Figure 2.9) has a hydrophilic/lipophilic balance similar to that of octanoic acid, but the influence of the two acids on amphiphilic association structures is entirely different, as shown in Figure 2.10 [93]. The octanoic acid causes the formation of a liquid crystal when added to a solution of water in hexylamine. The size of the lamellar liquid crystalline region is large (Figure 2.10a). Addition of the dicarboxylic acid, in contrast, gives no liquid crystal, and it may be concluded that its action in concentrated systems is similar

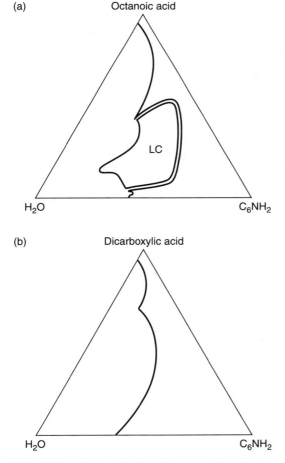

**FIG. 2.10** (a) Combination of water and hexylamine with octanoic acid (a) gives a very large area of a lamellar liquid crystal (LC). (b) Combination with the dicarboxylic acid of Figure 2.9 results in an isotropic liquid solution only.

to that of the common short-chain hydrotropes despite its long hydrocarbon chain (Figure 2.10b).

Activity in dilute systems was investigated using a model system from Unilever [96] in which octanol mimics the oily dirt. A lamellar liquid crystal is present at low concentration of the oily dirt [93], in the absence of the hydrotrope, because the formation of a lamellar liquid crystal on the addition of the octanol is the limiting factor in its solubilization into the micelles. After addition of the hydrotrope, the amount of model oily dirt solubilized into the aqueous micellar solution is greatly enhanced (Figure 2.11). Clearly, this hydrotrope functions not only as a destabilizer of liquid crystals in the formulation concentrate but also as a destabilizer of liquid crystals under the dilute conditions of the washing process [95].

The molecular mechanism behind the destabilization of liquid crystals was subsequently clarified [25]. The specific disordering promoted by the hydrotrope in the water–surfactant–oily liquid crystal was first determined, followed by an investigation into the conformation of the diacid molecule itself [92].

The order of the individual groups in the hydrocarbon chains in a liquid crystal is directly obtained from nuclear magnetic resonance (NMR) spectra using amphiphiles with deuterated chains. Each methylene group and the terminal methyl group give a NMR signal doublet, and the difference in frequency between the two signals is proportional to the order parameter [25]. Using a lamellar liquid crystal model system of "oily dirt," [96] surfactant, and water, the influence of the hydrotrope on the structure can be directly determined. Addition of the hydrotrope molecule results in a narrowing of the difference between the NMR signals due to

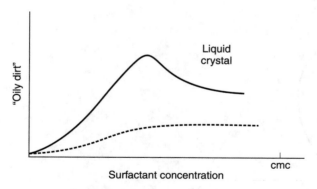

**FIG. 2.11** Solubilization of a model compound for oily dirt is small in a surfactant solution at concentrations below the critical micelle concentration (----) because of the formation of a liquid crystal. A combination of hydrotrope and surfactant gives an increased solubilization (—) caused by the hydrotrope destabilizing the liquid crystal.

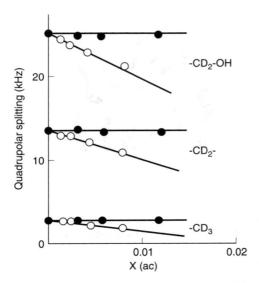

**FIG. 2.12** Addition of a hydrotrope (that of Figure 2.9) to a lamellar liquid crystal gives a reduction of the order parameter of the surfactant hydrocarbon chain (○); addition of a surfactant gives no change in order (●).

a disordering of the liquid crystal, as shown in Figure 2.12 [25]. It was assumed that this is the primary factor in the destabilization of the liquid crystal.

The diacid conformation was determined after it was added to the oily dirt liquid crystalline phase. Figure 2.13 shows two possibilities for the conformation of the hydrotrope in the liquid crystal. In one form of the diacid (Figure 2.13, right), both polar groups are located at the interface between the amphiphile polar groups and the water; the other possibility is that only the terminal carboxylic group is found at this site (Figure 2.13, left). The two conformations would result in different interlayer spacing (Figure 2.14) and a determination of this dimension can be used to distinguish between the two alternatives. Low-angle x-ray diffraction gives the interlayer spacing directly from the maxima in the diffraction pattern.

Interpretation of the results is straightforward. If addition of the diacid to a lamellar liquid crystal model dirt system does not increase the interlayer spacing, the conformation on the right in Figure 2.13 is correct; if an increase does take place, the situation on the left in Figure 2.13 would describe the structural organization of the diacid molecule.

The interlayer spacing with the diacid added [25] is very close to that of the host liquid crystal (Figure 2.14), and the conformation shown on the right in

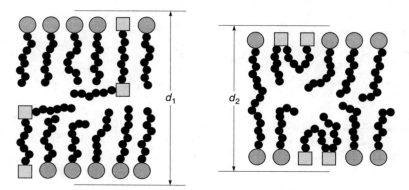

**FIG. 2.13** A hydrotrope (that of Figure 2.9) conformation with only one polar group at the water–amphiphile interface (left) results in an enhanced interlayer spacing, $d_1$, compared with the value, $d_2$, for a conformation with both polar groups at the interface (right).

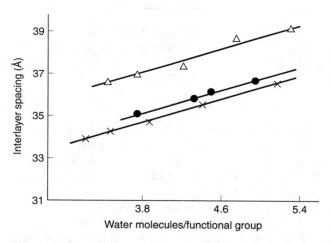

**FIG. 2.14** Low-angle x-ray values for interlayer spacing in a lamellar liquid crystal (X) show the spacing is unchanged with the addition of the hydrotrope (●) of Figure 2.9. Addition of a long-chain compound, oleic acid, gives the expected increase (Δ).

Figure 2.13 is obviously the one encountered in the liquid crystal. As a comparison, the addition of oleic acid with one polar group located at the interface gives the expected increase in interlayer spacing, as shown in Figure 2.14. Destabilization of the lamellar liquid crystal is not only affected by the diacid: it appears to be a general property shared by other hydrotropes, such as alkanols, short-chain quaternary ammonium salts, xylenesulfonates, and glycols, as shown by Pearson and Smith (11) and by Darwish *et al.* [97].

In some cases, the oily dirt is less polar than the model system of Kielman and Van Steen [96]. For less polar fatty oils the concept of hydrotropic breakdown of a liquid crystal is also useful [98].

## V. SUMMARY

The function of hydrotropes in detergency has been discussed as regards their interaction with surfactant colloidal association structures, especially lyotropic liquid crystals. The main activity of the hydrotrope as a part of a liquid detergent is to avoid gelation in both the concentrated package system and under the dilute conditions in the actual laundry process.

Both these activities are directly related to a detergent's phase equilibria with hydrophobic amphiphiles. These phase equilibria illustrate and explain the two basic characteristics of hydrotropes: their high association concentration and their pronounced solubilizing power.

## REFERENCES

1. Balasubramanian, D. and Friberg, S.E., in *Surface and Colloid Science*, Vol. 15, Matijevic, E., Ed., John Wiley, New York, 1993, pp. 197–220.
2. Friberg, S.E. and Brancewicz, C., in *Liquid Detergents*, Surfactant Science Series, Vol. 67, Lai, K.Y., Ed., Marcel Dekker, New York, 1997, pp. 21–33.
3. Matero, A., in *Handbook of Applied Surface and Colloid Chemistry*, Vol. 2, Holmberg, K., Ed., John Wiley, New York, 2002, pp. 407–420.
4. Neuberg, C., *Biochem. Z.*, 76, 107–176, 1916.
5. Neuberg, C., *J. Chem. Soc.*, 110(II), 555, 1916.
6. McKee, R.H., *J. Ind. Eng. Chem.*, 38, 382–384, 1946.
7. Lumb, E.C., *Trans. Faraday Soc.*, 47, 1049–1055, 1951.
8. Licht, W. and Wiener, L.D., *J. Ind. Eng. Chem.*, 42, 1538–1542, 1950.
9. Lawrence, A.S.C., *Nature* (London) 183, 1491–1494, 1959.
10. Friberg, S.E. and Rydhag, L., *Tenside Surf. Deterg.*, 7, 80–83, 1970.
11. Pearson, J.T. and Smith, J.M., *J. Pharm. Pharmacol.*, 26, 123–124, 1974.
12. Yu, W., Zou, A., and Guo, R., *Colloids Surf. A*, 167, 293–303, 2000.
13. Moaddel, T., Huang, T., and Friberg, S.E., *J. Dispersion Sci. Technol.*, 17, 459–475, 1996.
14. Guo, R., Yu, W.L., and Friberg, S.E., *J. Dispersion Sci. Technol.*, 20, 1429–1445, 1999.
15. Burns, R.L. and Duliba, E.P., *J. Surfactants Detergents*, 3, 361–368, 2000.
16. Nishihata, T., Rytting, J.H., and Higuchi, T., *J. Pharm. Sci.*, 70, 71–75, 1981.
17. Woolfson, A.D., McCafferty, D.F., and Launchbury, A.P., *Int. J. Pharm.*, 34, 17–22, 1986.
18. Jain, N.K., Patel, V.V., and Taneja, L.N., *Pharmazie*, 43, 194–196, 1988.

19. Animar, H.O. and Khalil, R.M., *Pharmazie*, 51, 490–493,1995.
20. Tavare, N.S. and Jadhav, V.K., *J. Chem. Ind. Eng. Data*, 41, 1196–1202, 1996.
21. Bary, A.A., Mahmoud, H.A., and Naim, O., *Bull. Fac. Pharm.*, 33, 51–57, 1995.
22. Gupta, G.D., Jain, S., and Jain, N.K., *Pharmazie*, 52, 709–712, 1997.
23. Suzuki, H. and Sunada, H., *Chem. Pharm. Bull.*, 46, 125–130, 1998.
24. Simamora, P., Alvarez, J. M., and Yalkowsky, S.H., *Int. J. Pharm.*, 213, 25–29, 2001.
25. Friberg, S.E., Rananavare, S.B., and Osborne, D.W., *J. Colloid Interface Sci.*, 109, 487–492, 1986.
26. Lee, S.C., Acharya, G., Lee, J., and Park, K., *Macromolecules*, 36, 2248–2255, 2003.
27. Matero, A., Mattsson, A., and Svensson, M., *J. Surf. Deterg.*, 4, 485–489, 1998.
28. Johansson, I., Strandberg, C., Karlsson, B., Karlsson, G., and Hammarstrand, K., *Prog. Colloid Polym. Sci.*, 116, 26–32, 2000.
29. Janakiraman, B. and Sharma, M.M., *Chem. End. Sci.*, 40, 2156–2158, 1985.
30. Godye, R., Smith, F., Westaway, K., Humera, A., Baldisera, L., Laberge, L., and Roussell, J., *Tetrahedron Lett.*, 27, 279–282, 1986.
31. Giguere, R.J., Bray, T.L., Duncan, S.M., and Majetich, G., *Tetrahedron Lett.*, 27, 4945–4948, 1986.
32. Khadilkar, B.M., Gaikar, V.G., and Chitnavis, A.A., *Tetrahedron Lett.*, 36, 8083–8086, 1995.
33. Khadilkar, B.M. and Madyar, V.R., *Org. Proc. Res. Develop.*, 5, 452–455, 2001.
34. Liu, J. and Wan, M., *J. Mater. Chem.*, 11, 404–407, 2001.
35. Larpent, C., Bernard, E., Brisse-le Menn, F., and Patin, H., *J. Mol. Catal. A: Chem.*, 116, 277–288, 1997.
36. Buurma, N.J., Blandamer, M.J., and Engberts, J.B.F.N., *Adv. Synth. Catal.*, 344, 413–420, 2002.
37. Friberg, S.E., Yang, J., and Huang, T., *Ind. Eng. Chem. Res.*, 35, 2856–2859, 1996.
38. Colonia, E.J., Dixit, A.B., and Tavare, N.S., *J. Cryst. Growth*, 166, 976–980, 1996.
39. Tavare, N.S. and Colania, E.J., *J. Chem. Eng. Data*, 42, 631–635, 1997.
40. Colonia, E.J., Dixit, A.B., and Tavare, N.S., *J. Chem. Eng. Data*, 43, 220–225, 1998.
41. Narayanan, K.V. and Ananth, M.S., *AICHE J.*, 40, 621–626, 1994.
42. Friberg, S.E., Fei, L., Campbell, S., Yang, H., and Lu, Y., *Colloids Surf.*, 127, 233–239, 1997.
43. Gandhi, N.N., Kumar, M.D., and Sathyamurthy, N., *J. Chem. Eng. Data*, 43, 695–699, 1998.
44. Kumar, M.D. and Gandhi, N.N., *J. Chem. Eng. Data*, 45, 419–423, 2000.
45. Tavare, N.S. and Jadhav, V.K., *J. Cryst. Growth*, 199, 1320–1325, 1999.
46. Bodhankar, S.S., Rajamani, V., and Gaikar, V.G., *J. Chem. Technol. Biotechnol.*, 71, 155–161, 1998.
47. Raman, G. and Gaikar, V.G., *Ind. Eng. Chem. Res.*, 41, 2966–2976, 2002.
48. Badwan, A.A., El-Khordagui, L.K., Saleh, A.M., and Khalil, S.A., *Int. J. Pharm.*, 13, 67–74, 1982.
49. Ali, B.A., Zughul, M.B., and Badwan, A.A., *J. Dispersion Sci. Technol.*, 16, 451–468, 1995.
50. Coffman, R.E. and Kildsig, D.O., *J. Pharm. Sci.*, 85, 848–853, 1996.
51. Srinivas, V., Rodley, G.A., Ravikumar, K., Robinson, W.T., Turnbull, M.M., and Balasubramanian, D., *Langmuir*, 13, 3235–3239, 1997.

52. Srinivas, V. and Balasubramanian, D., *Langmuir*, 14, 6658–6661, 1998.
53. Kahlweit, M., Busse, G., and Jen, J., *J. Phys. Chem.*, 95, 5580–5586, 1991.
54. Strey, R., Vilsanen, Y., Aratono, M., Kratohvil, J., Yin, Q., and Friberg, S.E., *J. Phys. Chem. B*, 103, 9112–9116, 1999.
55. Horwath-Szabo, G. and Friberg, S.E., *Langmuir*, 17, 278–287, 2001.
56. Koshinuma, M. and Sasaki. T., *Bull. Chem. Soc. Japan*, 48, 2755–2759, 1975.
57. Vikingstad, E., *J. Colloid Interface Sci.*, 72, 68–74, 1979.
58. Cutler, S.G., Meares, P., and Hall, D.G., *J. Chem. Soc., Faraday Trans.*, 174, 1758–1767, 1978.
59. Kale, K.M., Cussler, E.L., and Evans, D.F., *J. Sol. Chem.*, 11, 581–592, 1982.
60. Kale, K.M., Cussler, E.L., and Evans, D.F., *J. Phys. Chem.*, 84, 593–598, 1980.
61. Bhat, M. and Gaikar, V.G., *Langmuir*, 15, 4740–4751, 1999.
62. Bhat, M. and Gaikar, V.G., *Langmuir*, 16, 1580–1592, 2000.
63. Hassan, P.A., Raghavan, S.R., and Kaler, E.W., *Langmuir*, 18, 2543–2548, 2002.
64. Pal, O.R., Gaikar, V.G., Joshi, J.V., Goyal, P.S., and Aswal, V.K., *Langmuir*, 18, 6764–6768, 2002.
65. Hassan, P.A., Fritz, G., and Kaler, E.W., *J. Colloid Interface Sci.*, 257, 154–162, 2003.
66. Schubert, B.A., Kaler, E.W., and Wagner, N.J., *Langmuir*, 19, 4079–4489, 2003.
67. Acosta, E., Uchiyama, H., Sabatini, D.A., and Harwell, J.H., *J. Surf. Deterg.*, 5, 151–157, 2002.
68. Graciaa, A., Lachaise, J., Cucuphat, C., Bourrel, M., and Salager, J.L., *Langmuir*, 9, 669–672, 1993.
69. Salager, J.L., Graciaa, A., and Lachaise, J., *J. Surfact. Deterg.*, 1, 403–406, 1998.
70. Shinoda, K. and Arai, H., *J. Phys. Chem.*, 68, 3485–3490, 1964.
71. Gonzales, G., Nassar, E.J., and Zaniquelli, M.E.D., *J. Colloid Interface Sci.*, 230, 223–228, 2000.
72. Sharma, R. and Bahadur, P., *J. Surfact. Deterg.*, 5, 263–268, 2002.
73. Mansur, C.R.E., Spinelli, L.S., Oliveira, C.M.S., Gonzales, G., and Lucas, E.F., *J. Appl. Polym. Sci.*, 69, 2459–2468, 1998.
74. Mansur, C.R.E., Spinelli, L.S., Lucas, E.F., and Gonzales, G., *Colloids Surf. A*, 149, 291–300, 1999.
75. Mansur, C.R.E., Benzi, M.R., and Lucas, E.F., *J. Appl. Polym. Sci.*, 82, 1668–1676, 2001.
76. Varade, D. and Bahadur, P., *Tenside Surf. Det.*, 39, 48–52, 2002.
77. Friberg, S.E., Brancewicz, C., and Morrison, D.S., *Langmuir*, 10, 2945–2949, 1994.
78. Guo, R., Compo, M.E., Friberg, S.E., and Morris, K., *J. Dispersion Sci. Technol.*, 17, 493–507, 1996.
79. Rance, D.G. and Friberg, S. E., *J. Colloid Interface Sci.*, 60, 207–209, 1977.
80. Friberg, S.E., Yang, H.F., Fei, L., Rasmussen, D.H., and Aikens, PA, *J. Dispersion Sci. Technol.*, 19, 19–30, 1998.
81. Heldt, N., Zhao, J., Friberg, S.E., Zhang, Z., Slack, G., and Li, Y., *Tetrahedron*, 56, 6985–6990, 2000.
82. Banerjee, S. and Chatterjee, S.N., *Indian J. Biochem. Biophys.*, 19, 373–381, 1982.
83. Almog, S., Kushnir, T., Nir, S., and Lichtenberg, D., *Biochemistry*, 25, 2597–2605, 1986.

84. Murthy, K., Easwar, N., and Singer, E., *Colloid Polym. Sci.*, 276, 940–944, 1998.
85. Campbell, S.E., Yang, H., Patel, R., Friberg, S.E., and Aikens, P.A., *Colloid Polym. Sci.*, 275, 303–306, 1997.
86. Campbell, S.E., Zhang, Z., Friberg, S.E., and Patel, R., *Langmuir*, 14, 590–594, 1998.
87. Friberg, S.E., Hasinovic, H., Yin, Q., Zhang, Z., and Patel, R., *Colloids Surf. A*, 156, 145–156,1999.
88. Blute, I., Effect of hydrotropes on nonionic surfactant/water system. Phase and rheological behavior, manuscript in preparation.
89. Raney, K. H. and Miller, C.A., *J. Colloid Interface Sci.*, 119, 539–549, 1987.
90. Burns, R.L., *J. Surfact. Deterg.*, 2, 13–16, 1999.
91. Burns, R.L., *J. Surfact. Deterg.*, 2, 483–488, 1999.
92. Flaim, T., Friberg, S.E., *J. Colloid Interface Sci.*, 97, 26–37, 1984.
93. Cox, J.M. and Friberg, S.E., *J. Am. Oil Chem. Soc.*, 58, 743–745, 1981.
94. Ward, B.F., Jr., Force, C.G., Bills, A.M., and Woodward, F.E., *J. Am. Oil Chem. Soc.*, 52, 219–224, 1975.
95. Flaim, T., Friberg, S.E., Force, C.G., and Bell, A., *Tenside Detergents*, 20, 177–180, 1983.
96. Kielman, H.S. and Van Steen, P.J.F., *Surfactant Act. Agents Symposium*, Society of the Chemical Industry, London, 1979, pp. 191–198.
97. Darwish, I.A., Florence, A.T., and Saleh, A.M., *J. Pharm. Sci.*, 78, 577–581, 1989.
98. Friberg, S.E. and Rydhag, L., *J. Am. Oil Chem. Soc.*, 48, 113–115, 1971.

# 3
# Phase Equilibria

**GUY BROZE**   Advanced Technology Department, Colgate-Palmolive Research and Development, Inc., Milmort, Belgium

## I.  INTRODUCTION

All liquid detergents contain at least one surfactant in the presence of other materials, such as electrolytes, oily materials, and other impurities. Unlike the academic researcher, the formulator must work with industrial-grade raw materials containing significant amounts of different molecules, the properties of which may significantly differ from those of the main material. The understanding of how a given property of a "pure" system is affected by "impurities" is accordingly of essential practical importance. Understanding the principles by which a given product behaves (as is or under use conditions) allows us to replace counterproductive trial-and-error by more efficient methods with a broader range of potential applications. Phase diagrams are very useful tools to achieve this understanding.

**39**

## II.  WHAT IS A PHASE DIAGRAM?

A phase diagram is a graphic representation of the phase behavior of a system under study. The behavior of a single component as a function of temperature and pressure can be represented on a phase diagram, which will show the conditions under which a material is a solid, liquid, or gas. More complex phase diagrams may involve several components. Phase diagrams are very useful tools for formulation, as they allow one to define not only the acceptable composition range of a product but also enable one to optimize the order of addition of the different raw materials.

### A.  Two-Component Phase Diagrams
### 1.  Temperature and Composition

Whether a given proportion of two (liquid) ingredients will mix is defined by thermodynamics. Although in regular systems the entropy of mixing is always positive and accordingly favorable to mixing, the enthalpy of mixing can be positive or negative depending on the energy of formation of heterocontacts at the expense of homocontacts.

An exothermic mixture usually leads to mixing in all proportions. This is the case for water and ethanol. If the mixing is endothermic, the number of coexisting phases and their composition depend on temperature. Increasing the temperature usually results in an increase in the mutual solubility of the two compounds, eventually leading to complete miscibility above a critical temperature, the upper consolute temperature (UCT). Note that some abnormal systems can also have a lower consolute temperature (LCT). Both UCT and LCT are thermodynamic *critical points*. At a critical point, the compositions of the two phases in equilibrium become identical.

Figure 3.1 shows a schematic representation of a two-component phase diagram characterized by a UCT. The left axis corresponds to pure component A and the right axis to pure component B. The abscissa corresponds to different A–B compositions. It is very common to express the compositions in weight fraction. Mole fraction or volume fraction can also be used. The central, shaded area corresponds to the two-phase domain, also referred to as the *miscibility gap*. The clear zone surrounding it represents a single phase.

### 2.  Tie Lines and Lever Rule

When a mixture separates into two phases, it is important to know the compositions and the amounts of the two phases in equilibrium. A *tie line* links the two conjugated compositions in equilibrium. This means that any composition located on the same tie line will separate in the same two phases, the compositions of which are defined by the points of contact of the tie line with the phase boundary.

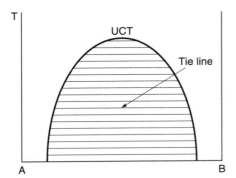

**FIG. 3.1** Phase diagram of two components (A and B) that are only partly miscible at low temperature and become fully miscible above the upper consolute temperature (UCT).

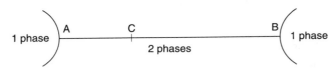

**FIG. 3.2** Lever rule allowing quantification of the proportion of two coexisting phases in a two-phase domain of a phase diagram.

The relative amounts of the two phases are determined according to the lever rule (Figure 3.2). If the compositions are expressed in weight fractions, the weight fraction of phase A is CB/AB and the weight fraction of phase B is AC/AB.

## B. Three-Component Phase Diagrams

Practical systems involve more than two components. A three-component system can be represented by an equilateral triangle (Figure 3.3). A corner of the triangle represents a pure component, a side represents the binary mixture of the components represented by the adjacent corners, and any point in the triangle represents one and only one three-component composition.

The weight fraction of component A in the composition represented by P in the triangle is given by the ratio of the lengths of the segments perpendicular to the sides: $P_a/(P_a + P_b + P_c)$. Similarly, the amount of B is given by $P_b/(P_a + P_b + P_c)$ and the amount of C by $P_c/(P_a + P_b + P_c)$.

Such a phase diagram is valid at one temperature. The effect of temperature on a three-component phase diagram can be visualized in three dimensions, with temperature on the elevation axis. The phase diagram looks like a triangular prism, with every horizontal slice corresponding to one temperature.

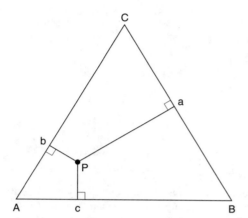

**FIG. 3.3**  Method of determining the composition of a three-component mixture.

## 1.  Fields and Densities

There is an important difference among the thermodynamic functions of state as far as phase equilibria are concerned. Some thermodynamic functions of state, such as temperature and pressure, have the same value in all the phases of a system under equilibrium conditions. They are actually the "forces" driving a system to its equilibrium. Such functions are referred to as *fields* [1].

The other thermodynamic functions of state generally have different values in the different phases of a system at equilibrium. Typical examples are the phase volumes, composition, enthalpy, etc. Such functions are known as *densities*.

A thermodynamic expression of functions of state can be expressed as a sum of field variables multiplied by their conjugated density. For example

$$G = U + PV - TS + \sum \mu_i n_i$$

where $U$ is the internal energy, $PV$ is the product of the field variable *pressure* and the density variable *volume*, $TS$ is the product of the field variable *temperature* and the density variable *entropy*; and $\mu_i n_i$ is the product of the field variable *chemical potential* of component $i$ and the density variable *number of moles* of component $i$. The chemical potentials are the field functions conjugated with the concentrations.

## 2.  Phase Rule

For a multicomponent system, the phase rule [2] allows us to know the number of independent variables necessary to define completely (from a compositional point of view) a system. This number is called the *number of degrees of freedom* or the *variance* of the system. The variance $f$ is given by

$$f = C - \Phi + 2$$

where $C$ is the number of chemically independent components in the system, $\Phi$ is the number of coexisting phases at equilibrium, and the last term takes care of temperature and pressure. Note that this definition of the variance supposes that the components do not react with each other. For systems at constant pressure, such as all systems under atmospheric pressure, the last term should be 1. Similarly, systems studied at constant temperature and pressure have 0 as the last term.

A direct implication of the phase rule is that a three-component system in one phase at atmospheric pressure and at 25°C has a variance equal to 2. This means that two dimensions are necessary to describe fully such a system. Another implication is that such a system could show a maximum of three coexisting phases. Indeed, a negative variance does not have any physical meaning.

A system based on five components will need, according to the phase rule, a four-dimension hyperspace to be completely described. To represent such a system, some variables are usually grouped. The accuracy of such a representation is, of course, imperfect.

A more accurate procedure is to set a variable to a constant value. This is impossible with a composition because it is a density and is usually different in each of the coexisting phases. The phase rule determines the number of independent variables a system needs to be represented but does not introduce any restriction on the choice of the independent variables. It is accordingly much better, whenever possible, to fix a field variable to reduce a system of one dimension. Instead of using concentrations (density variables), a representation as a function of the chemical potentials is easier to read and is more accurate. The problem is that, in practice, it is very complicated to work at defined chemical potentials.

## 3. Tie Lines and Critical Points

Let us consider two liquids A and B that are not very soluble in each other. Addition of liquid C increases the miscibility of B in A and of A in B. The addition of C has the same effect as increasing the temperature in the binary phase diagram. The major difference is that the tie lines are no longer necessarily parallel to the baseline, and the critical end point is no longer at the maximum of the miscibility gap (Figure 3.4). This is because C does not partition evenly between the two coexisting phases. In the present case, C goes preferably into B. The critical end point is located near the A corner. An isothermal critical end point is usually referred to as a *plait point*.

## 4. Three-Phase Domain

In some cases a three-phase region occurs (Figure 3.5). The coexistence of three phases in equilibrium in an isothermal three-component phase diagram is

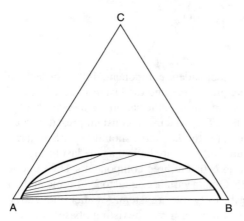

**FIG. 3.4**   A Winsor II ternary phase diagram.

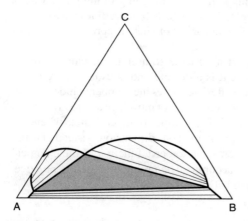

**FIG. 3.5**   A Winsor III ternary phase diagram.

a zero-variant situation. Of course, an infinity of different compositions fall inside the three-phase triangle, but the compositions of the three coexisting phases are the same for all the initial compositions falling in the three-phase triangle. They are represented by the three corners of the three-phase triangle. What changes are their respective amounts.

## C.   Recording Phase Diagrams

There are basically two methods for recording phase diagrams: the titration method and the constant composition method. Both have advantages and drawbacks.

## 1. Titration Method

In the titration method a mixture is titrated by another. Typically, a mixture of two of the components is titrated by the third. The weight of titrant to reach a phase boundary is carefully recorded and plotted on the phase diagram. The process is then repeated to cover the whole domain to be investigated. Such a method is relatively fast and can give a good idea of the phase boundaries.

There are two major drawbacks to this method. First, this method gives the phase diagram at one temperature only. To determine the phase diagram at another temperature, the process must be repeated. The temperature domain available with the titration method is limited for practical reasons, as all the components must be kept at the same temperature.

The second drawback is that the method is usually used in out-of-equilibrium conditions. In some systems, such as those involving lyotropic liquid crystals, the time required to reach equilibrium can be very long; metastable phases can also be encountered.

A phase diagram recorded by the titration method should be used as a guide only and should never be applied for long-term stability prediction.

## 2. Constant Composition Method

In the constant composition method a series of compositions covering the composition range to be studied are prepared in test tubes, which are sealed. The test tubes are shaken thoroughly and allowed to stand in a thermostatic bath. The test tubes containing turbid solutions are allowed to stand until they separate into two or more completely clear phases. The number of clear phases can be reported on the phase diagram, and the phase domains can be mapped.

This method is very time consuming, but it allows one to approach true equilibrium conditions, and the tubes can be used at other temperatures. Another advantage of this method is that, when a system gives more than one phase, it is possible to analyze the phases and accordingly know exactly where the phase boundaries are, as well as the orientation of the tie lines.

## III. PHASE DIAGRAMS FOR IONIC SURFACTANT-CONTAINING SYSTEMS
## A. Ionic Surfactant and Water
## 1. Krafft Point

The Krafft point can be defined as the temperature $T_k$ above which the amphiphile (surfactant) solubility in water greatly increases [3]. The reason is that the water solubility of the amphiphile, which increases with temperature, reaches the amphiphile critical micelle concentration ($C_M$ in Figure 3.6). When the solubility curve is above $C_M$ the dissolved amphiphile forms micelles and the amphiphile

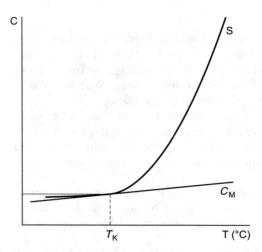

**FIG. 3.6** The Krafft point is the temperature at which the solubility of the amphiphile becomes higher than its critical micelle concentration ($C_M$).

activity in water solution no longer increases. There is accordingly no longer a limitation to solubilzation.

The Krafft point is a triple point because at this temperature three "phases" coexist [4]: hydrated solid amphiphile, individual amphiphile molecules in solution (unimers), and amphiphile molecules involved in micelles.

The value of $T_k$ increases as the amphiphile hydrophobic chain length increases. The Krafft points of the sodium salts of the classic amphiphiles (alkyl sulfates, sulfonates, and benzenesulfonates) are usually below room temperature. The Krafft point is a function of the counter-ion. Alkaline earth cations give higher Krafft points: for sodium laurylsulfate, $T_k = 9°C$; the values for the calcium, strontium, and barium salts are 50, 64, and 105°C, respectively.

Because the Krafft point imposes a limitation in formulation, the following rules to reduce $T_k$ are of interest:

- Chain branching and polydispersity reduce $T_k$.
- Complexation of Mg and Ca reduces $T_k$.
- The presence of unsaturation decreases $T_k$.

A very efficient way to reduce $T_k$ is to incorporate two or three oxyethylene monomers between the amphiphile hydrophobic chain and the polar head group (alcohol ethoxy sulfates). In each case other properties of the amphiphile, such as the surface activity, can be consequently modified.

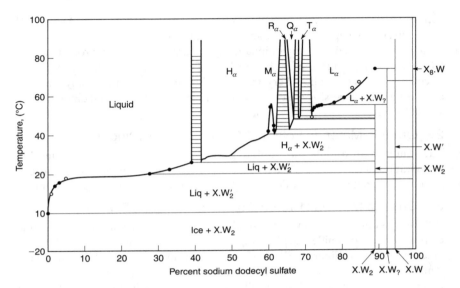

**FIG. 3.7** Typical phase diagram of a water–anionic surfactant system.

## 2. Phase Diagram

The phase diagram of sodium dodecyl sulfate–water is representative of many ionic systems (Figure 3.7) [5]. In Figure 3.7 "Liquid" is the aqueous micellar phase; $H_\alpha$ is the hexagonal lyotropic liquid crystal, sometimes called the middle phase; and $L_\alpha$ is the lamellar lyotropic liquid crystal, sometimes called the neat phase. On the surfactant-rich side, several hydrated solid phases are present.

As a general rule, in any (real) phase diagram, at any point representative of a region and on its boundaries, the number of phases and their nature are similar.

A tie line is the line joining the points representing two coexisting phases. If the total composition of a mixture falls in a two-phase region, it separates into the two phases located at both sides of the tie line that passes the formulation point. The weight distribution of the two phases is given by the lever rule.

## B. Ionic Surfactant, Water, and Organic Material Ternary Systems

### 1. Organic Material: Hydrocarbon

Let us consider an isotherm of a water–ionic amphiphile binary mixture above the Krafft point (for example, water–sodium octanoate) [6]. At an amphiphile concentration of 7% (the critical micellar concentration), the micellar isotropic solution L1 appears and lasts up to 41%. Between 41 and 46% is the miscibility gap between L1 and H1, the hexagonal phase, which lasts up to 52%. Above 52% is the miscibility gap between H1 and the hydrated crystal.

If a nonpolar component (aliphatic hydrocarbon or tetrachloromethane) is added, almost nothing happens (Figure 3.8a). The solubility of octane in either the micellar solution or the liquid crystal is very limited. This is true of any molecule exhibiting only dispersion cohesive forces (induced dipole–induced dipole van der Waals forces).

### 2.   Organic Material: Polar but Not Proton Donating

The solubility of a molecule exhibiting dipole–dipole cohesive forces and low H-bonding cohesive forces, such as methyl octanoate, is greater than that of a hydrocarbon, but nothing particular happens in the center of the phase diagram.

### 3.   Organic Material: Proton Donating

If the third component is a water-insoluble alcohol (five carbons or more), amine, carboxylic acid, or amide, the phase topography is profoundly modified. The phase diagram shown in Figure 3.8b [7] shows in addition to L1 and H1 a very large lamellar phase, a narrow reverse hexagonal phase H2, and, even more important, a "sector-like" area of reverse micelles L2. This means that the solubility of *n*-decanol in a sodium octanoate–water mixture containing between 25 and 62% amphiphile is far more important (30 to 36%) than pure water (4%) and pure sodium octanoate (almost zero). This phase is essential to obtain water-in-oil (w/o) microemulsions.

The solubility of *n*-decanol in the L1 phase is also important (up to 12% at the "end" of the L1 phase). The L1 phase is accountable for the observation of oil-in-water (o/w) microemulsions. The $L_\alpha$ domain, generally located in the middle of the diagram, points toward the water side for a critical surfactant-to-cosurfactant ratio. (A 1:2 sodium octanoate to *n*-decanol ratio leads to a lamellar phase with as little as 17% surfactant–cosurfactant mixture.) In some cases, such as for octyl trimethylammonium bromide (OTAB)–hexanol–water, the lamellar phase already exists for 3% hexanol + 3% OTAB!

The practical interest of a lamellar liquid crystal lies in its suspending capability. A lyotropic liquid crystal exhibits a viscoelastic behavior that allows suspension of solid particles for a very long time. The lamellar phase is additionally characterized by an ideal critical strain to provide the suspension with good resistance to vibrations and convections, without impairing its flowability with too great a viscosity.

## C.   Ionic Surfactant, Water, Proton-Donating Material, and Hydrocarbon Quaternary Systems

The "solubility" of an oil such as decane in the micellar isotropic solution L1 or in the reverse micellar isotropic solution L2 can be very important. L1 leads to w/o

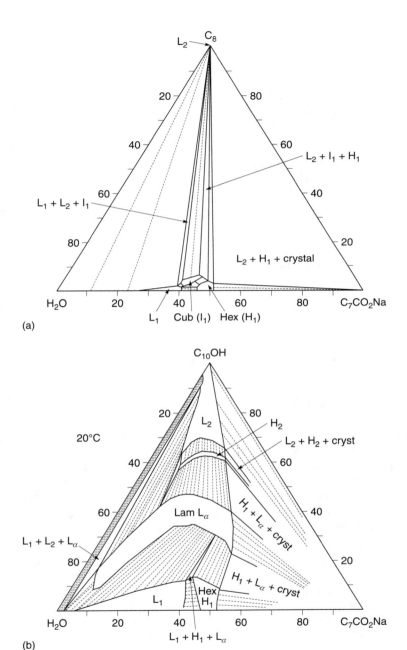

**FIG. 3.8** (a) Typical ternary phase diagram of water, an amphiphile (sodium octanoate), and a hydrocarbon (octane). (b) Typical ternary phase diagram of water, an amphiphile (sodium octanoate), and a co-amphiphile (decanol). This phase diagram was established by Ekwall in 1975.

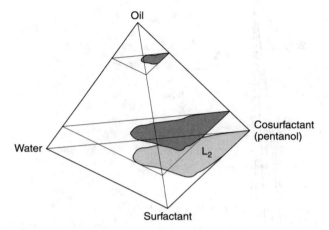

**FIG. 3.9**  Phase diagram of a water-in-oil microemulsion.

microemulsions and L2 to o/w microemulsions. Note that the cosurfactant is an amphiphile with (generally) a lower molecular weight than the main amphiphile, the surfactant.

## 1.  Water-in-Oil Systems

As shown in Figure 3.9, the L2 phase is able to solubilize a very large amount of a hydrocarbon such as decane or hexadecane. In fact, a composition containing up to 75% decane and water/surfactant/cosurfactant proportions corresponding to the L2 phase is still clear, fluid and isotropic, forms spontaneously, and is thermodynamically stable. The structure of this microemulsion can be (to some extent) regarded as a dispersion of tiny water droplets (reverse micelles) in a continuous phase of the hydrocarbon. The surfactant and cosurfactant are mainly located at the water/oil interface. This type of system is often referred to as a w/o microemulsion.

The term "microemulsion" to describe such systems is not well chosen: it conveys the idea of an actual emulsion characterized by submicrometer (below 0.1 μm) droplets. As is well known, an emulsion is not thermodynamically stable and cannot be represented by a single-phase domain in a thermodynamic phase diagram. The so-called microemulsions must be considered as real micellar solutions containing oil in addition to water and surfactants. These solutions, although very far from ideal in the thermodynamic sense, are nevertheless always real in the thermodynamic sense. Another important difference between microemulsions and emulsions is that, in general, a microemulsion requires significantly more surfactant than an emulsion.

These w/o microemulsions exhibit other important characteristics:

- The domain of existence is large. Significant compositional changes can occur without crossing a phase boundary. Such behavior is particularly important for manufacturing processes, because it provides robustness to the formulation.
- They are very stable in a large temperature range, usually from the Krafft point up to the boiling point. Moreover, the phase boundaries are almost insensitive to temperature.
- The phase topography remains almost unchanged even if up to 75% of the ionic amphiphile is replaced by a nonionic amphiphile.

To obtain a wide w/o microemulsion phase it is essential to adjust carefully the cosurfactant structure (usually its chain length) and its relative amount. Although trial and error is still the most commonly used method for obtaining microemulsions, a tentative rule is to combine a very hydrophobic cosurfactant (n-decanol) with a very hydrophilic ionic surfactant (alcohol sulfate) and a less hydrophobic cosurfactant (hexanol) with a less hydrophilic ionic surfactant (OTAB). For very hydrophobic ionic surfactants, such as dialkyl dimethylammonium chloride, a water-soluble cosurfactant, such as butanol or isopropanol, is adequate (this rule derives at least partially from the fact that an important feature of the cosurfactant consists of readjusting the surfactant packing at the solvent/oil interface).

## 2.  Oil-in-Water Systems

It was stated earlier that the solubility of decane in the L1 phase is almost zero. For a well-defined surfactant-to-cosurfactant ratio, very large quantities of decane (or any hydrocarbon) can be solubilized in the L1 phase. A thin, snake-like single-phase domain develops toward the oil vertex of the phase diagram (Figure 3.10). This phase can be regarded as amphiphile micelles swollen with oil.

Generally, the o/w microemulsion phases are only metastable systems. As with any metastable system, o/w microemulsions need an activation energy to separate; sometimes this activation energy is so large that the separation almost never occurs. Such systems are not thermodynamically stable and should accordingly not be considered in a phase diagram. However, they form spontaneously and are stable (because of the high activation energy for separation) for a very long time.

A typical example of a very stable metastable system is a mixture of one volume of oxygen with two volumes of hydrogen. The mixture is spontaneous and stable for a very long time, without being thermodynamically stable. The final thermodynamically stable state is obtained by adding a catalyst (platinum foam) or a flame to the mixture.

Although not thermodynamically stable, o/w microemulsions form spontaneously and are accordingly useful (ease of manufacture).

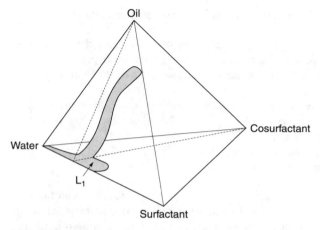

**FIG. 3.10**   Phase diagram of an oil-in-water microemulsion.

Thermodynamic instability implies some constraints on o/w microemulsions:

- Their position may depend on the order of addition of the raw materials and on the shear imposed on the system.
- Their domain of existence is generally narrow.
- They can be sensitive to freeze and thaw cycles.

## IV.  PHASE DIAGRAMS FOR NONIONIC SURFACTANT-CONTAINING SYSTEMS

The phase topography of a ternary system involving water, a hydrocarbon, and a polyethoxylated fatty alcohol depends on the hydrocarbon chain length, branching, degree of unsaturation, aromaticity, etc., on the amphiphile structure (hydrophobic and hydrophilic chain length), and also on temperature, which exerts a very strong influence on the configuration (and accordingly on the solubility) of the polyoxyethylene segments in water solution. A review has been presented in a series of papers [8–11].

The phase topography is strongly influenced by the more elementary behaviors of the binary amphiphile–oil and amphiphile–water systems.

## A.  Nonionic Surfactant and Oil

Polyethylene oxide is not soluble in hydrocarbons such as hexane or decane. If a fatty chain is attached to a short segment of polyethylene oxide (4 to 8 ethylene oxide units), the nonionic amphiphile obtained exhibits a solubility profile in oil

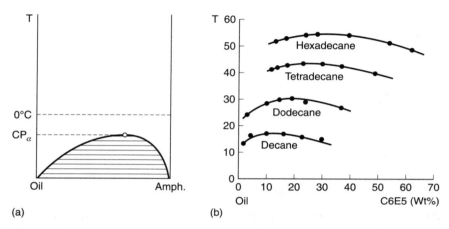

**FIG. 3.11** (a) Haze point temperature (CP, critical point; amph, amphiphile). (b) Haze point temperature dependence on oil structure.

depending on temperature. At low temperatures a miscibility gap is obtained, translating to the insolubility of the polyethylene oxide chain in the oil. At high temperatures the effect of the entropy is predominant and the amphiphile is soluble in all proportions in the oil.

As predicted by the Flory–Huggins theory, such a system shows a lower miscibility gap characterized by an upper critical point, at temperature $T_\alpha$, which depends on both the oil and the amphiphile structure (Figure 3.11a). The critical composition is usually not far from the pure oil side.

Figure 3.11b shows the lower miscibility gap between some $n$-alkanes and C6E5 (pentaethylene glycol monohexyl ether). The upper critical temperature $T_\alpha$ increases with increasing hydrocarbon chain length (hydrophobicity).

The critical temperature $T_\alpha$ is often referred to as the haze point temperature, and the miscibility gap between oil and amphiphile plays an essential role in the ternary phase diagram.

## B. Nonionic Surfactant and Water Cloud Point

The phase diagram of a nonionic amphiphile–water binary system is more complicated (see Figure 3.12). A "classic" upper critical point exists, but it is usually located below 0°C. At higher temperatures most nonionic amphiphiles show a miscibility gap, which is actually a closed loop with an upper as well as a lower critical point. The lower critical point $CP_\beta$ is often referred to as the cloud point temperature. The upper critical point often lies above the boiling temperature of the mixture (at 0.1 MPa). The position and the shape of the loop depend on

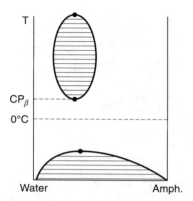

**FIG. 3.12**  Phase behavior of a water–nonionic amphiphile system (CP, critical point; amph, amphiphile).

the chemical structure of the amphiphile. The cloud point temperature plays an essential role in three-component phase diagram topography.

The closed loop can be regarded as a vertical section through a "nose" in the concentration–temperature–pressure space at constant pressure (see Figure 3.13a). When the pressure increases, the surface covered in the temperature–concentration phase by the phase separation loop decreases and vanishes at a critical pressure $P^*$.

The shrinking of the loop of the water–ethylene glycol butyl ether (C4E1) system with increasing pressure is shown in Figure 3.13b. The critical conditions for the loop to vanish are $T^* = 95°C$, $P^* = 80$ MPa, and $C^* = 28$ wt%.

To show the multidimensional nature of these phenomena, note that similar effects (shrinking of the loop, f.i.) can be achieved by the addition of "hydrotropic" electrolytes at constant pressure or by increasing the hydrophilicity of the amphiphile. Figure 3.13c shows the loop areas of butanol (C4E0), ethylene glycol butyl ether (C4E1), and diethylene glycol butyl ether (C4E2). The last does not exhibit a loop at 0.1 MPa (1 atm), but the system behaves actually as if the nose were "lurking."

Although no phase separation occurs in water, the lurking nose exerts some influence on the three-component phase diagram. Another way to look at the same phenomenon is to consider that, in conditions close to $T = 90°C$ and $C = 30$ wt%, the C4E2–water system is such that the mixing entropy is just high enough to maintain the molecules in a single phase, the enthalpic term being positive (endothermic). As soon as a third incompatible component (the oil f.i.) is incorporated, the entropy is no longer able to maintain the molecules in a single phase, and phase separation occurs.

In Table 3.1, the hydrophilic/lipophilic balance (HLB) is calculated according to the empirical equation HLB $= 20M_H/M$, where $M_H$ is the molar mass of the

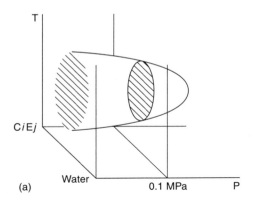

(a)

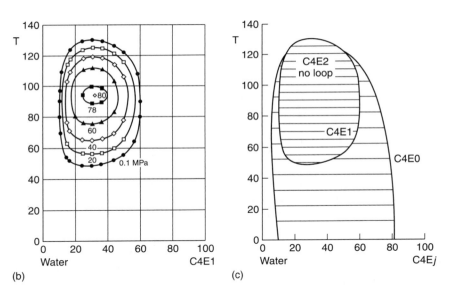

(b)                                    (c)

**FIG. 3.13** (a) Effect of pressure on the size of the closed loop. (b) Closed loop of the water–ethylene glycol butyl ether system at different pressures. (c) Effect of the hydrophilic group of the amphiphile on the shape of the closed loop. (From Schneider, G., *J. Phys. Chem.* (Munich), 37, 333, 1963. With permission.)

hydrophilic group and $M$ the total molar mass of the ethoxylated amphiphile. The parameter $\gamma_{min}$ is the minimum amphiphile concentration required for the homogenization of a 1:1 (wt%) mixture of water and $n$-decane at around 40°C and $T_\beta$ and $C_\beta$ are the coordinates of the lower critical points (cloud point). Although the HLB seems to be correlated with the cloud point, it cannot give any information on

**TABLE 3.1** Values of HLB, $\gamma_{min}$, $T_\beta$ and $C_\beta$ for Selected Amphiphiles

| Amphiphile | HLB | $\gamma_{min}$ (wt%) | $T_\beta$ (°C) | $C_\beta$ (wt%) |
|---|---|---|---|---|
| C4E1 | 10.3 | 58.9 | 48.7 | 29.0 |
| C6E3 | 12.7 | 47.4 | 45.4 | 13.5 |
| C8E4 | 12.6 | 29.6 | 39.6 | 6.9 |
| C10E5 | 12.5 | 19.7 | 40.3 | 3.5 |
| C12E6 | 12.4 | 10.6 | 48.0 | 2.2 |

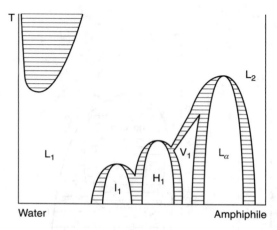

**FIG. 3.14** Binary phase diagram of a water–ethoxylated nonionic amphiphile system, including lyotropic liquid crystal domains. (From Kalhweit, M. and Strey, R., *Angew. Chem. Int. Ed. Engl.*, 24, 654, 1985. With permission.)

the amphiphile efficacy ($\gamma_{min}$). Even if the HLB remains constant, increasing both the polar and the nonpolar parts of a surfactant molecule significantly improves its efficacy (at least its water–oil coupling efficacy).

The closed loop is not the only characteristic of the nonionic surfactant–water binary phase diagram. Like the ionic surfactant–water mixture, nonionic surfactants, at higher concentration in water, exhibit lyotropic mesophases. Figure 3.14 shows a typical binary phase diagram exhibiting the full lyotropic mesophase sequence: I1, cubic isotropic phase; H1, direct hexagonal phase (middle phase); V1, special cubic ("viscous" phase); $L_\alpha$, lamellar phase (neat phase). Note the presence of the two-phase domains surrounding each mesophase, the critical point on top of each, and the zero-variant three-phase feature.

Although very difficult to determine with accuracy, the miscibility gaps always exist, as well as the three-phase situations. Of course, the critical temperatures and concentrations corresponding to each mesophase depend on the chemical nature of the amphiphile, the pressure, and the optional presence of an electrolyte.

Figure 3.15 shows some examples of real nonionic amphiphile–water binary phase diagrams [10,12]. As a rule, amphiphiles with a hydrocarbon chain length of eight or fewer carbon atoms exhibit only the loop (in a domain depending on the ethoxylation) and no mesophase.

Longer chain amphiphiles show one or more mesophases (usually one). The type of the main mesophase (the one having the highest critical temperature) depends on the relative volumes of the ethoxylate and hydrocarbon chains. If the volumes are similar, the lamellar phase is predominant. This is the case for C12E6. If the volume of the ethoxylate chain is significantly higher than that of the hydrocarbon chain, the hexagonal phase will melt at higher temperature (C12E7); if the volume of the ethoxylate chain is much higher than that of the hydrocarbon chain, the cubic phase I1 may appear.

In some cases, such as for C12E5, the lamellar phase $L_\alpha$ (or the H1 phase) interferes with the loop (with the cloud point curve) and induces the so-called critical phase L3. L3 is an isotropic, often bluish phase, exhibiting a zero-variant three-phase critical point at its lowest temperature of existence. The three phases present at the critical conditions are W (water with a minute amount of amphiphile), L3, and $L_\alpha$. The L3 phase seems to have a beneficial action on cleaning performance, maybe because of the presence of the critical point.

## C. Nonionic Surfactant, Water, and Oil

From the phase behavior of both binary mixtures (water–amphiphile and oil–amphiphile), it is now possible to account, at least qualitatively, for the three-component phase diagram as a function of temperature. The presence of a haze point on the oil–amphiphile phase diagram (critical point $\alpha$) at temperature $T_\alpha$ shows that the surfactant is more compatible with the oil at high than at low temperature. The presence of a cloud point on the water–amphiphile phase diagram (the lower critical point $\beta$) at temperature $T_\beta$ shows that (at least in the neighborhood of the temperature domain) the amphiphile is less compatible with water at high than at low temperature. As a consequence (the other parameters being kept constant), the amphiphile behavior depends on temperature.

At low temperature the amphiphile is more compatible with water than with oil. The phase diagram corresponding to this situation is shown in Figure 3.16 (al or a2). The tie line orientation is directly deduced from the partitioning of the amphiphile between water and oil: because under the current conditions the surfactant is more compatible with water than with oil, the majority of the amphiphile is in the water phase and only a limited amount of amphiphile is present in the oil. Accordingly, the tie lines point in the direction of the oil vertex. The phase diagrams al and a2 of Figure 3.16 are referred to as Winsor I (WI).

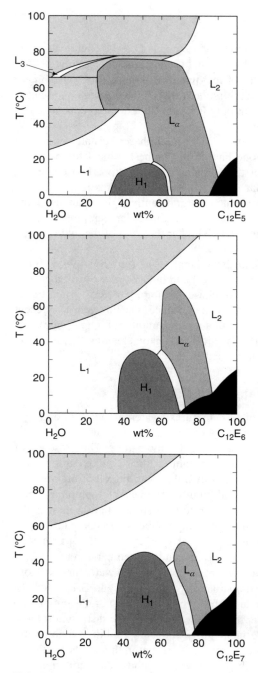

**FIG. 3.15** Examples of water–ethoxylated nonionic amphiphile binary phase diagrams. (From Broze, G., *Comm. J. Com. Esp. Deterg., Barcelona*, 20, 133, 1989. With permission.)

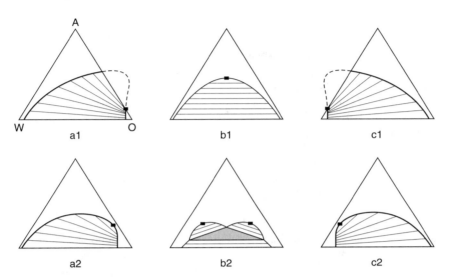

**FIG. 3.16** Evolution of water–ethoxylated nonionic amphiphile–oil ternary phase diagrams with temperature (temperature increasing from a to c).

If the temperature at which the phase diagram is recorded is above $T_\alpha$ (the haze point), a critical point $CP_\alpha$ is present near the oil vertex (although pure amphiphile and pure oil are miscible, the presence of a small amount of water "recalls" the lack of compatibility between amphiphile and oil). If the temperature is below $T_\alpha$, no critical point appears in the three-component phase diagram (it would be positioned at a negative water concentration).

At high temperatures these mixtures are more compatible with oil than with water. The phase diagram corresponding to this situation is shown in Figure 3.16 (cl or c2). The amphiphile partitioning now favors the oil, and the tie lines point in the direction of the water vertex. Phase diagrams cl and c2 of Figure 3.16 are referred to as Winsor II (WII). A critical point $CP_\beta$ occurs if the temperature is below the cloud point $T_\beta$, but more often the critical point lies outside the Gibbs triangle ($T > T_\beta$).

In WI and WII representations, the critical points $CP_\beta$ and $CP_\alpha$ are called plait points. If the difference between the temperature $T$ at which the phase diagram is recorded and the critical point of the binary mixture, $T_\beta$ or $T_\alpha$, increases, the distance from the plait point to the oil–amphiphile axis for $CP_\beta$ and the water–amphiphile axis for $CP_\alpha$ also increases. An important characteristic of a ternary system is the line that links the plait points as a function of temperature. The plait point curve is really the trace of the partitioning of the amphiphile between oil and water. The closer the plait point is to the oil, the more water soluble the amphiphile, and vice versa.

At low temperatures the amphiphile is more compatible with water because water interacts strongly with the hydrophilic head group. Accordingly, the hydrodynamic volume of the head group is greater than that of the hydrocarbon tail. At high temperatures head group hydration is reduced and so is the hydrodynamic volume, which becomes smaller than the hydrodynamic volume of the hydrocarbon tail. There is necessarily a temperature at which the hydrodynamic volumes of the two antagonistic parts of the amphiphile molecule are equal. This particular temperature, represented by $\widetilde{T}$, is the phase inversion temperature (also called the HLB temperature). The phase inversion temperature is a characteristic (and is accordingly a function) of the nature of the oil, the amphiphile, and the water solution (if electrolytes are present). If the pressure can vary (as in oil recovery), this also changes $\widetilde{T}$. It is important to realize that $\widetilde{T}$ can be higher than both $CP_\beta$ and $CP_\alpha$ when the amphiphile solubility is very small in water and oil.

The topography of the phase diagram at the phase inversion temperature depends on the mutual incompatibilities among oil–amphiphile, water–amphiphile, and water–oil. Even with a polar oil and water containing a chaotropic (hydrotropic) electrolyte, the water–oil incompatibility is enough to guarantee a miscibility gap from 0 to 100°C.

For the amphiphile the situation is not as simple. At $\widetilde{T}$ the amphiphile is equally compatible with water and oil, but no assumption is made about the degree of compatibility. Two limiting cases can occur:

1. The amphiphile is very compatible with both water and oil. The phase diagram will look like diagram b1 of Figure 3.16, with a plait point only for an equal amount of oil and water and with the lines parallel to the water–oil axis (equal partitioning). This plait point corresponds to the merging of the $CP_\alpha$ and $CP_\beta$ lines, and the projection of the plait point curves on the oil–water–temperature phase diagram should look like those shown in Figure 3.17a or Figure 3.17b.

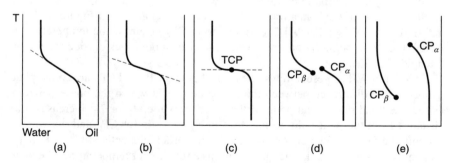

**FIG. 3.17** Transition from an infratricritical situation (a and b) to a supertricritical situation (d and e) through a tricritical point (TCP) (c). (From Kalhweit, M. and Strey, R., *Angew. Chem. Int. Ed. Engl.*, 24, 654, 1985. With permission.)

2.  The amphiphile is equally (and significantly) incompatible with both water and oil. The phase diagram will now look like diagram b2 of Figure 3.16. A three-phase triangle (3PT) appears.

Three phases are now in equilibrium:

1.  A water-rich phase (W)
2.  An oil-rich phase (O)
3.  An amphiphile-rich phase (S)

The amphiphile-rich phase is also called the surfactant phase or the middle phase. These terms, due to Shinoda, result from the physical appearance of a three-phase system:

1.  A dense, water-rich phase at the bottom
2.  A light, oil-rich phase at the top
3.  A phase containing most of the amphiphile in the middle

It is worth noting that with higher molar volume amphiphiles, such as C12E4, a significant amount of the amphiphile can be present in the oil phase, even at $\tilde{T}$. Here, too, the plait points $CP_\alpha$ and $CP_\beta$ will be inside or not be inside the Gibbs triangle depending on the relative positions of $\tilde{T}$, $T_\alpha$, and $T_\beta$.

If the phase diagram exhibits a 3PT it is called a Winsor III (WIII) system. In such a situation, the plait point curves do not merge but "cross" each other and stop at two terminal critical points (see Figure 3.17d or Figure 3.17e).

The sequence of the evolution of a three-component system when temperature is increased can be summarized as follows. If the amphiphile is strongly incompatible with oil and water the sequence is WI $\rightarrow$ WIII $\rightarrow$ WII. If the amphiphile is compatible or is weakly incompatible with oil and water the sequence is WI $\rightarrow$ WII.

A way to modify amphiphile compatibility with oil and water is to change the amphiphile molecular mass, keeping the appropriate balance between lipophobicity and hydrophobicity. A high-molecular-weight amphiphile like C12E6 will show a WI–WII–WII sequence; a low-molecular-mass amphiphile like C4E2 will show (with decanol acetate as the oil) a WI–WII sequence. By varying the amphiphile compatibility through the molecular mass, it is possible to pass from a WI–WII to a WI–WIII–WII sequence. At a certain point a situation as shown in Figure 3.17c will occur: the plait point curves just merge critically and the 3PT collapses. This situation corresponds to a *tricritical point*, an essential concept in theoretical thermodynamics.

When the system is such that a 3PT appears (by far the most common case), the 3PT is present from a temperature $T_l$ lower than $\tilde{T}$ to a temperature $T_u$ above $\tilde{T}$. To some extent the difference between $T_l$ and $T_u$ is a measure of how far the system is from the tricritical conditions. (Note that $\tilde{T}$ is not necessarily the mathematical average of $T_l$ and $T_u$, but it is close to it.)

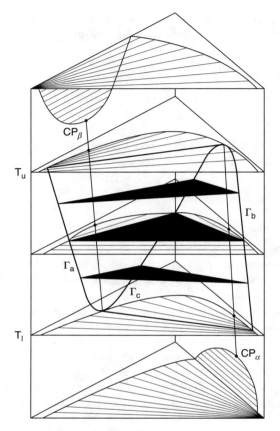

**FIG. 3.18** Detailed evolution of the phase diagram of a water–oil–ethoxylated nonionic amphiphile (low molecular weight) with temperature. (From Kalhweit, M. and Strey, R., *Angew. Chem. Int. Ed. Engl.*, 24, 654, 1985. With permission.)

The thermal evolution of a typical system, with broken critical lines (see Figure 3.18), can be summarized as follows:

For $T < T_1$, the phase diagram is a typical WI.

$T_1$ is the temperature of the critical end point of the $CP_\beta$ curve. At $T = T_1$, the phase diagram is still a WI, but with a critical tie line from which the 3PT will appear with the slightest increase in temperature.

For $T_1 < T < \tilde{T}$, the corner of the 3PT corresponding to the amphiphile phase remains close to the water side but moves "clockwise" in the Gibbs triangle.

For $T = \tilde{T}$, the amphiphile corner of the 3PT reaches "12 o'clock" (phase inversion temperature).

For $\widetilde{T} < T < T_\mathrm{u}$, the corner of the 3PT corresponding to the amphiphile phase keeps on "moving clockwise" to the oil phase.

$T_\mathrm{u}$ is the temperature of the critical end point of the $CP_\alpha$ curve. At $T = T_\mathrm{u}$, the amphiphile corner of the 3PT merges with the oil corner and the 3PT collapses in a critical tie line of a WII phase topography.

At $T > T_\mathrm{u}$, the phase diagram is a typical WII.

It is important to remark on the shape of the line joining the three corners of the 3PT triangle ($\Gamma$ lines). It is a single, continuous gauche line, with a minimum at $T_\mathrm{l}$ on the water side of the critical tie line and a maximum at $T_\mathrm{u}$ on the oil side of the critical tie line. The branches can be identified ($\Gamma_\mathrm{a}$, $\Gamma_\mathrm{b}$, and $\Gamma_\mathrm{c}$), each corresponding to the compositions of each corner of the 3PT. It is important to note that $\Gamma_\mathrm{a}$ has nothing to do with $CP_\alpha$ and that $\Gamma_\mathrm{b}$ has nothing to do with $CP_\beta$. $\Gamma i$ are composition curves, and $CP_i$ are critical point curves. At a critical end point, however, the critical point curve meets the extreme of the composition curve ($CP_\alpha$ meets $\Gamma_\mathrm{a}$ at $T_\mathrm{u}$ and $CP_\beta$ meets $\Gamma_\mathrm{b}$ at $T_\mathrm{l}$).

Another important characteristic of these systems is that the best compatibility capacity is achieved at $T = \widetilde{T}$, when the hydrodynamic volumes of both parts of the surfactant are equal. Under phase inversion conditions, the amount of amphiphile needed to make compatible a mixture of equal amounts of oil and water is minimal. The phase inversion conditions are accordingly looked for to minimize the amphiphile quantity needed to achieve a given task.

Winsor behavior is not the only characteristic of water–oil–nonionic amphiphile systems. The lyotropic mesophases appearing on the water–amphiphile binary phase diagrams expand to some extent in the Gibbs triangle (Figure 3.19).

Amphiphiles based on alcohols lower than C8 do not generate liquid crystals at all (amphiphiles based on alcohols of C4 and less do not even give micelles). Alcohol-based nonionic amphiphiles of C10 and above give lyotropic liquid crystals, at least usually up to $T_\mathrm{u}$. Figure 3.19 shows the typical and general behavior of a ternary system with an amphiphile giving liquid crystals. At a temperature below $T_\mathrm{l}$ each lyotropic mesophase appearing on the water–amphiphile binary phase diagram expands in the Gibbs triangle. At a temperature close to $\widetilde{T}$ generally only the lamellar liquid crystal phase is present, and points toward the amphiphile corner of the 3PT. Above $T_\mathrm{u}$ all the liquid crystals are molten.

## D. Effects of System Parameters on Phase Behavior

### 1. Nonionic Surfactant Structure

The parameters $T_\mathrm{u}$, $T_\mathrm{l}$, and $\widetilde{T}$ increase if more hydrophilic amphiphiles are used. This is easily explained by the HLB concept: a more hydrophilic amphiphile will remain in water up to a higher temperature.

Another fundamental effect of the amphiphile is a result of its molecular mass (or molar volume): increasing the molecular mass of an amphiphile at constant HLB

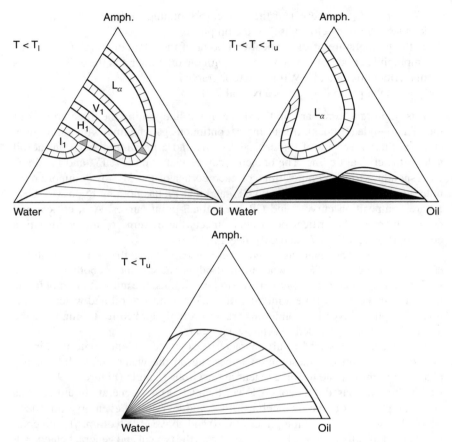

**FIG. 3.19** Ternary phase diagrams involving nonionic amphiphiles (amph) generating lyotropic liquid crystals with water. (From Kalhweit, M. and Strey, R., *Angew. Chem. Int. Ed. Engl.*, 24, 654, 1985. With permission.)

results in much less amphiphile being required to compatibilize equal amounts of water and oil, as illustrated in Figure 3.20.

## 2.   Effect of Oil

*(a)   Molar Volume.*   Increasing the oil molar volume results in an increase in $T_u$, $T_l$ (if they exist), and $\widetilde{T}$. This is illustrated in Figure 3.21 for aliphatic hydrocarbon, alkyl benzene, and alkanol acetate mixtures with water and diethylene glycol monobutyl ether (C4E2). Increasing the oil molar volume corresponds to increasing the oil–water incompatibility. Another result is an increase in the difference

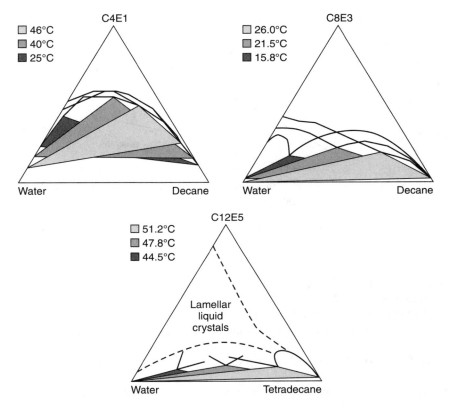

**FIG. 3.20** Effect of the molecular weight of an amphiphile on the shape of the ternary phase diagram. (From Kalhweit, M., Strey, R., and Haase, D., *J. Phys. Chem.*, 89, 163, 1985. With permission.)

between $T_u$ and $T_l$. This is illustrated by the alkyl benzene series in Figure 3.21, which presents a tricritical point for an alkyl chain length between six and seven.

*(b)  Polarity.*  Because increasing the oil polarity (by moving from aliphatic hydrocarbons to alkyl benzenes to alkanol acetates) decreases the water–oil incompatibility, it is not surprising that the Winsor transitions ($T_u$, $T_l$, and $\widetilde{T}$) occur at lower temperatures and the supertricriticality decreases.

The polarity of the oil can be estimated from Hansen's three-dimensional solubility parameters. Hansen separated Hildebrand's solubility parameter into three independent components: $\delta_d$ for the dispersion contribution, $\delta_p$ for the polar contribution, and $\delta_h$ for the H-bonding contribution. As an estimation of the oil polarity, we define $D_{ph}$ as the square root of the square of the polar component plus the

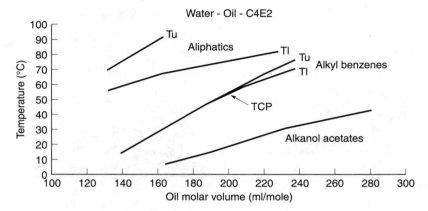

**FIG. 3.21** Effects of oil molar volume and polarity on the characteristics of ternary phase diagrams obtained with water and diethylene glycol butyl ether.

square of the H-bonding component of the solubility parameter, using the table published by Barton [13].

For ternary mixtures of alkanes, alkyl benzenes, or alkanol acetates with water and C4E2, the phase inversion temperature can be satisfactorily expressed as

$$\widetilde{T}(°C) = 1.016V - 0.00121V^2 - 44.72D_{ph} + 4.747D_{ph}^2 - 2.74$$

where $V$ is the oil molar volume in ml/mol and $D_{ph}$ is the oil polar character in MPa$^{1/2}$. The effects of the molar volume and of the polarity appear to be independent, at least in this specific case.

## 3. Effect of Electrolytes

It is possible to modify the behavior of water by adding electrolytes. Electrolytes usually reduce the solubility of uncharged organic components in water. Although the great majority of electrolytes exert a salting-out action, several exceptions exist, such as perchlorates and thiocyanates, which have a salting-in action.

The salting-out or salting-in characteristics of electrolytes were discovered at the end of the nineteenth century by Hofmeister [14] and essentially remain a mystery. It has been established that this effect has nothing to do with ionic strength: different salts at the same ionic strength have different salting-out or salting-in characteristics. Besides, unlike a classic electrostatic effect, it is almost linear with salt concentration (at least in the low concentration range).

In water, the effects of anions are much more pronounced than the effects of cations. The sequence of anions for increasing salting-out character in aqueous solutions is as follows: nitrate < chloride < carbonate < chromate < sulfate.

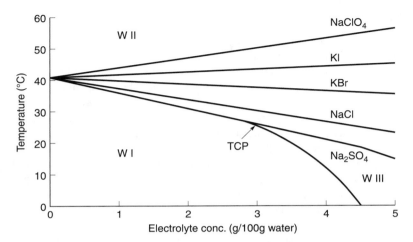

**FIG. 3.22** Effect of the nature and concentration of electrolytes on the characteristics of ternary phase diagrams obtained with water, diethylene glycol butyl ether, and tridecanol acetate.

Perchlorates and thiocyanates are salting-in electrolytes. A salting-out electrolyte strengthens the structure of water and makes it less available to hydrate organic molecules; salting-in electrolytes disrupt the structure of water, creating "holes." Salting-in electrolytes usually have a positive enthalpy of solubilization in water (endothermic solubilization).

In practice, salting-out electrolytes make water even more incompatible with oil. The result is a decrease in the Winsor transition temperatures and an increase in the supertricritical character. The amount of amphiphile necessary to compatibilize water and oil generally increases in the presence of a salting-out electrolyte. All these tendencies are reversed with a salting-in electrolyte. Figure 3.22 illustrates the effects of different electrolytes on a C4E2–C13 acetate–water system.

## 4. Effect of Low-Molecular-Mass, Water-Soluble Organic Molecules

Water can be made less incompatible with oil by adding small, uncharged, water-soluble organic molecules, such as amides and substituted ureas [15].

High salting-in performance is obtained with organic molecules with the following characteristics:

- Amphiphilic structure
- Hydrophobic segment short enough to prevent aggregation
- Concentration well below the solubility limit

Typical examples are urea derivatives, especially butyl urea.

## V. EMULSIONS

Unlike microemulsions, emulsions are not thermodynamically stable. When a mixture of oil (such as decane) and water is shaken, oil molecules come into contact with water. The area of contact between oil and water increases. As the two liquids are not miscible, increasing the area of contact results in an increased energy. In the case of decane, creating 1 $m^2$ of contact with water requires 0.050 J. This does not appear to be very much; however, if one considers one liter of a 50:50 volume mixture, creating a dispersion of 10 $\mu$m droplets requires 15 J. The net result is that water and oil will rapidly separate.

Adding a well-chosen surfactant can reduce the rate of separation between oil and water. Indeed, many of the surfactant molecules will be located at the oil/water interface, reducing significantly the interfacial tension to the order of 1 mN/m. The energy required to create the interface will be accordingly reduced. However, the role of surfactants is not limited to interfacial tension reduction. By adsorbing at the surface of the droplets, a surfactant will create a protecting barrier, which will significantly increase the droplet lifetime.

## A. Emulsion Instability and Breakdown

One can distinguish two different types of instabilities. There are instabilities resulting from thermodynamics, such as flocculation, coalescence, and Ostwald ripening, and there are those resulting from gravity, such as sedimentation, creaming, and coacervation.

## 1. Flocculation and Coalescence

Let us consider an emulsion of an oil (e.g., decane) in water. The oil droplets are constantly in motion due to thermal agitation or convection. In the course of their movements they may collide with each other. The collision may be perfectly elastic, in which case only the velocities of the droplets change, but their size or number do not. The collision, however, may not be elastic. The droplets may stick together, leading to flocculation. In a flocculation process the number of droplets remains constant, but they are no longer independent: they move together. In fact, flocculation involves many droplets, leading to large structures named flocs, which may even create a three-dimensional network. This can result in a viscosity increase, which may lead to a paste or a gel state.

Flocculation, by itself, can be reversible. Agitation may be sufficient to redisperse the droplets. However, when in close contact, the wall separating two droplets may break, allowing the droplets to merge into a single, bigger droplet. This phenomenon is referred to as coalescence and is not reversible. It results in a drift of the particle size distribution toward larger values, and may even lead to total phase separation.

Flocculation and coalescence can be avoided by preventing droplets approaching each other. In fact, flocculation (which may lead to coalescence) is the result of the van der Waals forces between two droplets. These forces act at relatively short distances, and are always attractive. There are two ways to counteract these forces: electrostatic repulsion and steric repulsion.

*(a) Electrostatic Repulsion.* The creation of electrical charges on droplets induces electrostatic repulsion forces that act at a longer distance than van der Waals forces. If strong enough, the electrostatic repulsion can offset van der Waals attraction, which means that the particles will repel each other. Ionic surfactants or polymers are commonly used to achieve electrostatic repulsion.

Electrostatic repulsion has a limitation. It works only for systems that do not contain large quantities of electrolytes. Indeed, the presence of electrolytes reduces the so-called Debye length, which is basically the distance at which electrostatic repulsion is effective. Electrolytes also compress the electrical double layer. The result is a reduction of the electrostatic repulsion, which may become weaker than van der Waals attraction.

*(b) Steric Repulsion.* Rather than creating electrical charges at the surface of droplets, it is possible to adsorb a water-soluble polymer. This polymer will expand in the aqueous phase to a certain distance. When two particles covered with polymer approach each other the polymer chains are compressed. This compression results in loss of configurational entropy. Besides, in the area where the polymer shells overlap, due to compression the local polymer concentration is higher. Osmotic pressure will pull the particles apart.

The selection of polymer is critical. If too water soluble, the polymer will not adsorb very well on the droplet surface. If not hydrophilic enough, the polymer will lie flat on the surface, so that van der Waals attraction can again take place.

The molecular weight of the polymer is also important. A polymer of too low a molecular weigh will not be efficient. The use of a high-molecular-weight polymer is better, but its concentration has to be high enough to cover the surface of all the droplets. If the amount of polymer is not high enough the same polymer molecule can anchor onto two different droplets, leading to the phenomenon known as bridging flocculation.

Properly selected nonionic surfactants can be good candidates to stabilize emulsions through steric repulsion.

## 2. Ostwald Ripening

Ostwald ripening also leads to a shift of particle size distribution toward higher values, but the mechanism is fundamentally different. Let us consider two droplets of different sizes located close together but not in contact. The Laplace pressure, which is equal to twice the interfacial tension divided by the droplet radius, is higher in the smaller particle. Now, if the oil has a nonzero solubility in water (which is the case for decane: its solubility in water is very small, but finite), diffusion will occur from the smaller particle toward the larger one through the continuous aqueous phase. The result is a growth of the larger particle at the expense of the smaller one.

Ostwald ripening obviously depends on particle polydispersity. If all the particles have the same size, there is no reason for one to grow at the expense of another. Ostwald ripening is also a function of the solubility of the oil in water and of the diffusion coefficient. This provides an excellent means to reduce its effect. The addition of a moderate amount of a water-insoluble oil such as a triglyceride is usually enough to reduce the impact of this destabilization mechanism.

## 3. Concentration Depletion

It may be logical to think that the higher the surfactant concentration, the more stable the emulsion. This is not always true. Indeed, a surfactant in excess forms micelles, which are significantly smaller than the emulsion droplets. Droplets are of the order of a micrometer and micelles are 5 to 100 nm. Droplets are surrounded by numerous micelles which bombard them constantly due to Brownian motion. When two droplets happen to be close to each other the collisions of the micelles are no longer isotropic. There are fewer micelles between the two droplets. The result from the unbalanced collisions is that the droplets are actually brought into contact.

This phenomenon is highly sensitive to droplet size and it is sometimes used as a method to prepare homodisperse emulsions by fractionation.

## 4. Sedimentation and Creaming

Sedimentation or creaming, depending on the relative densities of the oil and water phases, results from the action of gravity on the droplets. Under a gravitational field a spherical droplet will accelerate until it reaches a velocity for which the friction force balances the gravitational force. At this point the particle will move at a constant velocity $v$ predicted by Stokes' law:

$$v = \frac{\Delta \rho r^2}{6\pi \eta}$$

where $\Delta \rho$ is the density difference between the oil and water phases, $r$ is the radius of the supposed spherical droplet, and $\eta$ is the viscosity of the continuous phase.

Stability can be reached if the density of the dispersed phase exactly matches that of the continuous phase. This is difficult to achieve in practice as the volumic expansion coefficients are different for oil and water. Density matching accordingly holds only at one temperature. Decreasing the particle size reduces the separation rate but does not stop it. Moreover, particle size reduction commonly leads to a viscosity increase, and even to gelling. Increasing the viscosity contributes to separation reduction but it is not very easy to achieve in practice.

Good physical stability can, however, be obtained by developing a viscoelastic network in the continuous phase. The elastic component acts as a net that prevents the droplets from settling or creaming. Viscoelastic networks can be obtained with high-molecular-weight water-soluble polymers or lyotropic liquid crystals.

## 5. Coacervation

The electrostatic repulsion discussed above may be strong enough to prevent particles from coming into close contact but still be too weak to maintain the droplets far enough away from each other to counteract the effect of gravitation. The result is the sedimentation (or creaming) of some of the droplets. Unlike flocculation, coacervation is very easily reversible by simple agitation. Sometimes even convection currents are enough to redisperse the droplets.

Coacervation is usually observed in electrostatically stabilized systems in which the electrolyte concentration is slightly too high.

# VI. NANOEMULSIONS

The term "nanoemulsion" naturally creates confusion with the term "microemulsion." One may think that nanoemulsions have droplets smaller than microemulsions, since the prefix "nano" indicates a quantity three orders of magnitude smaller than a quantity indicated by the prefix "micro." As mentioned earlier in the chapter, the term microemulsion is not well chosen, but it is too well established to change it.

Microemulsions are thermodynamically stable phases, which can be represented by clear areas in equilibrium phase diagrams. Nanoemulsions are really small emulsions, with the main characteristics of emulsions: they are not thermodynamically stable and the way they are prepared has a great impact on their physical stability. The only difference with common emulsions is their very small droplet size, which ranges from 10 to 500 nm. Accordingly, nanoemulsions may look bluish, due to light diffusion (brown/yellow by transmission), just like microemulsions close to a critical point.

In contrast to thermodynamically stable microemulsions, nanoemulsions can be highly efficient in releasing oily materials. Indeed, they are highly metastable: the droplet size is small, but the interfacial tension is not so small. This results in the Laplace pressure inside the droplets being very high. Metastability is due to the activation energy required for two droplets to merge.

There are essentially two ways to prepare nanoemulsions. These are the phase inversion temperature (PIT) process and the high-pressure homogenization (HPH) process.

## A. PIT Process

A regular emulsion is prepared with a surfactant that is mainly water soluble, at a temperature lower than the PIT of the system. The emulsion is heated to the PIT.

At this temperature the interfacial tension is very small, and a very limited mechanical energy is required to mix thoroughly the ingredients. (Note that it is not necessary to add enough surfactant to reach the middle phase.) The temperature is then rapidly reduced to room temperature. The small droplets are accordingly "frozen" in their state before the rapid temperature reduction.

Unfortunately, the droplet size distribution of a nanoemulsion prepared by the PIT process is relatively large. Due to the high Laplace pressure, Ostwald ripening takes place rapidly, limiting the lifetime of the nanoemulsions to a few minutes to a few days. The addition of a water-insoluble component can significantly reduce the breakdown kinetics; however, long-term stability is rarely achieved with this process.

## B.  HPH Process

A high-pressure homogenizer is an instrument able to generate high-speed collisions between the droplets of a preformed emulsion. The result of these collisions is the production of very small (nanometric) droplets. If the process conditions are carefully optimized, narrow droplet size distributions can be obtained, and the addition of a water-insoluble oil can largely overcome Ostwald ripening.

The drawbacks of the HPH process are the expensive investment required, the constant attention of a highly skilled engineer during the operation, and the delicate cleaning and sanitization of the production line.

## REFERENCES

1.  Griffiths, R.B. and Wheeler, J.C., *Phys. Rev.*, A2, 1047–1064, 1970.
2.  Ricci, J.E., *The Phase Rule and Heterogeneous Equilibrium*, Van Nostrand, New York, 1951.
3.  Krafft, F., *Ber. Deutsche Chern. Ges.*, 32, 1596, 1899.
4.  Shinoda, K., *Solvent Properties of Surfactant Solutions*, Surfactant Science Series, Vol. 2, Marcel Dekker, New York, 1967, p. 12.
5.  Kekicheff, P., Gabrielle-Madelmont, C., and Ollivon, M., *J. Colloid Int. Sci.*, 131, 112, 1989.
6.  Ekwall, P., Mandell, L., and Fontell, K., *Mol. Cryst. Liq. Cryst.*, 8, 157, 1969.
7.  Friman, R., Danielsson, I., and Stenius, P., *J. Coll. Interface Sci.*, 86, 501, 1982.
8.  Kalhweit, M., Lessner, E., and Strey, R., *J. Phys. Chem.*, 87, 5032, 1983.
9.  Kalhweit, M., Lessner, E., and Strey, R., *J. Phys. Chem.*, 88, 1937, 1984.
10.  Kalhweit, M., Strey, R., and Haase, D., *J. Phys. Chem.*, 89, 163, 1985.
11.  Kalhweit, M. and Strey, R., *Angew. Chem. Int. Ed. Engl.*, 24, 654, 1985.
12.  Mitchell, D.J., Tiddy, G.I.T., Waring, L., Bostock, T., and McDonals, M.P., *J. Chem. Soc., Faraday Trans. 1*, 79, 975, 1983.
13.  Barton, A.F.M., *Handbook of Solubility Parameters and Other Cohesion Parameters*, CRC Press, Boca Raton, FL, 1983.
14.  Hofmeister, F., *Arch. Exp. Patal. Pharmackol.*, 24, 247, 1888.
15.  Broze, G., *Comm. J. Com. Esp. Deterg.*, Barcelona, 20, 133, 1989.

# 4
# Rheology of Liquid Detergents

**R.S. ROUNDS**   Fluid Dynamics, Inc., Flemington, New Jersey

## I. INTRODUCTION

Liquid detergent products span the entire rheology spectrum from low-viscosity Newtonian fluids to semisolid pastes, as demonstrated in Figure 4.1. Consumers can readily purchase the product form most in keeping with their preferences throughout the personal and household care product lines. For example, shampoos can be easily found that are low-viscosity Newtonian solutions, viscoelastic dispersions, or highly elastic gels. Similarly, laundry detergent products range in consistency and form from low-viscosity Newtonian liquids, to viscous pastes, to solid tablets. Developing these products to yield the desired shelf life and rheology stability is a complex task considering the number of components included in final commercial formulations.

Research scientists and engineers involved in successful development and manufacture of commercial products have different rheology needs. Advanced technology emphasis may be on fundamental studies of interactions of product

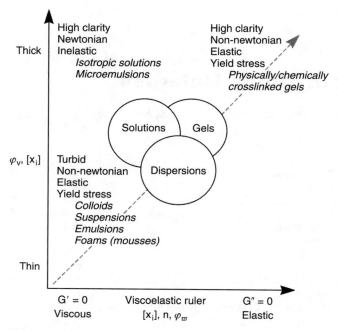

**FIG. 4.1** Liquid detergent product delivery forms and rheology spectrum. (Courtesy of Fluid Dynamics, Inc.)

ingredients and phase behavior of multicomponent surfactant systems, for example, while product formulators may need to benchmark rapidly the rheology of market prototypes and competitive products. Testing conditions for process engineers may extend rheology measurements to high shear rates and temperatures in keeping with process conditions for surfactants and/or product manufacture. Further, quality control professionals generally require test protocols for finished goods at factory locations. In response to these highly diversified needs, instrument manufacturers have produced a broad range of rheometers and viscometers.

From R&D to quality control, rheology measurements for each phase of the product development life cycle involve raw materials, premixes, solutions, dispersions, emulsions, and full formulations. Well-equipped laboratories with stress- and strain-controlled oscillatory/steady shear rheometers and viscometers can generally satisfy most characterization needs. When necessary, customized systems are designed to simulate specific user or process conditions. Rheology measurements are also coupled with optic, thermal, dielectric, and other analytical methods to further probe the internal microstucture of surfactant systems. New commercial and research developments are briefly discussed in the following sections.

Rheology is frequently cited in patents, often as a claim, and Section II provides examples of liquid detergent patents recently issued where rheology is cited as a key property. Instruments and methods used for product characterizations in several patents are included. Section III highlights new developments in rheology measurements, also listing several patents for new rheology measurement technology, since there have been dramatic advances in the science of rheology over the past decade. Examples of the flow behavior of commercial detergent products, including laundry detergents, shampoos, dishwashing liquids, and dentifrices, are included in Section IV and selections of fundamental rheology studies in surfactant systems are presented in Section V.

This chapter is intended to expand on the chapter of the same title in the first edition of this book [1]. An overview of the rheology of liquid detergents, including dispersions, suspensions, gels, and surfactant systems, is included in that chapter.

## II. PATENT SURVEY OF LIQUID DETERGENT FORMULATIONS

Personal and household care detergents have applications ranging from hard surface cleaners, to body washes, to dental pastes and gels. Products used by consumers need to be poured, pumped, squeezed, or sprayed while maintaining constant rheology profiles throughout the shelf life. Since rheology is a performance and consumer perceived property, patents describe compositions and manufacturing procedures needed to achieve desirable rheological properties and shelf life stability. This applies also to surfactants and admixtures during detergent manufacture, since these fluids need to be easily processed.

A representative catalog of recent patents is provided in Table 4.1 in chronological order. Each of these patents focuses attention on flow properties of key raw materials or full formulations. For high- and low-pH formulations, patents describe rheology modifiers meeting the demands of these difficult systems. Various rheology modifiers are also disclosed for the purpose of targeting specific rheology requirements. Several of the patents listed in Table 4.1 are discussed in the following sections.

### A. Home Care Products

Numerous patents have been granted for automatic dishwasher detergents, hard surface cleaners, and laundry detergents. Of special interest are patents concerning liquid compositions containing bleach, due to the problems encountered with the rheology and chemical and physical stability of these complex systems. Increasingly, patents describe "thickened" compositions and specifically cite viscoelastic behavior. A stable perfumed bleaching composition is described in U.S. Patent 6,248,705 for hard surface cleaning having a pH less than 2. The perfume cited is

**TABLE 4.1** Summary of Patents Relating to Rheology of Liquid Detergents

| Patent number and date | Assignee and inventors | Title |
|---|---|---|
| U.S. 6,576,602, June 10, 2003 | Procter & Gamble Company M.A. Smerznak, W.A.M. Broeckx, I. Goderis, R. Jones, D. Parry, J. Kahn, J. Wevers | Nonaqueous, particulate-containing liquid detergent compositions with surfactant-structured liquid phase |
| U.S. 6,506,716, January 14, 2003 | Procter & Gamble Company P. Delplancke, F. de Buzzaccarini, A. Fredj, P. Reddy, R. oswell, E. Sadlowski | Aqueous, gel laundry detergent composition |
| U.S. 6,331,291, December 18, 2001 | W.R. Glace, R.L. Ibsen, G.A. Skoler | Dentifrice gel/paste compositions |
| U.S. 6,313,085, November 6, 2001 | Cognis Deutschland GmbH C. Le Hen-Ferrenbach | High-concentration flowable anionic surfactant mixtures containing alkyl ether sulfates and alkyl sulfates |
| U.S. 6,306,916, October 23, 2001 | Henkel Kommanditgesellschaft auf Aktien A. Ansmann, R. Kawa, G. Strauss | Pearly luster concentrate with Newtonian viscosity |
| U.S. 6,294,511, September 25, 2001 | Clorox Company B.P. Argo, C.K. Choy, S.L. Nelson | Thickened aqueous composition for the cleaning of a ceramic surface and methods of preparation thereof and cleaning therewith |
| U.S. 6,274,539, August 14, 2001 | Procter & Gamble Company M.L. Kacher, D.P. Wallace, F.S. Allouch | Light-duty liquid or gel dishwashing detergent compositions having controller pH and desirable food soil removal, rheological, and sudsing characteristics |
| U.S. 6,274,546, August 14, 2001 | Henkel Kommanditgesellschaft auf Aktien D. Legel, J. Penninger, T. Voelkel | Stable high-viscosity liquid detergents |

| Patent/Date | Assignee / Inventors | Title |
|---|---|---|
| U.S. 6,271,187, August 7, 2001 | Ecolab Inc. C.A. Hodge, C.J. Uecker | Hand soap concentrate, use solution, and method for modifying a hand soap concentrate |
| U.S. 6,271,192, August 7, 2001 | National Starch &Chemical Investment Holding Co. E.W. Verstrat, J.S. Maxim, Jr., J. Rosie | Associative thickener for aqueous fabric softener |
| U.S. 6,277,798, August 21, 2001 | Procter & Gamble Company P.E. Russell, N.J. Phipps | Cleansing compositions containing water-soluble gel-forming nonionic surfactant |
| U.S. 6,258,859, July 10, 2001 | Rhodia, Inc. M. Dahayanake, J. Yang, J. G. Niu, P.-J. Derian, R. Li, D. Dino | Viscoelastic surfactant fluids and related methods of use |
| U.S. 6,248,705, June 19, 2001 | Procter & Gamble Company S. Cardola, L. Pieroni, R. Scoccianti | Stable perfumed bleaching compositions |
| U.S. 6,268,324, July 31, 2001 | Ecolab Inc. M.E. Besse, R.O. Ruhr, G.K. Wichmann, T.A. Gutzmann | Thickened hard surface cleaner |
| U.S. 6,241,812, June 5, 2001 | Pharmacia Corporation B.A. Smith, G.T. Colegrove, W.G. Rakitsky | Acid-stable and cationic-compatible cellulose compositions and methods of preparation |
| U.S. 6,221,827, April 24, 2001 | Henkel Kommanditgesellschaft auf Aktien M. Mendoza Cruz, E. de Jorge | Viscoelastic bleaching and disinfecting compositions |
| EP1088545, April 4, 2001 | Procter & Gamble Company G.N. McKelvey, K. Rigal | Hair care compositions |

*(continued)*

**TABLE 4.1** (Contd.)

| Patent number and date | Assignee and inventors | Title |
|---|---|---|
| U.S. 6,187,221 B1, February 13, 2001 | National Starch & Chemical Investment Holding Co. C.G. Gore, S.M. Steele | Controlled release bleach thickening composition having enhanced viscosity stability at elevated temperatures |
| U.S. 6,180,594 B1, January 30, 2001 | Witco Surfactants GmbH M. Fender, K. Hans-Jurgen, S. Schussler | Low-concentration, high-viscosity aqueous fabric softeners |
| U.S. 6,177,396 B1, January 23, 2001 | Albright & Wilson UK Limited R.M. Clapperton, J.R. Goulding, B.W.Grover, I.F. Guthrie, W.P. Haslop, E.T. Messenger, J.E. Newton, S.A. Warburton | Aqueous-based surfactant compositions |
| U.S. 6,150,320, November 21, 2000 | 3M Innovative Properties Company J. McDonell, J. Mlinar | Concentrated cleaner compositions capable of viscosity increase upon dilution |
| U.S. 6,150,445, November 21, 2000 | Akzo Nobel AV P. Bostrom, A. Myrstrom | Aqueous concentrate of an associative thickening polymer, and use of a nonionic surfactant for reducing the viscosity of the concentrate |
| U.S. 6,140,413, October 31, 2000 | Henkel Corporation L.N. Castles, S.C. James, J. Stewart | Silicone softener viscosity reducer |
| U.S. 6,126,922, October 3, 2000 | 3M Innovative Properties Company B. Wang, S.B. Mitra, S.M. Rozzi | Fluid-releasing compositions and compositions with improved rheology |
| WO0046331, August 10, 2000 | Procter & Gamble J.M. Clarks, G.K. Embleton, H.D. Hutton, J.D. Sadler, M.L. Kacher, D.P. Wallace | Diols and polymeric glycols in dishwashing detergent compositions |

| | | |
|---|---|---|
| U.S. 6,100,228, August 8, 2000 | Clorox Company B.P. Argo, C.K. Choy, A. Garabedian, Jr. | Bleaching gel cleaner thickened with amine oxide, soap, and solvent |
| U.S. 6,087,320, July 11, 2000 | Henkel Corporation A.D. Urfer, V. Lazarowitz, P.E. Bator, B.A. Salka, G. de Goederen, R.A. Alaksejczyk | Viscosity-adjusted surfactant concentrate compositions |
| U.S. 6,083,893, July 4, 2000 | Procter & Gamble Company D.R. Zint, T. Pace, R. Owens, M.L. Kacher | Shaped semisolid or solid dishwashing detergent |
| U.S. 6,083,854, July 4, 2000 | Procter & Gamble Company M.S. Bogdanski, U.C. Glaser | Wet wipes with low-viscosity silicone emulsion systems |
| WO0015180, March 23, 2000 | Hercules Inc. J.E. Brady | Rheology-modified compositions and processes thereof |
| U.S. 6,028,043, February 22, 2000 | Procter & Gamble Company R.W. Glenn, Jr., M.D. Evans, M.E. Carethers, S.C. Heilshorn | Liquid personal cleansing compositions which contain a complex coacervate for improved sensory perception |
| U.S. 6,008,261, December 8, 1999 | Condea Augusta SpA C. Genova, F. Montesion, E. Bozzeda | Aqueous surfactant compositions with a high viscosity |
| U.S. 6,008,184, December 28, 1999 | Procter & Gamble Company J.G.L. Pluyter, M.G. Eeckhout | Block copolymers for improved viscosity stability in concentrated fabric softeners |
| U.S. 5,997,764, December 7, 1999 | B F Goodrich Co. S.V. Kotian, H. Ambuter | Thickened bleach compositions |
| U.S. 5,939,375, August 17, 1999 | Th. Goldschmidt AG F. Muller, J. Peggau | Low-viscosity alkaline cleaning emulsion |

*(continued)*

**TABLE 4.1** (Contd.)

| Patent number and date | Assignee and inventors | Title |
|---|---|---|
| U.S. 5,932,538, August 3, 1999 | Procter & Gamble Company R.W. Glenn, Jr., M.D. Evans, M.E. Carethers, S.C. Heilshorn | Liquid personal cleansing compositions which contain an encapsulated lipophilic skin moisturizing agent comprised of relatively large droplets |
| U.S. 5,922,667, July 13, 1999 | Diversey Lever, Inc. E.C. van Baggem, N.J. Pritchard, G. de Goederen, R. Jakobs | Cleaning gels |
| U.S. 5,922,664, July 13, 1999 | Colgate-Palmolive Company H.C. Cao, P. Pagnoul | Pourable detergent concentrates which maintain or increase in viscosity after dilution with water |
| U.S. 5,851,979, December 22, 1998 | Procter & Gamble Company S. Scialla, S. Dardola, G.O. Boamcjetto | Pseudoplastic and thixotropic cleaning compositions with specifically defined viscosity profile |
| U.S. 5,981,457, November 9, 1999 | Kay Chemical Company F.U. Ahmed | Concentrated liquid gel warewash detergent |
| U.S. 5,965,502, October 12, 1999 | Huels Aktiengesellschaft D. Balzer | Aqueous viscoelastic surfactant solutions for hair and skin cleaning |
| U.S. 5,962,392, October 5, 1999 | Solvay Interox Limited | Thickened peracid compositions |
| U.S. 5,939,375, August 17, 1999 | Th. Goldschmidt AG F. Fuller, J. Peggau | Low-viscosity alkaline cleaning emulsion |
| U.S. 5,922,664, July 13, 1999 | Colgate-Palmolive Company H.C. Cao, P. Pagnoul | Pourable detergent concentrates which maintain or increase in viscosity after dilution with water |

| Patent | Company / Inventors | Description |
|---|---|---|
| U.S. 5,916,859, June 29, 1999 | Clorox Company<br>C.K. Choy, P.F. Reboa | Hexadecylamine oxide/counterion composition and method for developing extensional viscosity in cleaning compositions |
| U.S. 5,912,220, June 15, 1999 | S.C. Johnson & Son, Inc.<br>J.A. Sramek, H.A. Doumaux, T. Tungsubutra, P.J. Schroeder | Surfactant complex with associative polymeric thickener |
| U.S. 5,888,487, March 30, 1999 | Hankel Kommanditgesellschaft auf Aktien<br>G. Baumoeller, A. Wadle, C. Ansmann, H. Tesmann, T. Foerster | Low-viscosity opacifier concentrates |
| U.S. 5,851,979, December 22, 1998 | Procter & Gamble Company<br>S. Scialla, S. Cardola, G.O. Bianchetti | Pseudoplastic and thixotropic cleaning compositions with specifically defined viscosity profile |
| SK279419B, November 4, 1998 | Colgate-Palmolive Company<br>G.A. Durga, M. Prencipe | Viscoelastic dentifrice composition |
| U.S. 5,804,540, September 8, 1998 | Lever Brothers Company<br>L.S. Tsaur, M. He, M. Massaro, M.P. Aronson | Personal wash liquid composition comprising low-viscosity oils prethickened by nonantifoaming hydrophobic polymers |
| U.S. 5,811,383, September 22, 1998 | Dow Chemical Company<br>J. Klier, C.J. Tucker, G.M. Strandburg | High water content, low-viscosity, oil continuous microemulsions and their use in cleaning applications |
| U.S. 5,798,324, August 25, 1998 | S.C. Johnson & Son, Inc.<br>G.J. Svoboda | Glass cleaner with adjustable rheology |
| U.S. 5,776,883. July 7, 1998 | Lever Brothers Company<br>T.V. Vasudevan | Structured liquid detergent compositions containing nonionic structuring polymers providing enhanced shear thinning behavior |

*(continued)*

**TABLE 4.1** (Contd.)

| Patent number and date | Assignee and inventors | Title |
|---|---|---|
| U.S. 5,759,989, June 2, 1998 | Procter & Gamble Company S. Scialla, S. Cardola, G.O. Bianchetti | Stable aqueous emulsions of nonionic surfactants with a viscosity controlling agent |
| U.S. 5,733,861, March 31, 1998 | BASF Corporation S. Gopalkrishnan, K.M. Guiney, J.V. Sherman, D.T. Durocher, M.C. Welch | Hydrophilic copolymers for reducing the viscosity of detergent slurries |
| U.S. 5,728,665, March 17, 1998 | Clorox Company C.K.-M. Choy, B.P. Argo | Composition and method for developing extensional viscosity in cleaning compositions |
| U.S. 5,688,435, November 18, 1997 | Reckitt & Colman Inc. D.A. Chang, J.W. Cavanagh | Pigmented rheopectic cleaning compositions with thixotropic properties |
| U.S. 5,389,157, February 14, 1995 | Clorox Company W.L. Smith | Viscoelastic cleaning compositions with long relaxation times |
| U.S. 5,409,630, April 25, 1995 | Colgate-Palmolive Co. R. Lysy, M. Marchal | Thickened stable acidic microemulsion cleaning composition |
| U.S. 5,336,426, August 9, 1994 | J.E. Rader, W.L. Smith | Phase stable viscoelastic cleaning compositions |

a cyclic terpene/sesquiterpene compound, such as eucalyptol, added with cationic surfactants to yield the desired viscosity. An optimum rheology is claimed for vertical hard surface applications such as toilet bowl cleaners. Optimum viscosity is most preferably 250 or 900 cP at 20°C measured using a Brookfield viscometer at 60 r/min using Spindle no. 2 or with the Carri-Med viscometer at a shear stress of 50 dyn/cm$^2$. Thickened aqueous bleach cleaners containing hypohalite or peroxygen bleaches for hard surfaces are also the subject of U.S. Patent 5,997,764. Storage stability viscosity data at 20 r/min and 20°C are provided for example formulations containing viscosity stabilizers. A thickened bleach gel cleaner comprising hypochlorite-generating compounds, a ternary thickening system consisting of an alkali metal soap, hydrotrope, and bleach stable solvent, buffer/electrolyte stabilizer, and water is mentioned in U.S. Patent 6,100,228.

Viscoelastic non-Newtonian bleaching and disinfecting compositions are further cited in U.S. Patent 6,221,827. Brookfield RVT (Spindle no. 1, 60 r/min) viscosity data are included for several examples following storage for four weeks at 40°C. The compositions are noted to yield high stability in storage. A controlled release bleach thickening composition cited to have enhanced viscosity stability at higher temperatures is disclosed in U.S. Patent 6,187,221. The controlled release thickening composition contains halogen bleach, water, a crosslinked carboxylated polymer, and degradable crosslinking monomer. Brookfield viscosity of examples aged at 50°C is provided. Thickened peracid compositions are included in U.S. Patent 5,962,392 containing an aliphatic alcohol ethoxylate and an amine oxide cosurfactant. Brookfield viscosity data (Spindle no. 2, 50 r/min) are included in the patent text.

A rheopectic pigmented bleach (alkali metal hypochlorite) hard surface cleaner formulated with bentonite clay is disclosed in U.S. Patent 5,688,435. Examples of time-dependent shear effects determined from constant shear rate measurements at 1, 10, 50, and 100 sec$^{-1}$ are provided in the patent and shown in Figure 4.2 and Figure 4.3. The viscosity data show evidence of shear thickening as a function of time at constant shear rates of 1 and 10 sec$^{-1}$ and thixotropy occurs at 50 and 100 sec$^{-1}$. The formulation is rheopectic at 10 sec$^{-1}$. Dynamic mechanical data are also contained in the patent and the storage and loss modulus as a function of strain amplitude is shown in Figure 4.4, for one patent example.

Hypochlorite hard surface and drain cleaner compositions exhibiting enhanced extensional viscosity are mentioned in U.S. Patents 5,728,665 and 5,916,859. The viscoelastic compositions are intended for use with trigger sprayers and the hexadecyl amineoxide/organic counterion compositions provide low bleach odor and reduced spray misting. The patent contains extensional viscosity data in support of the claims. Viscosity as a function of shear rate at various $C_{16}$ diphenyloxide disulfonate concentrations is shown in Figure 4.5. Examples of steady shear and extensional viscosity as a function of shear rate and extensional rate are shown in Figure 4.6 and Figure 4.7.

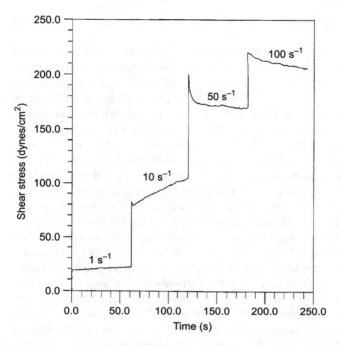

**FIG. 4.2** Step shear rate measurements of a bleach hard surface cleaner containing bentonite clay. (Reprinted from U.S. Patent 5,688,435.)

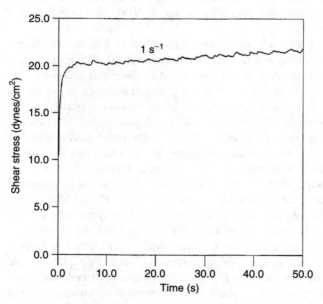

**FIG. 4.3** Step shear rate measurement of a bleach hard surface cleaner containing bentonite clay. (Reprinted from U.S. Patent 5,688,435.)

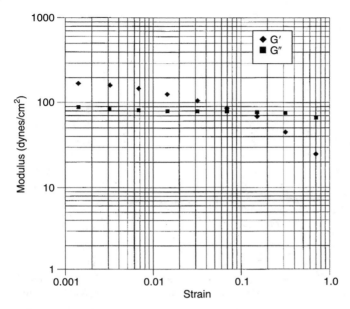

**FIG. 4.4**  Dynamic mechanical test results for bleach hard surface cleaner. (Reprinted from U.S. Patent 5,688,435.)

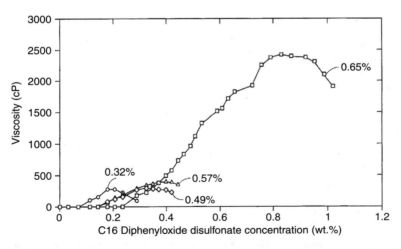

**FIG. 4.5**  Viscosity as a function $C_{16}$ diphenyloxide disulfonate concentration. (Reprinted from U.S. Patent 5,728,665.)

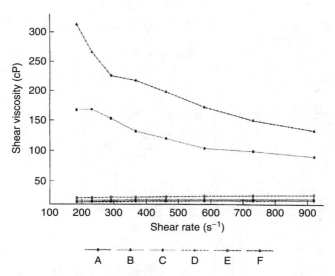

**FIG. 4.6** Viscosity differences between state-of-the-art Newtonian systems and the extensional system of an invention cited in a patent. (Reprinted from U.S. Patent 5,918,665.)

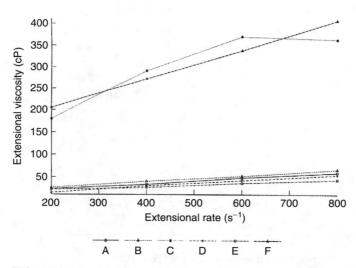

**FIG. 4.7** Extensional viscosity differences between state-of-the-art Newtonian systems and the extensional system of an invention cited in a patent. (Reprinted from U.S. Patent 5,918,665.)

Structured surfactant systems are also cited in patents. One example is a stable, pourable, spherulitic surfactant composition containing up to 80% by weight surfactant in U.S. Patent 6,177,396. This patent also contains a list of prior art worldwide patents using "structured surfactant" systems. Structured viscoelastic surfactant systems developed for suspending particles are cited in U.S. Patent 6,258,859, comprised of amphoteric/zwitterionic surfactants and their mixtures. The systems are also cited to increase effectively the average droplet size of sprays. The patent contains steady shear and dynamic mechanical data for disodium tallowiminodipropionate with phthalic acid, as well as disodium oleomidopropyl betaine solutions containing potassium chloride and phthalic acid. An acidic microemulsion cleaning composition, a viscoelastic gel, is described in U.S. Patent 5,409,630. The composition contains a sodium paraffin sulfonate, nonionic detergent, one aliphatic carboxylic acid, water-insoluble perfume, water, and an associative polymeric thickener. The pH is preferably in the range of 1 to 5 and dynamic mechanical data are cited in the patent as a function of strain and frequency.

U.S. Patent 6,268,324 describes low-viscosity cleaning compositions that increase in viscosity upon dilution. Thickening is attributed to rod-like micellar structuring. A shear thinning heavy-duty liquid containing 30 to 80% surfactants as lamellar drops dispersed in an aqueous medium is disclosed in U.S. Patent 5,776,883. Shear thinning behavior is cited using the Sisko model constants, $k$ and $n$, as well as pour viscosity ($21 \text{ sec}^{-1}$) and it states that $n$ should be less than 0.35, preferably less than 0.3. A nonaqueous liquid detergent composition containing a surfactant-structured continuous phase is disclosed in U.S. Patent 6,277,804. The "particulate containing detergent compositions" exhibit viscosity ranging from 300 to 5000 cP as measured with a Carrimed CSL2 rheometer at a shear rate of $20 \text{ sec}^{-1}$. Examples are cited in the patent for nonaqueous base systems giving yield values and pouring viscosity test results. The transformation of a detergent concentrate from micellar to lamellar phase in the presence of a water-soluble electrolyte produces an increase in viscosity upon dilution (U.S. Patent 5,922,664). The viscosity enhancement of an illustrative laundry detergent concentrate using potassium citrate as a function of concentration is provided in the patent. Enhancement is demonstrated with Brookfield viscosity measurements (Spindle nos. 1 and 2 at 25°C).

U.S. Patent 5,798,324 discloses a glass cleaner containing a synthetic polymer thickener in the presence of certain glycol ethers, nonionic surfactants, and linear alcohols that increases viscosity synergistically. Cited compositions exhibit optimal vertical cling and ease of use.

Light-duty or gel dishwashing compositions containing an alkyl ether sulfate-based anionic surfactant, polyhydroxy fatty acid amide nonionic surfactant, suds boosters/stabilizers, aqueous liquid carrier, pH control agent, and acrylic copolymer thickener are disclosed in U.S. Patent 6,274,539. Viscosity as determined using a Brookfield viscometer with RV no. 2 spindle at 1 r/min ranges from 800 to 1500 cP at 25°C.

A pseudoplastic and thixotropic cleaning composition is disclosed in U.S. Patent 5,851,979, which is suitable for both fabric and hard surface care. The viscosity values of the compositions range from 60 to 1500 cP at 12 r/min, 40 to 800 cP at 30 r/min, and 20 to 500 cP at 60 r/min (Spindle no. 2 and 20°C).

Concentrated compositions that increase in viscosity when diluted are described in U.S. Patent 6,150,320. A Bostwick consistometer is used for all viscosity measurements and equivalence is offered to Brookfield measurements using Spindle no. 1, at 60 r/min, and the Zahn viscometer, no. 1 cup. A low-viscosity hard surface cleaning emulsion, approximately 12 mPa s, is described in U.S. Patent 5,934,375 that increases in viscosity upon dilution with water to 800 to 1200 mPa s.

Pseudoplastic and thixotropic liquid detergents as emulsions are the subjects of U.S. Patent 5,851,979. Equilibrium viscosity values measured using a Brookfield viscometer with Spindle no. 2 at 20°C are cited at 12, 30, and 60 r/min. For one example containing hydrogen peroxide, the viscosity is 1020, 400, and 220 cP at 12, 30, and 60 r/min, respectively. Pseudoplasticity is clearly evident, as the viscosity decreases with increasing rotational speed.

## B.  Personal Care Products

Rheology is a product attribute frequently exploited in personal care products to create visual appeal to prospective consumers. Liquid products in transparent packaging may highlight the gel strength of the continuous phase with obvious suspension of the particulate phases. In certain instances, aeration may be introduced to emphasize the gel-like consistency of the product. Premium brand products may include stable suspensions captured within the gel matrix of encapsulated fragrances, moisturizers, exfoliating compounds, etc.

Because of the frequent use of personal care products by the consumer for hair, body, and skin care, rheological properties are designed to achieve product differentiation. Products are formulated to achieve efficacy within a definite matrix of rheological properties. Certain manufacturers lean toward lower viscosity systems, while others focus on a thicker, "richer" composition. Regardless of the rheology preference, formulators need to overcome obstacles to achieve robust product design, including variables such as pH and electrolyte concentration.

Aqueous viscoelastic surfactant solutions for hair and skin care are disclosed in U.S. Patent 5,965,502. Rheology conditions are specified for optimum flow behavior, in terms of the shear modulus as a function of temperature and pH. A representative graph of the storage and loss modulus as a function of angular frequency is presented in the patent and this is shown in Figure 4.8. Cited compositions contain anionic, betainic, and nonionic surfactants, electrolytes, and a water-soluble polymer. A nonionic gel personal cleanser is specified in U.S. Patent

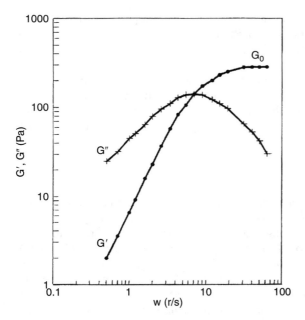

**FIG. 4.8**  Dynamic mechanical test results for a personal care product. (Reprinted from U.S. Patent 5,965,502.)

6,277,798. The high-viscosity composition has a viscosity (Helipath, Spindle A, 10 r/min and 25°C) ranging from 500 to 10,000 cP. Emulsion, moisturizing personal cleansing compositions containing a complex coacervate are disclosed in U.S. Patent 6,028,043 having a viscosity ranging from 2,000 to 100,000 cP and a yield point from about 5 to 50 dyn/cm$^2$. Patent rheology data cited are determined using the Carrimed CSL 100 controlled stress rheometer and the Wells-Brookfield cone/plate viscometer (2.4 cm cone). The instruments are used to determine the consistency and shear index, $k$ and $n$. The complex coacervate is characterized using the Stable MicroSystems Universal TA.XT2 texture analyzer. U.S. Patent 5,932,528 cites encapsulated lipophilic skin moisturizing agents. The Carrimed CSL 100 controlled stress rheometer is using to determine yield stress, as the amount of stress required to produce a strain of 1%.

Viscoelastic dentifrice compositions are disclosed in patents SK279419B and PL169998B. Hair care compositions comprising at least one associative polymer are disclosed in patent EP1088545. They are said to be easy to dispense and apply to the hair, having an excellent rheology profile.

A skin cleansing and moisturizing composition is disclosed in U.S. Patent 5,804,504. Low-viscosity oils having a viscosity less than 1000 cP prethickened with hydrophobic polymers having a low degree of crystallinity are used to deliver skin benefits without compromising foaming. A hand soap concentrate having a viscosity of 200 cP is disclosed in U.S. Patent 6,271,187 that increases in viscosity when mixed with an aqueous solution. Preferred thickeners for this application are polyalkylene ether diesters.

A nonaqueous dentifrice gel and/or paste composition is disclosed in U.S. Patent 6,331,291 B1, cited as a thixotropic and smooth-flowing substance. Brookfield viscosity values are reported in the patent for illustrative examples of the invention using Spindle no. 6 at 10 r/min and 23.5°C. The preferred viscosity range is 75,000 to 150,000 cP.

## C.   Actives

Modifying the rheological behavior of high-concentration surfactants is desirable for various reasons and this is also reflected in the patent literature. One example is U.S. Patent 6,313,085 involving high-concentration anionic surfactant mixtures of alkyl ether sulfate (60 to 90%) and alkyl sulfates (10 to 40% alkyl sulfate). This patent defines "flowable" by means of a Brookfield viscosity, as measured with an RVT instrument, 20°C, 10 r/min, Spindle no. 1. To be flowable, the patent states that the viscosity is less than 50,000 mPa s, preferably less than 10,000 mPa s. A pumpable, flowable, and pourable surfactant concentration is disclosed in U.S. Patent 6,087,320 for an aqueous blend of alkylpolyglycoside (70%) and anionic or amphoteric surfactants (30%) in the presence of inorganic and/or organic electrolytes. Viscosity determinations are included obtained using a viscometer at 25°C with Spindle no. 4, at 10 r/min.

Hydrophilic copolymers that reduce the viscosity of detergent slurries are disclosed in U.S. Patent 5,733,861. Viscosity-reducing properties are illustrated using data obtained from a Brookfield viscometer, Spindle no. 4, 20 r/min, at 25°C. The copolymer comprises an unsaturated hydrophilic monomer copolymerized with an oxyethylated monomer.

The process for producing detergent agglomerates from high active surfactant pastes is discussed in U.S. Patent 5,574,005. The surfactant paste is identified as having nonlinear viscoelastic properties, described as a power law fluid. An example of paste characterization is provided using a stress-controlled rheometer with truncated 2° cone (4 cm in diameter) and solvent vapor trap. A schematic of the rheometer tooling is shown in the patent with a shear stress–shear rate diagram for the paste where "shear fracture" is evidenced. Using a Carri-Med rheometer, a stress ramp from 5 to 5000 dyn/cm$^2$ is applied over a three-minute period.

It is very apparent that a great deal of effort is put in by research groups in defining the relevant mechanical properties of personal and home care liquid detergent formulations. Throughout the industry it is apparent that more rigorous characterization methods are being applied and included in product definitions comprising corporate patent portfolios.

## III.  RHEOLOGY MEASUREMENTS
## A.  Measurement Technology

During the past decade there has been a surge of technical developments in the rheometer industry and this is reflected in worldwide patents. Several examples are provided below:

WO0169231, Method of Fluid Rheology Characterization and Apparatus
U.S. 6,378,357, Method of Fluid Rheology Characterization and Apparatus
U.S. 5,456,105, Rheometer for Determining Extensional Elasticity
U.S. 5,532,289, Apparatus and Method for the Study of Liquid–Liquid Interfacial Rheology
U.S. 6,200,022, Method and Apparatus for Localized Dynamic Mechano-Thermal Analysis with Scanning Probe Microscopy
U.S. 5,543,594, Nuclear Magnetic Resonance Imaging Rheometer
U.S. 6,220,083, Elongational Rheometer and On-Line Process Controller
U.S. 5,520,042, Apparatus and Method for the Simultaneous Measurements of Rheological and Thermal Characteristics of Materials and Measuring Cell

Rheology is increasingly being coupled to other analytical test methods for more comprehensive material characterizations. Many of these developments are driven by research needs for broadened characterization capability. For fundamental studies of detergent systems this offers a broad suite of methods to probe surfactant mesophases and internal microstructure.

An overview of the viscometer and rheometer market from 1969 to 1999 is given by Barnes *et al.* [2] and examples of other advances are addressed in the technical literature [3–16]. Concurrent with new developments in rheology instrumentation are both introductory and advanced texts on rheology for industrial scientists [17–22].

Flow visualization, conductivity, turbidity, and light scattering can be simultaneously conducted with rheology measurements. Small-angle light scattering (SALS) coupled to rheology measurements is provided by Paar Physica (Rheo-SALS). Using a modular design concept, the SALS system is an add-on accessory to the research rheometer using concentric cylinder, parallel plate geometries. The laser wavelength is 658 nm and the maximum scattering angle for the concentric cylinder geometry is 11.3°.

GBC Scientific Equipment offers a Micro Fourier Rheometer, MFR 2100. The rheometer applies a squeezing motion to the sample, performing analyses on sample volumes less than 100 µl. An automated sample injection system is included in the instrument design. The rheometer is capable of handling low-viscosity fluids, 1 mPa s, with storage modulus measurement down to $10^{-4}$ Pa. Using a different measurement technique, both benchtop and in-line, the real-time ultrasonic rheometer and fluid characterization device uses spatially resolved ultrasonic

Doppler velocimetry techniques to monitor rheology of fluids and slurries (Pacific Northwest National Laboratory, Richland, WA).

Brookfield Engineering Laboratories has recently introduced a stress-controlled and yield stress viscometer, and the Thermo Haake RheoScope1 includes optical microscopy with video accessory for the cone and plate rheometer. With the Rheo-Scope1, rheology measurements are integrated with image/video acquisition. This accessory permits flow visualization during rheology measurements to observe and document shear-induced microstructural changes.

Although there have been steady advances in rheology measurement technology, not all areas have been equally addressed. One of the most difficult is the facile transition of characterization tools from R&D laboratories to the factory floor. For viscoelastic compositions, frequently encountered in personal and household care products, this presents a challenge to both R&D and production facilities. Further, rotational devices are limited in the shear rate and shear stress operating range. For process simulations, high-shear measurement tools are not readily available in the appropriate viscosity range. Several additional needs are discussed in the following sections.

## B.  Quality Control Metrics

"Simple" rheology measurements appear to be the measurements of choice in many industrial settings. Relative consistency indices are used routinely in both product development and manufacturing facilities as benchmarks, regardless of the complex nature of the fluids under consideration, obtained using analog or digital rotational devices. For structured detergents with yield stresses, the vane tool is more widely accepted and other characterization methods are frequently applied [23–26]. For many R&D and quality control (QC) applications, viscometry is still the principal characterization tool and several review articles discuss the use of rotational instruments in QC applications [27,28].

Texture analyzers are also used to assess deformability of a fluid, using penetration force vs. depth profiles, etc. These instruments in addition to Brookfield and Haake viscometers are common QC metrics. Other methods include viscosity flow cups and bubble or falling ball viscometers, and several relevant standard test methods include ASTM D1200, DIN/ISO 2431, ASTM D5125, BS3900:Part A6, ASTM D1545, and ASTM D1725.

For viscoelastic liquid detergent systems, oscillatory measurements may be more appropriate for QC applications. A benchtop, portable QC oscillatory instrument providing storage and loss modulus, complex viscosity values as a function of time and/or temperature, known as the T2SR® (Fluid Dynamics, Inc.) time/temperature scanning rheometer, has recently been introduced. The instrument uses a simple testing geometry in the shape of a flattened blade that is relatively noninvasive. The instrument, operating at 110 to 120 or 220 to 240 V

**FIG. 4.9** T2SR rheometer with temperature controller and high-temperature heating cell. (Courtesy of Fluid Dynamics, Inc.)

(50 to 60 Hz), is shown in Figure 4.9 [29]. A schematic of the instrument is shown in Figure 4.10. Originally designed at the University of Strathclyde as a rheometer for cure studies [30], the rheometer has been redesigned and electronically upgraded with modifications producing an oscillatory rheometer for general R&D and QC use for structured fluid systems.

Examples of time sweep test results at 2 Hz for an antibacterial hand soap are shown in Figure 4.11 and Figure 4.12. Figure 4.11 summarizes the complex modulus components, $G'$ and $G''$, and the complex viscosity, $n^*$, while Figure 4.12 shows the experimental variables of phase angle and amplitude obtained at 23 to 24°C.

For high-consistency viscoelastic personal and home care products, the T2SR provides a means of obtaining complex rheology information under constant frequency conditions in time sweeps, isothermally, or with temperature control in temperature sweeps. The viscosity range is listed as 10 to 10,000 Pa s, in the frequency range 0.5 to 5 Hz, and in the temperature range −20 to 400°C.

## C. High-Shear Viscometry

Most capillary rheometers are designed for high-viscosity materials such as polymer melts and have limited application to lower viscosity liquid detergent systems.

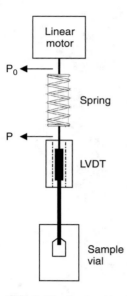

**FIG. 4.10** Design elements of the T2SR (time/temperature scanning rheometer). (Courtesy of Fluid Dynamics, Inc.)

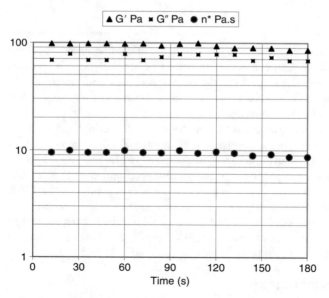

**FIG. 4.11** Measurement results for an antibacterial hand soap determined at 2 Hz using the T2SR rheometer at room temperature.

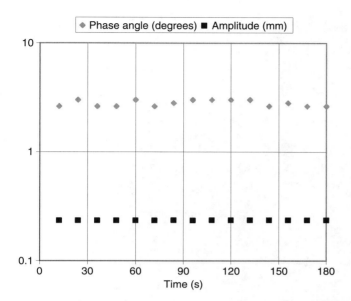

**FIG. 4.12**   Phase angle and amplitude measurement results for an antibacterial hand soap at 2 Hz using the T2SR rheometer.

There are notable exceptions, however, such as the CEP Lodge Stressmeter®. This rheometer is unique in that it measures both viscosity and first normal stress difference, N1. Both laboratory and in-line configurations are available. The instrument is shown in Figure 4.13. The Lodge Stressmeter is applicable to low-viscosity systems and available commercially from Chemical ElectroPhysics Corp. (Delaware, NJ).

The ACAV A4 (ACA Systems Oy, Finland), a pneumatic instrument, can also handle lower viscosity fluids, <50 mPa s. The ACAV A4, designed for coating applications, covers a broad shear rate range of $1 \times 10^3$ to $4 \times 10^6$ sec$^{-1}$ [31]. The benchtop instrument is shown in Figure 4.14. To demonstrate the low-viscosity range, Figure 4.15 provides viscosity vs. shear rate data for simple sugar/water solutions [32].

With design modifications, rotational rheometers can be used for high-shear-rate measurements. A high-shear rotational rheometer constructed with optically transparent parallel plates set up for simultaneous birefringence measurements on thin films is reported by Mriziq *et al.* [33]. Rheology and birefringence measurements for a perfluoropolyether lubricant are reported over a range of strain rates from $10^3$ to greater than $10^6$ sec$^{-1}$.

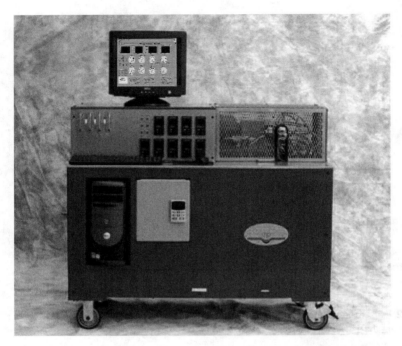

**FIG. 4.13**   CEP Lodge Stressmeter. (Courtesy of Chemical ElectroPhysics Corp.)

## D.   Extensional Viscosity Measurements

While dynamic mechanical and steady shear measurements are frequently used in rheology studies of surfactant systems, extensional viscosity measurements are lacking. This can be attributed to the difficulties associated with such measurements and the lack of commercial laboratory instrumentation since the discontinuance of the Rheometric Scientific RFX rheometer. For many detergent compositions, the relatively low viscosity further complicates such measurements. There appear to be very few data on extensional or elongation viscosity for detergent consumer products and actives in the technical literature at this time.

Filament stretching and capillary breakup rheometers are two experimental instruments used to impose uniaxial extension to fluids [34–39]. In both of these devices a fluid is placed between two surfaces or platens, and the spacing between the platens holding the sample is increased, as shown in Figure 4.16.

There are many practical situations in which extensional flow properties are important, both in processing detergent compositions and during consumer use. One of the most problematic can be the filling operation where a clean separation of the fluid and the filling nozzle does not occur. When extensional viscosity is

**FIG. 4.14** ACAV A4 capillary rheometer. (Courtesy of ACA Systems, Oy, Finland.)

high, a consumer will experience problems with "stringiness" in dispensing fluid from a pump or tube. This has been observed with various commercial personal care products. An example of one hair care product is shown in Figure 4.17, where the dispersion is quite stringy. This property is readily perceivable by the consumer and might not be an acceptable characteristic, since a clean break of the fluid from the dispensing orifice is generally desirable. Certainly this is true of fluids that are processed in high-throughput filling lines.

A commercial instrument for extensional viscosity measurements is currently offered by the Thermo Electron Corporation [40]. The device uses capillary breakup techniques and is called the Haake CaBER$^{TM}$. Vilastic Scientific, Inc. also offers an orifice attachment to their oscillatory rheometer for extensional viscosity determinations [41,42]. The principle of operation of the rheometer is oscillatory tube flow [43,44]. Dynamic mechanical properties can be determined

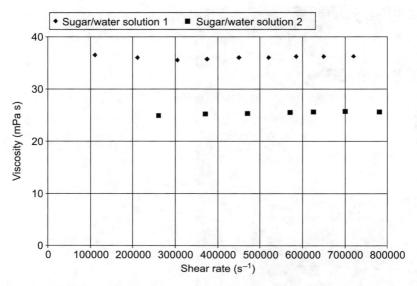

**FIG. 4.15** Example test results using the ACAV capillary rheometer. (Courtesy of ACA Systems, Oy, Finland.)

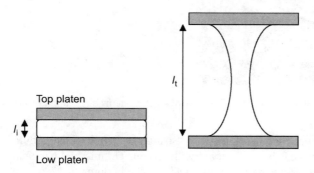

**FIG. 4.16** Example of an extensional viscosity measurement configuration using parallel plates.

as a function of frequency in the range 0.01 to 40 Hz. Shear stress and shear rates are 0.01 to 1000 dyn/cm$^2$ and 0.1 to 1000 sec$^{-1}$, respectively, for 0.4 to 90°C. This tube flow rheometer uses water as a calibration fluid and can handle very low-viscosity fluids for testing. With the orifice attachment, measurement of oscillatory pressure and flow through the converging channel allows extensional viscosity and elasticity to be calculated. The rheometer operates with small sample volumes (3 ml).

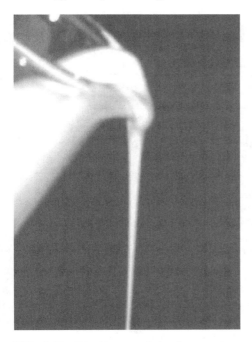

**FIG. 4.17**  Liquid soap solution in extension.

# E.  Interfacial Rheology

The properties of liquid/surface and liquid/liquid interfaces are fundamental to surfactant science. Surface tension measurements are quite common but interfacial rheology measurements are not and the rheology of these interfaces determines emulsion and foam stability, for example. Various experimental methods have been developed to determine interfacial rheology, including Gibbs elasticity and surface dilatational viscosity [45–52]. Common testing geometries for interfacial rheology measurements are shown in Figure 4.18.

A commercial stress-controlled interfacial rheometer is available from Camtel Ltd, CIR-100, equipped for use with a Langmuir trough accessory, CIR-LT. A schematic diagram of the CIR-100 drive mechanism and test sensor is shown in Figure 4.19 [53]. For the Camtel CIR-100, the platinum Du Nouy ring is the standard geometry. This rheometer applies an oscillating stress to a test sample and interfacial viscosity and elasticity are calculated from strain amplitude ($\gamma$) and phase angle ($\delta$), as shown in Figure 4.20. Measurement capabilities include time, strain, frequency, and temperature sweeps in simple shear and under changing surface pressure. Interfacial dilatational complex modulus can be determined at

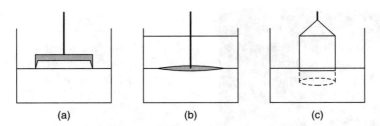

**FIG. 4.18**   Typical geometries for use in interfacial rheometers.

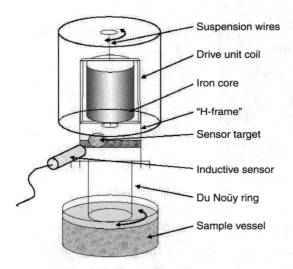

Suspension wires

Drive unit coil

Iron core

"H-frame"

Sensor target

Inductive sensor

Du Noüy ring

Sample vessel

**FIG. 4.19**   Schematic diagram of drive mechanism and sensor of CIR-100. (Courtesy of Camtel Ltd, U.K.)

the liquid/gas and liquid/liquid interfaces as a function of surface area and pressure, using the Langmuir trough. Application notes describing the operation of the CIR-100 are available from Camtel Ltd [54] and there are publications reviewing the principles and applications of surface/interfacial rheology [55–57].

Sinterface Technologies Profile Analysis Tensiometer (PAT1) can also be used for dynamic dilatational rheology measurements using a different testing method from the Camtel instrument [58]. The PAT1 is a sessile or pendant drop and drop/bubble oscillation instrument consisting of an automatic dosing system and video camera with framegrabber (Figure 4.21). Oscillations can be programmed to determine surface elasticity with temperature control in the range 10 to 350°C.

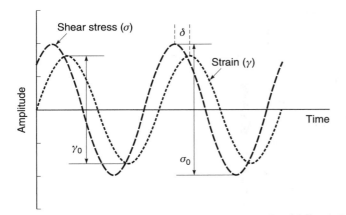

**FIG. 4.20**  Stress/strain relationship for oscillatory shear viscoelastic measurements.

**FIG. 4.21**  PAT1 instrument. (Courtesy of Sinterface Technologies.)

There is an extra oscillation module, based on direct measurements of the capillary pressure, which operates from 1 to 150 Hz. There is also an additional accessory for the PAT1 for low-frequency oscillations. The range of surface and interfacial tension is 1 to 1000 mN/m with a resolution of ±0.1 mN/m. The instrument allows for transient relaxation measurements, using perturbations such as ramp, square pulse, or trapezoidal area changes.

An Interfacial Shear Rheometer (ISR-1) is also offered by Sinterface Technologies for measuring interfacial shear properties, in the frequency range 0.02 to 0.2 Hz, dependent on the measurement system, in the temperature range 10 to 50°C. The measurement ranges of the rheometer include surface shear viscosity

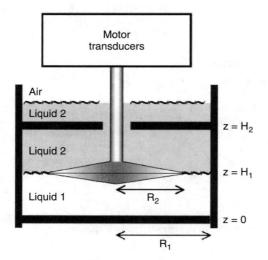

**FIG. 4.22** Schematic of biconical tool for surface rheology measurements using the Paar Physica interfacial rheometer. (Courtesy of American Institute of Physics.)

range of 1 $\mu$Ns/m to 100 mNs/m and surface shear elasticity range of 1 $\mu$Pa s to 100 mPa s.

A biconical disk interfacial rheometer is available from Anton Paar, known as the Physica Interfacial Rheology System (IRS). A schematic of the rheometer tool is shown in Figure 4.22. Current specifications of the instrument include a torque range of 0.02 $\mu$Nm to 150 mNm with temperature control from 5 to 70°C. All rheological test modes are available for the interfacial rheometer including oscillatory testing [59].

## IV. PRODUCT AND RAW MATERIAL CHARACTERIZATIONS

U.S. commercial products were selected for rheological characterization, demonstrating the breadth of rheology exhibited by current household and personal care products. Products include fabric softeners, dishwashing liquids, laundry detergents, shampoos, and dentifrices.

For dynamic mechanical and steady shear measurements, the Rheometric Scientific RFSII rheometer was used equipped with the sensitive range force rebalance transducer and couette geometry or parallel plate tooling.

Liquid detergent formulations covering the personal care and household care product categories exhibit a very wide range of rheological properties as shown in

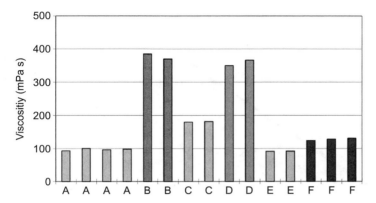

**FIG. 4.23** Newtonian viscosity of U.S. liquid laundry detergent products.

Figure 4.1. Within a single product category such as hair care one finds simple low-viscosity Newtonian fluids, non-Newtonian viscoelastic dispersions with time-dependent shear effects, and transparent highly elastic gels.

## A. Fabric Care Products

U.S. laundry detergents are typically Newtonian fluids and the viscosity of six commercial products is summarized in Figure 4.23. Several lots of each product, labeled A to F, were obtained and measurements completed at room temperature, 20 to 25°C, as a function of shear rate from 0 to 500 sec$^{-1}$. A typical shear stress–shear rate diagram is shown in Figure 4.24 for a product sampled from a 50 fl oz container. All six products tested are Newtonian with a viscosity less than 0.5 Pa s at room temperature, 21 to 23°C, with the shear rate ramped from 0 to 500 sec$^{-1}$ at an acceleration rate of 0.83 sec$^{-2}$. Newtonian behavior was confirmed through additional step shear rate measurements within the selected shear rate range.

Liquid fabric softeners are generally non-Newtonian and examples of the shear stress–shear rate relationship for two commercial products (A and B) is shown in Figure 4.25, determined at 22.5°C. In the shear rate range 0 to 250 sec$^{-1}$, we note non-Newtonian pseudoplastic behavior.

Dynamic mechanical strain-controlled measurements for both concentrated fabric softeners are shown in Figure 4.26. There are significant differences between the two products as regards the magnitude of the complex viscosity and complex modulus components and their strain dependence. Product B exhibits a higher viscosity and markedly longer linear region. The zero shear viscosity of product B is approximately 95 mPa s whereas that of product A is approximately half of this value at 50 mPa s.

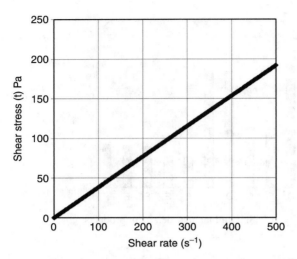

**FIG. 4.24**   Typical shear stress–shear rate relationship for Newtonian laundry detergent.

Both liquid fabric softeners exhibit time-dependent shear effects as shown in step shear rate measurements at room temperature. Figure 4.27 summarizes the steady shear viscosity as a function of time at shear rates of 0.1, 0.5, 1, and 5 $sec^{-1}$. Each shear rate is held for a period of 30 sec.

Fabric softeners, as demonstrated by these two commercial concentrated products, are more complex compositions due to their dispersion characteristics. The systems studied are non-Newtonian with time-dependent shear effects.

## B.  Personal Care Products

Shampoos, conditioning shampoos, body washes, and dentifrices cover a broad range of the rheology spectrum. This is a creative category where there are as many types of rheological fluids as there are containers. Examples of shear stress–shear rate profiles of randomly selected premium and value brand products obtained during thixotropic loop measurements, 0 to 25 $sec^{-1}$/60 sec, are shown in Figure 4.28. These products include clarifying and conditioning shampoos.

Several conditioning shampoos exhibit rheopectic behavior at low shear rates and examples for two commercial products are provided in Figure 4.29 at 0.05 and 0.1 $sec^{-1}$. The shear rates are applied sequentially for a time interval of 120 sec. Within this timeframe an equilibrium steady state shear stress is not reached. For product A, the shear rate is extended to 5 $sec^{-1}$ with similar results.

The oral care category has become a more complex product category, with the introduction of many products tailored to the youthful and senior consumer.

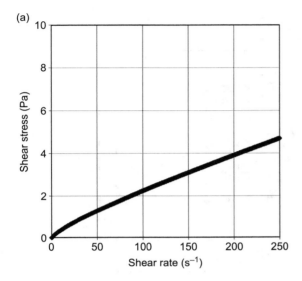

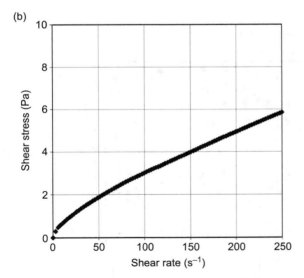

**FIG. 4.25** (a) Pseudoplasticity of product A (concentrated fabric softener). (b) Psuedo-plasticity of product B (concentrated fabric softener).

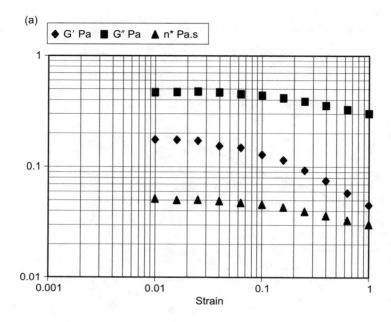

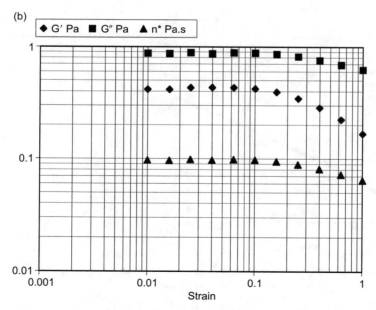

**FIG. 4.26** Strain sweep measurements at 10 rad/sec: (a) product A (concentrated fabric softener); (b) product B (concentrated fabric softener).

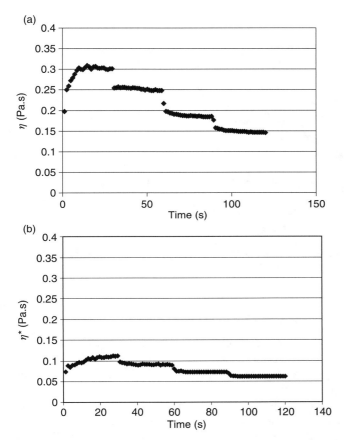

**FIG. 4.27** Viscosity during consecutive step shear rates of 0.1, 0.5, 1, and 5 $sec^{-1}$: (a) product A (concentrated fabric softener); (b) product B (concentrated fabric softener).

Different flavors, colorants, and "sparkling" additives, as well as packaging, are clearly adding complexity to the R&D product development venue for the junior market. In addition, the dentifrice product category has seen the recent introduction of many bleaching or "tooth whitening" compositions from many manufacturers. Some of these products are in the form of films for direct placement on the teeth, and others are in the conventional dentifrice form of pastes or gels. Rheology measurements show that the properties of these "whitening" compositions have broadened the spectrum of the rheology matrix, since some of these products appear markedly lower in consistency.

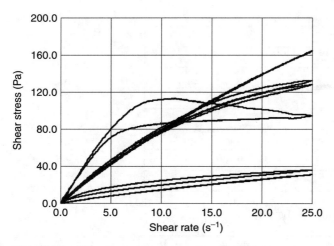

**FIG. 4.28**  Thixotropic loop measurement results for six commercial U.S. shampoos products.

## V.  FUNDAMENTAL RHEOLOGY STUDIES

Over the past few decades there has been an increase in the research tools for fundamental rheology studies of surfactant solutions and commercial detergent formulations. The coupling of rheometers with other methods has broadened the range of studies that can be completed, leading to a better understanding of solution properties, self-assembled mesophases, multiple-component dispersions, and gels.

An x-ray shear cell for studying the complex fluid of nematic surfactants in time-dependent shearing flows has been developed by Caputo *et al.* [60]. Shear aligning and director tumbling are cited for two surfactant systems, SDS/decanol and CPyCl/hexanol. A microscopic particle imaging velocimeter with a torsional shearing-flow cell has also been used to study the shear thickening of worm-like micelle solutions [61]. The effect of wall slip on the rheology of the micellar solutions as a function of shear rate is deduced from coupled flow visualization and rheology measurements. Particle image velocimetry of micellar solutions in unstable capillary flow has also been carried out [62]. At a critical stress found to be independent of strain rate, the worm-like micelle filaments rupture near the axial midplane. Filament failure is thought to occur from local scission of individual micellar chains.

Coupled controlled velocity, magnetic resonance imaging (MRI)/rheology measurements of thixotropic and yielding colloidal suspensions further demonstrate the importance of paired measurements [63]. Shear rate profiles obtained in laminar tube flow for both Newtonian and non-Newtonian fluids from MRI

(a)

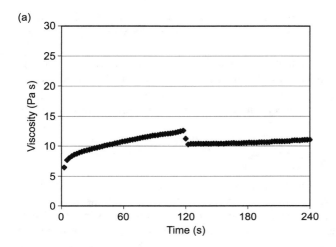

(b)

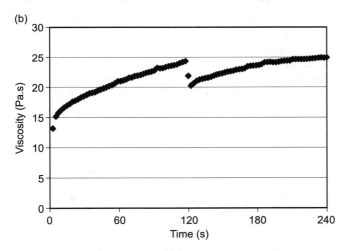

**FIG. 4.29** Time-dependent shear effects at step shear rates of 0.05 and 0.1 $\text{sec}^{-1}$: (a) conditioning shampoo A; (b) conditioning shampoo B.

and laser Doppler velocimetry data using Tikhonoz regularization is discussed by Yeow and Taylor [64]. NMR investigations during rheology measurement at rest and under shear for a nematic surfactant system (sodium dodecylsulfate, decanol, water) have been carried out. These measurements are used to determine the director orientations of the surfactant [65]. Shear-thickening self-assembling fluids have been studied using rheooptics, revealing unusual flow behavior at various flow conditions [66]. Measurements included stress growth, steady state viscosity, and stress relaxation for aqueous CTSAB/sodium salicylate solutions.

An overview of rheological measurements coupled with magnetic resonance is provided by Callaghan [67]. Rheo-NMR of emulsified systems has been studied, the systems including formulations with yield stress exhibiting wall slip [68]. Comparisons are provided between conventional rheological techniques and Rheo-NMR characterization.

## VI. CONCLUSIONS

The past decade has seen many advances in the science of rheology with applications to liquid detergent systems. This is in keeping with the progressive developments in detergent systems for personal, household, and industrial use. Very common household products exhibit remarkably rich rheology profiles and significant effort is being directed toward understanding how to expand, manipulate, and control these properties to generate high-performance products for consumer use. With new generations of raw materials, this continues to be a difficult field of rheology research.

Coupling of rheology measurements to other analytical techniques such as light scattering and NMR facilitates the study of micellar solutions and liquid crystalline phases, microemulsions, vesicles, etc., leading to the development of new surfactant systems. We anticipate continuing advances in rheology measurement technology with direct applications to the study of liquid detergent systems.

## REFERENCES

1. Rounds, R.S., in *Liquid Detergents*, chap. 4, Lai, K.Y., Ed., Marcel Dekker, New York, 1996, pp. 67–127.
2. Barnes, H.A., Schimanski, H., and Bell, D., *Appl. Rheol.*, 9, 69–76, 1999.
3. Fuller, G.G., *Optical Rheometry of Complex Fluids: Theory and Practice of Optical Rheometry (Topics in Chemical Engineering)*, Oxford University Press, New York, 1995.
4. Collyer, A.A. and Clegs, D.W., Eds., *Rheological Measurement*, 2nd ed., Chapman & Hall, London, 1998.
5. Tanner, R.I. and Walters, K., *Rheology: An Historical Perspective*, Elsevier, New York, 1998.
6. Melle, S. Calderon, O.G., Rubio, M.A., and Fuller, G.G., *J. Non-Newtonian Fluid Mech.*, 102, 135–148, 2002.
7. Gotz, J., Kreibich, W., Peciar, M., and Buggisch, H., *J. Non-Newtonian Fluid Mech.*, 98, 117–139, 2001.
8. Della Valle, D., Tanguy, P.A., and Carreau, P.J., *J. Non-Newtonian Fluid Mech.*, 94, 1–13, 2000.
9. Clasen, E. and Lukicke, W.-M., *Rheol. Acta*, 40, 67–73, 2001.

10. Ivanov, Y., Kavardjikov, V., and Pashkuleva, D., *Appl. Rheol.*, 11, 320–324, 2001.
11. Sjodahl, M., *Appl. Optics*, 13, 2875–2885, 1997.
12. Ouriev, B., *Appl. Rheol.*, 10, 148–150, 2000.
13. Adrian, R.J., *Annu. Rev. Fluid Mech.*, 23, 261–304, 1991.
14. Kotaka, T., Kojima, A., Kubo, H., and Okamoto, M., Society of Rheology Annual Meeting, Texas, 1997.
15. Lee, E.C., Colomon, M.J., and Muller, S.J., Society of Rheology Annual Meeting, Texas, 1997.
16. Brunn, P.O., Muller, M., and Harder, C., *Appl. Rheol.*, 7, 204–210, 1997.
17. Barnes, H.A., Hutton, J.F., and Walters, K., *An Introduction to Rheology*, Elsevier Science, New York, 2001.
18. Barnes, H.A., *A Handbook of Elementary Rheology*, Institute of Non-Newtonian Fluid Mechanics, University of Wales, 2000.
19. Goodwin, J.W. and Hughes, R.W., *Rheology for Chemists: An Introduction*, Royal Society of Chemistry, 2000.
20. Morrison, F.A., *Understanding Rheology*, Oxford University Press, New York, 2001.
21. R.G. Larson, *The Structure and Rheology of Complex Fluids*, Oxford University Press, New York, 1999.
22. Macosko, C.W., *Rheology: Principles, Measurements and Applications*, John Wiley & Sons, New York, 1994.
23. Junus, S. and Briggs, J.L., *Appl. Rheol.*, 11, 264–270, 2001.
24. Nguyen, Q.D. and Boger, D.V., *J. Texture Stud.*, 9, 335–347, 1985.
25. Glenn, T.A. and Keener, K.M., *Appl. Rheol.*, 10, 80–89, 2000.
26. Baravian, C., Lalante, A., and Parker, A., *Appl. Rheol.*, 12, 81–87, 2002.
27. Barnes, H.A., *Appl. Rheol.*, 11, 89–101, 2001.
28. Gleissle, W., *Appl. Rheol.* 5, 14, 1995.
29. T2SR rheometer brochure, application note FDI040810, Fluid Dynamics, Inc., New Jersey, August 2004.
30. Affrossman, S., Hayward, D., McKee, A., MacKinnon, A., Lairez, D., Pethrick, R.A., Batalis, A., Baker, F.S., and Carter, R.E., in *Rheology of Food, Pharmaceutical and Biological Materials With General Rheology*, Elsevier Science, New York, 1989, pp. 304–314.
31. ACA Systems, Applications, ACAV A4, August 2004.
32. Data courtesy of ACA Systems, Oy, Finland.
33. Mriziq, K.S., Dai, J.J., Dadmun, M.D., Jellison, G.E., and Cochran, H.D., *Rev. Sci. Instrum.*, 75, 2171–2176, 2004.
34. Anna, S.A., McKinley, G.H., Nguyen, D.A., Sridhar, T., Muller, S.J., Huang, J., and James, D.F., *J. Rheol.*, 45, 82–114, 2001.
35. Anna, S.L. and McKinley, G.H., *J. Rheol.*, 45, 115–138, 2001.
36. Tirtaatmadja, T. and Sridhar, T., *J. Rheol.*, 36, 277–284, 1993.
37. James, D.F., Walters, K., and Colyer, A.A., Eds., *Techniques of Rheological Measurement*, Elsevier, New York, 1994, pp. 33–53.
38. Bazilevsky, A.V., Entov, V.M., Rozhkov, A.N., and Oliver, D.R., Eds., *Third European Rheology Conference*, Elsevier, New York, 1990, pp. 41–44.
39. Tripathi, A., Whittingstall, P., and McKinley, G.H., *Rheol. Acta*, 39, 321–337, 2000.
40. Thermal Electron Corp. technical brief, Haake CaBER 1 extensional viscometer, 2004.

41. Henderson, N. and Thurston, G., Society of Rheology Annual Meeting, Viscoelastic Effects in Elongational Flow: A New Orifice Impedance Method, Minnesota, October, 2002.
42. Vilastic Scientific, Inc., Extensional Viscosity and Elasticity of Fluids, technical note 11, 2004.
43. Thurton, G.B. and Martin, C.E., *J. Acoust. Soc. Am.*, 25, 26–31, 1953.
44. Shrewsbury, P.J., Muller, S.J., and Lipemann, D., *Biomed. Microdevices*, 3, 225–238, 2001.
45. Tian, Y., Holt, R.G., and Apfel, R.E., *J. Colloid Interface Sci.*, 187, 1–10, 1997.
46. Burgess, D. and Sahin, N.O., *J. Colloid Interface Sci.*, 189, 74–82, 1997.
47. Nagarajan, R., Chung, S.I., and Wasan, D.T., *J Colloid Interface Sci.*, 204, 53–60, 1998.
48. Wantke, L.-D., Fruhner, H., Fang, J., and Lunkenheimer, K., *J. Colloid Interface Sci.*, 208, 34–48, 1998.
49. Myrvold, R. and Hansen, F.K., *J. Colloid Interface Sci.*, 207, 97–105, 1998.
50. Lopez, J.M. and Hirsa, A., *J. Colloid Interface Sci.*, 206, 231–239, 1998.
51. Wantke, K.-D. and Fruhner, H., *J. Colloid Interface Sci.*, 237, 185–199, 2001.
52. Nagarajan, R. and Wasan, D.T., *Rev. Sci. Instrum.*, 65, 2675–2679, 1994.
53. Courtesy of Camtel Ltd, U.K.
54. Camtel Ltd, Comparison of the Interfacial Rheological Properties of Commercial Shampoos, DS5 data sheet, April 2003.
55. Wilde, P., *Interfacial Rheology in Industry*, SISC Manual, Interfacial Rheology & Industry, Camtel Ltd, April 2003.
56. Wang, Z. and Narsimhan, G., Surface and Dilatational and Shear Rheological Properties of beta-Lactoglobulin at Air–Water Interface, AIChE Annual Meeting, Biomolecules at Interfaces, 2003.
57. Warburton, B., in *Techniques in Rheological Measurement*, Chapman & Hall, London, 1993, pp. 55–97.
58. Loglio, G., Pandolfinii, P., Miller, R., Makievski, A.V., Ravera, F., Ferrari, M., and Liggieri, L., in *Studies in Interface Science*, Vol. 11, Elsevier, Amsterdam, 2001, pp. 439–484.
59. Erni, P., Fischer, P., and Windhab, E.J., Rheology of Surfactant Assemblies at the Air/Liquid and Liquid/Liquid Interface, 3rd International Symposium on Food Rheology and Structure, Zurich, Switzerland, February 9–13, 2003.
60. Caputo, F.K., Ugaz, V.M., Burghardt, W.R., and Berret, J.-F., *J. Rheol.*, 46, 927–946, 2002.
61. Hu, H., Larson, R.G., and Magda, J.J., *J. Rheol.*, 46, 1001–1021, 2002.
62. Mendez-Sandez, A.F., Perez-Gonzalez, J., de Vargas, L., Castrejon-Pita, J.F., Castrejon-Pita, A.A., and Huelsz, G., *J. Rheol.*, 47, 1455–1466, 2003.
63. Raynaud, J.S., Moucheront, P., Baudez, J.C., Bertrand, F., Guilbaud, J.P., and Coussot, P., *J. Rheol.*, 46, 709–732, 2002.
64. Yeow, Y.L. and Taylor, J.W., *J. Rheol.*, 46, 351–365, 2002.
65. Thiele, T., Berret, J.-F., Muller, S., and Schmidt, C., *J. Rheol.*, 45, 29–48, 2001.
66. Hu, Y., Wang, S.Q., and Jamieson, A.M., *J. Rheol.*, 37, 521–546, 1993.
67. Callaghan, P.T., *Rep. Prog. Phys.*, 62, 599–670, 1999.
68. Hollingsworth, K.G. and Johns, M.L., *J. Rheol.*, 48, 787–803, 2004.

# 5

# Rheology Modifiers and Thickeners for Liquid Detergents

**PAUL REEVE**   Rohm and Haas France SAS, Sophia Antipolis, France

**TOM TEPE and JAN SHULMAN**   Rohm and Haas Company, Research Laboratories, Spring House, Pennsylvania

## I.  INTRODUCTION

Liquid detergents make a major contribution to the overall detergent and cleaners market, due in part to their handling characteristics and their ease of use. However, in numerous cases, the formulations would lack either consumer appeal or essential physical properties if they did not include additives to modify their viscosity or rheology. Generally speaking, viscosity build is required to improve the aesthetics of a formulation and meet the demands of consumers, for whom the concept "thicker is better" often remains valid [1]. This is especially true in those liquid formulations where the resulting viscosity without additives is barely above that of water itself. A simple increase in viscosity, though, is often not sufficient to meet the technical demands of a formulation. To address these needs, the rheology of the system has to be taken into consideration, and this is intimately bound up with the nature of the formulation, as well as its intended delivery system and its use.

**113**

Thus, a scouring cream will require suspending properties to prevent the finely divided abrasive from precipitating, yet it must be pourable or squeeze-dispensable from its package. Similarly, a spray cleaner will require a low viscosity under conditions of high shear in order to facilitate the passage through the spray or trigger mechanism. In the case of a wall or bathroom cleaner, a certain degree of "vertical cling" will be needed to maximize the contact time between the formulation and the surface. All these properties can be obtained through the appropriate choice of rheology modifiers [2].

The selection of the most suitable rheology modifier will depend on the type of flow, or rheology, required, based on considerations such as those indicated above. This will be largely, but not exclusively, an inherent property of the modifier itself. Selection will also depend to some extent on the nature of the formulation, as not all rheology modifiers are necessarily physically or chemically compatible with all other formulation components. For example, anionically charged rheology modifiers are often precluded from use in cationic-based systems such as fabric softeners in order to avoid incompatibilities with the surfactant. Even presuming compatibility, the other components of a formulation can alter the rheological properties of the additive being used, particularly if this is an associative rheology modifier. Clarity can be another issue with certain additives *per se*, or in combination with different ingredients. Bentonite, which is useful for contributing to the suspending properties of a formulation, gives opacity. Polyacrylic acids, whereas clear in aqueous solution, may show lack of clarity when certain surfactants are present.

In the case of liquid detergents, surfactants are almost always present. At low to intermediate concentration, most neat surfactant solutions have low viscosity and are close to Newtonian in flow. Only at higher surfactant concentrations, when structured micellar bilayers and other complex phases are formed, do systems tend to differ greatly from Newtonian. This behavior also helps drive the viscosity of finished formulations. In the great majority of liquid detergent formulations, concentrations of surfactant are such that little structure is developed by the surfactants themselves, resulting in formulations of low viscosity. As such, thickeners and/or rheology modifiers are often required to obtain the desired viscosity and flow characteristics.

In this chapter we survey the most common types of rheology modifiers that are used today in liquid detergents. This covers both natural and synthetic modifiers, with numerous subclasses in each, as illustrated in Table 5.1. Guidelines for the types of rheological profiles each modifier can provide, as well as general formulation issues, are presented.

## II.  RHEOLOGY

When a stress is applied to a liquid it will begin to deform, or flow. The deformation per unit of time, referred to as the shear rate, will increase as the applied stress

**TABLE 5.1**   Classification of Various Types of Organic and Inorganic Rheology
Modifiers

---

**Organic thickeners**
Nonassociative
  Naturally derived
    Nonionics (e.g., hydroxyethyl cellulose), anionic (e.g., carboxymethyl cellulose)
    Other polysaccharides (e.g., xanthan)
    Miscellaneous (e.g., alginates)
  Synthetic
    Nonionics (e.g., polyvinyl alcohol)
    Alkali swellables (e.g., crosslinked acrylics)
    Alkali solubles (e.g., noncrosslinked acrylics)
Associative
  Naturally derived
    Nonionics (e.g., hydrophobically modified hydroxyethyl cellulose, HMHEC)
  Synthetic
    Nonionics (e.g., hydrophobically modified ethoxylated urethanes, HEUR;
      hydrophobically modified nonionic polyols, HNP)
    Anionics (e.g., hydrophobically modified alkali-soluble emulsions, HASE)
**Inorganic thickeners**
  Salts (e.g., sodium chloride, magnesium chloride)
  Clays (e.g., bentonite, hectorite)

---

increases. If the relationship between increasing applied stress and increasing
shear rate is linear, then the liquid is defined as Newtonian, and the viscosity is the
slope of the plot of shear rate against applied stress. As the relationship is linear,
then the slope is constant, and so the viscosity is independent of the shear rate
(Figure 5.1).

In reality, few systems are Newtonian, and some of the other principal rheological profiles are also shown in Figure 5.1. In many cases a Newtonian behavior
is not desirable for a formulated product. This can be illustrated by the case of a
spray cleaner. A certain minimum viscosity is often required such that the material
appears to be "concentrated" in the bottle. The visual appearance is referred to in
this chapter as the "apparent viscosity" and is generally considered to correspond
to a shear rate of the order of 10 sec$^{-1}$ (reciprocal seconds). If the formulation is
Newtonian, then the viscosity will remain the same even at the relatively high shear
rates corresponding to spraying (Figure 5.2). This is not desirable, as the spray
pattern obtained varies considerably with the viscosity of the fluid in the spray nozzle, and better atomization is observed when the viscosity is low. Consequently,
an ideal profile for such a formulation is one in which the viscosity decreases as

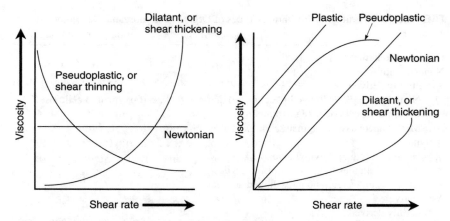

**FIG. 5.1**  Illustrative examples of various rheological responses.

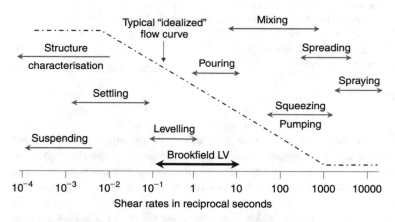

**FIG. 5.2**  Various flow events and their link to typical shear rate ranges.

the shear rate increases, as shown in Figure 5.1. Such a fluid is described as shear thinning, or pseudoplastic, and the degree of pseudoplasticity can be adjusted by the choice of rheology modifier. Figure 5.3 shows the viscosity response for aqueous solutions of four different rheology modifiers. Note the differences in the shear rate dependences of the solutions, with the Laponite being the most pseudoplastic of the four.

In the case of suspensions, be they opaque dispersions of abrasives found in certain scouring creams, or the suspension of visual cues or active ingredients now

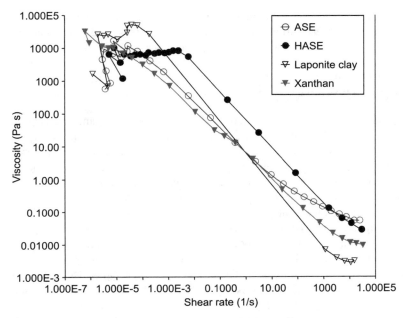

**FIG. 5.3** Measured flow curves for aqueous solutions of four common types of rheology modifiers. ASE is an alkali swellable/soluble emulsion-type rheology modifier and HASE is a hydrophobically modified alkali-soluble emulsion.

encountered in transparent hand, dish, and toilet cleaners, the properties required from the rheology modifier must ensure no settling of the components in the bottle. There are several techniques used to evaluate whether or not a formulation will adequately suspend ingredients, and it is not intended to evaluate the merits of each here. One of the simplest methods to use is the Brookfield yield value, which is a useful comparative tool, and in many cases is certainly sufficient to obtain a good approximation of the suspending properties of a system. This is evaluated by measuring the Brookfield viscosity at 0.5 and 1 rpm, and then calculating the yield value (YV) as:

$$YV = \frac{\text{viscosity (0.5 rpm)} - \text{viscosity (1 rpm)}}{100}$$

More sophisticated techniques use a controlled stress rheometer to evaluate the minimum stress necessary to obtain flow (the yield stress) or to calculate, by appropriate modeling, the zero shear viscosity of the system. Whichever technique is employed, the yield stress/value required to achieve a stable suspension will be greater as the size of the particle to be suspended increases, and as the density difference between the matrix and the particle increases.

## III. ORGANIC THICKENERS

Within the group of polymers classified under the heading of organic thickeners, there are both associative and nonassociative variants of several of the polymers. Associative rheology modifiers are those polymers that contain hydrophobic moieties at various levels in their composition. If this modification is made to a high-molecular-weight polymer, then it gives rise to an additional mechanism for modifying the rheological characteristics of the matrix. As well as the swelling and/or chain entanglement that occurs with polymers of a high molecular weight, inter- or intramolecular hydrophobic association can also take place in aqueous media. This is similar to the hydrophobic association that takes place in aqueous surfactant solutions, and which drives the surfactant molecules to form micelles. In the case of associative polymers, these interactions can take place between the polymer molecules, with other hydrophobes present in the matrix, including surfactants, or even with certain particle surfaces. By associating with other components in the system, additional structure can be developed which can modify rheology, and also contribute to the overall stability of the matrix.

## A.  Acrylics

A wide range of acrylic-derived polymers is available, and they can be classified in various ways. There are homopolymers and copolymers, and they can be emulsion polymerized (in water) or inverse polymerized (in an organic solvent). In addition they can be associative or nonassociative. Different acronyms and nomenclatures are used to describe the various classes of polymers. Some of the earliest acrylic rheology modifiers were the carbomers, which are crosslinked homopolymers of polyacrylic acid manufactured by inverse polymerization in a suitable solvent. They are generally recovered from the solvent by precipitation and are available as powders. A second class of nonassociative acrylic rheology modifiers are the alkali swellable/soluble emulsion (ASE) polymers which is subdivided into two categories. There are both crosslinked and noncrosslinked ASE polymers, which are essentially acrylic copolymers produced by aqueous (emulsion) polymerization and which are in the form of low-viscosity aqueous dispersions. The different product forms of the various classes of polymer can have an impact on the choice of the most suitable additive for a given situation. The equipment required for handling the aqueous-based emulsion polymers is simpler than that required for handling powders, in particular since polyacrylic acid-based powders are hygroscopic. The water-based emulsion polymers also show an advantage for high-throughput systems, as they can be integrated into continuous manufacturing processes, which is more difficult with a powdered additive.

Both of the types of polymer mentioned above can be modified by the incorporation of hydrophobic monomers onto the essentially hydrophilic acrylate backbone. The effect of this is to modify their characteristics by giving them so-called "associative" properties. These hydrophobes can interact or "associate" with other hydrophobes in the formulation (e.g., surfactants, oils, or hydrophobic particles) and thus build additional structures in the matrix [3–11]. These associative polymers are termed cross-polymers when they are based on carbomer-type chemistry [12] and hydrophobically modified alkali-soluble emulsions (HASEs) when based on ASE technology.

Although all of these additives are based on acrylic chemistry, both the behavior and the performance of the different categories of polymer vary considerably. One point in common, however, is that they are nearly all supplied in the acidic form and require neutralization to develop their thickening and rheological properties. The precise pH range over which these properties are obtained varies with the composition of the material, but in general the carbomers and the cross-polymers begin to develop their rheology-modifying behavior at a pH value of about 1 to 2 units below that of the ASE and HASE polymers.

The simplest system to consider is represented by a dispersion of the neutralized polymer in water. All show excellent clarity, but the carbomers (and the cross-polymers, not shown) are undoubtedly the most efficient in terms of their simple aqueous thickening properties, as indicated in Figure 5.4. The quantity of polymer required to achieve a given mid-shear rate viscosity is close to an order of magnitude less than that required for a crosslinked ASE thickener. The HASE polymers are generally found to have an efficiency between that of the ASE and the carbomer/cross-polymers.

The shape of the aqueous rheology curve also varies with the nature of the polymer. Both the carbomers and the cross-polymers show fairly similar behavior. Crosslinked ASE polymers, as shown in Figure 5.4, show a profile close to that of the carbomers, giving highly shear-thinning properties, although this is to some extent dependent on molecular weight. A noncrosslinked ASE polymer shows a more Newtonian profile. The HASE polymers tend to show a behavior between that of a noncrosslinked ASE thickener and the carbomers or crosslinked ASE polymers.

Build of significant low shear viscosity is used in many applications, and contributes significantly to formulation properties such as the vertical cling and the ability to suspend particles in a matrix. The obvious use of particle suspension is in slurries, where stability of the suspension over time is required. However, it is clear that this ability to suspend can also be applied to emulsions, which are simply suspensions of one immiscible liquid phase in another. As such, acrylic polymers

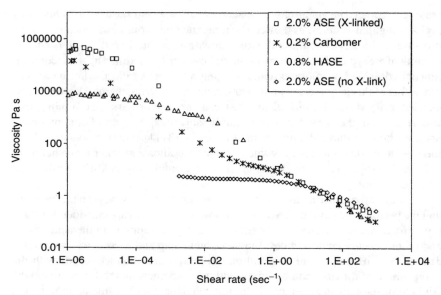

**FIG. 5.4**   Measured flow curves for aqueous solutions of four different acrylic thickeners.

showing low shear viscosity build can be very useful additives for stabilizing liquid–liquid emulsions and solid–liquid dispersions.

In HASE polymers the nature of the hydrophobe has a significant impact on both the efficiency and the pseudoplasticity of the resulting aqueous solution [13]. The longer the hydrophobe chain, and, within limits, the greater the number of hydrophobes on the polymer backbone, the greater the efficiency and the more pseudoplastic the polymer. This is illustrated in Figure 5.5, where the medium and high shear viscosities of the two polymers are equivalent. The two HASE polymers vary by the length of the hydrophobic moieties, C18 vs. C22. To obtain these results, a significantly lower content of the more hydrophobic HASE is required (0.55% vs. 0.8% for the less hydrophobic variant). In addition, the more hydrophobic C22 HASE polymer remains pseudoplastic over a wider shear rate range.

The above guidelines for the acrylic rheology modifiers are most useful in simple systems that are mainly water. However, in more complex matrices such as many finished formulations the situation can be very different. Both the carbomer and the cross-polymer type of rheology modifier are very sensitive to the presence of electrolytes, and this has a dramatic effect on the efficiency of the polymer, as well as on clarity. It is frequently found that in systems containing electrolytes, be they inorganic salts or anionic surfactants, the efficiency of a crosslinked ASE

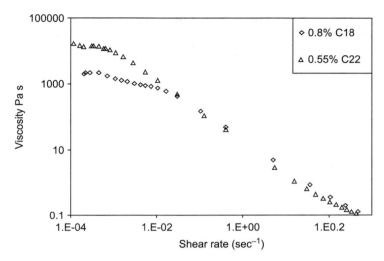

**FIG. 5.5** Measured flow curves for aqueous solutions of two different HASE polymers, illustrating the effect of hydrophobe size (see legend) on the low shear rate viscosity.

(x-ASE) or a HASE can often be as good if not better than a carbomer or a cross-polymer. This can also frequently be allied with a better formulation clarity.

Whereas in simple aqueous solutions the carbomers and the cross-polymers show a significantly better efficiency, Table 5.2 shows that in surfactant-based systems this no longer holds true. The use levels indicated in the table are the quantities of polymer required to obtain a given apparent Brookfield LV-60 viscosity of about 3000 mPa s, and in all the anionic surfactants the results are more equitable for the different classes of polymer compared with the situation in water. In the case of the nonionic surfactant, the difference is still maintained, however.

In terms of clarity, though, there is a notable advantage when using a HASE polymer in place of other acrylic polymers. Clarity is measured as the optical density, and from Table 5.2 the better clarity of the surfactants thickened with the HASE polymer is apparent. An optical density of 0.05 or less can be considered clear, and between 0.05 and 0.075 as showing a very slight haze. Above a value of 0.1 a loss of clarity becomes easily apparent.

In terms of the overall rheology profile of acrylic polymers when used in finished formulations, the behavior of the nonassociative thickeners is relatively easy to predict, as there is little interaction from a rheological point of view between the thickener and the matrix. Significantly higher polymer levels will be required if electrolytes are present, but the overall formulation rheology (e.g., pseudoplasticity, yield development) will remain similar. In most circumstances, though,

**TABLE 5.2**   Use Level (% Solids Based on Total) Required of Four Rheology Modifiers to Obtain a Surfactant Solution with Brookfield Viscosity of 3000 mPa s, and the Measured Clarity Values (Optical Density) of these Solutions

| Surfactant | HASE | x-ASE | Cross-polymer | Carbomer |
|---|---|---|---|---|
| 10% SLES |  |  |  |  |
| Use level | 1.1 | 4.5 | 1.2 | 1.6 |
| Clarity | 0.05 | 0.33 | 0.30 | 1.30 |
| 10% SLS |  |  |  |  |
| Use level | 1.35 | 3.0 | 1.1 | 1.4 |
| Clarity | 0.03 | 0.28 | 0.15 | 0.70 |
| 5% CAPB |  |  |  |  |
| Use level | 1.75 | 2.1 | 1.0 | 1.2 |
| Clarity | 0.05 | 2.36 | 0.16 | 0.83 |
| 5% SLES |  |  |  |  |
| Use level | 1.05 | 3.7 | 1.1 | 1.3 |
| Clarity | 0.03 | 0.18 | 0.13 | 0.81 |
| 5% APG |  |  |  |  |
| Use level | 1.7 | 2.0 | 0.4 | 0.5 |
| Clarity | 0.02 | 0.01 | 0.19 | 0.84 |

Surfactants: anionics sodium lauryl ether sulfate (SLES) and sodium lauryl sulfate (SLS); amphoteric cocamidopropyl betaine (CAPB); nonionic alkyl polyglucoside (APG).
HASE, hydrophobically modified alkali-soluble emulsion; x-ASE, crosslinked alkali swellable/soluble emulsion.

the clarity of the formulation will decrease as the electrolyte content increases, particularly in the case of the carbomers. Additionally, the ultimate stability of the formulation may be suspect.

The cross-polymers show rheology similar to that found in water, albeit at markedly reduced efficiency, and with some loss in clarity. However, the absence of clarity is not so great as that found with the carbomers. This improvement in clarity, but with little change in rheology, is attributed to the fact that these cross-polymers contain relatively small amounts of hydrophobe. The HASE polymers generally show good compatibility with electrolytes with little loss of clarity and efficiency. Nevertheless, the relatively high hydrophobe content of these polymers leads to strong associations between the polymer and the hydrophobes of the surfactant, and hence the rheology of the HASE polymers in surfactant solutions tends to differ significantly from the behavior found in water. This change depends to some extent on the surfactant, but also on the nature and quantity of the hydrophobe present on the polymer. It is sometimes found judicious to blend polymers of differing characteristics in order to achieve the required physical properties in a finished formulation.

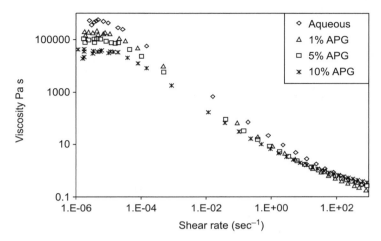

**FIG. 5.6** Measured flow curves for a crosslinked ASE polymer in solutions of increasing alkyl polyglucoside (APG) concentration.

In the following examples some of the matrix effects observed between rheology modifiers and different surfactants are illustrated. The case of a crosslinked ASE polymer in the presence of a nonionic alkyl polyglucoside (APG) surfactant is shown in Figure 5.6. The rheological profiles of the polymer in different concentrations of APG are very similar to those of the aqueous polymer results, indicating that the surfactant has very little effect on the rheological behavior of the ASE polymer.

The situation is a little different in the case of a crosslinked ASE with an anionic surfactant such as sodium lauryl sulfate (SLS), as shown in Figure 5.7. The overall shapes of the curves at the different SLS contents are similar to that of the aqueous solution. However, the curves are shifted downwards, illustrating a loss in efficiency of the polymer. This, however, is not a surfactant effect, but an electrolyte effect, showing how the ionic strength of the matrix is reducing the swelling of the polymer and reducing its efficiency.

The case of a HASE polymer is different, as illustrated in Figure 5.8 and Figure 5.9. Here the overall profiles of the rheological curves are significantly altered in the presence of the different surfactants, the system becoming less pseudoplastic as the surfactant concentration increases [14,15]. It is interesting to note that in this case the efficiency of the polymer varies with the shear rate. At low shear rates the viscosity shows a decline as the surfactant content increases, but at higher shear rates the surfactant–polymer solutions show a higher viscosity than the simple aqueous solution of the polymer. This change occurs with both

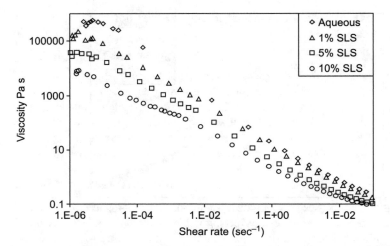

**FIG. 5.7** Measured flow curves for a crosslinked ASE polymer in solutions of increasing SLS concentration.

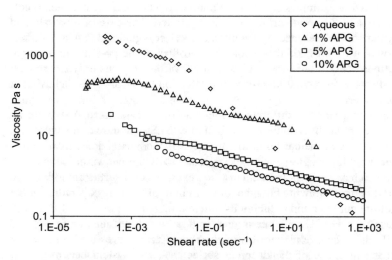

**FIG. 5.8** Measured flow curves for a HASE polymer in solutions of increasing alkyl polyglucoside (APG) concentration.

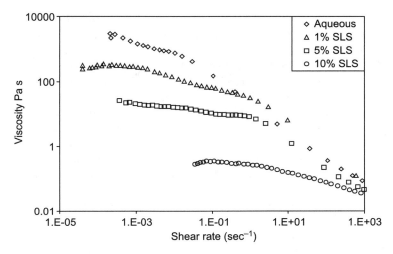

**FIG. 5.9** Measured flow curves for a HASE polymer in solutions of increasing SLS concentration.

anionic and nonionic surfactants, and is more marked with the less hydrophobic HASE rheology modifiers. As the polymer becomes more hydrophobic, either due to longer chain hydrophobes or to a greater number of hydrophobes (hydrophobe density), so this effect becomes somewhat attenuated.

The noncrosslinked ASE polymers are of interest due to their good electrolyte tolerance and their tendency to thicken formulations containing high levels of alkaline salts and builders. Thus, these types of polymer are often preferred in industrial and institutional cleaners.

## B. Synthetic Nonionic Polymers

Water-soluble synthetic nonionic polymers represent a large class of thickeners and rheology modifiers. Included in this group of commercial rheology modifiers are polymers based on polyacrylamide (pAm), polyethylene oxide or glycol (PEO or PEG), block copolymers (ethylene oxide [EO] and propylene oxide [PO]), polyvinyl alcohol (PVA), and polyvinyl pyrrolidone (PVP) (Figure 5.10). Nonionic polymers are generally compatible with anionic, nonionic, amphoteric, and cationic surfactants. They also have a much better tolerance for electrolytes than anionic polymers. Depending on the specific chemistry, nonionic polymers may exhibit a cloud point behavior, undergo base or acid hydrolysis, and may be unstable (certain types, e.g., pAm) to harsh environments such as peroxides, persulfates, or hypochlorite.

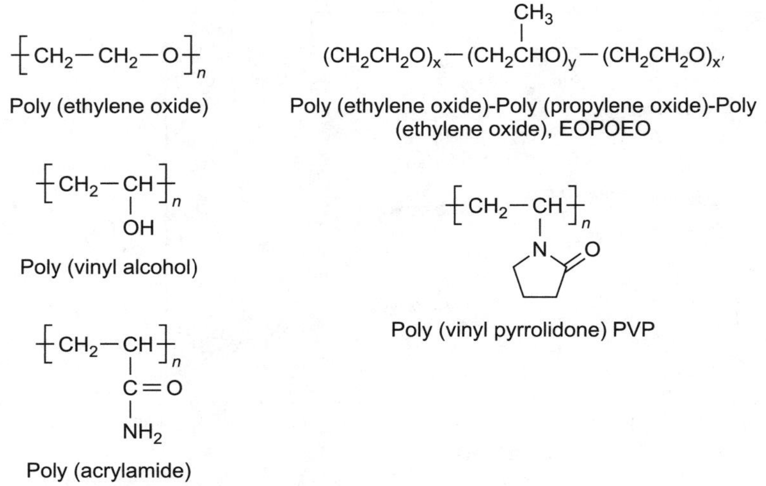

$$\left[ CH_2 - CH_2 - O \right]_n$$

Poly (ethylene oxide)

$$CH_3$$
$$|$$
$$(CH_2CH_2O)_x - (CH_2CHO)_y - (CH_2CH_2O)_{x'}$$

Poly (ethylene oxide)-Poly (propylene oxide)-Poly
(ethylene oxide), EOPOEO

$$\left[ CH_2 - CH \right]_n$$
$$\qquad\quad |$$
$$\qquad\quad OH$$

Poly (vinyl alcohol)

$$\left[ CH_2 - CH \right]_n$$

Poly (vinyl pyrrolidone) PVP

$$\left[ CH_2 - CH \right]_n$$
$$\qquad\quad |$$
$$\qquad\quad C = O$$
$$\qquad\quad |$$
$$\qquad\quad NH_2$$

Poly (acrylamide)

**FIG. 5.10**   Representative synthetic nonionic homopolymers.

The rheological behavior of aqueous pAm solutions is typically pseudoplastic. An example of the use of pAm polymer to thicken an acidic composition is disclosed in patent application WO 9419443 A1 [16].

EO-based polymers are classified as PEGs or PEOs depending upon their molecular weight [17]. Low-to-medium-molecular-weight (200 to 25,000) homopolymers of EO are referred to as PEGs and polymers with molecular weight range of 100,000 to 2,000,000 are classified as PEOs. PEG esters, and in particular the diesters, can also be used as thickeners in surfactants. The most common of this class is the PEG-6000 distearate, which is often referred to as a hydrophobically modified nonionic polyol, or HNP. Being an ester, it has a limited pH range over which it can be used. Materials thickened with this additive tend to be relatively Newtonian in their behavior. Heat is required to incorporate the thickener, which has a melting point of about 60°C. One of the limitations with this class of thickener is that in general the higher the surfactant content, the greater the concentration of thickener required to achieve the desired viscosity. As such, they are usually encountered in formulations that contain less than about 15% surfactant. The stearic acid diester of PEG, $(PEG)_n$–stearate ($n = 2$ to 175), is most often used to thicken shampoos. The esters are also used as thickeners in lotions, emulsions, cream deodorants, and hair conditioners [17].

Block copolymers of EO and PO such as EO–PO–EO, which are formed by condensing EO onto polypropylene glycol, are useful rheological additives with applications in household cleaners (toilet bowl cleaners, gels for cleaning vertical surfaces), personal care (shampoos, shaving creams, hair styling gels, antiperspirant gels, etc.), and pharmaceutical products (such as toothpastes and ointments). The viscosity build is the result of hydrogen bonding in aqueous systems, caused by the attraction of the polymer ether oxygen atoms to water protons. Alkylated EO–PO polymers are also suggested to thicken a liquid fabric softening composition [18].

Poly(*N*-vinyl-2-pyrrolidone), PVP, is available in various molecular weights (10,000, 40,000, 160,000, and 360,000) and can yield solutions of varying viscosities. PVP is best known for its unusual complexing ability toward many types of small molecules and for its physiological inertness [19]. As a thickener, it is used in biomedical, pharmaceutical, cosmetic, and personal care products (hair styling gels, shaving creams, shampoos, emollient creams and lotions, etc.).

## C.   Urethanes

The hydrophobically modified ethoxylated urethane (HEUR) rheology modifiers are intermediate molecular weight nonionic polymers which combine a hydrophilic backbone of varying chain length PEG with a hydrophobic, long-chain alcohol via a diisocyanurate linkage. The hydrophobicity of the long-chain alcohols can be adjusted by altering the alkyl chain length, grafting one or more hydrophobes onto the polyol chain, or attaching these hydrophobes either terminally or pendant to the polymer backbone. Due to the fact that HEUR rheology modifiers possess a relatively low molecular weight, particularly when compared to HASE polymers, thickening is achieved through the associative interaction between the hydrophobic portion of the molecule and other hydrophobic components in the formulation (surfactants, oils, pigments).

In the presence of surfactants, studies have indicated that the degree of association achieved with nonionic rheology modifiers tends to be greater when surfactants of lower hydrophilic–lipophilic balance (HLB) are employed [20,21]. This leads to significantly higher measured viscosities (using a fixed concentration of HEUR rheology modifier) with minor changes in the surfactant composition (Figure 5.11). The "size" of the surfactant hydrophobe at a given HLB also plays a role in determining the performance of these polymers, larger hydrophobes generally being preferred.

The rheology of HEUR-type polymers varies with a given formulation, but these polymers typically impart Newtonian behavior to the systems with which they are mixed, particularly at higher shear rates.

Due to the chemical nature of HEUR rheology modifiers, no neutralization is required to induce thickening. These materials are therefore compatible with anionic, nonionic, and cationic surfactant matrices, and are effective across a wide range of pH (3 to 13) [22]. HEUR rheology modifiers are used in cationic systems such as rinse-added fabric softeners (where anionic thickeners have

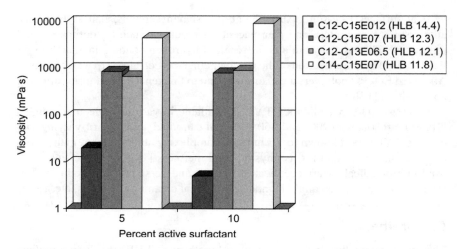

**FIG. 5.11**  Effect of nonionic surfactant HLB on viscosity of HEUR rheology modifiers.

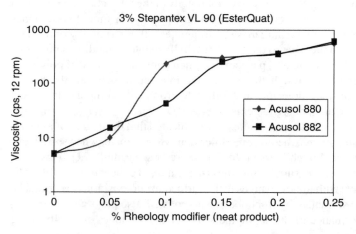

**FIG. 5.12**  Effect of rheology modifier concentration on cationic surfactant viscosity.

incompatibility problems). The new generation of highly biodegradable cationics (esterquats) deliver extremely low viscosities at use concentrations in aqueous fabric softeners. Incorporation of very low concentrations of the appropriate HEUR thickener can generate significant viscosity build and formula stabilization, providing consumer compositions with more acceptable product aesthetics (Figure 5.12).

Another area where HEUR rheology modifiers find utility is in acidic household cleaning products (typically formulated for the bathroom or kitchen). Many of the more commonly used rheology modifiers provide thickening benefits only upon neutralization (via chain–chain entanglement), and are ineffective in these types of applications. The addition of a small amount of the appropriate surfactant and HEUR rheology modifier to a system containing citric, sulfamic, or phosphoric acid can deliver acceptable viscosity, a transparent appearance, and good overall product stability. These polymers have been used successfully to thicken peroxide bleach formulations containing up to 25% hydrogen peroxide without inducing any appreciable loss of active oxygen.

## D. Synthetic Cationic Polymers

There are instances where a formulator may need to turn to cationic rheology modifiers, although these are less widely used. Synthetic cationic polymers are of three types: ammonium (primary, secondary, tertiary, and quaternary), sulfonium, and phosphonium compounds [23]. Of these, the ammonium-based polymers constitute a large class of materials with diverse applications such as additives for shampoos and soaps, as antistatic and thickening agents for rinse-added fabric softeners, in papermaking, mineral processing, and petroleum recovery, as stabilizers for emulsion polymerization, as biocides in waste water treatment, and in grease thickeners, hair sprays, and hair gels.

The copolymer of acrylamide and ammonium acrylate is used to build viscosity in rinse cycle fabric softeners. This polymer is compatible with nonionic and most cationic surfactants that are used in fabric softener formulations. The polymer is incompatible with anionic surfactants and strong oxidizing agents, and it is sensitive to electrolytes. An example of other cationic polymers useful as thickeners for aqueous acid solutions is described in patent application EP 395282 [24].

## E. Celluloses

Modified organic thickeners can be derived from naturally occurring water-insoluble polymers such as cellulose, chitin, and starch [25,26]. The most common derivatives include carboxymethyl, hydroxyethyl, hydroxypropyl, and methyl cellulose. Cationic, anionic (sulfate, phosphate), and zwitterionic derivatives have also been reported in the literature.

Sodium carboxymethyl cellulose (CMC, cellulose gum) is an anionic, water-soluble polymer (Figure 5.13). It is stable in a pH range of 4 to 10 and is compatible with most monovalent and divalent salts, as well as most anionic and nonionic materials. However, it is generally incompatible with cationic species due to its anionic nature. The structural stability of dispersions induced by CMC is highly dependent upon the concentration of the polymer. CMC is used as a thickener in toothpastes, skin creams, lotions, and food applications. The degree of

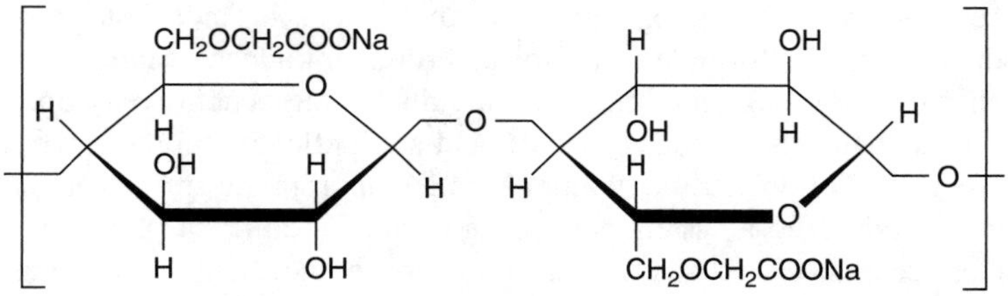

**FIG. 5.13** Idealized unit structure of CMC, with a DS of 1.0. (Reproduced with permission from Hercules Inc., Aqualon Division, Copyright 2004, Hercules Inc., Wilmington, DE.)

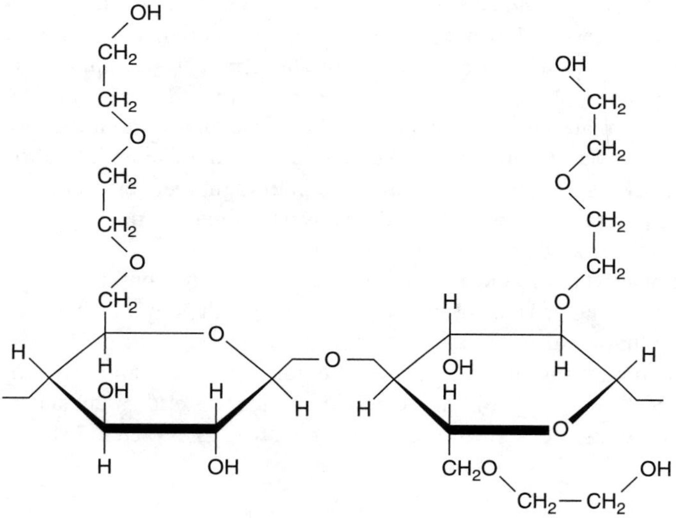

**FIG. 5.14** Idealized structure of hydroxyethyl cellulose. (Reproduced with permission from Hercules Inc., Aqualon Division, Copyright 2004, Hercules Inc., Wilmington, DE.)

substitution (DS) for a given CMC grade is the average number of substituted hydroxyl groups per ring. Therefore, the theoretical maximum DS is 3. The maximum substitution level of commercial CMC is a DS of 1.4. Thixotropy in CMC typically increases with decreasing DS.

Hydroxyethyl cellulose (HEC) is a water-soluble nonionic polymer having the general structure shown in Figure 5.14 [27]. The water solubility of HEC depends upon DS and the molar substitution (MS; also termed moles of substitution).

MS is the average number of moles of substitution (in the case of HEC, hydroxy-ethyl and ethoxy units) added per anhydroglucose ring. The MS value, unlike DS, can exceed 3 in the case of HEC, since side chains of PEO can form. Commercial water-soluble HEC samples have DS values in the range 0.85 to 1.35 and MS values in the range 1.3 to 3.4. HEC aqueous dispersions are pseudoplastic and thermally reversible. HEC is compatible with nonionic, cationic, and anionic materials (salts and surfactants). It is stable in the pH range 2 to 11. As a thickener, it is used in hair care products (conditioners, etc.), liquid soaps, shaving products, cationic lotions, antiperspirants, and deodorants.

Hydroxypropyl methyl cellulose and methyl cellulose are also water-soluble nonionic polymers [28]. They are compatible with inorganic salts and ionic species up to a certain concentration. Methyl cellulose can be salted out of solution when the concentration of electrolytes or other dissolved materials exceeds certain limits. Hydroxypropyl methyl cellulose has a higher tolerance for salts in solution than methyl cellulose. Both are stable over a pH range of 3 to 11. Commercial water-soluble methyl cellulose products have a methoxy DS of 1.64 to 1.92. A DS of lower than 1.64 yields material with lower water solubility. The methoxy DS in hydroxypropyl methyl cellulose ranges from 1.3 to 2. The hydroxypropyl MS ranges from 0.13 to 0.82. Methyl cellulose and hydroxypropyl methyl cellulose polymers have a number of applications and are used as thickeners in latex paints, food products, shampoos, creams and lotions, and cleansing gels. U.S. Patent 5,565,421 is an example of the use of hydroxypropyl methyl cellulose polymer to gel a light-duty liquid detergent containing anionic surfactants [29].

As is the case for HASE and cross-polymers, grafting of long-chain alkyl hydrophobes onto water-soluble cellulosic polymers leads to modified solution properties such as enhanced viscosity, surface activity, and unusual rheological properties [30–35]. Associative cellulosic thickeners build viscosity through two mechanisms: hydrogen bonding with water molecules (as with the unmodified cellulosic polymers) and micellar interactions that occur between the hydrophobic groups. The hydrophobic association can be viewed as pseudo-crosslinks which induce a three-dimensional network. The hydrophobic groups on the polymer can also interact more favorably with surfactant micelles to build viscosity in dispersions. The enhanced solution viscosity of C16 hydrophobically modified HEC is the result of intermolecular associations via the hydrophobic groups (Figure 5.15). Primary applications, sensitivity to electrolytes, and pH stability of this hydrophobically modified polymer are similar to those of unmodified HEC.

A polymeric quaternary ammonium salt of HEC, polyquaternium-24, in combination with certain primary surfactants, salts, and other viscosifying agents can be used as a thickener in personal care products such as shampoos, hair conditioners, creams, and aftershave gels.

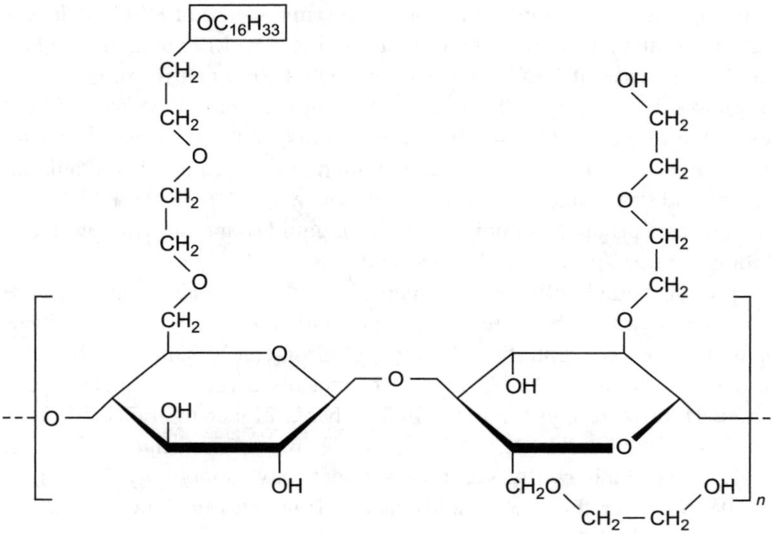

**FIG. 5.15** Idealized structure of hydrophobically modified HEC, with hydroxyethyl MS = 2.5. (Reproduced with permission from Hercules Inc., Aqualon Division, Copyright 2004, Hercules Inc., Wilmington, DE.)

## F.  Gums

Several natural biopolymers originally developed for use in the food industry, including xanthan, gum arabic, carrageenan, succinoglycan, gellan, locust bean gum, and alginates, have found use recently in the detergents industry. From a commercial point of view the most significant of these today is xanthan, a water-soluble polymer based on an anionic heteropolysaccharide, and produced by bacterial fermentation, followed by recovery of the resulting exopolymer. The organism employed during the fermentation process is a species of the bacterium *Xanthamonas campestris*. The polymer, due to its nature, is biodegradable, and it is necessary to ensure that formulations using this thickener are adequately preserved to prevent bacterial spoilage from taking place over the life of the product.

Xanthan is a slightly hygroscopic powder that requires hydrating prior to use. As a consequence, it is generally introduced into the water being used to prepare the formulation at the beginning of the processing, and stirred well to disperse completely and hydrate prior to addition of the remaining components. The behavior of xanthan is extremely pseudoplastic, with very high viscosities being developed under conditions of low shear, as shown in Figure 5.16. As shear is removed, the solution rebuilds structure almost instantaneously. Xanthan is thus a good

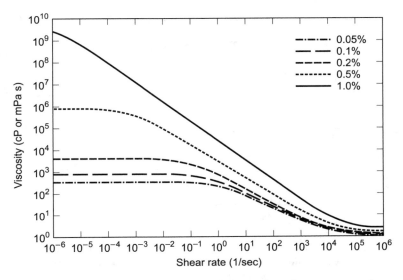

**FIG. 5.16**   Flow curves for aqueous solutions of xanthan gum at various concentrations. (Reproduced with permission from CP Kelco ApS, Copyright 2004, CP Kelco, San Diego, CA.)

candidate for providing suspending properties to formulations, as well as giving vertical cling properties. Slight turbidity is often observed in finished products.

Xanthan gum is one of the few rheology modifiers stable over a wide range of pH, including both acidic and alkaline ranges (Figure 5.17). It is thus suitable for acid cleaners and scale removers as well as the traditionally neutral detergents and alkaline hard surface cleaners. Xanthan gum can be quite tolerant of both monovalent and divalent metal salts, but the presence of trivalent metal ions ($Fe^{3+}$, $Al^{3+}$, and $Cr^{3+}$) often leads to marked crosslinking, in some cases causing precipitation. Sequestering such ions will ensure stability.

Figure 5.18 shows the viscosities of some common natural polymers over a range of shear rates. At low shear rates, solutions of xanthan gum have approximately 15 times the viscosity of guar gum and an even higher margin over the viscosity of CMC and sodium alginate. This further explains the strength of xanthan gum as a stabilizer for suspensions and for providing vertical cling.

We have already mentioned that xanthan gum solutions are tolerant to both acids and bases. Solutions of xanthan gum also have excellent compatibility with many surfactants, water-miscible solvents, and other thickeners. As an anionic polysaccharide, xanthan gum is most stable with anionic surfactants (up to 20% active), nonionic surfactants (up to 40% active), and amphoteric surfactants (up to

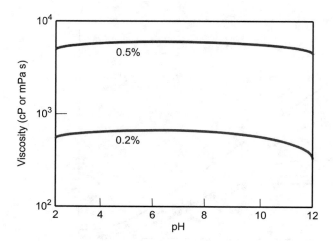

**FIG. 5.17** Viscosity as a function of pH for xanthan gum solutions at 0.2 and 0.5% solids. (Reproduced with permission from CP Kelco ApS, Copyright 2004, CP Kelco, San Diego, CA.)

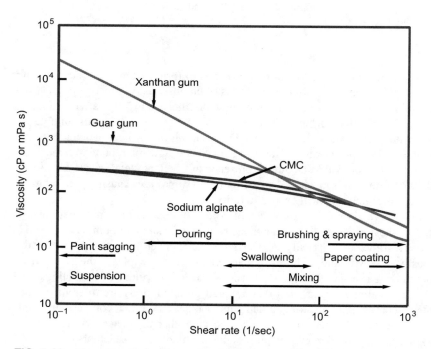

**FIG. 5.18** Flow curves of various natural polymers. (Reproduced with permission from CP Kelco ApS, Copyright 2004, CP Kelco, San Diego, CA.)

25% active) depending upon the composition of the surfactant. Generally, xanthan gum is not recommended for use with cationic surfactants, bleach solutions, and strong reducing environments.

Although encountered less frequently than xanthan, carrageenan is another of the gums sometimes used in liquid detergents. This polysaccharide is extracted from certain varieties of red seaweed [36]. The polysaccharide is made up of repeating galactose units and 3,6-anhydrogalactose, both sulfated and nonsulfated, joined by alternating $\alpha$1-3 and $\beta$1-4 glycosidic linkages, and is available as the kappa, iota, or lambda forms, with varying sulfate galactose ratios (1:2, 2:2, and 3:2 for the three versions, respectively). The lambda type is the most soluble and dissolves in cold water. The kappa and iota types of carrageenan form thermally reversible gels at low concentrations (1% by weight). Both kappa and iota types also form strong gels in the presence of specific ions ($K^+$, $Rb^+$, and $Ca^{2+}$) [37]. The carrageenans are stable in the pH range 3.5 to 9, with all three types undergoing hydrolysis at pH $< 3.5$. Carrageenans are used in toothpastes, skin creams, lotions, and food products (such as puddings, chocolate milk, and ice cream). One of the potential drawbacks is the fact that carrageenan systems have the tendency to show syneresis.

Alginates are extracted from brown seaweed, and can be thickeners or gellifiers depending on the type of alginate and the matrix. They have various conformations depending on the source of the seaweed. In the alginic acid form or as the Ca salt they have very low solubilities, but the Na or K salts are soluble. The Mg salt is the only soluble divalent salt. Even hard water will cause thickening of alginates, but the process can be controlled by the use of complexants and chelatants. Alginates show rheology much closer to Newtonian than most of the other gums.

Locust bean gum can be extracted from the European carob tree. As the extract, with impurities removed, it gives clear solutions. It shows significant synergy with carrageenan and xanthan, usually at about 50/50 levels, and this synergy tends to eliminate the syneresis often seen with carrageenan. Being insoluble in cold water, heat is required to obtain solutions.

Guar gum has a major processing advantage in that it is soluble in cold water. However, it is very much less pseudoplastic than xanthan and tends to give formulations with a long or "stringy" rheology.

## IV.  INORGANIC THICKENERS

The two most commonly used inorganic additives for rheology modification of liquid detergents are salts and smectite clays. Other inorganics such as silica and alumina have found more limited use in detergent cleaners. Each of these additive types has a distinctly different mechanism for modifying the rheology of a detergent, and very critical sensitivities within a formulation.

## A.  Inorganic Salts

Perhaps the greatest benefit of using inorganic salts to thicken a system is the low cost it adds to a formulation. It is not solely salt addition that drives thickening in a detergent, but rather the interaction that the salt has with other components of the system, most notably charged (anionic and cationic) surfactants. It is this interaction between salt and surfactant of which formulators typically attempt to take advantage, and this strategy works best in detergents with relatively high surfactant levels. Liquid laundry detergents, hand dishwashing (light-duty) liquids, and hand soaps are common examples of products thickened by simple salt addition.

However, the ability to thicken surfactant systems with salt addition is not universal, even in products with relatively high levels of surfactant. Not all anionic surfactants respond to added salt with increased viscosity, but the more commonly used alkylbenzene sulfonates (ABSs) and alkyl ether sulfates (AESs) do. In particular for AESs the shorter the EO segment, the more sensitive they are to salt addition, with the alkyl sulfate (no EO) at the most responsive extreme. The ABSs can be thickened with salt addition as well, but only over a narrow viscosity range before they salt out of solution. Addition of certain nonionics, especially alkyl polyglucosides (APGs) or alkanolamides, can enhance the thickening effect in anionic surfactants, often reducing the amount of salt needed to achieve a given viscosity.

Fundamental studies of salt effects on well-characterized anionic and cationic surfactant solutions have provided a mechanistic picture that links the surfactant structures and the rheology. In Chapter 4, rheology modification by neat surfactant solutions was detailed. This rheology is a complex result of, among other things, surfactant type(s) and concentration, mixture ratios, pH, and added solvents, all of which determine the structure of the surfactant aggregates, directly affecting the solution viscosity. Addition of salt provides another way of changing how surfactants develop structure. For example, Rybicki [38] showed decreasing viscosity of low-concentration sodium dodecylbenzene sulfonate solutions with addition of many different salts, while Wang [39] showed increased viscosity at higher surfactant concentration. In the former case, the salt is believed to screen the electrostatic field on the spherical micelle surface, reducing the effective volume of the micelle and thus the relative viscosity. In the latter case, the added salt drives a structural change from spherical to elongated, asymmetric micelles. The magnitude of the viscosity increase has also been observed to be dependent on the type of salt, and especially the type of counterion that is chosen (e.g., $Na^+$ vs. $Li^+$ vs. $Mg^{2+}$). Additionally, Gamboa and Sepulveda [40] have shown NaCl can increase the viscosity of anionic sodium dodecyl sulfate solutions and cationic cetyltrimethylammonium bromide solutions.

There are a number of potential downsides of this thickening technique. A formulator usually can control the viscosity only within a limited range, and often there is no added control over the shape of the flow curve. That is, a Newtonian product will remain Newtonian in flow rather than building in rheological features such as pseudoplasticity, thixotropy, or yield stresses. These are not restrictive limitations if one is only attempting to provide a little "body" to a formulation through small viscosity increases, but it is unlikely that salt will provide properties such as suspendability or vertical cling. Also, inorganic salts may interact negatively with other formulation components (e.g., electrolytes), resulting in a cloudy, or in the worst case unstable, product. Salts can promote irritation and possibly create corrosive conditions at high use levels. Finally, seemingly small differences in the byproducts that are formed during the surfactant synthesis can affect the salt thickening profile, resulting in potential for product variations if one does not use good process control.

## B.  Inorganic Clays

Smectite clays are naturally occurring water-swellable clays. Often, one finds the terms smectite, bentonite, hectorite, saponite, montmorillonite, and magnesium aluminum silicate (MAS) clays used interchangeably, leading to the potential for confusion. For clarity, smectite is the name of the subgroup of clays that encompass hectorite, saponite, and montmorillonite. Bentonite is the geological term commonly used to refer to smectite clays, the latter being a mineralogical term. The differences between hectorite, saponite, and montmorillonite clays lie in their chemical makeup and structure, with the latter two having MAS compositions [41]. All of the natural clays are mined and purified for use, typically being sold in powder form that requires a hydration step. The extent of purification of these clays can have an impact on the efficiency, clarity, and cost. The synthetic hectorite clays such as Laponite (Southern Clay Products) are typically sold as powders and similarly require hydration prior to use, but they are prepared free from impurities.

Both natural and synthetic clays are used as rheology modifiers for liquid detergents. They can stabilize emulsions and provide excellent particle suspension via the development of yield stresses. They can tolerate significant levels of water-miscible cosolvents like glycols and glycol ethers, thus finding greatest usage in surface cleaners of various types (e.g., toilet and oven cleaners). They are some of the few rheology modifiers stable to hypochlorite, and they can be formulated in products covering a wide pH range (roughly 3 to 12, but can be as low as 1), although they are not compatible with cationic species. Since these materials function via electrostatic interactions, the rheology modification they impart is essentially temperature-insensitive, and their mineral nature makes them resistant to biological degradation. After inorganic salts, natural clays tend to be the least expensive of the rheology modifiers, although the purification steps result in higher costs than less purified materials. The synthetic hectorites are the most expensive on a per weight basis, but they also tend to be significantly more efficient than the natural clays, resulting in potential for lower use levels in formulations. Additionally, these synthetic clays can form clear, aqueous solutions due to their small particle dimensions.

The hydration step of inorganic clays is vital to their effective usage in liquid formulations. In powdered form, all smectite clays (including the synthetic analogs) exist as aggregates of stacks of primary, disk-like clay platelets. When stirred in water, these aggregates break up toward the individual stacks, which can then hydrate, swell, and delaminate to the primary clay particle. This is shown schematically for the Laponite example in Figure 5.19. Energetic mixing and

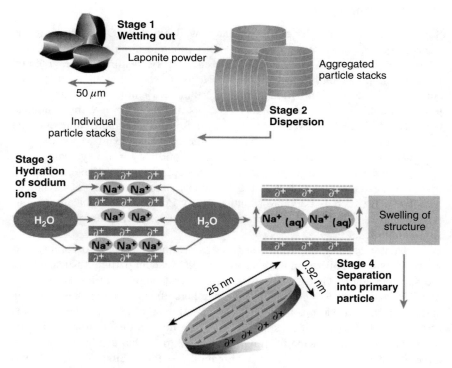

**FIG. 5.19**  Schematic of the wetting and delamination of inorganic clay particles. (Supplied by and used with the permission of Southern Clay Products.)

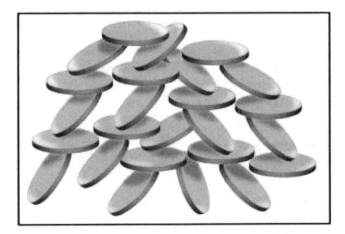

**FIG. 5.20**  Schematic of the "house-of-cards" structure derived from clay platelet edge–face interactions. (Supplied by and used with the permission of Southern Clay Products.)

sufficient free water is required for full hydration of the particles and for full effectiveness to be achieved. Completely delaminated, each primary clay particle has a thickness typically of the order of 1 to 3 nm with face dimension dependent on clay type, ranging from about 25 to 1000 nm. The flat, larger face holds an anionic charge, while the thin outer edge of the plate is slightly positive. At sufficient concentration, the clay particles align and fill space in what is known as a house-of-cards structure, with the positive edges of one particle interacting with the negatively charged face of another, as shown in Figure 5.20. This stacking provides structure in the aqueous system resulting in the creation of a yield stress.

These clays are unique in the fact that although they can build yield structures, they typically do not provide significant thickening. Once sufficient shear is applied to break the three-dimensional structure, the small clay particles provide minimal resistance, and thus the viscosity decreases to essentially that of the clay-free system. (This is one reason clays are usually used in conjunction with co-thickeners.) Once the fluid is brought again to rest, the clay platelets reorient and rebuild the yield stress. The yield value achieved and the time it takes to rebuild after breakdown are dependent on clay concentration and can be low and slow at low usage levels. Thus, clay-modified systems are examples of thixotropic, highly pseudoplastic, or yield-containing solutions. The result can be readily pourable systems that maintain a stable suspension or provide a degree of wall cling after spraying.

As described above, the ability of smectite clays to act as rheology modifiers is a result of their interparticle electrostatic interactions. Thus, their behavior and stability are very sensitive to relatively low levels of electrolyte in a formulation. For example, Mourchid and co-workers [42] have studied the phase behavior and stability of aqueous Laponite solutions in the presence of NaCl. For solutions above roughly 0.01 M, they find a flocculated clay state, independent of the concentration of the clay. In clay-modified detergent formulations such flocculation can lead to syneresis and instability, as well as increased opacity.

It is commonly recommended to utilize a second rheology modifier along with smectite clays in order to increase formulation stability and/or to provide more precise control to the overall rheology. Often, synergistic rheology is observed in these blends, resulting in viscosity increases greater than what would be expected based on the behavior of the two individual modifiers. This can lead to utilization of lower levels of each thickener, reducing total formulation cost and potentially eliminating some of the drawbacks of using the higher level of the co-thickener. Recommended co-thickeners include organic gums such as xanthan gum, cellulosics such as CMC and HEC, and polyacrylics such as HASE and carbomer polymers.

## V.  SUMMARY

There are many distinct types of rheology modifiers to which a formulator can turn so as to achieve the flow characteristics necessary for a liquid detergent formulation. Deciding where to start, or even who to contact for help, can be complicated even more by the numerous unique "flavors" that are available within each larger class of modifier. Initially, one can try to narrow the choices based on some general aspects of the formulation, such as pH or salt level. Table 5.3 provides a summary of each of the technology classes discussed in this chapter, their broad applicability under various formulation conditions, and some handling considerations.

It is clear that some classes of rheology modifiers will have more utility in certain detergent systems than in others. For example, the subset of modifiers that function in the neutral-to-alkaline range of pH would not be suitable for an acid-based surface cleaner at very low pH. Table 5.4 provides some guidance for choosing a rheology modifier based on different detergent applications.

These summary tables, along with the accompanying discussion throughout the chapter, provide formulators with much of the background knowledge needed to make an educated initial choice of rheology modifier for their specific formulation needs. However, with the on-going growth of liquid detergents in the consumer market, there is a corresponding need to differentiate these products to the consumer. Formulations will get more intricate, the demands on the rheology modifier system will become more complex, and the current stable of modifiers may not meet these demands. Solutions to these requirements may come from unique mixtures of available modifiers, or they may come from newly developed chemistries within the various classes of rheology modifiers. As many of the companies that

**TABLE 5.3** Summary of the Applicability of Various Rheology Modifiers, with their Common As-Supplied Formats

| Class | Electrolyte tolerance | pH range | Pseudoplasticity | Neutralization required | Typical form of raw material |
|---|---|---|---|---|---|
| ASE (crosslinked) | Moderate | Neutral to alkaline | High | Yes | Liquid |
| ASE (noncrosslinked) | Excellent | Neutral to alkaline | Low | Yes | Liquid |
| Carbomers | Poor | Neutral to alkaline | High | Yes | Powder |
| Cellulosics (ionic) | Poor | Neutral to alkaline | Medium | Yes | Powder |
| Cellulosics (nonionic) | Excellent | Acidic to alkaline | Medium | No | Powder |
| Cross-polymers | Poor | Neutral to alkaline | High | Yes | Powder |
| HASE | Good | Neutral to alkaline | High | Yes | Liquid |
| HEUR | Good | Acidic to alkaline | Low | No | Liquid |
| Synthetic nonionics/HNP | Moderate | Acidic to alkaline | Low | No | Powder |
| Inorganic clays | Poor | Acidic to alkaline | Very high | No | Powder |
| Xanthan | Good | Acidic to alkaline | High | No | Powder |

**TABLE 5.4** Summary of Preferred Rheology Modifiers by Application

| | ASE | HASE | Carbomer/cross-polymers | HEUR | Xanthan | Inorganic clays | Cellulosics | Synthetic nonionics/HNP | Salts |
|---|---|---|---|---|---|---|---|---|---|
| Hand dishwashing | + | + | | | | | + | + | + |
| Laundry | | + | | | | | | + | |
| Auto dishwashing | | | | | | | | | |
| Chlorinated | | | + | | | | | | |
| Nonchlorinated | + | | + | | + | | | | |
| Hand soap | | + | | | + | | | + | + |
| Surface cleaners | | | | | | | | | |
| Acidic | | | | + | + | + | | | |
| Alkaline | + | + | + | | | + | | | |
| Peroxide | | | + | + | | + | | | |
| Hypochlorite | | | + | | | + | | | |
| Fabric softeners | + | | | + | | + | + | | |
| Shampoos/conditioners | + | | | | | + | | + | + |

offer rheology modifiers have active development programs, formulators should be encouraged to contact the technical staff of these suppliers for further specific guidance for individual formulation needs.

## REFERENCES

1. Hunting, A.L.L., *Cosmet. Toilet.*, March, 53, 1982.
2. Farooq, A., in *Handbook for Detergents*, Surfactant Science Series, Vol. 82, Zoller, U. and Broze, G., Eds., Marcel Dekker, New York, 1999, chap. 22, p. 757.
3. Berg, J.M., Lundberg, D.J., and Glass, J.E., *Prog. Org. Coat.*, 17, 155, 1989.
4. Lundberg, D.J., Glass, J.E., and Eley, R.R., *Polym. Mater. Sci. Eng.*, 61, 533, 1989.
5. Sau, A.C. and Landoll, L.M., *Adv. Chem. Ser.*, 223, 343, 1989.
6. Glass, J.E., Lundberg, D.J., Ma, Z., Karunasema, A., and Brown, R.G., *Proceedings of the 17th Waterborne Coatings Symposium*, New Orleans, LA, 1990, p. 102.
7. Lundberg, D.J., Glass, J.E., and Eley, R.R., *J. Rheol.*, 35, 1255, 1991.
8. Aubry, T. and Moan, M., *J. Rheol.*, 40, 441, 1996.
9. Svanholm, T., Molenaar, F., and Toussaint, A., *Prog. Org. Coat.*, 30, 159, 1997.
10. Tirtaatmadja, V., Tam, K.C., and Jenkins, R.D., *AIChE J.*, 44, 2756, 1998.
11. Kaczmarski, J.P., Tang, X., Ma, Z., and Glass, J.E., *Colloids Surf.*, 147, 39, 1999.
12. Nagarajan, K. and Ambuter, H., in *Liquid Detergents*, Surfactant Science Series, Vol. 67, Lai, K.Y., Ed., Marcel Dekker, New York, 1996, chap. 5, p. 129.
13. Reeve, P., *Proceedings of the 5th World Surfactants Congress*, CESIO, Florence, 2000, Vol. 1, p. 367.
14. Reeve, P., *J. Com. Esp. Deterg.*, 28, 161, 1998.
15. Reeve, P., *J. Com. Esp. Deterg.*, 29, 151, 1999.
16. Janota, T.E., Krogh, J.A., and Miller, J.C., Patent WO 9419443 A1, 1997, to Exxon Chemical Patents, Inc.
17. Clarke, M.T., in *Rheological Properties of Cosmetics and Toiletries*, Laba, D., Ed., Marcel Dekker, New York, 1993, p. 55.
18. Ceulemans, R.A.A., DeBlock, F.J.M., and Hubesch, B.A.J., European Patent EP 799887 A1, 1997, to Procter & Gamble.
19. Molyneux, P., Ed., *Water-Soluble Synthetic Polymers: Properties and Behavior*, CRC Press, Boca Raton, FL, 1982, p. 147.
20. Jones, C., *Proceedings of the 4th World Surfactants Congress*, CESIO, Barcelona, 1996, Vol. 2, p. 439.
21. Reeve, P., *Proceedings of the XXVII CED Conference*, Barcelona, 1997, Vol. 27, p. 159.
22. Acusol 880 and 882 Detergent Grade Rheology Modifiers and Stabilizer, Rohm and Haas technical literature FC-397a, 1998.
23. Hoover, M.F., *J. Macromol. Sci. Chem.*, A4, 1327, 1970.
24. Hawe, M. and Farrar, D., European Patent EP 395282 A2, 1990, to Allied Colloids Ltd.
25. McCormick, C.L., in *Structural Design of Water-Soluble Copolymers, Water-Soluble Polymers*, Shalaby, S.W., McCormick, C.L., and Butler, G.B., Eds., American Chemical Society, Washington D.C., 1991, p. 2.

26. Hebeish, A., Waly, A., El Rafie, M.H., and El Sheikh, M.A., in *Book of Abstracts, 213th ACS National Meeting*, San Francisco, April 13–17, 1997, American Chemical Society, Washington D.C., 1997, p. 149.

27. Natrosol Hydroxyethylcellulose: Physical and Chemical Properties, Aqualon technical literature 250-11C, 1987.

28. Methocel Cellulose Ethers in Personal Care Products, technical literature, Dow Chemical Company, 1992.

29. Aszman, H., Gomes, G., and Lee, C., U.S. Patent 5,565,421, 1996, to Colgate-Palmolive.

30. Lundberg, D.J., Alahapperuna, Z.M.K., and Glass, J.E., in *Polymers as Rheology Modifiers*, Schulz, D.N. and Glass, J.E., Eds., American Chemical Society, Washington D.C., 1991, p. 234.

31. Wang, T.K., Iliopoulos, I., and Audebert, R., in *Structural Design of Water-Soluble Copolymers, Water-Soluble Polymers*, Shalaby, S.W., McCormick, C.L., and Butler, G.B., Eds., American Chemical Society, Washington D.C., 1989, p. 218.

32. Glass, J.E., Ed., *Polymers in Aqueous Media*, American Chemical Society, Washington D.C., 1989, p. 113 (giving examples of model associative thickeners and associative thickeners with commercial potential).

33. Tarng, M.-R., Kaczmarski, J.P., Lundberg, D.J., and Glass, J.E., *Adv. Chem. (Hydrophilic Polym.)*, 248, 305, 1996.

34. Hawe, M., *Nord Pulp Paper Res.*, J8, 188, 1993.

35. Bieleman, J.H., *Verfkroniek*, 68, 29, 1995.

36. The Carrageenan People: Introductory Bulletin A-1, FMC Corporation Marine Colloids Division, 1988.

37. Clark, A.H. and Ross-Murphy, S.B., *Adv. Polym. Sci.*, 83, 60, 1987.

38. Rybicki, E., *Tenside Surfactants Detergents*, 27, 5, 1990.

39. Wang, J., *Colloids Surf. A*, 70, 1, 1993.

40. Gamboa, C. and Sepulveda, L., *J. Colloid Interface Sci.*, 113, 2, 1986.

41. Ciullo, P.A., in *Industrial Minerals and Their Uses: A Handbook and Formulary*, Ciullo, P.A., Ed., William Andrews, Westwood, NJ, 1996, chap. 2, p. 17.

42. Mourchid, A., *Langmuir*, 14, 17, 1998.

# 6

# Nonaqueous Surfactant Systems

**MARIE SJÖBERG and TORBJÖRN WÄRNHEIM*** Institute for Surface Chemistry, Stockholm, Sweden

## I.  INTRODUCTION

There has been much interest in studying surfactant aggregation in polar solvents other than water over the last few decades. In a large number of studies various surfactant systems have been mapped and evidence for self-assembly of surfactants in some nonaqueous polar solvents has been published. During the last few years more detailed information on the structure of the aggregates and on the characteristics of the aggregation processes have been provided.

The research on aggregation of surfactants in nonaqueous, polar solvent systems can be motivated, mainly, with two different arguments. First, are the basic considerations of amphiphile aggregation involving a description of the hydrophobic interaction leading to, for example, micelle and liquid crystal formation. What can be learned from comparing water with other polar solvents? Much work has been performed to elucidate those properties of the solvent that are essential in order to obtain a hydrophobic (or "solvophobic") interaction. Comparisons of critical micelle concentrations in different solvents with parameters characterizing the solvent are numerous in the literature [1,2].

Second, there are technical applications where amphiphile aggregates and structures are needed to promote a specific effect, while circumstances may prevent the particular use of water due to certain reactions, corrosion, or other specific

---

*Current affiliation: ACO HUD AB Upplands Väsby, Sweden

interactions with water. Of particular interest in this context are, for example, alcohol-based systems for cleaning purposes [3]. Another highly relevant area is that of chemical reactions in an aprotic solvent such as formamide [4–6].

This review deals with the first, fundamental, point, in particular the formation of micelles and the mapping of phase equilibria. This is a logical starting point, since a prerequisite for most applied work in the field is some knowledge of aggregation processes and the relevant phase diagrams.

Methodological questions have often been raised when studying nonaqueous systems, since many early studies on micellization were performed using indirect methods for detecting aggregation [7–14]. This has caused considerable confusion due to apparently irreconcilable results. Also, several studies have pointed out the difference between a proper micellization process and ordinary aggregation. As is discussed in this chapter, depending on the combination of solvent and surfactant [15–18], it is possible to have a cooperative aggregation (micelle formation) as well as a more gradual aggregation process.

## II. MICELLAR AGGREGATION

Micellization has been studied in a large number of nonaqueous polar solvents, such as different alcohols, formamide, fused salts [19–26], hydrazine, hydrogen fluoride [27], and $N$-methylsydnone [28,29]. However, most of the early investigations used indirect methods such as surface tension measurements or conductimetry for the detection of surfactant aggregation. More recently, direct methods have been used to prove the existence of aggregates in the solution phase of polar solvent other than water. For example, PGSE-NMR [17], fluorescence spectroscopy [30], and SANS [31] have proven to be powerful methods for probing micelle formation in aqueous and nonaqueous systems.

The nonaqueous polar solvent that has been studied most extensively in this context is probably formamide. Lattes and co-workers have studied the aggregation of surfactants in formamide [32–35]. They have investigated the SDS–formamide and the $C_{16}$TABr–formamide systems. A sharp rise in solubility of the surfactant with increasing temperature was noted in these systems and interpreted as the Krafft point, i.e., the temperature where the monomeric solubility of the surfactant exceeds the c.m.c. The c.m.c. as well as the Krafft point were found to be considerably higher in formamide than in water for both surfactants. Other studies of surfactant aggregation in formamide, where aggregates were not found, have been performed at temperatures below the Krafft point [36–38]. The $C_{16}$TABr–formamide system has been widely studied with a number of different techniques, such as NMR relaxation and self-diffusion [18,35,39,40], small-angle x-ray scattering [34], positron annihilation [38], or Raman spectroscopy [41]. Most studies agree that aggregates start to form at considerably higher surfactant concentration than in water and that they are considerably smaller than in water. An aggregate radius of 9 Å was found at a concentration close to the c.m.c. [34], while it was

**TABLE 6.1**  Interfacial Tension Between Solvent and Hydrocarbon, and
Dielectric Constant for the Solvent

| Solvent | Interfacial tension$^a$ (mN/m) | Dielectric constant |
|---------|--------------------------------|---------------------|
| Water | $50^b$ | 78 |
| Glycerol | $29.7^c$ | 42 |
| Formamide | $27.3^c$ | 109 |
| Ethylene glycol | $17.2^c$ | 37 |
| N-methylformamide | $12.5^c$ | 182 |

$^a$At 20°C.
$^b$Against hexadecane.
$^c$Against dodecane.
*Source*: From Wärnheim, T. and Jönsson, A., *J. Colloid Interface Sci.*, 125, 627, 1988.

found to be about 15 Å at five times the c.m.c. [18]. This corresponds to an aggregation number of approximately one third of that in water. Aggregate growth with concentration has been verified both in a study of the solvent binding to the aggregates [39] and in a study of the counterion binding [40]. An increase in the micellar size with increasing surfactant concentration was also found in the SDS–formamide system [42].

An investigation of counterion binding of a cationic surfactant, $C_{16}$TAF, in formamide, ethylene glycol, and water showed that the degree of counterion binding is very different in the different solvents, depending on the dielectric constant of the solvent [40]: high in water and ethylene glycol but lower in formamide. Calculations confirmed that the effects of the dielectric constant (Table 6.1) could account for this trend [40]. This observation supports the study of Binana-Limbele and Zana [15], who found the micelles to be small and highly ionized in formamide.

The aggregation process of cationic and anionic surfactants in formamide has also been studied by SANS [31,43]. For N-alkylpyridinium halides, it was found that at an alkyl chain length of 12 carbons, only small, unstructured aggregates are formed while at a chain length of 16 to 20 carbons micelles are sole species. The micelles are smaller and with a higher charge than in water. Moreover, in a study of SDS in formamide it was found that micelles are formed but the mechanism of self-association is in agreement with a multiple equilibrium model rather than a pseudophase model. That is, the aggregates increase in size with surfactant concentration over a large region. The authors of the study conclude that the aggregation process in formamide is analogous to that of short-chained surfactants in water.

Micelles of cationic surfactants have been found to form both in glycerol [44] and in ethylene glycol [18]. The micelle formation of $C_{16}$PyBr in ethylene glycol and glycerol was studied with surfactant-selective electrodes [45,46]. The monomer concentration could in this way be measured at different total surfactant concentrations, and it was concluded that there is some premicellar aggregation and that the c.m.c. is not very well defined. The dissociation of $C_{16}PyBr$ micelles in mixtures of water and ethylene glycol has been studied and it was concluded that the degree of counterion association decreases with increasing amount of ethylene glycol in the solvent [47]. This is consistent with earlier estimations of the counterion binding in the water–ethylene glycol system, where conductivity measurements suggested a decrease in counterion association when ethylene glycol was added to the water [48].

For micelles of $C_{14}TABr$ in mixtures of water and ethylene glycol the solvent penetration in the micelles was investigated [30] through the fluorescence anisotropy of different probe molecules residing in different regions of the micelles. When the ethylene glycol–water ratio is increased the microviscosity in the hydrophobic regions of the micelles is constant while the microviscosity at the micellar surface increases. This indicates that the micellar interior does not change but the solvent penetration at the micellar surface increases upon addition of the ethylene glycol cosolvent.

Aggregation of nonionic surfactants in these nonaqueous solvents could, in principle, be more energetically favorable than that of ionic surfactants since, at least in water, the repulsive interaction between the polar head groups is smaller. The values of c.m.c. of different polyethylene glycol alkyl ethers ($C_iE_j$) have been determined in different nonaqueous solvents [17,49–55]. Different $C_iE_j$–formamide systems have been investigated using NMR self-diffusion [17]. Micelles are formed but are smaller than in water. In contrast to what is found in water, no micellar growth occurs at high temperatures, high surfactant concentration, or when approaching the lower consolute temperature. The same systems were later examined in a calorimetric study [16], and it was found that the enthalpies of micelle formation of $C_iE_j$ in formamide are much smaller and not as temperature dependent as in water. The aggregation numbers were found to be smaller, and for the $C_{12}E_j$ surfactants the smoothly bended titration curves indicate that the micelle formation extends over a significant concentration region.

Ruiz et al. have investigated the micellization of the nonionic surfactant Triton X-100 (p-tert-octyl-phenoxy(9.5)polyethylene ether) in mixed solvents of water and ethylene glycol [56], or water and formamide [57]. They found that for both solvent combinations there is a decrease in the micellar size, due to a decrease in the micellar aggregation number, with increasing cosolvent concentration. Moreover, the cloud point for the nonionic surfactant was found to increase with addition of formamide or ethylene glycol. This increase in the cloud point can be explained by the increased solubility of the EO chain in the solvent at high temperatures with increasing cosolvent content. A fluorescence study suggested that there is a considerable contact of the cosolvent with the inner region of the micelles for the Triton X-100 surfactant in water–formamide mixtures.

The effect of three alcohols (glycerol, propylene glycol, and 1-propanol) on the surfactant aggregation of $C_{12}E_8$ in water has been studied by Kunieda et al. [58].

They observed that addition of propylene glycol or 1-propanol results in smaller micelles and more solvent penetration into the palisade layer of the aggregates. In contrast, in the corresponding system with glycerol the micelles grow in size with increasing glycerol content. Moreover, the cloud point for the nonionic surfactant was found to decrease with increasing glycerol concentration. Both these observations are opposite to what has been found for the nonionic surfactant in formamide or ethylene glycol (higher cloud point and smaller micelles with more cosolvent). Both SAXS and PGSE-NMR results reveal that the addition of glycerol induces dehydration of the EO chain of the surfactant. The consequence of this dehydration is that the surfactant becomes increasingly hydrophobic the higher the glycerol content, which is consistent with larger micelles and lower cloud point upon glycerol addition. This is compared with a "salting-out" effect, i.e., when the added species are depleted from the surfactant film [59].

In a SANS study of the aggregation of nonionic surfactants in water mixed with glycerol or ethylene glycol, Penfold et al. [60] found similar differences comparing the two alcohols. With addition of ethylene glycol the cloud point of the surfactant increases, while addition of glycerol causes a reduction of the cloud point. They have also shown that the micellar aggregation number increases for $C_{12}E_8$ when the glycerol concentration increases. This increase in the micellar size is associated with the dehydration of the EO head group, similar to the observations made by Kunieda et al. [58].

The aggregation of amphiphilic poly(ethylene oxide)–poly(propylene oxide)–poly(ethylene oxide) block copolymers is in many ways similar to the aggregation of nonionic $C_iE_j$ surfactants. The phase behavior of these block copolymers in nonaqueous polar solvents was first reported by Samii et al. [61]. More recently, these systems have been investigated thoroughly by the group of Alexandridis [62–64]. This group has studied the micelle formation of the block copolymer Pluronic P105 ($EO_{37}PO_{58}EO_{37}$) both in pure formamide and in mixed solvents of water and formamide, ethanol, or glycerol. They conclude that micelles are formed in pure formamide but at higher concentration and temperature than in water [62]. Moreover, the enthalpy of micellization is lower in formamide and both the micelle radii and the association numbers are lower in formamide than in water. For the block copolymer in a mixed water and formamide solvent, it was concluded that the polymer volume fractions in both the micelle core and the micelle corona decreased with increasing formamide-to-water ratio [62]. Thus, addition of formamide causes an increased solvation of the micelle core and corona, thereby favoring the formation of smaller micelles.

Comparing micellization behavior in the cosolvents formamide, ethanol, and glycerol, some interesting trends were observed [64]. With formamide or ethanol as cosolvent the micelle formation of Pluronic P105 occurs at higher concentrations and temperatures compared to water without cosolvent. However, for glycerol the results show an opposite trend. The addition of glycerol promotes the formation of micelles and micellization starts at lower concentrations and temperatures. Also, the micelle association number increases and the polymer volume fraction in the corona increases when the glycerol content is increased (see Figure 6.1). This is similar to the differences found for nonionic surfactants comparing, for example,

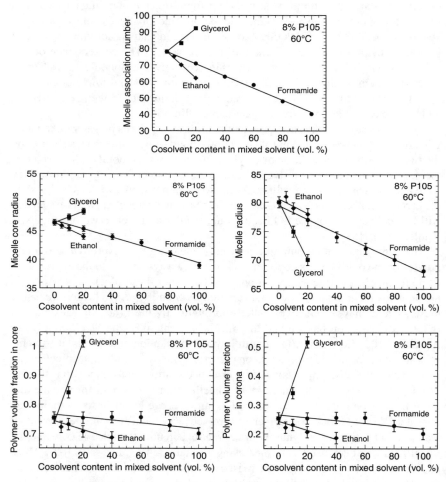

**FIG. 6.1** Structural information obtained from SANS for an 8 wt% $EO_{37}PO_{58}EO_{37}$ solution at 60°C plotted as a function of the cosolvent (glycerol, formamide, or ethanol) content in the mixed solvent. First row: micellar association number ($N_{association}$); second row: radii of core and core + corona ($R_{core}$ and $R_{micelle}$); third row: polymer volume fraction of core and corona ($\alpha_{core}$ and $\alpha_{corona}$). (From Alexandridis, P. and Yang, L., *Macromolecules*, 33, 5574, 2000.)

formamide and glycerol. Consequently, it is probable that the observed differences are due to the dehydration of the EO head groups.

The aggregation of fluorinated surfactants in nonaqueous solvents has also been studied. These surfactants form aggregates at lower concentrations than ordinary hydrogenated surfactants in water. Chrisment *et al.* have studied nonionic fluoro-alkyllipopeptides in DMSO and found progressive and very limited aggregation in this solvent as expected from the low polarity of the solvent [65]. In addition, the lithium salt of nonadecafluorodecanoic acid has been studied with $^{19}$F NMR in formamide, *N*-methylformamide, and ethylene glycol [66].

The thermodynamics of micellization in nonaqueous polar solvents have been studied by a number of authors. Important work has been published by Evans *et al.* using hydrazine as solvent [67,68] and later by Ruiz using ethylene glycol as solvent [69]. Both groups conclude that even though both the enthalpic and entropic contributions to the micellization differ substantially comparing water and the other solvent, these effects cancel out in the standard molar Gibbs free energy of micellization. Evans *et al.* could from their work challenge the conventional view that the structural properties of water would be necessary to obtain a driving force for aggregation [67,68].

The experimental work published so far on micelle formation in polar solvents other than water is clearly very extensive. Efforts to use theoretical models to predict the aggregation behavior have been more scarce. However, the group of Nagarajan has reported on theoretical thermodynamic treatment of these systems [70–72]. They could predict some trends that previously have been observed experimentally. For example, they predict an increase in the c.m.c., a decrease in the average micelle size, an increase in the aggregate polydispersity, and a stronger dependence of the aggregation number on the total surfactant concentration for nonaqueous solvents compared with water (see Figure 6.2). Also, they conclude that:

1. The high c.m.c. values in nonaqueous solvents are mainly due to the smaller magnitude of the surfactant tail transfer free energy to the nonaqueous solvent compared to water.
2. The small aggregation numbers in nonaqueous solvents originate mainly from the smaller magnitude of the hydrocarbon–solvent interfacial tension compared to water.
3. Neither the c.m.c. nor the micellar size is affected to any great extent by the lower dielectric constant of the nonaqueous solvent compared to water.

It is evident from all these experimental and theoretical investigations that micelles are formed in a selection of nonaqueous, polar solvents but that the aggregates, comparing the same surfactant, are generally smaller than in water. There is consequently a larger contact between the inner regions of the micelles and the solvent in these small aggregates. In the nonaqueous solvents investigated

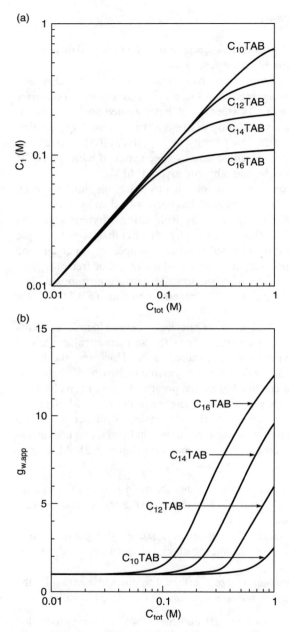

**FIG. 6.2** Calculated dependence of (a) the monomer concentration $C_1$ and (b) the weight average aggregation number $g_w$ on the total surfactant concentration $C_{tot}$ for decyl, dodecyl, tetradecyl, and cetyl trimethylammonium bromides in ethylene glycol solutions. (From Nagarajan, R. and Wang, C.-C., *J. Colloid Interface Sci.*, 178, 471, 1996.)

the interfacial tension between solvent and hydrocarbon is smaller (Table 6.1) and smaller aggregates are thus less energetically unfavorable from this point of view. Also, the micelles are generally formed at higher surfactant concentrations and the micelle formation extends over a significant concentration region. However, there is one exception to this trend, i.e., the aggregation of nonionic $C_iE_j$ surfactants or nonionic block copolymers ($EO_iPO_jEO_i$) in glycerol. In mixtures of glycerol and water the micellization starts at lower concentrations and temperatures, the aggregates grow in size, and the cloud point decreases with increasing glycerol content of the solvent. This has been explained by the dehydration of the EO groups with increasing glycerol content.

## III. CONCENTRATED SURFACTANT SYSTEMS
## A. Liquid Crystals

The first report on a nonaqueous lyotropic liquid crystal in a polar solvent appeared in 1979, where Friberg and co-workers revealed the existence of a lamellar (D) phase in the lecithin (dialkylphosphatidylcholine)–ethylene glycol system [73]. In a series of papers a large number of lecithin–diol systems have been characterized, and detailed structural properties of the systems have been elucidated [74–81]. The interlayer distance in the D-phase with ethylene glycol is shorter than in water, indicating an enhanced disorder in the lipid layers [73]. It was suggested that the primary solvation shell of the phosphatidylcholine group contains one bound solvent molecule per polar head group, with several more loosely associated, as determined by $^2$H NMR measurements [74,76]. Extensive phase studies reveal that lecithin readily forms D-phases with the homologous series of $\alpha,\omega$-diols, from ethylene glycol up to 1,7-heptanediol, although the swelling decreases with increasing molecular size of the solvent [75]. Oligomers and polymers of EO [79] and polyethylene glycol alkyl ether also form D-phases with lecithin, the latter as mixed lamellae containing the acyl part of the lecithin and the alkyl chain of the ethers [81].

The existence of a lamellar phase with lecithin has also been demonstrated for ethylammonium nitrate [82]. For lecithin and formamide, $N$-methylformamide, or $N,N$-dimethylformamide, the full phase diagrams have been determined [83] showing a gradual disappearance of liquid crystalline phases with increasing methylation of the solvent (Figure 6.3). The lamellar phase is stable with $N$-methylformamide, but disappears with $N,N$-dimethylformamide.

To summarize, the lecithin studies provide a qualitative picture of how the solvent affects the phase behavior for a zwitterionic surfactant with a large hydrophobic moiety. Lecithin forms lamellar lyotropic liquid crystals with a wide variety of solvents; a sufficiently hydrophilic solvent — and indeed even amphiphilic compounds with a hydrophilic moiety — stabilizes lamellar phases.

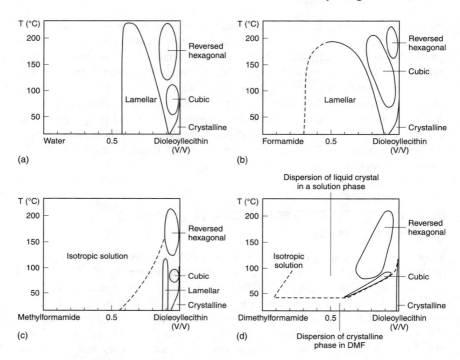

**FIG. 6.3**  Phase diagrams of dioleoyllecithin and (a) water, (b) formamide, (c) methyl-formamide, and (d) dimethylformamide. For notation, see Section V. (From Bergenståhl, B.A. and Stenius, P., *J. Phys. Chem.*, 91, 5944, 1987.)

Full and partial phase equilibria have also been determined for a large number of systems with ionic surfactants, and there are numerous reported combinations of surfactant and solvent that form liquid crystals. The first reported liquid crystal formed in a binary system containing a single-alkyl-chain ionic surfactant was for the $C_{16}TABr$–formamide system. $C_{16}TABr$ forms hexagonal, cubic, and lamellar phases in formamide, in analogy with the aqueous system [84–88]. The liquid crystals form at higher temperatures and melt at lower temperatures than in water. The melting point of solvated crystals can be lowered by addition of alcohol cosurfactants, as in the aqueous systems, which would be of importance in different technical applications [89]. $C_{16}TABr$ and the homologous series of alkyltrimethyl-ammonium bromides have been extensively characterized, and there are reports of liquid crystals formed in glycerol [86], ethylene glycol [86], mixtures of ethylene glycol and water [90], and $N$-methylsydnone [91,92] (Figure 6.4).

An extensive comparison between the aggregation of $C_{16}TABr$ and $C_{16}PyBr$ in a series of solvents, formamide, $N$-methylformamide, $N,N$-dimethylformamide,

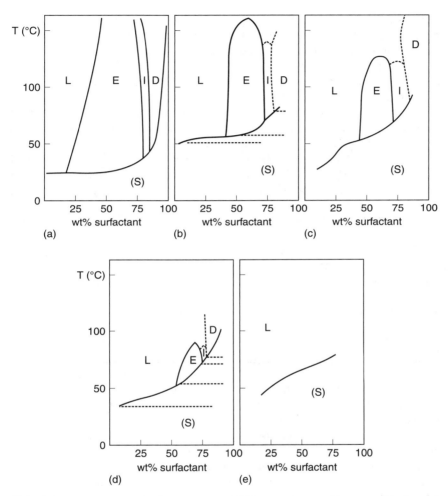

**FIG. 6.4** Phase diagrams of $C_{16}$TABr and different solvents: (a) water, (b) glycerol, (c) formamide, (d) ethylene glycol, and (e) methylformamide. For notation, see Section V. (From Auvray, X., Anthore, P., Petipas, C., Rico, I., and Lattes, A., *C. R. Acad. Sci. Paris*, 306, 695, 1988; Wärnheim, T. and Jönsson, A., *J. Colloid Interface Sci.*, 125, 627, 1988.)

glycerol, ethylene glycol, or *N*-methylsydnone, revealed some interesting differences between the surfactants. With formamide, ethylene glycol, and glycerol, the phase sequence was E → I → D for both surfactants, and with *N,N*-dimethylformamide, the least polar solvent, only D-phases formed. However, only $C_{16}$PyBr showed the sequence E → I → D with *N*-methylformamide and *N*-methylsydnone [91,92].

Considering solely the phase diagram, the main effect of exchanging water for a more weakly polar solvent in these systems is to decrease their existence regions. In addition to that, more subtle phenomena may occur in the nonaqueous systems. It has been demonstrated that for alkylpyridinium chlorides and bromides in glycerol and formamide, cubic phases can occur intermediate to the $L_1$- and E-phase region, where they are not stable in water [93].

E-, I-, and D-phases have been observed in the SDS–formamide system. In other solvents, ethylene glycol, glycerol, N-methylformamide, only a D-phase is formed at high surfactant concentrations with SDS [94].

Binary phase diagrams of the homologous series of potassium soaps (with alkyl chain length $C_{12}$–$C_{22}$) and ethylene glycol, butylene glycol, or glycerol have been determined [95]. Extensive structural investigations of the different phases using x-ray diffraction have been performed [96].

The nonaqueous systems also form liquid crystals analogous to aqueous systems in ternary systems with an added weakly hydrophilic component. SDS has been extensively employed in studies of ternary and quaternary systems with glycerol or formamide, a long-chain alcohol, and, sometimes, hydrocarbon [97–101]. In the SDS–glycerol–decanol system the lamellar phase swells extensively, even more so than in water [97]. While no liquid crystals form at room temperature in the binary systems, a D-phase occurs when decanol is added.

Aerosol OT (sodium diethylhexylsulfosuccinate) is another extensively studied ionic surfactant. This surfactant forms a lamellar and cubic phase with formamide [103] and glycerol [104], just as with water. With ethylene glycol and N-methylformamide, no liquid crystals except the inverse hexagonal occur [103].

In contrast, for a solvent such as propylene glycol, which has a less polar character, no liquid crystals are formed even for ionic surfactants with a reasonably large hydrophobic moiety such as didodecyldimethylammonium bromide [105].

The formation of liquid crystals by nonionic surfactants of the polyethylene glycol alkyl ether type, $C_iE_j$, has been much less considered. Phase diagrams of $C_{12}E_3$, $C_{12}E_4$, $C_{16}E_4$, $C_{16}E_6$, and $C_{16}E_8$ with formamide as solvent have been determined [17]. No liquid crystals are stable for the $C_{12}E_j$ surfactants; however, there is a clouding, a lower consolute temperature, in the $C_{12}E_3$ system [17]. The $C_{16}E_j$ series follow the same trend as the aqueous systems [107]: $C_{16}E_4$ gives a D-phase, $C_{16}E_6$ an E-phase, and $C_{16}E_8$ an I-phase, most likely of the $I_1$ type. As with ionic surfactants, the existence regions of the liquid crystalline phase are smaller, and there are fewer phases present, comparing formamide with water as solvent [17,107]. For $C_{12}E_8$, the phase diagrams with glycerol, propylene glycol, and propanol as solvents together with water show that no liquid crystalline phases are stable at volume fractions of polar cosolvent above 0.5 [108].

More recently, investigations of the solution behavior of block copolymers of the poly(ethylene oxide) (PEO)–poly(propylene oxide) (PPO) type have been extended to nonaqueous, polar solvent systems. The block copolymer Pluronic

P105 ($EO_{37}PO_{58}EO_{37}$) forms a variety of liquid crystalline phases (micellar cubic, hexagonal, bicontinuous cubic, and lamellar phase with increasing polymer concentration) in formamide [109] (Figure 6.5). Investigations on the aggregation behavior of Pluronic 105 in other solvents or solutes (ethanol, glycerol, propylene glycol) show that the formation of liquid crystals is limited to formamide.

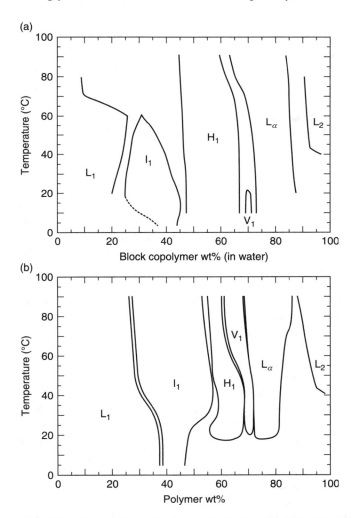

**FIG. 6.5** Binary phase diagrams of (a) Pluronic 105 ($EO_{37}PO_{58}EO_{37}$)–water and (b) Pluronic 105–formamide. For notation, see Section V. (From Alexandridis, P., *Macromolecules*, 31, 6935, 1998; Ivanova, R., Lindman, B., and Alexandridis, P., *Adv. Coll. Interface Sci.*, 89–90, 351, 2001.)

Some interesting features are found for the lattice parameters and aggregate dimensions derived from SAXS for the liquid crystalline phases when gradually exchanging water for another polar solvent. Cosolvents that are shown to have a smaller effect on the lattice parameters (e.g., propylene glycol) will maintain the microstructure and stability of the phase up to high cosolvent-to-water ratios [110]. The block copolymers could be of particular practical relevance for different applications, e.g., within the pharmaceutical area. This prompts investigations of cosolvent–water systems that are acceptable in this context [110,111].

Monoglycerides form an inverse hexagonal phase with glycerol, as in water [112]. Mixtures of triethanolamine and oleic acid form a nonaqueous lamellar liquid crystal with a surfactant bilayer of soap and acid with intercalated ionized and unionized alkanolamine as solvent [113,114]. Lamellar liquid crystals form analogously with dodecylbenzenesulfonic acid and triethanolamine [115].

These and other systems reported to contain a nonaqueous liquid crystalline phase are summarized in Table 6.2.

## B.  Microemulsions

The first reports on nonaqueous microemulsions, isotropic solutions containing a hydrophilic and a lipophilic component, stabilized by a surfactant, were made by Palit and McBain in 1946 [116] and by Winsor in 1948 [117]. They both used glycols as polar solvents. The microemulsion regions were only observed visually so no structural information could be obtained.

Three groups reported independently the observation of microemulsions with nonaqueous polar solvents in 1984. The group of Lattes found microemulsions in the $C_{16}$TABr–formamide–cyclohexane system with butanol as cosurfactant [118–120] while Friberg and Podzimek detected a narrow microemulsion region in the lecithin–ethylene glycol–decane system [121] (Figure 6.6). The third group investigated glycerol in heptane microemulsions, with AOT as surfactant [122], using dynamic light scattering to study the aggregation.

Lattes and co-workers have investigated the $C_{16}$TABr–formamide–butanol system with cyclohexane or isooctane as an oil component [118–120]. In both of these systems, conductivity measurements with varying composition were interpreted as indicative of percolation. When an x-ray scattering study was conducted on the latter system, no discrete droplets could be detected. When increasing the hydrocarbon volume of the surfactant by using didodecyldimethylammonium bromide $((C_{12})_2DABr)$, microemulsions form without cosurfactant in formamide and in ethylene glycol using dodecane and toluene as oil [102].

Fletcher et al. have investigated the glycerol-in-oil microemulsion stabilized by $C_{16}$TABr using a mixture of $n$-heptane and chloroform as oil with dynamic light scattering, giving a hydrodynamic radius of reverse micellar aggregate, glycerol droplets, and an area per surfactant head group [123].

**TABLE 6.2** Nonaqueous Lyotropic Liquid Crystalline Phases Reported in the Literature

| Surfactant | Solvent | Additive | Phases detected[a] | Ref. |
|---|---|---|---|---|
| Lecithin | EG | | D | 73 |
| Lecithin | 1,3-Propanediol | | D | 75 |
| Lecithin | 1,4-Butanediol | | D | 75 |
| Lecithin | 1,5-Pentanediol | | D | 75 |
| Lecithin | 1,6-Hexanediol | | D | 75 |
| Lecithin | 1,7-Heptanediol | | D | 75 |
| Lecithin | $(EG)_{1-4}$ | | D | 79 |
| Lecithin | PEG | | D | 79 |
| Lecithin | $C_{12}Ej$ | | D | 81 |
| Lecithin | EAN | | D | 82 |
| Lecithin | FA | | D, I, F | 83 |
| Lecithin | MFA | | D, I, F | 83 |
| Lecithin | DMF | | I, F | 83 |
| Lecithin | EG | Methanol | D | 80 |
| Lecithin | EG | Decanol | D | 80 |
| Lecithin | EG | Decane | D | 80 |
| $C_{16}$TABr | FA | | E, I, D | 84–87 |
| $C_{16}$TABr | G | | E, I, D | 86, 87 |
| $C_{16}$TABr | EG | | E, I, D | 86 |
| $C_{16}$TABr | MFA | | D | 91, 92 |
| $C_{16}$TABr | NMS | | D | 91, 92 |
| $C_{16}$TABr | EG | Decanol | D | 102 |
| $C_{16}$TACl | FA | | I, E... | 91, 92 |
| $C_{16}$TASO$_4$ | EG | | E, D | 86 |
| $C_{14}$TABr | G | | E, I, D | 86 |
| $C_{14}$TABr | EG | | E, D | 86 |
| $C_{20}$PyBr | FA | | ...I... | 93 |
| $C_{18}$PyBr | FA | | ...I... | 93 |
| $C_{16}$PyCl | FA | | I, E, I... | 91–93 |
| $C_{16}$PyCl | G | | E, I, D | 93 |
| $C_{16}$PyCl | MFA | | E, I, D | 91, 92 |
| $C_{16}$PyCl | DMF | | D | 91, 92 |
| $C_{16}$PyCl | EG | | E, I, D | 91, 92 |
| $C_{16}$PyCl | NMS | | E, I, D | 91, 92 |
| SDS | FA | | E, I, D | 94 |
| SDS | FA | Decanol | D | 97 |
| SDS | FA–$H_2O$ | Decanol | D (E) | 124 |
| SDS | FA | Decanol + toluene | D | 97 |
| SDS | G | | D | 94 |
| SDS | EG | | D | 94 |
| SDS | MFA | | D | 94 |
| SDS | G | Decanol | D | 97 |

*(continued)*

**TABLE 6.2**   (Contd.)

| Surfactant | Solvent | Additive | Phases detected[a] | Ref. |
|---|---|---|---|---|
| $KC_{22}$ | G | | E, D | 95 |
| $KC_{18}$ | G | | E, I, D | 95 |
| $KC_{16}$ | G | | E, D | 95 |
| $KC_{14}$ | G | | E, I, D | 95 |
| $KC_{12}$ | G | | E, D | 95 |
| $KC_{22}$ | EG | | E, D | 95 |
| $KC_{18}$ | EG | | E, I, D | 95 |
| $KC_{22}$ | BG | | D | 95 |
| $KC_{18}$ | BG | | I, D | 95 |
| AOT | FA | | D, I, F | 103 |
| AOT | G | Decanol | D... | 104 |
| AOT | G | Decane | D... | 104 |
| AOT | G | $p$-Xylene | D... | 104 |
| TEAOl | TEA | G, EG | D | 113, 114 |
| DBSA | TEA | G, EG, TEG | D | 115 |
| $C_{16}E_4$ | FA | | D | 17 |
| $C_{16}E_6$ | FA | | E | 17 |
| $C_{16}E_8$ | FA | | I, E | 17 |

*Note*: For notation, see Section V.
[a]Ellipses indicate that the entire phase diagram has not been investigated.

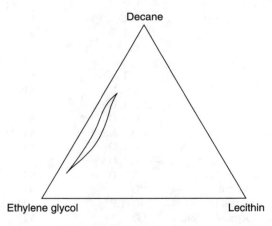

**FIG. 6.6**   Phase diagram of the lecithin–ethylene glycol–decane system at 25°C, showing a narrow microemulsion region. (From Friberg, S.E. and Podzimek, M., *Colloid Polym. Sci.*, 262, 252, 1984.)

Friberg and co-workers have studied a number of SDS-stabilized formamide microemulsions using different hydrocarbons and cosurfactants [124–126]. The system containing toluene and hexanol showed a microemulsion region but from light scattering studies it was concluded to be a nonstructured solution [125]. However, when increasing the alkyl chain length of the alcohol to decanol two isotropic solutions could be observed; the formamide solution gave no indication of an organized structure, but in the decanol solution the results were interpreted in terms of inverse micelles [126].

Ceglie and co-workers examined different microemulsions using SDS and formamide but changing the oil to $p$-xylene and using alcohols of different alkyl chain length, going from pentanol to octanol [37,127–129]. Both ternary and quaternary systems were investigated with two different NMR techniques: self-diffusion and frequency variable relaxation measurements. The self-diffusion study gave no indications of any organized structures, but from relaxation measurements a fraction of the SDS molecules was observed to aggregate into some interfacial domains when octanol was used as cosurfactant. Solution regions were observed with methylformamide and dimethylformamide as polar component and octanol as cosurfactant, but the phases were found to be nonaggregated solutions [37,127].

Another polar solvent that has been used in SDS-stabilized microemulsions is glycerol. Hexanol or decanol have been used as cosurfactants and systems both with and without oil have been studied. The ternary system with hexanol as cosurfactant was examined with SANS and NMR self-diffusion measurements by two different groups and both found the microemulsions to be structureless solutions [130,131]. Similar behavior was found from a self-diffusion study of the quaternary systems with $p$-xylene or decane as the oil component [131,132].

AOT–glycerol microemulsions have been carefully studied by two different groups: that of Friberg and that of Robinson. The former group studied ternary phase diagrams with decanol, decane, or $p$-xylene as the third component and isotropic solution phases were detected in all systems [104]. The latter group studied two other ternary systems with heptane or octane as the third component [122,133]. Reverse micelles with glycerol were found in both systems using dynamic light scattering for the heptane system and quasielastic neutron scattering for the octane system. AOT microemulsions have been the subject for additional investigations with formamide as polar solvent and isooctane or decane as an oil component [134–137]. Light scattering and steady-state adsorption spectra of the molecular probe Coumarin 343 were interpreted in terms of the formation of reverse micelles not only in formamide but also in a variety of less polar solvents such as methanol and acetonitrile [134]. The effects due to the molecular size of the probe have also been considered [135], effects that will be of importance in restricted systems [136].

Nonaqueous microemulsions with nonionic surfactants have been studied. The $C_{12}E_4$ surfactant was found to stabilize microemulsions of formamide and dodecane [138]. The ternary phase diagrams were studied at different temperatures and the solubilization of hydrocarbon was shown to be very temperature dependent (Figure 6.7). It was also observed that the temperature intervals of the three-phase regions are dependent on the hydrocarbon used; larger aliphatic hydrocarbons

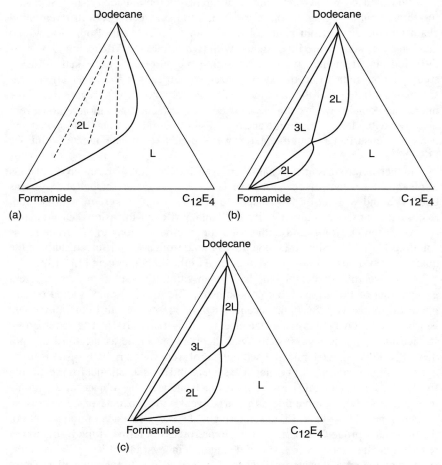

**FIG. 6.7** Phase diagrams of the $C_{12}E_4$–formamide–dodecane system at (a) 30°C, (b) 40°C, and (c) 50°C, showing the growth of the three-phase region where a solution phase is formed at minimum surfactant concentration. (From Wärnheim, T. and Sjöberg, M., *J. Colloid Interface Sci.*, 131, 402, 1989.)

give three-phase regions at higher temperatures, observations completely analogous to the corresponding aqueous systems. In a SANS study Schubert and Strey [139] investigated the order/disorder transition in microemulsions of nonionic surfactants, water–formamide mixtures, and oil. As the water content of the systems was decreased a more disordered microstructure was observed. However, even with pure formamide as polar solvent, the existence of internal interfaces, although uncorrelated, could be detected. Studies using NMR self-diffusion measurements and SANS in microemulsions formed with $C_{12}E_5$–propylene glycol and/or glycerol–alkane are interpreted in terms of droplet structures at low oil contents which through a percolation form an oil continuous structure [140].

As a brief conclusion it can be noted that many nonaqueous microemulsions reported do not seem to contain an organized structure, being simply molecular solutions. Since the degree of organization already in many aqueous microemulsion is low, in particular for quaternary systems containing ionic surfactant and cosurfactant, this is not really surprising.

## IV. CONCLUDING REMARKS

Evidently, the mapping of surfactant aggregation in nonaqueous polar solvents has grown to be very extensive, and investigations of many different combinations of surfactants and solvents are available in the literature. Furthermore, a wide range of experimental techniques have been used. The results from different studies are quite consistent and most of the authors agree on some basic trends.

Qualitatively, the general aggregation behavior is similar to water. That is, surfactant aggregation in the form of micelles, liquid crystals, or microemulsions is possible in polar solvents other than water. However, the Krafft temperatures of ionic surfactants and c.m.c. values are higher and the aggregation numbers of the micelles are lower in these nonaqueous solvents. The existence regions for liquid crystal phases in nonaqueous solvents are reduced and the phase diagrams are less complex than in water. Also, the microemulsions formed in nonaqueous solvents have often a more disordered microstructure than in water. It is tempting, in a qualitative manner, to ascribe these differences to the less extensive solvophobic interaction in the polar solvents used compared to water.

There are other ways of expressing and discussing this solvophobic interaction than, for example, comparing the interfacial tensions between solvent and hydrocarbon as in Table 6.1. The nonaqueous solvents that have been reported to promote aggregation of surfactant molecules have one property in common: they all have high cohesive energy. That is, the net attractive interactions between the solvent molecules are strong. Hildebrand *et al.* [141] have derived a cohesive energy parameter from the heat of vaporization of the solvent. Another measure of the cohesive energy is the Gordon parameter [142], $\gamma/V^{1/3}$ ($\gamma$ = surface tension,

$V$ = molar volume). This parameter has the advantage that it can be used for both liquids and fused salts.

Different authors [15,143] have tried to determine a limiting value for the cohesive energy of the solvent above which a certain solvent should be able to promote surfactant aggregation. However, it is curiously often overlooked that the hydrophobicity of the surfactant must also be taken into account in this context. A surfactant with a long hydrocarbon chain or a fluorinated hydrocarbon chain will be able to aggregate in solvents where a less hydrophobic surfactant will remain in monomeric form, just as in water.

## V. NOTATION

| | |
|---|---|
| AOT | Aerosol OT (sodium diethylhexylsulfosuccinate) |
| $(C_{12})_2DAB$ | Didodecyldimethylammonium bromide |
| $C_iE_j$ | Nonionic surfactant of the polyethylene glycol alkyl ether type; the alkyl chain contains $i$ carbon atoms and the polar group $j$ ethylene glycol units |
| c.m.c. | Critical micelle concentration |
| $C_xNH_3Br$ | Cationic surfactant of the alkylammonium bromide type; the alkyl chain contains $x$ carbon atoms |
| $C_xPyBr$ | Cationic surfactant of the alkylpyridinium bromide type; the alkyl chain contains $x$ carbon atoms |
| $C_xTABr$ | Cationic surfactant of the alkyltrimethylammonium bromide type; the alkyl chain contains $x$ carbon atoms |
| $C_xTASO_4$ | Cationic surfactant of the alkyltrimethylammonium sulfate type; the alkyl chain contains $x$ carbon atoms |
| CTbPB | Cetyltributylphosphonium bromide |
| D | Lyotropic liquid crystalline phase with lamellar structure |
| DBSA | Dodecylbenzene sulfonic acid |
| DMF | $N,N$-dimethylformamide |
| DMSO | Dimethylsulfoxide |
| E | Lyotropic liquid crystalline phase with hexagonal structure |
| EAN | Ethylammonium nitrate |
| EG | Ethylene glycol |
| EO | Ethylene oxide |
| F | Lyotropic liquid crystalline phase with reverse hexagonal structure |
| FA | Formamide |
| G | Glycerol |
| H | Lyotropic liquid crystalline phase with hexagonal structure (cf. E) |
| I | Isotropic liquid crystalline phase (used for structures with discrete aggregates) (cf. V) |

| | |
|---|---|
| $KC_x$ | Potassium soap with $x$ carbon atoms in the alkyl chain |
| $L_\alpha$ | Lyotropic liquid crystalline phase lamella structure (cf. D) |
| MFA | $N$-methylformamide |
| NaOl | Sodium oleate |
| NMS | $N$-methylsydnone |
| PEG | Polyethylene glycol |
| PG | Propylene glycol |
| PGSE NMR | Pulsed gradient spin echo NMR for self-diffusion measurements |
| PO | Propylene oxide |
| SANS | Small-angle neutron scattering |
| SAXS | Small-angle x-ray scattering |
| SDS | Sodium dodecyl sulfate |
| TEA | Triethanolamine |
| TEAOl | Triethanolammonium oleate |
| TEG | Triethylene glycol |
| V | Isotropic liquid crystalline phase (used for bicontinuous structures) |

# REFERENCES

1. Ray, A., *J. Am. Chem. Soc.*, 91, 6511, 1969.
2. Ramadan, M.S. Evans, D.F., Lumry, R., and Philson, S., *J. Phys. Chem.*, 89, 3405, 1985.
3. Ellis, B.N., *Cleaning and Contamination of Electronic Components and Assemblies*, Electrochemical Publications, Port Erin, U.K., 1986.
4. Lattes, A. and Rico, I., in *Microemulsion Systems*, Rosano, H. and Clausse, M., Eds., Marcel Dekker, New York, 1987, p. 377.
5. Rico, I., Couderc, F., Perez, E., Laval, J.P., and Lattes, A., *J. Chem. Soc., Chem. Commun.*, 15, 1205, 1987.
6. Samii, A., de Savignac, A., Rico, I., and Lattes, A., *Tetrahedron*, 41, 3683, 1985.
7. Gopal, R. and Singh, J.R., *Kolloid Z. Z. Polym.*, 239, 699, 1970.
8. Gopal, R. and Singh, J.R., *J. Indian Chem. Soc.*, 49, 667, 1972.
9. Gopal, R. and Singh, J.R., *J. Phys. Chem.*, 77, 554, 1973.
10. Singh, H.N., Singh, S., and Tewari, K.C., *J. Am. Oil Chem. Soc.*, 52, 436, 1975.
11. Singh, H.N., Saleem, S.M., Singh, R.P., and Birdi, K.S., *J. Phys. Chem.*, 84, 2191, 1980.
12. Varma, R.P., Bahadur, P., and Dayal, R., *Rev. Rom. Chim.*, 25, 201, 1980.
13. Ionescu, L.G. and Fung, D.S., *Bull. Chem. Soc. Jpn.*, 54, 2503, 1981.
14. Ionescu, L.G. and D. S. Fung, *J. Chem. Soc., Faraday Trans. I*, 77, 2907, 1981.
15. Binana-Limbele, W. and Zana, R., *Colloid Polym. Sci.*, 267, 440, 1989.
16. Olofsson, G., *J. Chem. Soc., Faraday Trans.*, 87, 3037, 1991.
17. Jonströmer, M., Sjöberg, M., and Wärnheim, T., *J. Phys. Chem.*, 94, 7549, 1990.
18. Sjöberg, M., Henriksson, U., and Wärnheim, T., *Langmuir*, 6, 1205, 1990.
19. Bloom, H. and Reinsborough, V.C., *Aust. J. Chem.*, 20, 2583, 1967.
20. Bloom, H, and Reinsborough, V.C., *Aust. J. Chem.*, 21, 1525, 1968.
21. Reinsborough, V.C. and Valleau, J.P., *Aust. J. Chem.*, 21, 2905, 1968.

22. Bloom, H. and Reinsborough, V.C., *Aust. J. Chem.*, 22, 519, 1969.
23. Reinsborough, V.C., *Aust. J. Chem.*, 23, 1473, 1970.
24. Evans, D.F. Chen, S.H., Schriver, G.W., and Arnett, E.M., *J. Am. Chem. Soc.*, 103, 481, 1981.
25. Evans, D.F., Yamauchi, A., Roman, R., and Casassa, E.Z., *J. Colloid Interface Sci.*, 88, 89, 1982.
26. Evans, D.F., Yamauchi, A., Wel, G.J., and Bloomfield, V.A., *J. Phys. Chem.*, 87, 3537, 1983.
27. Peters, D. and Miethchen, R., *Colloid Polym. Sci.*, 271, 91, 1993.
28. Beesley, A.H., Evans, D.F., and Laughlin, R.G., *J. Phys. Chem.*, 92, 791, 1988.
29. Auvray, X., Petipas, C., Perche, T., Anthore, R., Marti, M.J., Rico, I., and Lattes, A., *J. Phys. Chem.*, 94, 8604, 1990.
30. Ruiz, C.C., *J. Colloid Interface Sci.*, 221, 262, 2000.
31. Perche, T., Auvray, X., Petipas, C., Anthore, R., Perez, E., Rico-Lattes, I., and Lattes, A., *Langmuir*, 12, 863, 1996.
32. Escoula, B., Hajjaji, N., Rico, I., and Lattes, A., *J. Chem. Soc., Chem. Commun.*, 1233, 1984.
33. Rico, I. and Lattes, A., *J. Phys. Chem.*, 90, 5870, 1986.
34. Auvray, X., Petipas, C., Anthore, R., Rico, I., Lattes, A., Ahmah-Zadeh Samii, A., and de Savignac, A., *Colloid Polym. Sci.*, 265, 925, 1987.
35. Belmajdoub, A., ElBayed, K., Brondeau, J., Canet, D., Rico, I., and Lattes, A., *J. Phys. Chem.*, 92, 3569, 1988.
36. Almgren, M., Swarup, S., and Löfroth, J.E., *J. Phys. Chem.*, 89, 4621, 1985.
37. Das, K.P., Ceglie, A., Monduzzi, M., Söderman, O., and Lindman, B., *Prog. Colloid Polym. Sci.*, 73, 167, 1987.
38. Alfassi, Z.B. and Filby, W.G., *Chem. Phys. Lett.*, 144, 83, 1988.
39. Belmajdoub, A., Boubel, J.C., and Canet, D., *J. Phys. Chem.*, 93, 4844, 1989.
40. Sjöberg, M., Jansson, M., and Henriksson, U., *Langmuir*, 8, 409, 1992.
41. Amorim da Costa, A.M., *J. Mol. Struct.*, 174, 195, 1988.
42. Ceglie, A., Colafemmina, G., Della Monica, M., Olsson, U., and Jönsson, B., *Langmuir*, 9, 1449, 1993.
43. Perche, T., Auvray, X., Petipas, C., Anthore, R., Rico-Lattes, I., and Lattes, A., *Langmuir*, 13, 1475, 1997.
44. Fletcher, P.D.I. and Gilbert, P.J., *J. Chem. Soc., Faraday Trans. I*, 85, 147, 1989.
45. Gharibi, H., Palepu, R., Tiddy, G.J.T., Hall, D.G., and Wyn-Jones, E., *J. Chem. Soc., Chem. Commun.*, 115, 1990.
46. Palepu, R., Gharibi, H., Bloor, D.M., and Wyn-Jones, E., *Langmuir*, 9, 110, 1993.
47. Callaghan, A., Doyle, R., Alexander, E., and Palepu, R., *Langmuir*, 9, 3422, 1993.
48. Backlund, S., Bergenståhl, B.A., Molander, O., and Wärnheim, T., *J. Colloid Interface Sci.*, 131, 393, 1989.
49. McDonald, C., *J. Pharm. Pharmacol.*, 19, 411, 1967.
50. McDonald, C., *J. Pharm. Pharmacol.*, 22, 148, 1970.
51. McDonald, C., *J. Pharm. Pharmacol.*, 22, 774, 1970.
52. Ray, A., *Nature*, 231, 313, 1971.
53. Couper, A., Gladden, G.P., and Ingram, B.T., *Faraday Discuss.*, 63, 1975.
54. Cantu, L., Corti, M., Degiorgio, V., Hoffman, H., and Ulbricht, W., *J. Colloid Interface Sci.*, 116, 384, 1987.

55. Thomason, M.A., Bloor, D.M., and Wyn-Jones, E., *J. Phys. Chem.*, 95, 6017, 1991.
56. Ruiz, C.C., Molina-Bolivar, J.A., and Aguiar, J., *Langmuir*, 17, 6831, 2001.
57. Molina-Bolivar, J.A., Aguiar, J., Peula-Garcia, J.M., and Ruiz, C.C., *Molecular Phys.*, 100, 3259, 2002.
58. Aramaki, K., Olsson, U., Yamaguchi, Y., and Kunieda, H., *Langmuir*, 15, 6226, 1999.
59. Kabalnov, A., Olsson, U., and Wennerström, H., *J. Phys. Chem.*, 99, 6220, 1995.
60. Penfold, J., Staples, E., Tucker, I., and Cummins, P., *J. Colloid Interface Sci.*, 185, 424, 1997.
61. Samii, A.A., Karlström, G., and Lindman, B., *Langmuir*, 7, 1067, 1991.
62. Yang, L. and Alexandridis, P., *Langmuir*, 16, 4819, 2000.
63. Alexandridis, P. and Yang, L., *Macromolecules*, 33, 3382, 2000.
64. Alexandridis, P. and Yang, L., *Macromolecules*, 33, 5574, 2000.
65. Chrisment, J., Delpuech, J.J., Hamdoune, F., Ravey, J.C., Selve, C., and Stébé, M.J., *J. Chem. Soc., Faraday Trans.*, 89, 927, 1993.
66. Sjöberg, M., Ph.D. dissertation, Royal Institute of Technology, Stockholm, 1992.
67. Ramadan, M., Evans, D.F., and Lumry, R., *J. Phys. Chem.*, 87, 4538, 1983.
68. Evans, D.F. and Wightman, P.J., *J. Colloid Interface Sci.*, 86, 515, 1982.
69. Ruiz, C.C., *Colloid Polym. Sci.*, 277, 701, 1999.
70. Nagarajan, R. and Wang, C.-C., *Langmuir*, 11, 4673, 1995.
71. Nagarajan, R. and Wang, C.-C., *J. Colloid Interface Sci.*, 178, 471, 1996.
72. Nagarajan, R. and Wang, C.-C., *Langmuir*, 16, 5242, 2000.
73. Moucharafieh, N. and Friberg, S.E., *Mol. Cryst. Liq. Cryst. Lett.*, 49, 231, 1979.
74. Larsen, D.W., Friberg, S.E., and Christenson, H., *J. Am. Chem. Soc.*, 102, 6565, 1980; Larsen, D.W., Rananavare, S.B., El-Nokaly, M., and Friberg, S.E., *Finn. Chem. Lett.*, 96, 1982.
75. El-Nokaly, M.A., Ford, L.D., Friberg, S.E., and Larsen, D.W., *J. Colloid Interface Sci.*, 84, 228, 1981.
76. Rananavare, S.B., Ward, A.J., Friberg, S.E., and Larsen, D.W., *Mol. Cryst. Liq. Cryst.*, 133, 207, 1986.
77. Rananavare, S.B., Ward, A.J., Friberg, S.E., and Larsen, D.W., *Mol. Cryst. Liq. Cryst. Lett.*, 4, 115, 1987.
78. Larsen, D.W., Rananavare, S.B., and Friberg, S.E., *J. Am. Chem. Soc.*, 106, 1848, 1984.
79. El-Nokaly, M., Friberg, S.E., and Larsen, D.W., *J. Colloid Interface Sci.*, 98, 274, 1984; El-Nokaly, M., Friberg, S.E., and Larsen, D.W., *Liq. Cryst. Ordered Fluids*, 4, 441, 1984.
80. Friberg, S.E., Solans, C., and Gan-Zuo, L., *Mol. Cryst. Liq. Cryst.*, 109, 159, 1984.
81. Gan-Zuo, L., El-Nokaly, M., and Friberg, S.E., *Mol. Cryst. Liq. Cryst. Lett.* 72, 183, 1982.
82. Evans, D.F., Kaler, E.W., and Benton, W.J., *J. Phys. Chem.*, 87, 533, 1983.
83. Bergenståhl, B.A. and Stenius, P., *J. Phys. Chem.*, 91, 5944, 1987.
84. Belmajdoub, A., Marchal, J.P., Canet, D., Rico, I., and Lattes, A., *New. J. Chem.*, 11, 415, 1987.
85. Auvray, X., Anthore, P., Petipas, C., Rico, I., and Lattes, A., *C. R. Acad. Sci. Paris*, 306, 695, 1988.

86. Wärnheim, T. and Jönsson, A., *J. Colloid Interface Sci.*, 125, 627, 1988.
87. Auvray, X., Petipas, C., Anthore, R., Rico, I. and Lattes, A., *J. Phys. Chem.*, 93, 7458, 1989.
88. Ylihautala, M., Vaara, J., Ingman, P., Jokisaari, J., and Diehl, P., *J. Phys. Chem. B*, 101, 32, 1997.
89. Fuller, S., Li, Y., Tiddy, G.J.T., and Wyn-Jones, E., *Langmuir*, 11, 1980, 1995.
90. Backlund, S., Bergenståhl, B.A., Molander, O., and Wärnheim, T., *J. Colloid Interface Sci.*, 131, 393, 1989.
91. Auvray, X., Abiyaala, M., Duval, P., Petipas, C., Rico, I., and Lattes, A., *Langmuir*, 9, 444, 1993.
92. Auvray, X., Perche, T., Petipas, C., Anthore, R., Marti, M.J., Rico, I., and Lattes, A., *Langmuir*, 8, 2671, 1992.
93. Bleasedale, T.A., Tiddy, G.J.T., and Wyn-Jones, E., *J. Phys. Chem.*, 95, 5385, 1991.
94. Auvray, X., Perche, T., Anthore, R., Petipas, C., Rico, I., and Lattes, A., *Langmuir*, 7, 2385, 1991.
95. Dörfler, H.D. and Senst, A., *Colloid Polym. Sci.*, 271, 173, 1993; Dörfler, H.D. and Knape, M., *Tenside Surfactants Detergents*, 30, 196, 359, 1993.
96. Heike, A. and Dörfler, H.D., *Colloid Polym. Sci.*, 277, 494, 762, 777, 1104, 1109, 1999.
97. Friberg, S.E. and Liang, P., *Colloid Polym. Sci.*, 264, 449, 1986.
98. Friberg, S.E., Liang, P., Liang, Y.C., Greene, B. and van Gilder, R., *Colloid Surf.*, 19, 249, 1986.
99. Friberg, S.E., Ward, A.J.I., and Larsen, D.W., *Langmuir*, 3, 735, 1987.
100. Ward, A.J.I., Rong, G. and Friberg, S.E., *Colloid Surf.*, 38, 285, 1989.
101. Ward, A.J.I., Rong, G., and Friberg, S.E., *Colloid Polym. Sci.*, 267, 730, 1989.
102. Wärnheim, T., Jönsson, A., and Sjöberg, M., *Prog. Colloid Polym. Sci.*, 82, 271, 1990.
103. Bergenståhl, B., Jönsson, A., Sjöblom, J., Stenius, P. and Wärnheim, T., *Prog. Colloid Polym. Sci.*, 74, 108, 1987.
104. Friberg, S.E., and Liang, Y.C., in *Microemulsion Systems*, Rosano, H.L. and Clausse, M., Eds., Marcel Dekker, New York, 1987, p. 103.
105. Martino, A. and Kaler, E.W., *Colloid Surf. A*, 99, 91, 1995.
106. Wärnheim, T., Bokström, J. and Williams, Y., *Colloid Polym. Sci.*, 266, 562, 1988.
107. Mitchell, D.J., Tiddy, G.J.T., Waring, L., Bostock, T., and McDonald, M.P., *J. Chem. Soc., Faraday Trans. I*, 79, 975, 1983.
108. Aramaki, K., Olsson, U., Yamaguchi, Y., and Kunieda, H., *Langmuir*, 15, 6226, 1999.
109. Alexandridis, P., *Macromolecules*, 31, 6935, 1998.
110. Ivanova, R., Lindman, B., and Alexandridis, P., *Adv. Coll. Interface Sci.*, 89–90, 351, 2001.
111. Ivanova, R., Lindman, B., and Alexandridis, P., *J. Colloid Interface Sci.*, 252, 226, 2002.
112. Larsson, K., in *Lipid Handbook*, Gunstone, F.D., Harwood, J.L., and Padley, F.B., Eds., Chapman and Hall, New York, 1986.
113. Friberg, S.E., Liang, P., Lockwood, F.E., and Tadros, M., *J. Phys. Chem.*, 88, 1045, 1984.

114. Friberg, S.E., Wohn, C.S., and Lockwood, F.E., *J. Pharm. Sci.*, 74, 771, 1985.
115. Friberg, S.E., Wohn, C.S., Uang, Y.J., and Lockwood, F.E., *J. Disp. Sci. Technol.*, 8, 429, 1987.
116. Palit, S.R. and McBain, J.W., *Industr. Eng. Chem.*, 38, 741, 1946.
117. Winsor, P.A., *Trans. Faraday Soc.*, 44, 451, 1948.
118. Rico, I. and Lattes, A., *Nouv. J. Chim.*, 8, 429, 1984.
119. Rico, I. and Lattes, A., in *Surfactants in Solution*, Vol. 6, Mittal, K.L. and Bothorel, P., Eds., Plenum Press, New York, 1986, p. 1397.
120. Rico, I. and Lattes, A., in *Microemulsion Systems*, Rosano, H. and Clausse, M., Eds., Marcel Dekker, New York, 1987, p. 357.
121. Friberg, S.E. and Podzimek, M., *Colloid Polym. Sci.*, 262, 252, 1984.
122. Fletcher, P.D.I., Galal, M.F., and Robinson, B.H., *J. Chem. Soc., Faraday Trans. I*, 80, 3307, 1984.
123. Fletcher, P.D.I., Galal, M.F., and Robinson, B.H., *J. Chem. Soc., Faraday Trans. I*, 81, 2053, 1985.
124. Rong, G. and Friberg, S.E., *J. Disp. Sci. Technol.*, 9, 401, 1988.
125. Friberg, S. E. and Rong, G., *Langmuir*, 4, 796, 1988.
126. Friberg, S.E., Rong, G., and Ward, A.J.I., *J. Phys. Chem.*, 92, 7247, 1988.
127. Ceglie, A., Monduzzi, M., Söderman, O., and Lindman, B., *Prog. Colloid Polym. Sci.*, 76, 308, 1988.
128. Das, K.P., Ceglie, A., and Lindman, B., *J. Phys. Chem.*, 91, 2938, 1987.
129. Ceglie, A., Monduzzi, M., and Söderman, O., *J. Colloid Interface Sci.*, 142, 129, 1991.
130. Ranavare, S.B., Ward, A.J.I., Osborne, D.W., Friberg, S. E., and Kaiser, H., *J. Phys. Chem.*, 92, 5181, 1988.
131. Friberg, S.E. and Liang, Y.C., *Colloid Surf.*, 24, 325, 1987.
132. Das, K.P., Ceglie, A., Lindman, B., and Friberg, S.E., *J. Colloid Interface Sci.*, 116, 390, 1987.
133. Fletcher, P.D.I., Robinson, B. H., and Tabony, J., *J. Chem. Soc., Faraday Trans. I*, 82, 2311, 1986.
134. Riter, R.E., Kimmel, J.R., Undiks, E.P, and Levinger, N.E., *J. Phys. Chem. B*, 101, 8292, 1997.
135. Riter, R.E, Undiks, E.P., Kimmel, J.R., and Levinger, N.E., *J. Phys. Chem. B*, 102, 7931, 1998.
136. Raju, B.B. and Costa, S.M.B., *Spectrochim. Acta*, 56, 1703, 2000.
137. Laia, C.A.T., López-Cornejo, P., Costa, S.M.B., d'Oliveira, J., and Martinho, J.M.G., *Langmuir*, 14, 3531, 1998.
138. Wärnheim, T. and Sjöberg, M., *J. Colloid Interface Sci.*, 131, 402, 1989.
139. Schubert, K.V. and Strey, R., *J. Chem. Phys.*, 95, 11, 1991.
140. Martino, A. and Kaler, E.W., *Langmuir*, 11, 779, 1995.
141. Hildebrand, J.H., Prausuitz, J.M., and Scott, R.L., *Regular and Related Solutions*, Van Nostrand Reinhold, New York, 1970.
142. Gordon, J.E., *The Organic Chemistry of Electrolytes Solutions*, Wiley, New York, 1975.
143. Evans, D.F., *Langmuir*, 4, 3, 1988.

# 7
# Light-Duty Liquid Detergents

**JOAN GAMBOGI, EVANGELIA S. ARVANITIDOU, and
KUO-YANN LAI** Global Technology, Colgate-Palmolive Company,
Piscataway, New Jersey

## I. INTRODUCTION

Light-duty liquid detergents (LDLDs) are mixtures of surfactants dispersed in water and, as opposed to heavy-duty liquid detergents (HDLDs), are free of builders or alkaline inorganics. They are used primarily for hand washing of dishes, glasses, pots and pans, and other cooking and serving utensils. They are also used for washing hands, cleaning kitchen countertops, cutting boards, stove surfaces, and less often for washing delicate fabrics and general household cleaning.

Consumers expect LDLDs to clean, foam, and be mild to their hands. In addition, many consumers have come to want long-lasting foam, pleasing appearance and fragrance, ease of rinsing, safety for dishes, consumers, and the environment, convenient packaging and ease of dispensing, and good value. In the developed markets, LDLDs are now more and more concentrated, some are antibacterial for those concerned about family health, and some are more experiential. With the introduction of these "ultras," antibacterial, and sensorial variants of LDLDs in the 1990s, the face of the hand dishwashing liquid market in the developed markets has changed significantly. In the developing markets, LDLDs are in general more dilute with lower active levels and generally do not have the added benefits (such as antibacterial ability or the "aromatherapy" experience). However, the fundamental consumer need is still a dishwashing liquid that cleans fast, is convenient to use, and is not too expensive.

The LDLD market is worth over $900 million in the U.S. [1]. In a recent Habits and Practices study [2] conducted by Colgate-Palmolive in the U.S., it was found that an average household has on average 6.8 main meals per week. An LDLD is used to some extent after 94% of these meals. This is despite the fact that 60% of households have an automatic dishwasher (which is the highest incidence of these appliances in homes in the world). Even with the popularity of automatic dishwashers, a great deal of dishwashing is still performed manually; in particular the toughest to clean items are mostly washed by hand rather than in a dishwasher (86% vs. 16%).

The literature specifically devoted to the discussion of LDLDs is limited [3–11]. The advances in technology in this area are primarily documented in patents.

This chapter attempts to provide a thorough review of all aspects of LDLDs, including discussions on typical compositions and ingredients, the hand dishwashing process and the chemistry involved, test methods and performance evaluations, formulation technology, and new products and future trends. The LDLD chapter from the first edition [11] has been updated and sections rewritten to reflect the recent advances in technology and new products and future trends in the markets.

## II. TYPICAL COMPOSITION AND INGREDIENTS
## A. Typical Composition

LDLDs consist of a mixture of ingredients designed to provide cleaning, foaming, solubilization, preservation, fragrance, color, and in some cases antibacterial action. A typical light-duty liquid composition is detailed in Table 7.1.

Surfactants are the main active ingredients in an LDLD formulation and usually make up the bulk of the solids. Surfactants are surface-active agents and

**TABLE 7.1** Typical Light-Duty Liquid Composition

| Ingredient | Content (%) | Purpose |
|---|---|---|
| Surfactants | 1–50 | Cleaning, foaming |
| Hydrotrope | 0–10 | Phase stability, solubility |
| Salts | <3 | Viscosity control |
| Preservative | <0.1 | Micro stability |
| Fragrance | 0.1–1 | Aesthetics |
| Dye | <0.1 | Aesthetics |
| Other Additives | 0–3 | Chelant, antibacterial agent, enzymes, divalent ions, UV stabilizer |
| Water | Balance | |

**TABLE 7.2** Typical Physical and Chemical Characteristics of Light-Duty Liquid Detergents

| Characteristic | Typical value |
|---|---|
| Viscosity, cP | 100–500 |
| pH | 5–8 |
| Cloud point, °C | <5 |
| Clear point, °C | <10 |
| Solids level, % | 10–50 |
| Specific gravity | 1.0–1.1 |

their function is to penetrate and loosen soil, enhance water absorption and wetting of surfaces, suspend, disperse, and emulsify soil in water, and generate and stabilize foam.

Typical physical characteristics of LDLDs are summarized in Table 7.2. They are generally slightly viscous, Newtonian fluids with viscosities in the range 100 to 500 cP. The pH has always typically been near neutral (pH = 5 to 7) to match the natural pH range of the skin. Very recently more extremes in product pH have come onto the market with the relaunch of some of Procter & Gamble's products in the U.S. (pH = 8 to 8.5) and new antibacterial products by Colgate-Palmolive in Europe (pH = 3.5). LDLDs are usually between 10 and 50% solids in water.

## B. Ingredients

### 1. Surfactants

The primary cleaning ingredients in hand dishwashing liquids are surfactants. Surfactants are also responsible for providing the foaming, which is an important sensory indicator of efficacy for consumers. Another factor that is important to the consumer is mildness to the skin, as hand dishwashing is one cleaning task in which the hands are exposed to the cleaning solution for an extended time. Surfactants, since they are present in the formula in the largest amount, contribute the most to the irritation or lack of irritation of a product. The type of surfactants typically used in LDLDs are anionics and to a lesser degree nonionics and amphoterics. Cationics have not been used historically because of their lesser cleaning ability and incompatibility with anionic surfactants. Table 7.3 summarizes some of the surfactants and their structures falling into these three classes (anionic, nonionic, and amphoteric) that are found in LDLDs.

Anionic surfactants have been used predominantly because of their availability, good cleaning properties, excellent foaming properties, and low cost [12]. The anionics that have been most widely used either have a sulfonate ($SO_3^-$) or sulfate ($OSO_3^-$) head group. Some common sulfonates are linear alkylbenzene sulfonate (LAS), $\alpha$-olefin sulfonate (AOS), and paraffin sulfonate (PS, also referred to as secondary alkane sulfonate, SAS). Some typical sulfates are alkyl sulfate (e.g., sodium lauryl sulfate, SLS) and alkylethoxy sulfate (AEOS or more specifically sodium lauryl ethoxy sulfate, SLES), which differ only by the number of moles of ethylene oxide groups. Another anionic surfactant used on a limited scale is alpha sulfomethyl ester (ASME). In general, ASMEs have excellent detergency and are potential substitutes for LAS [13]. In the U.S. (and much of the rest of the world) LAS and AEOS are the most commonly used anionics, while in Europe PS is the major anionic used.

Nonionic surfactants have been used to a lesser extent because of their lower foaming performance and higher cost [14]. However, when used in combination with anionic surfactants they provide benefits to the overall formulation, such as mildness, improved wetting, foam boosting, and foam stabilization. Some nonionics that are found in LDLDs are ethoxylated alcohols, in particular 11-carbon chains with 9 moles of ethoxylation (e.g., Neodol 1-9 from Shell). Surfactants derived from sugar, such as alkylpolyglycosides (APG) [15] and fatty acid glucamides [16], are also used in many hand dishwashing formulations.

Another important class of nonionics are amine oxides, such as DMDAO (dimethyldodecyl amine oxide) and CAPAO (cocoamidopropyldimethyl amine oxide). This type of surfactant is nonionic at pH values above its $pK_a$ and cationic below that point. When functioning as a nonionic, amine oxides have many useful properties. They interact strongly with anionics which can result in performance benefits [17]. Amine oxides help to mitigate anionic surfactant irritation, act as foam stabilizers, and can also function to improve grease removal.

Amphoteric surfactants, in particular betaines, especially cocoamidopropyl betaine, typically provide synergistic benefits with anionic surfactants [18]. Similar to the benefits of amine oxides, they have been found to mitigate the inherent

**TABLE 7.3**   Surfactants Commonly Used in Light-Duty Liquid Detergents

| Chemical description | Chemical structure |
|---|---|

Anionic surfactants
    Alkylbenzene sulfonate

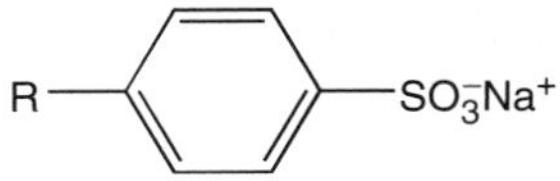

$$R-\bigcirc\!\!\!\!\bigcirc-SO_3^-Na^+$$

    Paraffin sulfonate

$$CH_3(CH_2)_m-CH-SO_3^-Na^+$$
$$|(CH_2)_nCH_3$$

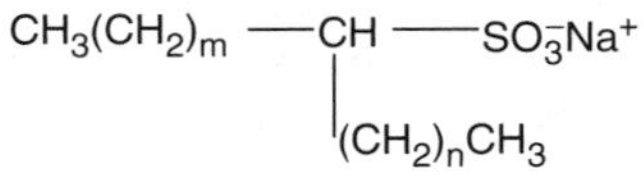

    α-Olefin sulfonate $\quad$ $R–CH_2–CH=CH–H_2–SO_3^-\ Na^+$

    Alkyl sulfate $\quad$ $R–OSO_3^-\ Na^+$

    Alkylethoxy sulfate $\quad$ $R–(OCH_2CH_2)_n–OSO_3^-\ Na^+$

    Alpha sulfomethyl ester

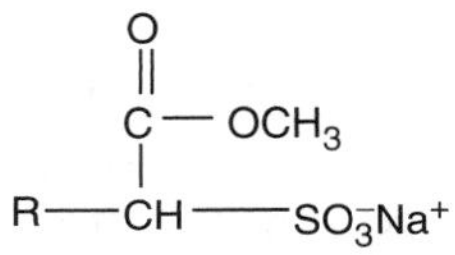

$$\overset{\displaystyle O}{\overset{\displaystyle \|}{C}}-OCH_3$$
$$R-CH-SO_3^-Na^+$$

Nonionic surfactants
    Alcohol ethoxylate $\quad$ $R(OCH_2CH_2)_nOH$

    Alkylpolyglycoside

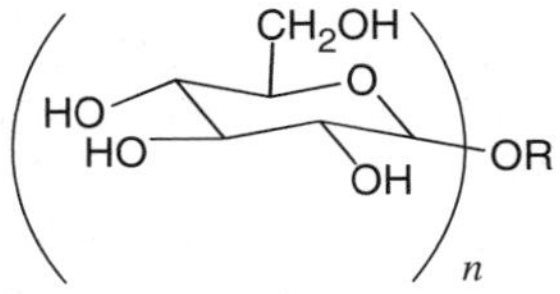

$$\left(HO\atop HO\right.\ \ {CH_2OH \atop \diagdown O}\ \left.{OR}\atop{OH}\right)_n$$

    Fatty acid glucamide

$$R-\overset{\displaystyle O}{\overset{\displaystyle \|}{C}}-NHCH_2CH_2OH$$

    Amine oxide

$$CH_3$$
$$R-N\longrightarrow O$$
$$CH_3$$

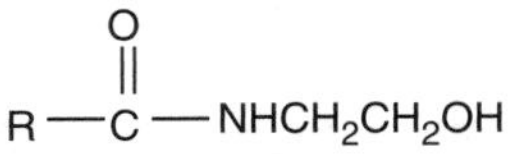

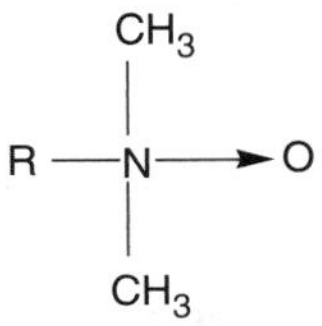

Amphoteric surfactants
    Cocoamidopropyl betaine

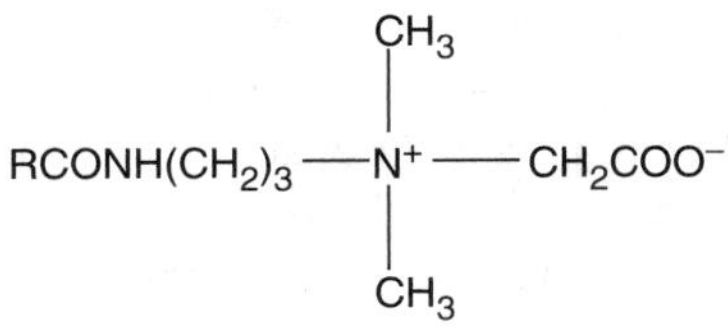

$$CH_3$$
$$RCONH(CH_2)_3-N^+-CH_2COO^-$$
$$CH_3$$

irritation of anionics, boost foaming and foam stability, and enhance grease removal. Taking proper advantage of positive surfactant interactions allows for the use of overall less total surfactant for similar performance benefits.

Surfactant suppliers [19] are concentrating their research on improving the cost/performance attributes of the surfactants. Their efforts have been and are focused on:

- The ability to remove and emulsify the suspended soil.
- Foaming and foam stability in the presence of soils.
- Solubility in the aqueous phase.
- The ability to coexist with other ingredients under extreme conditions as well as at room temperature.
- A good environmental profile.

## 2.  Foam Stabilizers

Foam is an important visual signal for LDLDs. While there is no direct correlation between foam and cleaning, consumers in general use foam volume and foam persistence to judge the performance of an LDLD. There is a wide variety of stabilizers for foam [20]. Among the most commonly used in LDLDs are the following:

- Fatty alkanol amides, such as LMMEA (lauric/myristic monoethanol amide), LMDEA (lauric/myristic diethanol amide), CDEA (cocodiethanol amide), and CMEA (cocomonoethanol amide). (Although recently there have been negative reports about DEA, diethanol amides [21].)
- Amine oxides, such as DMDAO (dimethyldodecyl amine oxide) and DMMAO (dimethylmyristyl amine oxide).

Details of fatty alkanol amides and amine oxides commonly used as foam stabilizers in LDLDs can be found in the literature [22].

## 3.  Hydrotropes

Hydrotropes are often added to an LDLD to help solubilize certain surfactants or other materials that are not easily soluble in water to ensure the stability of the formulation. The fundamental properties of hydrotropes and their hydrotropic action in liquid detergents are discussed in Chapter 2. The addition of a hydrotrope affects the formula viscosity and cloud/clear points.

The hydrotropes most widely used in LDLDs are sodium xylene sulfonate (SXS), sodium cumene sulfonate (SCS), sodium toluene sulfonate (STS), urea, and ethanol, as shown in Table 7.4. SXS, SCS, and ethanol are the most often used hydrotropes in LDLDs since they are nearly odorless and colorless. Urea is an effective and cheap hydrotrope; however, it has been found to raise the pH

**TABLE 7.4**  Hydrotropes Commonly Used in Light-Duty Liquid Detergents

| Hydrotrope | Chemical structure |
| --- | --- |
| Sodium xylene sulfonate (SXS) | |
| Sodium toluene sulfonate (STS) | |
| Sodium cumene sulfonate (SCS) | |
| Urea | |
| Ethanol | $CH_3CH_2OH$ |

of a formulation upon aging, which could potentially result in an ammonia odor. Urea is also a good nutrient for bacteria. With proper attention toward pH and preservation, urea can be utilized. Other molecules used as hydrotropes include isopropanol, propylene glycol, and polyethylene glycol ethers.

## 4.  Minor Ingredients

Many minor ingredients are added at the level of less than 1% and mainly to affect product aesthetics. Examples of these include fragrances, dyes, preservatives, chelators, viscosity modifiers, and pH modifiers. The fragrance and color of an LDLD are of critical importance to its success. The selection of these, together with packaging, creates the image for the product.

Preservatives are often needed to prevent microbial and fungal growth in LDLDs. Preservatives commonly used are formaldehyde, gluteraldehyde, benzoic acid, Kathon®, Dowicil®, Bronopol®, various esters of hydroxybenzoic acid, and others.

Chelants are used to ensure that no precipitation occurs on aging. The most common problem is iron, which is introduced as an impurity from surfactants and salts. The chelants most commonly used are EDTA, HEDTA, citrate salts, and disodium diethylene pentaacetate.

Viscosity modifiers are used to achieve the desired product viscosity. These include alcohols, salts, polymers, and hydrotropes. Viscosity adjustment with polymers, particularly in the high surfactant concentrations of "ultra" LDLD formulations, can be challenging. Acids, such as citric or sulfuric acids, or bases, such as hydroxide, are also added to bring the product pH to the desired level.

In certain products, specialty ingredients are also added for extra benefits, specific aesthetic effects, or for marketing claims. Some examples of these ingredients include antibacterial agents, enzymes, protein, lemon juice, opacifier, abrasives, polymers, bleach, and aloe. Antibacterial agents, such as triclosan, are popular in U.S. hand dishwashing liquids and provide a benefit to consumers of killing germs on hands. Recently, Procter & Gamble has introduced enzymes into two of its hand dishwashing liquids for cleaning or skin conditioning benefits. Polymers have been added to LDLDs in order to improve foaming, or grease release, or to enhance mildness.

## III.  HAND DISHWASHING
### A.  Variables
#### 1.  Mechanical Action

Mechanical action is very important in hand dishwashing. When people wash dishes they actively rub the surface. This mixes the surfactants with the soils and accelerates the cleaning. It also physically removes the soil.

The amount of mechanical action used in hand dishwashing is extremely variable and hard to quantify. This is typified by the large number of dishwashing performance tests that are used (see Section IV). Consumers may soak items that are difficult to clean in a low mechanical action environment. Under these conditions, surface chemistry is very important. Consumers may also scrub vigorously directly on the soiled area, break up the soil particles, and suspend them. At this point interfacial processes become important again. All individuals have their own techniques. Individuals vary the amount of effort they use depending on the type and distribution of the soil on the item. However, they usually do not use enough sustained mechanical action to make a stable oil-in-water emulsion.

Much of the cleaning occurs from water, heat, and mechanical action alone, as is true in automatic dishwashing machines. Large food particles, sugars, many starches, and some protein soils are readily removed by rinsing or soaking with plain hot water. However, some food soils, such as baked-on starches or polymerized fats, are extremely resistant. These require vigorous, direct mechanical action. In some cases vigorous chemical action is used, as in oven cleaners, but this mechanism is not within the scope of typical LDLDs.

## 2. Washing Methods

Dishwashing methods are extremely diverse and vary greatly according to geography, local tradition, and individual person, lifestyle, and diet.

*Neat dishwashing.* Neat dishwashing refers to the practice of placing the dishwashing liquid directly on the item to be washed or directly on the washing tool (sponge, brush, rag, etc.). The item is usually rinsed first. In neat dishwashing the surfactant concentration ranges from a few percent to 30% or more, depending on the active ingredient level of the product. In developed countries neat dishwashing is used when there are only a few dishes to wash or when one particular item is especially soiled. Neat dishwashing is also a common habit in places such as Brazil, India, and Japan [23].

*Dilute methods.* Dilute dishwashing is the widely used practice of filling a tub, sink, or pot with water and adding the dishwashing liquid (1 to 10 g) to make a solution [24]. The dishes are either submerged in the solution all at once or submerged one at a time and then washed. Typical surfactant concentrations range from 0.06 to 0.2% for U.S. consumers (using an "ultra" concentrated product) and from 0.06 to 0.3% for European consumers. The sinks range in size from 5 to 20 liters [25].

*Soaking.* The soaking method is used for hard-to-clean, baked-on grease and soils. One or two squirts of the product are added directly to the cookware, which is filled with hot water and left to soak for a period of time. After soaking, the items are cleaned with much less effort since the soils have been loosened. Typical soak concentrations are 0.2 to 0.5%, temperatures are usually those of domestic hot water supplies (40 to 50°C), and time is 10 to 15 minutes.

*Dip and dab methods.* The dip and dab method consists of adding product (10 to 100 g) to a small bowl and filling the bowl with water. The soiled item is then washed with this solution, but not submerged as in the dilute method. The dip and dab method generally has much higher concentrations (1 to 3%) than the dilute method and is generally only used in developing countries.

*Rinsing.* Once items are washed, they are generally rinsed with clean water. This is especially important when higher concentrations of dishwashing liquid are used. Some consumers in German-speaking countries do not rinse. They scrape the plates thoroughly so there is a minimum soil load, wash in very dilute solutions, and dry with a towel.

## 3. Soils

Many food soils are encountered in hand dishwashing, such as grease, carbohydrate, protein, dairy, and mixed soils. Baked-on soil requires more vigorous treatment, either mechanical or chemical [26]. The type of oily soil is almost exclusively triglycerides. The hydrocarbon chains in food triglycerides are predominantly C12 to C16, although higher and lower chains are also present.

When heated by cooking these fats and oils can undergo significant chemical changes due to oxidation and polymerization.

## 4. Surfaces

In dishwashing, one must consider soil and surfactant adsorption to both polar and nonpolar surfaces. Metals (aluminum, stainless steel, carbon steel, cast iron, silver, and tin), siliceous surfaces (china, glass, and pottery), and organics (polyethylene, polypropylene, polyethylene terephthalate (PET), polytetrafluoroethylene (PTFE), and wood) present a wide variety of surface characteristics. They span the range of high interfacial free energy (metals and many ceramics) to low interfacial free energy (hydrocarbon polymers) surfaces [27,28].

## 5. Water Temperature and Hardness

The water temperature used in hand dishwashing is highly variable and depends on the climate and the availability of hot water. In tropical places the ambient temperature of water is 30 to 37°C. Many edible fats are at least partly liquefied at this temperature. In some temperate zones the water can be close to 0°C, depending on its source and time of year. A source of hot water is essential under these conditions. A typical wash temperature is usually between 32 and 43°C. The maximum wash temperature is about 50°C due to the previously mentioned exposure of the consumer's skin during the washing process. Above this temperature the water becomes dangerously hot. Hotter water is used for tough jobs, but the items are left to soak and are not handled, unless perhaps gloves are used. Raising the washing temperature can markedly increase the amount of cleaning [29].

The amount of hardness ions in the water can vary greatly according to geographic location. Typical values in the U.S. are around 50 ppm (soft water) to 300 ppm (hard water). Water hardness can be beneficial or detrimental to performance, depending on the application and the product composition. In hand dishwashing, water hardness generally increases the efficacy of LDLDs. High water hardness increases grease removal and increases the number of items washed with a given amount of surfactant. Water hardness has a variable effect on foam depending on the range of hardness and the anionic or nonionic nature of the surfactants. Hard water can also increase the likelihood of spotting on articles when they are left to air dry.

## B. Mechanisms of Performance and Relevant Physical and Chemical Properties

### 1. Cleaning Mechanisms

The three main mechanisms for soil removal from hard surfaces are chemical, mechanical, and detergent action [30]. Cleaning of dishes by hand is accomplished primarily by mechanical action, warm water, and the detergent. The role of the

LDLD, through the use of surfactants, is to provide the detergency. Different surfactant physicochemical processes are relevant depending on whether the soil is a liquid or a solid.

Liquid soil is usually removed by roll-up, emulsification, direct solubilization, and possibly formation of microemulsion or liquid crystalline phases. The oil emulsification capability of the surfactant solution and the oil–water interfacial tension are relevant physicochemical parameters.

The first mechanism of cleaning hard surfaces is the roll-up of liquid soils or soils that have been liquefied by heat or surfactant penetration [31]. The mechanism involves successive steps. The first step is the wetting step in which the washing solution adsorbs on the grease and the substrate. The interfacial tensions of the grease–water and substrate–water interfaces are reduced to the same range as that of the substrate–grease interface. At this point, convection or mild mechanical agitation can be enough to detach the grease droplet from the surface. If the substrate–grease interfacial tension remains lower than the substrate–water tension, then all the grease cannot be removed. A portion is removed, however, and the remainder can be emulsified or solubilized.

Above the critical micelle concentration (CMC), surfactant molecules aggregate into structures called micelles. The hydrophobic portion of the surfactant molecule occupies the core of the aggregate and the hydrophilic head groups point toward the water phase. Solubilization is the spontaneous dissolving of grease in the hydrophobic core of the micelles. This results in swollen micelles, the size of which is still well below the wavelength of light, resulting in transparent systems. Swollen micellar systems are thermodynamically stable and can be considered oil-in-water microemulsions. The extent to which the LDLD can solubilize oily soil depends on the chemical structure of the surfactants, its use concentration, and the temperature. High concentrations of surfactants can accommodate much larger amounts of oil.

When insufficient surfactant is present to solubilize all of the oily soil, the remainder can be suspended in the bath by emulsification. An emulsion is a thermodynamically unstable suspension of liquid particles in a second liquid phase. Emulsion particles are much larger than micelles, about 500 nm or greater. The fact that emulsions are not thermodynamically stable is irrelevant since the dirty suspension is drained down the sink and the dishes are rinsed.

A recent review [32] describes several cases in which the maximum soil removal occurs when the soil is incorporated into an intermediate phase, such as a microemulsion or lamellar liquid crystals. These intermediate phases form at the interface between the soil and the washing bath. The phases grow up to a point and then, as a result of agitation, break off into the bath, where they are emulsified into the aqueous solution.

Solid inorganic soils, such as dust particles, are removed through a wetting and suspension mechanism. Solid organic soils, such as greases, are broken up and suspended in the bath by the LDLD. Penetration into the solid grease can cause it to swell and liquify. The relevant physicochemical processes are wetting and adhesion tension. The first step in cleaning hard surfaces is the adsorption of the surfactants at the soil interface. The cleaning solution first must effectively wet the surface. Wetting involves the interaction of a liquid with a solid. For dishwashing, it is usually the spreading of a liquid over a surface. Wettability can be measured by the contact angle ($\theta$), which is the angle that the liquid makes when it is in equilibrium with the other phases in contact with it. A low contact angle means high wettability and a high contact angle means poor wettability.

The work required to separate a unit area of liquid from a solid is called the work of adhesion. Adhesion tension is found from the product of surface tension and the cosine of the contact angle made by a drop of surfactant solution on a solid surface. Adhesion tension measurements have been conducted at a model grease surface–water interface and used to design superior consumer products [33]. Nonionic surfactants are useful in solid grease removal because they are efficient at covering the grease substrate and reducing the interfacial tension even at very low surfactant concentrations. Anionic surfactants are necessary, however, to disperse and emulsify the soil. It has been shown that by making adhesion tension measurements and using predictive diagrams the optimum ratio of anionic to nonionic surfactants can be formulated for superior cleaning [33].

## 2. Foaming

The foam of LDLDs is an important signal to the consumer of product efficacy. High sudsing signals good detergency and suds decrease signals detergency decrease to the consumer. Today, nearly all LDLDs are formulated to deliver long-lasting foam. Even though high foaming is not necessarily related to a product's actual cleaning ability (e.g., nonionic surfactants offer good cleaning but in general they do not foam well), modern products deliver both good cleaning and long-lasting foam to meet consumer expectation.

Consumers evaluate several different properties related to foam. The amount of foam formed when the water and product are first introduced into the wash basin or sink is referred to as the flash foam or the initial foam volume. The persistence of the foam in the presence of food soils is also judged. This is referred to as foam stability in the presence of soils. Foam must be present until the end of the dishwashing process. This performance measure is referred to as foam mileage or longevity. For some consumers, the quality of foam might also be important. This is particularly true for those consumers that use LDLDs to wash their hands. The quality or hand feel of the foam should be rich and thick; however, the foam must rinse quickly from the dishes.

The different properties of foam find their origin in different physicochemical phenomena. The amount of mechanical energy introduced can be the most important factor in determining foamability and in fact may play a greater role than the actual surfactant properties. Surfactant systems perform differently depending on whether the foam is generated with high shear or low shear. The relationship between a surfactant's chemical structure and its foaming ability can be quite complex. Also, there is not always a direct relationship between a surfactant's performance in foam volume and foam stability. A surfactant's foaming ability depends on its ability to lower the surface tension of the solution, its ability to diffuse to interfaces, the packing or structure it takes at the interface, the elastic properties it imparts on interfaces, and its ability to stabilize thin films.

In a homologous series of surfactants, the foam volume will generally go through a maximum as the chain length increases with C12 often providing the best foaming. This is due to a balance between conflicting effects. As the chain length increases, the CMC decreases and surface tensions are lowered. In contrast, as the chain length increases the solubility and speed of diffusion to interfaces is reduced.

In general, surfactants with lower CMCs are more *efficient* foamers [34]. Foam heights generally increase with an increase in concentration below the CMC and level off near the CMC. Thus the CMC of a surfactant system gives an indication of its foam efficiency. Above the CMC, different surfactants are able to foam to different heights, i.e. the effectiveness. The *effectiveness* of a surfactant system depends on both its ability to reduce the surface tension and the magnitude of intermolecular forces. Researchers from Exxon found that foam heights (effectiveness) correlate with the effectiveness of surface tension reduction at the CMC [35]. Foam stability correlated with the rate of surface tension reduction at the air–water interface [35]. Other researchers have also found correlations between dynamic surface tension and foam height generated in a kinetic manner [36]. In studies of SDS modified with long-chain alcohols it was found that foamability exhibits opposite behavior depending upon the rate of foam generation [37]. In slow gentle foam generation, more foam was generated with a system of lower equilibrium surface tension. In vigorous foam generation, more foam was generated with a system of lower dynamic surface tension.

Foams become unstable due to three basic mechanisms: drainage (gravity), film rupture (coalescence), and coarsening (Laplace pressure). Coarsening is driven by surface tension and causes gas to diffuse from small to larger bubbles. Effective methods for stabilizing foams attempt to overcome one or more of these mechanisms. All three phenomena depend on the elasticity and viscosity of the foam surface. Thus surface rheology measurements are becoming of increasing importance [38].

LDLDs are mixtures of surfactants. Surfactant mixtures often perform better than the sum of the individual surfactant contributions, or perform synergistically. The origin of this synergistic interaction is head group interactions and is dipolar in nature. The surfactant pairs having the greatest dipolar forces have the largest synergies: anionic–cationic > anionic–amphoteric > anionic–nonionic. However, anionic–cationic surfactant pairs tend to have solubility issues. For this reason, amphoteric surfactants, such as betaines, are often used in surfactant systems to improve foamability and foam stability. Rosen and Zhu have studied the relationship between synergism in foam height (by the Ross–Miles test) and synergism in surfactant adsorption properties [39]. The only correlation they were able to establish was between synergism in initial foam heights and surface tension reduction effectiveness. This term refers to the surface tension at the CMC of the mixture being lower than either pure component. However, no relationships for foam stability were found with equilibrium surface adsorption properties.

## 3.  Mildness

The mildness of an LDLD is important since consumers spend time with their hands exposed to the product either neat or soaking in diluted product. Negative signals can be a redness or roughness feeling. Skin feel can change with time. Surfactants are the ingredients primarily responsible for the lack of mildness in a product, although there is irritation to the skin just from soaking in hot water.

There are several known mechanisms of surfactant-induced skin irritation [40]. The first is by binding of surfactants to sites on the stratum corneum. The second is swelling of the membrane in response to binding. Stratum corneum swelling studies have shown that swelling increases over time, consistent with a diffusion process [41]. The rate of swelling changes with concentration after the CMC is reached. This contributes to the commonly held belief that it is surfactant monomers and not micelles that contribute to surfactant irritation. Swelling has been found to be a maximum for C12 or C14 isomers for various anionic surfactant series. Again this is probably due to a balance between opposing forces: decreasing CMC with increasing chain length but a decrease in water solubility. Nonionics have been found not to swell the stratum corneum significantly. The third mechanism is release of mediators of inflammation or removal of biomolecules. The release of inflammatory mediators is essentially from living cells, the Langerhans cells present in the epidermis and the keratinocytes, also in the epidermis. These mediators initiate a cascade of release and other cell types also participate in this release, such as macrophages, lymphocytes, leucocytes that are present in the dermis. The removal of biomolecules refers to the NMF (natural moisturizing factor) molecules that are extracted from the corneocytes and the stratum corneum and cause the stratum corneum to dehydrate. The fourth mechanism is the denaturation of proteins (discussed in Section IV.C). The fifth mechanism is permeability of surfactants through membranes. Finally, surfactants may be responsible for removal of skin lipids.

Surfactant mixtures have been shown to counter irritation. The explanation for this phenomenon is that the two or more surfactant types compete for a limited number of binding sites on the stratum corneum. Also, the two surfactants can form a complex that is less irritating. A mixture of surfactants very often has different solutions properties, as, for example, the CMC might be lowered. Adding an amphoteric surfactant, such as a betaine, to an anionic-based product can result in reduced skin irritation [42].

## 4. Rheology

Concentrated LDLDs, or "ultra" dishwashing liquids, typically contain 40 to 45% of surfactants and are isotropic, Newtonian liquids with viscosity in the range 100 to 500 cP. For most ionic surfactant systems, if the viscosity is plotted against the salt concentration the curve is found to increase to a maximum and then decrease again with higher salt concentration. Most concentrated LDLDs are already on the side of the salt curve where viscosity decreases with added salt. Therefore, salt can be used to reduce the viscosity of the concentrated surfactant system to the desired level. Alcohols such as ethanol are also used to reduce the viscosity. The addition of solvents such as alcohols or glycols also helps to solubilize and lower cloud/clear points. However, this needs to be done with care to avoid imparting undesirable odor to the product.

In some cases it has been observed that concentrated LDLDs thicken upon dilution. This would indicate that the micelles in the system are becoming elongated and entangled. This could be a potential problem for consumers who add water as the LDLD product bottle becomes empty in an attempt to wash out the remaining amounts of product.

## 5. Antibacterial Properties

There are two kinds of antibacterial effects: bacteriostatic, which means to stop bacteria growth; and bacteriocidal, which means to kill bacteria. In the U.S. the claims are bacteriocidal and specific to the hands. While there is no real signal of antibacterial action, consumers believe the claims that reputable dishwashing liquid manufacturers make. Claims are government regulated which helps justify consumer belief. The U.S. Food and Drug Administration (FDA) regulates claims regarding human skin (and therefore claims for hand washes), while the Environmental Protection Agency (EPA) regulates claims regarding surfaces (i.e., hard or porous). An FDA-regulated product must contain an antibacterial ingredient covered by an NDA (New Drug Application) of the Tentative Final Monograph, and efficacy of the product must be shown versus the vehicle (product without antibacterial ingredient). An EPA-regulated product must contain an active that is approved by the EPA and all other ingredients must be present on an EPA inert-ingredient list. Data must be presented on the product's efficacy and safety. In Europe there are different regulations, as discussed in Sections IV.C and VI.A.

## IV.  TEST METHODS AND PERFORMANCE EVALUATION

There are a number of laboratory tests used by formulators to evaluate the various performance aspects of an LDLD. These tests use a variety of soils and washing conditions (e.g., temperature, time, water hardness, mechanical action or not). A detailed description of various test methodologies is given in the literature [11]. Some of those methodologies are reviewed in this section. In addition, a subsection on antibacterial test methodology is included.

The test methods discussed below are classified in five categories: evaluation of cleaning performance; evaluation of foam performance (volume and stability); evaluation of mildness; evaluation of antibacterial efficacy; and other tests.

### A.  Cleaning Performance

Cleaning performance is the most important characteristics of a dishwashing liquid since consumers purchase the product for washing dishes, and their principal expectation is the removal of greasy soils. As stated previously, many food soils such as carbohydrates can be effectively removed sometimes with just plain hot water. However, this is not true of greasy soils which tend to pose more of a cleaning challenge. Consequently, the literature is replete with formulations directed at efficacious grease cleaning. Test methods for cleaning performance have mainly focused on greasy soils.

### 1.  Baumgartner Test

The Baumgartner test was originally developed by Hoechst AG Chemical Company [43]. This test attempts to mimic actual dishwashing by incorporating physical energy into the grease removal process. Polypropylene test tubes are coated with a thin layer of Armour lard (about 60 to 70 mg) and repeatedly dipped, at a controlled rate, in and out of the test detergent solution. The weight of the soil removed from each tube is measured. Test solutions contain 0.667 g LDLD/l of 150 ppm artificially hardened water for ultra formulas, corresponding to about 0.30 g/l surfactant for a 45 AI product. The experiment is executed at 300 or 600 dips and the temperature of the test solutions is held at 108°F (42°C).

Results are reported as % soil removed calculated using the following formula:

$$\% \text{ Soil removed} = \frac{\text{Amount of soil removed}}{\text{Original amount of soil}} \times 100$$

A higher percentage soil removed reflects better greasy soil cleaning ability.

### 2.  Cup Test

Another test used to assess the grease-removing ability of test solutions is the cup test. This test is based on Procter & Gamble's grease removal test [44] with a few modifications. The test consists of solidifying about 6.5 g of beef tallow in the bottom of a tripour cup. Test cleaning solutions are heated to 115°F (46°C) and then poured into the soiled cup and allowed to soak for 15 minutes. The cleaning solution is then poured out of the cup with any soil it has removed. It is important that the temperature of the solution in the cup be below that of the melt temperature of the greasy soil or combination of soils (e.g., chicken, beef, pork fat) otherwise a true reading of the product's performance will not be obtained, as the soil will melt.

The evaluation of the product can be done by either gravimetric or turbidity means. The gravimetric method is preferred since it is simpler. Cups are allowed to dry overnight and the final weight of the grease remaining measured and used to calculate the % grease removed. The amount of grease is determined before and after the solution is added. Results are reported as % soil removed, as described above for the Baumgartner test.

The test concentration is 2.67 g/l for an "ultra" dishwashing liquid, or 1.2 g/l of surfactant and 4.0 g/l for regular products. Six replicas are usually measured.

## 3. Hand Dishwashing Test (Plate Count)

Hand dishwashing tests provide the best performance information about the entire product as they use soils and wash conditions as close as possible to those encountered by consumers under normal conditions. The indicator of a product's performance is the number of plates washed. In this test [46] the product is placed directly onto a sponge and soiled plates are washed until they cannot be cleaned any further (greasy residue on the plate). Performance is equated to the number of plates washed: the greater the number of plates washed, the better the product.

Although this test provides a more accurate assessment of a product's performance, it has some drawbacks. This type of test usually takes a long time to complete. Another limitation of this test is that it is subjective, and thus can vary from operator to operator.

## 4. Static Soaking Test [46]

This test measures the amount of soil (Crisco®, a vegetable shortening derived from cotton seed and soy bean oils and manufactured by Smucker Co.) removed from a plate after soaking for 30 seconds in a test solution of 0.1% detergent, 150 ppm hardness, 100 ppm alkalinity, and at 50°C. The plates are transferred to an ice bath following immersion in the warm test solution to stop the soil-removing process. The plates are dried and weighed. The % percent soil removal is calculated as:

$$\% \text{ Soil removal} = \frac{\text{Amount of soil removed}}{\text{Original amount of soil}} \times 100$$

The higher the % soil removal, the better the performance of the dishwashing detergent.

## 5. Emulsion Stability Test [47]

This test measures a liquid detergent composition's ability to keep greasy soils emulsified. The test is performed by placing an oil, like corn oil, into a vial containing a solution of detergent. The vial is agitated for a fixed number of rotations at a fixed rate and let stand for a fixed time. Readings are taken with a turbidity meter or colorimeter at given time intervals. Higher turbidity values indicate more stable emulsions and lower colorimetry values indicate more stable emulsions.

## B.  Foam Performance

Foam volume and foam mileage tests are widely used for evaluating LDLDs. Foam volume tests measure the amount of foam a composition can generate with and without soil. Foam mileage, sometimes referred as foam stability, measures the ability of a detergent to maintain its foam with soil present or while it is introduced. This attribute is extremely important, because it constitutes a powerful signal to the consumer that it is time to add more detergent during the dishwashing process, and therefore influences consumers' estimation of the cost per use of the product.

## 1.  Foam Volume Tests

Foam volume is an important characteristic of light-duty dishwashing detergents. Higher foam heights are desirable, as consumers generally equate foaming with cleaning performance [4]. An ASTM method [48] for foam volume evaluations of light-duty dishwashing liquids is more commonly known as the Ross–Miles foam test. This method is widely used for evaluating the foaming ability of detergents or surfactants in general. Numerous other foam volume tests are cited in the patent literature that often differ significantly from the Ross–Miles test. These test methods are faster and easier for foam volume evaluation [49,50].

Foam volume tests can be conducted with soil or without soil. One foam volume test without soil [51], called the shake foam test or inverted cylinder test, is conducted by placing a solution of a composition into a cylinder. An amount of 100 g of LDLD solution (0.33 g/l concentration for an ultra formulation) is placed in a 500 ml graduated cylinder. The cylinder is shaken or inverted a fixed number of times or for a set amount of time (e.g., 40 rotations at 30 r/min). The foam height is measured in centimeters or milliliters, which are conveniently measured if graduated cylinders are used. A foam volume test with soil is conducted the same way but soil is placed into the cylinder. The soil can either be added with the solution initially or added after foam is generated. The cylinder is rotated or inverted the desired number of times and the resultant foam height is measured in milliliters or centimeters; usually at least three replicas are recorded.

## 2. Foam Stability Tests

Various types of foam mileage or foam stability tests are found in the literature. All use the same basic theme of generating foam under constant agitation and introducing soil until the foam collapses. Foam mileage tests measure a product's ability to resist foam depletion in the presence of soil. The amount of soil needed to break the foam is an indication of the product's ability to keep foaming while the consumer is washing dishes. The more soil it takes to break the foam the better the product's performance. Examples of some foam mileage tests are briefly described here.

*(a) Miniplate Test.* The Miniplate test [52] is an automated test designed to measure foam mileage. Foam is generated in a vessel and titrated to an endpoint with soil. The Miniplate apparatus consists of a computer system connected to an experimental chamber. The experimental chamber contains a thermostatically regulated water bath that is supplied with deionized water, a thermostatically controlled plate on which the reaction vessel sits, a motorized pump used to inject the soil, a brush with a system of gears to control its movement, a turbidity meter, and a photocell used to measure electronically the foam endpoint.

Concentrated LDLD solutions are prepared at 5% for regular and at 3.33% for ultra products in deionized water. A mixture of hard water concentrate, LDLD solution, and deionized water from the thermostat bath in the instrument is stirred by the motorized brush to generate the foam in the vessel. The final LDLD concentration is 0.083% for ultra products and 0.125% for regular products. The soil used in all Miniplate experiments is Crisco shortening (other soils can also be used) filled into plastic syringes. The amount of soil as well as the time required until the foam fully disappears is used to calculate the theoretical amount of plates washed. This method has been found to be correlated with a hand dishwashing procedure.

*(b) Shell Test.* The Shell test is a similar method to the Miniplate test, in that it is used to evaluate the foam performance of light-duty dishwashing liquid in the presence of soil [53]. Initial foam is generated in the test vessel by stirring a dilute solution of the test LDLD product. The soil is then titrated into the vessel, under constant-rate stirring, until the foam endpoint is reached. The quantity measured is the weight of soil added to reach the foam endpoint. This value can then be normalized to determine a foam performance rate (FPR). FPR [54] is the ratio of the weight of the soil used in the test formula to the average weight of the soil used in the standard LDLD formula under the same test conditions (e.g., temperature, test soil, water hardness):

$$\text{FPR} = \frac{\text{Weight of test product}}{\text{Weight of control}} \times 100$$

FPR can be used in comparing different products. A higher FPR value indicates better foam stability. The soil used is a mixed soil consisting of Crisco vegetable shortening, Pillsbury brand instant mashed potatoes, Progresso olive oil, homogenized milk, formaldehyde, and water. (The formaldehyde is added so that the soil mixtures are preserved from microbial degradation when a batch is stored for use over a period of days.) The solution concentration for ultra LDLDs is 0.0266% and for regular LDLDs is 0.04%.

*(c)  Tergotometer Test.*   The tergotometer test is commonly used in evaluating detergency of laundry products. It was modified [55] for the evaluation of LDLDs. The tergotometer provides constant agitation where a dilute solution of test composition is added and foam is generated. Planchets covered with soil are periodically added until the foam height is reduced to a foam endpoint or to a fixed height. The number of small metal plates added to the foam endpoint reflects the foam stability of a composition.

*(d)  Piston Plunger Test.*   Another foam mileage test is referred to as the piston plunger test [56]. In this test, similar to the tergotometer test, foam is generated and soil is added until the foam is depleted. The soil used in this test is a mixture of the commercial foods Crisco shortening and Ragu spaghetti. The amount of soil needed to break the foam is an indication of a product's efficacy.

*(e)  Dishwashing Test (Foam Endpoint).*   An ASTM method [57] exists for the evaluation of foam stability of hand dishwashing detergents that uses foam endpoint to indicate products' performance. The objective of this test is to provide a standard test for formulators and suppliers, a screening test for formulations, and quality control. One method described uses a dishcloth to wash the front and back of a series of soiled dishes at 30-second intervals. Dishes are continually washed to a foam endpoint as determined by an originally uniform layer of foam in a dishpan solution becoming a thin layer of foam covering half the surface. Standard soils are described, an example being a mixture of oils (Wesson and corn), grease (lard), oleic acid, gelatin, salt, flour, and water.

Another example of a plate wash test [56] consists of soiling dishes with a fixed amount of Crisco shortening. A basin of water is prepared where the temperature, water hardness, and product concentration are adjusted to meet the habits and practices for the intended market. Foam is generated by a mixer or by manual agitation. The plates are washed using standard washing implements like sponges or cloths. It is common to fix the time and number of strokes used to wash the plates to minimize variability. The cleaned plate is then stacked and the process continues until the foam endpoint is reached. Again, the foam endpoint is when half the surface of the water is covered with foam. The plates washed are counted and used as a relative measure of a product's performance.

*(f)  Modified Schlachter–Dierkes Test [58].*   In this test the product is placed in a graduated cylinder and inverted to generate foam. Soil increments are added at fixed intervals until the foam collapses. The results are recorded as the number of soil increments, with a higher value indicating better foam stability.

*(g) Total Suds and Suds Mileage [59].* Four graduated cylinders are charged with a test solution consistent with conditions for the intended market. The cylinders are stationed side-by-side, rotated to generate foam, and initial foam heights are measured. Soil is added and the cylinders are again rotated. The foam height is again recorded. Soil is repeatedly added to a low foam height. A control is tested with each run and compared to the test product. Total suds and suds mileage are determined as follows:

$$\text{Total suds} = \frac{\text{Overall suds of test product}}{\text{Overall suds of control}} \times 100$$

$$\text{Total mileage} = \frac{\text{Overall mileage of test product}}{\text{Overall mileage of control}} \times 100$$

## C. Mildness Evaluation

Mildness has become an important attribute of LDLDs. Assessments typically involve clinical and sensory evaluations of skin irritation. Mildness evaluations are usually conducted in both *in vivo* and *in vitro* testing.

### 1. *In Vivo* Tests

*(a) Frosch–Kligman Soap Chamber Test.* The Frosch–Kligman soap chamber test [60] is designed to evaluate the mildness of surfactant compositions (8% solution) for panelists with hypersensitive skin. It is one of the most popular tests and consists of a five-day test procedure. The exposure varies from 24 hours on the first day to 6 hours on days 2 to 5. The first application is four times longer than the others in order to induce damage on the skin barrier and allow the solution to start its irritating effect. The subsequent 6-hour applications are to allow the sites to develop scaling, flaking, and wrinkling. Evaluation of redness (erythema), scaling/flaking (dryness), and fissuring is carried out visually by a trained professional. Each of the parameters is given a score on a 0 to 4 (erythema) scale or a 0 to 3 scale (Table 7.5). The total score is determined by summing up the averages of the three parameters. If at any time a panelist experiences high irritation, the specific product is not reapplied.

The test is run in exaggerated nonrealistic conditions in order to ensure differentiation between products. The authors claim they can differentiate between two soaps with just 10 panelists. The big advantage is that multiple products (up to eight) can be evaluated at once, and it takes only eight days to complete the test.

*(b) Patch Tests.* There are a number of patch tests used to evaluate skin irritation. Table 7.6 lists the reaction and symptom rating scales for patch tests.

*Twenty-four-hour occlusive test.* This test is mainly used to screen potential compositions for a quick comparison [61,62]. A patch is used to keep the diluted product against the skin for 24 hours. The chambers are patched on the forearms

**TABLE 7.5**   Rating System for the Frosch–Kligman Soap Chamber Test

| Erythema | Scaling | Fissures |
|---|---|---|
| 1: slight redness, spotty or diffuse | 1: fine | 1: fine cracks |
| 2: moderate, uniform redness | 2: moderate | 2: single or multiple broader fissures |
| 3: intense redness | 3: severe with large flakes | 3: wide cracks with hemorrhage or exudation |
| 4: fiery red with edema | | |

**TABLE 7.6**   Reaction Rating Scale and Symptom Ratings for Patch Test

| Reaction | Rating | Symptom | Rating |
|---|---|---|---|
| No visible reaction | 0 | Vesicles | 5 |
| Reaction only just visible | 1 | Edema | 4 |
| Slight reaction | 2 | Redness | 3 |
| Moderate reaction | 3 | Flaking | 2 |
| Serious reaction | 4 | Dryness | 1 |
| | | Wrinkles | 1 |
| | | Semitransparency | 1 |
| | | Glasslike | 1 |

of volunteers. The arms are evaluated after 24 hours and in the case of a strong irritant response the product is likely to be irritating. If no irritation is observed then repeated exposures need to be performed before a safe conclusion on the irritation potential of the product is reached.

A variation of this test is that the arms are evaluated after another 24 hours without contact with the product have passed. Other variations include plastic or aluminum disks used for patching [63,64]. The overall rating is determined by multiplying the degree of reaction by the rating for the observed symptom.

*Four-hour patch test.* While the other patch tests include diluted products to simulate the most often encountered use conditions, in this test the product is used neat. As is known from Habits and Practices studies, a significant percentage of the population are neat product users. A procedure was developed by Dillarstone and Paye [65] where neat products are applied to the forearms of volunteers for a duration of 4 hours. The irritation potential of those products is compared to that of 10% SLS solution which is a "skin irritant." If a product is suspected to be an irritant, then several patches are performed and removed at 1, 2, 3, and 4 hours; if irritation occurs before 4 hours all patches are removed.

**TABLE 7.7** Twenty-one Day Cumulative Patch Test Numeric Score Index

| 0 | Negative reading (questionable erythema not covering entire patch area) |
|---|---|
| 1 | Definite erythema covering entire patch area |
| 2 | Erythema and induration |
| 3 | Vesiculation |
| 4 | Bullous reaction |

*Twenty-one-day cumulative patch test.* This test [66] is designed to investigate the irritation potential of an LDLD that comes in contact with the skin for a prolonged period of time. Patches with diluted product (use conditions or slightly exaggerated) are applied on panelists for 21 days. The patches are removed every 24 hours, read 30 minutes after, and the new patch placed immediately after. At the end of the study, skin irritation in terms of erythema, dryness, and edema is visually evaluated based on a 0 to 4 scale (Table 7.7). This cumulative test has the advantage of detecting even weak irritants, so false negatives are extremely rare.

*Human repeat insult patch test (HRIPT).* This is the test usually used to determine an allergic reaction, and subsequently to make "hypoallergenic" claims. The product at 5% concentration is patched onto skin, 3 times a week for 3 weeks, followed by a rest period of 2 weeks without patching. The product is again patched on sites not previously patched, and the new sites are evaluated for erythema and edema after 2- and 3-day exposures, and are compared to the initial patching. If the levels of redness and swelling are higher than the original one, the panelist is considered to have an allergic reaction.

*(c) Hand Soaking Tests.* In these type of tests panelists soak their hands in LDLD solutions for 15 to 20 minutes, 2 or 3 times a day for 5 consecutive days; the conditions are defined in such a way that mimic realistic use conditions. After soaking, the hands are rinsed with tap water, and patted dry with a paper towel. The skin is assessed for erythema and dryness before and after soaking; instrumental measurements often accompany the visual assessments. Hard water and temperature are carefully monitored.

Patel *et al.* [67] used hand immersion testing to measure the skin-smoothing properties of a dishwashing detergent that incorporates a skin-smoothing compound. In this test panelists immersed their hands into a 1% solution of test composition at 41°C for 60 seconds, rinsed under tap water for 60 seconds, and towel dried. The panelists examined their hands as they dried them evaluating the smoothness. All panelists evaluated the test composition as making their hands smoother than the control without the skin-smoothing compound.

These tests discriminate very well, and have the added benefit of being run under normal usage conditions.

**TABLE 7.8**  Irritation Potential Based on Zein Value

| Irritation classification | Zein value (mg N/100 ml) |
|---|---|
| Nonirritant | 0–200 |
| Slightly irritant | 200–400 |
| Severe irritant | >400 |

## 2.  *In Vitro* Tests

*In vitro* tests are designed to allow screening of the irritation potential of LDLD products before conducting the more expensive *in vivo* tests. Most of these *in vitro* tests show a good correlation with the *in vivo* tests. They are rather simplified models that attempt to simulate what is really happening in the skin. The most often used tests address the denaturation of the skin surface proteins. A detailed list of the tests is offered by Paye [68].

*(a)  Zein Test.*  The procedure was originally developed by Gotte [69,70]. The Zein test determines the extent of denaturation of the Zein corn protein, which is insoluble in water, by surfactants. An amount of 0.5 g of Zein (Sigma, St. Louis, MO) is incubated with a solution of the test product at pH 7.0 for one hour, at constant temperature and with slight agitation. The solution is filtered and processed in a centrifuge. The zein value is calculated as:

Zein value $= A - B$ mg/N$_2$/100 ml solution

where $A$ is the amount of nitrogen measured in the filtered solution using a micro-Kjeldahl method and $B$ is the amount of nitrogen in the surfactant solution without Zein. The proposed irritation potential based on zein value is shown in Table 7.8. The more Zein that is solubilized, the more irritating the product.

A disadvantage of the test is that it does not work in the presence of magnesium, so all test formulas have to be made specifically for Zein testing.

*(b)  Collagen Swelling Test.*  This test [71] is used to determine the irritation potential of LDLD solutions containing anionic surfactants. It is based on the denaturation and swelling of the collagen protein (stratum corneum can also be used). The collagen sheets are incubated for 24 hours at 50°C in the presence of titrated water. After the 24-hour period, the sheets are rinsed, digested, treated, and analyzed for radioactivity. This measurement is used to calculate the water volume uptake by the dry collagen.

The swelling of collagen does not only occur because of denaturation, but also because of negatively charged surfactant binding to the protein. When cationic, amphoteric, or nonionic surfactants are present, swelling is minimal because of lack of adequate binding to the protein. Therefore, this method is valuable for

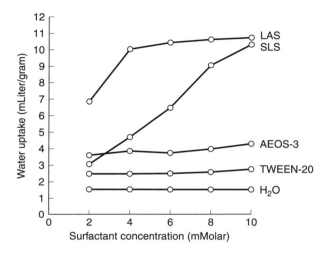

**FIG. 7.1** Water uptake as a function of surfactant concentration using the collagen swelling test. (From Blake-Huskins, J.C., Scala, D., and Rhein, L.D., *J. Soc. Cosmet. Chem.*, 37, 199, 1986. Reproduced with permission.)

assessing the irritation potential of anionic surfactants and anionic-surfactant-containing LDLD products.

Figure 7.1 shows that LAS and SLS have a high irritation potential as compared to AEOS-3, Tween-20, and water, since they produce a high level of swelling. Similar results for LAS and SLS are reported by Putterman *et al.* [72] for isolated stratum corneum.

*(c) Protein Denaturation.* Protein denaturation by surfactants is considered to be one of the major causes of skin irritation and roughness induced by surfactants [73]. Consequently, simple and reproducible methods for evaluating protein denaturation by surfactants have been developed. These tests work by measuring the amount of protein that has been denatured after the protein and test composition have been in contact with each other for a given period of time. Prottey *et al.* [74] developed a method that can be used to predict the mildness of surfactants and thus dishwashing liquids. They found a correlation between the changes observed in the stratum corneum phosphatase specific activity during normal dishwashing conditions and dryness and flakiness. Figure 7.2 shows the correlation of hand dryness and acid phosphatase specific activity after hand immersion in various surfactants.

A commercial manufacturer of light-duty dishwashing liquids used this method to show a liquid detergent composition to have good mildness [75].

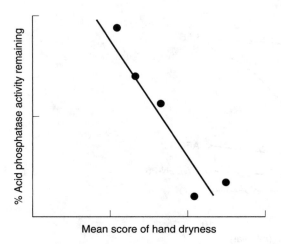

FIG. 7.2 Correlation of hand dryness and acid phosphatase activity. (From Prottey, C., Oliver, D., and Coxon, A.C., *Int. J. Cosmet. Sci.*, 6, 263, 1984. Reproduced with permission.)

In the protein denaturation test developed by Prottey *et al.* [74] the interaction of detergents with acid phosphatase enzyme is performed on skin. Others have developed a protein denaturation test that uses wheat germ [76]. Another protein denaturation method mixes an aqueous protein solution of egg yolk albumin and detergent composition [77]. The mixture is subjected to chromatography and compared to the results for only the egg yolk solution. Mildness is determined by the amount of denaturation, with the more mild the composition, the less the percent denaturation. Yet another method uses ovalbumin and human serum albumin and measures protein denaturation via gel permeation chromatography (GPC) [78]. Figure 7.3 shows an example of a chromatogram using this type of method. Percent denaturation is then calculated as:

$$\% \text{ Denaturation} = \frac{H_0 - H_t}{H_0} \times 100$$

*(d) Corneosurfametry.* This new test [68] combines protein denaturation and penetration through a barrier. Although it is an exaggerated test, it is useful to predict irritation by surfactant-based products. Superficial layers of the stratum corneum of human volunteers are collected by the use of an adhesive sheet; the diluted products are sprayed onto the skin samples, incubated, and rinsed off. The damage caused on the stratum corneum is determined by the degree of staining of the skin by a blue dye solution. With this method the direct interaction between

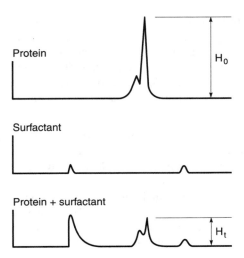

**FIG. 7.3** GPC elution curves for protein, surfactant, and mixtures of protein and surfactant. (From Miyazawa, K., Ogawa, M., and Mitsui, T., *Int. J. Cosmet. Sci.*, 6, 33, 1984. Reproduced with permission.)

surfactant solutions and the skin is measured, and no artifacts are caused by the presence of magnesium.

## D. Antibacterial Efficacy Evaluation

### 1. *In Vivo* Tests

These methods [79] help substantiate a claim of antibacterial activity on skin surfaces, where the efficacy of a product containing the active is compared to the efficacy of a placebo. Ideally, the test takes place once the formula is proven to be efficacious via *in vitro* testing.

*(a) Agar Patch.* The scope of the test is to assess the residual antibacterial activity of topical antibacterial products. Agar contact plates are inoculated with bacteria and are placed in contact with areas of the skin which have been washed with an antibacterial product. The surviving bacteria on the plate are enumerated and compared to the placebo. The objective is usually to achieve a 1 log reduction of organisms between the product containing the antibacterial ingredient and the placebo.

*(b) Skin Occlusion and Bacteria Recovery Test or Cup Scrub Test.* This test determines the immediate and/or residual antibacterial efficacy of topical preparations against a marker organism. The test is used for products which are applied

to skin and are either left on or rinsed from the skin. The antibacterial compounds are applied to the skin; the bacteria is applied to the treated area of the skin and occluded with a waterproof covering. After an appropriate period, the bacteria are harvested using a diluent, and the surviving bacteria are enumerated. The bacteria numbers from the sites treated with the placebo are compared to those treated with the active, and the difference determines the activity. At least 1 log reduction is desirable.

*(c)  Hand Imprint Test.*   This test is a qualitative test. It assesses the residual activity on hand products. The antibacterial and placebo products are applied to each hand, and the hands are washed individually, with product rinsed off as normal. The hands are placed on hand imprint seeded agar plates for a short contact time. After incubation, based on clearing (inhibition), a comparison is made between the residual activity of the active and the placebo. The objective is to observe a significant difference between the hand washed with the placebo product and the hand washed with the product containing the active.

*(d)  Health Care Personnel Hand Wash (HCPHW).*   This test measures the efficacy of antibacterial skin cleansing products to eliminate bacteria from the hands after a single contamination/wash cycle as compared to a baseline (untreated hand). The hands are artificially contaminated with a marker organism. The hands are prewashed with a nonantimicrobial soap, and bacteria are applied to hands and air-dried. Bacteria are then collected using the glove juice method to attain baseline counts. The hands are rewashed with a nonantimicrobial soap. Hands are recontaminated, air-dried, washed with the test product, and then rinsed off.

   The objective of this test is to observe a significant difference between the baseline bacterial counts and the posttreatment bacterial counts. Counts are converted to log cfu (colony forming unit), and a t-test compares the mean log (baseline counts) to the mean log (posttreatment counts).

## 2.   *In Vitro* Tests

These tests are used to test a new raw material or a complete formula. They are also used to support claim substantiation. They are easy, safe, and cheap tests and are designed to simulate actual use conditions. The first three of the tests discussed below can be used to screen products for either handwash products or for efficacy as surface disinfectants, as they test for bacteriocidal or bacteriostatic action without reference to a surface. The last two tests are specifically used to document surface disinfection claims.

*(a)  Minimum Inhibitory Concentration (MIC) Test.*   This test is used to screen possible antibacterial agents, or compare whole-formula inhibitory concentrations. Test tubes containing decreasing concentrations of the test agent are inoculated with the test organism, incubated, and examined for the presence or absence of growth. The MIC is determined from the last test tube in the dilution sequence where there

is no microbial growth. MIC is calculated based on the initial concentration of the active ingredient, and is given in ppm.

*(b)   Zone of Inhibition.*   The scope of the test is to determine the bacteriostatic activity of a compound against various organisms. Molten agar is inoculated with a pure culture and seed layer on the base agar. The compound is applied directly onto a disk or in a well to the seeded plates. After incubation the no-growth zones around the disks are measured. This is a qualitative test and is influenced by the diffusion rates of the actives. The results are given in millimeters, and they are comparable to the MIC values. The perimeter of the zone is where the diffusion has reduced that particular active ingredient to its MIC.

*(c)   Short Interval Kill Test (SIKT).*   The scope of the SIKT is to determine the bacteriocidal activity of compounds over a short period of exposure. Solutions are inoculated and neutralized at the end of a finite period. The surviving organisms are enumerated and compared to the control. The difference is expressed in log reduction.

*(d)   Use Dilution Test (UDT).*   The scope of the UDT [80] is to determine the germicidal efficacy of water-miscible disinfectants. Contaminated carriers are exposed to a diluted disinfectant solution. After a specific period of time, the carriers are transferred to a neutralizer broth (which stops the antibacterial action), incubated, and examined for growth. The test is usually given as 60 carriers or tubes, and growth must be absent from 59 to 60 of the tubes. Growth in two or more samples is failure in the test. This test is used for EPA registration of products as surface disinfectants. Depending on the organisms used, various levels of claim are recognized.

*(e)   European Requirements [81].*   In European products the antibacterial efficacy claim is substantiated through the EN series of quantitative suspension tests where a 5 log microorganism kill needs to be achieved for bacteria, 4 log for fungi, and 4 log for viruses.

*EN 1040: Bacteria.* The test germs are *Pseudomonas aeruginosa* and *Staphylococcus aureus* and the test conditions include a temperature of 20°C and contact times of 1, 5, 15, 30, 45, and 60 minutes. This test evaluates the basic efficacy of the product.

*EN 1279: Bacteria.* This is the same as above, but two additional bacteria, *Escherichia coli* and *Enterococcus hirae*, are included. The temperature of the test remains at 20°C, while the contact time is 5 minutes. The test is performed with hard water, to better simulate household conditions.

*EN 1275: Fungi.* The test organisms are *Candida albicans* and *Aspergillus niger* (spores). The test is conducted at 20°C, and the contact times are 5, 15, 30, and 60 minutes.

*EN 1650: Fungi.* This is the same as EN 1275 but hard water is used.

*DVV: Virus.* This test is to satisfy the German regulatory requirements. It is conducted at 20°C and the contact time is adjustable up to 60 minutes.

## E.   Other Tests

Various other tests are used by formulators to aid formulation or to evaluate some specific benefits. Drainage and rinsing tests are among the examples.

### 1.   Drainage Tests

The consumer benefits of liquid detergents that have good draining characteristics are faster and spot-free drying. A test method to measure the draining of light-duty liquids was described in U.S. Patent 5,154,850 [49]. In this test plates are immersed in a solution of product a fixed number of times, taken out, and air dried under ambient conditions. This cycle is repeated with the final dipping followed by rinsing. The plate is allowed to air dry and the water spots are counted. A product that provides good draining will give few to no spots on the plates.

In another drainage test [82] various regular kitchen utensils, such as drinking glasses, glass dinner plates, and ceramic dinner plates, are washed in test compositions under controlled conditions. The utensils are then rinsed and placed in a rack to dry. The time at which drainage begins and the percentage area dried by this drainage are recorded.

### 2.   Rinsing Tests

Although copious and long-lasting foams are desirable for LDLDs, consumers also want the foam to be easily rinsed away from the dishes so as not to leave a residue that could appear as spots. A test method was disclosed in U.S. Patent 5,154,850 for the evaluation of the rinsability of foam generated for an LDLD [49]. This involves making a solution of product, charging it to a container, and stirring. The solution is discharged from bottom of the container leaving residual foam in the container. Tap water is added to the container with residual foam and stirred again. The stirring and draining steps are repeated until no foam remains in the container. The product that needs fewer additions of water has better rinsing properties.

### 3.   Cloud/Clear Point Tests

It is important that a product does not turn cloudy either on the shelf at the point of sale or during the time of use or storage at home. To ensure this important product aesthetic, a cloud/clear point test is usually conducted. The product sample is put into a clear container and then immersed in some kind of cooling bath, such as a salt/ice bath. The turbidity of the sample is monitored as the temperature drops. The temperature at which the sample first becomes turbid is the cloud point. If the sample is removed from the cooling bath the temperature of the sample rises. The temperature increases slowly until the sample becomes clear again.

**TABLE 7.9**  Important Attributes of a
Hand Dishwashing Liquid Detergent

| Attribute |
| --- |
| Effective cleaning |
| Copious and long-lasting foam |
| Mildness to hands |
| Pleasant fragrance |
| Convenient to use |
| Safe to humans |
| Safe to dishes and tableware |
| Storage stability |
| Economic to use |

This temperature is the clear point. The acceptable cloud and clear temperatures
are set based on the conditions to which the products may be subject.

## V. FORMULATION TECHNOLOGY
### A. Formulation

Formulating an LDLD is both a science and an art. It requires a good balance
between product performance, aesthetics, safety, and cost. From the consumer
point of view, the important attributes for a hand dishwashing liquid are listed in
Table 7.9. Liquid dishwashing detergents are formulated to deliver against these
consumer-relevant attributes.

Formulation of LDLDs typically involves (1) selecting appropriate raw materials for the desired performance, (2) developing formulas and optimizing for
performance, (3) optimizing product aesthetics, (4) testing product safety, (5) optimizing product cost, (6) aging for product stability, (7) validating with consumers,
and (8) documenting advertising claims. These steps are usually not sequential,
but often take place in parallel.

The following sections present a review on formulation against these performance attributes with the intent of providing some guidelines.

### B. Guidelines and Examples
#### 1. Formulating for Effective Cleaning

The most important performance attribute of an LDLD is cleaning. As discussed in
an earlier section, cleaning of dishes with an LDLD primarily relies on the
interfacial properties provided by surfactants. Various surfactants exhibit different
interfacial properties and thus have varying ability in removing different soils from
various surfaces. In general, use of a combination of surfactants is necessary for

an LDLD to be effective against a wide spectrum of soils encountered on a variety of surfaces in the real world.

A significant number of patents on LDLDs have been issued in the U.S., Europe, and Japan in recent years. Listed in Table 7.10 are examples of LDLD patents formulating for effective soil removal/cleaning. The technology utilized in these patents ranges from special surfactants, surfactant mixtures, salts, and microemulsion to the use of special additives such as lemon juice and abrasives.

## 2. Formulating for High and Long-Lasting Foam

It is well recognized that foam is the most important visual signal that consumers use to judge the performance of an LDLD. This is in spite of the lack of direct correlation between foaming and cleaning properties of an LDLD, as discussed earlier. Therefore, it is critically important that an LDLD is formulated with copious and long-lasting foam.

Copious foam usually requires the use of high-foaming surfactants, typically anionic or amphoteric surfactants or a mixture of surfactants. Long-lasting foam often requires the use of foam stabilizers in addition to surfactant mixtures.

Table 7.11 summarizes recent LDLD patents formulating for good foaming properties. The technologies involved either use novel surfactants/surfactant blends or use novel foam stabilizers.

## 3. Formulating for Mildness

For some consumers, mildness to skin is an important attribute of an LDLD, especially for those who have sensitive skins.

There are essentially two approaches to formulate an LDLD for mildness: (1) use mild surfactants such as nonionic surfactants, amphoteric surfactants, or a combination of such surfactants; (2) use additives that are anti-irritants such as modified protein or polymers. Examples of recent LDLD patents with mildness benefit are listed in Table 7.12.

## 4. Formulating for Desirable Aesthetics

The aesthetic attributes of LDLDs are just as important as their performance. This includes color, fragrance, cloud and clear points, viscosity, and product stability. Color, fragrance, and viscosity are usually chosen based on consumer preference. The cloud and clear points have to be adequate for the temperature to which the product is likely to be exposed.

*(a) Cloud and Clear Points.* The cloud point is the temperature at which the product begins to turn cloudy or hazy upon cooling. The clear point is the temperature at which the cloudy product turns clear again upon warming. In North America and Europe it is desirable that the cloud point be below 4°C and the clear point not exceed 10°C.

**TABLE 7.10** Formulating LDLDs for Effective Cleaning: Patent Examples (1979–1995)

| Patent (year) | Inventor(s) (company) | Technology | Claimed benefit |
|---|---|---|---|
| U.S. 5,378,409 (1995) | Ofosu-Asante (Procter & Gamble) | Alkylethoxy carboxylate mixture and Ca ions; pH 8–10 | Good grease removal, mildness to skin, and good storage stability |
| U.S. 5,376,310 (1994) | Cripe *et al.* (Procter & Gamble) | Alkylethoxy carboxylate and Mg ions buffered at pH 8–10 | Good grease removal, mildness to skin, and good storage stability |
| U.S. 5,298,195 (1994) | Brumbaugh (Amway) | Amido amine oxide or alkylethoxylated carboxylate | Improved detergency and foam stability over range of water hardness |
| U.S. 5,269,974 (1993) | Ofosu-Asante (Procter & Gamble) | Alkylamphocarboxylic acid and Mg or Ca ions at pH 7–10 | Improved grease cleaning, sudsing, and stability |
| U.S. 5,230,823 (1993) | Wise *et al.* (Procter & Gamble) | Alkylethoxy carboxylate with minimal alcohol ethoxylate and soap byproducts; high pH and Mg ions | Good grease removal and mildness; high pH and Mg versions improve grease removal maintaining mildness |
| U.S. 5,236,612 (1993) | Rahman *et al.* (Lever Brothers) | Alkyl glycerates as coactive | Enhanced oil removal |
| U.S. 5,167,872 (1992) | Pancheri *et al.* (Procter & Gamble) | Polymeric and anionic surfactants forming complexes | High sudsing; improved grease efficacy |
| U.S. 5,096,621 (1992) | Tosaka *et al.* (Kao) | Dialkylamine oxides and anionics or nonionics | Permeability and efficacy |
| U.S. 4,992,212 (1991) | Corring *et al.* (Lever Brothers) | Organic base, zinc salt, and complexing agent | Superior cleaning without staining of aluminum utensils |
| U.S. 4,923,635 (1990) | Simion *et al.* (Colgate-Palmolive) | 0.5–1.8% Mg; triethanolammonium | Improved oily soil removal |
| U.S. 4,904,359 (1990) | Pancheri *et al.* (Procter & Gamble) | Polymeric and anionic surfactants forming complexes | High sudsing; improved grease efficacy |
| U.S. 4,919,839 (1990) | Durbut *et al.* (Colgate-Palmolive) | Microemulsion | Superior grease removal |
| U.S. 4,834,903 (1989) | Roth *et al.* (Henkel) | Low DP long-chain polyglucoside alkylene oxide adducts | Improved detergency |

*(continued)*

**TABLE 7.10** (Contd.)

| Patent (year) | Inventor(s) (company) | Technology | Claimed benefit |
|---|---|---|---|
| U.S. 4,797,231 (1989) | Schumann et al. (Henkel) | Abrasives | Neat — regular dishwashing; dilute – abrasive |
| U.S. 4,839,098 (1989) | Wisotzki et al. (Henkel) | Alkyl glucoside and dialkyl sulfosuccinate | Improved detergency and foam stability on proteinaceous soils |
| U.S. 4,853,147 (1989) | Choi (Colgate-Palmolive) | Cationic surfactant; $C_{21}$-dicarboxylic salt | Low-temperature detergency |
| U.S. 4,732,704 (1988) | Biermann et al. (Henkel) | Fatty alkyl C12–C14 monoglucosides | Enhanced foaming and detergency |
| U.S. 4,772,425 (1988) | Chirash et al. (Colgate-Palmolive) | Suspended low-density abrasives | Baked-on/dried-on food removal; improved stability with low viscosity |
| U.S. 4,772,423 (1988) | Pancheri et al. (Procter & Gamble) | Anionic-based with polymeric surfactant and betaine | Improved grease efficacy |
| U.S. 4,681,704 (1987) | Bernardino et al. (Procter & Gamble) | Mg alkylethoxy sulfate, C10–C14 alkyl amine oxide and amidoalkyl betaine | Greasy soil removal; good suds mileage |
| U.S. 4,614,612 (1986) | Reilly et al. (Lever Brothers) | Lemon juice | Effective against difficult soils |
| U.S. 4,492,646 (1985) | Welch (Procter & Gamble) | Highly ethoxylated drainage promoting nonionic surfactant | Complete drainage reducing spotting and filming on tableware |
| U.S. 4,430,237 (1984) | Pierce et al. (Colgate-Palmolive) | Nonionic mixture of alkyl glyceryl esters | Improved grease cleaning and foam stability |
| U.S. 4,368,146 (1983) | Aronson et al. (Lever Brothers) | Anionic/nonionic based plus polymers and alkali metal salt of casein | Rapid and uniform draining with no spotting or filming |
| U.S. 4,316,824 (1982) | Pancheri (Procter & Gamble) | Mg alkylethoxy sulfate and C10–C14 alkyl amine oxide | High sudsing; more effective detergency |
| U.S. 4,268,406 (1981) | O'Brien et al. (Procter & Gamble) | Reducing agent and nitrogen-containing protein denaturant | Superior cleaning of protein and carbohydrate soils |
| U.S. 4,133,779 (1979) | Hellyer et al. (Procter & Gamble) | Mg alkyl polyethoxy sulfate and C8–C16 alkyl amine oxide | Removal of greasy soils |

**TABLE 7.11** Formulating LDLDs for High and Long-Lasting Foam: Patent Examples (1981–1995)

| Patent (year) | Inventor(s) (company) | Technology | Claimed benefit |
|---|---|---|---|
| U.S. 5,352,387 (1994) | Rahman et al. (Lever Brothers) | Novel N-alkylglyceramide surfactants | Enhanced foam stability |
| U.S. 5,338,491 (1994) | Connor et al. (Procter & Gamble) | Glycerol amides [N-(1,2-propanediol) fatty acid amide] | Foaming and solubility benefits to comparable ethanolamides |
| U.S. 4,877,546 (1989) | Lai (Colgate-Palmolive) | Nonionic hydroxypropyl guar gum derivative | Enhanced foaming |
| U.S. 4,732,707 (1988) | Naik et al. (Lever Brothers) | Alkyl ether sulfates based on specific aliphatic carbon chain | Excellent foaming and detergency |
| U.S. 4,663,069 (1987) | Llenado (Procter & Gamble) | Alkyl polysaccharide-based surfactant composition | Stable foam, easily rinsed |
| U.S. 4,680,143 (1987) | Edge et al. (Lever Brothers) | Ternary system: dialkyl sulfosuccinate, LAS and/or SAS and alkyl ether sulfate | Improved performance and physical characteristics |
| U.S. 4,596,672 (1986) | MacDuff et al. (Lever Brothers) | Dialkyl sulfosuccinate and fatty acid dialkanolamide | Enhanced performance and physical characteristics |
| U.S. 4,599,188 (1986) | Llenado et al. (Procter & Gamble) | Alkyl polysaccharides and anionic surfactant blends | Stable foam, readily rinsed |
| U.S. 4,576,744 (1986) | Edwards et al. (Lever Brothers) | Polymers | Enhanced foam stability and increased viscosity |
| U.S. 4,528,128 (1985) | Naik (Lever Brothers) | Dialkyl sulfosuccinates of particular chain lengths | Enhanced foam stability in both soft and hard water |
| U.S. 4,537,709 (1985) | Edge et al. (Lever Brothers) | Particular chain length LAS and alkyl ether sulfates | Improved foaming performance |
| U.S. 4,556,509 (1985) | Demangeon et al. (Colgate-Palmolive) | Low-molecular-weight organic diamine diacid salt | Improved foam stability and degreasing activity in soft water |

*(continued)*

**TABLE 7.11** (Contd.)

| Patent (year) | Inventor(s) (company) | Technology | Claimed benefit |
|---|---|---|---|
| U.S. 4,536,318 (1984) | Cook et al. (Procter & Gamble) | Alkyl polysaccharide with anionic cosurfactants | Stable foam, readily rinsed |
| U.S. 4,434,087 (1984) | Hampson et al. (Lever Brothers) | Dialkyl sulfosuccinate (C6 and C8 alkyl) | Good foaming and cleaning |
| U.S. 4,434,089 (1984) | Billington et al. (Lever Brothers) | Dialkyl sulfosuccinate with protein | Enhanced foaming and cleaning in hard water |
| U.S. 4,434,090 (1984) | Hampson et al. (Lever Brothers) | Unsymmetrical dialkyl sulfosuccinate and dialkyl sulfosuccinate | Enhanced foaming and cleaning |
| U.S. 4,434,091 (1984) | Cox et al. (Lever Brothers) | Unsymmetrical dialkyl sulfosuccinate | Enhanced foaming and cleaning |
| U.S. 4,454,060 (1984) | Lai et al. (Colgate-Palmolive) | Novel cationic copolymer | Improved foam stability |
| U.S. 4,435,317 (1984) | Gerritsen et al. (Procter & Gamble) | Tertiary anionic surfactant mixture; Mg level corresponds to alkyl sulfate level | Maximum foam stability |
| U.S. 4,486,338 (1984) | Ootani et al. (Kao) | Succinic acid derivative with anionic and tertiary amine oxide | Superior foaming, detergency, and stability |
| U.S. 4,490,279 (1983) | Schmolka (BASF) | Nonionic block polymer surfactant with an amine oxide | High foaming and good foam stability |
| U.S. 4,277,378 (1981) | Tsujii et al. (Kao) | Partially neutralized succinic acid derivative | Good detergency and foaming; mild to skin |
| U.S. 4,235,752 (1980) | Rossall et al. (Lever Brothers) | Effective secondary alkyl sulfate isomers | Improved foam stability |

**TABLE 7.12** Formulating LDLDs for Mildness: Patent Examples (1980–1995)

| Patent (year) | Inventor(s) (company) | Technology | Claimed benefit |
|---|---|---|---|
| U.S. 5,387,373 (1995) | Naik (Unilever) | C10–C11 primary alkyl sulfate | Enhanced mildness |
| U.S. 5,284,603 (1994) | Repinec et al. (Colgate-Palmolive) | Nonionic-based surfactant system | Mildness |
| U.S. 5,340,502 (1994) | Palicka (Berol Novel AB) | Anionic surfactant; combination of three amphoteric compounds | Mildness maintaining good cleaning |
| U.S. 5,230,835 (1993) | Deguchi et al. (Kao) | Alkyl polyglucoside with polymers | Reduced irritation, increased foaming and detergency; rinsing and feel to hands |
| U.S. 5,084,212 (1992) | Farris et al. (Procter & Gamble) | C8 alkyl glyceryl ether sulfonate; foam enhancers | Ultramild surfactant |
| U.S. 5,139,705 (1992) | Wittpenn et al. | Blend of nonionics with low and high melting points | Mild, nonirritating composition |
| U.S. 5,154,850 (1992) | Deguchi et al. (Kao) | Alkyl glucoside | Reduced irritation and damage to hair and skin |
| U.S. 5,075,042 (1991) | Allison et al. (PPG Industries) | Alkyl polyethyleneoxy sulfonate; anionic/nonionic | Reduced primary skin irritation potential |
| U.S. 5,025,069 (1991) | Deguchi et al. (Kao) | Alkyl glucoside-based compositions; terpene and isothiazolone derivatives | Reduced irritation and damage to hair and skin; color and odor stability |
| U.S. 5,073,293 (1991) | Deguchi et al. (Kao) | Alkyl glucoside and dicarboxylic acid surfactant | Mild, pleasant feel to hands |
| U.S. 4,595,526 (1986) | Lai (Colgate-Palmolive) | Novel high-foaming nonionic surfactant-based composition | Superior skin mildness |

*(continued)*

**TABLE 7.12** (Contd.)

| Patent (year) | Inventor(s) (company) | Technology | Claimed benefit |
|---|---|---|---|
| U.S. 4,555,360 (1985) | Bissett (Procter & Gamble) | Anionic surfactants with betaine and amine oxide | Improved mildness |
| U.S. 4,554,098 (1985) | Klisch et al. (Colgate-Palmolive) | Alkyl ether sulfate, nonsoap anonionic, zwitterionic, and alkanoic acid alkanolamide | Reduced skin irritation; good cleaning and foaming |
| U.S. 4,526,710 (1985) | Fujisawa et al. (Kao) | Anionic phosphate surfactants in combination with amides and amine oxide | Mild to hands |
| U.S. 4,37,146 (1983) | Jones et al. (Procter & Gamble) | Surfactant with tertiary alcohol | Mild to skin |
| U.S. 4,247,425 (1981) | Egan et al. (Sherex Chemical) | Nonionic with polyoxyalkylene chain composed of randomly distributed oxyethylene and oxypropylene residues | Low eye and skin irritation |
| U.S. 4,256,611 (1981) | Egan et al. (Sherex Chemical) | Ethylene oxide adduct of partial glycerol esters of detergent-grade fatty acid and certain anionic surfactants | Low eye and skin irritation; adjust viscosity of aqueous solutions |
| U.S. 4,287,102 (1981) | Miyajima et al. (Lion) | Salt of olefin sulfonic acid (C12–C16) and tertiary amine oxide | Good detergency and little and roughening |
| U.S. 4,259,216 (1981) | Miyajima et al. (Lion) | α-Olefin sulfonate, alkylethoxy sulfate, and alkyl amine oxide with aryl sulfonate | No hand roughening and good for vegetables |
| U.S. 4,235,759 (1980) | Ohbu et al. (Lion) | Polyoxyalkylene alkyl ether sulfate and cationic surfactant | Mild to skin and good detergency |
| U.S. 4,195,077 (1980) | Marsh et al. (Procter & Gamble) | Modified protein | Protect keratinous material |

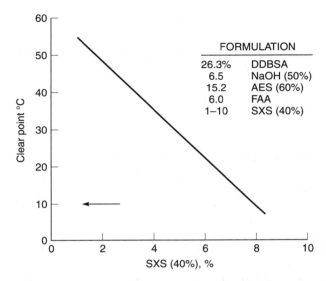

**FIG. 7.4**  Effect of sodium xylene sulfonate on the clear point of a premium LDLD blend. (From Drozd, J.C., *Chemical Times & Trends*, 8, 49, 1985. Reproduced with permission.)

The cloud and clear points of an LDLD can be adjusted using hydrotropes [83,84] such as sodium xylene sulfonate (SXS), sodium cumene sulfonate (SCS), alcohols, or urea. Figure 7.4 illustrates the significant effect of SXS on the clear point of an LDLD formulation.

*(b)  Viscosity.*  The viscosity of an LDLD is very important for its consumer acceptability and its dispersibility on dilution [85]. The viscosity of an LDLD is typically in the range 100 to 500 cP. In some markets such as Malaysia, Singapore, and Hong Kong, consumers prefer a much thicker product with viscosity in the range 2000 to 3000 cP. Consumers generally associate a thicker product with more "ingredients" in the product. However, technically, the viscosity of an LDLD is a strong function of not only its active ingredient level but also the isomer distribution in the surfactant, the relative amount of different surfactants, and the salt levels. Salt can be both a viscosity builder and a viscosity reducer, depending on where the formulation is on its salt curve, as mentioned above. An example of a simple system is sodium dodecylether sulfate (2.8EO) at 15% concentration. The viscosity first increases with the addition of NaCl and then decreases with further increase of the amount of NaCl [86]. AEOS with a narrow EO distribution thickens much more than AEOS with a conventional, broad EO distribution. Other factors that affect salt thickening are carbon chain length and carbon chain distribution [86].

Salt also has a significant effect on LAS. Depending on the cation of the LAS, salt in the range 0 to 2% can have a modest or large effect on the viscosity.

Fatty alkanolamides are mainly used as foam stabilizers, but they can also have a large effect on the viscosity of an LDLD formulation, usually increasing it. Other viscosity modifiers include hydrotropes such as alcohol, SXS, SCS, urea, and water-soluble polymers. However, not all of these have the same magnitude of effect, which depends on the surfactant system in the product.

*(c) Physical Stability.* Physical stability is another important product attribute that cannot be overlooked. Consumers would not want to purchase a product that changes physically over time. This may include precipitation, phase separation, or microbial contamination.

Aging studies are typically conducted to ensure physical stability of products at market age. Various aging conditions are necessary to simulate the conditions the product may encounter from warehousing, transportation, and storage in stores and at home. These conditions include elevated temperature such as 50°C and low temperature just above freezing point.

The other standard aging study normally conducted is to expose the product to sunlight to simulate storage of the product at home near a kitchen window to test for color and phase stability.

During aging, periodic examinations of products are made to check for effects to the key product characteristics such as pH, color, fragrance, product appearance; packaging is checked for any changes and deviations from room temperature samples. Any unacceptable changes and deviations need to be investigated to identify the cause and to determine the corrective measures. The entire series of aging studies need to be repeated when corrections are made to the formula.

To ensure a product's ability to withstand microbial contamination, adequacy of preservation studies need to be conducted. Consumers can contaminate products during use in the home. If the product is not able to control the growth of microorganisms, unsightly growth could result in the product affecting its quality. If the formulation is not self-preserving then the incorporation of a suitable preservative would be necessary.

## C.  Factors Affecting Performance

## 1.  Effect of Surfactant Type

Detergents in general do a better job of cleaning if they contain a mixture of surfactants. It is best if the mixture includes a variety of types of surfactants as well as a variety of carbon chain lengths. As discussed earlier, it is an advantage to mix anionic with nonionic or amphoteric surfactants. These mixtures can show a reduction in CMC and improvement in grease cleaning and foaming compared with individual surfactants due to the favorable interaction of head groups.

**TABLE 7.13**  Optimum Chain Length of LAS as a
Function of Hardness

| Hardness (ppm) | Optimum foam stability[a] |
|---|---|
| 0 | C13 |
| 50–150 | C11–C12 |
| >150 | C10–C11–C12 |

[a]Product formulation: 24% LAS, 6% AEOS, and 2% LMMEA.

## 2.  Effect of Carbon Chain Distribution

Several authors have described the effect of alkyl chain length on performance. For LAS, a widely used surfactant in LDLDs, the most important factor in the performance is the carbon chain distribution [87]. The optimum chain length of LAS for foam stability in a typical LDLD formulation varies depending on water hardness. This is illustrated in Table 7.13 for a formulated product of 24% LAS, 6% AEOS, and 2% LMMEA.

The location of the phenyl group on the alkyl chain (within the limits of commercially available LAS) has little effect on the performance [87]. For SAS, an anionic surfactant widely used in European LDLDs, more plates are cleaned in the C14–C15 range than at longer or shorter lengths. This is true for SAS alone and for SAS/AEOS mixtures (see Figure 7.5) [88].

## 3.  Effect of pH

The pH can do more than giving a stable formulation. Acidic pH can act as an antibacterial condition. Slightly acidic pH (around 5.5) is matched to the physiologic pH of skin. Alkaline pH can help clean greasy soils, although it is irritating to skin. Shifts in pH can improve preservation of the product improving the efficacy of preservatives or shifting the formulation to pH values more hostile to microbes. As a balance to give maximum grease cleaning while maintaining skin mildness, most commercial products have a pH close to neutral. Most surfactants are also most stable under neutral pH.

## 4.  Effect of Inorganic Ions

An important factor in the formulation of dishwashing liquids is the presence of inorganic ions. These can be present as impurities in all commercial surfactants. They are also present as calcium and magnesium carbonates, sulfates, and chlorides in hard water. They are deliberately added to products by some manufacturers as important performance boosters and viscosity modifiers. As performance boosters, divalent ions have a special function and are added as an inorganic salt, such as $MgSO_4$, or as the counterion of an anionic surfactant, such as $Mg(LAS)_2$.

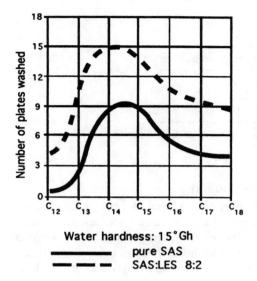

**Water hardness: 15°Gh**

————— pure SAS

— — — SAS:LES 8:2

**FIG. 7.5** Miniplate test of SAS (and mixtures) as a function of the C chain length. Concentration: 0.3 g/L of active substance. (*Source*: From Ref. 88. Reproduced with permission.)

One of the effects of increased electrolyte concentration is lowering of the CMC. This has direct implications for detergency. Electrolytes do this for ionic surfactants by screening the ionic head groups in a micelle from each other. Adding salts decreases the CMC in the following order for lauryl sufate: $Ca^{2+}$, $Mg^{2+}$ < $N(C_2H_5)_4^+$ < $N(CH_3)_4^+$ < $Cs^+$ < $K^+$ < $Na^+$ < $Li^+$. The larger the polarizability of the ion and the larger the valence, the greater is the decrease in the CMC. The smaller the radius of the hydrated ion, the greater is the decrease in the CMC [34].

A similar effect occurs in the electrical double layer that surrounds the surface of an object to be cleaned. Items with a relatively highly charged surface, such as glass, ceramics, and metals with oxide coatings, first repel surfactants with a like charge and attract surfactants with an opposite charge. Adding salts causes a decrease in the adsorption of the like-charged surfactant. In dishwashing liquids, many of the surfaces to be cleaned are negatively charged. Because the principal surfactants used in dishwashing detergents are anionic, adding salts increases the adsorption of anionics onto a negatively charged surface. For nonpolar items, such as many plastics, electrolytes also increase the adsorption of ionic surfactants because the mutual repulsion between the head groups is decreased.

Ionic strength also affects the stability of the emulsions that are formed as the dishes are cleaned. Emulsions in general decrease in stability with increasing electrolyte concentration. In the presence of electrolyte, the thickness of the electrical double layer is decreased. This decreases the stability of the emulsion. If the stability of the emulsion is too low, redeposition of soil can be a problem. Fortunately, the emulsion formed in dishwashing need not last very long. Also, many modern dishwashing formulations include substantial amounts of both anionic and nonionic surfactants. Nonionic emulsions are much less affected by ionic strength.

The most important effect of electrolytes is the effect of $Ca^{2+}$ and $Mg^{2+}$ ions on dishwashing performance. Numerous patents deal with the beneficial effects of both ions on dishwashing performance. It is thought that these divalent ions form a complex with anionic surfactants. This complex allows the anionic surfactant to adsorb more readily on surfaces with a negative surface charge.

A preferred mode of adding magnesium to a dishwashing liquid is to use the MgO or $Mg(OH)_2$ to neutralize the surfactants after they are formed by sulfonation or sulfation [89]. This method is preferred because adding magnesium from salt generally requires additional hydrotrope. Magnesium is typically preferred for use over calcium, since calcium surfactant salts are much less soluble. However, one series of patents discloses the use of calcium added to the formulation in the form of calcium xylene sulfonate for improved stability [90].

Recent patents describe the use of low-molecular-weight organic diamines in place of the Ca or Mg divalent ions [91,92]. The inventors claim that the diamines in hand dishwashing detergents are most effective when the pH is in the range 8 to 12. The diamines provide the same function as the divalent ions but with additional benefits. The reduction or elimination of the divalent ions leads to improved benefits in dissolution, rinsing, and low-temperature product stability, according to the patents.

## 5. Effect of Raw Material Variations

Some LAS mixtures appear to be self-hydrotroping; others need extra hydrotrope [93]. Commercial LAS contains various amounts of the 2-phenyl isomer and dialkyltetralin sulfonates. The 2-phenyl content significantly affects the solubility, and the tetralins considerably reduce the viscosity. This variation in 2-phenyl isomer and tetralin content is a result of the industrial process used to make the linear alkylbenzene. LAS is made from three primary processes that use different catalysts, $AlCl_3$, HF, and DETAL, resulting in different amounts of 2-phenylalkanes and dialkyltetralins. Material from the $AlCl_3$ process is considered high in 2-phenyl content and high in dialkyltetralins, material from the HF process is considered low in 2-phenyl content and low in dialkyltetralins, and material from the DETAL process is considered high in 2-phenyl content and low in dialkyltetralins.

Another consideration is the source of the hydrophobic end of the surfactant. It is well known that oleochemical fatty alcohols have even numbers of carbons and petrochemicals fatty alcohols have odd and even numbers. As discussed earlier, carbon chain length has an effect on dishwashing performance. However, when dishwashing compositions were made out of lauryl range (C12) hydrophobes, there was no difference in performance or physical properties between those from oleochemical and petrochemical sources [94].

## D.  Recent Patent Trends (1996–2003)

Over 160 U.S. patents were granted between 1996 and 2003 in the area of LDLDs or hand dishwashing liquids. Many of these can be classified as general composition of matter patents often utilizing mixed surfactant systems for beneficial overall performance. Some patents describe the use of microemulsion technology for improved grease removal. Many patents describe the use of novel surfactants or the use of nontraditional surfactants formulated in LDLDs, described in more detail below. There are a significant number of patents describing use of various polymers in LDLDs. An interesting new development is the use of enzymes in LDLDs. Several recent patents relate to enzymes in hand dishwashing. Disinfectant-type applications or formulations are also described. A few patents describe cleaning extra tough soils and other unique applications. All of these recent patent trends are described in more detail below.

### 1.  Novel Surfactants

Patent examples utilizing new or atypical surfactants are summarized in Table 7.14. The types of surfactants used in the examples are anionic, nonionic, amphoteric, and one cationic. Examples of novel anionic surfactants are mid-chain branched ethoxy sulfates. These surfactants are claimed to prove useful in the cleaning of heavily soiled dishware. The chelating surfactant ethylene diaminetriacetate is claimed to provide good foaming and grease cutting properties, particularly in hard water. Some novel nonionic surfactants include an ethoxylated/propoxylated nonionic surfactant, a gemini surfactant, and a bridged polyhydroxy fatty acid amide. A few patents list use of the amphoteric surfactant sultaine, which is new for use in LDLDs but has found previous use in personal care products.

### 2.  Polymers in LDLDs

Several recent patents describe the benefits of polymers in LDLDs (Table 7.15). Polymers are well known to interact with surfactants and provide many interesting properties. Some of the benefits claimed in the patents summarized in Table 7.15 are soil resistance due to amino acid copolymers, polyethylene glycol as a grease release agent, increased grease removal from polyoxyethylene diamine, enhanced foam volume and duration, increased solubility, and enhanced mildness by ethylene oxide–propylene oxide copolymers. As described in these various patents, the addition of polymers to LDLDs can aid performance in many important attributes of the product.

**TABLE 7.14** Formulating LDLDs with Novel Surfactants: U.S. Patent Examples (1996–2003)

| Number | Year | Inventors | Company | Title | Benefit |
|---|---|---|---|---|---|
| 6,617,303 | 2003 | Smith, Smadi | Huntsman | Surfactant compositions containing alkoxylated amines | Contains alkoxylated amines |
| 6,602,838 | 2003 | Koester, Behler, Neuss, Schmid, Elsner | Cognis | Hand dishwashing liquid comprising an alkoxylated carboxylic acid ester | Contains an alkoxylated carboxylic acid ester |
| 6,495,507 | 2002 | Arvanitidou | Colgate | High-foaming, grease-cutting light-duty liquid detergent | Contains an amphoacetate |
| 6,492,314 | 2002 | Jakubicki, Szewczyk | Colgate | High-foaming, grease-cutting composition containing a C12/C14 alkylamidopropyldimethyl amine oxide | Contains a C12/C14 alkylamidopropyldimethyl amine oxide |
| 6,423,678 | 2002 | Brumbaugh, Faber, Berube | Amway | Alcohol ethoxylate–PEG ether of glycerin | Hydrotrope provides increased foam generation |
| 6,281,181 | 2001 | Vinson, Cripe, Scheper, Stidham, Connor | Procter & Gamble | Light-duty liquid or gel dishwashing detergent compositions comprising mid-chain branched surfactants | Contain mid-chain branched ethoxy sulfates; useful for heavily soiled dishware at low temperature and high hardness |
| 6,268,331 | 2001 | D'Ambrogio, Connors | Colgate-Palmolive | Grease-cutting light-duty liquid detergent comprising lauryol ethylene diaminetriacetate | Contains ethylene diaminetriacetate — a chelating surfactant; good foaming and grease-cutting properties |
| 6,187,734 | 2001 | Erilli, Gallant | Colgate-Palmolive | High-foaming, grease-cutting light-duty liquid detergent comprising dialko sulfo succinates and zwitterionic surfactants | High-foaming, good grease-cutting properties |

*(continued)*

**TABLE 7.14** (Contd.)

| Number | Year | Inventors | Company | Title | Benefit |
|---|---|---|---|---|---|
| 6,187,733 | 2001 | Fabry, Weuthen | Henkel | Aqueous manual dishwashing composition containing a monoglyceride sulfate and at least two other surfactants | Contains monoglyceride sulfate |
| 6,127,328 | 2000 | D'Ambrogio, Jakubicki, Arvanitidou, Gambogi | Colgate-Palmolive | High-foaming grease-cutting light-duty liquid composition containing a C12 alkyl amido propyl dimethyl amine oxide | Contains a C12 alkylamidopropyldimethyl amine oxide for superior grease cutting |
| 6,066,755 | 2000 | Koch, Kwetkat | Huels | Amphiphilic compounds with a plurality of hydrophilic and hydrophobic groups based on carbonic acid derivatives | Amphiphilic compounds which have at least two hydrophilic and two hydrophobic groups |
| 6,022,844 | 2000 | Baillely, Perkins | Procter & Gamble | Cationic detergent compositions | Cationic esters to facilitate removal of greasy soils |
| 5,932,534 | 1999 | Gorlin, Gambogi, D'Ambrogio, Jakubick, Zyzyck | Colgate-Palmolive | Light-duty liquid containing sultaine surfactants for cleaning dishware and leaving a shiny appearance | Contains sultaine surfactants |
| 5,922,662 | 1999 | Thomas | Colgate-Palmolive | High foaming mild nonionic surfactant-based liquid detergents | Contains an alkyl succinamate |
| 5,888,955 | 1999 | Foley, Clarke, Yi-Change, Vinson | Procter & Gamble | Liquid dishwashing detergent compositions | Contains a bridged polyhydroxy fatty acid amide for improved sudsing |
| 5,872,111 | 1999 | Au, Harichian, Hung, Vermeer | Lever Brothers | Compositions comprising glycosylamide surfactants | Contain environmentally friendly carbohydrate surfactants |

| Patent number | Year | Inventors | Company | Description | Key feature |
|---|---|---|---|---|---|
| 5,811,384 | 1998 | Tracy, Li, Dahanayake, Yang | Rhodia | Compositions comprising at least one nonionic gemini surfactant – is useful, e.g., in personal care compositions, household cleaning products and industrial cleaners | Contain at least one nonionic gemini surfactant and are extremely effective emulsifiers and provide improved detergency |
| 5,780,417 | 1998 | Gorlin | Colgate-Palmolive | Light-duty liquid cleaning compositions | Contains ethoxylated/propoxylated nonionic surfactant |
| 5,739,092 | 1998 | Ofosu-Asante | Procter & Gamble | Liquid or gel dishwashing detergent containing alkyl ethoxy carboxylate divalent ok ions and alkylpolyethoxypolycarboxylate | Certain alkylpolyethoxypolycarboxylate surfactants prevent insoluble salt precipitation |
| 5,736,503 | 1998 | Vinson | Procter & Gamble | High-sudsing detergent compositions with specially selected soaps | Contain specially selected soap such as sodium 2-butyl-1-octanoate; provide spontaneous emulsification of grease |
| 5,489,393 | 1996 | Connor, Scheibel, Fu | Procter & Gamble | High-sudsing detergent with n-alkoxy polyhydroxy fatty acid amide and secondary carboxylate surfactants | Use of secondary carboxylate surfactants |
| 5,480,586 | 1996 | Jakubicki, McCandlish | Colgate-Palmolive | Light-duty liquid detergent composition containing a sulfosuccinamate-containing surfactant blend | Use of sulfosuccinamate |
| 5,393,466 | 1995 | Ilardi, Massaro, Rerek, Wenzel | Lever Brothers | New fatty acid ester compounds of polyoxyalkylene isethionate salts — are mild surfactants with good calcium tolerance for use in liquid and solid detergent compositions and personal care products | Superior to nonalkoxylated isethionates in mildness, performance, and calcium tolerance |

**TABLE 7.15** LDLDs Containing Polymers: U.S. Patent Examples (1996–2003)

| Number | Year | Inventors | Company | Title | Benefit |
|---|---|---|---|---|---|
| 6,645,925 | 2003 | Sivik, Bodet, Kluesener, Scheper, Yeung, Bergeron | Procter & Gamble | Liquid detergent compositions comprising quaternary nitrogen containing and/or zwitterionic polymeric suds enhancers | Contains a polymer claimed to be a suds enhancer, suds volume extender, and effective in preventing redeposition of grease |
| 6,509,306 | 2003 | Wisniewski, Thomas, Paye | Colgate | Light-duty liquid cleaning composition | Contains a silicon polymer; demonstrates improved sensory attributes and foam |
| 6,455,482 | 2002 | D'Ambrogio, Hassan, Dixit | Colgate | Light-duty liquid cleaning compositions comprising a crosslinked polymer | Contains a crosslinked polymer |
| 6,380,150 | 2002 | Toussaint, Oldenhove, Broze | Colgate | Light-duty liquid composition containing gelatin beads and polyacrylate thickener | Stably suspended oil containing gelatin beads with a polyacrylate polymeric thickener |
| 6,172,024 | 2001 | Arvanitidou | Colgate-Palmolive | High-foaming and grease-cutting light-duty liquid detergents comprising a polyoxyethylene diamine | Contain positively charged polymer to provide increased grease removal in compositions already exhibiting excellent disinfectant properties on hard surfaces |
| 6,172,023 | 2001 | Arvanitidou | Colgate-Palmolive | High-foaming grease-cutting light-duty detergent | Addition of polymers to high disinfectant acidic compositions to improve grease cutting |
| 6,172,022 | 2001 | Arvanitidou | Colgate-Palmolive | High foaming and grease cutting light-duty liquid detergents comprising a polyoxyethylene diamine | Contain positively charged polymer such as a poly(oxyethylene diamine) to provide increased grease removal |
| 6,160,110 | 2000 | Thomaides, Rodriques, Peterson | National Starch | Amino acid copolymers having pendant polysaccharide moieties and uses thereof | Copolymers assist in providing soil resistance |

| Patent | Year | Inventors | Assignee | Title | Notes |
|---|---|---|---|---|---|
| 6,133,217 | 2000 | Lewis, Lewis | Huntsman | Solubilization of low 2-phenyl alkyl benzene sulfonates | Addition of EO–PO block copolymers to detergents to increase solubility |
| 6,083,897 | 2000 | Lewis, Lewis | Huntsman | Solubilization of low 2-phenyl alkyl benzene sulfonates | Addition of polyethylene glycols to detergents to increase solubility |
| 5,985,813 | 1999 | Arvanitidou | Colgate-Palmolive | Liquid cleaning compositions based on cationic surfactant, nonionic surfactant and nonionic polymer | Addition of EO–PO block copolymers to cationic-based detergents to increase grease cutting |
| 5,977,275 | 1999 | Rodriques, Carrier, Furr | National Starch | Polymers having pendant polysaccharide and uses thereof | Polymer provides soil resistance to an article |
| 5,756,439 | 1998 | He, Fair, Massaro | Lever Brothers | Liquid compositions comprising copolymer mildness actives | Enhanced mildness by addition of EO–PO copolymers |
| 5,604,195 | 1997 | Misselyn, Erilli, Broze | Colgate-Palmolive | Liquid cleaning compositions with polyethylene glycol grease release agent | Contains a grease release agent |
| 5,552,089 | 1996 | Erilli, Mahieu, Misselyn, Yianakopoulos | Colgate-Palmolive | Aqueous light-duty liquid detergent compositions preventing grease buildup — comprising surfactants, solubilizing agent and grease release agent | Exhibits grease release effect, grease buildup prevented |
| 5,486,307 | 1996 | Misselyn, Mahieu, Erilli | Colgate-Palmolive | Liquid cleaning compositions with grease release agent and perfume | Contains grease release polymer |
| 6,207,631 | 2001 | Kasturi, Schafer, Sivik, Kluesener, Scheper | Procter & Gamble | Detergent compositions comprising polymeric suds volume and suds duration enhancers and methods for washing with same | Polymeric materials provide enhanced foam volume and duration |

## 3.   Use of Enzymes in LDLDs

Enzymes have been used widely in laundry detergents and automatic dishwashing detergents, but not previously in applications involving contact with the skin. Table 7.16 summarizes some recent patents describing the use of enzymes in hand dishwashing detergents. The enzyme supplier Novo Nordisk has patented modified polypeptides that have reduced allergenicity for use in hand dishwashing and personal care products [95]. Several formulations containing enzymes are patented by Procter & Gamble. The enzymes are added for skin conditioning, removing protein soils (protease), removing juice soils (pectinesterase), removing starch-based soils (amylase), and fat degrading (lipase).

## 4.   Disinfectants

The series of patents listed in Table 7.17 describe LDLD formulations that are disinfecting to hard surfaces. One technology used to obtain disinfectant properties is the use of acids such as salicyclic acid or alpha-hydroxy acids. The pH of these formulations are claimed to be between 3 and 6. The second type of disinfecting formulation utilizes quaternary ammonium compounds as the active antimicrobial agent. However, these formulations will need to avoid using common anionic surfactants so as not to form the inactive anionic–cationic complex. Also claimed as active ingredients are Zn salts, terpene alcohols, trichlorocarbanilide, hydrogen peroxide, and an iodophor.

## 5.   Enhanced Mildness and Skin Feel

Several patents listed in Table 7.18 describe the use of high levels of nonionic surfactants formulated with the specific purpose of increased mildness. As discussed earlier, this is one approach to formulating mild LDLDs. The remainder of the patents listed in Table 7.18 describe the addition of an ingredient to enhance mildness or improve the skin feel of the hands of the person doing the dishwashing. The ingredients claimed to benefit the skin feel attributes of the hands are an organosilane, monoalkyl phosphate ester, succinamate, and sucroglyceride surfactant.

## 6.   Heavy-Duty Cleaning

A few recent patents are listed in Table 7.19 that describe the cleaning of really tough, burnt-on soiled dishware. In general these products are alkaline (pH > 8), might be thickened with a polymer thickener, and may contain a soluble abrasive. In one case, the use of branched ethoxy sulfate surfactants proves useful in cleaning heavily soiled dishware.

## VI.   NEW PRODUCTS

Since 1993 an incredible wave of evolution has taken place in the hand dishwashing liquid market. New products not only include "smarter" surfactants

**TABLE 7.16** LDLDs Containing Enzymes: U.S. Patent Examples (1996–2003)

| Number | Year | Inventors | Company | Title | Benefit |
|---|---|---|---|---|---|
| 6,544,941 | 2003 | Lee, Ghatlia | Unilever | Dishwashing composition | Contains serine protease and a metalloprotease and shows protein soil removal |
| 6,201,110 | 2001 | Olsen, Hansen, Beck | Novo Nordisk A/S | Polypeptide with reduced respiratory allergenicity | Modified polypeptides with reduced allergenicity for use in dishwashing, personal care products, etc. |
| 6,162,778 | 2000 | McKillop, Foley, Crabtree, Burckett-St. Laurent, Clarke, Patil | Procter & Gamble | Light-duty liquid or gel dishwashing detergent compositions having beneficial skin conditioning, skin feel and rinsability aesthetics | Nonionic surfactant- and hydrotrope-containing skin feel/rinsability enhancing system and skin conditioning protease enzymes |
| 6,136,778 | 2000 | Kamiya | — | Environment safeguarding aqueous detergent composition comprising essential oils | Composition maximizes the decomposing action of an enzyme and minimizes the use of surfactant |
| 6,113,655 | 2000 | Tsunetsugu, Moese, Baeck, Herbots | Procter & Gamble | Detergent compositions comprising a pectinesterase enzyme | Contains a pectinesterase; useful for removal of body, plant, fruit, and vegetable juice soils |
| 5,952,278 | 1999 | Mao, Marshall, Visscher | Procter & Gamble | Light-duty liquid or gel dishwashing detergent compositions containing protease | Small amounts of protease |
| 5,851,973 | 1998 | Foley | Procter & Gamble | Manual dishwashing composition comprising amylase and lipase enzymes | Comprises surfactants, calcium or magnesium ions, enzymes, and polymer thickeners |
| 5,830,837 | 1998 | Bisgard-Frantzen, Borchert, Svendsen, Thellersen, Van der Zee | Novo Nordisk A/S | Amylase variants | Variant of the parent alpha-amylase enzyme having improved dishwashing performance |
| 5,786,316 | 1998 | Baeck, Busch, Verschuere, Katrien | Procter & Gamble | Cleaning compositions comprising xylanolytic enzymes | Compositions have xylanase activity; show an excellent boost in cleaning performance on fruit, vegetables, and mud or clay soils |

**TABLE 7.17**  Disinfectant LDLD Formulations: U.S. Patent Examples (1996–2003)

| Number | Year | Inventors | Company | Title | Benefit |
|---|---|---|---|---|---|
| 6,617,296 | 2003 | Connors, D'Ambrogio, Nascimbeni | Colgate-Palmolive | Antibacterial light-duty liquid detergent | Contains a stable zinc inorganic salt which provides antibacterial benefits |
| 6,258,763 | 2001 | Arvanitidou, Sandhu | Colgate-Palmolive | Light-duty liquid composition containing an acid | High-foaming liquid-disinfecting composition which has good grease-cutting properties; has a pH of 3–6 |
| 6,187,735 | 2001 | Gambogi, Durbut, Broze, Zyzyck | Colgate-Palmolive | Light-duty liquid detergent | High level of disinfectant properties, based on cationic and nonionic surfactants |
| 6,184,194 | 2001 | Arvanitidou, Suriano, Engels, Jakubicki | Colgate-Palmolive | High-foaming, grease-cutting light-duty liquid detergent having antibacterial properties comprising proton donating agent | High foaming and good grease-cutting properties, good mildness, as well as excellent disinfecting properties on hard surfaces |
| 6,152,152 | 2000 | Reynen, Aryana | Procter & Gamble | Antibacterial liquid dishwashing detergent compositions | Comprise surfactant, hydrotrope, and unsaturated terpene alcohol; disinfect dishware cleaning implements |
| 6,140,290 | 2000 | Gorlin | Colgate-Palmolive | High foaming nonionic surfactant-based liquid detergent | Contains an antibacterial agent (trichlorocarbanilide) which is soluble in polyethylene glycol |
| 6,140,289 | 2000 | Frank, McCandlish | Colgate-Palmolive | Cleaning composition for manual dishwashing comprises cationic, ethoxylated nonionic, amine oxide, and alkyl polyglucoside surfactants | High level of disinfectant properties, based on cationic and nonionic surfactants |

| Patent | Year | Inventors | Assignee | Title | Description |
|---|---|---|---|---|---|
| 6,113,933 | 2000 | Beerse, Morgan, Baier, Bartolo, Bakken Schuette | Procter & Gamble | Mild, rinse-off antimicrobial liquid cleansing compositions containing acidic surfactants | Antimicrobial cleansing compositions based on anionic and nonionic surfactants and a pH of 3.0–5.5 |
| 6,106,851 | 2000 | Beerse, Morgan, Baier, Bartolo, Bakken Schuette | Procter & Gamble | Mild, rinse-off antimicrobial liquid cleansing compositions containing salicylic acid | Antimicrobial cleansing compositions based on 0.15–2% salicylic acid |
| 6,103,683 | 2000 | Romano, Trani, Minervini, Brown | Procter & Gamble | Disinfecting compositions and processes for disinfecting surfaces | Disinfecting of surfaces with 0.1–15% hydrogen peroxide, antimicrobial essential oil or mixture |
| 5,968,539 | 1999 | Beerse, Morgan, Baier, Chen, Bakken | Procter & Gamble | Mild rinse-off antimicrobial liquid cleansing compositions which provide residual benefit versus gram-negative bacteria | Antimicrobial compositions based on organic acids and antibacterial ingredients |
| 5,728,667 | 1998 | Richter | Reckitt & Colman | Compositions containing organic compounds | Anionic formulation containing quaternary ammonium germicidal compound |
| 5,707,955 | 1998 | Gomes, McCandlish, Fischler | Colgate-Palmolive | High-foaming liquid detergents containing a nonionic surfactant and three anionic surfactants | Contains disinfecting agent (iodophor) complexed with nonionic |

**TABLE 7.18** Formulating LDLDs for Enhanced Mildness or Skin Feel: U.S. Patent Examples (1996–2003)

| Number | Year | Inventors | Company | Title | Benefit |
|---|---|---|---|---|---|
| 6,509,306 | 2003 | Wisniewski, Thomas, Paye | Colgate | Light-duty liquid cleaning composition | Contains a silicon polymer; demonstrates improved sensory attributes and foam |
| 6,214,781 | 2001 | Gambogi, Dalimier, Paye, Zocchi | Colgate-Palmolive | Light-duty liquid cleaning compositions comprising an organosilane | Improved sensory attributes for the hands |
| 6,013,611 | 2000 | Thomas, Gomes, Drapier, Church | Colgate-Palmolive | Light-duty liquid cleaning compositions especially for dishwashing | Contains a water-soluble nonionic surfactant |
| 5,874,394 | 1999 | Thomas, Gomes | Colgate-Palmolive | Light-duty liquid cleaning compositions containing monoalkyl phosphate ester | Enhanced mildness to the human skin |
| 5,869,439 | 1999 | Thomas, Gomes | Colgate-Palmolive | High-foaming nonionic surfactant-based liquid detergent | Contains alkyl succinamate; improved skin feel attributes |
| 5,856,291 | 1999 | Thomas, Gomes | Colgate-Palmolive | Light-duty liquid cleaning compositions containing alkyl sucroglyceride | Contains a sucroglyceride surfactant |
| 5,629,279 | 1997 | Erille, Repinec, Gomes | Colgate-Palmolive | High-foaming nonionic surfactant-based liquid detergent | High foaming nonionic based |

**TABLE 7.19** Formulating LDLDs for Tough Soils or Other Specialty Products: U.S. Patent Examples (1996–2003)

| Number | Year | Inventors | Company | Title | Benefit |
|---|---|---|---|---|---|
| 6,589,926 | 2003 | Vinson, Oglesby, Scheper, Kasturi, Ofosu-Asante, Clarke, Owens, Castro, Embleton | Procter & Gamble | Dishwashing detergent compositions containing organic diamines | Low-molecular-weight diamine-containing compositions have improved grease removal, sudsing, low-temperature stability and improved dissolution properties |
| 6,362,155 | 2002 | Kinscherf | Colgate-Palmolive | Thickened microemulsion cleaning compositions comprising xanthum gum | Superior cling to a vertical surface and effective in the removal of oily and greasy soil |
| 6,337,312 | 2002 | Kinscherf, Aszman, Thomas | Colgate-Palmolive | Liquid crystal compositions comprising an abrasive and magnesium sulfate heptahydrate | Effective as prespotting agent and for removing hard-to-remove soils from substrates |
| 6,281,181 | 2001 | Vinson, Cripe, Scheper, Stidham, Connor | Procter & Gamble | Light-duty liquid or gel dishwashing detergent compositions comprising mid chain branched surfactants | Contain branched ethoxy sulfate surfactants found especially useful for cleaning heavily soiled dishware |
| 6,274,539 | 2001 | Kacher, Wallace, Allouch | Procter & Gamble | Light-duty liquid or gel dishwashing detergent compositions having controlled pH and desirable food soil removal, rheological, and sudsing characteristics | Alkaline product thickened with acrylic copolymer useful for heavily soiled dishware |
| 6,228,832 | 2001 | Kinscherf, Thomas, Slezak, Psihoules | Colgate-Palmolive | Microemulsion cleaning composition | Alkaline, microemulsion composition effective in cleaning burnt-on greasy soils |

*(continued)*

**TABLE 7.19** (Contd.)

| Number | Year | Inventors | Company | Title | Benefit |
|---|---|---|---|---|---|
| 6,225,272 | 2001 | Giesen, Zaika, Middelhauve, Hofmann, Legel | Henkel | Dishwashing detergent with enhanced cleaning effect | Contains soluble abrasive component for excellent cleaning against dried-on and burnt-on food soils |
| 6,165,958 | 2000 | Arvanitidou | Colgate-Palmolive | High-foaming, grease-cutting light-duty liquid detergent comprising vinylidene olefin sulfonate | Addition of vinylidene olefin sulfonate to α-olefin sulfonate to improve rheology of AOS and increase detergent performance in hard water |
| 6,051,542 | 2000 | Pollack, Gomes | Colgate-Palmolive | Post foaming cleaning compositions comprising isopentane | Compositions sprayed on surface and then composition foams |
| 5,919,312 | 1999 | Wierenga, Weikel, Underwood | Procter & Gamble | Compositions and methods for removing oily or greasy soils | Composition for cleaning cooking surfaces has a pH no less than 8 |
| 5,891,836 | 1999 | Kacher | Procter & Gamble | Light-duty liquid or gel dishwashing detergent compositions which are microemulsions and which have desirable greasy food soil removal and sudsing characteristics | Microemulsion compositions for heavily soiled dishware |

and surfactant mixtures, but also address multiple consumer needs offering multidimensional benefits.

## A.  Antibacterial Products

The first liquid hand soap and hand dishwashing liquid product that offered long-lasting antibacterial protection on hands was introduced in 1994 by Colgate-Palmolive. The antibacterial ingredient was triclosan, and the product delivered the advertised efficacy as documented in clinical and laboratory testing. The color of the product was chosen carefully through extensive market research to be orange, conveying a strong antibacterial benefit. In addition, the fragrance had no strong "medicinal" connotation. A trend was established, and followed by other LDLD manufacturers in North America, Europe, and the rest of the world. In Europe, however, these products have not enjoyed the tremendous success they found in the U.S. The choice of triclosan as the antibacterial ingredient was unanimous in the U.S., while European countries chose either essential oils such as geraniol or *d*-limonene or used until recently buffering organic acids that impart a low pH to the system. The local regulatory requirements are always the driving force for the choice of the active ingredient.

### 1.  Typical Antibacterial Actives

Triclosan (2,4,4′-trichloro-2′-hydroxydiphenyl ether) is used in most antibacterial LDLD products in the U.S. It is usually known as Irgasan or Irgacare or DP300, and its major supplier is Ciba Geigy. It is almost insoluble in water (0.004%), and soluble in ethanol and Tween 20 or 80. Triclosan's thermal, hydrolytic, and light stability is very high. Triclosan has a broad spectrum of antimicrobial activity against gram-positive and gram-negative bacteria, as well as fungi. Depending upon the concentration, it exhibits both bacteriostatic and bactericidal activity. Studies indicate that the mode of action is nonspecific and the primary site of the antimicrobial activity of triclosan is the cytoplasmic membrane. At low bacteriostatic concentrations, triclosan interferes with the uptake of essential nutrients such as amino acids from the medium needed for biological activity. At bactericidal concentrations, triclosan disorganizes the cytoplasmic membrane causing leakage of low-molecular-weight compounds and other vital cellular constituents such as nucleic acids leading to the death of the cell. More details on the mode of action of triclosan were offered by Regos and Hitz [96].

Triclosan is highly substantive to many surfaces including skin, fabric, and hair, and can provide residual bacteriostatic activity. The residual antibacterial activity is summarized in Table 7.20 and Table 7.21.

Triclorocarban (3,4,4′-triclorocarbanilide or TCC) is an odorless solid compound that is sparingly soluble in water (50 to 100 ppb at room temperature) and

**TABLE 7.20**  Agar Patch Test Results on Skin for an LDLD Containing Triclosan

| Organism/bacteria count (log) | Placebo | LDLD with triclosan | Log reduction |
|---|---|---|---|
| Staphylococcus aureus | 1.59 | 0.05 | 1.53 |
| Escherichia coli | 1.70 | 0.00 | 1.70 |

**TABLE 7.21**  Zone of Inhibition Test Results on Cotton Fabric for a Fabric Softener Containing Triclosan

| Organism/bacteria count (log) | Control | Fabric softener with triclosan |
|---|---|---|
| Staphylococcus aureus | 0/20 | 20/20 |
| Escherichia coli | 0/20 | 20/20 |

fairly soluble in organic solvents and surfactants. It is highly stable in acidic pH and hydrolyzes under alkaline conditions and high temperature resulting in dichloro- and *para*-chloroanilines.

TCC is highly effective against gram-positive bacteria but not against gram-negative bacteria. Its mode of action is based on destabilization of the integrity of the cytoplasmic membrane through destruction of the semipermeability of the membrane, uncoupling of oxidative phosphorylation, and inhibition of transportation of the essential substances through the membrane [97,98]. TCC is mostly used today in personal care products, e.g., soap bars, liquid hand soap formulations, and shower gels, and it is not found in LDLD formulations in the U.S., probably because of its limited antibacterial activity. It could become a viable alternative for triclosan.

*Para*-chloro-*meta*-xylenol, or PCMX, is a phenolic antimicrobial agent. It is a white to off-white crystalline powder with a faint phenol odor. It is sparingly soluble in water, and soluble in alcohol.

PCMX acts by destroying the cell wall and by inactivating enzymes. It is highly effective against gram-positive bacteria but less effective against gram-negative bacteria, mycobacteria, fungi, and viruses. The pseudomonas species are resistant to PCMX. Its antibacterial activity is also reduced by nonionic surfactants due to incorporation into the micelles. Compared to triclosan it is less effective as an antimicrobial agent, but it is more effective than TCC against gram-negative bacteria. PCMX is currently used in private label LDLD formulations. The effectiveness (MIC values) of the previously discussed antibacterial ingredients is listed in Table 7.22. The regulatory status of these ingredients is summarized in Table 7.23.

**TABLE 7.22**  Effectiveness of Three Antibacterial Agents as Shown by their Minimum Inhibitory Concentration (MIC)

| Organism | MIC (ppm) of triclosan | MIC (ppm) of PCMX | MIC of TCC |
|---|---|---|---|
| *Staphylococcus aureus* | 0.01 | 20 | 0.1 |
| *Streptococcus pyogenes* | 3.0 | — | <0.05 |
| *Escherichia coli* | 0.3 | 75 | >10,000 |
| *Pseudomonas aeruginosa* | >1000 | — | ~10,000 |
| *Serratia marces* | >100 | — | ~10,000 |
| *Klebsiella pneumoniae* | 0.3 | 125 | — |
| *Staphylococcus epidermis* | 0.01 | 30 | 0.5 |
| *Candida albicans* | 3 | — | — |
| *Aspergillus niger* | 30 | — | — |

## 2.  Food and Drug Administration

Antibacterial LDLD compositions are regulated by the FDA in the U.S. The reason is because they are positioned as "antibacterial liquid hand soaps."

The proposed OTC Monograph in 1974 established seven categories for topical antimicrobial preparations, which are: antimicrobial soap, health care personnel hand wash, patient preoperative skin preparation, skin antiseptic, skin wound cleanser, skin wound protectant, and surgical hand soap. The reason for establishing seven categories was because not all antimicrobial products are used for the same purpose; therefore, the requirements for effectiveness should not be the same. By 1994 the tentative Monograph had eliminated antimicrobial soaps from being a separate category. Since that time, antimicrobial soaps must meet the efficacy requirements that equate to healthcare products. Antimicrobial body wash, hand wash, and food handling hand wash products are all treated "equally." The test method requirements are as follows:

- *In vitro:*

  MIC and SIKT (no criteria listed by FDA)

  Ingredient, product, and vehicle (product without active)

- *In vivo:*

  HCPHW test (2 log reduction after first wash, 3 log reduction after tenth wash)

  Patient preoperative test

  Surgical hand scrub test

**TABLE 7.23**  Regulatory Status of Three Antibacterial Agents in the U.S., Europe, and Japan

| | Triclosan | TCC | PCMX — chloroxylenol |
|---|---|---|---|
| U.S. (FDA) | Category III for Safety and effectiveness | Category I for safety, Category III for effectiveness | Short-term use (0.24–3.75%): Category I for safety, Category III for effectiveness; Long-term use: Category III for safety and effectiveness |
| Europe | EU directive on cosmetics: preservative for cosmetics up to 0.3%; higher level allowed as antibacterial agent CAS: 3380-34-5. Dangerous substance directive: classified in Annex I (29th atp) as: Xi, R36/38; N, R50/53; with specific concentration limits: 0.0025–0.025: R52-53; 0.025–0.25: R51-53; ≥0.25%: R50-53; ≥ 20%: R36/38, R50/53 | EU directive on cosmetics: preservative for cosmetics up to 0.2%; higher level allowed as antibacterial agent CAS: 101-20-2. Dangerous substance directive: *not* classified in Annex I; supplier classification: N, R50/53 | EU directive on cosmetics: preservative for cosmetics up to 0.5% CAS: 88-04-0. Dangerous substance directive: classified in Annex I as: Xn, R22; Xi, R36/38; Xi, R43; without specific concentration limits |
| Japan | Japanese standard on cosmetics, preservative for all cosmetics up to 0.1%. Japanese standard on quasi-drugs, active ingredient for toothpaste up to 0.02% | Japanese standard on cosmetics, preservative for cosmetics on use to mucous membrane and cosmetics on use without rinse up to 0.3% | This ingredient is not found in Japanese standard. It is necessary to register as new ingredient to MOH for new preservative for cosmetics with data on safety or for new active ingredient for quasi-drug (toothpaste) with data on safety and effectiveness |

*Note:* Category III means that insufficient data exist.

The industry has been active since 1997 (HealthCare Continuum Model) organizing meetings, gathering data, and discussing test methodologies in order to guide future FDA rulings.

The current expectation is that the Antimicrobial Handwash Monograph will be split into two sections. One section will be for professional hand wash (hospitals, food handlers, etc.) and the second will deal with consumer products. The best estimate for the FDA to take action on the consumer product section is around 2006.

## 3.  Current Antibacterial Dish Liquid Products

A list of some antibacterial products marketed in North America and Europe is given in Table 7.24 and Table 7.25.

## B.  Concentrated Products

Up to 1995 the highest surfactant level at which an LDLD was formulated was 32 to 34% worldwide. Procter & Gamble in 1995 introduced in the U.S. for the first time a more concentrated "ultra" LDLD that the consumer can use at a dose that is 2/3 of the previous dose. The products are formulated at higher surfactant level (about 48%) to provide the same performance, and the package has also shrunk from 22 oz to 14.7 oz to flag the latest change. Most manufacturers in the U.S.

**TABLE 7.24**  Antibacterial LDLD Products Marketed in North America

| LDLD product | Active | Company |
|---|---|---|
| *U.S.* | | |
| Ajax Antibacterial | Triclosan | Colgate-Palmolive |
| Ajax Fiesta | Triclosan | Colgate-Palmolive |
| Dawn Antibacterial | Triclosan | Procter & Gamble |
| Dawn Power Plus | Triclosan | Procter & Gamble |
| Dawn Fresh Escapes | Triclosan | Procter & Gamble |
| Joy Antibacterial | Triclosan | Procter & Gamble |
| Joy Escapes | Triclosan | Procter & Gamble |
| Palmolive Antibacterial | Triclosan | Colgate-Palmolive |
| Palmolive Original Antibacterial | Triclosan | Colgate-Palmolive |
| Palmolive Lemon Antibacterial | Triclosan | Colgate-Palmolive |
| Palmolive Spring Sensations Ocean Breeze | Triclosan | Colgate-Palmolive |
| Sunlight Antibacterial | Triclosan | Unilever |
| *Canada* | | |
| Ivory Antibacterial Orange | Triclosan | Procter & Gamble |
| Palmolive Antibacterial Orange | Triclosan | Colgate-Palmolive |
| Palmolive Lemon Antibacterial | Triclosan | Colgate-Palmolive |
| Spring Sensations Ocean Breeze | Triclosan | Colgate-Palmolive |
| Sunlight Antibacterial Orange | Triclosan | Unilever |

**TABLE 7.25**  Antibacterial LDLD Products Marketed in Europe

| LDLD product | Company | Country |
|---|---|---|
| Dawn von Fairy | Procter & Gamble | Germany, Benelux |
| Dreft | Procter & Gamble | Sweden, Finland |
| Fairy | Procter & Gamble | U.K., Greece, Portugal, Spain |
| Yes | Procter & Gamble | |
| Mir | Henkel | France |
| Nelsen | Procter & Gamble | Italy |
| Paic XL Antibacterial | Colgate-Palmolive | France |
| Palmolive Antibacterial | Colgate-Palmolive | France, Germany, Austria, Switzerland, Greece |
| Palmolive Spring Sensations Antibacterial Ocean Breeze | Colgate-Palmolive | France, Germany, Austria, Switzerland, Greece |
| Persil | Unilever | U.K. |
| Pril | Henkel | Germany |
| Pril 2-in-1 | Henkel | Germany |
| Super Pop antibacterial | Colgate-Palmolive | Portugal |
| Super Pop Spring Sensations Antibacterial — Ocean Breeze | Colgate-Palmolive | Portugal |
| Svelto | Unilever | Italy, Greece |
| Vel | Colgate-Palmolive | Denmark |

followed this trend, which became a major success with increased sales volumes. At the same time, they kept producing the regular sizes. This trend certainly was tested in Canada and later in Europe. The bulk of sales in these regions remain in the regular strength products.

## C.  Hand Care Products

The consumer need for a milder dishwashing liquid was met in the early 1990s with the successful introduction of Palmolive Sensitive Skin and the milder positioning of Ivory LDLD. The use of nonionic and amphoteric surfactants resulted in clinically milder and less irritating products. In the last decade, however, further improving mildness was a significant objective for the major players in the LDLD market. Dawn Hand Care was the first to introduce protease, an enzyme that is claimed to soften skin by exfoliation of the top layer.

## D.  High-Efficacy Products

These products are designed to offer an advantage on tough soils such as greasy, starchy, or cooked-on/burnt soils. The formulation of these products is based on a high anionic surfactant mix, an abrasive, polymers, and enzymes specific to target starch or grease. Examples of such products include Palmolive Pots & Pans, Dawn Power Plus, and Palmolive Max Power. In the high-pace environment of the developed world, consumers demand speed and high efficacy. Nonetheless, baked-on food and grease removal with manual dishwashing liquids still remains a challenge.

## E. Sensorial Products

Efficacy, hand care, and added benefits have been the fundamental elements of every LDLD product since the second half of the 1990s. With the dawn of the new millennium dishwashing moved into a new dimension: the experiential dimension, with different colors and fragrances introduced to add more fun to a cleaning task. Colgate-Palmolive introduced the Spring Sensations line in the spring of 2000. The line has not stayed stagnant but is constantly "rejuvenating" itself with new introductions such as Orchard Fresh and Green Apple. Procter & Gamble followed with Joy Invigorating Splash and Tropical Calm, and more recently in the spring of 2001 with Dawn Fresh Escapes featuring Citrus Burst Apple Blossom and Wildflower Medley. In the summer of 2001 Colgate-Palmolive also launched Ajax Fiesta, taking the nonpremium LDLD segment to the experiential dimension as well. The sensorial trend has already moved to Canada, Europe, Asia, and Latin America. The fragrance experience was taken a step further with the launch of Colgate-Palmolive's Aromatherapy variants in 2002, an extension of a trend that was gaining momentum in personal care products.

## VII. FUTURE TRENDS

Some consumers feel unsafe [99] because kitchen sites are contaminated and cross-contaminated through the use of sponges, dishcloths, cutting boards, and other cleaning or food preparation implements [100,101]. This cross-contamination is responsible for food poisoning incidents with salmonella and *Escherichia coli*, which are on the increase. Consumers remain germ-phobic, and guarding the family is their primary objective [102]. Therefore, the demand for antibacterial products will not only remain in the future, but the need for more effective ones will dominate. Efficacy is geared toward a broader range of microorganisms, and ideally with antibacterial actives that are naturally derived.

Washing the dishes is undoubtedly a chore and it is very difficult to make it sound appealing. Consumer product companies, however, have found a way to make it more fun by introducing pleasing and nontraditional fragrances. Experiential fragrance notes and appealing perfumes are here to stay.

These new fragrance trends came to complement the more traditional fragrances which are more associated with cleanliness and germ killing, such as lemon, lime, citrus bouquets, and mandarin orange accords [103].

In Japan, Family Pure dishwashing liquid by Kao is positioned as having a new deodorizing effect. Containing a new deodorizing agent "ASA" and herbal extract, it removes stubborn odors while cleaning the dishes. More deodorizing and air freshening benefits are expected to be added to the current experiential variants.

The experiential dimension does not have to stay in the fragrance or color arena. It is expanding to the beads, pearls, and other attractive elements seen in personal care products. A recent example is the Rainett aux Algues Marines LDLD introduced in France by Werner & Mertz. This product contains natural marine algae extracts encapsulated in beads. The extracts offer hand care benefits. Pearls or beads can be used as carriers of actives to enhance efficacy or mildness perceptions.

Aromatherapy has had an impact on the air care and candle segments of the household market, where products are often positioned as "soothing" and "relaxing." This trend is currently expanding with tremendous speed in body care products such as shower gels (Bed & Body Works, Bed & Bath stores) and soon in fabric care products such as fabric conditioners. With the recent launch of Ultra Palmolive Anti-Stress Aromatherapy Dish Liquid, Colgate-Palmolive has led the way of taking dishwashing to "a whole new sensation."

The trend toward natural products that use "natural" ingredients and "natural" fragrances are also growing. Natural and clean go together. This trend originated in Europe where consumers are more aware and passionate about environmental issues; soon the trend is expected to cross the Atlantic to North America. Garden Fresh and Fresh Rain are some examples [103].

Cleaning implements are also expected to enter the dish liquid market shortly. These implements could be either in the form of a wipe or dish tools like sponges/scourers recently introduced by Procter & Gamble in the UK. Wipes are now expanding from personal care applications to cleaning (floor or furniture wipes). The major advantage they offer is convenience for today's busy consumer. Dish tools can add speed to traditional dishwashing.

Can LDLD products be upscaled? There are already some manifestations of this trend; it remains to be seen if it will stay for a long time. Stores like the Good Home Company, The Thymes, Vermont Soapworks, Restoration Hardware, and Williams Sonoma include dish liquid in a series of home cleaning products, inspired by memories and nostalgia, designed to enhance life's daily duties and the overall kitchen experience. (According to the National Association of Professional Organizers a woman spends nearly 1100 hours a year in the kitchen!)

Enzymes made a splash in the automatic dishwashing detergents market first, and recently have been introduced in LDLDs for manual washing (e.g., Dawn Hand Care and Power Plus). The intention is to increase the cleaning power by introducing an ingredient like amylase that will break starchy food, or improve hand appearance by introducing enzymes such as protease. Kao also developed a genetically engineered amylase and expected to introduce it in its LDLD products.

Nowadays consumers are more sophisticated than they were 10 years ago, when they bought what was offered. Today, they demand choice. And choice can also be manifested through packaging and labels. This was especially evident in the original offering of the Method line of dishwashing products [104]. The bottles were designed in unusual shapes, and were intended to be stored upside down because they incorporated squeeze-activated valve to dispense the product. These products have gained significant market share in some large discount stores. Beautiful scenes and decorative bottles can definitely complement an efficacious product with a pleasing and appealing scent.

## REFERENCES

1. Euromonitor International (www.euromonitor.com), Dishwashing Products in United States, 2003.
2. Dishwashing Liquid Habits and Practices Study, Colgate-Palmolive, November 1999.
3. Kaiser, C., in *Detergent in Depth '80*, Symposium Series, Soap and Detergent Association, San Francisco, 1980, p. 30.
4. Chirash, W., *J. Am. Oil Chem. Soc.*, 58, 362A, 1981.
5. Drozd, J.C., *Chemical Times & Trends*, 7, 29, 1984.
6. Drozd, J.C., *Chemical Times & Trends*, 7, 41, 1984.
7. Drozd, J.C., *Chemical Times & Trends*, 8, 49, 1985.
8. Fernee, K.M., in *Proceeding of the Second World Conference on Detergents*, Baldwin, A.R., Ed., American Oil Chemists Society, 1987, p. 109.
9. Berth, P., Jeschke, P., Schumann, K., and Verbeek, H., in *Proceeding of the Second World Conference on Detergents*, Baldwin, A.R., Ed., American Oil Chemists Society, 1987, p. 113.
10. Heitland, H. and Marsen, H., in *Surfactants in Consumer Products: Theory, Technology and Application*, Falbe, J., Ed., Springer-Verlag, Heidelberg, 1987, chap. 5, pp. 318–319.
11. Lai, K.Y., McCandlish, E.F.K., and Aszman, H., in *Liquid Detergents*, Surfactant Science Series, Vol. 67, Lai, K.Y., Ed., Marcel Dekker, New York, 1996, chap. 7.
12. Linfield, W.M., Ed., *Anionic Surfactants*, Surfactant Science Series, Vol. 7, Marcel Dekker, New York, 1976.
13. Cohen, L. and Trujillo, F., *J. Surf. Det.*, 2, 363, 1999.
14. Schick, M.J., Ed., *Nonionic Surfactants: Physical Chemistry*, Surfactant Science Series, Vol. 23, Marcel Dekker, New York, 1987.
15. Balzer, D. and Luders, H., Eds., *Nonionic Surfactants: Alkylpolyglucosides*, Surfactant Science Series, Vol. 91, Marcel Dekker, New York, 2000.
16. Fu, Y.C. and Scheibel, J.J., International application WO 92/06157 to Procter & Gamble Co., 1992.
17. Durbut, P. and Gambogi, J.E., AOCS Conference, May 2002.
18. Domingo, X., in *Amphoteric Surfactants*, Surfactant Science Series, Vol. 59, Lomax, E.G., Ed., Marcel Dekker, New York, 1996, chap. 3, pp. 75–190.
19. Sajic, B., Ryklin, I., and Frank, B., *Household and Personal Products Industry*, 35, 94, 1998.

20. Lai, K.Y. and Dixit, N., in *Foams: Theory, Measurements and Applications*, Surfactant Science Series, Vol. 57, Prud'homme, R. and Khan, S.A., Eds., Marcel Dekker, New York, 1995, chap. 8, pp. 315–338.
21. Schoenberg, T., *Household and Personal Products Industry*, 36, 7, 1999.
22. McCutcheon's *Emulsifiers and Detergents*, North American and International Edition, 1999.
23. Branna, T., *Household and Personal Care Products*, 31, 86, 1994.
24. *Soap, Cosmet. Chem. Specialties*, 70, 53, April 1994.
25. Fu, Y.C., and Scheibel, J.J., International application WO 92/06157 to Procter & Gamble Co., 1992.
26. Day, D.M., *J. Am. Oil Chem. Soc.*, 52, 461, 1975.
27. Carroll, B.J., *Colloids Surf. A*, 74, 131, 1993.
28. Myers, D., in *Surfactant Science and Technology*, VCH, New York, 1988, p. 277.
29. Power, N.C., *J. Am. Oil Chem. Soc.*, 40, 290, 1963.
30. Cox, M.F., *J. Am. Oil Chem. Soc.*, 63, 559, 1986.
31. Adam, N.K., *J. Soc. Dyers Colour*, 53, 121, 1937.
32. Miller, C. A. and Raney, K.H., *Colloids Surf. A*, 74, 169, 1993.
33. Durbut, P. and Broze, G., *Comun. J. Ecom. Esp. Deterg.*, 28, 115, 1998.
34. Rosen, M.J., *Surfactants and Interfacial Phenomena*, 2nd edition, Wiley-Interscience, New York, 1989.
35. Varadaraj, R., Bock, J., Valint, P., Zushma, S., and Brons, N., *J. Colloid Interface Sci.*, 140, 31, 1990.
36. Engels, Th., von Rybinsk, W., and Schmiedel, P., *Prog. Colloid Polym. Sci.*, 111, 117, 1998.
37. Patist, A., Axelberd, T., and Shah, D.O., *J. Colloid Interface Sci.*, 208, 259, 1998.
38. Edwards, D.A. and Wasan, D.T., in *Foams: Theory, Measurements and Applications*, Surfactant Science Series, Vol. 57, Prud'homme, R. and Khan, S.A., Eds., Marcel Dekker, New York, 1995, chap. 3, pp. 189–215.
39. Rosen, M.J. and Zhu, Z.H., *J. Am. Oil Chem. Soc.*, 65, 663, 1988.
40. Rhein, L., in *Surfactants in Cosmetics*, Surfactant Science Series, Vol. 68, Reiger, M.M. and Rhein, L.D., Eds., Marcel Dekker, New York, 1999, chap. 18, pp. 397–426.
41. Rhein, L.D., Robbins, C.R., Fernee, K., and Cantore, R., *J. Soc. Cosmet. Chem.*, 37, 125, 1986.
42. Lomax, E.G., Ed., *Amphoteric Surfactants*, Surfactant Science Series, Vol. 59, Marcel Dekker, New York, 1996.
43. Jakubicki, G.J. and Warschewski, D., EP 0487169A1 to Colgate-Palmolive Co., 1991.
44. Pancheri, E.J. *et al.*, U.S. Patent 4,904,359 to Procter & Gamble Co., 1990.
45. Tosaka, M., Haayakawa, Y., and Deguchi, K., GB 2219594A to Kao Corp., 1989.
46. Lai, K.Y., U.S. Patent 4,595,526 to Colgate-Palmolive Co., 1986.
47. Putkik, C.F. and Borys, N.F., *Soap Cosmet. Specialties*, 86, 34, 1986.
48. ASTM 1173-53, Vol. 15.04, *Annual Book of ASTM Standards*, American Society for Testing and Materials, Philadelphia, PA, 1992.
49. Deguchi, K., Saito, K., and Saijo, H., U.S. Patent 5,154,850 to Kao Corp., 1992.
50. Jakubicki, G.J. and Warschewski, D., EP 0487169 A1 to Colgate-Palmolive Co., 1991.

51. Rubin, K.F., Van Blarcom, D., and Lopez, J.A., WO 93/03120 to Unilever, 1993.
52. Anstett, R.M. and Schuck, E.J., *J. Am. Oil Chem. Soc.*, 43, 576, 1966.
53. Customer report by Shell Development Company, January 1987.
54. Nguyen, C., Riska, G., Hawrylak, G., and Malihi, F., Characterization of Foam Properties in Light Duty Liquid Dishwashing Products, presented at the AOCS Symposium, Cincinnati, OH, May 4–7, 1989.
55. Duliba, E.P., Nguyen, C., Riska, G.D., Hawrylak, G.W., and Bala, F.J., WO 90/02164 to Colgate-Palmolive Co., 1990.
56. Massaro, M. and Retek, M.E., EP 0373851 A2 to Unilever, 1989.
57. ASTM D 4009-92, Vol. 15.04, *Annual Book of ASTM Standards*, American Society for Testing and Materials, Philadelphia, PA, 1992.
58. Naik, A.R., U.S. Patent 4,528,128 to Lever Brothers Co., 1985.
59. Dyet, J.A. and Foley, P.R., WO 92/06171 to Procter & Gamble Co., 1992.
60. Frosch, P. and Kligman, A., *J. Am. Acad. Dermatol.*, 1, 35, 1979.
61. Tronnier. H. and Heinrich, U., *Parf. Kosmet.*, 76, 314, 1995.
62. Tausch, I., Bielfeldt, S., Hildebrand, A., and Gassmuller, J., *Parf. Kosmet.*, 75, 28, 1996.
63. Massaro, M. and Perek, M.E., EP 0373851 A2 to Unilever, 1989.
64. Smid, J.K. and Meljer, H.C., EP 0215504 A1 to Stamicarbon BV, 1986.
65. Dillarstone, A. and Paye, M., *Cont. Derm.*, 30, 314, 1994.
66. Berger, R.S. and Bowman, J.P, *J. Tox. Cutaneous Occular*, 1, 109, 1982.
67. Patel, A.M., *et al.*, EP 0410567 A2 to Colgate-Palmolive Co., 1990.
68. Paye, M., in *Handbook of Detergents, Part A: Properties*, Surfactant Science Series, Vol. 82, Broze, G., Ed. Marcel Dekker, New York, 1999, pp. 470–479.
69. Gotte, E., *Chem. Phys. Surface Active Subst., Proc. Int. Congr.*, 4, 83, 1964.
70. Gotte, E., *Aesthet. Med.*, 10, 313, 1966.
71. Blake-Huskins, J.C., Scala, D., and Rhein, L.D., *J. Soc. Cosmet. Chem.*, 37, 199, 1986.
72. Putterman, G.J., Wolejsza, N.F., Wolfram, M.A., and Laden, K., *J. Soc. Cosmet. Chem.*, 28, 521, 1977.
73. Imokawa, G., Sumura, K., and Katsumi, M., *J. Am. Oil Chem. Soc.*, 52, 484, 1975.
74. Prottey, C., Oliver, D., and Coxon, A.C., *Int. J. Cosmet. Sci.*, 6, 263, 1984.
75. Naik, A.R. and Coxon, A.C., EP 0232153A2 to Unilever NV, 1987.
76. Tanaka, H., Horiuchi, Y., and Konishi, K., *Anal. Biochem.*, 66, 489, 1975.
77. Kanekiyo, T., Tanaka, N., and Sano, S., EP 0356784A2 to Mitsubishi Petrochemical Co., 1989.
78. Miyazawa, K., Ogawa, M., and Mitsui, T., *Int. J. Cosmet. Sci.*, 6, 33, 1984.
79. Sandhu, S., *et al.*, Review of Antimicrobial Agents for Use in Consumer Products, Colgate-Palmolive, 1998.
80. Engler, R., Disinfectants, *Official Methods of Analysis of the Association of Official Analytical Chemists*, Williams, S., Ed., Association of Official Analytical Chemists, Arlington, VA, 1984, p. 67.
81. Cremieux, A., Freney, J., and Davin-Regli, A., Methods of testing disinfectants, *Disinfection, Sterilization and Preservation*, Block, S.S., Ed., Lippincott Williams & Wilkins, New York, 2001, p. 1318.
82. Aronson, M.P., Larrauri, E.A., and Hussain, Z.J., EP 0013585 to Unilever Ltd, 1980.

83. Matson, T.P. and Berretz, M., *Soap Cosmet. Chem. Specialties*, 55, 41, 1979.
84. Drozd, J.C., *Chemical Times & Trends*, Jan., 49 and 57, 1985.
85. Drozd, J.C. and Gorman, W., *J. Am. Oil Chem. Soc.*, 65, 398, 1988.
86. Smith, D.L., *J. Am. Oil Chem. Soc.*, 68, 629, 1991.
87. Matheson, K.L. and Matson, T.P., *J. Am. Oil Chem. Soc.*, 60, 1693, 1983.
88. Quack, J.M. and Trautman, M., *Ann. Chim.*, 77, 245, 1987.
89. Jakubicki, G., Riska, G., Uray, A., and Nguyen, C., U.S. Patent 5,565,146 to Colgate-Palmolive Co., 1996.
90. Ofosu-Asante, K. and Smerznak, M.A., U.S. Patent 5,415,814 to Procter & Gamble Co., 1995.
91. Vinson, P.K., Oglesby, J.L., Scheibel, J.J., Scheper, W.M., Ofosu-Asante, K., Clarke, J.M., and Owens, R.N., U.S. Patent 5,990,065 to Procter & Gamble Co., 1999.
92. Vinson, P.K., Oglesby, J.L., Scheibel, J.J., Scheper, W.M., Kasturi, C., McKenzie, K.L., Ofosu-Asante, K., Clarke, J.M., and Owens, R.N., U.S. Patent 6,069,122 to Procter & Gamble Co., 2000.
93. Cohen, L., Vergara, R., Moreno, A., and Berna, J.L., *J. Am. Oil Chem. Soc.*, 72, 115, 1995.
94. Condon, B.D. and Matheson, K.L., *J. Am. Oil Chem. Soc.*, 71, 53, 1994.
95. Olsen, A.A., Hansen, L.B., and Beck, T.C., U.S. Patent 6,201,110 to Novo Nordisk A/S, 2001.
96. Regos, J. and Hitz, H.R., *Zbl. Bakt. Hyg. I. Abt. Orig.*, A226, 390, 1974.
97. Beaver, D., Roman, D.P., and Stoffel, P.J., *J. Am. Chem. Soc.*, 79, 1236, 1957.
98. Baichwal, R.S., Baxter, R.M., Kandel, S.I., and Walker, G.C., *Can. J. Biochemistry*, 38, 245, 1960.
99. Laske, Ch., *et al.*, *Comun. J. Com. Esp. Deterg.*, 29, 173, 1999.
100. Josephson, K.L., Rubino, J.R., and Pepper, I.L., *J. Appl. Microbiol.*, 83, 737, 1997.
101. Bloomfield, S.F. and Scott, E., *J. Appl. Microbiol.*, 83, 1, 1997.
102. Branna, T., *Household and Personal Products Industry*, 36, 75, Dec. 1999.
103. Hickley, J., *Household and Personal Products Industry*, 36, 95, Jan. 1999.
104. Walker, R., *New York Times Magazine*, Section 6, Feb. 29, 2004, p. 42.

# 8

# Heavy-Duty Liquid Detergents

**AMIT SACHDEV**   Research and Development, Global Technology, Colgate-Palmolive Company, Piscataway, New Jersey

**SANTHAN KRISHNAN**   Research and Development, GOJO Industries, Inc., Akron, Ohio

**JAN SHULMAN**   Research Laboratories, Rohm and Haas Company, Spring House, Pennsylvania

## I. INTRODUCTION

Heavy-duty liquid detergents (HDLDs) were introduced into the laundry market many years after the introduction of powder detergents. The first commercial heavy-duty liquid appeared in the U.S. in 1956. Liquid detergents were introduced in the Asia/Pacific region and Europe as recently as the 1970s and 1980s, respectively. A number of commercial heavy-duty liquids from the U.S., Europe, and Asia/Pacific are depicted in Figure 8.1.

HDLDs have several advantages when compared to powder detergents. Liquid detergents readily dissolve in warm or cold water, leaving no detergent residue on dark fabrics. They can be easily dispensed from the bottle or refill package, and their dispensing caps allow for the unused liquid to flow back into the container without spilling. In addition, liquids do not suffer from adverse effects after exposure to moisture (powders can "cake" in storage when exposed to high humidity). Furthermore, liquid detergents lend themselves to a pretreatment regimen at full strength by pouring directly on soils and stains, providing a convenient way to facilitate the removal of tough stains.

A typical heavy-duty liquid consists of all or some of the following components: surfactants, builders, enzymes, polymers, optical brighteners, and fragrance. In addition, it may contain other special ingredients designed for specific functions.

Both anionic and nonionic surfactants are used in the formulation of liquid detergents. Surfactants are primarily responsible for wetting the surfaces of fabrics as well as the soil (reducing surface and interfacial tension), helping to lift the stains off the fabric surface, and stabilizing dirt particles and/or emulsifying grease droplets [1–4]. The main anionic surfactants are sodium alkylbenzene sulfonates, alkyl sulfates, and alkylethoxylated sulfates. The nonionic surfactants used to formulate heavy-duty liquids are primarily ethoxylated fatty alcohols. Other surfactants are also used in HDLDs and are discussed in a subsequent section.

Builders are formulated into detergents mainly to sequester hardness ions ($Ca^{2+}$, $Mg^{2+}$) found in water, as well as to disperse the dirt and soil particulates in the wash water. Common builders used in liquid detergents are sodium and potassium polyphosphates (except in the U.S.), carbonates, aluminosilicates (zeolite A), silicates, citrates, and fatty acid soaps [5].

**FIG. 8.1**   Commercial North American (above) and European and Asia/Pacific HDLDs.

HDLDs usually incorporate a protease and an amylase enzyme. In addition, a premium liquid detergent may also utilize lipase and cellulase enzymes to enhance performance. The function of the protease enzyme is to digest protein soils such as blood and proteinaceous food stains, while the amylase selectively acts on starchy soils (e.g., gravy). Lipase attacks fatty chains in greasy soils and facilitates the breakdown of these soils during the wash cycle. Cellulase is an enzyme that acts on cellulose, and is used in detergents for removing pills from cotton fabrics, thereby restoring the reflectance of the fabric surface and making colors look brighter [6].

Polymers now play an increasingly important role in heavy-duty liquids. Low-molecular-weight, water-soluble polyacrylate dispersants prevent clay/particulate soils from redepositing on fabrics. Dye transfer inhibitors (polyvinylpyrrolidone) help keep fugitive dyes well dispersed in the aqueous bath, delivering part of the color care benefit often found in premium liquids. Soil release polymers facilitate the removal of oily/greasy soils from synthetic fabrics and blends. Deflocculating (hydrophobically modified) polycarboxylates have found utility in "coupling" structured liquids. Recent patents depict novel polymeric technologies for reducing wrinkling (ease of ironing) and preventing fiber abrasion/wear ("liquifiber"). Rheology modifiers have also made dramatic inroads in recent years. These thickening agents can do more than simply increase viscosity, including stabilization of duotropic systems and suspending actives or visual cues to improve consumer acceptance.

Optical brighteners are colorless fluorescent whitening agents that absorb ultraviolet radiation and emit bluish light, making fabrics look whiter and brighter to the human eye. Most detergents contain optical brighteners in their composition. Their content is adjusted more or less to reflect regional consumer preferences and marketing claims.

Liquid laundry detergents may be classified into two main types: unstructured liquids and structured liquids. Unstructured liquid detergents typically are isotropic, have a large and continuous water phase, and are the most widespread type of liquid detergent sold on the U.S. market. Structured liquid detergents are those consisting of multilamellar surfactant droplets suspended in a continuous water phase. These structured liquids are capable of suspending insoluble particles such as builders (phosphates, zeolites). These liquids have had some commercial utilization in Europe and in Asia/Pacific and were formerly sold in the U.S. in the early to mid-1990s. A third type of liquid detergent is one where the continuous phase is nonaqueous. These products have seen limited distribution throughout the world, but remain a topic of interest for many detergent manufacturers.

This chapter first describes the physical characteristics of heavy-duty liquids, which is followed by a detailed description of typical formulation components and their functions. This is followed by a brief discussion of evaluation methodologies. Finally, the emerging trends in the formulation of heavy-duty liquids are reviewed. A comprehensive listing of the patents relevant to HDLDs is given in the Appendix.

## II.  PHYSICAL CHARACTERISTICS OF HDLDs

The physical form and appearance of laundry liquids can vary greatly between different regions of the world. These variations in liquid types from region to region are largely dictated by the laundry habits and personal choices of the consumers in that particular market. HDLDs can be broadly classified into two main types: structured and unstructured liquids. A third category, nonaqueous liquids, has been actively studied and is discussed in this chapter.

Structured liquids are opaque and usually possess a moderate viscosity. These products are formed when surfactant molecules arrange themselves as liquid crystals [7–9]. This form of liquid detergent is largely marketed in Europe and the Asia/Pacific region. Unstructured liquids, on the other hand, are usually thin, clear or translucent, and are formed when all ingredients are solubilized in an aqueous media. Nonaqueous liquids, where the continuous medium consists of an organic solvent, can be either structured or unstructured.

## A. Structured Liquids

### 1. Introduction

The general tendency of liquids containing high levels of anionic surfactants and electrolytic builders is to form liquid crystalline surfactant phases [7–12]. This trend can be accelerated with the use of longer or branched-chain alkyl groups and by using a higher electrolyte level [13]. The resulting liquid is opaque, extremely thick, unpourable, and frequently physically unstable. It may also subsequently separate into two or more layers or phases: a thick, opaque surfactant-rich phase containing the flocculated liquid crystals and a thin, clear electrolyte-rich phase. The challenge, therefore, in developing such a liquid is to not only to prevent phase separation of the product but also to reduce the viscosity to a "pourable" level. A pourable level depends, of course, on the preferences, requirements, and convenience of the consumer. Viscosities of commercially available structured liquids vary from 500 to 9000 cP.

### 2. Lamellar Structures

The liquid crystalline phase in a structured liquid is frequently in the form of spherical lamellar bilayers or droplets [14–18]. The internal structure of these droplets is in the form of concentric alternating layers of surfactant and water. This configuration is often compared to the structure of an onion, which also has a similar concentric shell-like structure (Figure 8.2). It has been previously determined that the physical stability of these types of liquids is achieved only when the volume fraction of these bilayer structures is high enough to be space-filling. This corresponds to a volume fraction of approximately 0.6 [7,8,19]. An excessively high value of this volume fraction, however, will lead to flocculation, high viscosity, and an unstable product. A stable dispersion of the lamellar droplets makes it possible to suspend solids and undissolved particles between the lamellae and in the continuous electrolyte phase. This allows the use of relatively high builder/electrolyte levels [20]. Many patents have been issued for structured liquids that have the capability of suspending undissolved solids. The suspended solids include bleaches [21,22], builders such as zeolites [23], and softeners [24,25].

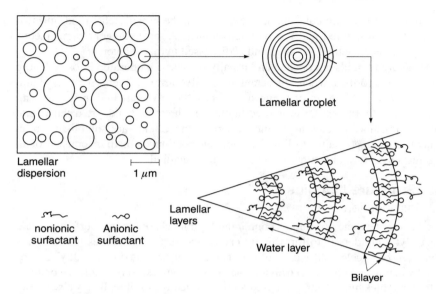

Lamellar droplet

Lamellar dispersion

1 μm

nonionic surfactant

Anionic surfactant

Lamellar layers

Water layer

Bilayer

**FIG. 8.2**  Schematics of a nonflocculated lamellar dispersion, a lamellar droplet, and the internal structure of a lamellar droplet. (Reproduced from Sein, A., Engberts, J.B.F.N., Vanderlinden, E., and van de Pas, J.C., *Langmuir*, 9, 1714, 1993. With permission.)

There are a number of factors that determine whether or not a lamellar droplet can form. As a general rule these bilayer structures will develop if the surfactant head group is smaller than twice the trans cross-sectional area of the alkyl chains of the surfactants [8,13]. This ratio of the areas of the alkyl chain and the surfactant head group is referred to as the packing factor of the surfactant system. Among the factors that can accelerate the formation of these structures is the use of longer alkyl chains, branched alkyl groups, dialkyl groups, and higher levels of electrolytes. Conversely, by using short, straight-chain alkyl groups, lower electrolyte levels, or hydrotropes, the onset of the liquid crystalline phase can be delayed.

A lamellar droplet is held together by an intricate balance of various inter- and intradroplet forces [10,11]. Any alteration or imbalance in these forces can have a direct impact on the stability of the structured liquid. Electrostatic repulsion between the charged head groups of anionic surfactants is compensated for by attractive van der Waals forces between the hydrophobic alkyl chains of the anionic and nonionic surfactants. In addition, there are also osmotic and steric forces between the hydrated head groups of nonionic surfactants. These particular interactions can be either attractive or repulsive depending on the "quality" of the solvent [8]. The resultant force has a direct influence on the size of the water layers, the size of the droplet, and eventually the stability of the liquid.

## 3. Stability of Structured Liquids

The balance of attractive forces between the surfactant layers and the compressive/repulsive forces due to steric/osmotic interactions makes highly concentrated formulations possible. However, a single-phase structured liquid, by its very nature, is never in a state of complete equilibrium. For practical purposes a stable structured liquid is achieved when the inter- and intralamellar forces are manipulated in such a way that phase separation is minimized or avoided. Depending upon the extent of concentration of the ingredients, various methods can be employed to stabilize these structured liquids (Figure 8.3).

The most basic means of stabilization and viscosity reduction is by the addition of electrolytes. The addition of cations in the form of electrolytes such as sodium citrate has the effect of screening out some of the repulsive forces between the negatively charged anionic head groups. Also, the electrolytes in the continuous layer provide an element of stability by giving it ionic strength. This screening out process reduces the size of the intralamellar water layer and consequently the size of the entire droplet. This reduction of the lamellae size frees up some

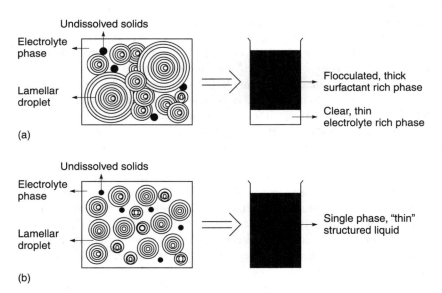

**FIG. 8.3** Schematics depicting the stability of (a) unstable and (b) stable structured HDLDs.

extra volume in the continuous phase and therefore provides an additional element of stability. Increasing the amounts of citrate works only up to a certain point beyond which there is a greater amount of undissolved salt which will be suspended *between* the lamellar droplets and can lead to excessive thickening. Another consequence of adding large amounts of electrolyte is the further erosion of the intralamellar water layer. This water layer has to be maintained at a level that is sufficient to hydrate the head groups of the nonionic surfactants. Salting-out electrolytes [26], of which sodium citrate is an example, also hydrate and therefore compete with the nonionics and other ingredients for the water. Excessive shrinkage of the water layer can, therefore, result in product instability.

*(a)  Free Polymers.*  The addition of electrolytes assists in lowering viscosities and in stabilizing a structured liquid only up to certain degree [27–29]. Polyethylene glycol and polyacrylates are examples of "free" polymers. These polymers are nonstructuring, and consequently they do not have the capability of adsorbing onto the lamellar dispersions. Instead, they function by means of osmotic compression which results in a shrinkage of the lamellar droplet. The consequence of this reduction in the volume of the individual droplets is a higher void fraction in the liquid. The polymer can therefore be used only up to the point at which the optimum void fraction is achieved. Further increases in the free polymer concentration often lead to depletion flocculation, which is also accompanied by large increases in viscosity as well as phase separation.

Concentrating the structured liquid by merely forming thinner lamellar layers and increasing the volume fraction of the lamellae can have implications for the rheology and pourability of the product. The best pourability characteristics are obtained when the volume fraction of the lamellar phase is as low as possible and the size of the lamellae is relatively large. A compromise between these two pathways/strategies has to be achieved in order to formulate a stable, concentrated liquid with acceptable rheological traits. This task becomes increasingly difficult at even higher concentrations. With only a limited void fraction available, the lamellar droplets, even though they are reduced in size, begin flocculating.

*(b)  Deflocculating Polymers.*  Free polymers are effective in reducing the sizes of the lamellar dispersion and thereby imparting stability. However, at ever increasing concentrations of surfactants and builders, simply reducing the intralamellar water layer is not sufficient to prevent flocculation. The problem was successfully addressed by researchers at Unilever who were able to prevent flocculation by altering the interlamellar forces [13,19,30–33]. This was achieved by means of a deflocculating polymer which can be considered as bifunctional. These polymers consist of a hydrophilic backbone that is attached to a hydrophobic side chain. The hydrophilic component is fundamentally like a free polymer or copolymer in structure as well as function. The hydrophobic side

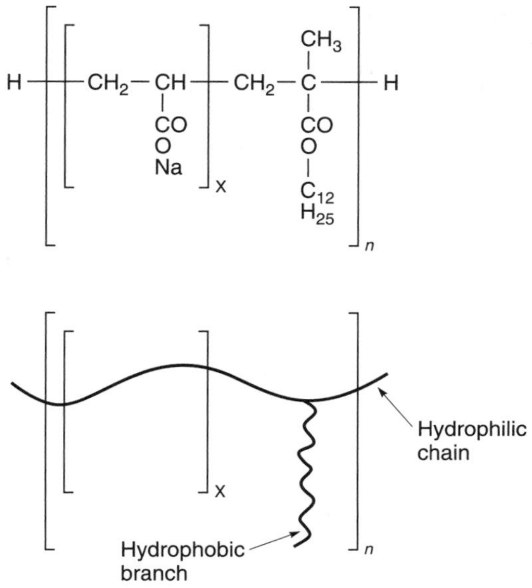

Acrylate–laurylmethacrylate copolymer

**FIG. 8.4** Example of a deflocculating polymer. (Reproduced from Broekhoff, J.C.P. and van de Pas, J.C., presented at American Oil Chemists Society Conference, Anaheim, CA, April 1993. With permission.)

chain is typically a long alkyl group. Figure 8.4 shows a schematic of deflocculating polymer — an acrylate–laurylmethacrylate copolymer. The unique aspect of this polymer is its ability to not only utilize its hydrophilic component to induce osmotic compression within the lamellar bilayers, but also to employ its hydrophobic side chain to adsorb onto the surfactant layers. This hydrophobicity also permits the deflocculating polymer to attach itself to the outer surface of the lamellar droplet and consequently be able to influence the interlamellar interactions. This trait prevents or at least reduces the likelihood of flocculation occurring.

The stability of these structured liquids, therefore, is obtained when the lamellae are not only smaller in size but also well separated (Figure 8.4). This results in not only a single-phase, stable liquid but also a product with good flow properties. Table 8.1 lists the ingredients typically found in a structured HDLD.

## B. Unstructured Liquids

### 1. Introduction

The current U.S. market for HDLDs is predominantly low-viscosity, clear, isotropic compositions. Besides the obvious differences in the physical appearance

**TABLE 8.1**  Example of a Structured HDLD Formulation

| Ingredient | Function | % |
|---|---|---|
| Sodium Linear Alkylbenzene Sulfonate | Anionic Surfactant | 0–30 |
| Sodium Alkyl Ether Sulfate | Anionic Surfactant | 0–10 |
| Alcohol Ethoxylate | Nonionic Surfactant | 0–10 |
| Sodium Carbonate | Builder | 0–25 |
| Zeolite | Builder | 0–25 |
| Sodium Perborate | Bleach | 0.0–10.0 |
| Polymer | Stabilizer | 0.0–1.0 |
| Protease | Enzyme | 0.0–1.5 |
| Fluorescent Whitening Agent | Brightener | 0.0–0.5 |
| Boric Acid | Enzyme Stabilizer | 0.0–5.0 |
| Preservative | | 0.05–0.2 |
| Fragrance | | 0.0–0.6 |
| Colorant | | 0.00–0.2 |

and properties between the structured and unstructured liquids, there are
other dissimilarities in the formulation of these liquids which can have a direct
impact on the cleaning performance of the product. Unstructured liquids are commonly formulated with higher amounts of surfactants in conjunction with lower
builder levels (see Table 8.2). This is in contrast to structured liquids, which utilize

**TABLE 8.2**  Example of an Unstructured HDLD Formulation

| Ingredient | Function | % |
|---|---|---|
| Sodium Linear Alkylbenzene Sulfonate | Anionic Surfactant | 0–15 |
| Sodium Alkyl Ether Sulfate | Anionic Surfactant | 0–15 |
| Alcohol Ethoxylate | Nonionic Surfactant | 0–15 |
| Sodium Citrate | Builder | 0–10 |
| Monoethanolamine | Buffer | 0–5 |
| Soap | Defoamer | 0–5 |
| Protease | Enzyme | 0.0–1.5 |
| Fluorescent Whitening Agent | Brightener | 0.0–0.5 |
| Boric Acid | Enzyme Stabilizer | 0.0–5.0 |
| Ethanol | Solvent | 0.0–5.0 |
| Sodium Xylene Sulfonate | Hydrotrope | 0.0–10.0 |
| Preservative | | 0.05–0.2 |
| Fragrance | | 0.0–0.6 |
| Colorant | | 0.0–0.2 |

more builders and electrolytes to sustain the structured phase. The physical appearance and stability of structured liquids are very dependent on surfactant ratios, whereas the clear, unstructured liquids allow far greater flexibility in choosing surfactant types/ratios as long as a single phase is maintained. The main advantage in structured liquids is their ability to suspend undissolved and insoluble solids. The unstructured clear liquids, on the other hand, by their very nature, typically do not permit the use of insoluble materials. This results in the use of only water-soluble builders at relatively low levels, and precludes the use of other useful builder ingredients such as zeolites. In the past few years there have been numerous research programs aimed at synthesizing modifiers with novel rheology that allow a formulator to suspend insoluble actives or visual cues in an unstructured (isotropic) liquid detergent matrix. These novel polymer chemistries are discussed in greater detail in Chapter 5.

It cannot be said that one form of liquid has a distinct advantage over the other. The formulation and marketing of either form may be dependent on such factors as efficacy targets, consumer preferences and habits, choice and availability of raw materials, as well as cost considerations.

## 2. Stability of Unstructured Liquids

Unlike structured liquids, these unstructured, low-viscosity, clear liquids can be developed only if the onset of the formation of liquid crystals is hindered or they are broken up. This can be accomplished by two different methods: by the addition of hydrotropes and solvents which can disrupt or prevent any liquid crystal formation as well as aid in solubilizing the other components in the formulation or by increasing the water solubility of the individual components. More than likely a combination of both these techniques is used to develop a stable liquid. The respective costs of these approaches ultimately determine their usage in the final formulation. Some of the methods used to formulate stable, single-phase, clear unstructured liquids are summarized below.

Compounds such as sodium xylene sulfonate (SXS), propylene glycol, and ethanol are useful in disrupting and preventing the formation of lamellar structures which can opacify and thicken a liquid. SXS is especially useful in solubilizing linear alkylbenzene sulfonate (LAS). Propylene glycol and ethanol also have the additional benefit of contributing to enzyme stability. The main drawback of using these compounds is that they do not contribute to the detergency performance of the product. Their principal function is to aid in achieving the low viscosity and clear appearance by solubilizing various ingredients and preventing precipitation/phase separation.

It is possible to form concentrated liquid detergents that do not require additional ingredients to assist in the maintenance of a clear appearance. This is usually accomplished by minimizing the use of LAS and electrolytes and maximizing the use of nonionic surfactants.

The use of ingredients with increased water solubility is probably the most effective tool for producing a single-phase, low-viscosity clear liquid. Potassium salts generally tend to be more soluble than their sodium cation counterparts. In these formulations, a higher level of potassium citrate (as opposed to sodium citrate) can be successfully incorporated. Detergency performance is not affected by replacing the $Na^+$ cation with $K^+$.

Citrate compounds are salting-out electrolytes — they tie up water molecules in the liquid and as a result help force the formation of liquid crystals or lamellar structures. It is sometimes possible to reverse this trend by the addition of "salting-in" electrolytes, compounds with high lyotropic numbers (>9.5) which can raise the cloud point of a liquid formulation [26]. This permits increased concentration without the onset of structuring.

Ethanolamines such as monoethanolamine (MEA) and triethanolamine (TEA) can also be invaluable in enhancing the solubility of ingredients. These compounds are bifunctional in that they have characteristics common to both alcohols and amines. As a result, salts of MEA and TEA are more soluble than those prepared with $Na^+$. Neutralizing sulfonic acid with MEA is a very effective way of freeing up additional water to allow for higher surfactant concentrations. In addition, any free alkanolamine that is not tied up as a salt behaves in a similar fashion to an alcohol and can aid in solubilizing other ingredients. These compounds also provide detergency benefits by buffering the wash water on the alkaline side.

## C.  Nonaqueous Liquids

Nonaqueous liquids may be classified as structured or unstructured depending on the level of surfactants and other components in their formulation [34]. These detergents have several advantages over aqueous formulations. Nonaqueous detergents can contain all the primary formulation components, including those that are not compatible with or difficult to formulate in aqueous systems. The liquid matrix is a nonionic surfactant or a mixture of nonionic surfactants and a polar solvent such as a glycol ether [35–38]. Builders such as phosphates, citrates, or silicates can be incorporated, although zeolites containing about 20% water are generally not recommended [39]. Phosphate-free formulations have also been reported [40]. Bleach systems such as TAED (tetraacetylethylene diamine) and activated sodium perborate monohydrate can be included in these formulations. Since these formulations do not contain water, enzymes may be added with minimal need for stabilizers. Softening ingredients can also be included [41,42].

Excellent flexibility in the concentration of the detergent can be attained since only the active cleaning ingredients are included in the formulation. The density of the finished product can be as high as 1.35 g/ml for these liquids, requiring lower dosages for equivalent cleaning. However, the two major challenges facing this technology are physical stability and dispensability of the product and its rapid solubilization in the washing machine.

## III. COMPONENTS OF HDLDs AND THEIR PROPERTIES

Heavy-duty liquid laundry formulations vary enormously depending upon the washing habits and practices of consumers in a given geographic region. The degree of complexity can range from formulations that contain minimal amounts of cleaning ingredients to highly sophisticated compositions consisting of superior surfactants, enzymes, builders, and polymers. This section describes the ingredients found in typical HDLD formulations.

### A. Surfactants

Surfactants are the major cleaning components of HDLD formulations throughout the world. Unlike powder detergents, physical and phase stability considerations greatly limit the usage of other cleaning ingredients, chiefly builders. Surfactants contribute to the stain removal process by increasing the wetting ability of the fabric surface and stains and by assisting in the dispersion and suspension of the removed soils.

A HDLD formulator has a vast array of surfactants from which to choose [43]. A comprehensive listing and description of these surfactants are beyond the scope of this discussion. The choice and levels of surfactants used in commercial HDLD products depend not only on their performance and physical stability characteristics but also on their cost effectiveness.

This section briefly describes the anionic and nonionic surfactants commonly used in commercial HDLD formulations. Cationic surfactants, although used on a large scale, are found predominantly in rinse-added fabric softener products. LAS, alcohol ethoxylates, and alkyl ether sulfates are three of the most widely used surfactants in liquid laundry detergents [44]. Recently, various external considerations, such as environmental pressures, have prompted manufacturers to change their surfactant mix to include newer natural-based surfactants [45–47], including alkyl polyglucosides (Henkel) [48].

### 1. Linear Alkylbenzene Sulfonate

The excellent cost–performance relationship of LAS makes it the dominant surfactant used in laundry detergents [49]. Recent trends in Europe and North America indicate a gradual reduction in its usage in HDLDs. Nevertheless, its use in laundry liquids globally is still substantial, especially in the developing regions of the world.

De Almeida *et al.* [50] and Matheson [51] provide a comprehensive examination of the processing, production, and use of linear alkylbenzene in the detergent

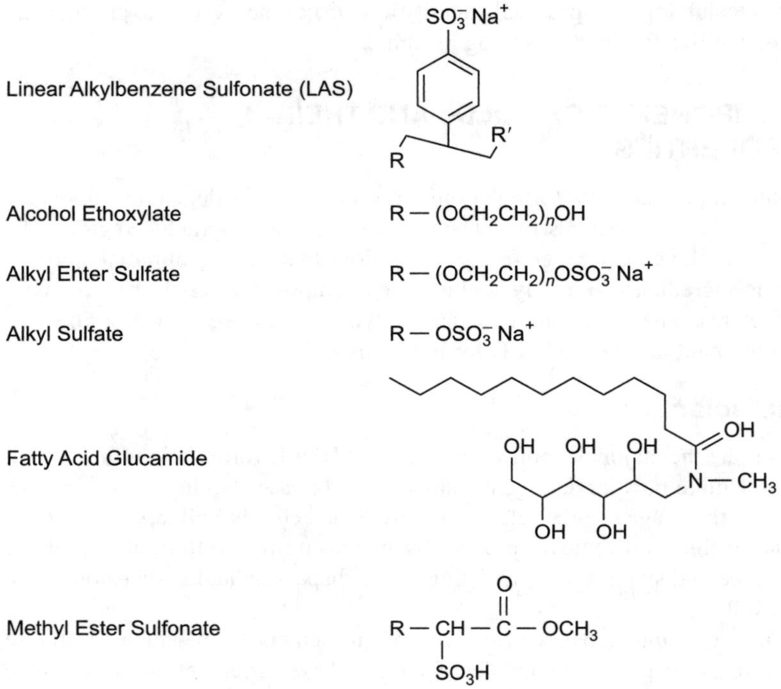

**FIG. 8.5**  Structures of typical HDLD surfactants.

industry. LAS are anionic surfactants and are prepared by sulfonating the alkyl-benzene alkylate and subsequently neutralizing it with caustic soda or any other suitable base. The alkylate group is typically a linear carbon chain of length ranging from C10 to C15, with a phenyl group attached to one of the secondary carbons on the alkyl chains (Figure 8.5). The alkylate portion of the molecule is hydrophobic whereas the sulfonate group provides the water solubility and the hydrophilicity. Most commercial alkylates are mixtures of various phenyl isomers and carbon chain homologs [52]. The position of the phenyl group depends on the manufacturing method. Systems using $AlCl_3$ or HF catalysts are the most common.

The length of the carbon chain and the isomeric distribution strongly influence the ease of formulation and performance of the surfactant. It has been determined that the surface activity of this surfactant increases with longer carbon chain lengths [53]. A longer alkyl chain increases the hydrophobicity of the molecule, lowers the critical micelle concentration (CMC), and generally provides better

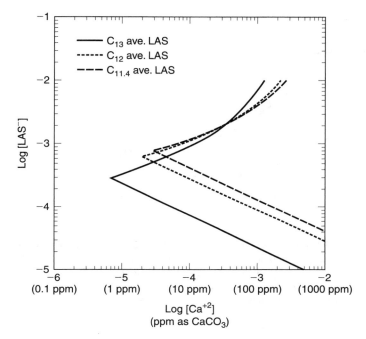

**FIG. 8.6** $Ca^{2+}$–LAS precipitation boundary diagrams. (Reproduced from Matheson, K.L., Cox, M.F., and Smith, D.L., *J. Am. Oil Chem. Soc.*, 62, 1391, 1985. With permission.)

soil removal characteristics [54–57]. LAS offers superior and very cost effective detergency performance, especially on particulate soils. However, due to its high sensitivity to water hardness, it is best utilized only when used with an accompanying builder [58]. Figure 8.6 shows the increased sensitivity to hardness ions for LAS with longer carbon chain lengths. Without the assistance of builders, the soil removal efficacy of LAS drops rapidly with increasing water hardness [4,59] (Figure 8.7).

The amount and type of LAS in HDLDs depends largely on the physical form of the laundry liquid — unstructured or structured. In unstructured liquids, solubility considerations require the use of smaller carbon chain lengths ($\sim$C11). In addition, the choice of cations can also enhance solubility. Potassium and amine cations such as MEA and TEA can be used instead of sodium ions to improve stability [60]. An increased ratio of the 2-phenyl isomer in the LAS can also increase solubility [61] and sometimes improve the hardness tolerance of the surfactant [62]. In structured liquids, a longer alkyl chain can be more desirable for the formation

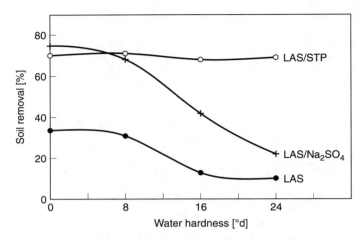

**FIG. 8.7**    Soil removal data for LAS as a function of water hardness. Results are shown for surfactant with builder (sodium tripolyphosphate, STPP) and electrolyte. (Reproduced from Coons, D., Dankowski, M., Diehl, M., Jakobi, G., Kuzel, P., Sung. E., and Trabitzsch, U., in *Surfactants in Consumer Products: Theory, Technology and Application*, Falbe, J., Ed., Springer-Verlag, New York, 1987. With permission.)

of surfactant lamellae. The choice of the counterion can also affect stability since ions such as $Na^+$ and $K^+$ have different electrolytic strengths which can also have an impact on phase stability.

A disadvantage of using LAS in HDLDs is their detrimental effect on enzymes. With the increasing use of enzymes it becomes necessary to devote a sizable portion of the formulation space and cost to enzyme stabilization. Alternative approaches using surfactants that are more compatible with enzymes can be employed.

## 2.   Alcohol Ethoxylates

Figure 8.5 shows the general structure of a nonionic alcohol ethoxylate surfactant. Its hydrophobic group is linear with the carbon chain length typically ranging from C10 to C15. The hydrophilic ethoxylate group can vary in size from an average of 5 to 12 moles of ethylene oxide [63–65]. Alcohol ethoxylates are marketed commercially under the trade names Neodol (Shell Chemical Co.), Bio-Soft (Stepan), Genapol (Clariant), Tergitol (Dow), Surfonic (Huntsman), and Alfonic (Sasol). The feedstock for the alcohol can be derived from natural coconut oil sources as well as from petroleum feedstock. These surfactants are usually sold at a 100% actives concentration and range in state from fluid liquids to soft solids.

Alcohol ethoxylate usage in HDLDs depends on the type or the physical form of the liquid detergent. The high aqueous solubility of alcohol ethoxylates makes them a useful ingredient in unstructured liquids. This solubility can be further enhanced by increasing the degree of ethoxylation and decreasing the carbon chain length. However, these modifications can sometimes have negative ramifications on the cleaning performance. The choice of carbon chain length and the degree of ethoxylation depends on the physical stability and cleaning requirements of individual formulations. Structured liquids, in contrast, can only tolerate a limited amount of the nonionic alcohol ethoxylate surfactant since the stability of these liquids is dependent upon the optimum distribution of the size and packing configuration of lamellar droplets. Excessive use of nonionic surfactants can disturb this somewhat delicate equilibrium and cause phase separation of the HDLD.

Nonionic surfactants like alcohol ethoxylates demonstrate superior tolerance to hard water ions. This characteristic is especially useful in unstructured HDLD formulations because solubility constraints limit the amount of builder that can be incorporated. They also provide excellent cleaning benefits and are commonly used in conjunction with LAS in HDLD formulations [57,66]. Studies have shown that in LAS-containing products, alcohol ethoxylates can lower the critical micelle concentration (Figure 8.8) as well as provide improvements in the detergency [66]. Superior cleaning is observed, especially on oily soils such as sebum (body sweat) on polyester fabrics [67]. The presence of alcohol ethoxylates

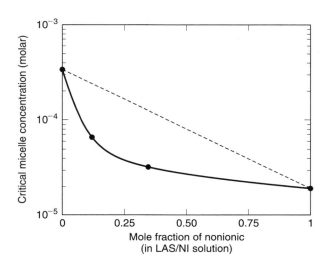

**FIG. 8.8** Critical micelle concentration (CMC) as a function of nonionic surfactant content in a LAS/NI solution. (Reproduced from Cox, M.F., Borys, N.F., and Matson, T.P., *J. Am. Oil Chem. Soc.*, 62, 1139, 1985. With permission.)

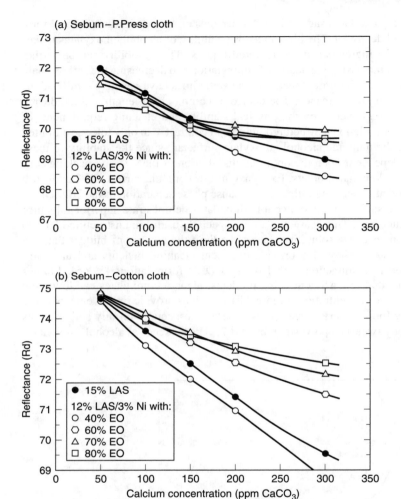

**FIG. 8.9**  Detergency performance at 100°F (38°C) of LAS and LAS/alcohol ethoxylate blends. The formulation also contained 25% sodium tripolyphosphate, 10% silicate, and 35% sodium sulfate. (Reproduced from Cox, M.F., Borys, N.F., and Matson, T.P., *J. Am. Oil Chem. Soc.*, 62, 1139, 1985. With permission.)

in an LAS-containing formulation is found to improve detergency, especially at higher hardness levels (Figure 8.9). Improvements have also been detected when narrow distribution ethoxylate surfactants (Figure 8.10) are used [68].

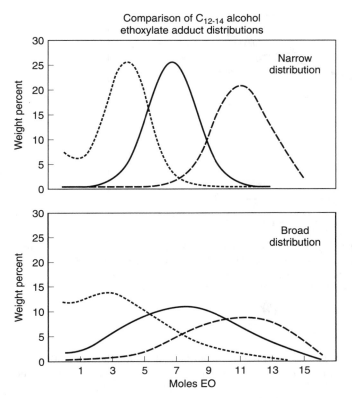

**FIG. 8.10** Typical ethoxylate adduct distributions in narrow range and broad range C12–C14 alcohol surfactants with similar cloud points. (Reproduced from Dillan, K.W., *J. Am. Oil Chem. Soc.*, 62, 1144, 1985. With permission.)

The insensitivity of nonionic surfactants to calcium ions also provides a very important benefit in the stabilization of enzymes (see Section III.C). It has been shown that these surfactants are not as detrimental to the preservation of enzymes in HDLDs as sodium LAS. With increasing reliance on the use of enzymes in the laundry cleaning process, nonionic surfactants like alcohol ethoxylates play an important role in enhancing enzyme stability.

## 3. Alkyl Ether Sulfates (AEOS)

These are also anionic surfactants which are manufactured by sulfating alcohol ethoxylate surfactants [69]. Figure 8.5 shows the structure of the molecule which consists of the alcohol ethoxylate connected to a sulfate group. The EO groups typically range in size from 1 to 3 moles.

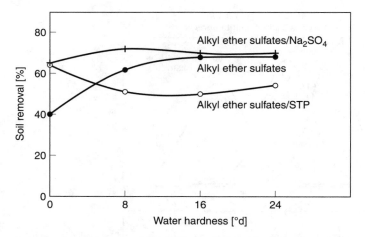

**FIG. 8.11**   Data showing the hardness tolerance of alkyl ether sulfate surfactants. (Reproduced from Coons, D., Dankowski, M., Diehl, M., Jakobi, G., Kuzel, P., Sung. E., and Trabitzsch, U., in *Surfactants in Consumer Products: Theory, Technology and Application*, Falbe, J., Ed., Springer-Verlag, New York, 1987. With permission.)

These surfactants provide numerous benefits that make them an attractive option to HDLD formulators. They are commonly used in both structured and unstructured liquids. Their high water solubility makes it possible to use a wide range of levels in unstructured liquids. They can also be successfully incorporated in structured liquids.

Unlike sodium LAS, alcohol ether sulfates are more tolerant to hardness ions and as a result do not require an accompanying high level of builder in the formulation. Figure 8.11 shows the relative insensitivity of AEOS to hardness ions. Small amounts of AEOS when added to LAS were found to improve the interfacial properties. They are more compatible with enzymes, which can also reduce the cost of enzyme stabilizers in the formulation. They are also milder to the skin and, as a result, are frequently used in hand dishwashing applications. The superior detergency performance of this surfactant is demonstrated by its superior efficacy in most stain categories.

## 4.  Alkyl Sulfates

These anionic surfactants (Figure 8.5) are used primarily in Europe as a substitute for LAS [45]. Environmental considerations have prompted manufacturers to use surfactants of this type, which can be derived from oleochemical sources. The carbon chain length can range from C10 to C18. Tallow alcohol sulfate is the common form used in HDLDs. It provides excellent detergency, good foaming properties, and favorable solubility characteristics.

## 5.  Polyhydroxy Fatty Acid Amides (Glucamides)

Polyhydroxy fatty acid amides (Figure 8.5) are currently used in light- and heavy-duty laundry liquids. Recent advances in the technology for the manufacture of these surfactants has made their use economically feasible [70–72]. The use of natural or renewable raw materials improves their biodegradation characteristics. Several patents have been filed for detergent formulations containing glucamides which claim superiority in cleaning efficacy for oily/greasy and enzyme-sensitive stains [73–76]. Synergies with other anionic and nonionic surfactants have also been reported [75,76]. Their improved skin mildness qualities can be useful in light-duty liquid applications [77]. Enzyme stabilization characteristics in glucamide formulations are also enhanced relative to LAS-containing HDLDs.

## 6.  Methyl Ester Sulfonates

This anionic surfactant (Figure 8.5) is also derived from oleochemical sources and has a good biodegradability profile. It is currently used in only a limited number of markets (primarily in Japan), but has recently gained some prominence in the U.S. market. Its good hardness tolerance characteristics (Figure 8.12) and its ability to also function as a hydrotrope make this surfactant a good candidate

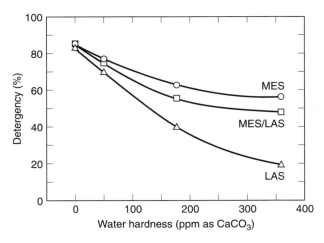

**FIG. 8.12**  Detergency as a function of water hardness in methyl ester sulfonate/LAS formulations. Conditions: 25°C, surfactant 270 ppm, $Na_2CO_3$ 135 ppm, silicate 135 ppm. (Reproduced from Satsuki, T., in *Proceedings of the 3rd World Conference on Detergents: Global Perspectives*, Cahn, A., Ed., AOCS Press, Champaign, IL, 1994, p. 135. With permission.)

for liquid detergents [78]. It has also been found to be a good cosurfactant for LAS-containing formulations. It can only be used in products with low alkalinity due to the likelihood of hydrolytic cleavage of the ester linkage under high pH conditions.

## 7.  Other Surfactants

Once used as a major surfactant in detergent formulations, soap is now used only as a minor ingredient in HDLDs. Its function is primarily to provide foam control in the washing machine. European liquid formulations contain higher soap levels than their counterparts in North America because of increased foaming tendencies in European machines. Soap also aids in the cleaning process by chelating divalent cations (such as calcium and magnesium). When employed at high concentrations it behaves as a precipitating builder and can leave behind an organic complex (of soap scum) on fabric surfaces. A variety of other surfactants are also used primarily for specialty applications [79]. They include amine oxides, amphoterics, and betaines.

## B.  Builders

The primary function of builders in the detergency process is to tie up the hardness ions, $Ca^{2+}$ and $Mg^{2+}$, which are naturally found in water. They also provide other valuable benefits including maintaining the alkalinity of the wash solution, functioning as antiredeposition and soil dispersing agents and, in some cases, as corrosion inhibitors [80–84].

The level of builder usage in liquid formulations depends largely on three main criteria: (1) the aqueous solubility of the builder, (2) the physical form of the liquid, and (3) the cost effectiveness of the ingredient. Due to inherent solubility constraints in formulating stable liquid detergents, the usage level of builders in HDLDs is significantly lower than in granulated detergents. This is especially true in the case of unstructured liquids where the solubility limitations of the builder largely dictate its level in the formulation. In structured liquids, however, a certain amount of electrolytic builder is necessary to induce structuring which allows the incorporation of significantly higher amounts of builder. Insoluble builders can also be added by suspending them in the liquid. Builder ingredients such as zeolites, phosphates, silicates, or carbonates can account for 20% or more of the total formulation.

## 1.  Mechanisms

Builder compounds decrease the concentration of the wash water hardness by forming either soluble or insoluble complexes with calcium and magnesium ions. The mechanisms by which these ingredients function can be broadly classified into three classes: (1) sequestration, (2) precipitation, and (3) ion exchange. All three methods have the ultimate effect of lowering the concentration of hardness ions that could interfere with the cleaning process by rendering the surfactants less effective.

In sequestration (chelation) the hardness ions are bound to the builder in the form of soluble complexes. Phosphates, citrates, and nitrilotriacetic acid (NTA) are examples of this class of builder compound. Table 8.3 lists the calcium binding capacities of various builders. Other strongly chelating compounds exist, such as phosphonates and EDTA, but they are generally not extensively used in HDLDs. The most efficient builder is sodium tripolyphosphate (STPP). Unfortunately, tripolyphosphate has been identified as a possible cause of eutrophication in lakes and rivers. It is severely controlled and even banned in several countries. As a result, most countries in North America and Europe have converted to nonphosphate formulations. Other regions are also gradually imposing restrictions on the use of phosphates.

Carbonates are examples of builders that precipitate out the calcium ions in the form of calcium carbonate. Precipitating builders can leave behind insoluble deposits on clothes and washing machine components. Aluminosilicates such as zeolites are ion exchange compounds; they remove (predominantly) calcium and magnesium ions and exchange them with sodium ions.

Most builders also contribute significantly to detergency by providing alkalinity to the wash water. A high pH ($>10$) solution aids in the removal of oily soils such as sebum by saponification. The insoluble fatty acids found in oily soils are converted to soluble soaps under alkaline conditions, facilitating their removal during the washing process.

## 2. Builder Classes

*(a) Inorganic.* In regions where phosphorous compounds are still permitted in detergent products, polyphosphates such as tripolyphosphates ($P_3O_{10}$) and pyrophosphates ($P_2O_7$) are unsurpassed in their cost effectiveness and cleaning ability. These ingredients are not only very good chelating agents but they also provide a soil suspending benefit. Stains, once removed from the fabric, can be suspended in the wash water by electrostatic repulsion, thereby preventing soils from redepositing onto clothing. To a certain extent phosphates also buffer the wash water. The solubility of tripolyphosphates can be enhanced by using the potassium salt. This would be more appropriate for unstructured liquids. In structured liquids, the sodium salt can be incorporated and at much higher levels.

Carbonate compounds offer an economical means of reducing the calcium content and also raising the alkalinity of the wash water. They lower the concentration of the calcium by precipitating it in the form of calcium carbonate. This could lead to fabric damage in the form of encrustation, which becomes especially apparent after repeated washing cycles under high water hardness conditions. Fortunately, this is not a major problem in unstructured HDLDs since the amount of carbonate used in the formulation is limited due to solubility restrictions. Compounds such

**TABLE 8.3** Sequestration Capacity of Selected Builders. Table Reproduced with permission from Jakobi, G. and Schwuger, M.J., *Chem. Z.*, 182, 1975.

| Structure | Chemical name | Calcium binding capacity (mg CaO/g) | |
|---|---|---|---|
| | | 20°C | 90°C |
| NaO–P–O–P–ONa with O double bonds and ONa, ONa groups | Sodium diphosphate | 114 | 28 |
| NaO–P–O–P–O–P–ONa with O double bonds and ONa, ONa, ONa groups | Sodium triphosphate | 158 | 113 |
| HO–P–C–P–OH with OH, CH₃, OH, O groups | 1-Hydroxyethane-1,1-diphosphonic acid | 394 | 378 |
| N(CH₂–PO₃H₂)₃ | Amino tris methylenephosphonic acid | 224 | 224 |

| Structure | Name | | |
|---|---|---|---|
| $$\begin{array}{c} O \\ \| \\ CH_2-C-OH \\ \| \\ N-CH_2-C-OH \\ \| \quad \| \\ CH_2 \quad O \\ \| \\ C-OH \\ \| \\ O \end{array}$$ | Nitrilotriacetic acid | 285 | 202 |
| $$\begin{array}{c} O \\ \| \\ CH_2-C-OH \\ \| \\ N-CH_2-C-OH \\ \| \quad \| \\ CH_2-CH_2-OH \quad O \end{array}$$ | N-(2-Hydroxyethyl)imino diacetic acid | 145 | 91 |
| $$\begin{array}{c} O \quad CH_2-C-OH \\ \| \quad \| \\ HO-C-H_2C \quad \quad O \\ N-(CH_2)_2-N \\ HO-C_2-H_2C \quad \quad CH_2-C-OH \\ \| \quad \quad \| \\ O \quad \quad O \end{array}$$ | Ethylenediamine tetraacetic acid | 219 | 154 |
| $$\begin{array}{c} HO \quad O \\ \backslash \| \\ C-OH \quad H \\ \| \quad \| \\ H \quad C-OH \\ \| \quad \| \\ O=C \quad O \\ \| \quad H \\ HO-C \quad H \\ \| \\ O \end{array}$$ | 1,2,3,4-Cyclopentane tetracarboxylic acid | 280 | 235 |

*(continued)*

**TABLE 8.3**  (Contd.)

| Structure | Chemical name | Calcium binding capacity (mg CaO/g) | |
|---|---|---|---|
| | | 20°C | 90°C |
| $CH_2$—$C$—$CH_2$ ; $C$=$O$ / HO ; $C$=$O$ / $C$—OH ; $C$—OH ; $C$=$O$ | Citric acid | 195 | 30 |
| O=$C$—OH ; $CH$—O—$CH_2$—$C$—OH (O) ; $C$=$O$ / $C$—OH | $O$-Carboxymethyl tartronic acid | 247 | 123 |
| O=$C$—$CH_2$—$CH$—$C$—OH (O) ; O—$CH_2$—$C$=$O$ / HO / OH | Carboxymethyl oxysuccinic acid | 368 | 54 |

as sesquicarbonates and bicarbonates that are less likely to lead to the formation of calcium carbonate precipitates have better solubility characteristics and can be used to a larger extent in unstructured liquids. Structured liquids offer the potential of incorporating much higher amounts of these compounds. Carbonates are also good wash water buffers and can provide the alkalinity needed for improved efficacy.

Another class of ingredient that is effective at providing alkalinity is sodium silicates [86]. Although they can also be good sequestrants, and are used as such in powder formulations, they provide this benefit only at higher concentrations. Once again, solubility restrictions prevent the incorporation of any substantial amounts in unstructured liquids. At the low levels at which they can be used, they are valuable as alkaline buffers. The use of sodium silicates in HDLDs is typically limited to the liquid silicates which have $SiO_2/Na_2O$ ratios from 3.2 to 1.8. Clariant recently introduced SKS-6, a pure sodium disilicate ($Na_2Si_2O_5$) that softens tap water more efficiently than amorphous sodium silicates [87]. Although initially positioned for powder detergents (laundry and automatic dishwashing), its performance profile may entice a few manufacturers to look at potential utilization in liquid formulations.

Aluminosilicates [$M_z(zAlO_2:ySiO_2)$] are another type of builder of which zeolite A is a common example [88]. Zeolite A is a sodium aluminosilicate, with an Al/Si ratio of 1:1 and a formula of $Na_{12}(SiO_2 \times AlO_2)_{12} \times 27H_2O$. It acts as a builder by exchanging sodium ions inside the lattice with calcium ions from the wash water. Zeolites are not effective in providing alkalinity and are normally used in conjunction with carbonates. They are insoluble in water and are not suitable for formulating unstructured liquids. In structured liquids, zeolites are suspended as solid particles.

*(b)   Organic.*   The restrictions placed on the use of phosphate compounds in detergent formulations have led to a variety of organic compounds that could function as builders but which also must be readily biodegradable. Although some of these compounds do approach the sequestration level of phosphates, they are not as cost effective [90].

Various polycarboxylate compounds, those with at least three carboxylate groups, have now become widely used as replacements for phosphates as the builder component of HDLDs. In liquid detergent formulations citrate compounds have become commonplace. Although their chelating ability is relatively low (Figure 8.13), citrate is used in HDLDs for a variety of reasons. Citrate's high aqueous solubility makes it useful in unstructured liquids, whereas in structured liquids its high electrolytic strength can aid in salting out and stabilizing the formulation. In addition, it is also used in enzyme-containing formulations where the maintenance of the pH at less than 9.0 is crucial to the stability of the enzyme.

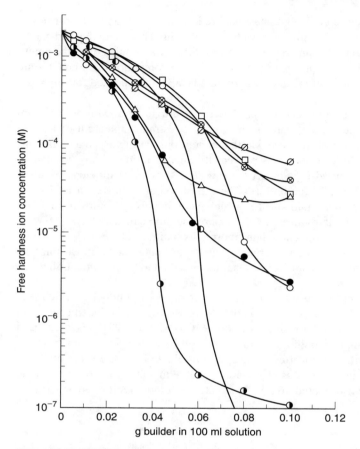

**FIG. 8.13** Sequestration of water hardness ions by detergent builders. Sodium polyacrylate $M_w = 170,000$, STPP (sodium tripolyphosphate), NTA (nitrilotriacetic acid), EDTA (ethylenediamine tetraacetic acid), sodium citrate, CMOS (sodium carboxymethoxysuccinate), sodium carbonate, zeolite A. (Reproduced from Nagarajan, M.K., *J. Am. Oil Chem. Soc.*, 62, 949, 1985. With permission.)

Citric acid itself has also been patented as an ingredient in protease stabilization systems [91].

Ether polycarboxylates have been found to provide improvements over the calcium and magnesium chelating ability of citrates. In a series of patents assigned to Procter & Gamble, it has been claimed that a combination of tartrate monosuccinates and tartrate disuccinates (Figure 8.14) delivers excellent chelating

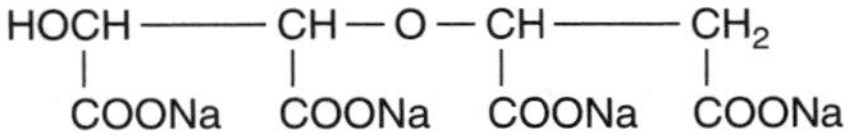

HOCH———CH—O—CH———CH$_2$
⎪      ⎪      ⎪      ⎪
COONa   COONa   COONa   COONa

Sodium Tartrate Monosuccinate (TMS)

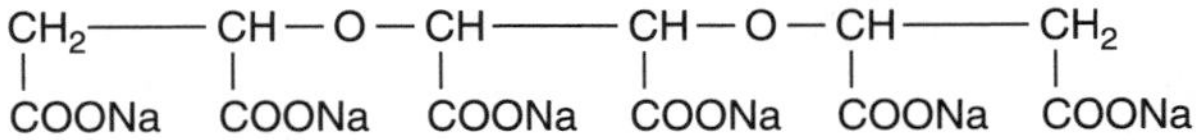

CH$_2$———CH—O—CH———CH—O—CH———CH$_2$
⎪    ⎪    ⎪    ⎪    ⎪    ⎪
COONa   COONa   COONa   COONa   COONa   COONa

Sodium Tartrate Disuccinate (TDS)

**FIG. 8.14** Ether polycarboxylate builders.

performance [92–94]. Data represented in Figure 8.15 indicate a high calcium binding capacity.

Salts of polyacetic acids, e.g., ethylenediamine tetraacetic acid (EDTA) and nitrilotriacetic acid (NTA), have long been known to be very effective chelating agents [95]. The chelating ability of NTA has been found to be comparable to that

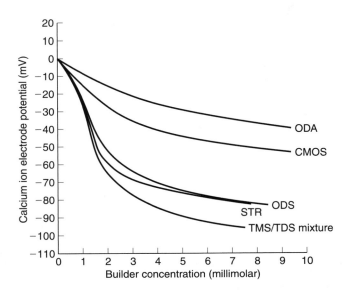

**FIG. 8.15** Effect of builder level on calcium ion concentration. CMOS, sodium carboxymethoxysuccinate; ODS, sodium oxydisuccinate; STP, sodium tripolyphosphate; TMS, tartrate monosuccinate; TDS, tartrate disuccinate. (From Bush, R.D., U.S. Patent 4566984 to Procter & Gamble Co., 1986.)

of STPP. Unfortunately, questions regarding the toxicity of this compound have all but prevented any large-scale use in HDLDs. Currently, NTA usage is primarily limited to a few powder and liquid detergent formulations in Canada. The high chelation power of EDTA has been used in compositions where metal impurities of iron and copper can be detrimental to the product stability, for example, in peroxygen bleach-containing liquids.

Polymeric polyelectrolytes have also found applications as alternative builder ingredients [96,97]. High-molecular-weight polyacrylate homopolymers and acrylic–maleic copolymers can be very effective in tying up calcium ions in the

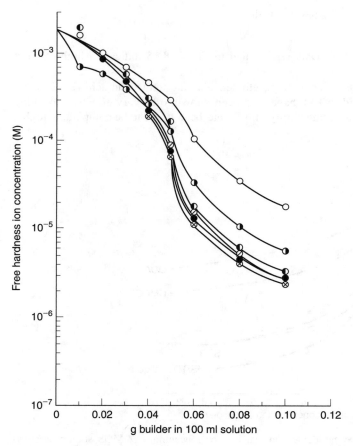

**FIG. 8.16** Sequestration of water hardness ions by sodium polyacrylate polymers. $M_W = 2,100, \ 5,100, \ 20,000, \ 60,000, \ 170,000, \ 240,000$. (Reproduced from Nagarajan, M.K., *J. Am. Oil Chem. Soc.*, 62, 949, 1985. With permission.)

wash bath (Figure 8.16). However, concerns about their biodegradability and limited compatibility in most surfactant matrices have significantly limited their use in liquid formulations. The lower molecular weight polymer analogues can also aid in soil dispersancy and clay/particulate soil antiredeposition. In products containing moderate to high levels of carbonate, these polymers disrupt the nucleating calcium carbonate crystal structure, preventing the deposition of inorganic precipitates on fabrics.

Fatty acids, such as oleic and coco fatty acid (saturation level) can serve a multifunctional role when added to HDLDs. Although they primarily provide a foam suppression capability, they can also precipitate out some of the calcium ions in the wash by forming calcium soap. This could, however, pose a problem, since soap scum, commonly known as lime soap, is insoluble and may have an impact on the overall cleaning result.

## C. Enzymes

Enzymes have become integral components of most liquid detergent compositions as they continue to play an increasing role in the stain removal process. This has come about due to many recent advances in enzyme technology and has resulted in more efficient and effective strains. The ability of these enzymes to target specific classes of stains can provide the formulator with the flexibility to tailor the development of products for consumers with different requirements and preferences. In addition, enzymes are especially effective when the liquid detergent is used as a prespotter.

There are four types of enzymes currently used in HDLDs: protease, lipase, cellulase, and amylase [6,98]. They are all proteins and are derived from various living organisms. Their role is to catalyze the hydrolysis of large biological molecules into smaller units which are more soluble and as a result are washed away relatively easily. The optimum conditions for the functioning of these enzymes depend on individual strains or types. Generally, the rates for enzymatic reactions rise with increasing temperatures, and are usually optimum within an alkaline pH range of 9 to 11.

Protease is by far the most widely used of all detergent enzymes. Introduced in the 1960s, it has since become one of the more important components of detergent formulations [6]. Proteases aid in the removal of many soils commonly encountered by the consumer, such as food stains (cocoa, egg yolk, meat), blood, and grass. This enzyme hydrolyzes or breaks up the peptide bonds found in proteins resulting in the formation of smaller and more soluble polypeptides and amino acids. Since most enzymes have to function under high pH conditions, subtilisin, a bacterial alkaline protease, is commonly used in laundry detergents. This particular protease does not hydrolyze any specific peptide bond in proteinaceous stains but cleaves bonds in a somewhat random manner.

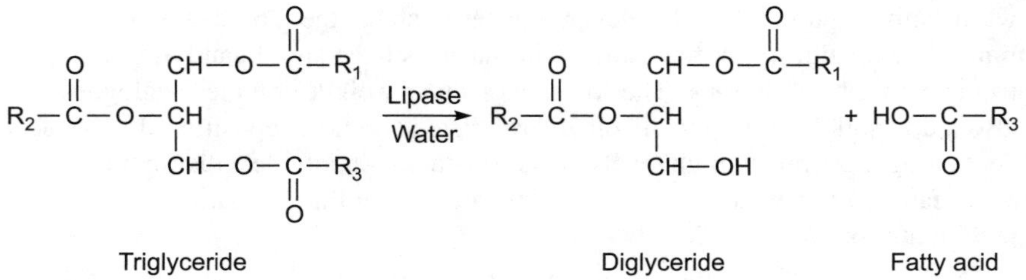

**FIG. 8.17**   Lipase-catalyzed conversion of insoluble oily (triglyceride) soils.

Amylase enzymes also work on food stains containing starches, such as rice, spaghetti sauce, potatoes, oatmeal, and gravy. These enzymes hydrolyze the 1–4 glucosidic bonds in starch, which leads to the formation of smaller water-soluble molecules. α-Amylase randomly hydrolyzes the bonds in the starch polymer to form dextrin molecules. β-Amylase, in contrast, cleaves the maltose units that are situated at the end of the starch polymer.

The use of lipase in detergents is a relatively recent occurrence. The first commercial detergent lipase was introduced in 1988 [6,98]. These enzymes target oily/greasy soils encountered with body sweat/collar soils, foods (butter, tallow, and sauces), and select cosmetics (lipstick, mascara) which are typically some of the most difficult stains to remove. The major components of most oily stains encountered in households are triglycerides. Lipases catalyze the hydrolysis of mostly the $C_1$ and $C_3$ bonds in the triglyceride molecule yielding soluble free fatty acids and diglyceride (Figure 8.17). In practice, it has been determined that lipases work best subsequent to the first wash (Figure 8.18). It is believed that the temperatures encountered in a typical drying process are needed to activate the enzyme. Although most oily stains can also be cleaned using traditional surfactant methods, the main benefit of lipases is their ability to perform at relatively low concentrations and low temperatures.

With greater emphasis being given to the care of fabrics, cellulase enzymes have become increasingly important in detergent products [98]. Repeated washing often leads to cotton fabrics looking faded and worn. This appearance is attributed to the damaged cellulose microfibrils on the fabric surface. Cellulase enzymes are able to hydrolyze the β(1–4) bonds along the cellulose polymer, resulting in smaller units which are carried away in the wash (Figure 8.19). The removal of these damaged microfibrils or "pills" gives the clothing a less faded appearance (Figure 8.20).

## 1.   Enzyme Stabilization

Enzymes are highly susceptible to degradation in heavy-duty laundry liquids. With the increasing emphasis on the use of enzymes as cleaning agents, it becomes all

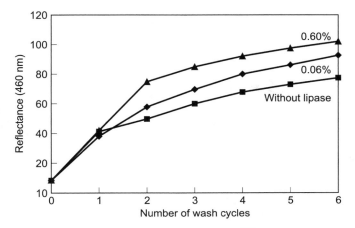

**FIG. 8.18**   Effect of lipase enzyme (Lipolase™) on lard/sudan red stains as a function of the number of wash cycles. Conditions: powder detergent, temperature = 30°C, ter-gotometer, pH = 9.7. (Reproduced from Gormsen, E., Roshholm, P., and Lykke, M., in *Proceedings of the 3rd World Conference on Detergents: Global Perspectives*, Cahn, A., Ed., AOCS Press, Champaign, IL, 1994, p. 198. With permission.)

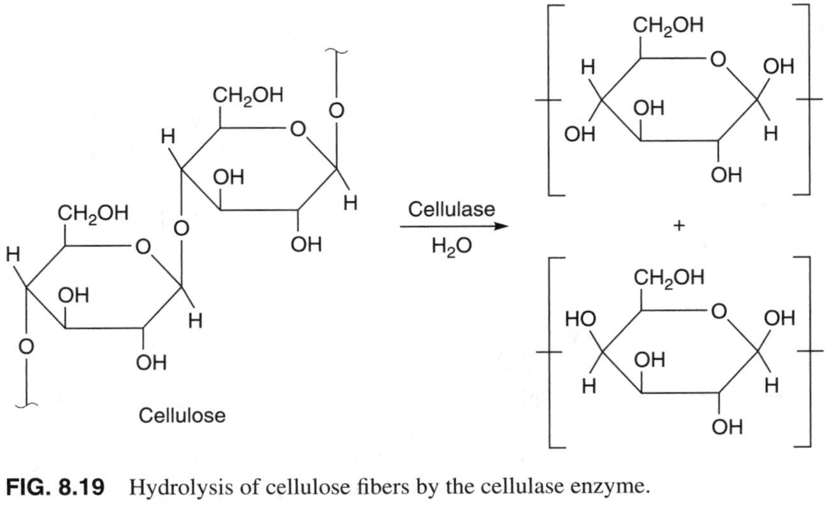

Cellulase
H$_2$O

+

Cellulose

**FIG. 8.19**   Hydrolysis of cellulose fibers by the cellulase enzyme.

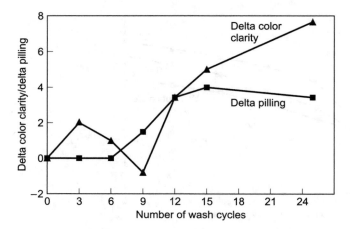

**FIG. 8.20**   Effect of cellulase on the color clarity and pilling tendency of a cotton fabric. European machine at 40°C using a new black cotton fabric. (Reproduced from Gormsen, E., Roshholm, P., and Lykke, M., in *Proceedings of the 3rd World Conference on Detergents: Global Perspectives*, Cahn, A., Ed., AOCS Press, Champaign, IL, 1994, p. 198. With permission.)

the more important that these enzymes are protected against premature degradation or at least maintain their performance throughout the shelf life of the product.

Many factors contribute to the denaturation of enzymes in HDLDs. They include free water, alkalinity, bleaches, and calcium ion concentration. The presence of free water in the formulation is a major cause of enzyme degradation. This process is greatly accelerated at increasingly alkaline conditions. Generally, enzyme-containing commercial HDLDs are maintained within a pH range of 7 to 9 (Figure 8.21). However, this constraint can affect the detergency, as most enzymes attain their optimum efficacy at pH ranges from 9 to 11. Certain additional ingredients, especially bleaches, can also have a major detrimental effect on enzyme stability.

It is believed that those ingredients that are capable of depriving an enzyme's active site of calcium ions are detrimental to enzyme stability. It is hypothesized that calcium ions bind at the bends of the polypeptide chain, resulting in a more stiff and compact molecule [100–102]. Builders and surfactants that have affinities toward calcium ions are examples of such ingredients. The degree of stability also varies greatly with the type of surfactant or builder used. LAS and alkyl sulfate surfactants have been found to be more detrimental to enzymes than alcohol ethoxylates or alkyl ether sulfates [99]. The degree of ethoxylation also affects the status of the enzyme. In ether sulfates, improved stability is observed with increasing EO groups up to 5 to 7 EO groups [100]. LAS is more likely to bind

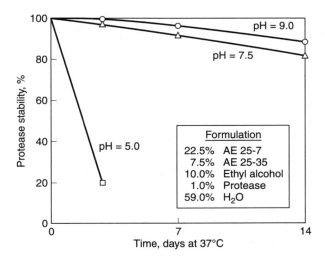

**FIG. 8.21** Effect of product pH on protease stability in a HDLD containing alcohol ethoxylate and alcohol ethoxy sulfates. (Reproduced from Kravetz, L. and Guin, K.F., *J. Am. Oil Chem. Soc.*, 62, 943, 1985. With permission.)

with the calcium ions in the product than other, more hardness-tolerant surfactants such as alkyl ether sulfates or the nonionic alcohol ethoxylate surfactants. This has been considered as a possible cause for faster enzyme degradation in LAS-containing HDLDs (Figure 8.22). Similarly, in formulations with builders or chelants, additional calcium is sometimes added to shift the equilibrium to favor the enzyme's active sites and prevent premature deactivation.

Other mechanisms for enzyme denaturation in the presence of surfactants have also been proposed. One hypothesis is that the high charge densities of ionic surfactants increase the probability of them binding strongly to protein sites. This causes conformational changes of the enzyme which subsequently leads to further enzyme deactivation [99,103].

The task of stabilizing enzymes is further complicated by the fact that HDLD formulations increasingly contain more than one enzyme (e.g., protease, lipase, and cellulase) system. In such systems, not only do the enzymes have to be protected against denaturation, but enzymes such as lipase and cellulase, which are themselves proteins, have to be shielded from the protease.

*(a) Protease-Only HDLDs.* All stabilization systems function by either binding to the active site of the enzyme or by altering the equilibrium of the formulation to favor the stable active sites. The system is effective in protecting the enzyme only if the stabilizing molecule binds strongly to the enzyme while in a formulation,

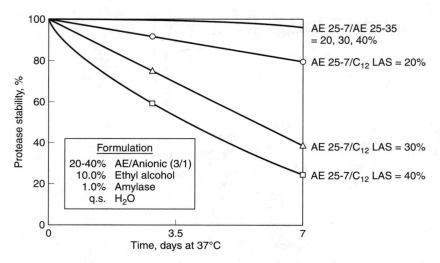

**FIG. 8.22** Effect of surfactant type on protease stability (AE, alcohol ethoxylate; AE25-3S, alcohol ethoxy sulfate; LAS, linear alkylbenzene aulfonate). (Reproduced from Kravetz, L. and Guin, K.F. *J. Am. Oil Chem. Soc.*, 62, 943, 1985. With permission.)

but easily dissociates from the enzyme's active sites when it encounters the dilute conditions in the wash.

Letton and Yunker [104] and Kaminsky and Christy [105] describe protease stabilization systems comprised of a combination of a calcium salt and a salt of a carboxylic acid, preferably a formate. These ingredients are moderately effective in enzyme stabilization and are relatively inexpensive. Care has to be taken, however, when adding divalent ions such as calcium to HDLDs to prevent the possibility of precipitation.

An improvement over this earlier system was attained with the addition of boron compounds such as boric acid or borate salts [106–108]. It has been hypothesized that boric acid and calcium form intramolecular bonds which effectively crosslink or "staple" an enzyme molecule together [107,108]. The use of polyols such as propylene glycol, glycerol, and sorbitol in conjunction with the boric acid salts further enhances the stability of these enzymes [109–111]. The patent literature contains numerous examples of enzyme stabilization systems that utilize borates, polyols, carboxylate salts, calcium, and ethanolamines, or combinations thereof [91,112–115].

*(b)   Mixed Enzyme HDLDs.*   In HDLD formulations with additional enzymes, it becomes increasingly difficult to stabilize all the enzymes. Amylases, lipases, and cellulases are themselves proteins and hence are susceptible to attack from

the protease. Various approaches to stabilizing a mixed enzyme system have been documented in the patent literature. One approach attempts to extend the stabilization techniques developed to stabilize protease-only formulations and apply them to mixed enzyme liquids [116–118].

Compounds that bind even more tightly to the protease active sites and as a result inhibit this enzyme's activity in the product during storage on the shelf have been identified. However, this method is effective only if this enzyme inhibition can be reversed under the dilute conditions of the wash water. Various boronic acids [119–123], such as arylboronic acids and alpha-aminoboronic acids, and peptide aldehyde [124], peptide ketone [125], and aromatic borate ester [126] compounds have been found which deliver this type of performance. It is believed that boronic acids inhibit proteolytic enzymes by attaching themselves at the active site. A boron-to-serine covalent bond and a hydrogen bond between histidine and a hydroxyl group on the boronic acid apparently are formed [122]. The patent literature also describes methods of stabilizing the cellulase enzymes in mixed enzyme systems with hydrophobic amine compounds such as cyclohexylamine and *n*-hexylamine [127].

Recently, alternative methods have also been developed to stabilize these complex enzyme systems. The technique of microencapsulation [128] is designed to prevent physically the protease enzyme from interacting with the other enzymes (Figure 8.23). This is accomplished by a composite emulsion polymer system which has a hydrophilic portion attached to a hydrophobic core polymer. The protease is stabilized by trapping it within the network formed by the hydrophobic polymer.

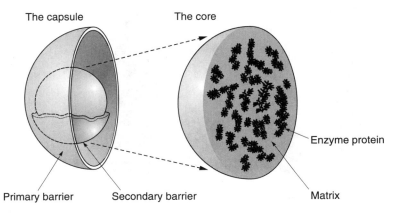

**FIG. 8.23** Enzyme microencapsulation. (Reproduced from Gormsen, E., Roshholm, P., and Lykke, M., in *Proceedings of the 3rd World Conference on Detergents: Global Perspectives*, Cahn, A., Ed., AOCS Press, Champaign, IL, 1994, p. 198. With permission.)

## D. Detergent Polymers

### 1. Polyacrylate Dispersants

Low- to moderate-molecular-weight polymeric dispersants have been utilized in powder laundry formulations for over 20 years. These polymers possess multiple benefits: (1) they act as crystal growth modifiers and prevent the formation and subsequent deposition of inorganic scale (i.e., calcium carbonate) on fabrics during the wash cycle, (2) they prevent the deposition of clay/particulate soils on fabrics, (3) they provide a modest increase in primary detergency on select soil/fabric combinations, (4) they reduce the viscosity of a high crutcher solids slurry, allowing easier processing, and (5) they improve particle integrity and the dissolution profile of spray-dried solids. In most isotropic liquid detergents, solubility limitations restrict the amount of inorganic builders incorporated into the formulation (so minimal inorganic scale is produced), and improvements in processing parameters are not warranted. The one measurable improvement has been attributed to clay soil antiredeposition, or enhanced whiteness maintenance. Incorporation of 3 to 5 ppm of active polymer into a liquid detergent can dramatically improve the whiteness index of a low- to moderate-cost formulation (Figure 8.24). These benefits are far less pronounced in systems utilizing high levels of surfactant with a modest builder system (citrate/fatty acid soap). Although these polymers are known to have modest sequestration properties for divalent metal cations (340 to 450 mg $CaCO_3$/g polymer), their use level in liquids is generally

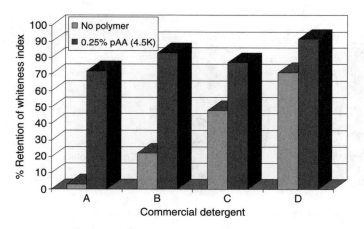

**FIG. 8.24**  Clay/oily soil redeposition of typical HDLDs. A, B, C, and D represent commercial liquid detergents, ranging from low-cost to premium brands, with and without the addition of a low-molecular-weight polyacrylate (pAA) homopolymer.

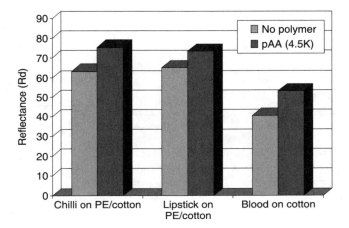

**FIG. 8.25** Effect of polymer incorporation in a commercial laundering application (35 lb machine, 100 ppm polyacrylate (pAA) homopolymer).

restricted to a fraction of a percent of the finished product. (Polyacrylates typically have limited compatibilities in surfactant matrices [129,130].) Studies have shown that these polymers are also effective in enhancing the primary detergency of select clay/oily soils, but this benefit is far more pronounced in industrial and institutional cleaning formulations than in heavy-duty liquids (Figure 8.25).

Polymer properties can be adjusted by modifying the polymer backbone through the introduction of alternative (hydrophilic/hydrophobic) monomers, adjusting the charge density (mono- versus dicarboxylic acids), varying the synthetic pathway (which can have an impact on polymer morphology), adjusting the molecular weight, or by utilizing different chain transfer (terminating) agents to cap the end functionality of the polymer chain, which changes the polymer's affinity and binding capacity for different substrates (Figure 8.26).

Carboxymethylcellulose (CMC) has also been widely utilized as an antiredeposition agent for cotton fabrics. It functions by forming a protective layer on the surface of the cellulosic fibers. However, the low aqueous solubility greatly limits its use in unstructured liquids.

## 2. Dye Transfer Inhibition: Polyvinylpyrrolidone (PVP)

Color-safe detergents are becoming a significant portion of the detergent market worldwide. Few technologies have been able to demonstrate real color safety in the washing cycle. Inhibition of dye transfer is one way whereby the color freshness of fabrics may be maintained after repeated washing. Polymers such as PVP are employed which inhibit transfer of fugitive dyes from colored fabrics onto other items in the washing machine [131,132].

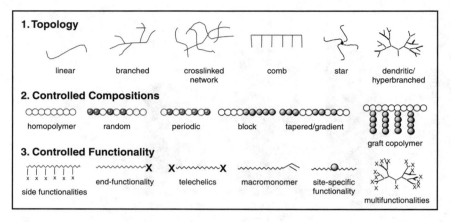

**FIG. 8.26**  Polymer morphology: production of polymers via alternative synthetic pathways.

PVP is a nonionic water-soluble polymer that interacts with water-soluble dyes to form water-soluble complexes with less fabric substantivity than the free dye. Additionally, PVP inhibits soil redeposition and is particularly effective with synthetic fibers and synthetic cotton blends. The polymer comprises hydrophilic, dipolar imido groups in conjunction with hydrophobic, apolar methylene and methine groups. The combination of dipolar and amphiphilic character make PVP soluble in water and organic solvents such as alcohols and partially halogenated alkanes, and will complex a variety of polarizable and acidic compounds. PVP is particularly effective with blue dyes and not as effective with acid red dyes.

## 3.  Soil Release Agents

Greasy and oily soils on polyester or polyester-containing fabrics are among the more difficult stains to displace. Removal of these soils from cotton fabrics, however, is far easier. This difference in cleaning can be attributed to the hydroxyl and carboxyl groups on the cellulosic fibers which give the surface of the cotton a hydrophilic nature and subsequently permit surfactants and water to more easily wash away adsorbed soils. The surface of polyester fabrics, in contrast, is hydrophobic since they are essentially composed of copolymers of terephthalic acid and ethylene glycol. This hydrophobicity not only creates an affinity towards oily soils, but makes them more difficult to remove.

The use of soil release agents in liquid laundry detergents is meant to address some of these issues [133]. These ingredients are usually polymeric and are composed of hydrophilic as well as hydrophobic segments. The hydrophobic functionality of the polymer allows it to be deposited and remain adsorbed onto the

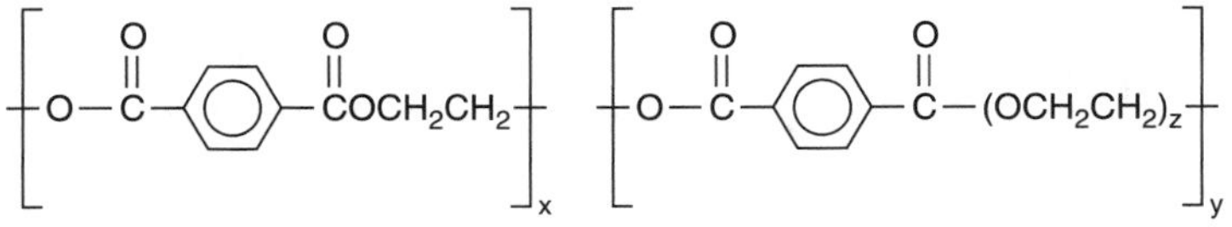

Poly (Ethylene Terephthalate, Polyoxyethylene Terephthalate)
(PET) (POET)

**FIG. 8.27** Polyethylene terephthalate (PET)–polyoxyethylene terephthalate (POET) soil release polymer. (Reproduced from Grime, K., in *Proceedings of the 3rd World Conference on Detergents: Global Perspectives*, Cahn, A., Ed., AOCS Press, Champaign, IL, 1994, p. 64. With permission.)

hydrophobic fibers of the polyester fabric during the washing and rinsing cycles. Once adsorbed the polymer can change the surface characteristics of the fabric to a more hydrophilic environment. In this way the polymer can not only make it easier for water and other cleaning agents to diffuse into the soil/fabric interface, but also prevents soils from adsorbing strongly onto the fabric. Consequently, soil release polymers are most effective only after the fabric has been treated with it. For this reason the largest impact is observed after several wash cycles.

A typical soil release polymer consists of segments of hydrophobic ethylene terephthalate and hydrophilic polyethylene oxide (Figure 8.27). A major drawback initially with these polymers was the limited aqueous solubility. This, in turn, led to lower performance, especially in particulate soil removal, due to weaker adsorption on fabric surfaces. The low solubility also limited the use of soil release polymers in unstructured liquids and sometimes led to instability of the product. Some of these obstacles have been overcome by using lower molecular weight polymers and by introducing end-capping groups [98,134]. These changes have led to improved solubility and better soil removal characteristics (Figure 8.28).

## 4. Rheology Modifiers

In recent years the use of rheology modifiers has taken on a more significant role in the formulation of unstructured liquid detergents. These polymers can improve product aesthetics, making the formulation more appealing to the consumer, and can enhance product stability by interacting with hydrophobic particles or pigments. The suspension of visual cues or actives has become a means of differentiating like products, and the combination of polymeric rheology modifiers and inorganic clays have produced systems with a yield stress. (This is discussed at greater length in Chapter 5.) Typically utilized rheology modifiers not only thicken the formulation but have an impact on liquid flow properties (i.e., pseudoplasticity, or shear thinning). These materials can be derived from organic compounds (natural or synthetic), can be nonassociative or associative (interact with surfactants or

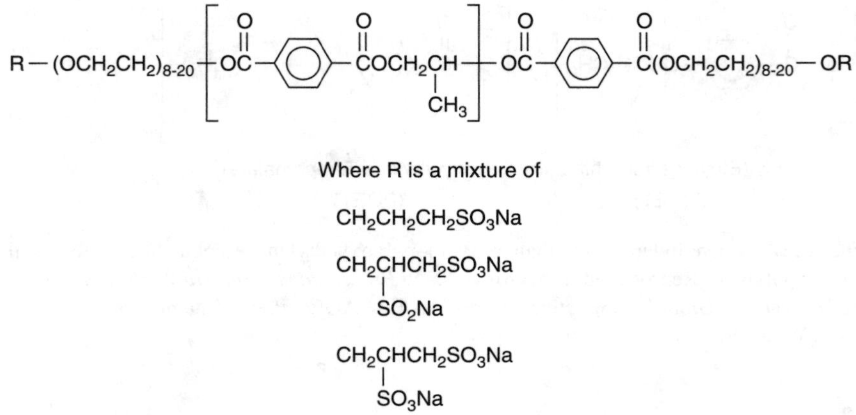

Where R is a mixture of

$$CH_2CH_2CH_2SO_3Na$$

$$CH_2CHCH_2SO_3Na$$
$$|$$
$$SO_2Na$$

$$CH_2CHCH_2SO_3Na$$
$$|$$
$$SO_3Na$$

**FIG. 8.28**  Soil release polymer with end-capping groups. (Reproduced from Grime, K., in *Proceedings of the 3rd World Conference on Detergents: Global Perspectives*, Cahn, A., Ed., AOCS Press, Champaign, IL, 1994, p. 64. With permission.)

other hydrophobic moieties), or can be (relatively) simple inorganic compounds such as salts or clays.

## E.  Bleaches

Bleaches play a significant role in detergent formulations since they can affect the cleaning efficacy which is easily consumer perceptible. Bleaching action involves the whitening or lightening of stains by the chemical removal of color. Bleaching agents chemically destroy or modify chromophoric systems as well as degrade dye compounds resulting in smaller and more water-soluble molecules which are easily removed in the wash. Typical bleach-sensitive stains include food/vegetable products, coffee, tea, fruits, red wine, and some particulate soils. Bleaches can also aid in minimizing "dinginess" which gives clothes a grayish/yellow tint caused by a combination of fabric fiber damage and dirt buildup. Bleaches also perform other functions in the wash liquor, namely improved sanitization and color protection (through oxidation of dyes and stains that have been solubilized during the cleaning process).

There are two types of bleaches used in the laundry process: hypochlorite and peroxygen bleaches. Although hypochlorite bleaches by themselves are effective bleaches, they lead to color fading and fabric damage and are difficult to incorporate into detergent formulations. Peroxygen bleaches, although not as effective, can be formulated into detergents and cause minimal color fading or fabric damage. They also bleach out food stains since the chromophores found in these soils are susceptible to peroxide bleaches. Fabric dyes, however, are not as active as food dyes and are not easily affected by the peroxygen compounds.

Most detergents with bleach formulations are in the powder form. Unfortunately, the aqueous nature of HDLDs does not easily permit the formulation of bleach components. This is especially true in unstructured liquids where the stability of the peroxygen components is severely compromised. Nevertheless, attempts have been made to produce HDLDs that also contain bleaches.

## 1. Peroxygen Bleaches

Peroxide bleach-containing detergent formulations contain either hydrogen peroxide or compounds that react to form hydrogen peroxide in the wash. The most direct source for peroxide bleaching is hydrogen peroxide. Numerous attempts have been made to develop stable hydrogen peroxide-containing HDLDs [135–137]. The stability of this ingredient in aqueous formulations, however, is of concern. Hydrogen peroxide is very susceptible to decomposition in aqueous environments largely because trace impurities of metal ions such as iron, manganese, and copper can catalyze its decomposition [138]. Alkalinity also accelerates this process. For these reasons HDLDs containing hydrogen peroxide are maintained at an acidic pH and usually also contain a strong chelating agent to sequester metal ions. A free radical scavenger can also be added to further enhance stability. Polyphosphonate compounds, polycarboxylates, and butylated hydroxytoluene (BHT) are examples of chelating agents and free radical scavengers, respectively, which are used in hydrogen peroxide-containing formulations [139]. Still, the bleaching performance of these products is inadequate, particularly at low temperatures ($<40°C$). These limitations have prompted manufacturers to look at other methods to develop bleach-containing HDLDs.

Inorganic peroxygen compounds such as sodium perborate tetrahydrate or monohydrate and sodium percarbonate can also be used as sources for hydrogen peroxide. These insoluble compounds release hydrogen peroxide on contact with the wash water. The challenge is to stabilize them within a HDLD formulation. The ability of structured liquids to suspend solids between the surfactant lamellae or spherulites can be made use of in these products [140–143]. It is possible to suspend sodium perborate in highly concentrated structured liquids. The minimization of contact with water prevents the peroxygen compounds from decomposing prematurely. It has also been found that the use of solvents further improves the stability [144]. Hydrophobic silica can enhance stability in unstructured liquids [145]. The most effective method of formulating with perborates and percarbonates is with nonaqueous liquids. The complete absence of water and a high level of solvents significantly enhance the stability of bleaches in the product.

## 2.  Peracid and Activated Peroxygen Bleaches

Peroxycarboxylic acids or peracids are far more effective bleaching compounds than peroxygen compounds especially at low and ambient temperatures. The high reactivity and low stability of these compounds have so far prevented them from being used in commercial detergent bleach formulations. Peracids are somewhat stable in aqueous solutions of neutral pH and they equilibrate with water in acid pH to form hydrogen peroxide and carboxylic acids. However, in alkaline conditions these compounds undergo accelerated decomposition. Nevertheless, attempts have been made to develop HDLDs that take advantage of the ability of structured liquids to suspend insoluble solids. Patents have been issued for liquid detergent formulations which incorporate peroxy acids such as diperoxydodecanedioic acid [146,147] and amido and imido peroxy acids [148,149]. These product formulations are also maintained at an acidic pH to reduce the premature reaction of the peracid.

An alternative and more desirable method of bleaching is by forming the peracid in the wash water. This is accomplished by reacting a bleach activator compound with a source of hydrogen peroxide such as perborate, percarbonate, or hydrogen peroxide itself in an aqueous environment. In this reaction, referred to as perhydrolysis (Figure 8.29), the bleach activator undergoes nucleophilic attack from a perhydroxide anion generated from hydrogen peroxide, resulting in the formation of percarboxylic acid. This formulation strategy is effectively utilized in powder detergents where perborate and percarbonate compounds are used along with activators such as tetraacetylethylene diamine (TAED) and sodium nonanoyloxybenzene sulfonate (NOBS).

In liquid detergents the challenge is to incorporate hydrogen peroxide and an activator compound in an aqueous formulation and prevent these two components from reacting prematurely in the product itself. A novel method utilizes emulsions formed by nonionic surfactants with varying hydrophilic/lipophilic balance (HLB) values to protect a soluble activator, acetyltriethyl citrate, from other ingredients, including hydrogen peroxide, in the product [150–152]. An acidic pH and the addition of a strong chelating compound aid in product stability. Other patents using NOBS [153], glycol and glycerin esters [154], and a lipase–anhydride combination [155] have been issued.

## F.  Optical Brighteners

It has been found that fabrics, especially cotton, begin to appear yellowish after repeated washing cycles. Virtually all modern liquid detergent formulations contain very small amounts (<1.0%) of optical brighteners or fluorescent whitening agents which absorb ultraviolet light (300–430 nm) and reemit it as fluorescent visible blue light (400–500 nm) [156–159]. This visible blue light offsets some of

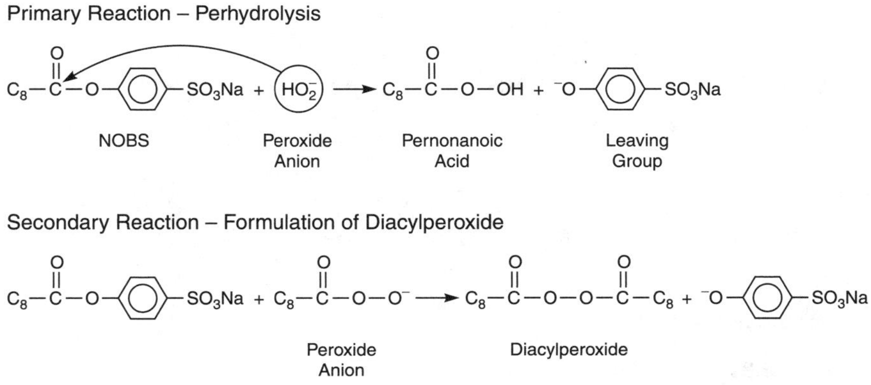

Primary Reaction – Perhydrolysis

NOBS    Peroxide Anion    Pernonanoic Acid    Leaving Group

Secondary Reaction – Formulation of Diacylperoxide

Peroxide Anion    Diacylperoxide

**FIG. 8.29** Bleach activator (nonanoyloxybenzene sulfonate) reactions. (Reproduced from Grime, K., in *Proceedings of the 3rd World Conference on Detergents: Global Perspectives*, Cahn, A., Ed., AOCS Press, Champaign, IL, 1994, p. 64. With permission.)

the yellow hues from the fabric and as a result provides a whitening/brightening effect as seen by the human eye.

The most widely used class of fluorescent whitening agents (FWAs) are known as CC/DAS types or cyanuric chloride diaminostilbene. The chromophore in these molecules is triazinylaminostilbene (Figure 8.30). The basic molecule can

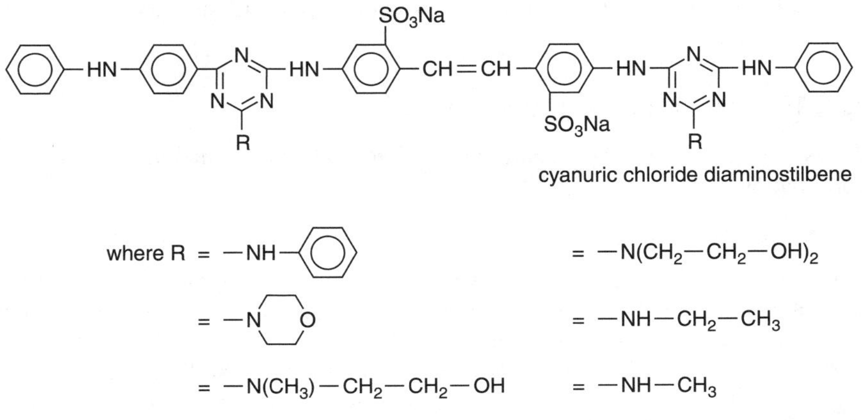

cyanuric chloride diaminostilbene

where R = $-NH-$ ⃝    = $-N(CH_2-CH_2-OH)_2$

= $-N$⟮O⟯    = $-NH-CH_2-CH_3$

= $-N(CH_3)-CH_2-CH_2-OH$    = $-NH-CH_3$

**FIG. 8.30** CC/DAS or DASC: bis-triazinyl derivatives of 4,4′-diaminostilbene-2,2′-disulfonic acid. (Reproduced from Whalley, G., *HAPPI*, Nov., 82, 1993. With permission.)

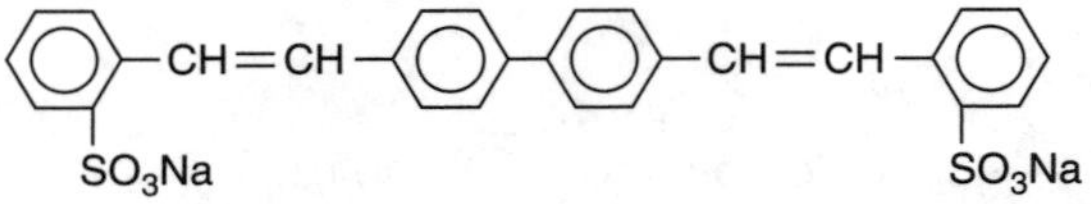

Distyrylbiphenyl type

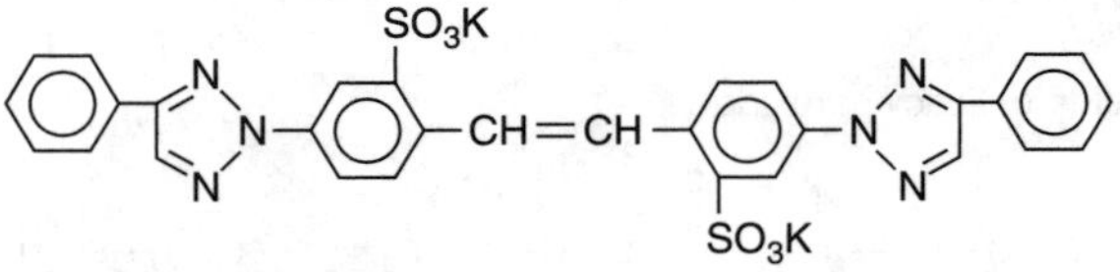

Triazolylstilbene type

**FIG. 8.31** Chlorine-stable fluorescent whitening agent: disodium 4,4′-bis(2-sulfostyryl)biphenyl and a triazolylstilbene type. (Reproduced from Whalley, G., *HAPPI*, Nov., 82, 1993. With permission.)

be altered by the addition of various substituent groups including alkoxy, hydroxy, or amino groups. The choice of these substituent groups can control the solubility, substantivity, and overall performance of the brightener. These brightener molecules are highly susceptible to electrophilic attack. As a result, they are ineffective and unstable in the presence of chlorine bleach. For this reason, it is common for detergent manufacturers to recommend that users add chlorine bleach at least five minutes into the wash. Another class of brighteners have been developed that are more resistant to chlorine bleach (Figure 8.31). These include distyrylbiphenyl derivatives such as disodium 4,4′-bis(2-sulfostyryl)biphenyl and triazolylstilbene.

## G. Miscellaneous Ingredients

A variety of other ingredients also serve valuable functions in HDLD formulations.

### 1. Buffers

An alkaline pH in the wash water can greatly improve the cleaning ability of the detergent. Certain oily/greasy soils can be removed from the fabric surface by saponification at high pH values. In addition, enzymes reach an optimum performance level within a pH range of 8 to 11. Examples of buffering compounds used in HDLDs include carbonates, liquid silicates, borates, and amines such as monoethanolamine (MEA) and triethanolamine (TEA). However, in enzyme-containing formulations care should be taken to prevent the product pH from exceeding a level that could lead to enzyme degradation.

## 2. Defoamers

Soap is used extensively to minimize excessive foaming in washing machines, particularly in European or horizontal axis machines where agitation is considerably higher and foam generation can become a significant problem. Long-chain fatty acids (palmitic/stearic) or their alkali metal salts (sodium or potassium) are typically incorporated into liquid formulations. Under high water hardness conditions the formation of lime soap (calcium fatty acids) presents a soil redeposition issue, and garments can become dingy and acquire an off odor. The problem can be mitigated by adjusting the surfactant system or incorporating a hydrophobically modified polymer to thoroughly disperse lime soap [160].

Silicone-based defoamers are also highly effective foam control agents. Marketed by Dow Corning, silicone foam control agents have a lower surface tension than most fluids, readily dispersing through the liquid film constituting a bubble, collapsing the soap bubble, and dramatically reducing the foam [161].

## 3. Hydrotropes

It is sometimes necessary to use hydrotropes to solubilize all ingredients in an unstructured liquid. Examples of hydrotropes include sodium xylene sulfonate (SXS), sodium cumene sulfonate (SCS), and sodium toluene sulfonate (STS). Alcohols (ethanol) and glycols (propylene glycol) are also commonly used to couple insoluble or incompatible actives to produce a homogeneous, isotropic liquid. An extensive description of these ingredients is presented in Chapter 2.

## 4. Minors

A number of ingredients, although used in small amounts, serve very important roles. Preservatives are needed to inhibit the growth of microorganisms in aqueous products. Included in this long list of actives are formaldehyde or formaldehyde donors (imidazolidinyl compounds or dimethylhydantoin), isothiazolones, combinations of dimethylhydantoin and iodopropylbutylcarbamate (IPBC), and select cationic surfactants (when properly formulated with nonionics) [162]. Fragrances and dyes help cover the odor and color of the base liquid and enhance the aesthetics of the product. A fragrance is comprised of a blend of volatile organic chemicals (typically 20 to 100 ingredients), containing one or more functional groups such as aliphatic hydrocarbons, alcohols, aldehydes, terpenes, and natural extracts. Liquid detergents inherently possess a straw or amber color. Dyes are used to produce a brighter, more attractive appearance that will appeal to the eye of the consumer. In some cases, the detergent manufacturer will change the appearance of a liquid detergent by intentionally adding an opacifier into a clear, isotropic liquid. This can be done to provoke the concept of mildness (as evidenced in rinse-added fabric softeners) or to depict a highly concentrated formulation. The opacifiers that are commonly used in these systems are styrene/(meth)acrylate emulsion polymers (sold by Rohm and Haas) with a particle size between 150 and 400 nm [163].

## IV.  PRODUCT EVALUATION METHODS
## A.  Physical Properties

The acceptability and formulation success of a laundry liquid can depend on several important physical properties. Foremost among these characteristics is the physical stability of the product. The consumer expects the final product to be homogeneous and single phase. Accelerated aging tests at temperatures the product is likely to encounter in its lifetime are conducted to test the storage stability. In addition, a test to check the ability of the liquid to withstand repeated freeze–thaw cycles and still remain a single phase is generally carried out. Rheology and flowability tests are especially crucial in characterizing structured liquids. Chapter 4 provides a good description of the basic rheological concepts. HDLDs have to maintain desirable pouring characteristics in order to be considered acceptable by the consumer. The stability of the fragrance is also of utmost importance to the product. Fragrance characteristics are analyzed in conjunction with the accelerated physical stability tests.

The importance of preserving enzymes has been discussed earlier in this chapter. The activity of enzymes is usually measured by performing an assay specific to the enzyme being tested. It is also recommended to conduct washing tests over an extended time period on enzyme-sensitive stained fabrics and determine the extent of enzyme loss as a function of time.

## B.  HDLD Detergency Evaluation

The mechanisms underlying the detergency and soil removal process have been reviewed by many authors [164–172]. This section briefly summarizes the test methods used to characterize the performance of liquid laundry detergents. There are typically three stages of testing during product development: (1) laboratory evaluation, (2) practical evaluation, and (3) consumer tests.

## 1.  Laboratory Tests

*(a)  Soil Removal.*    Soil removal testing on a laboratory scale is conducted using specialized equipment, typically a tergotometer, which is designed to simulate the actual laundry process. The tergotometer consists of a series of 1 l stainless steel buckets, each with an agitation mechanism. Soiled fabric swatches are added into the wash solution in order to measure stain removal and other attributes of the detergent product. These instruments offer the advantage of providing a controlled environment in which the effects of various variables can be measured. The effects of water temperature, water hardness, agitation rate, detergent concentration, etc., can then be determined.

Table 8.4 shows the categories and typical examples of soils that are artificially deposited on a range of fabrics such as cotton, polyester, and blends. These soils

**TABLE 8.4** Listing of Laundry Soils Used in Detergent Evaluations (Table reproduced with permission from Coons, D., Dankowski, M., Diehl, M., Jakobi, G., Kuzel, P., Sung. E., and Trabitzsch, U., in *Surfactants in Consumer Products: Theory, Technology and Application*, Falbe, J., Ed., Springer-Verlag, New York, 1987.)

| Water-soluble soils | Particulate soils | Fats/oils | Proteins source | Carbo-hydrates | Bleachable dye source |
|---|---|---|---|---|---|
| Inorganic salts | Metal oxides | Animal fats | Blood | Starches | Fruit |
| Sugar | Carbonates | Vegetable fats | Eggs | | Vegetables |
| Urea | Silicates | Sebum | Milk | | Wine |
| Perspiration | Humus | Mineral oils | Cutaneous scales | | Coffee |
| | Carbon black | Waxes | | | Tea |

represent stains that can be removed by physical as well as by chemical mechanisms. The removal of stains in the water-soluble and particulate categories are largely dependent upon mechanical agitation and the interfacial forces created by the detergent surfactants. Bleachable or oxidizable and enzymatic stains are examples of soils which are more responsive to the chemical nature of the wash liquor.

Detergency can be evaluated either visually by an expert experienced panel which rates the degree of soil removal or by instrumental techniques [173–175]. In the latter method, the stain removal $R$ is expressed as:

$$R = 100 - \left[(L_c - L_w)^2 + (a_c - a_w)^2 + (b_c - b_w)^2\right]^{1/2}$$

where $L$ = reflectance, $a$ = redness/greenness, $b$ = yellowness/blueness, c = unstained fabric washed in treatment conditions, and w = stained fabric washed in treatment conditions.

*(b) Brightening.* The brightening performance of detergent formulations is determined by washing a set of large unsoiled swatches representing various fabric types. These can include swatches of cotton, nylon, cotton blends, terry towels, and polyester fabrics. These swatches are washed for one to three cycles, and their brightness is subsequently measured using a reflectometer. The $b$ component (blue/yellow) of the light reading is measured for this test.

*(c) Soil Release.* Clean swatches are first prewashed, typically for one to three cycles, using the detergents to be tested. After the prewash, the swatches are stained with a variety of soils, usually greasy particulate and food soils. Subsequently, the swatches are washed again. At each stage, the reflectance reading of the fabric swatches is recorded.

*(d)   Antiredeposition.*   Once soil removal has occurred, the possibility exists that this soil can redeposit onto the fabric. The likelihood or the extent of this occurring for a particular detergent product can be measured by an antiredeposition test. In this test clean swatches along with typical soils are added into the wash solution. Representative particulate soils include vacuum cleaner dirt and various colored clays (grayish-brown, reddish-brown, orange); a mixture of triolein and mineral oil as well as an artificial sebum (body sweat) composition serve as oily/greasy soils [176,177]. The washing experiments can be conducted using a laboratory-scale tergotometer or a commercial washing machine. Generally, one to three cycles for a laboratory scale and five to ten wash cycles for a washing machine are necessary for the entire test. The degree of soil deposited is measured instrumentally by reflectance readings.

*(e)   Dye Transfer or Color Loss.*   Colored fabrics tend to lose color and fade after repeated detergent washes. In addition, white or undyed fabrics can sometimes acquire a small degree of color from the transfer of dyes when washed with colored fabrics. Tests methods have been developed to measure the contribution of detergents to this color loss. Detergent manufacturers perform these tests on a laboratory scale as well as on a practical level.

In a typical laboratory-scale test [178,179] conducted using a tergotometer, a nylon fabric dyed red and cotton fabrics dyed different shades of blue are added to the wash bucket along with a clean undyed white cotton fabric. This white cloth is meant to be a scavenger of dyes lost in the wash and provides an indication of color transfer. Various detergents can be tested and ranked according to their color loss properties. After the wash, the fabric swatches are instrumentally evaluated with a colorimeter. The $E$ value provides the degree of colorfastness of the fabric in a particular detergent:

$$E = \left[ (L_w - L_o)^2 + (a_w - a_o)^2 + (b_w - b_o)^2 \right]^{1/2}$$

where $L$ = reflectance and the subscripts w and o signify after the wash and before the wash, respectively.

Practical evaluations utilize commercial washing machines and use actual colored clothing materials in the test. These clothes are then washed repeatedly with detergent over a 10- to 50-cycle range. The clothes are then evaluated visually using an expert panel. Instrumental measurements can also be taken.

## 2.  Practical Evaluation

Additional performance evaluations are also conducted in commercial washing machines. Although this method does not permit testing in as controlled a fashion as the launderometers, they do predict consumer-relevant behavior. Soiled fabric swatches when added to commercial washing machines also need to be accompanied by additional clothing as ballast to provide more realistic clothes-to-detergent and bath-to-fabric ratios.

Fabric bundle tests are also an effective and realistic method of evaluating detergency of HDLDs. In this test, clothing articles are distributed among volunteers to be used in their normal manner. They are periodically returned to be washed and evaluated. The evaluation can be conducted visually or instrumentally.

### 3. Consumer Tests

The final and most important component of the evaluation of laundry detergents is the consumer test. A select group of consumers is provided with a product and instructed to use it with their normal laundry loads. Their feedback on various aspects of cleaning and product aesthetics is collected and analyzed. These data play a significant role in any decision regarding the composition of the liquid laundry detergent.

## V. RECENT PATENT TRENDS

Several thousand patents have been issued over the past 7 or 8 years covering HDLDs. Many of these patents describe improvements in detergent efficacy via more conventional or accepted approaches, including optimized surfactant systems and "bleach alternative systems" (optical brightener/enzyme packages). In recent years there has been a greater emphasis on adjuvants or additives that are incorporated into detergent formulations at relatively low concentrations that deliver significant, consumer-perceptible benefits. High on that list are fabric and color care compositions, with the ultimate goal being focused on preserving fabric appearance after multiple launderings ("looks newer longer") [180–185]. There has been a continuous effort to find/commercialize novel polymers (other than PVP and its associated derivatives) that reduce dye transfer in the wash or rinse cycle. Novel enzymes (peroxidases/oxidoreductases) are beginning to find utility in liquid detergents as alternative dye transfer inhibitors [186], with recent efforts aimed at reducing enzyme allergenicity by polypeptide modification [187]. Several patents on novel soil release technologies have also been granted, with the focus being shifted from synthetics and fabric blends to 100% cotton garments [189–193].

The market has also migrated towards consumer-friendly products that reduce fabric wrinkling and dramatically cut/eliminate ironing time [180,182,194,195]. Polymers have also been employed to build rheology (or structure) in various liquid formulations, with the objective of improved product aesthetics (through the suspension of actives or visual cues) [196]. Fragrance encapsulation and additives for masking malodors have also become topics of much interest [197,198].

Water-soluble packages or sachets have been launched commercially in Europe and the U.S., and several patents have also been published in this area [199,200].

These recent patent trends are described in greater detail in the following sections. A series of tables summarizing the principal advances in HDLDs are listed in the Appendix.

## A.  Polymers

The incorporation of polymers into HDLDs is one of the areas where patent activity and manufacturer interest remain quite high. Polymers can attach to a fabric and facilitate the release of various soils (soil release polymers) [188–193], prevent the transfer of fugitive dyes onto clean fibers in the wash bath (dye transfer inhibitors) [201–205], improve the cleaning efficacy/antiredeposition properties of oily/particulate soils [206,207], act as suds enhancers/foam boosters [208,209], increase fabric yarn strength [210], reduce wrinkling/ease of ironing [180,194,195], and eliminate fabric pilling/fuzz on select garments [181–185].

### 1.  Soil Release Polymers

The concept of polymeric soil release agents has been around for well over 25 years. The initial polymer chemistries (polyethylene terephthalate–polyoxyethylene terephthalate, PET–POET) were designed to deposit on fabrics and facilitate oily soil removal upon subsequent washing [98,133,134]. The limitation of this chemistry was its effectiveness on synthetics (polyester) alone, with limited benefits being observed on cotton and synthetic blends. In recent years the focus has shifted to delivering soil release on cotton. Two classes of polymer chemistries have been disclosed in the recent patent literature for cotton soil release: one based on hydrophobically modified polycarboxylates derived from acrylic acid and hydrophobic comonomers at defined molar ratios [188] and the other based on modified polyamines [189–193].

### 2.  Fabric Care

During the laundering process, fabrics undergo mechanical and chemical changes that damage the fibers comprising a garment and adversely affect fabric integrity. These changes can be observed in the form of fiber deterioration (pilling/fuzz formation), fabric wrinkling, fabric stiffness, color fading, and poorer fabric appearance. There have been several patents/applications that have utilized cellulose-based polymers and their derivatives to negate this effect. Cellulose-based polymers or oligomers have been found to impart fabric appearance and integrity benefits (improved abrasion resistance) to textiles/fabrics washed in a conventional laundry composition without promoting any reduction in cleaning efficacy [183]. Cationic celluloses have also been employed as enhancing agents or deposition aids designed to facilitate the deposition of insoluble fabric care benefit agents such as dispersible polyolefins and latexes through the laundering cycle [211]. Fabric care benefits have also been achieved via incorporation of fiber reactive additives (polyamide–polyamines) that provide several cleaning benefits (soil removal, soil dispersancy, and dye transfer inhibition) [184,185]. Polymers and prepolymers derived from polyoxyalkylene amines have also been utilized in a process to shrinkproof wool [212]. Yarn strength-enhancing agents selected from a group consisting of polysaccharides, clays, starches, chitosans, and mixtures of these materials have also been claimed [210]. The authors claim that strengthening the yarn reduces pilling and "wear and tear" by retarding/inhibiting fiber migration from within the yarn to the surface of the yarn.

Fabric care benefits have also extended to areas that simplify or reduce extra labor encountered by the consumer. The crease recovery of fabrics can be improved by using C20–C40 saturated or unsaturated aliphatic hydrocarbons which have melting points below 0°C [194]. These additives can be adopted into products for use in a tumble dryer (flexible sheets), sprayable formulations, or in fabric care compositions (such as rinse-added fabric softeners). Ease of ironing benefits have also been promoted by utilization of silicone gel compositions for ease of ironing (improved glide, wrinkles more readily removed) and improved appearance after ironing (less visible wrinkles) [195]. A recent patent application discloses the use of nanoparticles or nanolatexes (10 to 500 nm) as a crease resistance agent or additive for helping in the ironing of textiles in an aqueous or wet medium [180].

## 3. Color Care (Dye Transfer Inhibition)

PVP has been used for many years to inhibit dye deposition by complexing fugitive dyes during the washing of colored fabrics. The performance of PVP is adversely affected by the presence of anionic surfactants in the wash bath. Analogues of PVP, including poly(4-vinylpyridine-$N$-oxide) (PVPNO), polyvinylimidazole (PVI), copolymers of polyvinylpyridine and polyvinylimidazole (PVP–PVI), and polysulfoxide polymers, have also been employed in detergents to prevent dye redeposition onto garments [201,202]. Bleach-stable, modified polyamine additives have been found to inhibit dye transfer between fabrics during laundering [203], and poly(vinylpyridine betaines) containing a quaternary nitrogen and a carboxylate salt are also claimed [204]. A recent patent application has claimed that laundry detergent compositions with select, cationically charged dye maintenance polymers or oligomers (having a net positive charge) impart appearance and integrity benefits to fabrics and textiles [205]. These polymers (oligomers) or copolymers are comprised of one or more linearly polymerizing monomers, cyclic polymerizing monomers, or mixtures thereof. These additives associate with the fibers of the fabric, minimizing the natural tendency of the laundered garments to deteriorate in appearance over multiple wash cycles.

## 4. Enhanced Cleaning Efficiency

Polymers are not typically known for their cleaning prowess in detergent systems. Most of the observed benefits focus on preventing the formation or subsequent redeposition of inorganic scale or clay soils back onto fabric. Premium liquid detergents often utilize a combination of citric acid/citrate and fatty acid soaps to chelate calcium ions in the wash bath. The presence of high concentrations of fatty acid soap and divalent metal ions ($Ca^{2+}$, $Mg^{2+}$) leads to the formation of lime soap, or the insoluble calcium salt of fatty acids. Incorporation of selected surfactant blends and polymeric additives (maleic/olefin copolymers) act as lime soap dispersants, controlling the formation/deposition of an organic "scale" on fabrics, resulting in improved whiteness maintenance [160]. Other polymeric additives have been shown to deliver enhanced whitening benefits. Blends of soil release copolymers and conventional dye transfer inhibitors (PVP, PVPNO) provide improvements in soil antiredeposition [207], and ethoxylated/propoxylated polyalkyleneamine polymers (e.g., polyethyleneimine, with a degree of substitution of 1.0) deliver enhanced soil dispersancy in fabric laundering, dishwashing, and hard surface cleaning applications [206].

## 5. Product Aesthetics

There are several advantages offered by polymer chemistries in this area which cover many different properties. Liquid detergent compositions comprising quaternary nitrogen-containing and/or zwitterionic polymeric suds enhancers and suds volume extender have been claimed [208,209]. These polymers also exhibit an increased effectiveness for preventing the redeposition of greasy soils during the cleaning process. Aqueous compositions (liquid detergents, shampoos, personal care products, etc.) designed to impart a "thick or rich" appearance perceived by the consumer can be achieved by incorporation of specialized synthetic or natural polymers. Systems exhibiting pseudoplasticity, or shear thinning behavior, are usually quite desirable. At high surfactant concentrations, many of these rheology modifiers fail to deliver sufficient viscosity build. The combination of a lipophilically modified copolymer (based upon acrylic acid residues) and a colloidal inorganic clay have been found to expand that surfactant range [111]. Other approaches have focused on polymer gums (carrageenans, gellans, and agars) capable of suspending relatively large size particles that remain pourable (with good shear thinning properties) [196].

## B. Enzymes

Enzyme cocktails are now commonly used in liquid detergents. Most products contain a minimum of a protease (for removal of proteinaceous soils) and an amylase (to facilitate starchy food-based soil removal) to assist cleaning. Several other formulations contain lipases (for degrading fatty/oily soils) and cellulases (to improve fabric appearance by cleaving the pills/fuzz formed on cotton and synthetic blends). Numerous patents were granted on enzyme stabilization packages during the 1980s and 1990s, but the recent focus has shifted toward novel enzymes for improved cleaning (xylogluconases, neopullulanase, mycodextranase, and oxidoreductases) [213–216], reduced allergenicity [187], and the use of cellulose binding domains [217,218].

Currently, the generation of allergic responses to polypeptides is controlled by immobilizing, granulating, or coating the enzymes. The ability to reduce an allergic response by modifying the polypeptide (conjugating the enzyme with a polymer) allows the formulator to use enzymes in a wider range of products (light-duty liquids, personal care products).

The concept of utilizing cellulose binding domains to improve detergent efficacy is quite interesting. Cellulase enzymes are comprised, in part, of cellulose binding domains that have a high affinity for the surface of cotton. By linking a chemical entity to a cellulose binding domain, one can enhance/ensure deposition of actives (perfumes, polymers, bleaching or hygienic agents) onto the substrate through the wash.

## C. Fragrance Encapsulation/Odor Elimination

Although utilized in liquid detergents at relatively low concentrations, fragrances have become a critical selling feature of many cleaning products. The use of cyclodextrin molecules to reduce malodors generated by nitrogenous compounds (diamines) is just one example [197]. Instead of attempting to mask malodors with excess fragrance, the perfume is admixed with the cyclodextrin, which possesses an internal cavity capable of forming complexes. This technology delivers a fragrance to the wash cycle while the off odor emanating from other (undesirable) components is suppressed. Other mechanisms have been used in recent years to improve fragrance delivery or provide for controlled release. Fragrance raw materials can be delivered onto fabrics through the wash by way of a fragrance delivery system comprised of a single precursor pro-accord or pro-fragrance compound ($\beta$-ketoester) having higher fabric substantivity [219]. Compounds comprising at least one $\beta$-oxy or $\beta$-thio carbonyl moiety are capable of liberating an active molecule ($\alpha,\beta$-unsaturated ketone, aldehyde, or carboxylic ester) into the surrounding environment [220]. Pro-perfume compositions comprising an amino functional component and a benefit agent (which is suspended in the liquid matrix) providing enhanced deposition and long-lasting fragrance release are also disclosed [221]. Lastly, polymeric nanoparticles including olfactive molecules having a defined glass transition temperature are also claimed [198]. The fragrance is contained inside the polymeric nanoparticle (core/shell technology), protecting the perfume and ensuring slow release during the washing process.

## D. Fabric Protection/Optical Whiteners

Consumers experience color damage to their personal garments from prolonged exposure to the sun (from line drying and everyday wear). The effect is more pronounced in tropical and subtropical climates around the globe. Visible light is the largest contributor to fabric fading, and the incorporation of nonstaining, light-stable antioxidants into fabric care compositions has been achieved by addition of C8–C22 hydrocarbon fatty organic moieties [222]. A second approach involves novel compounds that are useful as ultraviolet absorbing agents and fluorescent whitening agents designed to improve the sun protection factor (SPF) of textile fibers, especially cotton, polyamide, and wool [223].

## E.  Sachets/Unit Dose

Unit dose products have made impressive gains into the European marketplace, particularly the automatic dishwashing detergent market, in the form of compressed tablets. Polyvinyl alcohol (PVA) sachets containing powders and liquids have appeared in Europe and Asia/Pacific, but their presence has not been observed in the U.S. liquid detergent market. (Automatic dishwashing compositions containing liquids are being sold commercially in automatic dishwashing applications in the U.S.). The pouches must dissolve rapidly in the wash liquor, independent of water temperature and mechanical agitation, yet have enough structural integrity not to rupture upon handling by the consumer. Detergent components capable of crosslinking the water-soluble PVA film may alter the properties of the container, adversely affecting the sachet's solubility in water. The preparation of an essentially anhydrous liquid detergent composition (<5% free water by weight) allows the formulator more freedom to practice a wider variety of technologies [199].

## F.  Surfactants

A series of patents covering novel anionic surfactants described as mid-chain branched surfactants deliver improved efficacy due to enhanced solubility (in cold water) and greater water hardness tolerance [224]. These surfactants could find greater utility in the U.S. with the growth of front loader washing machines and reduced washing temperatures. A second category of surfactants covers detergent compositions containing $\alpha$-sulfofatty acid esters and a process for synthesizing these materials [225]. These surfactants have performance comparable to other anionics in soft water, but exhibit better cleaning efficacy as water hardness increases.

## G.  Builders

The use of builders in liquid laundry formulations has effectively been restricted to citrates and fatty acid soaps (in isotropic compositions sold in the U.S. and Europe). Although liquid laundry products can contain tripolyphosphate, soda ash, NTA, and other common builders found in powder laundry, limitations on solubility (in a surfactant matrix) or regulatory concerns have restricted their utilization. There is a finite number of patents covering zeolite-built HDLDs, but problems relating to long-term stability of the slurry have always been a concern. A recent patent on the production of colloidal particles under $0.1\ \mu m$ (35 nm zeolite A) has rekindled interest in this application [226]. The drive is to mill zeolite particles even finer and reach a 15 nm particle size, where the zeolite suspension will become transparent. Theoretically, one could produce a "clear" zeolite slurry at a modest viscosity with a cost profile that is considerably more favorable than sodium citrate. Several articles/patents on nanoparticle technology are concerned with ultrafine particles (nanolatexes) capable of being suspended over time in a liquid detergent composition [190].

## VI.  NEW PRODUCTS

As the twentieth century drew to a close, liquid detergents in the U.S. achieved a milestone to which no other global laundry market comes close — they surpassed the sales of powder detergents. This trend has continued unabated for the past 5 to 10 years, and in 2004 liquids accounted for close to 70% of detergent market sales.

A number of formulation revisions and advanced technologies have contributed to the rapid growth in liquids. In some cases the changes were directed at consumer needs/requirements and were relatively "low tech;" in other cases the market was driven by novel technologies developed by the detergent manufacturers in concert with other suppliers.

## A.  High-Efficiency Detergents

In 1996 Frigidaire introduced a new extra large capacity, low-energy front loader clothes washing machine. The new machines, often referred to as horizontal axis washers, use considerably less water than comparable top loaders, and have performance features much like their European counterparts. The variation in mechanical agitation and fabric-to-water ratio necessitated the reformulation of the surfactant package to limit suds generation and prevent soil redeposition. Wisk HE and Tide HE were launched specifically to accommodate these new machines. Although these machines have been commercially available for close to eight years, the relatively high cost associated with front loaders has resulted in considerably less sales than anticipated, and the range of detergents manufactured for these types of machines remains limited.

## B.  Line Extensions

An inherent weakness concerning liquid detergents is the inability to stabilize cost effectively an active oxygen bleach species in the liquid detergent matrix. Nearly every major detergent manufacturer has launched a line extension of its formulation with a *bleach alternative* variant. The consumer receives a product with an improved cleaning and whiteness profile through the incorporation of an advanced enzyme/optical brightener package. All Free Clear, an enzyme-, dye-, and fragrance-free version of All, is a second example of targeting a niche market successfully without overhauling the existing formulation. Unilever was able to design an enzyme-linked immunosorbent assay to prove that All Free Clear effectively rids fabrics of dust mite matter, an allergen commonly found in the home. Procter & Gamble was able to leverage their cyclodextrin technology (first employed as a fragrance delivery vehicle) in their Febreze for the Wash laundry additive. Cyclodextrins are ring structures that are comprised of six, seven, or eight glucose monomers that can entrap small molecules [197]. Unwanted odors can be effectively removed from carpets and various textiles as opposed to masking the odor with perfumes that gradually lose their fragrance over time. (Henkel has recently introduced Fresh Magic, claimed to be the first detergent to neutralize unpleasant odors in synthetic garments as they are worn by the individual. The active ingredient, Neutralin, is a proprietary combination of a malodor absorber and a unique fragrance designed to deliver odor elimination and a long-lasting smell.)

## C.   Color/Fabric Care

Several of the formulation improvements over the past few years have strong roots in research and development. ISP introduced a new dye transfer inhibiting polymer (Chromabond S-100) based upon poly(4-vinylpyridinium betaine) in a European color-safe detergent. This next generation dye transfer inhibitor is less likely than PVP, the industry standard, to interact with other ingredients (linear alkylbenzene sulfonate) in the detergent formulation. Procter & Gamble strengthened its claim as the market leader by releasing Tide Clean Rinse, a formulation designed to deliver enhanced cleaning, better stain removal, and brighter clothes. The product utilizes an alkoxylated polyalkyleneimine polymer to prevent soil from redepositing [206] onto fabrics and a proprietary mannanase enzyme designed to remove carbohydrates that physically attract/retain other soils. Procter & Gamble also introduced a novel, hydrophobically modified cellulosic material ("Liquifiber") into its Cheer liquid detergent with Colorguard. The product contains multiple ingredients to help protect and maintain the original color of garments [183–185]. The cellulosic polymer binds to the fabric and helps prevent cotton fibrils from breaking loose during mechanical agitation. This reduction in fabric abrasion prevents garments from prematurely looking worn after multiple wash cycles. Cheer with Colorguard also utilizes a peroxide to scavenge chlorine, a cationic polymer based upon imidazole and epichlorohydrin to fix dyes in place, and the vinylimidazole derivative of PVP to inhibit dye transfer from fugitive dyes in the wash bath. Henkel now markets a product called Black Magic that contains a dye fixative that keeps black clothes black longer by preventing color fading.

## D.  Wrinkling Reduction/Ease of Ironing

Most consumers dread the idea of ironing their clothes after laundering. Unilever launched Wisk with Wrinkle Reducer in 2000 in an effort to reduce or eliminate the need for ironing, but product acceptance was not overwhelming. Procter & Gamble markets Downy Wrinkle Releaser as an auxiliary product that sprays directly onto garments. Although some benefits are realized on select fabrics, the premise of no ironing has not been achieved. (In Europe Procter & Gamble is marketing Bold Easy Iron with fabric softener. The product utilizes a combination of softening technologies and the hydrophobically modified cellulosic (Liquifiber) technology to reduce wrinkling concerns.)

## E.  Unitized Dose Products

In the past three to four years alternative product forms have taken center stage. The reintroduction of detergent tablets has met with great success in Europe (in particular, the automatic dishwashing category), but has generated little interest in the U.S. Liquid-containing unitized dose sachets are successfully marketed in the U.S. automatic dishwashing category (Electrasol gelpacs and Cascade 2-in-1 action pacs, a powder encased in a liquid over wrap), but the market has not evolved to liquid-filled sachets for laundry. In Europe Persil Liquits (Henkel) is a water-free liquid detergent packaged in a polyvinyl alcohol wrap. The polyvinyl alcohol sachet is water soluble and dissolves in the wash bath within a few minutes, leaving no residue on clothing. The unitized dose concept has not met with overwhelming interest among consumers in the laundry area, so predicting its acceptance in the U.S. is anything but simple.

## VII.  FUTURE TRENDS

During 2003 and 2004 the price of oil and natural gas increased dramatically. This in turn has resulted in significant increases in detergent raw material costs, and considerable emphasis has been placed on controlling product spend. With the prospect of higher prices persisting for the next few years, where will the detergent manufacturers focus their research efforts? Will the consumer be content with the status quo, or continue to demand additional benefits without incurring added costs?

## A.  Detergent/Fabric Softener Combinations: Return of the Two-in-Ones

One concept that appears to be of interest is the formulation of "two-in-one" or softergent liquids. This idea has seen its popularity rise and fall over the past two decades. The prospect of delivering fabric softening in a liquid detergent has often met with consumer pessimism. Most liquid detergent/fabric softener combinations are based upon cationic surfactants (to provide fabric softening and antistatic properties). In the past these products often had associated negatives with respect to primary detergency (removal of clay/particulate soils) and problems with clay soil redeposition during the wash cycle, leading to poor whiteness maintenance of cotton garments. In addition, the (ion pair) complex formed between the anionic and cationic surfactants resulted in less than desirable softening properties. In August 2004 Procter & Gamble introduced Tide with a Touch of Downy, a "pseudo" two-in-one composition. This product is not targeted as a replacement for rinse-added fabric softeners (as was the case with earlier attempts), but is being launched as a line extension to the current Tide formulation. The consumer gets a minimal enhancement of softening/antistatic properties, and the high level of cleaning performance anticipated from the market leader is maintained without any noticeable downsides in detergent efficacy. Unilever recently filed a U.S. patent application with its own version of a two-in-one formulation based upon a very high-molecular-weight cationic homo/copolymer as the active softening agent [227]. These cationic polymers have been used extensively in personal care applications (hair and skin), and have a high affinity for negatively charged surfaces (like cotton). It will be interesting to watch how this market evolves over the next 6 to 12 months from a technical and consumer perspective.

## B.  Suspension of Visual Cues

One of the recent trends in personal care products is to deliver a "visual cue," or stated in a more definitive way, a consumer-perceptible point of differentiation. For the most part, this has been accomplished by the introduction of a polymeric rheology modifier (organic gums, cellulosics, hydrophobically modified alkali-soluble emulsions (HASE), polyacrylates, or carbomers) and/or an inorganic (clay) to clear, gel formulations to establish a yield point. These thixotropic, highly pseudoplastic systems allow the suspension of various "actives" which are visible to the consumer. The list of actives can encompass substances ranging from air particles to insoluble builders to moisturizing beads to stabilized (micro)emulsions. This concept could be a way to promote a unique advantage offered by a cleaning product, even if the visual cue is actually inert in reality. The majority of liquid detergents sold in the U.S. are currently homogeneous, clear, single-phase products, so adapting this technology to a HDLD, although not trivial, is potentially viable.

## C.  Fabric/Color Care

Fabric and color care are areas that have received considerable emphasis over the past decade. The introduction of hydrophobically modified cellulosics (Liquifiber), multiple variants based upon derivatives of PVP technology for dye transfer inhibition, and a wealth of recent patent applications on polymeric additives to improve fabric appearance (looking newer longer, reduced pilling, fuzz reduction) will set the tone for the next round of product improvements.

## D. Ultraviolet Protection

Another area that has received considerable attention, although mainly outside of the U.S. and Europe, is ultraviolet light protection for garments (and skin). Ciba Specialties has marketed ultraviolet absorbing products under the Tinosorb tradename that boost the sun protection factor (SPF) of a typical cotton T-shirt from a value of 5 to 8 to at least 15 after 5 wash cycles. Although this technology has not been embraced in the more temperate regions of the world, climate changes and the impact of global warming will make these materials more popular, with rinse-added fabric softeners the likely delivery vehicle.

## E. Enhanced Detergency: Several Potential Approaches

Improvements in detergent efficacy will continue to capture the undivided attention of detergent manufacturers. There are multiple approaches that encompass a wide variety of current and novel technologies. Examples of these types of approaches range from unique combinations of enzymes (pectate, lyase, and mannanase) to facilitate the removal of food soil residues, to ethoxylated quaternized amines to improve soil suspension and cleaning of outdoor soils/stains, to nanoparticle technologies to deliver crease resistance properties in tumble dry additives and aqueous ironing formulations [180].

As discussed in Section V, detergent manufacturers have begun utilizing cellulose binding domains to enhance the deposition of actives onto fabric surfaces. Will this open a unique window into alternate biotechnology advances/ approaches? Can the emergence of water-soluble silicate builders (e.g., SKS-6) as replacements for zeolites (in Europe) translate to liquid detergent formulations? HDLDs have been formulated with zeolite 4A as the builder of choice, but hard packing and viscosity build upon aging have made these products very difficult to formulate successfully, and these liquids have not gained consumer acceptance. Despite the wealth of patent activity on soil release polymers for cotton, will these technologies become commercially viable (from a performance and cost/use standpoint)? Can encapsulation technologies be expanded beyond fragrances/odor protection to include actives such as bleaches?

## F. New and Developing Markets

Most of the emphasis in this section has focused on the expanding boundaries of patent art and technology, and ascertaining its impact on future (liquid) detergent compositions. The laundry detergent markets in the U.S., Europe, and Japan are well established, so where will the next influx of new products occur? The answer will probably be in emerging countries, where consumers will gradually upgrade from generic, low-cost domestic products to more high-tech offerings. Although liquids will not be the principal product form utilized by the consumer, limitations on phosphate levels (environmental pressures) and the implementation of alternative, poorer performing builders may lead the market back toward liquids. Eastern Europe (Poland, Hungary, the Czech Republic, and Slovakia), China, India, and parts of Latin America (Mexico, Brazil) will likely spur rapid growth and new product entries, further diversifying the variety and compositions marketed globally.

## APPENDIX

**TABLE A**   Recent HDLD Patents (1994–2004) Related to Fabric Care

| Patent no. | Issue date | Inventor/company | Technology |
|---|---|---|---|
| U.S. 2004/0038851 | 02/2004 | Aubay *et al.* | Deliver crease resistance/ease of ironing properties to fabrics by a treatment comprising nanoparticles |
| U.S. 6696405 | 02/2004 | Mooney/Unilever | Improved crease recovery of fabrics utilizing C20–C40 saturated or unsaturated aliphatic hydrocarbons. For use in fabric conditioning formulations used in the dryer cycle (sheets, liquids for rinse-added fabric softeners) |
| U.S. 2002/0016276 | 02/2002 | Spendel/Procter & Gamble | Yarn strength-enhancing agents suitable for laundry and/or fabric care compositions. The additives that improve yarn strength can include polysaccharides, clays, starches, chitosans, and mixtures thereof |
| U.S. 5336419 | 08/1994 | Coffindaffer *et al.*/ Procter & Gamble | Silicone gel for ease of ironing and improved fabric after treatment |

*(continued)*

**TABLE A**  (Contd.)

| Patent no. | Issue date | Inventor/company | Technology |
|---|---|---|---|
| U.S. 2004/0121930 | 06/2004 | Wang *et al.*/ Procter & Gamble | Cationic celluloses for enhanced delivery of fabric care benefits (softening, color protection, pill/fuzz reduction, antiabrasion, antiwrinkle). One water-insoluble fabric care benefit agent in combination with at least one delivery-enhancing agent (dispersible polyolefins and latexes) |
| WO 2003/027219 | 04/2003 | Cooke *et al.*/ Unilever | Polymeric material comprising one or more poly(oxyalkylene) amine groups and an epihalohydrin-derived terminal group that acts as a lubricant to reduce fabric abrasion during the tumble dryer/wash cycle, decreasing fabric wear and color loss on collars and cuffs |
| GB 2360792 | 04/2003 | Hopkinson *et al.*/ Unilever | Fiber rebuild polymers (cellulosics/polysaccharides with pendant ester groups) that impart unique properties to the fabric. These properties include replacing lost fiber weight (on cellulosics), repair/rebuild fiber strength, enhance fabric body/ smoothness, reduce fading, improve appearance and fabric comfort, control dye transfer, and can increase fiber stiffness, deliver antiwrinkling benefits, and ease of ironing properties. The polymers undergo a chemical change in the wash bath (hydrolysis of ester groups) that enhances their affinity for the fabric surface |

*(continued)*

**TABLE A**   (Contd.)

| Patent no. | Issue date | Inventor/company | Technology |
|---|---|---|---|
| WO 98/29530 | 07/1998 | Randall *et al.*/ Procter & Gamble | Laundry detergent compositions containing fiber reactive additives (polyamide–polyamines) to improve fabric appearance and integrity |
| WO 97/42287 | 06/1994 | Pramod *et al.*/ Procter & Gamble | Laundry detergent compositions containing fiber reactive additives (modified polyamines) to improve fabric appearance and integrity |
| WO 99/14245 | 03/1999 | Leupin *et al.*/ Procter & Gamble | Laundry detergent compositions containing cellulosic-based polymers to improve fabric appearance and integrity Laundering of fabric/textiles with the additive leads to overall improvements in fabric appearance, pill/fuzz reduction, antifading properties, improved abrasion resistance, and enhanced softening |
| U.S. 5571286 | 11/1996 | Connell *et al.*/ Precision Process Ltd | Polymers and prepolymers derived from polyoxyalkylene amines and their use in a process for shrinkproofing wool |

**TABLE B**   Recent Patents (1993–2004) on HDLDs with Enzymes

| Patent no. | Issue date | Inventor/company | Technology |
|---|---|---|---|
| U.S. 2003/ 0022807 | 01/2003 | Wilting *et al.*/ Novozymes North America, Inc. | Removing or bleaching soils/stains derived from xyloglucan-containing food or plants, select binding of soils on cellulosic fabrics |
| U.S. 6015783 | 01/2000 | Von der Osten *et al.*/ Novo Nordisk A/S | Removal or bleaching of soils/stains from cellulosics with an enzyme hybrid comprised of a catalytically activated amino acid sequence from a noncellulytic enzyme |

*(continued)*

**TABLE B** (Contd.)

| Patent no. | Issue date | Inventor/company | Technology |
|---|---|---|---|
| | | | linked to an amino acid comprising a cellulose binding domain. The enzyme hybrid in combination with a surfactant in a detergent formulation |
| WO 99/32594 | 07/1999 | Duval *et al.*/ Procter & Gamble | Cleaning compositions containing a neopullulanase for improved stain removal, enhanced overall cleaning, and sanitization of treated surface |
| WO 98/13457 | 04/1998 | Ohtani *et al.*/ Procter & Gamble | Cleaning compositions containing a mycodextranase for improved stain removal, enhanced overall cleaning, and sanitization of treated surface |
| EP 0603931 | 07/1993 | Pramod/Procter & Gamble | Liquid laundry detergents containing stabilized glucose/ glucose oxidase as a hydrogen peroxide generation system |
| U.S. 6734155 | 05/2004 | Herbots *et al.*/ Procter & Gamble | Cleaning compositions containing an oxidoreductase to facilitate the removal of colored and/or everyday body stains/soils |
| U.S. 6114509 | 09/2000 | Olsen *et al.*/ Novo Nordisk A/S | Modified polypeptides with reduced allergenicity |
| WO 00/18897, EP 1119613, EP 1117770 | 04/2000 | Smets *et al.*/ Procter & Gamble | Detergent compositions containing a chemical component linked to a cellulose binding domain. These materials deliver a higher effective concentration of the active to the fabric surface |
| U.S. 5981718 | 11/1999 | Nielsen *et al.*/ Novo Nordisk A/S | 4-Substitiuted phenyl boronic acids as enzyme stabilizers |
| U.S. 5834415 | 11/1998 | Nielsen *et al.*/ Novo Nordisk A/S | Naphthalene boronic acids |

**TABLE C**  Recent Patents (1996–2004) on HDLDs with Polymers

| Patent no. | Issue date | Inventor/company | Technology |
|---|---|---|---|
| U.S. 6372708 | 04/2002, | Kasturi et al./ Sivik et al./ Procter & Gamble | Polymeric additives (possessing a cationic charge) that deliver enhanced suds duration and suds volume |
| U.S.6645925 | 11/2003 | | |
| U.S. 5854197 | 12/1998 | Duccini et al./ Rohm and Haas | Cleaning compositions containing a lime soap dispersant (maleic/hydrophobe) that delivers improved whiteness maintenance in liquid detergents built with modest levels of fatty acid soaps |
| U.S. 6451756 | 09/2002 | Shulman et al./ Rohm and Haas | Hydrophobically modified polycarboxylates that deliver soil release benefits on cotton and cotton-containing blends. These polymers are effective on oil/greasy soils through the wash or during the rinse cycle |
| U.S. 6291415 | 09/2001 | Gosselink et al./ Procter & Gamble | Cotton soil release polymers from modified polyamines having functionalized backbones and improved stability to bleach. Laundry detergent compositions comprising these polymers possessing enhanced hydrophilic soil removal benefits |
| U.S. 6191093 | 10/2001 | | |
| U.S. 6087316 | 07/2000 | | |
| U.S. 6071871 | 06/2000 | | |
| U.S. 6057278 | 05/2000 | | |
| EP 1402877 | 03/2004 | Tepe/Rohm and Haas | Composition containing at least one lipophilically modified copolymer (acrylic residues) and a colloidal inorganic clay. Thickener for high surfactant concentrations ($>18\%$) |

*(continued)*

**TABLE C**  (Contd.)

| Patent no. | Issue date | Inventor/company | Technology |
|---|---|---|---|
| U.S. 5565145 | 10/1996 | Watson *et al.*/ Procter & Gamble | Cleaning and soil dispersing compositions comprising ethoxylated/propoxylated polyalkyleneamine polymers |
| U.S. 6369018 | 04/2002 | Hsu *et al.*/ Unilever | Easy pouring (high shear thinning), transparent liquid capable of suspending particles in the presence of high levels of surfactant and electrolyte. Polymer gum solution selected from carrageenans, gellans, and agars |
| U.S. 2003/0186832 | 10/2003 | Padron *et al.*/ Unilever/Procter & Gamble | Isotropic liquid detergents containing a soil release polymer and an antiredeposition enhancer (PVP, PVPNO) delivering a synergistic improvement in soil antiredeposition |
| U.S. 6664223 | 12/2003 | Zappone *et al.*/ Colgate-Palmolive | Fabric treatment composition that contains a polyfunctional molecule, such as derived from polyacrylic acid, in combination with a urea-derived compound. During pressing or ironing of the fabric, the urea-derived compound is said to crosslink the polyfunctional molecule and thereby provide crease resistance to the fabric |

**TABLE D**  Recent Patents (1998–2003) on HDLDs with Optical
Brighteners/Antioxidants/Fabric Protection

| Patent no. | Issue date | Inventor/company | Technology |
|---|---|---|---|
| U.S. 6482241 | 11/2002 | Metzger *et al.*/Ciba Specialty Chemicals | Method of improving the sun protection factor (SPF) of textile fabrics (cotton, polyamide, wool) using asymmetric stilbene derivatives |
| U.S. 5854200 | 12/1998 | Severns *et al.*/ Procter & Gamble | Rinse-added fabric softeners containing antioxidants for sun fade protection of fabrics. Nonfabric-staining, light-stable antioxidant compounds comprising C8–C22 hydrocarbon fatty organic moieties |
| U.S. 6015504 | 01/2000 | Reinehr *et al.*/Ciba Specialty Chemicals | New compounds (triazinyl diaminostilbenes) that are useful as ultraviolet absorbing agents and increase the SPF of textile fibers |
| U.S. 6613340 | 09/2003 | Koshti *et al.*/ Galaxy Surfactants | Substantive hydrophobic cationic UV absorbing compounds |

**TABLE E**  Recent Patents (1995–2004) on HDLDs with Dye Transfer Inhibition

| Patent no. | Issue date | Inventor/company | Technology |
|---|---|---|---|
| U.S. 5855621 | 01/1999 | Damhus *et al.*/ NovoNordisk A/S | Reduce DTI through the addition of a peroxidase/oxidase during the wash/rinse cycle |
| WO 99/15614 | 04/1999 | Shih *et al.*/ISP Investments | Poly(vinylpyridine betaines) |
| WO 97/42291 | 11/1997 | Panandiker *et al.*/ Procter & Gamble | Modified polyamines |
| U.S. 5880081 | 03/1999 | Gopalkrishnan *et al.*/ BASF | Hydrophilic copolymer (unsaturated "philic" copolymer with an oxyalkylated monomer) |

*(continued)*

**TABLE E** (Contd.)

| Patent no. | Issue date | Inventor/company | Technology |
|---|---|---|---|
| U.S. 2004/0038852 | 02/2004 | Brown *et al.*/ Procter & Gamble | Chlorine scavenger in concert with a polymeric DTI and less than 0.02% of a triazinylaminostilbene optical brightener |
| U.S. 6733538 | 05/2004 | Panandiker *et al.*/ Procter & Gamble | Dye maintenance polymer comprising one or more linearly polymerizing monomers, cyclically polymerizing monomers and mixtures thereof |
| U.S. 5466802, WO 95/27038 | 11/1995 | Panandiker *et al.*/ Procter & Gamble | PVPNO and PVP–VI |
| EP 664335 | 07/1995 | Abdennaceur *et al.*/ Procter & Gamble | Polysulfoxide polymers |

**TABLE F**  Recent HDLD Patents (1997–2003) Related to Surfactants

| Patent no. | Issue date | Inventor/company | Technology |
|---|---|---|---|
| WO 97/39089 | 10/1997 | Connor *et al.*/Procter & Gamble | Liquid cleaning compositions containing selected mid-chain branched surfactants and cosurfactants. These surfactants deliver enhanced cleaning in cold water and in the presence of hard water |
| WO 01/90293 | 11/2001 | Libe *et al.*/Huish Detergents | Compositions containing α-sulfofatty acid ester surfactants and hydrotropes and method of manufacture |
| U.S. 6596680 | 07/2003 | Kott *et al.*/Procter & Gamble | Specific alkylbenzene surfactant mixtures to improve detergency |

**TABLE G**   Recent HDLD Patents (2001–2004) on Perfume Adjuvants

| Patent no. | Issue date | Inventor/company | Technology |
|---|---|---|---|
| U.S. 6184188 | 02/2001 | Severns *et al./* Procter & Gamble | Fragrance delivery system for liquid detergent compositions comprising a β-ketoester |
| U.S. 2004/0018955 | 01/2004 | Wevers *et al./* Procter & Gamble | Pro-perfume composition comprising an amino functional component and a benefit agent that is stably suspended in a liquid detergent. Provides enhanced deposition and a long-lasting release on the treated fabric |
| WO 02/077150 | 10/2002 | Pashkovski *et al./* Colgate-Palmolive | Fragrance-containing gel delivering enhanced deposition and retention of said fragrance (from structured liquids) |
| EP 1146057 | 10/2001 | Quellet *et al./* Givaudan | Polymeric nanoparticles that incorporate olfactive components into an emulsion polymer and act as an efficient delivery system for these fragrances. Perfume is gradually released over a period of time, preventing "top notes" from volatizing too quickly |
| WO 03/049666 | 06/2003 | Fehr *et al./* Firmenich | Compounds comprising one β-oxy or β-thio carbonyl moiety capable of liberating a perfume molecule (α,β-unsaturated ketone, aldehyde, or carboxylic ester) |
| WO 01/23516 | 04/2001 | Foley *et al./* Procter & Gamble | Compositions that are particularly effective at masking malodors or odor suppression. Complexing agents (cyclodextrins) that have an internal cavity, forming complexes that incorporate the malodor |
| WO 2003015736 | 02/2003 | Ness *et al./*Quest International | Aqueous fabric care composition comprises surfactant, silicone insoluble in water, and perfume having a solubility parameter (SP) not exceeding about 20. By using a perfume with a low SP value, the invention enables good partitioning of perfume into the silicone of the composition, which means that the perfume will be associated with the silicone and deposited onto fabric in use |

**TABLE H**  Recent Patents (2000–2004) on Unit Dosed HDLDs

| Patent no. | Issue date | Inventor/company | Technology |
|---|---|---|---|
| WO 02/097026 | 12/2002 | Fregonese/Reckitt Benckiser | Liquid detergent compositions encapsulated in a polymer (especially compositions that contain ingredients capable of crosslinking a water-soluble polymer) |
| U.S. 6448212 | 07/2000 | Holderbaum *et al.*/ Henkel KgaA | Laundry detergent portion for use in a washing/dishwashing machine for a program taking place in an aqueous phase |
| WO 02/16541 | 02/2002 | Kaiser *et al.*/Reckitt Benckiser | Aqueous liquid detergent packaged in a water-soluble or water-dispersing package having an improved stability. Composition contains 20–50% water, at least one polyphosphate, and potassium and/or sodium ions |
| EP 1378564 | 01/2004 | Bonastre *et al.*/ Cognis Iberia | Laundry detergent portion |
| EP 1319706 | 06/2003 | Ramcharen *et al.*/ Unilever | Dispersed solid in a liquid detergent in a water-soluble pouch |

**TABLE I**  Recent Patents (1993–2004) Related to Builders

| Patent no. | Issue date | Inventor/company | Technology |
|---|---|---|---|
| U.S. 5704556 | 01/1998 | McLaughlin/DevMar | Process for rapidly producing finely divided aluminosilicate particles by media grinding techniques |
| U.S. 6699831 | 03/2004 | Takano *et al.*/Kao Corporation | Liquid detergent composition comprising an aluminosilicate or crystalline silicate |
| U.S. 5252244 | 10/1993 | Beaujean *et al.*/ Henkel KgaA | Aqueous zeolite-containing liquid detergent stabilized with an electrolyte mixture |

**TABLE J**   Recent HDLD Patents (1995–2004) with Nonaqueous Liquids

| Patent no. | Issue date | Inventor/company | Technology |
|---|---|---|---|
| U.S. 2003/0100468 | 05/2003 | Smerznak *et al.*/ Procter & Gamble | Nonaqueous particulate-containing liquid laundry detergents comprising a peroxygen bleaching agent and an organic detergent builder |
| U.S. 6770615 | 08/2004 | Aouad *et al.*/Procter & Gamble | Nonaqueous liquid laundry detergent compositions with a suspended solid particulate phase comprised of low-density particles (binding agent, alkalinity source, a chelant, and builder or mixtures thereof). Ingredients that are insoluble in the surfactant-rich phase can be incorporated into the liquid phase without segregation or separation |
| U.S. 5441661 | 08/1995 | Beaujean *et al.*/ Henkel KgaA | Nonaqueous liquid detergent containing a hydrated zeolite stabilized by a polar deactivating agent |

**TABLE K**   Recent Patents (2004) with Fabric Softeners

| Patent no. | Issue date | Inventor/company | Technology |
|---|---|---|---|
| U.S. 2004/0152617 | 08/2004 | Murphy *et al.*/ Unilever | Cationic polymers and anionic surfactants that provide optimal cleaning and fabric softening properties. High-molecular-weight cationic polymers are used in place of conventional quats (cationic surfactants) to provide the softening benefit on cotton cloth |

# REFERENCES

1. Rosen, M.J., *Surfactants and Interfacial Phenomena,* 2nd ed., Wiley-InterScience, New York, 1989.
2. Lynn, J.L., *Kirk-Othmer Encyclopedia of Chemical Technology*, 4th ed., Vol. 7, John Wiley, New York, 1993, p. 1072 (and references therein).
3. Schick, M.J., Ed., *Nonionic Surfactants: Physical Chemistry*, Surfactant Science Series, Vol. 23, Marcel Dekker, New York, 1987.
4. Coons, D., Dankowski, M., Diehl, M., Jakobi, G., Kuzel, P., Sung. E., and Trabitzsch, U., in *Surfactants in Consumer Products: Theory, Technology and Application*, Falbe, J., Ed., Springer-Verlag, New York, 1987.
5. Cahn, A., *INFORM*, 5, 70, 1994.
6. Gormsen, E., Roshholm, P., and Lykke, M., in *Proceedings of the 3rd World Conference on Detergents: Global Perspectives*, Cahn, A., Ed., AOCS Press, Champaign, IL, 1994, p. 198.
7. Jurgens, A., *Tenside Surf. Det.*, 26, 226, 1989.
8. van de Pas, J.C., *Tenside Surf. Det.*, 28, 158, 1991.
9. Gray, G.W. and Winsor, P.A., in *Liquid Crystals and Plastic Crystals*, Vol. 1, Ellis Horwood, Chichester, U.K., 1974, p. 223.
10. Sein, A., Engberts, J.B.F.N., Vanderlinden, E., and van de Pas, J.C., *Langmuir*, 9, 1714, 1993.
11. Schepers, F.J., Toet, W.K., and van de Pas, J.C., *Langmuir*, 9, 956, 1993.
12. Haslop, W.P., Allonby, J.M., Akred, B.J., and Messenger, E.T., U.S. Patent 4618446 to Albright and Wilson Ltd, 1986.
13. Broekhoff, J.C.P. and van de Pas, J.C., presented at American Oil Chemists Society Conference, Anaheim, CA, April 1993.
14. Hales, S.G., Khoshdel, E., Montague, P.G., van de Pas, J.C., and Visser, A., WO 9109109 to Unilever, 1991.
15. Machin, D., Naik, A.R., Buytenhek, C.J., and van de Pas, J.C., EP 0328176 to Unilever, 1989.
16. Montague, P. and van de Pas, J., GB 2237813 to Unilever, 1991.
17. van de Pas, J.C., Schepers, F.J., and Verheul, R., WO 9105844 to Unilever, 1991.
18. Montague, P. and van de Pas, J.C., EP 346995 to Unilever, 1989.
19. Montague, P.G. and van de Pas, J.C., U.S. Patent 5147576 to Lever Brothers Co., 1991.
20. Krishnan, S.V., U.S. Patent 5318715 to Colgate-Palmolive Co., 1994.
21. Liberati, P., McCown, J.T., Aronson, M., and van de Pas, J.C., U.S. Patent 5073285, 1991.
22. Boutique, J.P. and Surutzidis, A., EP 5727223 A1 to Procter & Gamble, 1993.
23. Yanaba, S., Shiobara, M., Masamizu, K., Morohara, K., and Abe, S., EP 0459077 to Lion Corp., 1991.
24. van de Pas, J.C., WO 9113963 to Unilever, 1991.
25. Cao, H.-C., U.S. Patent 5364553 to Colgate-Palmolive Co., 1994.
26. van de Pas, J.C., van Voorst, F., and Toet, W.K., EP 079646 A2 to Unilever, 1983.
27. Machin, D. and van de Pas, J.C., U.S. Patent 5006273 to Lever Brothers, 1991.
28. van de Pas, J.C. and Buytenhek, C.J., *Colloids Surf.*, 68, 127, 1992.
29. Machin, D. and van de Pas, J.C., EP 301883 to Unilever, 1989.

30. Buytenhek, C., Mohammadi, M., van de Pas, J.C., Schepers, F., and Leonardus, C., WO 9109107 to Unilever, 1991.
31. van de Pas, J.C., WO 9403575 AI to Unilever, 1994.
32. Dawson, P.L. and van de Pas, J.C., WO 9109108 AI to Unilever, 1991.
33. Repinic, S.T., Zappone, M., Fuller, R.L., and Krishnan, S.V., U.S. Patent 5602092 to Colgate-Palmolive Co., 1997.
34. Broze, G., presented at CSMA Conference, Chicago, April 1992.
35. Adams, R. and Crossin, M.L., U.S. Patent 4744916 to Colgate-Palmolive Co., 1988.
36. Dixit, N.S., Rhinesmith, R.J., Sullivan, J.J., Barone, C.A., Adams, R.P., and Lai, K.Y., U.S. Patent 4892673 to Colgate-Palmolive Co., 1990.
37. Russell, S.W. and Tomlinson, A.D., U.S. Patent 5102574 to Lever Bros., 1992.
38. van der Hoeven, P.C. and Prescott, J., WO 9425562 to Unilever, 1994.
39. Hormann, J.M., Houghton, M. and Verheul, R.C., WO 9423009 to Unilever, 1994.
40. Broze, G., Bastin, D., Laitem, L., and Ouhadi, T., U.S. Patent 4655954 to Colgate-Palmolive Co., 1987.
41. Julemont, M., Zocchi, G., Mineo, N., and Fonsny, P., U.S. Patent 5004556 to Colgate-Palmolive Co., 1991.
42. Broze, G. and Bastin, D., U.S. Patent 4806260 to Colgate-Palmolive Co., 1989.
43. *McCutcheon's Detergents and Emulsifiers*, North American Edition and International Edition, 1994.
44. Matson, T.P., *J. Am. Oil Chem. Soc.*, 55, 66, 1978.
45. Hövelmann, P., in *Proceedings of the 3rd World Conference on Detergents: Global Perspectives*, Cahn, A., Ed., AOCS Press, Champaign, IL, 1994, p. 117.
46. Satsuki, T., in *Proceedings of the 3rd World Conference on Detergents: Global Perspectives*, Cahn, A., Ed., AOCS Press, Champaign, IL, 1994, p. 135.
47. Brancq, B., in *Proceedings of the 3rd World Conference on Detergents: Global Perspectives*, Cahn, A., Ed., AOCS Press, Champaign, IL, 1994, p. 147.
48. Plantaren, *A New Generation of Surfactants*, Henkel Corporation, 1992.
49. Bryan, R., Trends in the Alkylbenzene Industry, presented at the World Petrochemical Conference, Houston, TX, 1994.
50. de Almeida, J.L.G., Dufaux, M., Bentaarit, Y., and Naccache, C., *J. Am. Oil Chem. Soc.*, 71, 675, 1994.
51. Matheson, K.L., in *Anionic Surfactants: Organic Chemistry*, Surfactant Science Series, Vol. 56, Stache, H.W., Ed., Marcel Dekker, New York, 1995, p. 109.
52. Matheson, K.L. and Matson, T.P., *J. Am. Oil Chem. Soc.*, 60, 1693, 1983.
53. van Os, N.M., Kok, R., and Bolsman, T.A.B.M., *Tenside Surf. Det.*, 29, 175, 1992.
54. Cox, M.F. and Matheson, K.L., *J. Am. Oil Chem. Soc.*, 62, 1396, 1985.
55. Smith, D.L., Matheson, K.L., and Cox, M.F., *J. Am. Oil Chem. Soc.*, 62, 1399, 1985.
56. Cohen, L., Moreno, A., and Berna, J.L., *J. Am. Oil Chem. Soc.*, 70, 73, 1993.
57. Aronson, M.P., Gum, M.L., and Goddard, E.D., *J. Am. Oil Chem. Soc.*, 60, 1333, 1983.
58. Matheson, K.L., Cox, M.F., and Smith, D.L., *J. Am. Oil Chem. Soc.*, 62, 1391, 1985.
59. Jakobi, G. and Schwuger, M.J., in *Waschmittelchemie*, Henkel & Cie, Heidelberg, 1976, p. 91.
60. Moreno, A., Cohen. L., and Berna, J.L., *J. Am. Oil Chem. Soc.*, 67, 547, 1990.
61. Cohen, L., Vergara, R., Moreno, A., and Berna, J.L., *J. Am. Oil Chem. Soc.* 72, 115, 1995.

62. Moreno, A., Cohen, L., and Berna, J.L., *Tenside Surf. Det.*, 25, 216, 1988.
63. Himpler, H.A., *INFORM*, 6, 22, 1995.
64. McKenzie, D.A., *J. Am. Oil Chem. Soc.*, 55, 93, 1978.
65. Dillan, K.W., *J. Am. Oil Chem. Soc.*, 62, 1144, 1985.
66. Cox, M.F., Borys, N.F., and Matson, T.P., *J. Am. Oil Chem. Soc.*, 62, 1139, 1985.
67. Dillan, K.W., Goddard, E.D., and McKenzie, D.A., *J. Am. Oil Chem. Soc.*, 57, 230, 1980.
68. Matheson, K.L., Matson, T.P., and Yang, K., *J. Am. Oil Chem. Soc.*, 63, 365, 1986.
69. Domingo, X., in *Anionic Surfactants: Organic Chemistry*, Surfactant Science Series, Vol. 56, Stache, H.W., Ed., Marcel Dekker, New York, 1995, p. 223.
70. Connor, D.S., Scheibel, J.J., and Kao, J.N., U.S. Patent 5338486 to Procter & Gamble, 1994.
71. Connor, D.S., Scheibel, J.J., and Segverson, R.G., WO 9206073 to Procter & Gamble, 1992.
72. Caswell, D.S., Murch, B.P., and Mao, M., WO 9206072 to Procter & Gamble, 1992.
73. Surutzidis, A., Boutique, J.P., Fu, Y.C., Murch, B.P., Connor, D.S., and Scheibel, J.J., U.S. Patent 5318728 to Procter & Gamble, 1994.
74. Mao, M., Cook, T., Panandiker, R., and Wolff, A., WO 9206154 to Procter & Gamble, 1992.
75. Murch, B.P., Morrall, S.W., and Mao, M., WO 9206162 to Procter & Gamble, 1992.
76. Collins, J.H. and Murch, B.P., WO 9206160 to Procter & Gamble, 1992.
77. Ofosu-Asante, K., Willman, K.W., and Foley, P.R., WO 9305132 to Procter & Gamble, 1993.
78. Smith, N.R., *Soap Cosmet. Chem. Specialties*, 48, 1989.
79. Friedli, F.E., Watts, M.M., Domsch, A., Tanner, D.A., Pifer, R.D., and Fuller, J.G., in *Proceedings of the 3rd World Conference on Detergents: Global Perspectives*, Cahn, A., Ed., AOCS Press, Champaign, IL, 1994, p. 156.
80. Houston, C.A., in *Proceedings of the 2nd World Conference on Detergents: Global Perspectives*, Baldwin, A.R., Ed., AOCS Press, Champaign, IL, 1986, p. 161.
81. Schwuger, M.J. and Smulders, E.J., in *Detergency, Theory and Technology*, Surfactant Science Series, Vol. 20, Cutler, W.G. and Kissa, E., Eds., Marcel Dekker, New York, 1987, p. 371.
82. Nagarajan, N.K. and Paine, H.L., *J. Am. Oil Chem. Soc.*, 61, 1475, 1984.
83. Rieck, H.P., in *Proceedings of the 3rd World Conference on Detergents: Global Perspectives*, Cahn, A., Ed., AOCS Press, Champaign, IL, 1994, p. 161.
84. Hollingsworth, M.W., *J. Am. Oil Chem. Soc.*, 55, 49, 1978.
85. Jakobi, G. and Schwuger, M.J., *Chem. Z.*, 182, 1975.
86. Schweiker, G.C., in *Proceedings of the 3rd World Conference on Detergents: Global Perspectives*, Cahn, A., Ed., AOCS Press, Champaign, IL, 1994, p. 63.
87. SKS-6, The Performance Builder for a Bright Future, Clariant, 2004.
88. Denkewicz, R.P., Jr. and Borgstedt, E.U.R., in *Proceedings of the 3rd World Conference on Detergents: Global Perspectives*, Cahn, A., Ed., AOCS Press, Champaign, IL, 1994, p. 213.
89. Crutchfield, M.M., *J. Am. Oil Chem. Soc.*, 55, 58, 1978.
90. Nagarajan, M.K., *J. Am. Oil Chem. Soc.*, 62, 949, 1985.
91. Dormal, M. and Noiret, J., U.S. Patent 4529525 to Colgate-Palmolive, 1985.

92. Bush, R.D., Connor, D.S., Heinzman, S.W., and Mackey, L.N., U.S. Patent 4663071 to Procter & Gamble Co., 1987.
93. Bush, R.D., U.S. Patent 4566984 to Procter & Gamble Co., 1986.
94. Bush, R.D., U.S. Patent 4654159 to Procter & Gamble Co., 1987.
95. Gresser, R., in *Proceedings of the 3rd World Conference on Detergents: Global Perspectives*, Cahn, A., Ed., AOCS Press, Champaign, IL, 1994, p. 153.
96. Perner, J., in *Proceedings of the 3rd World Conference on Detergents: Global Perspectives*, Cahn, A., Ed. AOCS Press, Champaign, IL, 1994, p. 168.
97. Wirth, W.J., in *Proceedings of the 3rd World Conference on Detergents: Global Perspectives*, Cahn, A., Ed., AOCS Press, Champaign, IL, 1994, p. 138.
98. Grime, K., in *Proceedings of the 3rd World Conference on Detergents: Global Perspectives*, Cahn, A., Ed., AOCS Press, Champaign, IL, 1994, p. 64.
99. Kravetz, L. and Guin, K.F., *J. Am. Oil Chem. Soc.*, 62, 943, 1985.
100. Kravetz, L. and Guin, K.F., presented at 2nd World Surfactant Congress, Paris, 1988.
101. Mathews, B.W., Weaver, L.H., and Kester, W.R., *J.Biol. Chem.*, 249, 8030, 1974.
102. Dahlquist, F.W., Long, J.W., and Bigbee, W.L., *Biochemistry*, 15, 1103, 1976.
103. Mozhaev, V.V. and Martinek, K., *Enzyme Microb. Technol.*, 6, 50, 1984.
104. Letton, J.C. and Yunker, M.J., U.S. Patent 4318818 to Procter & Gamble, 1982.
105. Kaminsky, G.J. and Christy, R.S., U.S. Patent 4305837 to Procter & Gamble, 1981.
106. Hora, J. and Kirits, G.A.A., U.S. Patent 4261868 to Lever Bros., 1981.
107. Severson, R.G., U.S. Patent 4537706 to Procter & Gamble, 1985.
108. Severson, R.G., U.S. Patent 4537707 to Procter & Gamble, 1985.
109. Tai, H.T., U.S. Patent 4404115 to Lever Bros., 1983.
110. Boskamp, J.V., U.S. Patent 4462922 to Lever Bros., 1984.
111. Boskamp, J.V., U.S. Patent 4532064 to Lever Bros., 1985.
112. Inamorato, J.T. and Crossin, M.C., U.S. Patent 4652394 to Colgate-Palmolive, 1987.
113. Crutzen, A., U.S. Patent 4842758 to Colgate-Palmolive, 1989.
114. Ramachandran, P. and Shulman, J.E., U.S. Patent 4900475 to Colgate-Palmolive, 1990.
115. Cao, H.-C., U.S. Patent 5221495 to Colgate-Palmolive, 1993.
116. Hessel, J.F., Cardinali, M.S. and Aronson, M.P., U.S. Patent 4908150 to Lever Bros., 1990.
117. Arnson, M.P., Cardinali, M.S., and McCown, J.T., U.S. Patent 4959179 to Lever Bros., 1990.
118. Panandiker, R.K., Thoen, C., and Lenoir, P., WO 9219709 to Procter & Gamble, 1992.
119. Bjorkquist, D.W. and Panadiker, R.K., U.S. Patent 5354491 to Procter & Gamble, 1994.
120. Bjorkquist, D.W. and Panandiker, R.K., WO 9404654 AI to Procter & Gamble, 1994.
121. Labeque, R., Lenoir, P., Panandiker, R.K., and Thoen, C., EP 583536 AI to Procter & Gamble, 1994.
122. Bjorkquist, D.W. and Panandiker, R.K., U.S. Patent 5442100 to Procter & Gamble, 1995.
123. Panandiker, R.K. and Bjorkquist, D.W., U.S. Patent 5431842 to Procter & Gamble, 1995.
124. Johnston, J. P., Lenoir, P., Thoen, C., Labeque, R., and McIver, J., EP 583534 AI to Procter & Gamble, 1994.

125. Labeque, R., McIver, J., and Thoen, C., EP 583535 AI to Procter & Gamble, 1994.

126. Panandiker, R.K., Thoen, C.A. J.K., and Lenoir, P.M.A., U.S. Patent 5422030 to Procter & Gamble, 1995.

127. Herbots, I. and Jansen, M., EP 633311 AI to Procter & Gamble, 1995.

128. Tsaur, L.S., Arnson, M.P., Morgan, L.J., Hessel, J.F., McCown, J.T., and Gormky, J.L., U.S. Patent 5281356 to Lever Bros., 1994.

129. Shulman, J.E. and Jones, C.E., U.S. Patent 5409629 to Rohm and Haas, 1995.

130. Amick, D.R., Jones, C.E., and Hughes, K.A., U.S. Patent 4797223 to Rohm and Haas, 1987.

131. Grifo, R.A. and Berni, R.P., *Soap Chem. Specialties*, Sept. 1968.

132. Hornby, J.C., *HAPPI*, Jan., 88, 1995.

133. Kissa, E., Schwuger, M.J., and Smulders, E.J., in *Detergency, Theory and Technology*, Surfactant Science Series, Vol. 20, Cutler, W.G. and Kissa, E., Eds., Marcel Dekker, New York, 1987, p. 333 (and references therein).

134. Gosselink, E.P., U.S. Patent 4702857 to Procter & Gamble, 1987.

135. Schwadtke, K. and Jung, E., U.S. Patent 5271860 to Henkel, 1993.

136. Aoyagi, M. and Takanashi, K., U.S. Patent 5118436 to Kao Corp., 1992.

137. Akabane, Y., Tamura, T. and Fujiwara, M., U.S. Patent 4820437 to Lion Corp., 1989.

138. Kirchner, J.R., in *Kirk-Othmer Encyclopedia of Chemical Technology*, Vol. 13, Wiley, New York, 1981, p. 12.

139. Mitchell, J.D. and Farr, J.P., U.S. Patent 4900468 to Clorox Co., 1990.

140. Boutique, J.P. and Depoot, K., WO 9424247 to Procter & Gamble, 1994.

141. Gray, R.L., Peterson, D., Chen, L., and Buskirk, G.V., U.S. Patent 5019289 to Clorox Co., 1991.

142. Donker, C.B., Hull. M., and van de Pas, J.C., EP 385522 A2 to Unilever, 1990.

143. Boutique, J.P., U.S. Patent 5264143 to Procter & Gamble, 1993.

144. de Buzzaccarini, F. and Boutique, J.P., U.S. Patent 5250212 to Procter & Gamble, 1993.

145. Gazeau, D. and Thoen, C., EP 482275 to Procter & Gamble, 1992.

146. Rerek, M.E. and Aronson, M.P., U.S. Patent 4824592 to Lever Bros. Co., 1989.

147. Emmons, S.A. and Hale, P., U.S. Patent 4929377 to Lever Bros. Co., 1990.

148. Coope, J., Madison, S., Hessel, J., Kuzmenka, D. and Humphreys, R., EP 0524250 to Unilever, 1993.

149. Nicholson, W., WO 9321295 to Procter & Gamble, 1993.

150. Scialla, S. and Scoccianti, R., EP 0629690 to Procter & Gamble, 1993.

151. Scialla, S., Cardola, S., and Bianchetti, G., WO 9502667 to Procter & Gamble, 1995.

152. Scialla, S. and Cardola, S., EP 0598973 to Procter & Gamble, 1993.

153. Showell, M., Kong-Chan, J., Sliva, P., Kinne, K., and Hunter, K., EP 0624640 to Procter & Gamble, 1994.

154. Hardy, F., Scialla, S., Scoccianti, R., and Manfredi, B., EP 0563460 to Procter & Gamble, 1992.

155. Trani, M., WO 9424257 to Procter & Gamble, 1993.

156. Whalley, G., *HAPPI*, Nov., 82, 1993.

157. Schuessler, U., in *Proceedings of the 2nd World Conference on Detergents*, Baldwin, A.R., Ed., American Oil Chemists Society, 1986, p. 187.

158. Eckhardt, C., Kaschig, J., Franke, K., Lee, F., and Ergenc, F., in *Proceedings of the 3rd World Conference on Detergents: Global Perspectives*, Cahn, A., Ed., AOCS Press, Champaign, Champaign, IL, 1994, p. 193.
159. Siegrist, A.E., *J. Am. Oil Chem. Soc.*, 55, 114, 1978.
160. Duccini, Y., Keenan, A.C., and Shulman, J.E., U.S. Patent 5854197 to Rohm and Haas Company, 1998.
161. Silicone Foam Control: Proven Solutions from the Silicone Experts, product selection guide, Dow Corning Corporation, 2001.
162. Kabara, J.J., *Cosmetic and Drug Preservation, Principles and Practice*, Marcel Dekker, New York, 1984.
163. Acusol Opacifiers for Home and Fabric Care, bulletin CS-728, Rohm and Haas Company, 2002.
164. Schick, M.J., in *Nonionic Surfactants: Physical Chemistry*, Surfactant Science Series, Vol. 23, Schick, M.J., Ed., Marcel Dekker, New York, 1987, p. 753.
165. Cutler, W.G. and Davis, R.C., Eds., *Detergency: Theory and Test Methods*, Part 1, Marcel Dekker, New York, 1972.
166. Cutler, W.G. and Davis, R.C., Eds., *Detergency: Theory and Test Methods*, Part 2, Marcel Dekker, New York, 1975.
167. Cutler, W.G. and Davis, R.C., Eds., *Detergency: Theory and Test Methods*, Part 3, Marcel Dekker, New York, 1981.
168. Lucassen-Reynders, E.H., Ed., *Anionic Surfactants: Physical Chemistry of Surfactant Actions*, Surfactant Science Series, Vol. 11, Marcel Dekker, New York, 1981.
169. Cutler, W.G. and Kissa, E., Eds., *Detergency: Theory and Technology*, Surfactant Science Series, Vol. 20, Marcel Dekker, New York, 1987.
170. Krüssmann, H., *J. Am. Oil Chem. Soc.*, 55, 165, 1978.
171. Galante, D.C. and Dillan, K.W., *J. Am. Oil Chem. Soc.*, 58, 356A, 1981.
172. Cahn, A., *HAPPI*, June, 77, 1995.
173. ASTM D 3050-87, *Annual Book of ASTM Standards*, ASTM, Philadelphia, 1993.
174. ASTM D 4265-83, *Annual Book of ASTM Standards*, ASTM, Philadelphia, 1993.
175. ASTM D 2960-89, *Annual Book of ASTM Standards*, ASTM, Philadelphia, 1993.
176. ASTM D 4008-89, *Annual Book of ASTM Standards*, ASTM, Philadelphia, 1993.
177. AATCC Test Method 152-1990, American Association of Textile Chemists and Colorists, technical manual, 1995, p. 270.
178. AATCC Test Method 151-1990, American Association of Textile Chemists and Colorists, technical manual, 1995, p. 270.
179. ASTM Test D 5548, *Annual Book of ASTM Standards*, ASTM, Philadelphia, 1993.
180. Aubey, E., Labeau, M.P., and Harrison, I., U.S. Patent Application 2004/0038851, 2004.
181. Cooke, D.J., Felton, J., and Parker, A.P., Patent Application WO 2003/027218 to Unilever, 2003.
182. Hopkinson, A., Jarvis, A.N., and Kukulj, D., GB 2360792 to Unilever, 2003.
183. Leupin, J.A., Boyer, S.L., Gosselink, E.P., Wang, J., Hunter, K.B., and Washington, N., WO 99/14245 to Procter & Gamble, 1999.
184. Randall, S.L. and Panandiker, R.K., WO 98/29530 to Procter & Gamble 1998.
185. Pramod, K., Panandiker, R.K., Ghosh, C.K., Watson, A.R., Kong-Chan, J.L., and De Buzzaccarini, F., WO 97/42287 to Procter & Gamble, 1997.

186. Damhus, T., Kirk, O., Pedersen, G., and Venegas, M.G., U.S. Patent 5855621 to Novo Nordisk, 1999.
187. Olsen, A.A., Hansen, L.B., and Beck, T.C., U.S. Patent 6114509 to Novo Nordisk, 2000.
188. Shulman, J.E., Kirk, T.C., Swift, G., Schwartz, C., Creamer, M.P., and Falcone, B.A., U.S. Patent 6451756 to Rohm and Haas Company, 2002.
189. Gosselink, E.P. and Price, K.N., U.S. Patent 6071871 to Procter & Gamble, 2000.
190. Gosselink, E.P. and Price, K.N., U.S. Patent 6057278 to Procter & Gamble, 2000.
191. Watson, R.A., Gosselink, E.P., and Price, K.N., U.S. Patent 6291415 to Procter & Gamble, 2001.
192. Watson, R.A. and Gosselink, E.P., U.S. Patent 6191093 to Procter & Gamble, 2001.
193. Watson, R.A., Gosselink, E.P., and Manohar, S.R., U.S. Patent 6087316 to Procter & Gamble, 2000.
194. Mooney, W., U.S. Patent 6696405 to Unilever Home & Personal Care, 2004.
195. Coffindaffer, T.W., Bartolo, R.G., and Belfiore, K.A., U.S. Patent 5336419 to Procter & Gamble, 1994.
196. Hsu, F.L.G., Kuzmenka, D.J., Murphy, D.S., Neuser, K.M., Bae-Lee, M., Garufi, K,. and Coccaro, D., U.S. Patent 6369018 to Unilever Home & Personal Care, 2002.
197. Foley, P.R., Kaiser, C.E., Sadler, J.D., Burckhardt, E.E., and Liu, Z., WO 01/23516 to Procter & Gamble, 2001.
198. Quellet, C., EP 1146057 to Givaudan SA, 2000.
199. Fregonese, D., WO 02/097026 to Reckitt Benckiser, 2002.
200. Holderbaum, T., Richter, B., Nitsch, C., and Haerer, J., U.S. Patent 6448212 to Henkel KgaA, 2002.
201. Panandiker, R.K., Wertz, W.C., and Hughes, L.J., U.S. Patent 5466802 to Procter & Gamble, 1995.
202. Abdennaceur, F. and Boutique, J.P., EP 664335 to Procter & Gamble, 1995.
203. Panandiker, R.K., Wertz, W.C., and Ghosh, C.K., WO 97/42291 to Procter & Gamble, 1997.
204. Shih, J.S., Srinivas, B., and Hornby, J.C., WO 99/15614 to ISP Investments, 1999.
205. Panandiker, R.K., Randall, S.L., Littig, J.S., Gosselink, E.P., and Bjorkquist, D.W., U.S. Patent 6733538 to Procter & Gamble, 2004.
206. Watson, R.A., Gosselink, E.P., and Zhang, S., U.S. Patent 5565145 to Procter & Gamble, 1996.
207. Padron, T., Binder, D.A., Diaz, N., Meyer, F.F., and Murphy, D.S., U.S. Patent Application 2003/0186832 to Unilever, 2003.
208. Sivik, M.R., Bodet, J.F., Kluesener, B.W., Scheper, W.M., Yeung, D.W.K., and Bergman, V., U.S. Patent 6645925 to Procter & Gamble, 2003.
209. Kasturi, C., Schafer, M.G., Sivik, M.R., Kluesener, B.W., and Scheper, W.M., U.S. Patent 6372708 to Procter & Gamble, 2002.
210. Spendel, W.U., U.S. Patent Application 2002/0016276 to Procter & Gamble, 2002.
211. Wang, J., Panandiker, R.K., Kindel, P.F., and Leyendecker, M.R., U.S. Patent Application 2004/0121930 to Procter & Gamble, 2004.
212. Tepe, T.R., EP 1402877 to Rohm and Haas, 2004.
213. Wilting, R., Bjornvad, M.E., Kaappinen, M.S., Schulein, M., and Dela, H., U.S. Patent Application 2003/0022807 to Novozymes North America, 2003.

214. Duval, D.L., WO 99/32594 to Procter & Gamble, 1999.
215. Ohtani, R. and Matsushita, N., WO 98/13457 to Procter & Gamble, 1998.
216. Herbots, I.M. and Busch, A., U.S. Patent 6734155 to Procter & Gamble, 2004.
217. Smets, J., Bettiol, J.L.P., Laudamiel, C., Boyer, S.L., Baeck, A.C., and Herbots, I.M., WO 00/18897 to Procter & Gamble, 2000.
218. von der Osten, C., Cherry, J.R., Bjornvad, M.E., Vind, J., and Rasmussen, M.D., U.S. Patent 6015783 to Novo Nordisk, 2000.
219. Severns, J.C., Sivik, M.R., Costa, J.B., Hartman, F.A., and Morelli, J.P., U.S. Patent 6184188 to Procter & Gamble, 2001.
220. Fehr, C., WO 03/049666 to Firmenich, 2003.
221. Wevers, J., Smets, J., Rosaldo, R.J., and Steenwinckel, P.C., U.S. Patent Application 2004/0018955 to Procter & Gamble, 2004.
222. Severns, J.C., Sivik, M.R., Baker, E.S., and Hartman, F.A., U.S. Patent 5854200 to Procter & Gamble, 1998.
223. Metzger, G., Reinehr, D., Eckhardt, C., and Cuesta, F., U.S. Patent 6482241 to Ciba Specialty Chemicals, 2002.
224. Connor, D.S., Crupe, T.A., Vinson, P.K., and Foley, P.R., WO 97/39089 to Procter & Gamble, 1997.
225. Libe, P.B., Huish, P.D., and Jensen, L.A., WO 01/90293 to Huish Detergents Inc., 2001.
226. McLaughlin, J.R., U.S. Patent 5704556 to DevMar Associates, 1998.
227. Murphy, D.S., Orchowski, M., Tartakovsky, A., and Binder, D.A., U.S. Patent Application 2004/0152617 to Unilever, 2004.

# 9
# Liquid Automatic Dishwasher Detergents

**LEN ZYZYCK, PHILIP A. GORLIN, NAGARAJ DIXIT, and
KUO-YANN LAI** Global Technology Research and Development,
Colgate-Palmolive Company, Piscataway, New Jersey

**319**

## I. INTRODUCTION

The concept of using mechanical devices for dishwashing was documented as early as 1865 with the issuance of a U.S. Patent to J. Houghton [1]. Subsequently several companies tried to manufacture and market automatic dishwashing machines for home as well as for institutional use [2,3]. However, it was not until the early 1950s that both detergents and mechanical dishwashers became widely available to consumers. By the early 1990s, an estimated 50% of the households in the U.S., 25% in Europe, and 8% in Japan had automatic dishwashers [4].

The focus of this chapter is on the U.S. market, as this is still the major market for liquid automatic dishwasher detergents (ADDs). However, we will touch briefly on the global situation that prevailed in the late 1990s. During that period, the global dishwashing market was divided between hand dishwashing and automatic machine dishwashing. The split was about 70% for hand dishwashing, and 30% for machine dishwashing. About 90% of ADD sales worldwide were concentrated in five countries: the U.S., France, Germany, the U.K., and Italy. The greatest incidence of households with dishwashing machines was in the U.S., with more than 50%, followed by France with more than 30%. In Europe the predominant form of ADD was unit dose, followed by powder, then liquid. In contrast, in the U.S. powder dominated at about 60%, followed by liquid at about 30%, and unit dose at about 10%. By 2004, the market in the U.S. had begun to shift more toward unit dose, although liquids continued to rise. The form distribution was about 40% powder, 40% liquid, and 20% unit dose.

The ADDs originally introduced into the U.S. market were in the powder form. These products have subsequently undergone major changes in composition to deliver better cleaning performance. Typical compositions of powder ADDs sold in the U.S. and Europe are shown in Table 9.1.

Attempts have been made to produce and market phosphate-free detergents with minimal success. Two proposed compositions, which might also be suitable for tablet making, are shown in Table 9.2.

In North America enzyme-based powder formulas have come to dominate the powder market.

Today, powders no longer dominate the U.S. market. While ten years ago they commanded nearly 70% of the market share [6], today they represent only about

**TABLE 9.1**   Conventional Machine Dishwashing Powder Formulation [4]

| Ingredient | % by weight |
|---|---|
| Sodium tripolyphosphate (STPP) | 15–45 |
| Sodium silicate | 15–60 |
| Sodium carbonate | 0–25 |
| Chlorocyanurates | 0–7 |
| Nonionic surfactant | 0–6 |
| Sodium sulfate | 0–40 |
| Water | Balance |

**TABLE 9.2**   Nonphosphate Machine Dishwashing Formulation [5]

| Ingredient | Composition 1 (wt %) | Composition 2 (wt %) |
|---|---|---|
| Na citrate 2 $H_2O$ | 35 | 35 |
| Na disilicate (granular) | 30 | 30 |
| Na perborate 1 $H_2O$ | 10 | 10 |
| TAED[a] | 3 | 3 |
| Acrylic-salt D | 10 | 0 |
| Acrylic-salt G | 0 | 10 |
| Na carbonate | 10 | 10 |

[a]Tetraacetylethylene diamine.

40% of the U.S. market. About 20 years ago the marketplace saw the advent of liquid automatic dishwasher detergents (LADDs). Clarification of what constitutes a "liquid" is worth mentioning here. Commercially, the products are marketed under different names such as: liquids, liquigels, liquid gels, or gels. Technically, these products are concentrated suspensions. The liquid matrix predominantly comprises an aqueous phase and thickening or structuring agents. The latter are typically either swellable clays or water-dispersible polymers and optionally a cothickener. The mechanical properties of these products are such that the product can be dispensed from the container without prior shaking.

ADDs in the liquid form offer the following advantages over powders:

1.  They offer convenience in dispensing and dosing.
2.  They dissolve quickly in the wash water, providing a residue-free wash.
3.  They are free from lumping or caking during storage.
4.  They do not release irritating dust upon handling.

The market entry of first-generation automatic dishwasher liquids dates back to 1986 in the U.S. and 1987 in Europe with the introduction of Palmolive Automatic and Galaxy, respectively, by Colgate-Palmolive [7]. Soon afterwards, Procter & Gamble and Lever Brothers introduced similar liquid products. The category has since been steadily growing. Today LADDs account for about 40% and 15% of total sales of the automatic dishwasher detergent market in the U.S. and Europe, respectively [8].

There has been an evolution of LADD technology over the years. This consists of clay hypochlorite bleach form, gel hypochlorite bleach form, and enzyme no bleach form. The first-generation LADDs were essentially powder compositions in the liquid form, in which functional components were suspended or dispersed in a structured liquid matrix. The liquid matrix consisted of water and the common structuring additives used were bipolar clays and a cothickener comprising a metal salt of a fatty acid or hydroxy fatty acid. These liquid products, although minimizing some of the shortcomings of the powders, suffered from two major disadvantages. First, the rheological properties of these products were such that the bottle needed to be shaken prior to dispensing of the product. This was due to phase separation and the production of "free" bleach solution. Second, the shelf life stability of the products did not meet consumer expectations. This problem was shortly recognized by the manufacturers and aesthetically superior, nonshake, stable, and translucent products were introduced to the market in 1991 as "gels." All the liquid products marketed today in U.S. are essentially in gel form using polymeric thickeners.

A new form of ADD (Electrasol), as a tablet, appeared in the U.S. in 1997. This was the first tablet for use in dishwashing machines. It was a pressed powder. It contained a disintegrant to help the tablet break apart and dissolve in the wash water. This form was based on both enzyme/oxygen bleach technology as well as chlorine bleach technology. In 2002 the next generation of unit dose products appeared (Electrasol). This was a "gel pac" that was based on enzyme technology. It consisted of a gel suspension in a water-soluble sachet. This was soon followed by a dual-compartment, water-soluble sachet. One compartment contained a typical enzyme/oxygen bleach cleaning system, while the second contained a liquid. The liquid was typically a nonionic or solvent system that contributed to cleaning and acted as a humectant to control moisture in the powder compartment.

By 2004, the ADD market was segmented in three forms: powder, gel, and unit dose. Approximate market share for these forms in the U.S. was 40, 40, and 20% respectively. While liquid gel systems continued to grow, they were challenged by the unit dose segment that consisted of solid, liquid, or hybrid products.

The discussion of this chapter focuses on the technology behind the development of currently marketed LADDs in either gel or unit dose form. Powder ADDs, which in most areas utilize analogous technology, will not be discussed except for comparison.

## II.  MECHANICS AND CHEMISTRY OF AUTOMATIC DISHWASHING

Cleaning in an automatic dishwasher is accomplished by a combination of three types of energy: mechanical, thermal, and chemical. In general, a combination of machine (mechanical), hot water (thermal), and detergent (chemical) is necessary for complete cleaning of dishware. Before these effects are discussed in more detail, the basic components and mechanics of an automatic dishwasher are described.

## A.  Components of Automatic Dishwashing Machines

Concurrent with the evolution and advancement in dishwasher detergents, automatic dishwashers have been improving over time in order to offer the consumer greater convenience and performance. European machines are similar to North American machines except for some minor but important differences. These differences allow for the use of different detergent technologies in the two markets. In recent years the machine technologies as well as the chemical technologies of Europe and North America have become closer and may merge. It is more cost effective for manufactures to offer machines with common components.

Ten years ago the European market was dominated by machines that heated the water and had stainless steel interiors. The North American market was dominated by machines that received hot water from the house plumbing, did not heat the water, and had plastic interiors. Today, high- and middle-tier North American machines are sensor controlled (water temperature and wash time) and come in stainless steel options (Figure 9.1).

Automatic dishwashing machines typically contain two racks that hold the items to be washed. All machines contain at least one spray arm, which spins due to the pressurized wash solution being pumped through the arm nozzles. This allows for an even distribution of the wash liquor over all the items being washed. The main spray arm is located underneath the bottom rack and directs the water pumped through it upwards. In some machines, a second and even a third spray arm are located below and above the top rack, respectively. Water can be delivered to the spray arms either through a "tower" which extends during the wash cycle, or through a "pipe" that runs along the back of the machine to deliver water to the upper spray arms. This is typical of a three-spray-arm machine. The wash liquor is recycled throughout the complete cycle by means of a pump which continually circulates it through the spray arms. A strainer in the wash tank removes large soil particles throughout the recirculation process. For each program cycle (wash or rinse) a new water supply is introduced. The amount of water introduced is a function of machine type and wash setting. As machines have become more energy and environmental friendly, less water is used. The cost to operate the machine is also lower.

**FIG. 9.1**   Stainless steel inside of an automatic dishwashing machine. (Courtesy of the General Electric Co.)

Machines generally contain two detergent dispensing cups, one of which has a lid and is therefore closed through part of the machine cycle. Newer machines have the "open" and "closed" cups side by side with one gasketed cover. The part of the cover that is over the open cup has vents to allow the dish liquid to be washed into the prewash cycle. As machines have become "smarter" the size of the second or prewash cup seems to be shrinking. With the new unit dose technologies, it may disappear altogether. A significant difference between American and European machines used to be that the latter contained a gasketed closed cup. This prevents the detergent from prematurely leaking out of the cup. The detergent is formulated as a viscous gel which will not flow unless stressed. As American machines evolve toward a more common structure with their European counterparts, this difference will probably disappear.

All European machines, and some American ones, also contain a rinse aid dispenser, which provides a dose of rinse aid during the final cycle. These dispensers only have to be filled about once a month. Rinse aids are especially useful in hard water areas (see Section VII).

A typical dishwashing program consists of several cycles of differing function. The number of each type of cycle and their order depends on the brand of machine and the wash program selected. Today's machines offer a wide selection of washing programs. The simpler machines may offer a few programs, such as heavy wash, normal wash, short wash, rinse only, plate warmer, and hot start. More elaborate machines offer a greater number of programs, such as antibacteria, cookware (pots and pans), normal wash, speed cycle, china crystal, glasses, plastics cycle, and rinse only. Additionally, options such as heated dry (on/off) delay hours, added heat (wash), prewash, control lock, and reset may be available. All washing programs consist of at least a rinse, a main wash, one rinse after the main wash, and a drying cycle, which is optionally heated. The rinse cycles are mechanically similar to the wash cycles, except that no detergent is present. In programs containing two wash cycles (heavy or pot-scrubber cycle), both dispensing cups are filled with detergent. If only a main wash cycle is to be used (normal or light cycle), only the closed cup is filled. In either case, it is essential that the detergent be structured so that it will not leak out of the closed cup before the cup opens. Otherwise, the main wash will be under dosed and will not result in effective cleaning. Proper structuring of gel detergents is discussed in Section IV.

The most important differences between European and North American machine designs are in the condition of the incoming wash water. As discussed later in this chapter, the presence of hardness ions ($Ca^{2+}$ and $Mg^{2+}$) in the wash decreases the overall effectiveness of cleaning. This is a problem especially in hard water areas and can only be overcome by softening of the water. Water hardness varies by country. The United States and Japan have water that would generally be considered soft. Most European countries, in contrast, have water that is considered to be hard [9].

European machines circumvent the problem of hard water by softening the water before it is introduced into the wash. This is accomplished by a built-in ion exchanger which works by replacing the hardness ions with sodium ions. Regeneration of the ion exchanger with sodium chloride is required periodically. In contrast, North American machines contain no water softening device and must therefore rely on the detergent for sequestration of the hardness ions. The presoftening of the wash water by European machines allows for detergents to be formulated that contain lower builder levels.

Another difference between North American and European dishwashing programs concerns the temperature of the incoming water. In North America the wash water is preheated by the household water heater and is introduced into the machine at temperatures of 110 to 140°F, typically 120°F (49°C). Newer machines may then heat the water in the various wash and rinse cycles. In contrast, European machines receive cold water and heat it via machine heating coils during the wash and rinse cycles. The final temperature that the water reaches in this manner is therefore higher than American machines, reaching 115 to 170°F (46 to 77°C), depending on the program selected. Schematic profiles of water temperature vs. time for U.S. and European machines are shown in Figure 9.2 and Figure 9.3, respectively.

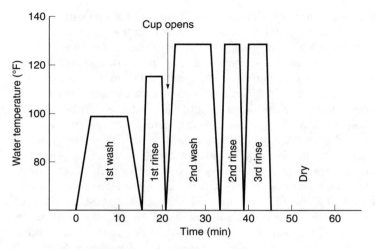

**FIG. 9.2**    Typical U.S. dishwashing cycle.

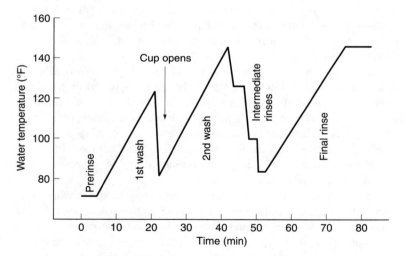

**FIG. 9.3**    Typical European dishwashing cycle.

## B. Mechanical Cleaning

Probably the most important component of cleaning in automatic dishwashing is the mechanical component, which primarily arises from the kinetic energy of the water being pressurized through the rotating spray arm nozzles. As the wash water is pumped through the spray arms, the water jets sprayed from the rotor arms help to dislodge soils adhered to the dishware. Better machines with three spray arms and a more efficient water feed will naturally be more efficient. Therefore, an increase in kinetic energy, which can be accomplished by either more water or higher pressure, will result in more efficient removal of soils from substrates [10]. It has been suggested that mechanical energy of the machine itself is responsible for 85% of the soil removal during the cleaning cycle; the detergent contributes the other 15% [11]. Thermal energy is in a sense a secondary effect, contributing to the effectiveness of both the mechanical and chemical components of cleaning.

## C. Thermal Cleaning

Most soil removal during an automatic dishwashing cycle is positively affected by thermal energy, which is delivered by hot water. In the past the wash water temperature was dictated by the incoming water temperature. Modern machines sense the wash water temperature and adjust it according to the program selected. In contrast, the wash temperature in European machines approaches 170°F (77°C). This is because the machine itself contains a heating element that heats the water.

Elevated temperatures have some advantageous effects on chemical processes involved in cleaning. For example, the solubility of slightly soluble salts increases with temperature. Otherwise, the deposition of such salts on glassware is the cause of much of the inorganic filming observed in hard water regions. Thermal energy also aids in the removal of fatty soils from items being washed. Above their melting points, fats are more easily removed since the interfacial forces binding them to the dishware and the cohesive forces of the soil are both lower. The activity of most oxidizing agents also increases with temperature. Conversely, higher temperatures can have a negative effect on fine china and crystal, causing etching. Some machines compensate for this by having a glass or crystal wash program.

## D. Chemical Cleaning

A typical wash load consists of several types of soils which must be solubilized or degraded by a combination of the detergent and machine. Detergents must control both food soils introduced by the items washed and inorganic scale produced by hardness ions in the wash water. Food soils consisting of either proteinaceous, starchy, or fatty materials must not only be effectively removed from the dishware, but they must also be prevented from redepositing on items being washed.

The composition of soils on the washware surface may vary both before and during machine washing. Heat during cooking or machine washing may cause a redistribution of fats on the surfaces, for example. Redeposition on glasses is of particular importance as it leads to undesirable spotting and filming which is easily consumer perceivable. Soils that produce stains are also a problem in automatic dishwashing. For example, coffee, tea, wine, and tomato sauce leave highly visible stains on wash items unless properly treated during the wash.

Removal of food soils from wash surfaces can only be accomplished once the attractive forces between surface and soil are overcome. Generally, dehydrated soils interact more strongly with surfaces due to stronger van der Waals interactions between soil and surface [10]. In contrast, the soil–surface interaction decreases as the surface becomes more hydrophobic due to less available polar sites. Exceptions are hydrophobic soils such as fats, which interact strongly with hydrophobic surfaces [10].

## III.  COMPONENTS OF LIQUID AUTOMATIC DISHWASHER DETERGENTS AND THEIR FUNCTIONAL PROPERTIES

LADDs are chemically complex mixtures, consisting of a variety of components working in unison to clean the items placed in the dishwashing machine. Each component performs a vital function, either alone or in conjunction with others. LADD ingredients can be broadly classified into two categories: those that perform a cleaning function and those that modify the rheology or aesthetics of the liquid. In this section the typical components found in LADDs are discussed.

A typical LADD composition and the functions of the individual components are shown in Table 9.3. Two examples of specific gel LADD formulas are shown in Table 9.4 [12].

## A.  Builders

Builders as a class perform several essential functions in the automatic dishwashing process. Ideally, a builder should possess the following properties: a high and rapid sequestration capacity for hardness ions in the wash water, soil dispersing properties, chemical stability and compatibility with other detergent components, low toxicity, high biodegradability, and low cost. The chemical structures of some common builders are shown in Figure 9.4.

The presence of alkaline earth ($Ca^{2+}$ and $Mg^{2+}$) ions during an automatic dishwashing cycle can lead to undesirable spotting and filming on items being washed. This occurs through the formation of insoluble metal complexes with proteinaceous soils, fatty acids, anionic surfactants, and carbonate. Research in

**TABLE 9.3** Typical Compositions and Functions of LADD Products

| Component | Typical | Amount (%) | Function |
|---|---|---|---|
| Builder | Sodium tripolyphosphate (STPP) | 5–30 | Sequestration |
| | Low-molecular-weight acrylate | | Soil suspension, alkalinity, emulsification |
| | Citrate | | |
| | Silicate | 3–15 | Anticorrosion, alkalinity, sequestration |
| Surfactant | Dowfax | | Spot/film prevention, sheeting action, soil dispersion |
| Bleach | Sodium hypochlorite | 0.5–2 | Stain removal, soil removal, disinfectant |
| | Perborate | | |
| Caustic | NaOH | 1–5 | Alkalinity |
| | KOH | | |
| Defoamer | | 0–1 | Foam prevention |
| Thickener | Carbopol | 0.5–2 | Gel structure |
| | Polygel | | |
| Enzymes | Protease | <1.0 | Food removal |
| | Amylase | <1.0 | |
| Color/fragrance | | <0.5 | Aesthetics |
| Water | | Balance | Solubilizer, flow properties |

**TABLE 9.4** Examples of Gel LADD Formulas [12]

| Ingredient | Formula A (%) | Formula B (%) |
|---|---|---|
| Sodium tripolyphosphate | 6 | 6 |
| Sodium disilicate | 12 | 12 |
| Potassium hydroxide | 3.89 | 3.89 |
| Sodium hydroxide | 0.87 | 0.87 |
| Acusol 445N | 1.92 | 1.92 |
| Carbopol 617 | 0.7 | 0.7 |
| Dowfax 3B2 | 0.23 | 0.23 |
| LPKn 158 | 0.16 | 0.16 |
| Sodium hypochlorite (13% solution) | 9.2 | 9.2 |
| Stearic acid | 0.11 | 0.16 |
| Perfume | 0.1 | 0.1 |
| Water | Balance | Balance |

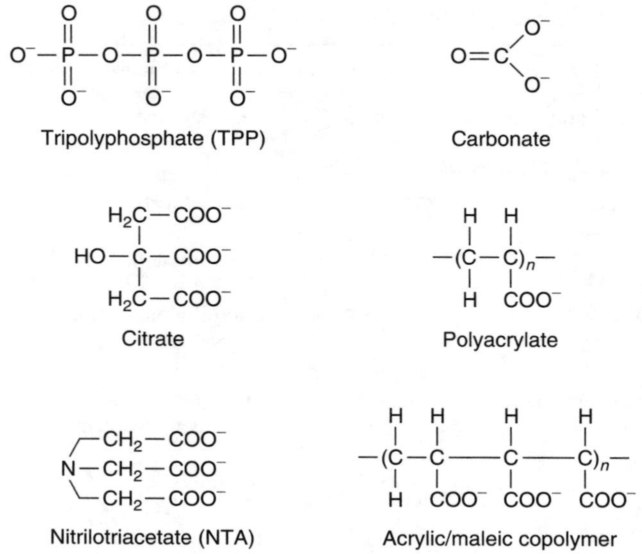

$$O^- - \overset{\overset{\displaystyle O}{\|}}{\underset{\underset{\displaystyle O^-}{\|}}{P}} - O - \overset{\overset{\displaystyle O}{\|}}{\underset{\underset{\displaystyle O^-}{\|}}{P}} - O - \overset{\overset{\displaystyle O}{\|}}{\underset{\underset{\displaystyle O^-}{\|}}{P}} - O^-$$

Tripolyphosphate (TPP)

$$O = C \big\langle \genfrac{}{}{0pt}{}{O^-}{O^-}$$

Carbonate

$$\begin{array}{c} H_2C - COO^- \\ | \\ HO - C - COO^- \\ | \\ H_2C - COO^- \end{array}$$

Citrate

$$-(\overset{\overset{\displaystyle H}{|}}{\underset{\underset{\displaystyle H}{|}}{C}} - \overset{\overset{\displaystyle H}{|}}{\underset{\underset{\displaystyle COO^-}{|}}{C}})_n-$$

Polyacrylate

$$N \genfrac{}{}{0pt}{}{}{} \begin{array}{l} -CH_2 - COO^- \\ -CH_2 - COO^- \\ -CH_2 - COO^- \end{array}$$

Nitrilotriacetate (NTA)

$$-(\overset{\overset{\displaystyle H}{|}}{\underset{\underset{\displaystyle H}{|}}{C}} - \overset{\overset{\displaystyle H}{|}}{\underset{\underset{\displaystyle COO^-}{|}}{C}} - \overset{\overset{\displaystyle H}{|}}{\underset{\underset{\displaystyle COO^-}{|}}{C}} - \overset{\overset{\displaystyle H}{|}}{\underset{\underset{\displaystyle COO^-}{|}}{C}})_n-$$

Acrylic/maleic copolymer

**FIG. 9.4**  Chemical structures of selected builders.

the field has led to a better understanding as to the causes of spotting and filming on items being washed, especially glassware [13,14]. Fatty soils, in combination with calcium ions, result in filming of glasses. However, organic soils do not deposit as film in soft water. In addition, inorganic filming (as $CaCO_3$) is also a problem in hard water conditions. Both types of film can be controlled or even prevented by efficiently sequestering the hardness ions in the wash water. This problem is obviously of greater concern in hard water areas. Spotting, in contrast, cannot be controlled by sequestration of hardness ions. Both fatty and proteinaceous soils can deposit as spots under all water conditions.

Because of the different mechanisms responsible for soil deposition on glassware, formulators of LADD products rely on two different approaches to control this problem. First, compounds that sequester the hardness ions (builders) are used. In this manner, the calcium and magnesium form water-soluble complexes and are removed with the wash water. In general, $Ca^{2+}$ sequestration is of greater importance than that of $Mg^{2+}$. This is not only due to the fact that $Ca^{2+}$ is generally present in higher concentrations in water supplies, but also because Ca–soil and $CaCO_3$ complexes are more stable and less soluble than the magnesium analogs. An important factor in the selection of builders is therefore their calcium binding affinity, $K_{Ca}$, usually reported as the $pK_{Ca}$ [15].

The equilibrium present in solution is as follows:

$$Ca^{2+} + L^{n-} \Longleftrightarrow CaL^{2-n}$$

$$K_{Ca} = [CaL^{2-n}]/[Ca^{2+}][L^{n-}]$$

$$pK_{Ca} = -\log K_{Ca}$$

where L is the complexing ligand. The $pK_{Ca}$ of builders typically used in LADDs are listed in Table 9.5. Also listed are the calcium binding capacities in terms of mg CaO per g of builder [16]. For two molecules with equal binding constants, the one with the lower molecular weight will be more efficient.

The second approach to spot/film prevention involves the use of compounds that either inhibit the deposition and precipitation of $Ca^{2+}$ complexes by slowing down crystallization or reduce the growth of existing crystallites, known as the threshold effect.

Generally, combinations of builders are used in LADD products, the reason being that some builders are more effective sequestrants of calcium, others more effectively bind magnesium, and yet others provide soil dispersion. Studies have also shown synergistic effects in many cases. For example, Lange has reported that interactions between silicate and phosphate show surface activity greater than that which each component alone would contribute [17]. Experiments have shown that combinations of low-molecular-weight polyacrylates and soda ash tolerate higher concentrations of $Ca^{2+}$ in the wash water than equal amounts of either sequestrant individually [15,18]. Builders also work synergistically with surfactants to increase the detergency of the liquid. By tying up free hardness ions, they prevent the formation of insoluble Ca–surfactant complexes.

**TABLE 9.5**  Values of $pK_{Ca}$ and Calcium Binding Capacity for Typical ADD Builders

| Builder | $pK_{Ca}$ | Ca binding capacity (mg $Ca^{2+}$/g) |
|---|---|---|
| Sodium tripolyphosphate | 6.0 | 198 |
| Nitrilotriacetic acid | 6.4 | 448 |
| Sodium citrate | 3.6 | 400 |
| Sodium carbonate | ? | 286 |
| Low-molecular-weight polyacrylate | 4.5 | 440 |
| Acrylate–maleic copolymers | 4.5 | 480 |
| Zeolite A | | 198 |

## 1. Phosphates

Details of the chemistry of phosphorous compounds pertinent to detergent applications are discussed in a comprehensive review by van Wazer [19]. Phosphates are the universal builder in LADDs due to their high performance-to-cost ratio. The term "phosphate" when used in reference to LADDs actually refers to oligophosphate ions, most common being the tripolyphosphate pentaanion (TPP), the pyrophosphate tetra-anion, and the cyclic trimetaphosphate trianion. The structure of TPP is shown in Figure 9.4. All polyphosphates hydrolyze to simple orthophosphate ($PO_4^{3-}$) over time. The rate of hydrolysis increases with increasing temperature or decreasing pH [19]. For tripolyphosphate under typical LADD conditions, the hydrolysis half-life is very long, on the timescale of years [18], and is thus of little concern for typical alkaline LADD compositions.

These builders are generally available as sodium, potassium, or mixed-metal salts, the latter being more soluble in water but also more costly. The widespread use of sodium TPP (STPP) in LADD formulations can be attributed to the many functions it performs during the wash cycle. Besides its efficient sequestration of hardness ions, STPP works to disperse and suspend soils, enhance the surface action of anionic surfactants, solubilize proteinaceous soils, and provide alkalinity and buffering action. Pyrophospates have been included in some LADD formulations because of its better solubility properties relative to tripolyphosphate [20].

Phosphates are currently the primary builder in all LADD products sold in North America, although various regulations limiting their use have been in place for the past 20 years. The concern with phosphates is that large amounts of them in waste water results in the eutrophication of lakes and ponds, leading to excessive algal growth. Because of this concern, government regulations banning their use in laundry detergents and limiting their use in ADD products were passed in the early 1970s.

## 2. Silicates

After phosphates, silicates are the most ubiquitous builders used in LADD formulations. Like tripolyphosphates, silicates are multifunctional. They are, however, better sequestrants of magnesium ions. A combination of phosphate and silicate is therefore generally used in ADD formulations. In addition to their sequestering properties, silicates provide alkalinity, soil suspension, and anticorrosion properties. A detailed treatment of the synthesis, chemistry, and applications of silicates has been undertaken by Iler [21]. Other reviews are also available [22,23].

Silicates used in detergents vary according to the $SiO_2:Na_2O$ ratio present. They are synthesized by the reaction of sand and sodium carbonate at elevated temperatures. Commercially, ratios of 0.5 to 4 are available, depending on the ratios of starting materials used [24]. An important transition occurs at a mole ratio of about two. Below an $SiO_2$:$Na_2O$ mole ratio of two, monomeric or dimeric silicate tetrahedra exist [23]. In contrast mole ratios greater than two result in higher molecular weight silicates due to polymerization. The equilibrium between monomeric and polymeric silicate is affected by the pH of the solution. As the solution becomes more alkaline, the amount of monomeric species increases. For LADD formulation purposes, disilicates with an $SiO_2$:$Na_2O$ ratio of 1:2 to 1:3 are generally used. At lower ratios, metasilicates form which render the detergent too alkaline, at which point safety problems due to their corrosiveness may become a concern.

An interesting study was reported in 2003 examining the cause of a cloudy, milky ring that can form in soda-lime-silicate glassware [25,26]. This ring was particularly noticed in the bowl of wineglasses that had undergone many washings in an automatic dishwasher. Under microscopic examination grooves and scratches were noted in the unwashed glassware that were not visible to the unaided eye. The glassware was then washed 100 times in an automatic dish washing machine. Glassware washed with solutions containing no sodium disilicate remained clear and transparent. Glasses washed with harsher, sodium disilicate solutions (0.7–1.5 g/l) became visibly corroded around the bowl where the microscopic scratches were noted. Interestingly, the most aggressive dishwashing solutions did not appear to produce this ring. This was attributed to a higher, more uniform rate of corrosion, which left the glass clear but thinner. These glasses may be more prone to breakage in subsequent washings.

## 3. Zeolites

A related class of siliceous builders are the aluminosilicates (zeolites). Zeolite A, in particular, has been studied as a builder in detergent formulations, generally as a phosphate replacement. It consists of alternating $SiO_2$ and $AlO_2^-$ building blocks forming a three-dimensional cage structure, with sodium ions balancing the charge. It works by ion exchange, replacing $Ca^{2+}$ in solution with $Na^+$. This is a different mode of action compared to the other builders, which form soluble complexes with calcium and magnesium. Because of the relatively small pore size of zeolite A (4.2 Å), the larger, hydrated $Mg^{2+}$ dications cannot be efficiently exchanged.

Studies have shown zeolites to be slower than STPP at removing $Ca^{2+}$ from the wash solution [27]. In addition, their insolubility in aqueous solutions has limited their use to powder detergents, especially in the laundry industry. A comprehensive review of the chemistry of aluminosilicates in detergent compositions is available [28]. Patents for LADD products utilizing zeolites have appeared in the literature [29–31]. The attraction is due to their low cost and low toxicity.

## 4. Carbonate

Sodium carbonate ($Na_2CO_3$) has been used in detergent formulations for many years both to sequester calcium ions (at pH > 9) in the wash water and as an alkalinity source. The structure of the carbonate dianion is shown in Figure 9.4. Currently, all powder ADDs contain high levels of sodium carbonate (15 to 40%), mainly as a source of alkalinity. In LADDs, where caustic can be incorporated into the compositions, the need for sodium carbonate is less important.

Because complexation with $Ca^{2+}$ results in insoluble $CaCO_3$, which deposits on items being washed, carbonate alone is not an effective builder system. In contrast, tripolyphosphate forms a soluble calcium salt, preventing deposition of insoluble salts. Sodium sesquicarbonate ($Na_2CO_3 \cdot NaHCO_3 \cdot 2H_2O$) and sodium bicarbonate ($NaHCO_3$) have not been used in LADD formulations, except where buffering action is needed.

## 5. Citrate

A builder that has been studied as a possible phosphate replacement is sodium citrate. The structure of the citrate trianion is shown in Figure 9.4. Several properties of sodium citrate restrict its use in LADD formulations. First, it is incompatible with hypochlorite, precluding its use in most LADD compositions. It is also inferior to STPP in its sequestration efficiency for calcium ions. Finally, it is almost three times more expensive than STPP. This combination renders citrate unsuitable as a replacement for STPP. In Europe, where the different wash conditions allow for the use of milder peroxygen bleaches or enzymes and the water is presoftened, citrate-built products are possible.

## 6. Organic Polymers

Polymers of carboxylic acids are being increasingly used in LADD formulations since they can perform several functions necessary of ADD components. Low-molecular-weight polycarboxylates (<250,000 a.m.u.) are useful both as builders and soil dispersing agents. Since these latter two functions in LADD formulations are usually performed by STPP, much effort has been directed toward the use of low-molecular-weight polycarboxylates as phosphate replacements. In particular, the soil dispersing properties of low-molecular-weight polycarboxylates makes them very attractive in LADD formulations since other builders such as carbonate, silicate, aluminosilicates, nitrilotriacetate, citrate, and even tripolyphosphate act mainly as sequestrants.

The most often used polycarboxylates consist of acrylic, maleic, and olefinic monomers, either as homo- or copolymers. Modifications to the side groups can be made in order to alter the hydrophobicity of the polymer. In addition, the relative ratios of monomers in copolymer structures can be varied in order to control the properties of the polymer. Both acrylic acid homopolymers and acrylic acid–maleic anhydride copolymers are commercially available in a wide range of molecular weights (1,000 to 250,000 a.m.u.). The structures of both types are shown in Figure 9.4. They are sold under the trade names Acusol® (Rohm & Haas) [32] and Sokalan® (BASF) [33], among others.

As calcium sequestering agents the copolymers are generally better than analogous acrylic homopolymers due to their greater concentration of $COO^-$ groups per monomeric unit. The differences are rather small, though. In addition, the calcium binding capacity increases with increasing molecular weight [18].The degree to which the carboxylate groups are ionized also affects the polymer properties. Polycarboxylates that exhibit only partial ionization in the wash have been found to be better sequestrants [34]. The degree of ionization determines the amount of association between $COO^-$ groups and the degree to which the polymer coils. Two mechanisms have been proposed for sequestration of $Ca^{2+}$ and $Mg^{2+}$ by polycarboxylic acids [35]. Through electrostatic binding, the hardness ions interact with electrostatic fields created by the polymer. In contrast, site binding relies on preferential binding of large cations (e.g., $Ca^{2+}$) over smaller ones (e.g., $Na^+$ or $K^+$) along specific binding sites [35].

Ca–polycarboxylate complexes are not very soluble, and will precipitate and deposit on glassware under high $Ca^{2+}$ concentrations. In fact, research at BASF [15] has shown that a combination of soda ash and low-molecular-weight polycarboxylates works better at preventing filming on glassware than either soda ash or the polymer alone. This is because soda ash will bind the calcium but will not be deposited as $CaCO_3$ due to the dispersing properties of the polymer. In this manner, a higher concentration of $Ca^{2+}$ can be tolerated.

In addition to their role as calcium sequestrants, polycarboxylates perform two other major functions [15]. In fact, because they are usually used in combination with other sequestrants (phosphate, carbonate, citrate) which behave more as sequestrants of hardness ions, it is this property that makes them so important. First, they prevent crystal growth of calcium precipitates, especially $CaCO_3$ (threshold effect). They also reduce the growth of crystallites already formed. In general, low-molecular-weight polycarboxylates ($<10,000$ a.m.u.) are more effective than high-molecular-weight ones. Second, and equally important, polycarboxylates are effective dispersants of particulate soils. In this role, low-molecular-weight polymers are again more efficient than high-molecular-weight analogs.

## 7.  Other Builders

A very effective STPP replacement not used in LADD compositions is nitrilotriacetic acid (NTA), commonly manufactured as the monohydrate. Its structure is shown in Figure 9.4. It is superior to STPP in sequestration of hardness ions, contribution to the alkalinity of the detergent, and solubility at high pH conditions. Because of these properties, it was once considered for use in phosphate-free LADD products. However, concerns about its toxicity have prevented widespread use. There is some controversy regarding its possible role as a carcinogen [36,37]. Several states have classified this material as a potential carcinogen, discouraging its use. Only sporadic use in the laundry detergent category is now observed in Canada and parts of Europe.

The last class of builder to be mentioned here is the ether carboxylates, or oxydicarboxylates. These have not found use in marketed LADD products mainly due to their high cost and incompatibility with hypochlorite. Nonetheless, some favorable characteristics have been identified for effective $Ca^{2+}$ binding [38]:

1. A $pK_{Ca} \geq 5$ is desired with minimum molecular weight.
2. Carboxylate and ether oxygens on $\alpha$-carbons are preferred.
3. Ether oxygens are preferred over ketal oxygens.
4. The order of preference is substituted malonate > malonate > succinate > acetate > propionate.
5. The total ionic charge should be >2 but <5, except for 2:1 complexes.
6. A close steric fit of $\geq 4$ donor oxygens around $Ca^{2+}$ is desired.
7. A minimum number of degrees of freedom in the carbon backbone is desired.

## B. Surfactants

The role of surfactants in the automatic dishwashing process is very different from their role in the hand dishwashing process. In contrast to light-duty liquid detergents (LDLDs) (see Chapter 7), which consist mainly of and rely on a combination of surfactants for foaming, grease cutting, and soil removal, LADDs contain surfactants only as a minor additive. The grease and soil removal is instead accomplished by the high alkalinity and bleach present. In addition, the high wash temperature helps to melt fatty soils and to denature proteinaceous soils. Surfactants do provide sheeting action on the items being washed to prevent soils from depositing as spots or film. However, their tendency to produce a large amount of foam is detrimental to the automatic dishwashing process. The reason is that dishwashing machines work by pumping the wash solution through spray arms, which spin because of the water pressure. Foam decreases the water pressure and therefore the cleaning efficiency of the machine (see Section II). In surfactant-containing compositions, defoamers are often used which reduce foaming by interfering with the formation of micelles (see Section III.C).

Surfactants suitable for LADD formulations must be low foaming. In addition, if the product is to contain chlorine bleach, the surfactants must be resistant to oxidation. In LADD formulations only anionic and nonionic surfactants have found use. Cationic or amphoteric surfactants have not been used because of their relatively high cost and/or incompatibility with other detergent ingredients. Anionic surfactants are available as alkali metal salts which dissociate in aqueous solution. Typical hydrophilic "head groups" useful in LADD formulations include carboxylates, sulfonates, sulfates, phosphates, and phosphonates. These surfactants are bleach stable unless the hydrophobic tail is oxidizable. A drawback to the use of anionics in LADDs is that they produce too much foam. Nonetheless, because of their bleach stability, alkyl ether sulfonates such as the Dowfax® line (Dow Chemicals) have been used at low concentration in LADD products. The bleach compatibility of a series of anionic surfactants is reported by Rosen and Zhu [39] and in a Dow Chemical bulletin [40].

Nonionic surfactants have also been investigated as components of LADD compositions. The hydrophilicity of nonionic surfactants arises from the polar linkages within the molecule. Typical nonionic surfactants include polyethylene glycol ethers, fatty acid alkanol amides, amine oxides, and ethylene oxide/propylene oxide (EO/PO) block polymers. The last mentioned are especially effective in LADD formulas, although their incompatibility with chlorine bleach has restricted their use to powder or bleach-free gel formulas. The relative hydrophobicity of the EO/PO polymeric surfactants can be controlled by variations in the EO:PO ratio, with hydrophobicity increasing with PO content. Nonionic surfactants possess several advantages over anionics as far as automatic dishwashing is concerned. At temperatures above their cloud points, foaming is reduced to nearly zero. In addition, nonionics are much more effective at lowering the surface tension of water and are therefore much better detergents.

## C. Defoamers

The formation of foam during the automatic dishwashing cycle is detrimental to the mechanical washing efficiency of the machine. To prevent foaming, LADD products often contain defoamers. The most commonly used hypochlorite-stable defoamers in LADDs are anionic surfactants such as alkyl phosphate esters and ethoxylated esters [41–43] and silicone oils [44]. The structures of the first two types are shown in Figure 9.5. Also described in the patent literature is the use of aliphatic alcohols or acids as defoamers [45].

The wash temperature has a significant effect on foaming due to its effect on the solubility of the defoaming surfactant. Nonionic surfactants are only effective above their cloud points. Careful consideration must therefore be taken when formulating with nonionics. Water hardness also plays a role in foam formation. In hard water, $Ca^{2+}$ and $Mg^{2+}$ form complexes with fats or anionic surfactants. These insoluble complexes interact with the foam, breaking up the micellar structure.

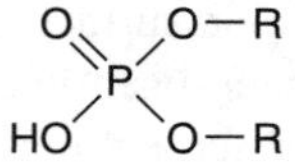

Phosphate Diesters

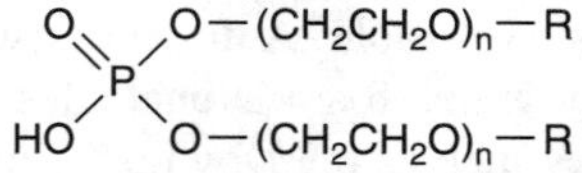

Ethoxylated Phosphate Diesters

**FIG. 9.5**  Chemical structures of phosphate ester defoamers.

## D.  Oxidizing Agents

The primary role of bleach in automatic dishwashing is in the removal of food-based stains by oxidation of the responsible molecules. These include antho-cyanins (berries), humic acid polymers (coffee, cocoa), tannins (tea, red wine), and carotinoid dyes (carrots, tomatoes). In addition, bleaches contribute to overall cleaning and disinfecting of dishware. They accomplish this by oxidizing food soils, thereby solubilizing them. Several factors contribute to the bleaching activity in the automatic dishwashing process. The most important ones are listed below [46]:

1.  All bleaches work faster with increasing temperature.
2.  Trace amount of metal impurities dramatically increase the rate of catalytic decomposition of bleaches.
3.  The rate of bleaching increases with increasing bleach concentration. However, bleach self-decomposition also increases.
4.  The bleach activity is often dependent on pH.

Two general types of bleaches have been used in ADDs: hypochlorite and peroxygen bleaches. Chlorine bleaches refer here to complexes that either contain or deliver hypochlorite ($OCl^-$) in solution. The compounds used in detergent products are typically alkali metal or alkaline earth hypochlorites, with liquid sodium hypochlorite being the most common. Solid calcium hypochlorite has also received attention, although its low solubility and its contribution of $Ca^{2+}$ to the wash liquor has limited its use in LADD products. Generally, NaOCl is added in amounts resulting in 1 to 2% available chlorine.

One drawback to the use of chlorine bleaches is that in addition to oxidizing soils the bleach also oxidizes some ingredients often used in the formulations. Because of these unwanted side reactions, care must be taken when formulating with chlorine bleaches. For example, enzymes, nonionic surfactants, sodium citrate, and other useful LADD components are easily and rapidly oxidized by hypochlorite. A study of the stability of sodium hypochlorite in the presence of surfactants has been published by Rosen and Zhu [39]. Even when using only bleach-stable detergent ingredients, the bleach present will decrease over time due to autodecomposition. Two different pathways contribute to bleach degradation [46]:

$$2NaOCl \rightarrow 2NaCl + O_2$$

$$3NaOCl \rightarrow NaClO_3 + 2NaCl$$

The first pathway accounts for about 95% of the consumption of $OCl^-$ in the absence of metallic catalysts [47]. The decomposition is affected by pH, concentration, temperature, salt content, and photolysis [39,47] and can lead to loss of product performance upon prolonged storage.

The peroxygen bleaches, typically perborate or percarbonate salts, produce peroxide in aqueous solutions. The effectiveness of peroxide as an oxidizing agent at relatively low wash temperatures (120 to 140°F) is unfortunately minimal. Peracids (RC(O)OOH), which can be generated by the activation of carboxylic acids by peroxide or can be added directly, have not found use in LADD compositions in spite of their relatively high oxidation potentials because of their instability in aqueous solution.

As mentioned earlier, European wash conditions are somewhat different from those encountered in the U.S. In particular, the water temperature during the wash cycles is higher in Europe. This difference has important consequences regarding the formulation of detergents. The higher wash temperatures of European dishwashing also allow for the use of milder oxygen bleaches in place of chlorine bleaches. As bleaching agents, the activity of peroxides and peracids increases with temperature. Under typical North American conditions, they are not suitable replacements for hypochlorite. However, they are more effective under European conditions. The use of oxygen bleaches allows for more freedom in formulations since many LADD components are incompatible with hypochlorite but not with peroxide. Because of the instability of peroxide in alkaline aqueous systems, their use is limited to nonaqueous LADDs. Unfortunately, these are not currently economically feasible.

## E. Enzymes

The use of enzymes in detergent formulations became a practical matter in the mid-1960s when Novo Industry began production of proteases by microbial fermentation. These enzymes were stable at the high temperatures and alkalinity encountered in dishwashing.

Several types of enzymes have found uses in LADD compositions [4,48]. Most common are proteases, amylases, and lipases, which attack proteinaceous, starchy, and fatty soils, respectively. Proteases work by hydrolyzing peptide bonds in proteins. Proteases differ in their specificity toward peptide bonds. The typical protease used in LADD formulations, bacterial alkaline protease (subtilisin), is very nonspecific. That is, it will attack all types of peptide bonds in proteins. In contrast to proteases, amylases catalyze the hydrolysis of starch. They attack the internal ether bonds between glucose units, yielding shorter, water-soluble chains called dextrins. Lipases work by hydrolyzing the ester bonds in fats and oils. Often, combinations are used because of the specificity of each kind to one type of soil. The commercially available enzymes are listed in Table 9.6.

While enzymes have played a role in powder detergents in the U.S. market over the years, more recently they have started to emerge as additives in LADD formulations. Both proteases and amylases have appeared in LADDs in spite of the fact that, unlike powders, oxygen bleaches could not be incorporated for stain removal. They first appeared in bottled liquid/gel products in 2000 in the U.S. These were introduced as premium products, with dual enzymes, and captured about 5% of the total automatic dishwashing market.

The first automatic dish gel pac was introduced in 2002. This was a dual enzyme-based formula that was packaged in a water-soluble sachet. This form of unit dose followed on the relative success of pressed powder (tablet) unit dose products in the market. This also captured about 5% of the total automatic dishwashing market.

**TABLE 9.6**  Commercially Available Detergent Enzymes

| Type | Trade name | Manufacturer |
|---|---|---|
| Protease | Esperase | Novo |
| | Savinase | |
| | Alcalase | |
| | Everlase | |
| | Kannase | |
| | Purfect® L | Genencor |
| | Purafect® L | |
| | Properase® L | |
| Amylase | Termamyl | Novo |
| | Duramyl | |
| | Purastar® HPAm L | Genencor |
| Lipase | Lipolase | Novo |
| | Lipolase Ultra | |
| | Lipo Prime | |

As enzyme technology improves, the share of enzyme-containing products can be expected to grow in the liquid LADD market. New forms will probably help support this move, and may ultimately lead to a dual enzyme, oxygen bleach technology.

## F.  Structuring Agents

LADD products currently in the marketplace are not real liquids, but are gels thickened by a structuring agent. Two main kinds of materials are used in LADD products currently available to consumers: water-swellable, high-molecular-weight ($>1,000,000$) crosslinked polyacrylates and clays such as bentonite or laponite. Often, cothickeners are added to improve the stability of the gel. For example, colloidal alumina or silica [49–51], fatty acids or their salts [50,52–63], and others have been described in the patent literature as being effective additives. It has also been found that incorporated air bubbles can also contribute to the stability of polyacrylate-thickened gels [71–74]. The effect of such thickening systems is to make the gel viscoelastic and thixotropic. This means that the gel is viscous when unstressed, but the viscosity decreases due to shear thinning upon application of an applied stress. The rheological properties of LADD products are discussed in Section IV. When LADD products contain bleach, potential structuring agents are very limited. They must be chemically stable to bleach for the life of the product.

Currently, products in the market are thickened mainly by polyacrylic acids, although some use a mixture of polymer and a clay. This is in contrast to the early LADD products, which used exclusively clay thickeners and therefore suffered from stability problems [7]. The products, being clay suspensions in an aqueous medium, tended to separate in a relatively short period of time. The introduction of bleach-compatible polymeric thickeners in the early 1990s overcame this problem [75].

## IV.  RHEOLOGY AND STABILITY OF LADDs

As discussed earlier, LADDs are complex, multicomponent mixtures consisting of both organic and inorganic compounds dispersed in a liquid matrix. Such compositions can exhibit a broad range of rheological characteristics from simple Newtonian to complex pseudoplastic flow. Shown in Figure 9.6 and Figure 9.7 are flow and viscosity profiles of Newtonian and non-Newtonian fluids as a function of applied shear rate. A number of mathematical models have been proposed [76] to describe the flow characteristics of various systems. These equations are called constitutive equations and are used to predict flow behavior in complex systems.

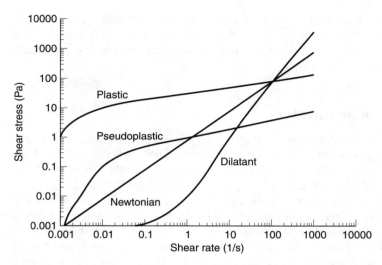

**FIG. 9.6**   Generic examples of shear stress vs. shear rate plots for different rheological systems. (Reproduced from Laba, D., Ed., *Cosmetic Science and Technology Series*, Vol. 13, Marcel Dekker, New York, 1993. With permission.)

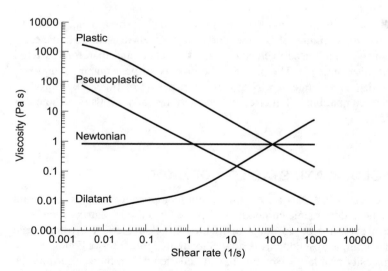

**FIG. 9.7**   Generic examples of viscosity vs. shear rate plots for different rheological systems. (Reproduced from Laba, D., Ed., *Cosmetic Science and Technology Series*, Vol. 13, Marcel Dekker, New York, 1993. With permission.)

Clearly, the response of fluids to an applied shear can be linear or nonlinear and depends on two major factors: shear rate and structural or mechanical properties of the system, which in turn depend upon the interaction between the components including the rheological additives. It is the latter that primarily determines the flow properties of LADDs. The intent of this section is to discuss various rheological properties and test methods pertinent to LADDs. Interested readers are referred to Chapter 4, which deals with the rheology of complex liquids and suspensions, and other books [76–79] and review articles [80–82] covering this subject. Heywood [83] discusses the criteria for selecting various commercial viscometers.

The ideal rheological characteristics of LADDs from the consumer point of view are:

1. The product must be homogeneous and phase-stable.
2. The flow properties of the product should be such that it can be readily dispensed to the dishwasher cup from the bottle without dripping.
3. Once the product is in the dishwasher cup it should recover its structure/viscosity such that the product stays in the cup until the wash cycle begins.

The third attribute is the most important for dishwashers sold in the U.S. since a majority of dishwasher cups are primarily designed for powders. They are therefore ungasketed and will not prevent improperly structured gel products from prematurely leaking out.

## A. Flow Properties

## 1. Viscosity

This is a fundamental rheological property that is universally measured and reported for all liquid products. Viscosity is resistance to flow. Different types of viscometers are used to measure this property. A major disadvantage of determining the viscosity of complex fluids is that the value obtained depends on the instrument, type of spindle, and rotational speed. Therefore, one should be extremely careful when comparing the viscosities of liquid products described in the literature.

## 2. Yield Value

Another rheological attribute of LADDs commonly referred to in the patent literature is yield value (also referred to as yield stress). The significance of the yield value is that it indicates shear stress or shear rate necessary to induce flow which is a characteristic of the system. This is only an apparent yield stress, since everything will flow even at zero stress if given enough time.

Several methods based on viscometers [84,85] and stress rheometers [86,87] have been proposed and described for measuring this parameter. One of the methods described [88] utilizes a Brookfield RVT model viscometer and a T-bar spindle.

The viscometer readings are recorded at 0.5 and 1.0 r/min after 30 seconds or after the system is stable. Shear stress at zero shear is equal to two times the 0.5 r/min reading minus the reading at 1.0 r/min. The yield value is then calculated as follows:

Yield value = 18.8 × stress at zero shear

According to this patent, the yield value for LADDs should be in the range 50 to 350 dyn/cm$^2$.

Another method described in the literature [85] is based on the Brookfield viscosity measurements at two different shear rates and the yield value is calculated according to the following equation:

Yield value = $2r_1(\eta_1 - \eta_2)/100$

where $r$ is the shear rate and $\eta_1$ and $\eta_2$ are viscosities at shear rates of $r_1$ and $2r_1$, respectively.

## 3. Thixotropy

The most desirable rheological behavior for LADDs is thixotropy. Such systems follow a shear-thinning pattern very similar to pseudoplastic systems, but when the shear is removed the structure rebuilds in a time-dependent manner instead of instantaneously. A rheological profile of a thixotropic liquid shows a characteristic hysterysis loop, the size of which is related to the degree of thixotropy and structure recovery time. A typical thixotropic loop shows the relationship between the viscosity or stress vs. shear rate. An example of such a plot for a commercial LADD is shown in Figure 9.8. Such flow curves are typically obtained by plotting viscosity or shear stress with increasing shear rate (up curve) followed by decreasing shear rate (down curve).

Some systems exhibit flow behavior opposite of thixotropic systems, that is, viscosity increases with increasing shear rate. Such fluids are referred to as dilatant or rheopectic. This type of behavior is not common for liquid products containing a low concentration of the dispersed phase.

## 4. Dynamic Properties

The vast majority of concentrated dispersions, such as LADDs, exhibit both viscous and elastic properties. These systems are therefore referred to as viscoelastic. The flow properties discussed in the previous section are not sufficient for complete rheological characterization of viscoelastic fluids. Dynamic mechanical properties, characterized by the storage modulus ($G'$) and loss modulus ($G''$), are normally

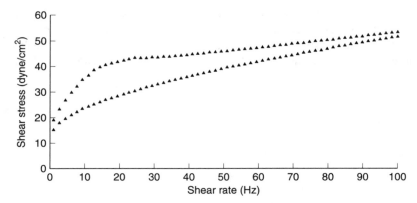

**FIG. 9.8**   Plot of shear stress vs. shear rate for a commercial LADD. The rheograms were recorded on a Carri-Med CSL 100 stress-controlled rheometer using a 4 cm acrylic parallel plate configuration with gap setting of 1000 μm.

measured to quantify the viscoelastic properties. The storage modulus represents the mechanical energy stored and recovered and is a direct measure of the elasticity of the fluid. The loss modulus is a measure of mechanical energy lost thermodynamically as heat. This type of energy dissipation occurs when the sample is undergoing viscous flow. For predominately elastic materials $G' > G''$, and for predominately viscous materials $G'' > G'$.

Both $G'$ and $G''$ are related to the complex modulus $G*$ and complex viscosity $\eta*$ by the following relationships:

$$G* = G' + iG''$$

$$\eta* = [(G'/\omega)^2 + (G''/\omega)^2]^{1/2}$$

where $\omega$ is the angular frequency of the oscillation.

The relative magnitudes of the two moduli provide significant information regarding strength of internal association or structure in fluids and dispersions. These moduli are measured as a function of strain, frequency, or time. For some dispersions, the magnitudes of $G'$ and $G''$ may remain constant as a function of either frequency or strain. Such materials are referred to as linearly viscoelastic.

Plots of $G'$ and $G''$ vs. % strain for the three major commercial gel LADDs sold in the U.S. are shown in Figure 9.9, Figure 9.10, and Figure 9.11. For these plots, a strain is applied to the sample and the stress response is measured.

The elasticity and viscosity of a gel are essential criteria for ease of dispensing and cup retention in the dishwasher. For example, a patent [89] claims that viscosities of 1000 to 20000 cP under 5/sec shear, 200 to 5000 cP under 21/sec shear, and a steady-state viscoelastic deformation compliance value of at least 0.01 are ideal for product dispensability and cup retention (as measured on a Haake Rotovisco RV-100 viscometer). A series of patents issued to Dixit and co-workers

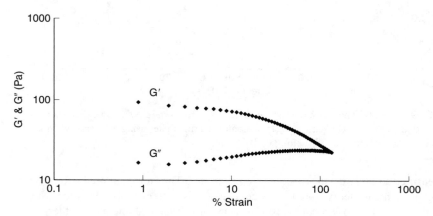

**FIG. 9.9**   Plots of $G'$ and $G''$ vs. % strain for a commercial LADD. The rheograms were recorded on a Carri-Med CSL 100 stress-controlled rheometer using a 4 cm acrylic parallel plate configuration with gap setting of 1000 μm.

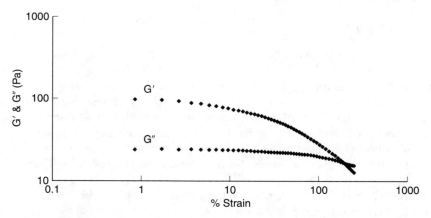

**FIG. 9.10**   Plots of $G'$ and $G''$ vs. % strain for a commercial LADD. The rheograms were recorded on a Carri-Med CSL 100 stress-controlled rheometer using a 4 cm acrylic parallel plate configuration with gap setting of 1000 μm.

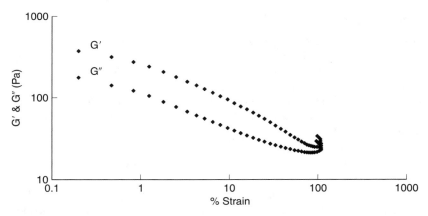

**FIG. 9.11** Plots of $G'$ and $G''$ vs. % strain for a commercial LADD. The rheograms were recorded on a Carri-Med CSL 100 stress-controlled rheometer using a 4 cm acrylic parallel plate configuration with gap setting of 1000 $\mu$m.

[55,56,59,71,72] emphasize the importance of linear viscoelasticity as defined by tan(delta) as an important rheological characteristic for LADDs.

## V. EVALUATION OF THE PERFORMANCE OF LADDs

Essential to the development of new LADD products is the evaluation of their performance. It is important that the test conditions closely reproduce the conditions encountered by the consumer under typical household situations; therefore, prototype formulas are usually tested in actual dishwashing machines using real food soils.

Typically, performance evaluation tests are run using a variety of soiled items, since a combination of soils is typically encountered by consumers. For example, spotting and filming tests on standard glass tumblers are run in conjunction with plates and cutlery soiled with egg, oatmeal, spinach, tomato sauce, and various other common food soils. In addition, the ability of the ADD product to remove tough stains such as coffee, tea, or blueberry is assessed on a variety of substrates during these multi-soil test cycles.

Ideally, testing should be conducted using machines from all major manufacturers, as differences exist in detergent dosing amounts, water fill amounts, and lengths and orders of cycles. It is also important that performance tests be carried out at different water hardnesses, since regional variations exist. Water hardness for testing purposes can be controlled by the addition of simple salts of calcium and magnesium to deionized water. A 2:1 mole ratio is commonly used since that is the $Ca^{2+}:Mg^{2+}$ ratio encountered in most water supplies.

## A.  Spotting and Filming

The standard method for the testing of ADD products as regards spot/film prevention on glassware is described in ASTM D 3556 [90]. The test entails the washing of glass tumblers using commercial automatic dishwashing machines. A soil consisting of margarine and powdered, nonfat milk (40 g, in an 8:2 ratio) is added prior to starting the cycle. The soil can be introduced by spreading onto dinner plates or by adding directly into the wash. This combination of soils provides the fats and proteins generally present during typical household wash cycles. Optionally, a similar soil mixture which also contains cooked cereal can be added, providing a source of starch. For completeness, dinner plates, dessert plates, and silverware are placed in the machine for ballast.

It is recommended that at least five complete cycles be run during a test, adding the soil at the beginning of each cycle. This is to ensure that product performance remains acceptable after repeated cycles. Detergents that are underbuilt, for example, will lead to heavy filming on the glasses only after several cycles. The test parameter that is typically varied during this spot/film test is the hardness. Spotting tends to occur under lower water hardness conditions, while film formation tends to happen more under harder water conditions.

To discriminate between similar products, it is often useful to run the tests under stress conditions; that is, low temperature and high water hardness. In contrast, to see how a detergent might perform under normal household conditions, higher temperatures and lower water hardness should be explored.

The ASTM procedure also describes a rating method for judging the spotting and filming on glasses. The readings are done visually, using a fluorescent light box to highlight spotting and filming on the glasses being inspected. The performance of a product is rated on a 1 to 5 scale for both spotting and filming, as shown in Table 9.7.

The use of photometry to rate the spotting and filming on glassware eliminates the possibility of subjective judging by humans. Several systems have been

**TABLE 9.7**  Rating Scale for Spotting and Filming of Glassware

| Rating | Spotting | Filming |
|---|---|---|
| 1 | No spots | None |
| 2 | Random spots | Barely perceptible |
| 3 | ¼ of surface covered | Slight |
| 4 | ½ of surface covered | Moderate |
| 5 | Virtually completely covered | Heavy |

described in the literature that take advantage of this technique [91]. As of this writing, the ASTM D 3556 test is under review.

## B. Soil Removal

When testing the performance of ADD products it is often useful to determine the ability of the detergent to remove tough food soils from items being washed. In selecting the soils to be used, the following criteria must be met. The soils must be representative of those encountered by consumers, but they must not be removed so easily as to render all products equal in cleaning efficiency. The soils can be roughly divided into two classes: water-soluble and water-insoluble soils. Examples of the former are sugars, starch, flour, or egg white, while the latter might be animal or vegetable fats.

Egg yolk has proven to be an especially useful soil for performance testing of ADD products, provided that it is prepared in a way that gives meaningful and reproducible results. This soil contains a very high protein content, and thus is not saponified by the heat and alkalinity during the wash cycle as fatty soils are. The efficient removal of egg yolk from dinner plates can only be accomplished by detergents that target proteinaceous soils. Often, $CaCl_2$ is mixed into the egg mixture to make a cohesive–adhesive egg complex capable of remaining on the dishes throughout the complete cycle [92]. Otherwise, the mechanical energy from the water jets alone will remove the egg from the plate. Typically, 2 to 3 g of $CaCl_2$/25 g of egg yolk are added. This more closely simulates typically soiled plates, which might derive a small amount of calcium from milk or salts present. A known weight of the egg soil should be spread onto dinner plates. The degree of egg removal after the cycle is determined either visually or by weight difference. An added bonus of this test is that it provides a proteinaceous soil to the wash liquor, making the spotting and filming scores obtained more realistic. Studies have shown proteinaceous soils to be a source of spotting on glasses [13]. It is important to place consistently the egg plates in the same position in the dishwasher in order to minimize spray arm effects.

The Association of Home Appliance Manufacturers (AHAM) has published a set of standards for testing of automatic dishwashers [93]. Although the focus of this test is the dishwasher performance, it can also be applied to the product performance.

The test procedure also describes how to load the items into the dishwasher and other test parameters. Variations in the types and number of soils and items used for a multi-soil test are permissible. It is desirable, though, to use a soil combination that contains the major types of soils (proteinaceous, fatty, and starchy), as some ADD products might effectively remove some but not others. This is especially true for formulas that contain enzymes.

## C. Stain Removal

Besides the prevention of spots and film and the removal of food soils from items being washed, LADD products must also effectively remove stains. For this purpose, LADDs often employ bleaching agents which act by oxidizing the chromophore responsible for the stain color. Generally, as the oxidizing potential of the bleach increases, its effectiveness also increases. The common method employed to gauge stain removal is to allow either coffee or tea to dry in a porcelain, plastic, or glass cup. The cups are then run in a standard soil test [93]. Additionally, coffee- and/or tea-stained melamine tiles can be used in evaluating stain removal. These types of tiles can be obtained from companies such as Test Fabrics in West Pittston, PA.

## D. Foaming

The minimization of foam generated by detergent and food soils during the wash cycles in the automatic dishwashing process is a prerequisite for efficient mechanical cleaning by the machine spray arms. Foam decreases the water pressure pumped through the rotors, decreasing the kinetic energy of the water jets. As stated in Section II, mechanical action has been estimated to be responsible for 85% of the cleaning in a machine dishwashing cycle. A lowering of the wash pressure therefore will have a noticeable effect on overall cleaning. Because of this, ADD formulations, especially those containing anionic surfactants, often contain defoamers.

A standard test method has been developed for the measurement of foam during an automatic dishwashing cycle. This method is described in the CSMA Compendium, Method DCC-001 [94]. The test involves the measurement of the machine spray arm rotational velocity (in r/min) at one-minute intervals over a ten-minute wash cycle. The rotor r/min will decrease in the presence of foam since the water pressure being pumped through the spray arm will be lower. In order to measure the revolutions per minute, a magnetically activated reed switch is used and the spray arm is fitted with a magnet. The detergent is dosed normally, and a high-foaming soil is added before the cycle begins. The recommended soil is nonfat powdered milk (10 g) or a powdered milk and egg white combination (1:1).

The efficiency score for a particular detergent is obtained by taking the average of the arm speed readings and dividing by the average reading for the control. In this experiment, the control is the same experiment without the detergent or soil. As in other performance tests, it is good practice to use the same machine for all comparative experiments in order to eliminate variations due to the use of different machines. Different machines will have slightly different motors which will produce slightly different rotor velocities.

## E.   Fine China Overglaze and Glass Corrosion

Fine china is often decorated with colored patterns made from various metal salts or oxides. Two methods are commonly used: underglaze, in which the color is put on before the glaze, and overglaze, in which it is applied after the glaze. The overglaze pattern on fine china is incompatible with the high alkalinity characteristic of ADD products. Unless the china is somehow protected from hydroxide ions during the wash, the overglaze and the color will be attacked and destroyed. To protect against this, LADDs are generally formulated with sodium silicate, which acts by coating the china with a protective siliceous layer, preventing the alkali from coming in contact with the overglaze itself. Silicate works in a similar manner to protect metal machine parts from corrosion.

A procedure has been developed to test the effectiveness of ADD products in fine china overglaze protection. This method is described in ASTM D 3565 [95]. Segments of a china plate are soaked in a 0.3% ADD solution held at 96.0 to 99.5°C in a steam bath. Two controls are also set up consisting of a sodium carbonate solution and water only. The segments of the china plate are placed on wire mesh supports to avoid contact with the bottom of the steel beakers used. The solutions are heated for six hours, after which the china segments are removed and rubbed vigorously with a 1.5 inch square of muslin. The plate segment is then washed, dried, and visually inspected for fading. An effective detergent should prevent any sign of wear.

Additional information about glass corrosion can be obtained from the research done by Sharma and Jain [25,26]. Although they evaluated the visible effect of wine glass corrosion over the course of 100 machine washings, a similar visible irreversible clouding of the wineglasses can be demonstrated by a modification of the ASTM D 3565 glaze test. The test can be modified by using soda-lime-silicate wine glasses in place of the china plate segments. The ring effect can be seen in as little as three hours.

## VI.   FORMULATION TECHNOLOGY
## A.   Cleaning

Consumers have several criteria for accepting a LADD. Some of these are listed in Table 9.8. The formulator must strike a balance among performance, aesthetics, convenience, and cost to meet consumer requirements. While cleaning dishware is the most important element, shine and spot and film prevention are also strongly desired. Corrosion protection, particularly related to glasses and china, is also an important attribute for a successful LADD. Formulators use a combination of components to achieve these goals. These include, but are not limited to, alkalinity, bleach, silicates, phosphates, polymers (both low and high molecular weight), and enzymes.

**TABLE 9.8**  Important Attributes of LADDs

---

Effective cleaning

Shine

Spot and film prevention

No powder residue

Convenient to use

Safe to dishes and tableware

Safe to dishwasher

Stable upon storage

Safe to humans

Economic to use

---

Aside from mechanical action, the bulk of the food removal from soiled dishes is accomplished by two mechanisms. Fatty soils are removed by a combination of the high temperature and alkalinity present, which melts and saponifies fats. Proteinaceous and starchy soils, in contrast, are solubilized by oxidation and hydrolysis. More effective soil removal is accomplished by the use of enzymes, which selectively and efficiently attack fats, starches, or proteins. Stain removal is another aspect of cleaning that is addressed by different technologies. A recent technology, based on benzoyl peroxide, has been shown to be effective against red tomato stains on plastic. The removal of tough stains such as coffee or tea can only be accomplished by strong bleaching agents such as hypochlorite. Hypochlorite also has a sanitizing effect on the washed items. Listed in Table 9.9 are relevant patents that disclose novel cleaning technologies in LADDs.

## B.  Thickeners

Equally important when formulating LADDs is making a stable and easy-to-handle product. LADDs are concentrated suspensions which must be properly structured so as to prevent separation upon storage. The product must also be thickened for two reasons: to prevent it from prematurely leaking out of the machine dispenser cup and to make it easier to control when dosing. However, the product must be shear thinning so that it will easily flow under an applied stress. Early LADDs were thickened by clay thickeners. More recently, the use of high-molecular-weight polymeric thickeners, optionally with fatty acid or other surfactant cothickeners, has solved the separation problems. Composition patents in this area are listed in Table 9.14. Patents that disclose processing or manufacturing methods used in LADD production are listed in Table 9.10.

**TABLE 9.9** Patents Relating to Cleaning Technology in LADDs

| Patent no. and year | Inventor(s) and company | Technology | Claimed benefits |
|---|---|---|---|
| U.S. 6602837 (2003) [96] | Patel (Procter & Gamble) | Nonaqueous; diacyl peroxide | Plastic stain removal especially carotenoids |
| U.S. 5929008 (1999) [98] | Goldstein (Procter & Gamble) | Perborate bleach | Improved cleaning at pH > 11.4 |
| U.S. 5858944 (1999) [105] | Keenaqn et al. | Polycarboxylates | Low film/spot |
| U.S. 5698507 (1997) [99] | Gorlin et al. (Colgate-Palmolive) | Dual enzyme | Soil removal |
| U.S. 5618465 (1997) [100] | Durbut et al. | Dual enzyme | Soil removal |
| U.S. 5597789 (1997) [101] | Sadlowsk | Polymer | Hard water film performance |
| U.S. 5545344 (1996) [102] | Durbut et al. | Enzymes, no phosphate, nonaqueous | Concentrated |
| U.S. 5527483 (1996) [103] | Kenkare et al. | Nonaqueous, enzymes | Soil removal |
| U.S. 5372740 (1994) [104] | Fair et al. | In situ TPP, Na/K balance | No added silicate |
| U.S. 5318715 (1994) [106] | Krishnan (Colgate-Palmolive) | Nonaqueous, dual enzymes | Improved cleaning at pH < 10.5 |
| WO 94/25557 (1994) [107] | Sadlowski (Procter & Gamble) | Modified polyacrylate copolymer | Enhanced hard water filming |
| WO 94/17170 (1994) [108] | van Dijk et al. (Unilever) | Itaconic acid–vinyl alcohol copolymer | Improved scale prevention; biodegradable |
| U.S. 5308532 (1994) [109] | Adler et al. (Rohm & Haas) | Water-soluble carboxylate terpolymers | Reduced spotting and filming |

*(continued)*

**TABLE 9.9** (Contd.)

| Patent no. and year | Inventor(s) and company | Technology | Claimed benefits |
|---|---|---|---|
| U.S. 5240633 (1993) [110] | Ahmed et al. (Colgate-Palmolive) | Nonaqueous; protein-engineered enzymes | Improved protein and carbohydrate soil removal; improved low-temperature cleaning |
| U.S. 5164106 (1992) [111] | Ahmed et al. (Colgate-Palmolive) | Nonaqueous; dual bleach system | Improved cleaning against proteinaceous and starchy soils |
| U.S. 5094771 (1992) [112] | Ahmed et al. (Colgate-Palmolive) | Nonaqueous; silica, alumina, or titanium dioxide | Readily dispersible in water; improved spotting and filming in hard water |
| U.S. 5076952 (1991) [113] | Ahmed et al. (Colgate-Palmolive) | Dual bleach sysactem | Improved cleaning |
| U.S. 4971717 (1990) [30] | Dixit (Colgate-Palmolive) | Aluminosilicate | Improved filming and spotting |
| U.S. 4970016 (1990) [114] | Ahmed et al. (Colgate-Palmolive) | Alumina or titanium dioxide | Improved filming in hard water |
| U.S. 4968446 (1990) [115] | Ahmed et al. (Colgate-Palmolive) | Alumina or titanium dioxide | Improved filming |
| U.S. 4968445 (1990) [116] | Ahmed et al. (Colgate-Palmolive) | Silica | Improved filming |
| U.S. 4931217 (1990) [117] | Frankena (Lever Bros.) | Quaternary ammonium salts | Enhanced fat removal at low temperature |
| EP 271155 (1988) [118] | van Dijk (Unilever) | Bacterial lipases | Improved spotting and filming |
| U.S. 4753748 (1988) [119] | Laitem et al. (Colgate-Palmolive) | Nonaqueous; stable polyphosphate builder suspension | Improved rinse properties |
| U.S. 4597886 (1986) [120] | Goedhart et al. (Unilever) | Enzymes and layered clay | Improved spotting and filming |
| U.S. 4539144 (1985) [121] | de Ridder et al. (Lever Bros.) | Modified polyacrylate; phosphate-free | Improved filming, spotting, streaking |
| U.S. 4306987 (1981) [122] | Kaneko (BASF Wyandotte) | Block polyoxyalkylene nonionic surfactant | Effective foam control; spot-free wash items; effective against encrusted protein soils |

**TABLE 9.10** Patents Relating to Thickeners in LADDs

| Patent no. and year | Inventor(s) and company | Technology | Claimed benefits |
|---|---|---|---|
| U.S. 5691292 (1997) [53] | Marshall et al. (Procter & Gamble) | Thixotropic liquid; chlorine bleach free | Thickener for enzyme-based ADDs; enzyme stabilizing |
| U.S. 5427707 (1995) [52] | Drapier et al. (Colgate-Palmolive) | Thixotropic thickener; long-chain fatty acids | Thixotropy with lower thickener levels |
| U.S. 5413727 (1995) [54] | Drapier et al. (Colgate-Palmolive) | Thixotropic thickener; long-chain fatty acids | Thixotropy with lower thickener levels |
| U.S. 5368766 (1994) [71] | Dixit (Colgate-Palmolive) | High-molecular-weight crosslinked polymer thickener; K/Na ratio > 1:1; incorporated air bubbles | Physical stability; low bottle residue; low cup leakage; improved cleaning |
| U.S. 5298180 (1994) [72] | Dixit (Colgate-Palmolive) | High-molecular-weight crosslinked polymer thickener: K/Na ratio > 1:1; incorporated air bubbles | Physical stability; low bottle residue; low cup leakage; improved cleaning |
| U.S. 5336430 (1994) [123] | Bahary et al. (Lever Bros.) | Polysaccharide thickener; encapsulated bleach and chlorine scavenger | Biodegradable structurant |
| U.S. 5252241 (1993) [55] | Dixit et al. (Colgate-Palmolive) | High-molecular-weight crosslinked polymer thickener; long-chain fatty acid; incorporated air bubbles | Physical stability; low bottle residue; low cup leakage; improved cleaning |
| U.S. 5252242 (1993) [56] | Dixit et al. (Colgate-Palmolive) | High-molecular-weight crosslinked polymer thickener; long-chain fatty acid; incorporated air bubbles | Physical stability; low bottle residue; low cup leakage; improved cleaning |
| U.S. 5229026 (1993) [57] | Dixit et al. (Colgate-Palmolive) | High-molecular-weight crosslinked polyacrylic acid; K:Na ratio > 1:1; incorporated air bubbles; fatty acid or salt | Exceptional physical stability; low bottle residue; low cup leakage; improved cleaning |

*(continued)*

**TABLE 9.10**  (Contd.)

| Patent no. and year | Inventor(s) and company | Technology | Claimed benefits |
|---|---|---|---|
| U.S. 5229026 (1993) [31] | Dixit (Colgate-Palmolive) | High-molecular-weight crosslinked polyacrylic acid; K:Na ratio > 1:1; incorporated air bubbles; aluminosilicate | Exceptional physical stability; low bottle residue; low cup leakage; improved cleaning |
| U.S. 5205953 (1993) [73] | Dixit (Colgate-Palmolive) | High-molecular-weight crosslinked polyacrylic acid; K:Na ratio > 1:1; incorporated air bubbles; polymeric chelating agent | Exceptional physical stability; low bottle residue; low cup leakage; improved cleaning |
| U.S. 5188752 (1993) [58] | Prencipe et al. (Colgate-Palmolive) | Crosslinked methyl vinyl ether–maleic anhydride copolymer; fatty acid/salt | Linear, viscoelastic |
| U.S. 5169552 (1992) [64] | Wise (Procter & Gamble) | Polymeric thickener; stabilizing agent; buffering agent | Improved bleach stability; improved rheology; shear thinning |
| U.S. 5135675 (1992) [65] | Elliot et al. (Lever Bros.) | Swellable clay, sulfonated polymer, and multivalent cations | Good salt tolerance; bleach stability; shear thinning |
| U.S. 5130043 (1992) [66] | Prince et al. (Procter & Gamble) | Polycarboxylate and phosphate esters | Enhanced stability and cohesiveness |
| U.S. 5098590 (1992) [59] | Dixit et al. (Colgate-Palmolive) | Thixotropic thickener; long-chain fatty acid/salt; equal specific gravities for bulk and liquid phases | Improved stability |
| U.S. 5057237 (1991) [60] | Drapier et al. (Colgate-Palmolive) | Clay thickener; polyvalent metal salts of long-chain fatty acids | Improved stability |
| U.S. 4950416 (1990) [50] | Baxter (Vista Chemical) | Thixotropic liquid; aqueous alumina dispersion; long-chain fatty acid or salt and mixtures | Nonpolymeric and nonclay compositions |

| Reference | Inventor (Company) | Description | Benefit |
|---|---|---|---|
| U.S. 4941988 (1990) [51] | Wise (Procter & Gamble) | Polyacrylate thickener and alkali metal silica colloid; optimized thickening system | Improved stability against phase separation |
| WO 89/04359 (1989) [67] | Donker (Unilever) | Thixotropic gel with alkyl phosphate, phosphonate, sulfate, or sulfonate | Lower clay levels needed |
| U.S. 4836946 (1989) [61] | Dixit (Colgate-Palmolive) | Clay thickener; alkali metal fatty acid salt | Improved stability |
| U.S. 4824590 (1989) [74] | Roselle (Procter & Gamble) | Shear-thinning gels with incorporated air | Improved stability against phase separation; increased viscosity; lower clay levels needed |
| U.S. 4801395 (1989) [62] | Chazard et al. (Colgate-Palmolive) | Clay thickener and long-chain fatty acids | Thixotropic liquid structured by surfactants only |
| EP 304328 (1989) [124] | Kreischer (Unilever) | Nonsoap anionic surfactant and soap; electrolyte level > 20% | Phase stability for more than 12 weeks; lower clay levels needed |
| U.S. 4752409 (1988) [63] | Drapier et al. (Colgate-Palmolive) | Inorganic colloid-forming clays or other thixotropic thickener; polyvalent metal salts of long-chain fatty acids | Improved dispensability and processing; improved cup retention |
| U.S. 4740327 (1988) [68] | Julemont et al. (Colgate-Palmolive) | Clay thickener and anionic surfactant | Low yield value; phase stability due to the filler material |
| GB 2168377 (1985) [69] | Taraschi (Procter & Gamble) | Water-insoluble abrasive colloidal clay; water-insoluble low-density particulate filler | Viscosity 3000–6000 cP; yield point 250–825 |
| WO 83/03621 (1983) [70] | Kolodny et al. (American Home Prod.) | Xanthan gum thickener; mixture of anionic and nonionic surfactants | Good viscosity and flow control |
| U.S. 4260528 (1981) [125] | Fox et al. (Lever Bros.) | Natural or synthetic gum thickener, urea and polyhydric alcohol | Reduced cup leakage; low foaming |
| U.S. 4226736 (1980) [126] | Bush et al. (Drackett) | Thixotropic gel | |

**TABLE 9.11** Patents Relating to LADD Processing

| Patent no. and year | Inventor(s) and company | Technology | Claimed benefits |
|---|---|---|---|
| U.S. 5624892 (1997) [128] | Angevaare et al. (Lever Bros.) | Aluminum sequestrant | Inhibits lead corrosion |
| U.S. 5395547 (1995) [129] | Broadwell et al. (Colgate-Palmolive) | Fatty acid/surfactant/defoamer predispersion | Improved stability |
| U.S. 5366653 (1994) [127] | Corring (Lever Bros.) | Dry-blending thickening polymer and sodium trimetaphosphate | Concentrated liquid |
| U.S. 5246615 (1993) [130] | Broadwell et al. (Colgate-Palmolive) | Preneutralization of polymeric thickener with alkali metal silicate or builder | Pumpable polymer gel premix |
| WO 93/21298 (1993) [131] | Gabriel et al. (Procter & Gamble) | Premix made, mixed aerated to increase density and aesthetics | Increased product density; enhanced aesthetics; improved rheological efficiency of polymer |
| U.S. 5075027 (1991) [132] | Dixit et al. (Colgate-Palmolive) | High shear dispersing to entrain air; in-line mixing | Stable product |
| U.S. 4927555 (1990) [133] | Colarusso (Colgate-Palmolive) | Wet grinding with high-speed disperser | Manufacture of thixotropic gel |

## C. Processing

The manufacture of stable, rheologically built products is not as simple as mixing the ingredients. Proper processing is required to provide the right rheology and density. Both of these attributes are required for a successful LADD. Table 9.11 provides a list of patents related to this subject.

## D. Corrosion

Other important considerations must be taken into account when formulating LADDs. Because of the corrosiveness of typical LADDs, ingredients must be added which protect both the machine itself and fine items such as china and silverware. Generally, silicate is added for this purpose, but its inherent alkalinity in aqueous solutions is not always desired. In Table 9.12 several patents are listed which describe other anticorrosion agents.

**TABLE 9.12**  Patents Relating to Corrosion Inhibition in LADDs

| Patent no. and year | Inventor(s) and company | Technology | Claimed benefits |
|---|---|---|---|
| U.S. 6448210 (2002) [135] | Keyes *et al.* (Johnson Diversey) | Zinc (gluconate) | Glassware protection |
| U.S. 6083894 (2000) [138] | Keyes *et al.* (S.C. Johnson) | Zinc (gluconate) | Glassware protection |
| U.S. 5731277 (1998) [136] | Gary *et al.* (Lever Bros.) | Aluminum tetrahydroxide | Tableware protection |
| U.S. 5783539 (1998) [137] | Angevaare *et al.* (Lever Bros) | Aluminum complex | Tableware protection |
| U.S. 5698506 (1997) [139] | Angevaare *et al.* (Lever Bros.) | Aluminum salt | Tableware protection |
| U.S. 5374369 (1994) [134] | Angevaare *et al.* (Lever Bros.) | Cyanuric acid and chloroamines, etc. | Prevents tarnishing of silver items |
| U.S. 4992195 (1991) [140] | Dolan *et al.* (Monsanto) | Sulfamic acid or water-soluble salts | Protection of silver items; stabilizes bleach present |
| U.S. 4933101 (1991) [141] | Cilley *et al.* (Procter & Gamble) | Insoluble inorganic zinc salts | Inhibits glassware corrosion |
| U.S. 4859358 (1989) [142] | Gabriel *et al.* (Procter & Gamble) | Long-chain hydroxy fatty acid salts | Protection of silver items |

## E.  Surfactants

Surfactants can provide greasy soil removal, and spot and film benefits to LADDs. The main restriction of surfactants is their ability to create foam. While this is a desirable property in hand dishwashing, it can create significant problems in automatic dishwashing products. They can cause foam which reduces the amount of energy delivered by the spray arms. Additionally, the foam can rise above the machine lip and cause external flooding. The foam can be controlled by using defoaming agents in conjunction with the surfactants, or selecting appropriate nonionic surfactants. Patent examples of these technologies are listed in Table 9.13.

## F.  Stability

As previously mentioned, a hurdle in formulating bleach-containing LADD products is that many useful ingredients are not stable toward hypochlorite. Table 9.13 lists patents that claim bleach-stable nonionic surfactants and Table 9.14 contains patents relating to the stabilization of LADD components.

**TABLE 9.13**   Patents Relating to Surfactants in LADDs

| Patent no. and year | Inventor(s) and company | Technology | Claimed benefits |
|---|---|---|---|
| U.S. 6034044 (2000) [97] | Scheper *et al.* | Nonionic surfactants | Superior grease cleaning |
| WO 94/22800 (1994) [143] | Bunch *et al.* (Olin) | Epoxy-capped poly(oxyalkylated) alcohol | Reduction in spotting and filming; biodegradable; low foaming |
| U.S. 4988452 (1991) [144] | Kinstedt *et al.* (Procter & Gamble) | Capped polyakylene oxide nonioncs | Bleach stable |
| EP 337760 (1989) [145] | Gabriel *et al.* (Unilever) | Capped polyalkylene oxide nonioncs | Bleach stable |
| U.S. 5073286 (1989) [146] | Otten *et al.* (BASF) | Sterically hindered polyether polyol nonionics | Bleach stable |
| U.S. 4438014 (1984) [147] | Scott (Union Carbide) | EO/PO adducts of alkoxylates | Bleach compatible; enhanced low foaming and wetting |
| U.S. 4436642 (1984) [148] | Scott (Union Carbide) | EO/PO adducts of alkylphenols | Bleach compatible; enhanced low foaming and wetting |

## G. New Forms

The last few years have seen the introduction of new forms for LADDs. These include both a solid tablet form and, perhaps more importantly, a liquid form delivered in a water-soluble sachet. Table 9.15 lists patents relating to some of these novel products.

## VII. AUXILIARY PRODUCTS FOR DISHWASHERS
## A. Liquid Prespotters

The automatic dishwasher detergents in today's marketplace, both liquid and powder versions, deliver cleaning performance that is acceptable to most consumers. However, one area in which consumers would like to see product improvement is in cleaning of baked-on, cooked-on, and dried-on food soils. These soils are tenaciously stuck to surfaces and are hard to remove unless strong mechanical forces are applied. To make the cleaning task of such hard-to-remove soils easier, special liquid formulations, often referred to as "prespotters," have been developed.

The idea of a prespotter is to apply the product onto the soiled surface and allow it to stand at ambient temperature over a period of 30 to 60 minutes before cleaning in the dishwasher. This process allows the soil to soften and debond or the adhesive forces between the soil and the substrate to loosen. The conditioned soiled substrate can be easily removed by mechanical and chemical forces in the dishwasher.

**TABLE 9.14** Patents Relating to Stabilization of LADD Components

| Patent no. and year | Inventor(s) and company | Technology | Claimed benefits |
|---|---|---|---|
| U.S. 5384061 (1995) [149] | Wise (Procter & Gamble) | Chlorine bleach ingredient, phytic acid, rheology stabilizing agent | Enhanced chemical and physical stability |
| U.S. 5258132 (1993) [150] | Kamel et al. (Lever Bros.) | Wax encapsulation of bleach, enzymes, or of bleach catalysts | Allows for stabilization of incompatible materials in liquid products |
| U.S. 5230822 (1993) [151] | Kamel et al. (Lever Bros.) | Wax encapsulation of bleach, enzymes, or of bleach catalysts | Allows for stabilization of incompatible materials in liquid products |
| U.S. 5225096 (1993) [152] | Ahmed et al. (Colgate-Palmolive) | Alkali metal iodate | Improved hypochlorite stability and efficacy |
| U.S. 5185096 (1993) [153] | Ahmed (Colgate-Palmolive) | Alkali metal iodate | Improved hypochlorite stability and efficacy |
| U.S. 5229027 (1993) [154] | Ahmed (Colgate-Palmolive) | Water-soluble iodide/iodine mixture | Improved hypochlorite stability and efficacy |
| U.S. 5200236 (1993) [155] | Lang et al. (Lever Bros.) | Solid core particles encapsulated in wax coating | Coat prolongs time that encapsulated particles remain active in water |
| EP 533239 (1993) [156] | Tomlinson (Unilever) | Encapsulated bleach | Excellent enzyme stability; reducing agent |
| U.S. 5141664 (1992) [157] | Corring et al. (Lever Bros.) | Encapsulation of bleach, bleach precursor, enzymes, surfactants, or perfumes | Uniformly dispersed active particles in a clear gel |
| EP 414282 (1991) [158] | Behan et al. (Quest) | Encapsulation of perfume in microorganism cells | Protects perfume from oxidizing agents |
| U.S. 4919841 (1990) [159] | Kamel et al. (Lever Bros.) | Blend of hard and soft waxes for coating bleach, perfumes, enzymes, or surfactants: process for encapsulation | Capsules useful for cleaning compositions |

Several prespotter formulations are disclosed in the patent literature. Acidic compositions that contain a mixture of nonionic surfactants and hydrotropes [163], thickened alkaline products with hypohalite bleaches [164,165], and enzyme-containing formulas [166] have all been developed for use as prespotters. Although

**TABLE 9.15**  Patents on New Forms of LADD

| Patent no. and year | Inventor(s) and company | Technology | Claimed benefits |
| --- | --- | --- | --- |
| U.S. 6632785 (2003) [161] | Pfeiffer et al. (Unilever) | New form: sachet | Water-soluble sachet; encapsulated bleach |
| U.S. 6605578 (2003) [160] | Fleckenstein et al. (Colgate-Palmolive) | New form: sachet | Water-soluble container; liquid composition |
| U.S. 6228825 (2001) [162] | Gorlin et al. (Colgate-Palmolive) | New form: sachet | Water-soluble package; nonaqueous |

these products are effective at cleaning tough soils, no product has made it to the marketplace, possibly due to economic factors. However, LADD products can be used as prespotters by applying directly to soiled items.

## B.  Rinse Aids

As discussed before, consumers judge the performance of automatic dishwasher detergents based on the overall cleaning, filming, and spotting on dishware, glasses, and utensils. Spots and film on glass surfaces are readily noticeable due to the differences in the refractive indices of the glass and the deposits. It is generally recognized that the deposits on glasses are predominantly water-soluble minerals such as salts of alkaline earth metal ions present in the water with proteins and fats of the soil as minor components. Clearly, the condition of the final rinse cycle water largely determines the degree of spotting and filming on glasses. Laboratory assessment indicates that water hardness exceeding 200 ppm as $CaCO_3$ results in poor performance on glasses and silverware unless the calcium ions are sequestered.

To minimize the mineral deposits and surface-active soil components on articles cleaned in the dishwasher, special formulations called rinse aids are often used for both home and institutional dishwashers. In the U.S. an estimated 40% of households use rinse aids [167]. Rheologically, rinse aid liquids are Newtonian type with viscosities in the range 50 to 200 cP. Their role is to reduce the interfacial tension between the dish- and glassware and the wash water. In this way, a uniformly draining film of the wash water is achieved on the items. Otherwise, uneven wetting will result in spotting and filming or on the items being dried.

Typically, rinse aid formulations for household dishwashers are composed of aqueous solutions containing a nonionic surfactant(s), a complexing agent such as citric acid or polyphosphate, hydrotropes (also known as coupling agents), fragrance, and color. Suitable preservatives are also added to the formulations to prevent the product from bacterial and fungal growth [168]. The pH of the formulations ranges from acidic to alkaline. Normally the rinse aid solution is injected during the final rinse cycle of dishwashing. A typical dosage of the product per rinse is about 0.3 to 1.0 g/l depending upon the level of nonionic surfactant in the formulation. Most rinse aids contain between 20 and 40% surfactant levels. Excessive or underdosage may have an adverse effect on the spotting and filming on glasses.

## 1. Ingredients for Rinse Aids

*(a) Nonionic Surfactants.* The heart of the rinse aid formulation is the surfactant and virtually all formulations contain nonionic surfactants. The primary function of the nonionic surfactant is to produce rapid sheeting action, achieving a quick and uniformly draining film which prevents the nonuniform drying of the hard water minerals on the utensil surfaces. Systematic studies have shown that the nonionic surfactants must satisfy the criteria for rinse aid applications, the most important ones of which are discussed below:

1. The nonionic surfactant must be an efficient wetting agent with low foaming characteristics, since excessive foaming not only influences the cleaning but also the rinsing effectiveness.
2. The foaming properties of the nonionic surfactants depend upon the temperature because of their inverse solubility temperature relationship. Above the cloud point they are nonfoamers and some nonionic surfactants may even function as defoamers above their cloud point temperature. Therefore, the nonionic surfactant selected for rinse aid formulations must have a cloud point below the temperature of the rinse water.
3. The aqueous surface tensions of the surfactant solutions must be low, in the range 30 to 40 dyn/cm$^2$. The surface tensions should be preferably measured at temperatures close to rinse water temperatures.

The nonionic surfactants commonly used in rinse aid formulations along with their structures are shown in Table 9.16.

*(b) Sequestering Agents.* Sequestering agents such as polyphosphates are added to the rinse formulations in order to condition the rinse cycle water (deactivation of alkaline earth metal ions) and to prevent or delay the formation of water-insoluble compounds like calcium bicarbonate or carbonate. These insoluble precipitates may deposit on glasses and appear as spots or film.

The addition of acidic additives such as citric acid is very popular in European formulations. The theory behind their use is that if sufficient acid is present in the final rinse solution, the acid converts the carbonate and bicarbonate ions into water and carbon dioxide, preventing the formation of insoluble salts. Citric acid formulations may also keep the dishwasher surfaces and nozzle of the spray arms free of limescale deposits. It is also believed that citrate may contribute to the brilliancy or shiny appearance of siliceous surfaces [175].

**TABLE 9.16**  Nonionic Surfactants for Rinse Aids

| Surfactant | Structure |
|---|---|
| Alkylphenoxy polyethenoxyethanol [169] | $R(C_6H_4)O(CH_2CH_2O)_nCH_2CH_3$ |
| Block polymers of ethylene and propylene oxide [170–172] | $RO(CH_2CH_2O)_n(CH_2CH[CH_3]O)_mH$ |
| Alkylphenoxy polyethenoxybenzyl ethers [173] | $R(C_6H_4)O(CH_2CH_2O)_nCH_2C_6H_5$ |
| Alkyl polyethenoxybenzyl ethers [174] | $R(CH_2CH_2O)_nCH_2C_6H_5$ |
| Ethoxylated alcohols | $R(OCH_2CH_2)_nOH$ |

*(c)  Hydrotropes.*   Hydrotropes (Chapter 2) or coupling agents play an important role in formulating rinse aid products. Their main functions include increasing the solubility of the nonionic surfactant in water and thus maintaining the clarity of the formulations. Judicious selection of hydrotropes is important since they may contribute to the foaming and potentially reduce the sheeting efficiency of the nonionic surfactant. Most effective are certain alkylnaphthalene sulfonates and sulfosuccinate esters, since they increase the solubility of the nonionic surfactants without leading to excessive foaming. Other hydrotropes utilized in rinse aid formulations include propylene glycol, isopropanol, and urea. In general, alcohols are not effective solubilizers in rinse aid formulas [175].

## 2.  Typical Rinse Aids

Typical examples of rinse aid formulations are shown in Table 9.17 [176]. Several patents describing rinse aid compositions have been issued. These are listed in Table 9.18.

**TABLE 9.17**   Examples of Rinse Aid Formulations [176]

| Ingredient | I (wt %) | II (wt %) |
|---|---|---|
| Plurafac® RA 30[a] | 50 | 17.5 |
| Plurafac® RA 40[a] | 1 | 17.5 |
| Isopropanol | 24 | 12 |
| Citric acid, dehydrated | — | 25 |
| Deionized water | 16 | 28 |

[a]BASF Corporation.

**TABLE 9.18**  Patents Relating to Rinse Aids

| Patent no. and year | Inventor(s) and company | Technology | Claimed benefits |
|---|---|---|---|
| WO 94/07985 (1994) [177] | De Smet et al. (Procter & Gamble) | Lime soap dispersant; lipase enzyme | Improved spotting and filming |
| U.S. 5294365 (1994) [178] | Welch et al. (BASF) | Hydroxypolyethers | Low foaming; improved spotting and filming |
| U.S. 5104563 (1992) [179] | Anchor et al. (Colgate-Palmolive) | Low-molecular-weight polypropylene which interacts with anionic or nonionic surfactants | Improved spotting and filming |
| GB 2247025 (1992) [180] | van Dijk et al. (Unilever) | Phospholipase $A_1$ and/or $A_2$ | Improved cleaning, filming, and spotting |
| EP 252708 (1988) [181] | van Dijk et al. (Unilever) | Nonplate-shaped colloids such as silica | Reduced drying time; improved spotting |
| U.S. 4443270 (1984) [182] | Biard et al. (Procter & Gamble) | Ethoxylated nonionic; organic chelating agent; water-soluble Mg, Zn, Sn, Bi, or Ti salts | Improved spotting and filming |
| U.S. 4416794 (1983) [183] | Barrat et al. (Procter & Gamble) | Ethoxylated nonionic; organic chelating agent; aminosilane | Improved spotting and filming |

# VIII.  FUTURE TRENDS

The majority of LADDs marketed today deliver performance that is acceptable to consumers. However, economic, environmental, and regulatory pressures necessitate formulators of LADDs to continue to improve the products. Over the last dozen years, machine redesign has consistently improved performance. Recent developments have led to smarter and more efficient machines. Microprocessors and sensors determine most of the washing conditions in modern machines. This has been driven by the need for more energy- and water-efficient machines. Both Europe and the U.S. have seen an explosion of unit dose products that are essentially concentrates, as they are only single-cup products. Exploration in the area of phosphate-free products continues, as state legislatures continue to be under pressure to ban their use in automatic dish detergents. Legislative pressure has also encouraged the exploration of nonchlorine bleaches. The use of enzyme-based formulas has continued to grow, and they now account for about 10% of the LADD market in the U.S.

## A. Reduction in Dose Size

Recommended inlet water temperature has dropped as much as 20°F, from 140 to 120°F (60 to 50°C). Sensors then adjust the water temperature, monitor the cleaning, and extend the cycles as needed. The amount of water used has also been reduced. Dishwashers have two cups for dosing detergent: a main wash cup and a prewash cup. The prewash cup has been reduced in size over recent years. All of these changes have led to a more consistent performance for the consumer, and the ability to use less automatic dishwasher detergent. This has also presented the product formulator with a challenge.

One solution for this has been the advent of unit dose. This generally consists of a tablet or sachet form. The sachet form may contain a liquid gel [184]. This is a water-based, enzyme-containing system in a water-soluble polyvinyl alcohol pouch. Recently a new variant has emerged. This is a combination liquid and powder in two separate compartments in a polyvinyl alcohol pouch [185,186]. The powder portion contains traditional enzyme, oxygen bleach chemistry, while the liquid portion contains an organic solvent that enhances "baked-on, burnt-on" cleaning. The liquid can also act as a humectant to control the stability of the active components of the powder portion. These provide about one half the weight of product that would be used for a single closed-cup wash. In the last few years these forms have grown to about 20% of the U.S. market.

The development of more sophisticated unit dose forms continues. An aqueous gel containing encapsulated bleach has been developed [187–189]. It is claimed that these have excellent cleaning properties with minimal spot and film formation.

## B. Nonphosphate Products

The low cost-to-performance ratio of phosphates, especially alkali metal tripolyphosphates, makes them the "work-horse" of detergents. Although phosphate builders are safe for humans, they are unfortunately beneficial to algae growth. Therefore, large amounts of phosphate in waste streams lead to eutrophication of lakes and streams. For this reason there have been concerns, especially in Europe, about the heavy use of phosphates in detergent products. In certain regions of the U.S. the quantity of phosphates used in LADDs is regulated by local governments. Therefore, one of the challenges that the automatic dishwasher detergent manufacturers face today is finding a substitute for polyphosphate. Ideally, the phosphate substitute must satisfy several criteria such as:

1. Free from phosphorous and nitrogen.
2. Soluble in water with readily biodegradable characteristics.
3. Chemically stable and compatible with oxygen and/or chlorine bleach
4. Cleaning performance characteristics equal to phosphates.

5. Safety to humans.
6. Economically practical.

Although many builder systems meet most of these criteria, they provide inferior performance. In particular, filming and spotting on glasses is a concern in hard water areas. This is probably the primary reason that nonphosphate liquid products have not found their way into the marketplace.

The search for an alternative to phosphates will continue as evident from the patent activity in the last 5 to 10 years. Many reviews have appeared which compare the characteristics of the more heavily studied alternatives [18,29,38,190–206]. Table 9.19 lists the recent patents on phosphate-free LADD formulations.

## C. Hypochlorite-Free Products

Another trend appears to be the formulation toward products containing no hypochlorite bleach. The reason is that the strong oxidizing nature of hypochlorite makes it incompatible with other easily oxidized components, such as nonionic surfactants, fragrances, and enzymes. There is also some concern regarding the possible formation of chlorinated organics [8]. Oxygen bleaches, such as perborates or percarbonates, which liberate peroxide in solution, are being studied to replace chlorine bleach. Being weaker oxidizing agents than halogen bleaches, they are compatible with some oxidizable LADD components. However, there are two main disadvantages of oxygen bleaches. First, they display acceptable performance only at elevated temperatures. Second, they are difficult to formulate in liquid products due to chemical stability problems.

The search for chlorine bleach alternatives as well as development of technologies for stabilizing peroxide bleaches will continue. Microencapsulation technology for formulating LADDs containing chlorine bleaches has shown limited success [150,151,155,159].

## D. Enzymatic Products

Although enzyme-based systems comprise the bulk of the powder market, it has only been in recent years that they have emerged in automatic dishwasher gels. Currently enzyme-based liquid products comprise about 10% of the ADD market. This comprises both bottled gels and unit dose sachets. Unlike the powder formulas, liquid enzyme formulas with bleach have not been successfully delivered. Efforts in this area include the development of a gel containing encapsulated bleach [187–189]. New enzymes are continuously being developed that are more bleach stable. As these become commercially viable, new aqueous gel products that provide the benefits of both bleach and enzyme cleaning will be available to the consumer.

**TABLE 9.19** Patents Relating to Phosphate-Free LADDs

| Patent no. and year | Inventor(s) and company | Technology | Claimed benefits |
|---|---|---|---|
| U.S. 6602837 B1 (2003) [191] | Patel (Procter & Gamble) | Diacyl peroxide | Stain removal from plastics |
| U.S. 5545344 (1996) [192] | Durbut et al. (Colgate-Palmolive) | Nonaqueous; enzyme | Long-term stability |
| U.S. 5510048 (1996) [193] | Durbut et al. (Colgate-Palmolive) | Enzyme; nonaqueous | Concentrated; stability |
| EP 0703974 B1 (1998) [194] | Ambuter et al. (Procter & Gamble) | Enzyme | Concentrated; free of bleach and silicate |
| EP 0625567 B1 (2001) [195] | Beck et al. (Stockhausen) | Anhydrous | Free of bleach |
| EP 0530635 B1 (1997) [196] | Golz et al. (Benkisser) | Neutral; carboxylic acid | Free of bleach and silicate |
| EP 0518721 B1 (1995) [197] | Burbut et al. (Colgate-Palmolive) | Enzyme; nonaqueous | Free of bleach; stability |
| WO 94/29428 (1994) [190] | Ambuter et al. (Procter & Gamble) | Concentrated; enzymes and stabilizing system | Long-term stability; free of chlorine bleach and silicate |
| WO 94/05763 (1994) [198] | Rattinger et al. (Unilever) | Pyridine carboxylates | Hypochlorite resistant; biodegradable |
| EP 561452 (1993) [207] | van Dijk et al. (Unilever) | Biodegradable polyamino acid | Improved scale prevention |
| U.S. 5169553 (1992) [208] | Durbut et al. (Colgate-Palmolive) | Binary mixture of protease and amylase; nonaqueous | Free of phosphates |
| WO 91/03541 (1991) [29] | Beaujean et al. (Henkel) | Aluminosilicate; stabilizing electrolyte system | Stable during storage and transport; do not sediment between 5 and 60°C |
| EP 476212 (1990) [199] | Boutique et al. (Procter & Gamble) | Citrate; C10–C16 alkyl- or alkenyl-substituted succinic acid | Physically stable; good building capacity |
| DE 3832478 (1988) [209] | Dixit (Colgate-Palmolive) | Aluminosilicate; polycarboxylates | Free of phosphates |

# REFERENCES

1. Houghton, J., U.S. Patent 7,365, 1865.
2. Meeker, D.A., *The Story of the Hobart Manufacturing Co.*, Newcomer Publications, Princeton University Press, Princeton, NJ, 1960.
3. Morrish, D.H., *History of Dishwashers*, General Electric Co., Louisville, KY, 1967.
4. Cater, D.H., Flynn, M.H., and Frank, P.F., *Inform*, 5, 1095, 1994.
5. Gauthier, F., Schneegans, W., Bernard, L., and Reeve, P., Rohm and Haas France, Sepawa, October 2003.
6. Branna, T., *HAPPI*, 18, 56, 1990.
7. Whalley, G.R., *HAPPI*, 23, 71, 1995.
8. Lake, R.F., in *Proceedings of the 3rd World Conference on Detergents: Global Perspectives*, Montreux, Switzerland, 1993, pp. 108–110.
9. Werdelmann, B., *Soap Cosmet. Chem. Specialties*, 50, 36, 1974.
10. Mizuno, W.G., Detergency, Theory and Test Methods, in *Surfactant Science Series*, Vol. 5, Part III, Marcel Dekker, New York, 1981.
11. Oberle, T.M., *Detergents in Depth*, The SDA, New York, 1974.
12. Gorlin, P., U.S. Patent 6,258,764 to Colgate-Palmolive, 2001.
13. Shulman, J.E., *HAPPI*, 20, 130, 1992.
14. Shulman, J. E. and Robertson, M.S., *Soap Cosmet. Chem. Specialties*, 68, 46, 1992.
15. Richter, F.H., Winkler, E.W., and Baur, R.H., *J. Am. Oil Chem. Soc.*, 66, 1666, 1989.
16. McGrew, G.T., *HAPPI*, 14, 66, 1986.
17. Lange, K.R., *J. Am. Oil Chem. Soc.*, 45, 487, 1968.
18. Hudson, A.P., Woodward, F.E., and McGrew, G.T., *J. Am. Oil Chem. Soc.*, 65, 1353, 1988.
19. van Wazer, J.R., in *Phosphorous and its Compounds*, Interscience, New York, 1958.
20. Shen, C.Y., *J. Am. Oil Chem. Soc.*, 45, 510, 1968.
21. Iler, R. K., *The Chemistry of Silica*, John Wiley, New York, 1979, p. 21.
22. Falcone, J.S., Ed., *Soluble Silicates*, ACS Symposium Series 194, Washington D.C., 1982.
23. Coffey, R. and Gudowicz, T., *Chem. Ind.* 6, 169, 1990.
24. The PQ Corp., Multi-Functional Characteristics of Soluble Silicate, 17-101/1291, 1991.
25. Sharma, A. and Jain, H., *J. Am. Ceram. Soc.*, 86, 1669, 2003.
26. Sharma, A. and Jain, H., *J. Surfactants Detergents*, 7, 111, 2004.
27. The PQ Corp., New Dimensions in Zeolite Detergent Chemistry, VAL-100/1192, 1992.
28. Schwuger, M.J. and Smulders, E.J., in *Surfactant Science Series*, Vol. 20, Marcel Dekker, New York, 1987, p. 371.
29. Beaujean, H.F., Bode, J., Paasch, S., Schwadtke, K., Smulders, E., and Sung, E., WO Patent 91/03541 to Henkel, 1991.
30. Dixit, N.S., U.S. Patent 4,971,717 to Colgate-Palmolive, 1990.
31. Dixit, N.S., U.S. Patent 5,229,026 to Colgate-Palmolive, 1993.
32. Rohm & Haas, Acusol 445 Series Detergent Polymers for Machine Dishwashing, FC-131, 1993.
33. BASF Specialty Chemicals, Sokalan Polymeric Dispersing Agents, 49.
34. Schaffer, J.F. and Woodhams, R.T., *Tenside Det.* 16, 240, 1979.

35. Hunter, M., da Motta Marques, D.M.L., Lester, J.N., and Perry, R., *Environ. Technol. Lett.*, 9, 1, 1988.
36. Dwyer, M., Yeoman, S., Lester, J.N., and Perry, R., *Environ. Technol.*, 11, 263, 1990.
37. Grilli, M.P. and Capucci, A., *Toxicol. Lett.*, 25, 137, 1985.
38. Crutchfield, M.M., *J. Am. Oil Chem. Soc.*, 55, 58, 1978.
39. Rosen, M.J. and Zhu, Z.H., *J. Am. Oil Chem. Soc.*, 69, 667, 1992.
40. Dow Chemicals, Dowfax Anionic Surfactants for High-Performance Products, 1994.
41. Scardera, M. and Scott, R.N., U.S. Patent 4,070,298 to Olin, 1978.
42. Jeschke, R., Schmidt, K., Lange, F., and Koren, K., WO Patent 92/20768 to Henkel, 1992.
43. Dahanayake, M. and Hashem, M., *Soap Cosmet. Chem. Specialties*, 65, 39, 1989.
44. Surutzidis, A. and Fisk, A.A., European Patent 593841 A1 to Procter & Gamble, 1994.
45. Kurzendörfer, C., Seiter, W., Beaujean, H., Block, C., and Holderbaum, T., WO Patent 94/12603 to Henkel, 1994.
46. Coons, D.M., *J. Am. Oil Chem. Soc.*, 55, 104, 1978.
47. Church, J.A., *Ind. Eng. Chem. Res.*, 33, 239, 1994.
48. Dalgaard, L.H., Kochavi, D., and Thellersen. M., *Inform*, 2, 532, 1991.
49. Farooq, A., Mehreteab, A., Broze, G., Dixit, N., and Hsu, D., *J. Am. Oil Chem. Soc.*, 72, 843, 1995.
50. Baxter, S.L., U.S. Patent 4,950,416 to Vista Chemical, 1990.
51. Wise, R.M., U.S. Patent 4,941,988 to Procter & Gamble, 1990.
52. Drapier, J., Gallant, C., Laitem, L., Delsignore, M., Shevade, M., Rounds, R., Kenkare, D., Davan, T., and Dixit, N.S., U.S. Patent 5,427,707 to Colgate-Palmolive, 1995.
53. Marshall, J.L., Hall, D.L., Ambuter, H., and Fitch, E.P., U.S. Patent 5,691,292 to Procter & Gamble, 1997.
54. Drapier, J. and Dixit, N.S., U.S. Patent 5,413,727 to Colgate-Palmolive, 1995.
55. Dixit, N.S., Shevade, M., Rounds, R., and Delsignore, M., U.S. Patent 5,252,241 to Colgate-Palmolive, 1993.
56. Shevade, M., Delsignore, M., Dixit, N.S., and Kenkare, D., U.S. Patent 5,252,241 to Colgate-Palmolive, 1993.
57. Dixit, N.S., Farooq, A., Rounds, R.S., and Shevade, M., U.S. Patent 5,232,621 to Colgate-Palmolive, 1993.
58. Prencipe, M., McCandlish, E.F., and Loprest, F.J., U.S. Patent 5,188,752 to Colgate-Palmolive, 1993.
59. Dixit, N.S., and Davan, T., U.S. Patent 5,098,590 to Colgate-Palmolive, 1992.
60. Drapier, J., Gallant, C., Wouters, F., and Laitem, L., U.S. Patent 5,057,237 to Colgate-Palmolive, 1991.
61. Dixit, N.S., U.S. Patent 4,836,946 to Colgate-Palmolive, 1989.
62. Chazard, G., Drapier, J., Gallant, C., and van De Gaer, D., U.S. Patent 4,801,395 to Colgate-Palmolive, 1989.
63. Drapier, J., Gallant, C., van De Gaer, D., and Delvenne, J., U.S. Patent 4,752,409 to Colgate-Palmolive, 1988.
64. Wise, R.M., U.S. Patent 5,169,552 to Procter & Gamble, 1992.
65. Elliot, D.L., and Sisco, R.M., U.S. Patent 5,135,675 to Lever Brothers, 1992.

66. Prince, M.J., and Glassco, T.H., U.S. Patent 5,130,043 to Procter & Gamble, 1992.
67. Donker, C.B., WO Patent 89/04359 to Unilever, 1989.
68. Julemont, M. and Marchal, M., U.S. Patent 4,740,327 to Colgate-Palmolive, 1988.
69. Taraschi, F.A., U.K. Patent GB 2168377 A to Procter & Gamble, 1986.
70. Kolodny, E.R. and Liebowitz, E., WO Patent 83/03621 to American Home Products, 1983.
71. Dixit, N.S., U.S. Patent 5,368,766 to Colgate-Palmolive, 1994.
72. Dixit, N.S., U.S. Patent 5,298,180 to Colgate-Palmolive, 1994.
73. Dixit, N.S., U.S. Patent 5,205,953 to Colgate-Palmolive, 1993.
74. Roselle, B.J., U.S. Patent 4,824,590 to Procter & Gamble, 1989.
75. Lockhead, R.Y., Sauer, C.E., and Nagarajan, M.K., Hypochlorite Tolerant Polymeric Rheology Modifiers, presented to the American Oil Chemists Society, Baltimore, MD, 1990.
76. Laba, D., Ed., *Cosmetic Science and Technology Series*, Vol. 13, Marcel Dekker, New York, 1993.
77. Ferry, J.D., in *Viscoelastic Properties of Polymers*, 3rd ed., John Wiley, New York, 1980, p. 40.
78. Sherman, P., in *Industrial Rheology*, Academic Press, London, 1970.
79. Walters, K., in *Rheometry: Industrial Applications*, Wiley, New York, 1980.
80. Tadros, T.F., *Colloids Surf.*, 18, 137, 1986.
81. Russell, W.B., *J. Rheology*, 24, 287, 1980.
82. Goodwin, J.W., in *Surfactants*, Tadros, T.F., Ed., Academic Press, New York, 1984.
83. Heywood, N., *Chem. Eng.*, 415, 16, 1985.
84. Dzwy, N.Q. and Boger, D.V., HaakeBuchler Instruments Inc., technical bulletin PB-856.
85. Bowles, R.L., *et al.*, *Modern Plastics*, 32, 142, 1955.
86. Casson, N., in *Rheology of Dispersed Systems*, Miles, C.C., Ed., Pergamon Press, New York, 1959.
87. Asbeek, A.W., *Official Digest*, 33, 65, 1961.
88. Gabriel, S.M. and Roselle, J., U.S. Patent 4,859,358 to Procter & Gamble, 1989.
89. Corring, R. and Gabriel, R., U.S. Patent 5,141,664 to Lever Brothers, 1992.
90. ASTM method D 3556-85 (Reapproved 1995), 2000, V15.04, p. 367.
91. Mausner, M. and Schlageter, M., *Household Personal Prod. Ind.*, Jan., 59, 1982.
92. Peart, M.V.K., Ph.D. thesis, Purdue University, 1969.
93. Association of Home Appliance Manufacturers, ANSI/AHAM DW-1-1992.
94. Chemical Specialties Manufacturers Associations, Inc., Detergents Division Test Methods Compendium, 1985.
95. ASTM method D 3565-89, 2000, V15.04, p. 372.
96. Patel, R.N., U.S. Patent 6,602,837 to Procter & Gamble, 2003.
97. Scheper, W.M., U.S. Patent 6,034,044 to Procter & Gamble, 2000.
98. Goldstein, A.S., U.S. Patent 5,929,008 to Procter & Gamble, 1999.
99. Gorlin, P., Kenkare, D., and Phillips, S., U.S. Patent 5,698,507 to Colgate-Palmolive, 1997.
100. Durbut, P., Fahim, F.U., and Drapier, J., U.S. Patent 5,618,456 to Colgate-Palmolive, 1997.
101. Sadlowski, E., U.S. Patent 5,597,789 to Procter & Gamble, 1997.

102.  Burbut, P., Ahmed, F., and Drapier, J., U.S. Patent 5,545,344 to Colgate-Palmolive, 1996.
103.  Kenkare, D., Dixit, N., and Durbut, P., U.S. Patent 5,527,483 to Colgate-Palmolive, 1996.
104.  Fair, M.J. and Nicholson, J.R., U.S. Patent 5,372,740 to Lever Brothers Co., 1994.
105.  Keenan, A.C., Kirk, T.C., McCallum, T.F., III, Shulman, J., Tallent, R.J., and Weinstein, B., U.S. Patent 5,858,944, 1999.
106.  Krishnan, S., U.S. Patent 5,318,715 to Colgate-Palmolive, 1994.
107.  Sadlowski, E.S., WO Patent 94/25557 to Procter & Gamble, 1994.
108.  van Dijk, W.R. and Khoshdel, E., WO Patent 94/17170, 1994.
109.  Adler, D.E., McCallum, T.F., Shulman, J.E., and Weinstein, B., U.S. Patent 5,308,532 to Rohm & Haas, 1994.
110.  Ahmed, F.U., Durbut, P., and Drapier, J., U.S. Patent 5,240,633 to Colgate-Palmolive, 1993.
111.  Ahmed, F.U. and Bochis, K., U.S. Patent 5,164,106 to Colgate-Palmolive, 1992.
112.  Ahmed, F.U., Buck, C.E., and Jakubicki, G., U.S. Patent 5,094,771 to Colgate-Palmolive, 1992.
113.  Ahmed, F.U. and Bochis, K., U.S. Patent 5,076,952 to Colgate-Palmolive, 1991.
114.  Ahmed, F.U. and Buck, C.E., U.S. Patent 4,970,016 to Colgate-Palmolive, 1990.
115.  Ahmed, F.U. and Buck, C.E., U.S. Patent 4,968,446 to Colgate-Palmolive, 1990.
116.  Ahmed, F.U. and Buck, C.E., U.S. Patent 4,968,445 to Colgate-Palmolive, 1990.
117.  Frankena, H., U.S. Patent 4,931,217 to Lever Brothers, 1990.
118.  van Dijk, W.R., European Patent 271155 A2 to Unilever, 1988.
119.  Laitem, L., Delvaux, M., Broze, G., and Bastin, D., U.S. Patent 4,753,748 to Colgate-Palmolive, 1988.
120.  Goedhart, M., Gortemaker, F.H., Kemper, H.C., and Kielman, H.S., U.S. Patent 4,597,886 to Lever Brothers, 1986.
121.  de Ridder, J.J.M., Hollingsworth, M.W., and Robb, I.D., U.S. Patent 4,539,144 to Lever Brothers, 1985.
122.  Kaneko, T.M., U.S. Patent 4,306,987 to BASF Wyandotte, 1981.
123.  Bahary, W.S. and Hogan, M.P., U.S. Patent 5,336,430 to Lever Brothers, 1994.
124.  Kreischer, P.H., European Patent 304328 A2 to Unilever, 1989.
125.  Fox, D.J., van Blarcom, D., and Rubin, F.K., U.S. Patent 4,260,528 to Lever Brothers, 1981.
126.  Bush, W.G. and Braun, V.D., U.S. Patent 4,226,736 to Drackett, 1980.
127.  Corring, R., U.S. Patent 5,366,653 to Lever Brothers, 1994.
128.  Angevaare, P.A. and Gary, R.G., U.S. Patent 5,624,892 to Lever Brothers, 1997.
129.  Broadwell, R. and Shevade, M., U.S. Patent 5,395,547 to Colgate-Palmolive, 1995.
130.  Broadwell, R., Shevade, M., and Kenkare, D., U.S. Patent 5,246,615 to Colgate-Palmolive, 1993.
131.  Gabriel, S.M., Glassco, T.H., Ambuter, H., and Fitch, E.P., WO Patent 93/21298 to Procter & Gamble, 1993.
132.  Dixit, N.S. and Davan, T., U.S. Patent 5,075,027 to Colgate-Palmolive, 1991.
133.  Colarusso, R.J., U.S. Patent 4,927,555 to Colgate-Palmolive, 1990.
134.  Angevaare, P.A. and Gary, R.G., U.S. Patent 5,374,369 to Lever Brothers, 1994.
135.  Keyes, G.B., Seaman, C.E., and Kasson, J.K., U.S. Patent 6,448,210 to Johnson Diversey, 2002.

136. Gary, R.G., Angevaare, P.A.J.M., Jensen, A.O., and Van Gorkom, L., U.S. Patent 5,731,277 to Lever Bros., 1998.
137. Angevaare, P.A. and Gary, R.G., U.S. Patent 5,783,539 to Lever Bros., 1998.
138. Keyes, G.B., Seaman, C., and Kassen, J.K., U.S. Patent 6,083,894 to S.C. Johnson Commercial Markets, Inc., 2000.
139. Angevaare, P.A. and Gary, R.G., U.S. Patent 5,698,506 to Lever Bros., 1997.
140. Dolan, M.J. and Jakse, F.P., U.S. Patent 4,992,195 to Monsanto, 1991.
141. Cilley, W.A. and Wise, R.M., U.S. Patent 4,933,101 to Procter & Gamble, 1991.
142. Gabriel, S.M. and Roselle, B.J., U.S. Patent 4,859,358 to Procter & Gamble, 1989.
143. Bunch, H.S., Groom, T., Grosser, F.R., Scardera, M., Targos, T.S., and Vanover, A.R., WO Patent 94/22800 to Olin, 1994.
144. Kinstedt, G.C. and Myers, S.L., U.S. Patent 4,988,452 to Procter & Gamble, 1991.
145. Gabriel, R., Aronson, M.P., and Steyn, P.L., European Patent 337760 A2 to Unilever, 1989.
146. Otten, J.G., Parker, E.J., and Kinnaird, M.G., U.S. Patent 5,073,286 to BASF, 1989.
147. Scott, R.J., U.S. Patent 4,438,014 to Union Carbide, 1984.
148. Scott, R.J., U.S. Patent 4,436,642 to Union Carbide, 1984.
149. Wise, R.M., U.S. Patent 5,384,061 to Procter & Gamble, 1995.
150. Kamel, A.A., Lang, D.J., Hanna, P.A., Gabriel, R., Theiler, R., and Goldman, A.S., U.S. Patent 5,258,132 to Lever Brothers, 1993.
151. Kamel, A.A., Lang, D.J., Hanna, P.A., Gabriel, R., and Theiler, R., U.S. Patent 5,230,822 to Lever Brothers, 1993.
152. Ahmed, F.U. and Shevade, M., U.S. Patent 5,225,096 to Colgate-Palmolive, 1993.
153. Ahmed, F.U., U.S. Patent 5,185,096 to Colgate-Palmolive, 1993.
154. Ahmed, F.U., U.S. Patent 5,229,027 to Colgate-Palmolive, 1993.
155. Lang, D.J., Kamel, A.A., Hanna, P.A., Gabriel, R., and Theiler, R., U.S. Patent 5,200,236 to Lever Brothers, 1993.
156. Tomlinson, A.D., European Patent 533239 A2 to Unilever, 1993.
157. Corring, R. and Gabriel, R., U.S. Patent 5,141,664 to Lever Brothers, 1992.
158. Behan, J.M., Birch, R.A., and Perring, K.D., European Patent 414282 A1 to Quest International, 1991.
159. Kamel, A., Hurckes, L.C., and Morelli, M.M., U.S. Patent 4,919,841 to Lever Brothers, 1990.
160. Fleckenstein, M. and Zyzyck, L., U.S. Patent 6,605,578 to Colgate-Palmolive, 2003.
161. Pfeiffer, N., Ghatlia, N.D., and Secemski, I.I., U.S. Patent 6,632,785 to Unilever Home & Personal Care USA, 2003.
162. Gorlin, P., Calabro, D., Martin, E., Fiore, M., and Canady, V., U.S. Patent 6,228,825 to Colgate-Palmolive, 2001.
163. Altenschoepfer, T., Jeschke, P., and Wisotzki, K., U.S. Patent 4,818,427 to Henkel, 1989.
164. Rupe, L.A., Tuthill, L.B., and Leikhim, J.W., U.S. Patent 4,116,851 to Procter & Gamble, 1978.
165. Leikhim, J.W., U.S. Patent 4,116,849 to Procter & Gamble, 1978.
166. McCandlish, E., Canadian Patent 2,093,783 to Colgate-Palmolive, 1993.
167. Roberts, G. and Welch, M.C., *Soap Cosmet. Chem. Specialties*, 71, 58, 1995.
168. Stache, H., in *Tensid-Taschenbuch*, Stache, H., Ed., Hanser, Munich, 1981, p. 484.

169. Wilson, J.L., Mizuno, W.G., and Crecelius, S.B., *Soap Cosmet. Chem. Specialties*, 34, 48, 1958.
170. Wyandotte Chemical Pluronic Grid, bulletin 189-61.
171. Reich, H.E., Patton, J.T., and Francis, C.V., *Soap Cosmet. Chem. Specialties*, 37, 55, 1961.
172. Fischer, W. K., in *Fatty Alcohols*, Henkel, Dusseldorf, 1982, p. 187.
173. Niederhauser, W. D. and Smialkowski, E.J., U.S. Patent 2,856,434 to Rohm & Haas, 1958.
174. Rue, L.M., Brunelle, T.E., and Mizuno, W.G., U.S. Patent 3,444,242 to Economics Laboratory, 1969.
175. Parker, E. J. and Schoene, K.F., *HAPPI*, 25, 83, 1988.
176. Heitland, H. and Marsen, H., in *Surfactants in Consumer Products*, Falbe, J., Ed., Springer-Verlag, Heidelberg, 1987, p. 321.
177. De Smet, B.L.A., Pluyter, J.G.L., and Jones, L.A., WO Patent 94/07985 to Procter & Gamble, 1994.
178. Welch, M.C., Otten, J.G., and Schenk, G.R., U.S. Patent 5,294,365 to BASF, 1994.
179. Anchor, M.J. and Roelofs, R.R., U.S. Patent 5,104,563 to Colgate-Palmolive, 1992.
180. van Dijk, W., Gerardus, A., and Bastein, T.M., British Patent GB 2247025 to Unilever, 1992.
181. van Dijk, J., Kielman, H.S., Los, L., and Verheul, R.C.S., European Patent 252708 A2 to Unilever, 1988.
182. Biard, D. and Lodewick, R., U.S. Patent 4,443,270 to Procter & Gamble, 1984.
183. Barrat, C.R., Walker, J.R., and Wevers, J., U.S. Patent 4,416,794 to Procter & Gamble, 1983.
184. Kaiser, R., Guzmann, M., and Wiedmann, R., WO Patent 02/16541 to Reckitt Benckiser, 2002.
185. Sharma, S., Kinloch, J.I., Greener, S.J., and Lynde, K.R., U.S. Patent Application 2002/0137648 A1 to Procter & Gamble, 2002.
186. Smith, D.J., Sharma, S., Kinloch, J.I., and Greener, S.J., U.S. Patent Application 2002/0142930 A1 to Procter & Gamble, 2002.
187. Pfeiffer, N., Ghatlia, N.D., and Secemski, I.I., U.S. Patent 6,475,977 B1 to Unilever, 2002.
188. Pfeiffer, N., Ghatlia, N.D., and Secemski, I.I., U.S. Patent 6,492,312 B1 to Unilever, 2002.
189. Pfeiffer, N., Ghatlia, N.D., and Secemski, I.I. U.S. Patent 6,632,785 B2 to Unilever, 2003.
190. Ambuter, H. and Alwart, T.S., WO Patent 94/29428 to Procter & Gamble, 1994.
191. Patel, R.H., U.S. Patent 6,6,02,837 B1, to Procter & Gamble, 2003.
192. Durbot, P., Ahmed, F.U., and Drapier, J., U.S. Patent 5,545,344 to Colgate-Palmolive, 1996.
193. Durbot, P., Ahmed, F.U., and Drapier, J., U.S. Patent 5,510,048 to Colgate-Palmolive, 1996.
194. Ambulter, H. and Todd, S., EP 0703974 B1 to Procter & Gamble, 1998.
195. Beck, R., Krause, F., and Shouenkaes, U., EP 0625567 B1 to Stockhausen, 2001.

196. Golz, K., Hertling, L., Magg, H., and Washenbach, G., EP 0530635 to Benkisser, 1997.
197. Durbut, P., Ahmed, F., and Drapier, J. EP 0518721 B1 to Colgate Palmolive, 1997.
198. Rattinger, G.B., Cotter, B., and Fair, M.J., WO Patent 94/05763 to Unilever, 1994.
199. Boutique, M.J., and Depoot, K.J.M., European Patent 476212A1 to Procter & Gamble, 1992.
200. Matzner, E.A., Crutchfield, M.M., Langguth, R.P., and Swisher, R.D., *Tenside Deterg.*, 10, 239, 1973.
201. Niewenhuizen, M.S., Kieboom, A.P.G., and van Bekkum, A.P.G., *J. Am. Oil Chem. Soc.*, 60, 120, 1983.
202. Kemper, H.C., Martens, R.J., Nooi, J.R., and Stubbs, C.E., *Tenside Deterg.*, 12, 47, 1975.
203. Santhanagopalan, S., Raman, H., and Suri, S.K., *J. Am. Oil Chem. Soc.*, 61, 1267, 1984.
204. Madden, R.E., Edwards, T.G., Kaiser, C.B., and Jaglowski, R.G., *Soap Cosmet. Chem. Specialties*, 50, 38, 1974.
205. Trulli, F. and Santacesaria, E., *Chimoggi*, 8, 13, 1993.
206. Gauthier, F., *Comun. J. Com. Esp. Deterg.*, 24, 109, 1993.
207. van Dijk, W.R., Rocourt, A.P.A.F., and van Drunen, R.W.P., European Patent 561452 A1 to Unilever, 1993.
208. Durbut, P., Ahmed, F., and Drapier, J., U.S. Patent 5,169,553 to Colgate-Palmolive, 1992.
209. Dixit, N.S., German Patent DE 3832478 A1 to Colgate-Palmolive, 1988.

# 10

# Shampoos and Conditioners

**JIASHI J. TARNG and CHARLES REICH**   Advanced Technology/Hair Care, Global Technology, Colgate-Palmolive Company, Piscataway, New Jersey

## I.  INTRODUCTION

The primary function of shampoos is to clean the hair, thereby improving its appearance through the removal of dulling deposits that can weigh the hair down and even cause it to stick together. The main function of conditioners is to reduce the magnitude of the forces associated with combing or brushing hair. This latter benefit is also provided, to differing extents, by conditioning shampoos, while both products can provide important secondary benefits such as dandruff control, hair moisturization, flyaway reduction, and improvement of shine.

In formulating shampoos and conditioners to provide the above benefits, several unique factors must be considered. The products must act quickly, of the order of minutes, and at relatively low temperatures between 20 and 40°C. The viscosity of the formulations must also be sufficiently high to avoid runoff from the hand while still spreading easily on the hair. In addition, a shampoo must generate a rich and stable lather that can be rinsed easily. Finally, since shampoos and conditioners will be used in contact with skin and eyes, they must exhibit low toxicity and irritation.

In this chapter the effects of these and other factors on product form and development are discussed. The first section describes general shampoo and hair conditioner compositions. Subsequent sections then discuss hair-cleaning mechanisms and product performance and efficacy, followed by methods of evaluating the cosmetic attributers of shampoos and conditioners. Finally, a brief discussion of damage to hair from shampooing and grooming is presented.

## II.  TYPICAL COMPOSITION AND INGREDIENTS

The basic ingredient composition of shampoos is summarized in Table 10.1. Some additional ingredients are listed in Table 10.2.

**TABLE 10.1** Basic Ingredient Composition for Cleansing Shampoos

| Basic ingredient | Activity (%) |
|---|---|
| Primary surfactants — anionic (e.g., ALES, ALS, AOS) | 8–20 |
| Secondary surfactants/foam boosters — nonionic and amphoteric (e.g., betaine, amine oxide, amphoacetate, sultaine, sulfosuccinate, APG) | 0–10 |
| Foam stabilizers (e.g., CDEA, CMEA) | 0–5 |
| Thickeners (e.g., salt, gum, polymer) | 0–5 |
| Other minors (preservatives, fragrance, acid, dye) | QS |
| Water | Balance |

**TABLE 10.2** Additional Ingredients for Various Shampoos

| Additional ingredient | Activity (%) |
|---|---|
| Conditioning agents (e.g., silicones, cationic polymers, cationic surfactants, oils) | 0–8 |
| Pearlizing agents (e.g., EGDS, EGMS) | 0.2–2.5 |
| Opacifiers (e.g., cetyl stearyl alcohol) | 0–3 |
| Rheology modifiers (e.g., gum, polymer) | 0–3 |
| Emulsifiers | 0–5 |
| Clarifying agents | 0–3 |
| Antioxidants | 0–2 |
| Antidandruff/antifungal agents | 0.5–5 |
| Promotional additives (UV absorbers, natural oils, botanic extracts, protein hydrolytes) | 0–2 |

## A. Surfactants in Shampoos

To provide adequate cleaning, lather, and viscosity, shampoos generally contain surfactants at concentrations between 8 and 20%, along with fragrances, color additives, and preservatives. Other possible ingredients include conditioning agents, opacifiers, clarifying agents for solubilization, thickeners for viscosity control, and antidandruff agents. Many shampoos also contain special ingredients such as vitamins, pro-vitamins, antioxidants, and herbal and marine extracts. These special additives are employed to support innovative claims involving repair, revitalization, nourishment, and color protection of hair.

The surfactants in shampoos can be classified according to whether or not they carry a charge. With the exception of baby shampoos, most primary surfactants are anionic; other surfactants are generally used in a secondary capacity.

## 1. Primary Surfactants

The main function of the primary surfactants in a shampoo is to provide a cleaning benefit. Primary surfactants are also necessary for adequate foam and viscosity control. As stated above, levels of surfactant between 8 and 20% are generally employed in shampoos. These levels are chosen primarily to provide acceptable lather and viscosity, since many common soils, e.g., sebum, are adequately cleaned at lower surfactant concentrations.

*(a) Alkyl and Alkyl Ether Sulfates.* The most common primary surfactants used today in shampoos are the lauryl and lauryl ether (laureth) sulfates. These materials were first introduced into the U.S. market more than 50 years ago [1], and since then the lauryl (sodium or ammonium, triethanolamine, diethanolamine) and laureth (sodium or ammonium) sulfates have dominated the market, in large part because their properties represent an excellent balance of cost, mildness, cleaning efficacy, lather, and viscosity control.

The lauryl and laureth sulfates are used either alone or in combination. The most commonly used variants are ammonium lauryl sulfate (ALS) and sodium or ammonium laureth sulfates (SLES or ALES) with an average of 2 or 3 moles of ethylene oxide [2,3]:

$$CH_3(CH_2)_{11}OSO_3^-NH_4^+ \quad CH_3(CH_2)_{11}(OCH_2CH_2)_xOSO_3^-Na^+$$
Ammonium lauryl sulfate    Sodium laureth sulfate

The lauryl sulfates are produced by sulfation of a mixture of synthetically prepared $C_{12}$–$C_{14}$ fatty alcohols or a mixture of coconut fatty alcohols (approximately 50% $C_{12}$). Depending on the manufacturer, the commercial lauryl and laureth sulfates contain different mixes of mostly $C_{12}$ and $C_{14}$ surfactants. These are chosen to improve the foam and surface activity of the species.

Schwuger [4] has investigated the effects of ether groups on the solubility, surface properties, and detergency of alkyl ether sulfates. Addition of ethylene oxide groups to the alkyl surfactants increases solubility, thus reducing the formation of precipitates and maintaining foam volume in the presence of $Ca^{2+}$ and $Mg^{2+}$ ions. The use of ether sulfates would be preferred over that of alkyl sulfates for a clear formulation.

The sulfate group is attached to the lauryl and laureth surfactants through an ester linkage. These detergents are therefore subject to hydrolysis at extreme values of pH. As a result, shampoos containing these surfactants are generally formulated with a pH between 5 and 9.

Minimal irritation is another essential property for shampoos because the products can easily come into contact with sensitive parts of the body, including the eyes, during the hair washing process. Studies on skin irritation by surfactants show that irritation is usually not a problem with the long-chain alkyl sulfates [5–9]. The presence of ethylene oxide groups reduces the irritation of these materials.

*(b) Alpha Olefin Sulfonates.* The alpha olefin sulfonates (AOS) rank second in use behind alkyl and alkyl ether sulfates. Nevertheless, because of the predominance of the latter surfactants, AOS has been confined to limited use in nonpremium shampoos. The detergent is actually a mix of four species in roughly equal quantities that can be represented by the following structures:

$$R-CH_2-CH=CH-CH_2-SO_3^- Na^+$$
$$R-CH=CH-CH_2-CH_2-SO_3^- Na^+$$
$$R-CH_2-CHOH-CH_2-CH_2-SO_3^- Na^+$$
$$R-CHOH-CH_2-CH_2-CH_2-SO_3^- Na^+$$

Commercial AOS is 14 to 16 carbons in chain length, so that R in the above structures represents a hydrocarbon chain length of 10 or 12.

AOS is fairy stable at low pH because the $SO_3$ attachment does not involve an ester linkage. It is therefore suitable for use in low pH formulations. It is also more soluble in hard water than SLS. Foaming of AOS has been reported to be comparable to SLS and SLES under various conditions especially in the presence of sebum [10,11]. Other distinctive properties that AOS exhibits include low cloud point, good solubilizing properties, and light color and odor.

Viscosity building with AOS is more difficult than with alkyl sulfates, although it can be done with the same types of materials, such as monoalkanolamides and salt [11–13]. In addition, the detergent has been reported to leave a harsher feel than lauryl and laureth sulfates [13].

## 2.  Secondary Surfactants

Secondary surfactants, which include nonionics, amphoterics, and some of the less widely used anionics, are often employed in a formulation to improve foam quality and stability, to provide additional detergency, and to enhance viscosity. Some of them are also used to reduce eye irritation in mild or baby shampoos.

*(a) Nonionic Surfactants.* Although the detergency of nonionic surfactants is equal to, or in many cases better than, that of anionic surfactants [14,15], nonionic detergents have not been used as primary hair cleansers due to their inferior foaming properties. This is because of the large surface area per molecule and the lack of charge on the surface films of nonionic foam [16].

An example of a nonionic for baby shampoos is Polysorbate 20, which is the monoester of lauric acid and anhydrosorbitol condensed with approximately 20 moles of ethylene oxide. Another example is PEG-80 sorbitan laurate, an ethoxylated sorbitan monoester of lauric acid with an average of 80 moles of ethylene oxide.

Fatty alkanolamides are another class of commonly employed nonionic surfactants. These are used in shampoos to enhance lather and viscosity. The most frequently used alkanolamides are cocoamide DEA (diethanolamide) and cocoamide MEA (monoethanolamide):

R–CO–N–(CH$_2$CH$_2$OH)$_2$   R–CO–NH–CH$_2$CH$_2$OH
Alkanolamide DEA        Alkanolamide MEA

where R is a hydrocarbon chain that, in the case of lauramide, for example, would contain 11 carbons. The monoethanolamides are reported to enhance viscosity more effectively than the corresponding diethanolamides [17,18].

The viscosity-building effect of the long-chain amides is a result of the building of ordered structures between detergent and amide molecules. This effect is promoted by the linear alkyl chain in the surfactant, which lines up easily to form ordered arrangements [1,3].

Amine oxides are also employed to improve foam characteristics and stabilize lather, especially at moderately acidic pH values. CAP (cocamidopropyl) amine oxide is one of the most commonly used amine oxides. These materials act as nonionics at the near-neutral pH encountered in shampoos but are easily protonated at acidic pH. As a result, they sometimes behave as cationics and act as conditioning and antistatic agents as well in a properly formulated system [13,19].

*(b) Amphoteric Surfactants.* Amphoteric surfactants are often used in conventional and baby shampoos to improve mildness and lather. Examples include lauroamphocarboxy glycinate [20], CAP betaine, and CAP hydroxysultaine:

CH$_3$(CH$_2$)$_{10}$CO–NH–CH$_2$CH$_2$–N–(CH$_2$CH$_2$OH)$_2$–CH$_2$COOH
Lauroamphocarboxy glycinate
R–CO–NH–(CH$_2$)$_3$–N–(CH$_3$)$_2$–CH$_2$COOH
Cocamidopropyl betaine
R–CO–NH–(CH$_2$)$_3$–N–(CH$_3$)$_2$–CH$_2$–CHOH-CH$_2$SO$_3$H
Cocamidopropyl hydroxysultaine

Betaines are very soluble over a wild pH range. Their charge nature changes with pH. At high pH the surfactant is anionic as a result of ionization of the carboxyl group, while at low pH the nitrogen is protonated, resulting in a cationic species. At the intermediate pH values normally found in shampoos, the carboxyl group is partially ionized and the nitrogen is partially in the protonated form.

In general, betaines are compatible with anionic surfactants. The simpler alkyl betaines are found to be less compatible with alkyl sulfates than the alkylamidopropyl betaines, especially when the concentration of betaine is about one half that of the lauryl sulfate [1]. This incompatibility is related to the pH of the system as well as the nature of the anionic species present.

Betaines act as foam modifiers, changing the loose and lacy foams normally generated by lauryl and laureth sulfates to thick and creamy lathers. They also help to thicken shampoo formulations and lower eye and skin irritation [21,22].

*(c) Miscellaneous Anionic Surfactants.* There are many other anionic surfactants that have been used at low concentration as secondary surfactants in shampoos or other specialty products. The materials on the list include paraffin sulfonate, alkylbenzene sulfonate, sulfosuccinates, linear alkylbenzene sulfonates,

$N$-acyl methyltaurates, $N$-acyl sarcosinates, acyl isethionates, $N$-acyl polypeptide condensates, polyalkoxylated ether glycolates, monoglyceride sulfates, and fatty glyceryl ether sulfonates [1,23].

## B. Conditioning Agents in Conditioners

Hair conditioners are used primarily to improve the appearance and manageability of hair. They can be either rinse-off or leave-in types having the form of emulsions, solutions, or creams, and having a wide range of viscosities. There is also a wide range of claimed benefits associated with different conditioners, such as improved ease of combing, reduction of damage from grooming, prevention of flyaway hair, and increased hair softness and shine.

Several major types of conditioning agents have been employed in conditioners including cationic quaternary ammonium compounds, cationic polymers, long-chain fatty alcohols, and silicones such as dimethicone and its derivatives. These conditioning agents provide various cosmetic benefits to hair and may exhibit different drawbacks. For example, some silicones may leave a greasy feel on dry hair. Use of high concentrations of cationic polymers, which bind strongly to the hair surface at multiple sites, may lead to over-conditioning and excessive buildup after repeated application. Therefore, combinations of different types of conditioning agents are often used to provide the best overall conditioning performance.

Most products, however, contain the same general classes of conditioning agents with differences mainly in concentrations, numbers of different agents, and the particular members of a conditioning class employed. Table 10.3 lists a formula example from U.S. Patent 6,287,545 for a rinse-off conditioner.

The major classes of conditioning agents are described in the following sections.

**TABLE 10.3**  Formula Example from U.S. Patent 6,287,545 for a Rinse-Off Conditioner

| Ingredient | Activity (%) |
| --- | --- |
| Stearyl alcohol | 1.00 |
| Cetyl alcohol | 3.00 |
| Stearamidopropyl dimethylamine | 1.00 |
| Distearyldimonium chloride | 0.75 |
| Dimethicone | 0.75 |
| Mineral oil | 0.55 |
| Cyclomethicone | 0.75 |
| Propylene glycol | 0.50 |
| Fragrance and preservative | 0.50 |
| Water | QS to 100 |

## 1. Cationic Surfactants

Cationic surfactants, primarily quaternary ammonium compounds (quats), are the most widely used conditioning agents in current commercial products [24–26]. Important reasons for this include effectiveness, availability, versatility, and low cost.

Examples of commonly used quats are stearalkonium chloride, cetrimonium chloride, and dicetyldimonium chloride:

$CH_3(CH_2)_{16}CH_2-N^+(CH_3)_2(C_6H_5CH_2)Cl^-$
Stearalkonium chloride
$CH_3(CH_2)_{14}CH_2-N^+(CH_3)_3Cl^-$
Cetrimonium chloride
$CH_3(CH_2)_{14}CH_2]_2-N^+(CH_3)_2Cl^-$
Dicetyldimonium chloride

Because of the positive charges on quaternary ammonium compounds, such as the above, they are substantive to hair, binding to negative sites on the hair surface. Treatment with these materials, therefore, results in a hydrophobic coating on the hair fiber that not only renders the hair softer and easier to comb [27] but also greatly reduces the buildup of static charge (flyaway) on the hair surface [28].

Deposition of conditioning quats has been found to increase with increasing negative charge on the hair surface. Thus, deposition is greater on chemically treated hair (bleached, permed, or dyed), which is oxidized as part of the treatment process and therefore carries a greater negative charge. Deposition is also greater on the tips of the hair, which are older and therefore subject to greater sunlight oxidation. This can be seen in Table 10.4, which lists deposition from a solution of stearalkonium chloride on the roots and tips of bleached and virgin hair [29].

Many conditioning properties of quaternary ammonium compounds are related to the degree of hydrophobicity of the lipophilic portion of the surfactants. Thus quat deposition increases with increasing alkyl chain length and also with an

**TABLE 10.4**  Binding of Radiolabeled Stearalkonium Chloride (SAC) to Human Hair[a]

| Hair type | mg SAC bound/g hair (root area)[b] | mg SAC bound/g hair (tip area)[b] |
|---|---|---|
| Albino virgin hair | 0.789 | 0.649 |
| Albino bleached hair | 1.62 | 1.83 |

[a]Test procedure: 0.67 g of 1% [$^{14}$C]SAC (30% ethanol:water) was applied to a 2 g tress and rubbed into the hair for 1 minute. Tresses were then rinsed in a beaker of tap water for 45 seconds, followed by rinsing in a second beaker for 15 seconds, and finally rinsed under running 38°C tap water for 1 minute. Portions of hair taken from different parts of the tress were then dissolved in $2M$ NaOH at 80°C, oxidized with $H_2O_2$, then mixed with Aquasol-2 LSC cocktail and perchloric acid and counted.
[b]Each number represents an average of 5 replicates.

increase in the number of alkyl chains [30–35]. As a result, tricetylmonium chloride deposits to a greater extent than does dicetyldimonium chloride, which, in turn, is more substantive than the monocetyl quat.

Increased hydrophobicity has also been found to correlate with increased conditioning. Thus, Garcia and Diaz [36] have reported greater improvements in wet combing from heavier conditioning quats, even when present on hair in lower amounts than less hydrophobic species.

The dependence of deposition on degree of hydrophobicity indicates that van der Waals forces play an important role in deposition of quaternary ammonium conditioners [35]. This conclusion is consistent with the entropy-driven deposition demonstrated by Ohbu et al. [37] for a monoalkyl quat and by Stapleton [38] for a protonated long-chain amine.

## 2. Lipophilic Conditioners

In commercial products quaternary ammonium surfactants are almost never used alone. They are often used in combination with long-chain fatty conditioners, especially cetyl and stearyl alcohols [25], which serve to boost the conditioning effects of the quats [39]. The addition of cetyl alcohol to cetrimonium bromide was found to reduce combing forces by nearly 50% [25]. In another study, Fukuchi et al. [40] found a significant decrease in surface friction with the combination of cetyl alcohol and behentrimonium chloride.

Another consequence of the addition of fatty alcohols to cationic surfactants is the formation, under the right conditions, of liquid crystal and gel networks [41–45] that can greatly increase viscosity and confer stability upon the emulsion. Formation of such liquid crystals has been observed even at low concentrations [44,45]; the ready formation of these structures, along with low cost, improved stability, and compatibility with cosmetic ingredients are important reasons why long-chain alcohols are so ubiquitous in conditioning formulations.

Other lipids found in conditioners include glycol, triglycerides, fatty esters, waxes of triglycerides, and liquid paraffin.

## 3. Quaternized Polymers

Quaternized polymers have been found to improve wet combing and reduce static charge. In general, they can be formulated with anionic surfactants; greater deposition occurs with a mixture of amphoteric and nonionic surfactants. Two of the most important examples are Polyquaternium-10, a quaternized hydroxyethylcellulose polymer, and Polyquaternium-7, a copolymer of diallyldimethylammonium chloride and acrylamide. These are the two most frequently used polymeric conditioning agents in commercial shampoos [46,47].

Other important polymers are Polyquaternium-11, a copolymer of vinylpyrrolidone and dimethylaminoethyl methacrylate quaternized with dimethyl sulfate.

Also used are Polyquaternium-16, a copolymer of vinylpyrrolidone and quaternized vinylimidazole; and Polyquaternium-6, a homopolymer of diallyldimethyl-ammonium chloride.

By virtue of their cationic nature, the above polymers are very substantive to hair. As a result of multiple points of electrostatic attachment to the hair fiber, they are also difficult to remove completely, especially when charge density is high [29,46]. It has been reported that deposition of many polymers on hair is inversely proportional to cationic charge density [48,49]. This was explained by the observation that smaller quantities of high-charge-density polymers would be needed to neutralize all of the negative charge on a hair fiber.

In a shampoo formula, Polyquaternium-10 and Polyquaternium-7 form negatively charged complexes with excess anionic surfactant [48,50]. These complexes are repulsed by the negatively charged hair surface, resulting in reduced deposition. The magnitude of this effect is determined by the particular anionic employed and the anionic surfactant/polymer ratio.

Despite the reduction in deposition, it has been reported that polyquaterium-SLS complexes resist removal from hair [51]. Therefore, care must be taken in formulating polyquats into both conditioners and conditioning shampoos to avoid excessive buildup and a heavy-coated feel on the hair with repeated use.

## 4.  Silicones

The use of silicones in hair care products has increased considerably in the past two decades due to the pleasing aesthetic properties they impart to the hair. They are used in a wide variety of products, including conditioners, shampoos, hairsprays, mousses, and gels [52]. The low surface free energy of these materials results in rapid formation upon deposition of a thin, uniform coating on the surface of hair [53].

Silicones have been claimed to improve combing, enhance feel, reduce flyaway, increase shine, reduce drying time, and lock in color [54–56]. The most frequently used silicone is dimethicone, which is a polydimethylsiloxane. Other important silicones are dimethiconol, which is a dimethylsiloxane terminated with hydroxyl groups, amodimethicone, which is an amino-substituted silicone, dimethicone copolyol, which is a dimethylsiloxane containing polyoxyethylene and/or propylene side chains, and cyclomethicone, which refers to a class of cyclic dimethyl polysiloxanes ranging from trimer to hexamer:

$CH_3-SiO(CH_3)_2-[SiO(CH_3)_2]_x-Si(CH_3)_2-CH_3$
Dimethicone
$OH-SiO(CH_3)_2-[SiO(CH_3)_2]_x-[SiO(CH_3)(CH_2)_3-NHCH_2CH_2NH_2]_yH$
Amodimethicone
$CH_3-SiO(CH_3)_2-[SiO(CH_3)_2]_m-[SiO(CH_3)(CH_2)_3-O(C_2H_4O)a]_n$
   $-Si(CH_3)_2-CH_3$
Dimethicone copolyol

The presence of amino groups in silicones was found by Wendel and Disapio [57] to greatly increase the substantivity of these materials. This is a result of the positive charge formed by these groups at the pH commonly found in commercial products.

In another experiment, Berthiaume *et al.* [53] found that deposition from a prototype conditioner formulation, as well as conditioning, softness, and detangling, increased with increasing amine content in a series of amodimethicones.

The relative conditioning efficacy of silicones and a series of cationic surfactants was compared [58]. It was found that dimethicone lowered frictional coefficients and surface energy of virgin hair to a greater extent than most cationic surfactants including a very effective conditioning agent, distearyldimemonium chloride. Dimethicones with molecular weight greater than 20,000 were found to be the most effective in reducing surface tension.

Synergistic effects have been observed when silicones are used in combination with a particular quat [59,60]. Deposition of silicones (30-second exposure followed by drying without rinsing) was found to nearly double if tricetyldimonium chloride was present in the silicone treatment solution. Reduction in combing forces was also almost doubled when silicones were deposited in the presence of the quat.

Dimethicone copolyols provide lighter conditioning effects due to their solubility in water and low level of substantivity. Because of that, they are less effective in rinse-off products. They are used, however, in leave-on products, such as hairsprays, styling mousses, and gels.

Cyclomethicone is volatile and does not remain on dry hair, especially after blow-drying. It helps other conditioning agent disperse, however, and form films on hair. It also helps improve wet combing and provides transient shine.

Silicones, especially dimethicone, have been employed as the primary conditioning agents in "two-in-one conditioning shampoos since the latter part of the 1980s. The level of conditioning from these types of shampoos is generally lower than that from stand-alone conditioners, especially on treated hair, which is more negatively charged and, therefore, has a lower affinity for hydrophobic substances like dimethicone.

Yahagi [58] studied the performance of dimethicone, amodimethicone, and dimethicone copolyols in two-in-one shampoos. Dimethicone and amodimethicone were found to provide hair with a similar degree of ease of combing. Unsurprisingly, soluble dimethicone copolyols did not perform well. Effects of silicones on foam volume were also investigated. A significant reduction in foam volume was observed with a model shampoo formula containing dimethicone, while amodimethicone and dimethicone copolyol showed a minimal effect on lather.

Table 10.5 lists various patents on hair conditioners granted since 1981. The development of more effective formulations with multiple benefits using novel

**TABLE 10.5** U.S. Patents Related to Hair Conditioners and Conditioning Agents for Hair Care

| Patent no. and year | Inventor(s) and company | Technology | Claimed benefits |
|---|---|---|---|
| U.S. 6730641 (2004) | Verboom et al. (Alberto-Culver) | Cetrimonium chloride/stearalkonium chloride = 0.65–2.0; ≤1% total amount of cetrimonium chloride + stearalkonium chloride | Synergistic effect; rinse-off or leave-in conditioner; provides silk wet feel and inhibits flyaway at 34% RH |
| U.S. 6726903 (2004) | Rutherford et al. (Unilever) | Monoalkyl quat (≥$C_{14}$, $C_{16}$–$C_{22}$)/dialkyl quat = 15:1–2:1; $C_{16}$–$C_{16}$ dialkyl/$C_{18}$–$C_{18}$ dialkyl = 1:3–3:1 | Foaming hair conditioner |
| U.S. 6723309 (2004) | Deane | Mixtures of conditioners, cooling agents, humectants, botanicals, and vitamins; No harsh chemicals and surfactants | Clean hair without removing essential oils; leave hair shiner, more body, and more manageable |
| U.S. 6645480 (2003) | Giles (Unilever) | Cationic surfactant; hydrophilically substituted cationic surfactant; lipid | An anionic surfactant-free formula providing acceptable cleaning and lathering; give wet slippy feel to hair |
| U.S. 6613316 (2003) | Sun et al. (Unilever) | Monoalkyl quat (≥$C_{14}$, $C_{16}$–$C_{22}$)/dialkyl quat = 15:1–2:1; monoalkyl quat (≥$C_{14}$, $C_{16}$–$C_{22}$)/dialkyl quat = 15:1–2:1; $C_{16}$–$C_{16}$ dialkyl/$C_{18}$–$C_{18}$ dialkyl = 1:3–3:1; fatty alcohol opacifier | Opacifying hair conditioner |
| U.S. 6602494 (2003) | Jahedshoar et al. (Wella) | Silicone surfactant; hydrophobic, nonsurfactant silicone; basic or cationic N-containing conditioning compound; Polyhydric alcohol | Optically clear, transparent or translucent hair conditioner; leave-in or rinse-off |

| Patent no. (year) | Inventor (assignee) | Composition | Benefit |
|---|---|---|---|
| U.S. 6569414 (2003) | Bernecker et al. (Henkel) | Lipid-soluble ester alcohol or ester polyol; water-soluble compound (panthenol and derivatives, sugar, polyvinyl pyrrolidine or mixtures) | Reduce split ends |
| U.S. 6537533 (2003) | Alvarado (Unilever) | (a) $C_{20}$–$C_{24}$ quaternary ammonium having ethosulfate or methosulfate as an anion; (b) $C_{20}$–$C_{24}$ quaternary ammonium having chloride or bromide as an anion; (a)/(b) = 1:10–10:1; a solid at room temperature containing a fatty alcohol, ester, amine, amide, acid, or a water-soluble polymer | Nonirritating to eyes; rinse-out conditioner |
| U.S. 6376455 (2002) | Friedli et al. (Gold-schmidt) | Quaternary fatty acid ($C_6$–$C_{22}$) amino alcohol esters of methylethanol isopropanolamine (HEIPA) | Improved biodegradability; as effective as the conventional dialkylammonium |
| U.S. 6287545 (2001) | Su (Colgate-Palmolive) | Combination of low HLB (2–9) and high HLB (10–19) ethoxylated branched fatty alcohol ethers or esters as stabilizers; emulsion has a pH 2.0–5.5 | Improved freezing and freeze-thaw stability; leave-in or rinse-off |
| U.S. 6235275 (2001) | Chen et al. (Unilever) | 0.1–10% cationic surfactant capable of forming lamellar dispersion; 0.5–30% oil; 0.1–20% silicone surfactant | Improved wet and dry combing and leave a soft dry feel; easy to rinse out |
| U.S. 6149899 (2002) | Pyles (Helene Curtis) | (a) $C_{16}$–$C_{22}$ monoalkyl quat; (b) $C_{16}$–$C_{22}$ dialkyl quat; (a)/(b) ≥ 4.0; 1–4% fatty alcohol | Low solid formulation providing substantial conditioning benefits with compromising viscosity to users |
| U.S. 6147038 (2000) | Halloran (Dow Corning) | Aminofunctional silicone microemulsions having at least one long-chain quaternary amine salt | Optically clear hair conditioner; increased beneficial effects |

*(continued)*

**TABLE 10.5** (Contd.)

| Patent no. and year | Inventor(s) and company | Technology | Claimed benefits |
|---|---|---|---|
| U.S. 5989533 (1999) | Deegan et al. (Revlon) | 0.1–20% cationic conditioning agent; 0.1–20% ester of α- or β-hydroxyl acids; 0.1–30% fatty alcohol; 0.001–10% nonionic surfactant | Improved substantive conditioning, shine, body, combing, and fullness; no greasy or tacky feel |
| U.S. 5750097 (1998) | Leidreiter et al. (Goldschmidt) | Diacetyl tartrate esters of $C_8$–$C_{18}$ fatty acid glycerides | A conditioning agent for hair rinses and conditioning shampoos |
| U.S. 5616758 (1997) | McCarthy et al. (Karlshamns) | Cationic quaternary aminosilicones | A substantive conditioning agent suitable for a variety of environments such as skin and hair conditioner, fabric softener, and fiber lubricant |
| U.S. 5552137 (1996) | Manning (Witco) | $(R^1)(R^2)N^+(CH_2CH_2OC(O)R)_2X^-$ wherein $R^1$ is $C_1$–$C_6$ alkyl, or $C_1$–$C_6$ hydroxyalkyl; $R^2$ is $C_1$–$C_6$ alkyl, or benzyl; R is $C_{12}$–$C_{22}$ alkyl having 0–3 C=C, provided that at least 2 different chain lengths R are present and 0, 1, and 2 C=C are present | A biodegradable conditioning agent; exhibit exemplary performance as a conditioner |
| U.S. 5393452 (1995) | Raleigh et al. (General Electric) | High-molecular-weight, high-viscosity silicone-polyether copolymer | Improved antistatic properties; conditioning shampoo |
| U.S. 5334376 (1994) | Robbins et al. (Colgate-Palmolive) | Particulate barium sulfate combined with one or more coreactants from the following groups: silicone free of amino group, long-chain fatty alcohol, long-chain fatty acid amide | Improved body, manageability, and style retention to hair |

| Patent (year) | Inventor (company) | Description | Features |
|---|---|---|---|
| U.S. 5332569 (1994) | Wood et al. (Alberto-Culver) | Silicon oil in an organic solvent-based carrier comprising PEG and an anionic-cationic emulsifier complex formed from: anionic (phosphate or sulfate) copolymer of dimethylpolysiloxane (20–40 units) and polyoxyethylene (3–15 units) and cationic conditioning compound having at least one quaternary nitrogen or amido amine and one hydrophobic aliphatic or silicone polymer chain | Improved stability of silicone emulsion |
| U.S. 5213793 (1993) | Moses et al. (Gillette) | ≤1% solid; 0.01–0.5% cationic conditioning agent; 5–10% volatile oil; ≤3% hydrophobic emulsifying agent | Rinse-on or leave-in; no visible residual by a leave-in application |
| U.S. 5120531 (1992) | Wells et al. (Procter & Gamble) | 0.2–20% hair styling polymer; 0.2–20% nonaqueous solvent; 0.05–25% conditioning agent | Rinse-off conditioner; styling + conditioning |
| U.S. 5100657 (1992) | Ansher-Jackson et al. (Procter & Gamble) | A mixture of conditioning agents: silicone, cationic surfactant, and fatty alcohol; nonionic long-chain alkylated cellulose ether as the primary thickener; water-insoluble surfactant as a secondary thickener | Provide cleaner hair conditioning; does not have the dirty hair feel and quick resoiling of hair associated with quaternary ammonium |
| U.S. 4978526 (1990) | Gesslein et al. (Inolex Chem) | Alkyl or alkylamido dimethyl 2,3-dihydroxypropyl ammonium | Clear and stable formulations; do not leave buildup on hair |

*(continued)*

**TABLE 10.5** (Contd.)

| Patent no. and year | Inventor(s) and company | Technology | Claimed benefits |
|---|---|---|---|
| U.S. 4973476 (1990) | Krzysik (Dow Corning) | Volatile silicone; a functional silicone | Leave-in conditioner |
| U.S. 4940576 (1990) | Walsh (Chesebrough-Pond's) | A lyotropic liquid crystal phase formed by oppositely charged polymer and surfactant upon dilution | Rinse-on conditioner; more effective on improving the ease of wet combing than a complex not in the form of a liquid crystal |
| U.S. 4913828 (1990) | Caswell et al. (Procter & Gamble) | Ion pair/wax composites formed by alkyl amine and anionic surfactant | Effective conditioning agent for rinse-off conditioner and fabric conditioners |
| U.S. 4886660 (1989) | Patel et al. (Colgate-Palmolive) | $C_{14}$–$C_{22}$ alkyltrimethyl quaternary ammonium, $C_{14}$–$C_{22}$ alkanol, cellulose polymer, copolymer of PVP/VA | Provide shiny, smooth, manageable hair, ease to comb, easy to style, long holding power |
| U.S. 4868163 (1989) | Takei et al. (Kao) | Monoalkyl phosphate having a β-branched alkyl group | Highly safe; transparent or semitransparent jelly-like composition; good moisturizing effect |
| U.S. 4726945 (1988) | Patel et al. (Colgate-Palmolive) | Distearyl or ditallow quaternary ammonium; $C_8$–$C_{18}$ amido $C_2$–$C_3$ alkyl di-$C_1$–$C_2$ alkylamine; $C_{14}$–$C_{18}$ alcohol; mineral oil; cyclomethicone; propylene glycol | Easy to wash out with SLES-based shampoos; provide softness, shine, ease of combing, flyaway control, and manageability; rinse-off conditioner |
| U.S. 4725433 (1988) | Matravers (Neutrogena) | 0.5–3% Laureth-4; 1–4% Choleth-24; 0.1–0.8% hydroxyethylcellulose; 0.4–0.8% Polyquaternium 10 | Free of oil and fatty alcohol; nonirritating; no buildup |

| Patent | Inventor | Composition | Comments |
|---|---|---|---|
| U.S. 4719104 (1988) | Patel (Helene Curtis) | 0.25–4% static-reducing agent; 0.05–1% cationic film-forming polymer; 0.25–4% distearyldimethylammonium; static-reducing agent/catonic polymer = 2:1–20:1 | Rinse-on; reduce static |
| U.S. 4714610 (1987) | Gerstein (Revlon) | Amine oxide; acid to provide a pH of 2.4–3.8 | Rinse-on |
| U.S. 4659565 (1987) | Smith et al. (Ethyl Corp.) | Di $C_6$–$C_{18}$ alkyl methylamine oxide | Improve hair body and flyaway reduction |
| U.S. 4610874 (1986) | Matravers (Neutrogena) | 0.1–6% ethoxylated/acetylayed lanolin derivative; 0.1–1% ionic polymer; 0.5–1% hydroxyethyl cellulose | A clear freely pourable conditioner |
| U.S. 4551330 (1985) | Wagman et al. (Helene Curtis) | Oil-in-water emulsion inverts to water-in-oil emulsion at the hair surface when being rubbed onto the hair; unctuous oleaginous; water-dispersible polyvalent metal salt having a cation selected from $Al^{3+}$, $Ce^{3+}$, $F^{3+}$, $Zr^{4+}$, and aluminum zirconium coordination complex; acid or alkali to give pH 1.5–7.5 | Rinse-on or rub-on |
| U.S. 4275055 (1981) | Nachtigal et al. (Conair) | 2.5–7.5% stearamidopropyl dimethylbenzylammonium chloride; 2–5% stearyl dimethylbenzylammmonium chloride; 0.25–0.75% NaCl | Conditioning agents also function as pearlizing agents to form a stable pearlescent conditioner |

ingredients and compositions has been claimed. More applications on the use of conditioning agents in conditioning (two-in-one) shampoos are discussed in Section III.C.

## C. Other Common Auxiliary Ingredients in Shampoos and Conditioners

### 1. Emulsifying Agents

Most conditioning products are oil-in-water emulsions requiring emulsifying agents for stability. As discussed above, fatty alcohols in combination with quats can confer stability on emulsions of this type. If necessary, other emulsifiers may also be added to improve stability [61,62]. Many emulsifiers used in conditioners are nonionic, including fatty alcohols, ethoxylated fatty alcohols, ethoxylated fatty esters, and ethoxylated sorbitan fatty esters [63].

Many silicones used in hair care products are insoluble in water and must therefore be stabilized in emulsions. To make the manufacturing process easier, many suppliers offer silicones as preformed emulsions.

### 2. Viscosity/Rheological Modifiers

A sufficiently high product viscosity, usually between 2,000 and 5,000 cP for a shampoo and 3,000 and 12,000 cP for a conditioner, is an important requirement for consumer acceptability.

Salts are the least expensive ingredients to thicken shampoos. The two most commonly used salts are sodium and ammonium chlorides. Salts increase the viscosity of the products by interacting with the long-chain surfactants, converting the small spherical micelles of the surfactants into larger aggregates or even liquid crystal structures [3]. It should be noted that viscosity of a system generally reaches a maximum during salt addition; adding more salt after this point results in a viscosity decrease [3,64].

Alkanolamides, betaines, and amine oxides build viscosity in shampoos by increasing structure (Sections II.A.2(a) and Section II.A.2(b)). This is the same mechanism by which shampoos are thickened by salts [3,4].

Polymeric gums are also important compounds for building viscosity in shampoos and conditioners. They are easily dispersed in water at common use levels of 0.5 to 1.5%. The most commonly used cellulose polymer is hydroxyethylcellulose, which is compatible with anionic and cationic surfactants and stable over a wide pH range [24]. Other cellulose polymers in use include methylcellulose and hydroxypropylmethylcellulose.

Associative thickeners, such as PEG-55 propylene glycol oleate and PEG-120 methyl glucose dioleate, have been employed in shampoos to not only thicken but also introduce advantageous rheological properties [65,66]. These materials combine the properties of surfactant and polymer in one molecule. They tend to self-associate or interact with other solids in the formula and form a shear-sensitive network structure.

Synthetic polymers are also effective thickeners. These materials produce plastic rheological properties, imparting a yield value to the continuous phase that helps to suspend permanently oil droplets, thus providing excellent stability against creaming or coalescence during storage [67]. The shear-thinning structures they produce can permit easy dispensing of the product from its container.

Examples of synthetic polymers include crosslinked acrylate copolymers (carbomers) [67–69] and modified acrylate derivatives (acrylates/alkyacrylate crosspolymers) [70].

For conditioners, as discussed previously in Section II.B.1, mixtures of fatty alcohols and quaternary ammonium compounds form liquid crystals and gel networks that can greatly increase viscosity.

It should be noted that the use of different thickening agents in shampoos and conditioners can result in varying rheological characteristics, which affects the choice of a particular thickening agent.

## 3. Foam Modifiers

An important reason besides cleaning for using combinations of primary and secondary surfactants is to improve the quality and volume of foam. As discussed in Section II.A.2, some secondary surfactants such as betaines, amine oxides, and fatty alkanolamides also act as foam modifiers. They change the foams from a loose lacy structure generated by lauryl and laureth sulfates to rich and creamy foams.

## 4. Opacifying and Clarifying Agents

Opacity or pearlescence in shampoos and conditioners can be generated by a number of raw materials. Cetyl or stearyl alcohol, ethylene glycol stearate, and glyceryl monostearate are frequently used with alkyl sulfates [3,63]. Recently it has been claimed that behenyl alcohol will provide an improved pearlescent appearance (Table 10.6, U.S. Patent 6,608,011). These materials are incorporated into surfactant solutions above their melting points; they then crystallize upon cooling, producing a pearlescent appearance. The degree of opacity depends on the size, distribution, shape, and reflectance of the precipitated crystals. Table 10.6 lists some patents relating to pearlescent shampoos.

In cases where clearness is desired, solubilizing agents are used to improve and stabilize the clarity of a shampoo. Typical clarifying agents include ethanol, isopropanol, propylene glycol, butylene glycol, and sorbitol. Phosphates and short-chain polyethoxylated alcohols and esters have also been used [63].

**TABLE 10.6**    U.S. Patents Related to Pearlescent/Opacified Shampoos

| Patent no. and year | Inventor(s) and company | Technology | Claimed benefits |
|---|---|---|---|
| U.S. 6608011 (2003) | Patel et al. (Colgate-Palmolive) | Behenyl alcohol | Improved pearlescence and stability |
| U.S. 6365168 (2002) | Ansmann et al. (Henkel) | Dialky ether having $C_{12}$–$C_{22}$ linear or branched alkyl and/or alkenyl; Cationc polymer | Mild and excellent conditioning properties; brilliant pearlescence; very good stability |
| U.S. 6309628 (2001) | Ansmann et al. (Henkel) | Dialky ether having $C_{12}$–$C_{22}$ linear or branched alkyl and/or alkenyl; silicone | Synergistic improvement in pearlescent effect and conditioning properties; capable of stabilizing silicones in aqueous formulations |
| U.S. 6165955 (2000) | Chen et al. (Rhodia) | Fatty acid-based compound | Mild, cold pearlizing concentrate; good high-temperature stability |
| U.S. 6106816 (2000) | Hitchen (Chesebrough-Pond's) | Insoluble, nonvolatile silicone; suspending polymer; titanium dioxide-coated mica | Improved stability for silicone and pearlizing particles |
| U.S. 5925604 (1999) | Chen et al. (Rhodia) | Cold pearlizing concentrates do not require CDEA for stabilization and heat for making | Ultramild and CDEA free; energy saving cold making process; excellent pearlescent effects and cleaning |
| U.S. 5562898 (1996) | Dowell et al. (Helene Curtis) | Long-chain amine ($\geq C_{16}$) or fatty amidoamine ($\geq C_{13}$), and an acid | Excellent opaque or pearlescent aesthetic properties; resist phase separation |
| U.S. 5529721 (1996) | Salka et al. (Henkel) | Liquid pearlizing composition: alkyl polyglycoside, glycol distearate, and betaine | Easier processing |

*(continued)*

**TABLE 10.6** (Contd.)

| Patent no. and year | Inventor(s) and company | Technology | Claimed benefits |
|---|---|---|---|
| U.S. 5384114 (1995) | Dowell *et al.* (Helene Curtis) | Long-chain amine (at least one carbon chain $\geq C_{16}$) or fatty amidoamine ($\geq C_{13}$); neutralized with a suitable acid | A new class of opacifier/pearlizer for water-based compositions; good stability and suspending ability |
| U.S. 4741855 (1988) | Grote *et al.* (Procter & Gamble) | Better suspending agents: long-chain esters of ethylene glycol, esters of long-chain fatty acids, long-chain amine oxides | Better suspending stability |
| U.S. 4654207 (1987) | Preston (Helene Curtis) | Fatty acid ester derived from a $C_{16}$–$C_{22}$ fatty acid and a saturated $C_{14}$–$C_{22}$ fatty alcohol; an anhydrous solubilizing agent to presolubilize the ester | Improved capability of making a consistently predictable pearlescence |
| U.S. 4438096 (1984) | Preston (Helene Curtis) | 0.25–1% myristyl myristate | Stable pearlescent shampoo |

## 5. Antioxidants/Free Radical Scavengers

Product stability and performance can be affected by exposure to several oxidative sources, including oxygen, free radicals, UV radiation, oxidative enzymes, catabolic oxidation, and chemical oxidation. Many antioxidants are also good UV absorbers due to their conjugated chemical structure. Typical antioxidants found in cosmetic products are flavonoids, polyphenols, carotenoids, thiols, tocopherol (vitamin E) and ascorbic acid (vitamin C) [71,72]. According to Black [73], a combination of antioxidants from different classes is more effective than a single antioxidant due to an antioxidant cascade mechanism.

## 6. Photofilters/UV Absorbers

UV absorbers have been used in shampoos for many years and mainly serve to improve color stability against prolonged sunlight exposure in clear packages.

The common UV absorbers available for product protection are benzophenone, methylbenzyledine camphor, and *para*-aminobenzoic acid (PABA).

Use of UV absorbers or sunscreens in skin care products to prevent photo damage has been widely accepted. The common absorbers for UVB (280 to 320 nm) found in cosmetic products include PABA, salicylic acid derivatives, octocrylene, and phenylbenzimidazole sulfonic acid. Effective UVA (320 to 400 nm) absorbers include methyl anthranilate and avobenzone. Zinc oxide and titanium dioxide can physically block the radiation [74].

The concept of using UV absorbers for hair has been gaining greater attention in recent years. Degradation from UV radiation has been shown to occur in many keratinous materials, including wool [75–77] and hair [78–83]. This process is mediated by oxygen and accelerated by water. Damage to hair from UV exposure includes reduced elastic strength, excessive drying, and discoloration or photo fading of natural or artificial color.

Nonsubstantive photofilters, such as salicylic acid derivatives and octyl methoxycinnamate or benzophenone derivatives, have been included in leave-on formulations for skin and hair. Revlon possesses a patent describing the use of benzophenone and PABA derivatives with cationic surfactants and nonionic film formers in a mousse for hair protection [84]. L'Oreal has included camphorbenzalkonium sulfate as a proprietary photofilter in hair and skin care formulations [85–92].

Recently a number of cationic photofilters with improved substantivity to hair have been developed for rinse-off products. Croda, Inc. has developed a more substantive cinnamido amine cationic quaternary salt (Crodasome UV-A/B) [93–95]. ISP has also marketed a cationic sunscreen dimethylparamidopropyl laurdimonium tosylate (Escalol HP-610) with improved substantivity and mildness to hair [96].

The combination of a UV filter and antioxidants can provide a greater effect on photo protection since the antioxidants can eliminate free radicals that are generated by the UV light.

## 7. Sequestrants

Sequestrants can bind Ca or Mg ions present in hard water, thus blocking the formation of insoluble soaps or other salts during washing and rising. They can also improve, to a lesser extent, product stability by preventing catalytic decomposition of coloring agents and perfumes in the presence of trace metal ions [97]. Citric acid, EDTA and its salts, and polyphosphates are commonly used sequestrants [63].

## 8. Preservatives

Preservatives are necessary in products to ensure microbiological robustness. Alkyl sulfates and alkyl ether sulfates, for example, are subject to degradation by esterases produced by bacteria or fungi at the concentrations normally employed in shampoos. Therefore, a suitable preservative system is required to maintain product stability [13].

Preservatives can be classified into two types: compounds that release formaldehyde and compounds that do not release formaldehyde. Formalin, an aqueous solution of formaldehyde, is a commonly used preservative in shampoos and conditioners [3]. Although formaldehyde has been known as a sensitizer, it is not a problem if used at 0.1% or lower. The use of formaldehyde in baby shampoos is not recommended. Other preservatives that fall in the formaldehyde-releasing group are diazolidinyl urea, imidazolidinyl urea, and DMDM (dimethyloldimethyl) hydantoin.

The most commonly used preservatives that do not release formaldehyde are parabens, quaternium-15, and a mixture of chloromethyl isothiazolinone and methyl isothiazolinone (Kathon CG) [3,63,98,99].

The most frequently used antimicrobials found in commercial shampoos are parabens, methylparaben and propylparaben. Mixtures of different preservatives provide broader protection against a wider spectrum of microorganisms and have been proven to be the most effective method to ensure product robustness [100]. The level of preservative necessary depends on the composition of ingredients, total alcohol level, pH value, and water activity.

Care should be taken in the selection of preservative systems to avoid interaction between preservatives and ingredients or packaging material that could inactivate the preservative, cause product instability, or irritation to skin [101].

## 9. Fragrances and Colorants

Fragrances and colorants are added to hair care products to mask any undesirable base odor and enhance product aesthetics. These materials can play a crucial role in a consumer's decision to purchase a product on the shelf.

Due to the many oils in fragrances, changes in fragrance may result in dramatic changes in rheological characteristics and phase stability of a formula.

## III. TYPES OF SHAMPOOS
## A. Cleansing Shampoos

Cleansing shampoos provide basic cleansing of hair. They can be formulated with a variety of primary and secondary surfactants depending on the desired cleaning efficacy and aesthetics of the product (Table 10.7). They usually contain higher levels of anionic surfactants to permit effective cleaning, including removal of substantive residues from styling and other hair care products.

**TABLE 10.7**  Typical Formula Composition for a Cleansing Shampoo

| Ingredient | Activity (%) |
|---|---|
| Primary surfactants — anionic | 8–20 |
| Secondary surfactants/foam boosters — nonionic and amphoteric | 0–10 |
| Thickeners — salt, gum, polymer | 0–5 |
| Clarifying agents | 0–3% |
| Antioxidants/sequestrants/UV absorbers | 0–5 |
| Fragrance, preservative, dye | QS |
| Water | Balance |

## B.  Mild and Baby Shampoos

The most important requirement for baby shampoos is minimal irritation to scalp, hair, skin, and, especially, eyes. These products are often formulated with levels of nonionic and amphoteric surfactants higher than those found in basic cleaning shampoos.

As mentioned in Section II.A.1(a), the presence of ethylene oxide moieties in alkyl ether sulfates reduces surfactant irritation. Polysorbate 20 and PEG-80, which have 20 and 80 moles of ethylene oxides, respectively, are often incorporated into baby shampoos as anti-irritants. Magnesium salts of these surfactants are also milder than the sodium salts. Table 10.8 shows an example of the formulation of a mild baby shampoo from U.S. Patent 3,928,251. More developments in mild and nonirritating shampoo formulas are listed in Table 10.9.

## C.  Conditioning (or Two-in-One) Shampoos

Two-in-one shampoos provide conditioning benefits, such as softness, ease of combing, and manageability, in addition to basic cleaning. These products, which were developed in the late 1980s, provide significantly more conditioning than

**TABLE 10.8**  Formula Example from U.S. Patent 3,928,251 for a Mild Shampoo

| Ingredient | Activity (%) |
|---|---|
| Sodium ethoxylated (3EO) coco sulfate | 6.6 |
| Sultaine | 4.9 |
| PEG-20 sorbitan monolaurate | 14.0 |
| Water | Balance |
| pH | 7.0 |

**TABLE 10.9** U.S. Patents Related to Mild and Nonirritating Shampoos

| Patent no. and year | Inventor(s) and company | Technology | Claimed benefits |
|---|---|---|---|
| U.S. 6514918 (2003) | Librizzi (Johnson & Johnson) | Fatty ($C_6$–$C_{30}$)amides with $EO_0$–$EO_{20}$ or/and $PO_0$–$PO_{40}$ groups | Capable of viscosity building and foam boosting without compromising the mildness and safety properties |
| U.S. 6503873 (2002) | Crudden et al. (Hampshire Chem) | N-acyl ethylenediaminetriacetic acid (ED3A) (Na or K lauroyl) | A novel chelating surfactant product ultra-mild detergent compositions in combination with alkyl sulfates; excellent lather stability |
| U.S. 6461598 (2002) | O'Lenick et al. (Biosil Res. Ins.) | Salt complexes formed by a fatty ammonium compound and an anionic compound | Extremely mild to the eyes; outstanding conditioning effects; suitable for baby shampoos and body washes |
| U.S. 6056948 (2000) | Baust et al. (Benckiser NV) | Alkyl polyglycol ether carboxylate with 2–5 EO; alkylether sulfate; Fatty acid amidopropylbetaine | Extra-mild formulation; low tenside concentration |
| U.S. 6013616 (2000) | Fabry et al. (Henkel) | Monoglyceride (ether) sulfate; fatty acid condensations (isethionates, taurates, or sarcosinates) | Mild detergent mixtures |
| U.S. 5981450 (2000) | Fabry et al. (Henkel) | Monoglyceride (ether) sulfate; amino acid derivatives (acyl gluamates, vegetable protein hydrolyzates, or vegetable protein fatty acid condensates) | Mild detergent mixtures |
| U.S. 5968496 (1999) | Linares et al. (Procter & Gamble) | (a) Imidazolinium amphoteric surfactant; (b) polyol alkoxy ester; (a)/(b) = 15:1–1:1 | Excellent cleaning performance and mildness; improved foam stability |
| U.S. 5922671 (1999) | Tracy et al. (Rhodia) | Bis-alkyphenol alkoxylated Gemini surfactants | Improved surfactant properties; mild and environmentally benign |

*(continued)*

**TABLE 10.9** (Contd.)

| Patent no. and year | Inventor(s) and company | Technology | Claimed benefits |
|---|---|---|---|
| U.S. 5792737 (1998) | Gruning and Weitemeyer (Goldschmidt AG) | Amodipropyl betaines with C7–C12 alkyl group, especially coconut oil fatty acids | Exceptional mild and low irritating properties |
| U.S. 5756439 (1998) | He et al. (Lever Brothers) | 0.1–25 wt% EO/PO polymers: HLB $\geq$ 12, EO portion $\geq$ 50%, mol wt 6,000–25,000; anionic:EO/PO polymer = 1:1–10:1 | Significantly enhanced mildness |
| U.S. 5753600 (1998) | Kamegai et al. (Kao) | Saccharide nonionic surfactant | Low irritation to skin and scalp; improve scalp/skin resistance to external stimuli like antimicrobial contagion; improved antibacterial effect |
| U.S. 5679330 (1997) | Matsuo et al. (Kao) | $C_6$–$C_{36}$ alkylene oxide adduct type | Mild to skin and hair; excellent lathering, detergency, and conditioning effects |
| U.S. 5514369 (1996) | Salka et al. (Henkel) | Alkyl polyglycosides, betain, and polymeric slip agents | Mild to skin and eyes; more efficient deposition of antidandruff agents; anionic surfactant free; compatible with cationic materials such as conditioners and colorants |
| U.S. 5478490 (1995) | Russo et al. (Lonza) | Polyglyceryl esters | Meet baby shampoo criteria without the need of using ethylene oxide derivatives; viscosity and clarity can be adjusted by tailoring the polyglyceryl ester |
| U.S. 5372744 (1994) | Kamegai et al. (Kao) | Alkyl saccharide; sucrose fatty acid ester; anionic and amphoteric surfactants | Low irritation to skin and eyes; creamy and abundant foam; excellent slippery feel |
| U.S. 5310508 (1994) | Subramanyam et al. (Colgate-Palmolive) | $C_4$–$C_{24}$ alcohol EO(1–10) glyceryl sulfonate | Reduced skin irritancy and superior cleaning ability |

| Patent | Inventor (Company) | Composition | Properties |
|---|---|---|---|
| U.S. 5234618 (1993) | Kamegai et al. (Kao) | 0.1–95% saccharide nonionic surfactant; antibacterial agent | High antibacterial effect; restrain the early occurrence of dandruff; does not weaken the cutaneous metabolic and barrier function of the skin |
| U.S. 5073293 (1991) | Deguchi et al. (Kao) | (a) Alkyl glycoside; (b) dicarboxylic acid; (a)/(b) = 600:1–1:1; (a) + (b) = 1–60% | Excellent foaming power and detergency; easily rinsed out; provides a pleasant feeling to hands during use |
| U.S. 5035832 (1991) | Takamura et al. (Kao) | Alkylsaccharide nonionic surfactant; silicone derivative | Fine, slippery, creamy foam; very mild to skin and hair; a tense, slippery feeling to the hair; a light, refreshing feeling to the skin |
| U.S. 4946136 (1990) | Fishlock-Lomax (Amphoterics) | Combination of two amphoteric surfactants (alkylamion or alkoxyalkylamino type, and acylamino type) and one anionic surfactant | Mild shampoos |
| U.S. 4426310 (1984) | Verunica (Colgate-Palmolive) | Na lauryl ether diethoxy sulfate; polyethoxylated (78EO) glyceryl monoeater of coconut oil fatty acids; polyoxyethylene (20EO) sorbitan monooleate; $N,N$-dimethyl-$N$-lauryl betaine; disodium lauryl diethoxy sulfosuccinate | Neutral pH and clear; low irritating and does not cause a burning sensation in contact with children's eyes |
| U.S. 4181634 (1980) | Kennedy et al. (Johnson & Johnson) | $C_{10}$–$C_{26}$ alkyleneoxylated (2–3EO) bisquatemaryammonium | Reduced irritant properties of anionic and amphoteric surfactant |
| U.S. 4154706 (1979) | Kenkare et al. (Colgate-Palmolive) | Amine oxide; alkyl glycoside; polyoxyethylene sorbitan monolaurate; cocoethanolamide; polyacrylamide | Mild nonionic shampoos; free of ionics, allow washing of hair without destroying or adversely affecting the disulfide bonds of the keratin without changing the isoelectric point of the hair |
| U.S. 3928251 (1975) | Bolich Jr. et al. (Procter & Gamble) | Ethoxylated anionic; zwitterionic (betaines, sultaines); polyethoxylated nonionic | Mild shampoo without stinging eyes |

was previously available from shampoos. As such, they represent a major advance in hair care technology (Table 10.10 and Table 10.11).

The primary conditioning agent used in most two-in-one shampoos is dimethicone. Other related silicones have also been used, either in a primary or secondary capacity, including dimethiconol, amodimethicone, and dimethicone copolyol. Because most of these materials are not soluble in water, two-in-one shampoos are generally oil-in-water emulsions, requiring the use of a suitable stabilizer or emulsifying agent.

As stated previously, the surface of chemically treated hair is more negatively charged than that of virgin hair. As a result, hydrophobic conditioning agents like dimethicone bind to treated hair to a lesser extent than to untreated fibers. As a result, some two-in-one shampoos incorporate cationic polymers to increase conditioning on more hydrophilic damaged hair. However, the conditioning performance of cationic polymers in two-in-one shampoos may be no better than dimethicone as a result of formation of complexes in shampoos with high levels of anionic surfactant (Section II.B.3).

It is important to note that cationic quaternary compounds (monoquats) are rarely used alone in two-in-one shampoos because their substantivity to hair is either greatly reduced in the presence of anionic detergents [102] or else they form undesired precipitates with anionic surfactants. Therefore, combinations of anionic, nonionic, amphoteric, and zwitterionic surfactants are often employed to minimize the formation of insoluble complexes.

**TABLE 10.10** Formula Example from U.S. Patent 6,007,802 for a Conditioning Shampoo

| Ingredient | Activity (%) |
| --- | --- |
| ALES (3EO) | 14.00 |
| CAP betaine | 2.70 |
| Polyquaternium-10 | 0.15 |
| B8/C10 diester of adipic acid | 0.30 |
| Cocamide MEA | 0.80 |
| Cetyl alcohol | 0.42 |
| Stearyl alcohol | 0.18 |
| Carbapol 981 | 0.50 |
| Dimethicone | 1.00 |
| Fragrance | 0.70 |
| DMDM hydantoin | 0.37 |
| Color solution (ppm) | 64 |
| Water and minors | QS to 100 |

**TABLE 10.11**  U.S. Patents on Conditioning (Two-in-One) Shampoo Formulas

| Patent no. and year | Inventor(s) and company | Technology | Claimed benefits |
|---|---|---|---|
| U.S. 6627184 (2003) | Coffindaffer et al. (Procter & Gamble) | $C_4$–$C_{16}$ polyalphaolefin | Improved clean hair feel; improved fullness and body |
| U.S. 6592856 (2003) | Giles et al. (Unilever) | Combination of conditioning agents: emulsified silicones, cationic polymers, fatty acid polyesters of cyclic polyols and/or sugar derivatives | Improved hair softness and ease of combings, especially for damaged hair, through environmental or harsh mechanical or chemical treatments |
| U.S. 6506372 (2003) | Dubief et al. (L'Oreal) | Amphoteric polymer with at least one monomeric unit from (meth)acrylate or (meth)acrylamide having at least one fatty chain ($C_8$–$C_{30}$) | Improved disentanglement and softness |
| U.S. 6489286 (2002) | Lukenbach et al. (Johnson & Johnson) | Nonionic/amphoteric/anionic surfactants; at least two conditioning agents selected from cationic celluloses, sugar derivatives, and homopolymers or copolymers | Nonirritating, suitable for children and adults having sensitive skin and eyes; imparts wet and dry detangling, and manageability |
| U.S. 6436383 (2002) | Murray (Unilever) | Amino-functionalized silicone; emulsified nonamino-functionalized silicone having average particle size $\leq 2$ $\mu$m | Improved conditioning performance; softer and more manageable hair |
| U.S. 6432393 (2002) | Bergmann et al. (Helene Curtis) | Elastomeric resinous material | Increase in hair body without scarifying conditioning attributes |
| U.S. 6387855 (2002) | De La Mettrie (L'Oreal) | Hydrophobic guar (galactomannan) gum | Improved suspending stability of silicone; good detergent and foaming properties; very good homogeneity and viscosity |

(*continued*)

**TABLE 10.11** (Contd.)

| Patent no. and year | Inventor(s) and company | Technology | Claimed benefits |
|---|---|---|---|
| U.S. 6375939 (2002) | Dubief et al. (L'Oreal) | Amphoteric polymer with at least one monomeric unit from (meth)acrylate or (meth)acrylamide having at least one fatty chain ($C_8$–$C_{30}$) | Improved deposition of antidandruff agent; improved softness and disentangling of the hair |
| U.S. 6335024 (2002) | Philippe et al. (L'Oreal) | Aminoalcohol derivatives containing a urea functional group | A conditioning agent for hair and a moisturizing agent for skin |
| U.S. 6306805 (2001) | Bratescu et al. (Stepan) | Anionic–cationic bridging surfactant blends; bridging surfactants selected from ethoxylated alkanolamide, semipolar nonionic, amphoteric, zwitterionic | Clear solutions at a variety of concentrations in water; impart cleaning, foaming, and conditioning properties to hair |
| U.S. 6264931 (2001) | Franklin et al. (Akzo Nobel) | ≤4% fatty aliphatic ($C_{11}$–$C_{24}$) quaternary ammonium having ester linkages | Comparable with anionic surfactants; biodegradable conditioning agent |
| U.S. 6162423 (2000) | Sebag et al. (L'Oreal) | Dialkylether with $C_{12}$–$C_{30}$ alkyl radicals, same or different, linear or branched, saturated or unsaturated | Improved homogeneity and stability of silicone; sufficient foaming power |
| U.S. 6156297 (2000) | Maurin et al. (L'Oreal) | Nonvolatile vegetable oil as a conditioning agent; anionic sulfate/alkyl glycoside ≤2 | Excellent cosmetic properties: disentangling, softness, sheen, and body of the hair; good washing and foam power |
| U.S. 6110450 (2000) | Bergmann (Helene Curtis) | Ceramide and/or glycoceramide and phytantriol | Especially advantageous wet disentangling; synergistic effect |
| U.S. 6106816 (2000) | Hitchen (Chesebrough-Pond's) | Suspending polymer (e.g., polyacrylic acid, carylates copolymer) for silicones; titanium dioxide-coated mica | Improved phase/suspending stability |

| Patent | Inventor (Company) | Composition | Benefit |
|---|---|---|---|
| U.S. 6051214 (2000) | Isbell et al. (U.S. Secretary of Agriculture) | Fatty acid estolides | Enhanced rinseability, wet feel, detangling, dry comb feel, style management, shine and/or body to the hair |
| U.S. 6051213 (2000) | Beauquey et al. (L'Oreal) | At least 2 wt% alpha-hydroxylated carboxylic acids and theirs derivatives | Enhanced lightness, smoothness, shine, and mechanical strength |
| U.S. 6048519 (2000) | Hiraishi et al. (Helene Curtis) | A particular combination of silicone compounds; silicone gum with a viscosity of ≥1 MP; silicone fluid with viscosity of ≤100 kP; amino-functionalized silicone | Excellent conditioning benefits |
| U.S. 6007802 (1999) | Coffindaff et al. (Procter & Gamble) | Ethoxylated alkyl sulfate; amphoteric surfactant; insoluble conditioning agent; cellulosic cationic polymers; synthetic esters | Excellent cleaning performance and improved conditioning; minimum buildup |
| U.S. 5997854 (1999) | von Mallek (Henkel) | Quaternary ammonium; amphoteric and anionic surfactants; alkyl polyglycoside | Enhanced conditioning in the absence of silicones |
| U.S. 5990059 (1999) | Finel et al. (Helene Curtis) | 0.01–10 wt% microemulsion of high viscosity, slightly crosslinked silicone with a particle size of ≤0.15 μm; 0.01–10 wt% cationic deposition polymer | Excellent mechanical stability; excellent conditioning ability; high optical transparency or translucency |
| U.S. 5980877 (1999) | Baravetto et al. (Procter & Gamble) | Nonvolatile conditioning agent having a dual particle size range of 2 μm and 5 μm | Excellent cleaning in combination with improved conditioning; minimize adverse side effects associated with excess buildup |
| U.S. 5977038 (1999) | Birtwistle et al. (Helene Curtis) | Cationic conditioning polymer having a charge density of ≤+3.0 meq/g | Selectively enhances the wet feel and ease of wet comb, while reducing the ease of dry combing to allow an easy styling |
| U.S. 5888489 (1999) | von Mallek (Henkel) | Quaternary ammonium, amphoteric, and anionic surfactants, alkyl polyglycoside, emollient, amide | Enhanced conditioning in the absence of silicones |

*(continued)*

**TABLE 10.11** (Contd.)

| Patent no. and year | Inventor(s) and company | Technology | Claimed benefits |
|---|---|---|---|
| U.S. 5776871 (1998) | Cothran et al. (Procter & Gamble) | An insoluble silicone stably suspended with a low level of cationic polymer | A stable conditioning antidandruff shampoo without the need for crystalline suspending agents or polymeric thickening agents; do not need the costly heating and cooling steps |
| U.S. 5747436 (1998) | Petal et al. (Colgate-Palmolive) | Anionic and amphoteric surfactants; complex acid:amine (1:1 mole ratio); mixture of monoalky and dialkyl ammonium | Improved antistatic properties; fee of silicone |
| U.S. 5733536 (1998) | Hill et al. (WhithHill Oral Tech.) | Ultramulsion comprises: a dispersed silicone with particle size of 0.1–10 $\mu$m, viscosity $1.5$–$4 \times 10^6$ cP, a EO/PO copolymer surfactant with mol wt 1,100–150,000; surfactant/silicone = 400:1–1:2 | Method for making the ultramulsion; distinctive conditioning, moisturizing, protecting, etc.; the silicone phase functions as a reservoir for various treatment substances |
| U.S. 5726137 (1998) | Petal et al. (Colgate-Palmolive) | 1–16% long-chain ($C_{24}$–$C_{50}$) aliphatic alcohol with 0–40 EO; an alkyl amine having a $pK_a$ of at least 7.5; a water-soluble, film-forming polycationic polymer | A low-silicone conditioning shampoo has a high degree of conditioning properties; a nonsilicone conditioning shampoo possesses a high degree of styling control properties |
| U.S. 5665267 (1997); U.S. 5587154 (1996) | Dowell et al. (Helene Curtis) | Long-chain fatty ($\geq C_{16}$) amine; water-insoluble hair-treating compounds (silicones, antidandruff agent); suitable acid | Improved suspending ability without a thickening agent; effectively cleanse hair and deliver hair-treating compounds |
| U.S. 5656258 (1997) | Cauwet et al. (L'Oreal) | A mixture of conditioning polymers: (a) quaternary polyammonium polymer and (b) polymer containing 70–90 wt% diallydialkylammonium units, (a)/(b) $\leq 1.0$ | Improved the disentanglement of hair (especially wet hair) and the softness of the hair and skin; reduced buildup after repeated applications |

| Patent/Year | Inventor | Description |
|---|---|---|
| U.S. 5641480 (1997) | Vermeer (Lever Brothers) | Heteroatom containing alkyl aldonamides and conditioning agents | Enhanced stability and viscosity; improved foam and clarity; improved conditioning characteristics |
| U.S. 5612025 (1997) | Cauwet-Martin et al. (L'Oreal) | Mixture of conditioning polymers: a quaternary polyammonium polymer with mol wt $\leq 100,000$, a polymer with 70–90% $C_1$–$C_{18}$ diallyldialkylammonium units and 30–10% acrylic or methacrylic units | Synergistic cosmetic effect; improved softness of hair and skin; improved disentangling |
| U.S. 5580494 (1996) | Sandhu et al. (Colgate-Palmolive) | High charge density (>200) cationic polymers with anionic surfactants | Improved stability; reduce the use of water-insoluble silicone; less harsh to hair protein |
| U.S. 5415857 (1995) | Robbins et al. (Colgate-Palmolive) | 0.3–5% aminosilicone; 0.1–5% cationic surfactant conditioning agent | Improved hair conditioning characteristics; reduced buildup after repeated applications |
| U.S. 5346642 (1994) | Patel et al. (Colgate-Palmolive) | Long-chain ($C_{25}$–$C_{45}$) saturated primary alcohol or a derivative | Improved emulsion stability; desirable pearlescent appearance; improved hair conditioning effect |
| U.S. 5213716 (1993) | Patel et al. (Colgate-Palmolive) | Long-chain ($C_{25}$–$C_{45}$) saturated primary alcohol or a derivative | Improved emulsion stability; desirable pearlescent appearance; improved hair conditioning effect |
| U.S. 5211883 (1993) | Yamashina et al. (Kao) | Amidoamine type amphoteric surfactant; silicone polymer | Excellent soft and smooth feeling during washing and rinsing; superb natural hair-set effect; no oily stickiness and roughness; easy passage of comb through the hair; mild to skin, eyes, and mucous |

*(continued)*

**TABLE 10.11**  (Contd.)

| Patent no. and year | Inventor(s) and company | Technology | Claimed benefits |
|---|---|---|---|
| U.S. 5145607 (1992) | Rich (Takasago) | Anionic surfactants (Na, K, NH$_4$ alkyl sulfate); cationic conditioning surfactant | Clear conditioning shampoo |
| U.S. 5106613 (1992) | Hartnett et al. (Colgate-Palmolive) | C$_6$/C$_8$/C$_{10}$ alkyl and alkyl lower alkoxylated sulfates; aminosilicones; microcrystalline waxes | Better conditioning properties than detergents that contain longer chain alkyl groups |
| U.S. 4997641 (1991) | Patel et al. (Colgate-Palmolive) | Poly-lower alkylene (e.g., polyethylene and polypropylene); hydrocarbon solubilizer | Improved wet and dry combing, manageability; reduced static charge and flyaway |
| U.S. 4997641 (1991) | Hartnett et al. (Colgate-Palmolive) | C$_6$/C$_8$/C$_{10}$ alkyl and alkyl lower alkoxylated sulfates | Improved hair conditioning properties; reduced buildup after repeated applications |
| U.S. 4741855 (1988) | Grote et al. (Procter & Gamble) | Long-chain acyl derivatives: long-chain esters of ethylene glycol, esters of long-chain fatty acid, long-chain amine oxides | Stable silicone-containing shampoo |
| U.S. 4728457 (1988) | Fieler (Procter & Gamble) | Heated premixes (silicone) are added to a main mix (surfactants and suspending agents) at ambient temperature | Improved process for making silicone-containing shampoos |
| U.S. 4704272 (1987) | Oh et al. (Procter & Gamble) | Tri long-chain alkyl quaternary ammonium, or tri long-chain amine; nonvolatile silicone | Good hair conditioning and stable products |

Conditioning from two-in-one shampoos is expected to occur primarily at the rinsing stage during which time the shampoo emulsion breaks, releasing the silicone for deposition on hair. This separation of cleaning and conditioning stages permits the shampoo to perform efficiently both of its functions: removal of soils and deposition of conditioning agents.

## D. Antidandruff Shampoos

In the U.S. hair care products containing an antidandruff ingredient are considered as over-the-counter (OTC) drugs. However, different regulations are applied in other regions. An antidandruff shampoo is treated as a quasi-drug in Japan, a therapeutic product in Australia, and may be a cosmetic or an OTC product in Europe, depending upon the claims.

Water-insoluble anti-inflammatory agents or antidandruff particulates are more effective than water-soluble particulates. This is because antidandruff particulates come out of suspension when diluted by application to wetted hair, and deposit on the hair and scalp. When the composition is rinsed from the hair, many particulates of the agent remain on the hair and scalp to provide an effective amount for treatment. A soluble agent is washed away for the most part during rinsing, providing only an ineffective amount remaining on the scalp.

Water-insoluble antidandruff agents, such as zinc pyrithione, selenium sulfide, climbazole, coal tar derivatives, and powder sulfur, have been used in many products for treating dandruff (Table 10.12). Although these materials are effective in controlling dandruff, several difficulties can occur in formulating these materials into a stable product. In general, most of these ingredients have high specific density, which makes it hard to suspend them in liquid shampoos. Selenium sulfide is also sensitive to pH, and begins to break down and form toxic sulfides during storage when the pH becomes greater than 6.5.

A great deal of work in formulating antidandruff shampoos has been performed in the past decade (Table 10.13), resulting in claims of improved

**TABLE 10.12** Antidandruff Agents for Shampoos [2]

| Ingredient | Concentration (%) |
|---|---|
| Zinc pyrithione | 1.0–2.0 |
| Climbazole | 0.1–2.0 |
| Selenium sulfide | 0.1–4.0 |
| Salicylic acid | 1.5–3.0 |
| Powder sulfur | 2.0–5.0 |
| Coal tar derivatives | 0.5–5.0 |

**TABLE 10.13** U.S. Patents on Shampoo Formulas with Antidandruff Efficacy

| Patent no. and year | Inventor(s) and company | Technology | Claimed benefits |
|---|---|---|---|
| U.S. 6663875 (2003) | Glauder et al. (Clariant) | Oxiconazole (Z-1-(2,4-dichlorophenyl)-2-(1H-imidazol-1-yl)-O-(2,4-dichlorobenzyl)ethanone oxime and salt) | Broad antimicrobial spectrum and low toxicity; free of cytotoxic agents |
| U.S. 6649155 (2003) | Dunlop et al. (Procter & Gamble) | 0.02–5% cationic guar gum derivative (mol wt from 50,000 to 700,000, charge density from 0.05 to 1.0 meq/g); 0.1–4% antidandruff particulate (pyridinethione salts, selenium sulfide, or sulfur); nonvolatile conditioning agent | Improved coacervate formed between the cationic polymer and anionic surfactant upon dilution of the shampoo; superior combination of antidandruff efficacy and conditioning (three-in-one) |
| U.S. 6515007 (2003) | Murad | Acidic component of a hydroxy acid or tannic acid or salts in combination with vitamin E component and an antigrowth agent for the antidandruff effect, or with niacin component and a 5-α reductase inhibitor for the antihair-thinning effect | Antidandruff; antihair thinning (three-in-one) |
| U.S. 6514490 (2003) | Odds et al. (Janssen Pharmaceutica) | Ergosterol biosynthesis inhibitor; 10'-undecen-3-oyl-aminopropyl trimethylammonium methylsulfate | Effective antifungal and antidandruff composition |
| U.S. 6451300 (2002) | Dunlop et al. (Procter & Gamble) | 0.005–1.5% polyalkylene glycol; nonvolatile conditioning agent; cationic polymer | Superior combination of antidandruff efficacy and conditioning (three-in-one) |

| Patent | Inventor (Company) | Composition | Description |
|---|---|---|---|
| U.S. 6410593 (2002) | De Mesanstourne et al. (CECA) | 1.5–3.5% Amphoram U (undecylenamidopropylbetaine, an amphoteric surfactant) | A multifunctional amphoteric surfactant serves as a cleaning and foaming agent and an antifungal agent to treat or prevent dandruff |
| U.S. 6333027 (2001) | Hopkins et al. (Johnson & Johnson) | 0.1–15% active ingredient selected from undecylenic acid, undecylenamidopropylbetaine, and mixtures thereof | A composition for treating and/or ameliorating dandruff, seborrheic dermatitis, psoriasis, and eczema; nonstinging to the eyes |
| U.S. 6323166 (2001) | Kamiya (Kamiya) | 0.05–5% essential oil selected from terpene esters and terpene hydrocarbons; 3–20% $N$-acylamino acid salt; 0.1–15% sucrose fatty acid ester or $C_6$–$C_{18}$ fatty acid alkylolamide | Soften the hair and give a rinsing effect and controlling dandruff and itch on the scalp; remove minute chemicals deposited on the skin responsible for atopic dermatitis |
| U.S. 5900393 (1999); U.S. 5834409 (1998) | Ramachandran et al. (Colgate-Palmolive) | Climbazole or a mixtures of climbazole and one or more cotherapeutics such as salicylic acid; ratio of amphoteric to anionic surfactant from 0.75 to 1.25 | Mild detergent composition; therapeutic effect on scalp disorders (itch, irritation, and dryness) encountered in warm weather and in tropical regions |
| U.S. 5723112 (1998) | Bowser et al. (Chesebrough-Pond's) | Metal pyrithione with at least 90% of particles $\leq 5$ μm; water-soluble cationic polymer as a deposition aid | Improved mechanical stability; excellent antidandruff ability |
| U.S. 5624666 (1997) | Coffindaffer et al. (Procter & Gamble) | 0.1–5% particulate antidandruff agent having an average particle size from 0.35 to 5 μm; 0.01–1% soluble cationic polymer as a stabilizing agent | Suspend without crystalline suspending agents or hydrophilic polymeric thickeners; have good, nonslimy feel |

*(continued)*

**TABLE 10.13** (Contd.)

| Patent no. and year | Inventor(s) and company | Technology | Claimed benefits |
|---|---|---|---|
| U.S. 5302323 (1994) | Hartung et al. (Abbott Lab.) | 0.5–2.5% selenium sulfide; two suspending agents: di(hydrogenated) tallow phthalic acid amide and one selected from hydroxypropylbmethylcellulose and Mg Al silicate; pH 4.0–6.5 buffer system with Na citrate and citric acid; at least 18% anionic surfactant | Improved pH stability, suspension stability, lathering, and conditioning (three-in-one) |
| U.S. 5154847 (1992) | LaPetina et al. (Helene Curtis) | Ethylene–maleic anhydride resin or polyacrylic acid resin, and alkanolamide and/or wax ester | Improved suspending stability; improved foaming properties |
| U.S. 4867971 (1989) | Ryan et al. (Colgate-Palmolive) | 0.1–2% 1-imidazolyl-1-(4-chlorophenoxy)-3,3-dimethylbutan-2-one (climbazole); pH 4–5.5 | A stable homogeneous liquid antidandruff shampoo; increased deposition of climbazole; enhanced antidandruff efficacy |
| U.S. 4854333 (1989) | Inman et al. (Procter & Gamble) | 0.005–0.9% peroxy oxidizing agent, e.g., hydrogen peroxide and Na percarbonate; 0.1–5% selenium sulfide having average particle size <25 μm | A selenium sulfide shampoo with a neutral pH (no need for a buffer system); improved color stability |
| U.S. 4379753 (1983) | Bolich (Procter & Gamble) | Metal salt of pyridinethione | Improved product aesthetics |

antidandruff efficacy, stability, and mildness. More complicated formulas, called three-in-one shampoos, have been developed to provide additional benefits, such as conditioning, temporary styling, and antihair-thinning, in addition to cleaning and dandruff control.

## E.   Shampoos with Specific Cosmetic Benefits

Shampoos providing specific cosmetic benefits other than conditioning have being developed and have gained popularity in recent years. Cosmetic benefits such as volume increase, shine increase, ease of styling, moisturizing, replenishing, etc., have been claimed. Table 10.14 lists some examples of shampoo formulas that provide specific cosmetic benefits.

## IV.   BASIC MECHANISM OF HAIR CLEANING

The primary function of shampoos is to clean hair. In order to assess the effectiveness of different detergents in cleaning various soils, it is important to understand the different cleaning mechanisms operating during the shampooing process.

## A.   Nature of the Hair Surface

Before discussing specific detergency mechanisms, it is necessary to consider the nature of the hair surface. The structure of hair is described in detail in Robbins' book [3]. Figure 10.1 shows a scanning electron microscopy (SEM) image of the root end of a typical, virgin hair fiber. The fiber consists of a hydrophilic central portion, the cortex, covered by 5 to 10 overlapping layers of cells termed the cuticle [3]. Compared to the cortex, the cuticle contains a large percentage of cystine residues, resulting in a highly crosslinked structure. The surface of the cuticle consists of a monolayer of covalently bound fatty acids that impart a hydrophobic nature to a healthy, undamaged hair fiber [103–105].

An important component of the cuticle structure is the cell membrane complex, or CMC, which consists of a ($\delta$) proteinaceous layer, sandwiched by two ($\beta$) lipid layers. The CMC is the only continuous structure in hair. It acts as a cement between different layers or components of the hair fiber and is responsible for the physical integrity of the hair structure.

Recently, a unique anteiso methyl-branched saturated fatty acid of 21 carbons, 18-methyl eicosanoic acid or 18-MEA, was identified in the outermost portion of the epicuticle, which is part of the CMC [104,106–110]. 18-MEA is the predominant fatty acid in the epicuticle. It makes up approximately 40% of the surface lipid layer of wool and human hair [106,107,109]. In addition to 18-MEA, other fatty acids have been isolated in smaller amounts from the epicuticle including

**TABLE 10.14** U.S. Patents on Shampoo Formulas with Cosmetic Benefits other than Conditioning

| Patent no. and year | Inventor(s) and company | Technology | Claimed benefits |
|---|---|---|---|
| U.S. 6689347 (2004) | Barbuzzi et al. (Unilever) | Water-insoluble particles having a layered structure comprising O atoms and silicone and/or P atoms, and organic functional groups bonded to silicone or P atoms by covalent bonds | Impart body attributes such as root lift, volume, bounce, and manageability in the absence of a styling polymer which leads to stickiness or dry feel |
| U.S. 6440907 (2002) | Santora et al. (Johnson & Johnson) | 0.01–3% humectant (cationically changed polyols, $C_6$–$C_{22}$ sugar derivatives) | Exceedingly mild to skin and eyes; leave skin and hair feeling moist but without feeling excessively oily and slippery; good dispersibility and foamability |
| U.S. 6432393 (2002) | Bergmann et al. (Helene Curtis) | Elastomeric resinous materials having a $G'$ modulus between $1 \times 10^2$ and $1 \times 10^5$ dyn/cm$^2$ | Increase in hair body without sacrificing conditioning attributes |
| U.S. 6348439 (2002) | Rousso et al. (Bristol-Myers Squibb) | Nonionic and/or cationic polymers; pH 8–10 when cationic polymer is present; pH 8–14 when cationic polymer is not present | Provide body, fullness, and texture to fine or very fine hair |
| U.S. 6231843 (2001) | Hoelzel et al. (Wella) | 5–50% anionic, nonionic, amphoteric surfactants; 2–10% fruit acids, at least two acids selected from lactic, citric, maleic, tartaric, gluconic, fumaric, and succinic acid; 0.2–2% pantothenol, pantothenic acid, and esters of pantothenic acid | Hair cleaning composition free of oily and greasy ingredients |
| U.S. 6046145 (2000) | Santora et al. (Johnson & Johnson) | 0.01–3% humectant (cationically charged polyols, $C_6$–$C_{22}$ sugar derivatives) | Extra mild to skin and eyes; leave skin and hair feeling moist but without feeling excessively oily and slippery; good dispersibility and foamability |

| U.S. 6024952 (2000) | Story et al. (Andrew Jergens) | Cationic polymer (polyquat-6); anionic emollient (sulfated castor oil) | A sparingly soluble moisturizing complex is formed; the anionic/cationic complex deposits strongly on the hair and is difficult to remove |
| U.S. 5994280 (1999) | Giret et al. (Procter & Gamble) | 3–40% insoluble, nonionic oil or wax or mixture; 0.1–8% $C_{10}$–$C_{18}$ fatty acid | Superior physical, viscosity, and foam stability; provide moisturizing effect |
| U.S. 5858340 (1999) | Briggs et al. (Procter & Gamble) | 0.5–20% polyhydric alcohol humectant; water-soluble polyglycerylmethacrylate lubricant; polyethyleneglycol (EO 2–200) glyceryl fatty ($C_5$–$C_{25}$) ester; hydrophilic gelling agent | Improved moisturization and skin feel; reduced tack and residue characteristics; excellent visual clarity and absorption characteristics |
| U.S. 5661189 (1997) | Grieveson et al. (Unilever) | Anionic, cationic, amphoteric, zwitterionic surfactants; thickening agents; benefit agents (silicones, fats and oils, vitamins, plant extracts, sunscreens, alkyl lactate, essential oils, etc.); small amount of soap | Effective and stable detergent system for delivering a wide variety of benefit agents |
| U.S. 4839162 (1989) | Komori et al. (Kao) | Diglyceride (liquid at room temperature); polyol type humectant and/or hydrophilic humectant | Provide high and long-lasting moisturizing effect; preserve the moisturizing function after a lapse of time |
| U.S. 4452989 (1984) | Deckner et al. (Charles of the Ritz) | 2-Pyrrolidone-5-carboxylic acid and salts | Nonirritant to skin and eye; provide unique moisturizing properties |
| U.S. 4374125 (1983) | Newell (Helene Curtis) | 0.01–1% sodium-2-pyrrolidone-5-carboxylate; 0.05–5% glycerine; 0.05–5% collagen protein | Restore the proper moisture level in initially moisture deficient hair; maintain the proper moisture level in hair initially having a normal moisture content |
| U.S. 4220168; U.S. 4220166 (1980) | Newell (Helene Curtis) | 0.01–1% sodium-2-pyrrolidone-5-carboxylate; 0.05–5% glycerine; 0.05–5% collagen protein | Maintain the proper moisture level in hair initially having a normal moisture content |

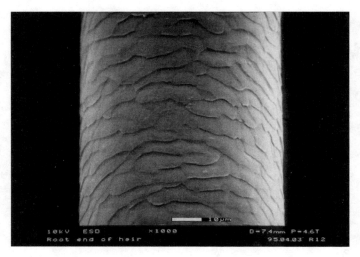

**FIG. 10.1**    SEM image of a typical root section of a virgin hair fiber.

lauric (C12), myristic (C14), palmitic (C16), oleic (C18:1), stearic (C18), arachidic (C20), behenic (C22), lignoceric (C24), and cerotic (C26) acids [107–110].

$$CH_3CH_2CH-(CH_2)_{16}-COOH$$
$$|$$
$$CH_3$$

18-Methyl Eicosanoic Acid (18-MEA)

Untreated hair comprises proteins that exhibit an isoelectric point near pH 3.67 [111]. As a result, despite its hydrophobic surface, hair carries a negative charge at the normal pH levels of hair care products. Loss and damage of the surface lipid layer also reduce the hydrophobicity of the hair. This combination of negative charge and hydrophobicity affects the types of soils that bind to hair as well as the ease with which different soils can be removed from the fiber surface.

This situation is further complicated by the fact that the concentrations of negative charges increase from the root to the tip of the virgin hair fiber [105,112,113]. This is primarily a result of weathering from sunlight exposure, which oxidizes the hair, converting cystine to cysteic acid and cystine *S*-sulfonate. The tips, being the oldest portion of a hair fiber will have been subjected to the greatest degree of stress and will carry the greatest degree of negative charge.

In addition to surface energy, the physical condition of the hair fiber also affects the types of soils attracted to and removed from the hair surface. Figure 10.2 shows an SEM image of the tip region of a hair fiber. The uplifting at the scale edges

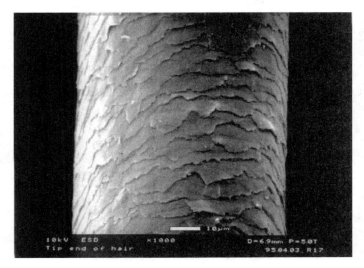

**FIG. 10.2**   SEM image of the tip end of the same hair fiber as in Figure 10.1.

caused by weathering and grooming exposes normally inaccessible areas in which soils can become physically entrapped.

If hair is further damaged to the point that the cuticle has split (Figure 10.3), the exposed hydrophilic cortex could strongly adsorb hydrophilic soils that would not bind as strongly to the intact hydrophobic cuticle layers.

## B.   Cleaning of Particulate Soil

The soil found on hair can be classified into two types: solid particulate and liquid or oily soil. Solid soils can come from hair care products or from the environment. Examples of the former might be polymeric resins or antidandruff agents, while the latter includes airborne particles carried by air currents, dust, carbon particles in the form of soot or clays, or rubber abraded from automobile tires [113–116].

Solid particles usually adhere to the hair surface through van der Waals or ionic forces [116–118]. In water, the ease of removing these soils from a surface depends upon the relative affinities for each other of the water, soil, and substrate. These affinities are expressed as $W_a$, the work of adhesion, which is defined as the free energy change per unit area involved in removing an adhered solid particle from a surface (in this case the hair fiber) to which it is adhered. In water, $W_a$ can be expressed as

$$W_a = \gamma_{PW} + \gamma_{FW} - \gamma_{PF} \qquad (1)$$

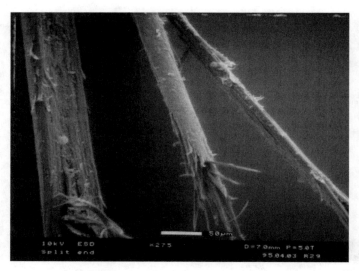

**FIG. 10.3**    SEM image of a split end.

where $\gamma_{AB}$ represents the interfacial tension between any two surfaces A and B. In this case, P represents the soil particle, W represents water, and F represents the fiber surface [119].

An examination of the above equation indicates that, in water, hydrophobic particles are quite difficult to remove from a hydrophobic hair surface because such systems result in high values for $\gamma_{PW}$ and $\gamma_{FW}$ and low values for $\gamma_{PF}$, resulting in a high work of adhesion.

Anionic and nonionic surfactants can reduce the work required to remove solid particles. This is because these surfactants adsorb to hair or hydrophobic soils with their hydrophobic tails in contact with the hydrophobic surfaces and their hydrophilic heads oriented toward the bulk solution. This has the effect of reducing $\gamma_{FW}$ and $\gamma_{PW}$ and, thus, $W_a$. Even more importantly, anionic surfactants remove particulates as a result of the increase of negative potentials on soil and hair upon anionic adsorption to these surfaces. This increases mutual repulsion between particulate and fiber, thus facilitating soil removal.

The ease of removing particulate soil from the hair surface is also dependent upon particle size. As size decreases, the surface area per unit weight of the particle, and consequently the area of actual contact per unit weight between particle and substrate, increases. As a result, more force per unit area is required to remove the particle [120]. In a normal cleaning process, particulates that are less than 0.1 $\mu$m in size cannot be effectively removed from fibrous substrates [121].

## C. Cleaning of Oily Soil

The most important type of oily soil found on hair is sebum, which is produced by the sebaceous glands on the scalp. Sebum is a mixture of fatty materials that is solid at room temperature, but almost completely molten at body temperature [122].

Various mechanisms are involved in removing the oily materials. These include roll-up, emulsification, liquid crystal formation, and solubilization. In the following sections the mechanisms and their relative importance in the hair cleaning process are discussed.

### 1. Roll-Up Mechanism

Figure 10.4 shows oil droplets adsorbed to a solid substrate. The equilibrium contact angle formed by the droplet is determined by Young's equation

$$\lambda_{FW} = \lambda_{FO} + \lambda_{OW} \cos \theta \tag{2}$$

where $\lambda$ is the interfacial tension between two phases, O represents the oil phase, W the water phase, and F the fiber, and $\theta$ is the contact angle.

According to Eq. (2), the adsorption of a surfactant to the fiber surface (oriented with the hydrophilic head pointing toward the aqueous phase) increases the contact angle, $\theta$, as a result of a reduction in $\lambda_{FW}$. If the reduction in $\lambda_{FW}$ is sufficiently large, the contact angle will increase to 180°, and the oil droplet will spontaneous separate from the fiber surface. At this point, then, the surfactant has increased the affinity of the fiber surface for water to such an extent that the water simply displaces the oil droplet and rolls it up. This process was first described by Adam [123] and by Kling [124] and is termed roll-back.

In practice, a thin film of oily soil can often form on the hair surface with a contact angle of zero. In this case, it may not be possible for a surfactant to roll back completely the hydrophobic soil without additional mechanical action such as rubbing and flexing.

A major hurdle to the roll-back process is high soil viscosity. Applying mechanical work in this case can increase soil removal, as does increasing temperature,

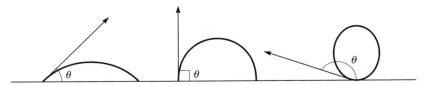

**FIG. 10.4** Different stages of the roll-up process. Note the increase in the contact angle as the oil droplet is rolled back from the substrate.

a step that increases rates of diffusion and surfactant adsorption, while reducing viscosity.

## 2.  Micellar Solubilization

Surfactants tend to form colloidal-sized association structures, or micelles, when the concentration is above a particular level termed the critical micelle concentration (CMC). In such structures, the hydrophobic portions of the surfactants are buried in the micelle interior, and the hydrophilic heads are oriented toward the bulk solution [125].

Because of their makeup, micelles can solubilize insoluble soils, such as fatty acids and hydrocarbons, in the hydrophobic interior of the micelle structure. Nonpolar soils are incorporated deep in the interior, while more polar materials are found closer to the hydrophilic heads [126,127].

The kinetics by which micelles solubilize fatty acid soils have been described as (1) adsorption of micelles on the soil surface; (2) incorporation of soil into the micelles; and (3) desorption of the soil-containing micelles. Diffusion of micelles to and away from the soil surface precedes and completes this solubilization process [128,129].

Solubilization is particularly important for hair cleansing because under normal shampooing conditions a significant concentration of micelles is expected to be present. Thus, final surfactant concentration in the lather during shampooing has been estimated to be 1 to 2%, a value that is 5 to 10 times the CMC of SLS [1]. This is an important value in view of reports that, for several surfactants, maximum detergency in cleaning of various fats and oils occurs at 6 to 10 times the CMC [130,131].

Addition of salts or other ingredients that shield negative charges decreases repulsion between the charged head groups of ionic surfactants, thus permitting closer packing of the surfactant molecules, leading to larger and rod-like micelles [132–136]. Such a size or shape change increases the volume of the inner core available for solubilizing hydrocarbons and long-chain polar compounds. Some ingredients in shampoos, such as long-chain amides, long-chain fatty alcohols, and betaines, form mixed micelles with the anionic detergents. The formation of mixed micelles also reduces repulsion between ionic head groups and results in larger micelles with more solubilizing capacity. As is the case with roll-up, soil removal by solubilization is greatly enhanced by increasing temperature [137] and application of mechanical work through rubbing and flexing of hair and through rinsing.

More work is needed to determine the exact contribution of solubilization to the cleaning of different soils. However, solubilization is undoubtedly a highly significant cleaning mechanism and, quite likely, is the most important means by which shampoos remove soils from human hair.

## 3. Emulsification, Penetration, and Mesophase (Liquid Crystal) Formation

Emulsification and liquid crystal formation are two important mechanisms by which detergents can effectively remove soil from substrates. Emulsification involves the breaking of a large oily mass into smaller droplets that can remain suspended for long enough in the cleaning medium to be washed away during rinsing. This process is accelerated in the presence of amphiphilic compounds in the soil, such as fatty acids or fatty alcohols, which can interact with detergents to produce spontaneous emulsification of the soil [120,138,139].

Oily soils containing amphiphilic species, such as fatty acids or fatty alcohols, can also be removed from substrates as a result of the formation of liquid crystal or mesomorphic phases between the amphiphile and a detergent. The liquid crystals are then broken up by subsequent osmotic penetration by water [140–142].

Removal of solid soils by mesophase formation can be accelerated by increasing the temperature. This has been reported for stearyl alcohol [143] and for lauric, palmitic, and stearic acids [128, 129] and is likely due at least in part to the increased penetration of the soils at higher temperatures [128,129,143].

Removal of solid soils by penetration without liquid crystal formation has been reported for tripalmitin, octadecane, and tristearin [143–145]. In these cases penetration of detergents occurred at crack and dislocation sites of soils.

## V.  ASSESSMENT OF CLEANING EFFICACY OF SHAMPOOS
### A.  Cleaning of Sebum

The most common soil found on human hair is sebum, a natural oily substance secreted onto the scalp by the sebaceous glands [146,147]. This material, which is a mixture of lipid components (Table 10.15), is distributed more or less uniformly over the hair surface as a result of contact with sebum-filled follicles by

**TABLE 10.15**  Average Sebum Composition in Adults [115]

| Ingredient | Percent of total |
| --- | --- |
| Cholesterol | 8.65 |
| Free fatty acids | 23.39 |
| Triglycerides | 32.71 |
| Wax and cholesterol esters | 19.53 |
| Squalene | 10.31 |
| Paraffins | 5.42 |

hair fibers [148], followed by mechanical actions such as combing, brushing, and rubbing against pillows.

Sebum, when not present in excess, lubricates hair, giving it a smooth and moisturized feel. However, too much deposited sebum causes the hair to become limp and clumped together. Under such conditions, the hair is perceived by panelists as dirty, greasy, and dull. In addition, because sebum is sticky, it tends to bind airborne particles and other materials with which it comes into contact [115,149,150], thus increasing soiling.

Most lipids found on the hair surface come from sebum. Hair also contains internal lipids, which are partly extractable [2,3]; much of this extractable material also appears to originate from the sebaceous glands. Robbins [2,3] has reported that the total extractable lipid can be as much as 9% of the total weight of hair that has not been shampooed for a week. The external and internal lipids are divided roughly equally among this extractable material.

During the relatively short period of time required for shampooing, detergents do not penetrate the hair fiber to any great extent. Shampoos, therefore, largely clean only surface lipid. The internal lipids left behind, however, do not contribute appreciably to consumer perceptions such as soiling, dulling, and feel of hair. Robbins [2,3], for example, found that there is no difference in the quantity of internal lipid extracted from oily or dry hair indicating that only surface lipids were responsible for the oily state of the hair.

An examination of the composition and physical state of sebum suggests that several cleaning mechanisms can operate during its removal from hair. Since sebum is completely molten at body temperature [122], it can be effectively removed by the roll-back mechanism. Also, the presence of approximately 25% free fatty acids in sebum indicates, as discussed in Section IV.C.3, that it is subject to removal by emulsification and mesophase formation. Finally, because the concentration of detergents during shampooing is well above their critical micelle concentrations, sebum can also be cleaned from hair by solubilization.

A number of studies have concluded that anionic surfactants are very effective at cleaning sebum at normal shampoo concentrations [2,3,151–155]. Shaw [152], for example, on the basis of SEM results, concluded that almost complete removal of surface lipid could be effected by anionic surfactants in a single application. Experiments involving extraction of wool swatches and hair clippings led Robbins [2] to reach the same conclusion for two surfactant applications.

The work done by Thompson [153] has shown that sodium laureth-2 sulfate (SLES-2) is superior to ammonium lauryl sulfate (ALS) in cleaning sebum, results that are consistent with those of Clarke and co-workers [152,154]. One reason for this finding is that SLES-2 has a lower CMC and a larger micellar aggregation number than ALS under the same conditions [155,156]. Thus, at a given concentration above the CMC, a solution of SLES-2 is likely to solubilize more sebum than ALS simply because more of its molecules are involved in micelle formation and it contains larger micelles having increased capacity for solubilizing nonpolar materials.

It should be noted that the results reported above were conducted, for the most part, on virgin hair having a hydrophobic surface. Chemically treated (permed, bleached, straightened, etc.) hair, however, has a more hydrophilic surface than virgin hair, resulting in a lower affinity for sebum and an increased ease of removal for these types of hydrophobic deposits. This reduction in sebum affinity is consistent with the increased dryness and need for conditioning reported for chemically treated and damaged hair.

## B. Cleaning of Quaternium Compounds from Conditioner-Treated Hair

As discussed in Section II.B.1, the most commonly used conditioning agents are long-chain quaternium compounds or quats. Two of the most widely used quats are cetyltrimonium chloride (CTAC) and stearalkonium chloride (SAC).

Removal of quats such as SAC and CTAC from hair can be expected to be more difficult than the removal of sebum. One reason for this is that the positively charged quats tend to deposit onto hair as a film having a strong electrostatic attraction for the negatively charged fiber surface. Because of the solid nature of these soils, the roll-back mechanism fails to apply, while the positive charge on the quats interferes with the cleaning mechanism discussed in Section IV.B: the mutual repulsion between solid soil and substrate resulting from adsorption of anionic surfactant onto the two surfaces.

Solubilization by anionic surfactants is another possible mechanism for cleaning quats. However, Reich and co-workers [157,158] found that solubilization of CTAC and SAC by lauryl and laureth sulfates (1 to 5 EO) was ineffective owing to formation of surfactant–quat complexes that were insoluble in ALS or SLES and, thus, difficult to remove from hair. In this case, reducing the carbon chain length of the quat to 12 or the chain length on the anionic surfactant to 10 resulted in more soluble complexes and more effective removal of the cationic soil.

## C. Cleaning of Cationic Polymers

As discussed in Section II.B.3, Polyquaternium-10 and Polyquaternium-7 (Merquat 550) are two of the most important polymers found in conditioning products.

Polyquaternium-10, also known as Polymer-JR, has a positive charge density of 670 (residue weight per unit of charge) [46,159] and is available with a molecular weight range from 250,000 to almost 1,000,000. Polyquaternium-7, also known as Merquat 550, has a reported positive charge density of 197 [47], which is more than three times the density of Polymer-JR, and a molecular weight of about 500,000.

Many studies on the binding of Polyquaternium-10 to hair have been published utilizing a variety of techniques including radiotracer methods, ESCA, and streaming potential measurements [51,159–161]. It has been shown that Polyquaternium-10 is quite substantive to hair, resisting complete removal by SLS even after 30 minutes of exposure to this detergent.

In radiotracer experiments conducted in the authors' laboratories, deposition of Polymer-JR was found to be almost 2.3 times greater for bleached hair than for virgin hair, indicating a greater number of negative binding sites on the former (Table 10.16).

Similar results were found in experiments run in the authors' laboratories utilizing radiolabeled Polyquaternium-10 (Table 10.17). Under conditions modeling actual hair washing (mechanical rubbing, short treatment times of one minute followed by rinsing), only 43% of bound Polyquaternium-10 could be removed in a single washing with SLS. The removal was increased to 75% for cleaning of Polymer-LR, a polymer with a similar structure to Polyquaternium-10 (JR), but a lower cationic charge density.

Fewer adsorption studies have been published for Polyquaternium-7. These include an ESCA study on hair [113] and a study on adsorption of Polyquaternium-7 from different surfactant solutions [46]. From the experiments conducted in the authors' laboratories (Table 10.16), deposition of Merquat 550 from solution was found to be somewhat lower, although this polymer has a higher reported charge density than Polymer-JR and was at a slightly higher concentration in the test solutions. However, the percent removal from wool was found to be similar for the two polymers. The reason that Merquat 550 is not harder to remove from wool than Polymer-JR remains unclear, despite its higher charge density. It may be due to steric reasons or may be related to the findings of Goddard and Harris [113], who reported that treating hair fibers with equal concentrations of Polquaternium-10

**TABLE 10.16**  Deposition of Radiolabeled Polymer-JR 400 on Hair[a]

| Substrate | mg JR bound/g hair[b] |
|---|---|
| Virgin hair | 0.424 |
| Bleached hair | 0.962 |

[a]Test procedure: 0.67 g of 1.76% [$^{14}$C]Polymer-JR in water was applied to 2 g tresses and rubbed into the hair for 1 minute. Tresses were then rinsed in a beaker of tap water for 45 seconds, followed by rinsing in a second beaker for 15 seconds, and a final rinse under 38°C running tap water for 1 minute. Portions of hair taken from the tress were then dissolved in 2 M NaOH at 80°C, oxidized with $H_2O_2$, then mixed with Aquasol-2 LSC cocktail and perchloric acid and counted.
[b]Each number represents an average of 5 replicates.

**TABLE 10.17** Deposition and Cleaning of Polycationic Conditioners[a]

| Treatment | mg polymer bound/g wool[b] | Cleaning (%) |
|---|---|---|
| 1.76% Polymer-JR | 5.77 | — |
| 1.76% Polymer-JR/5% SLS | 3.27 | 43 |
| 1.5% Polymer-LR | 4.63 | — |
| 1.5% Polymer-LR/5% SLS | 1.16 | 75 |
| 2.5% Merquat 550 | 4.78 | — |
| 2.5% Merquat 550/5% SLS | 2.81 | 41 |

[a]Test procedure: 0.15 ml of [$^{14}$C] polymer solution in water was rubbed into 0.15 g wool swatch for 1 minute, followed by rinsing in a beaker of tap water for 45 seconds, another rinse in a second beaker for 15 seconds, and a final rinse under 30°C running tap water for 1 minute. The same procedure was followed with SLS. Following this, swatches were dissolved in 2 M NaOH at 80°C, then mixed with Aquasol-2 LSC cocktail and perchloric acid and counted.
[b]Each number represents an average of 5 replicates.

and Polyquaternium-7 resulted in 25% surface coverage by the former material and only 10% coverage by the latter.

Deposition of Polyquaternium-7 and Polyquaternium-10 from anionic shampoos has been reported to be greatly decreased as a result of the formation of negatively charged polymer/surfactant complexes that are repulsed by negatively charged keratin surfaces [152,157]. As was stated earlier, however, these association complexes are still resistant to removal from hair [51].

## D. Cleaning of Fixative Residues

Neutral or negatively charged polymeric resins are commonly employed to provide styling benefits in products such as mousses, gels, hairsprays, and setting lotions. Typical examples in use today are the copolymer of vinyl acetate and crotonic acid, the copolymer of polyvinyl pyrrolidone and vinyl acetate (PVP/VA), the ethyl ester of the copolymer of polyvinyl methyl ether and maleic anhydride (PVM/MA), and the copolymer of octylacrylamide/acrylates/butylaminoethyl methacrylate (Amphomer).

Very few studies have been reported on the cleaning of hairspray resins from hair. In general, these resins are expected to be removed easily from hair due to their noncationic nature. This is consistent with a study by Sendelbach and co-workers [162] who reported 80 to 90% removal of different neutral or neutralized fixatives using a novel gravimetric study. Similar results were found in an experiment in the authors' laboratories employing a radiolabeled ethyl ester of PVM/MA. In this experiment it was found that 89% of the resin was removed from wool swatches with a single washing with 10% ALS.

The above experiments cover a limited selection of hairspray resins. More work needs to be done, therefore, to gain a more complete picture of fixative cleaning from hair, especially for polymers having differing degrees of neutralization.

## E. Cleaning of Dimethicone Residues

Dimethicone is the major conditioning agent used in two-in-one shampoos. As with fixative resins, little research has been done on the ease of removal of this material.

Rushton *et al.* [163] have studied the buildup and cleaning of dimethicone using ESCA and atomic absorption measurements. They observed a roughly 35% increase in dimethicone deposition on virgin hair after five washings with a commercial two-in-one shampoo compared to a single wash. No further deposition was observed, however, between 5 and 60 washes. They also found that a single wash with commercial shampoos removed more than 90% of deposited dimethicone. These latter experiments, however, were performed on solvent extracts of treated hair; no evidence was presented to show that all of the deposited dimethicone could be recovered in the solvent extract. More work is needed to determine unequivocally the ease of cleaning of dimethicone deposits.

## VI. IMPORTANT ATTRIBUTES OF SHAMPOOS AND CONDITIONERS

## A. Viscosity and Spreadability

Spreadability is an important product attribute that is related to viscosity and consistency of a formulation. The desired viscosity for a shampoo is generally between 2000 and 5000 cP: high enough for the product to be held in the palm without dripping but not too high to be difficult to spread over the hair. Conditioners, in general, have a higher viscosity than shampoos, of the order of 3,000 to 12,000 cP. A number of additives including alkanolamides, salts, and quaternary polymers can be used to control the consistency of shampoos and conditioners as discussed in Sections II.A.2 and II.C.1.

## B. Lather and Foam

Although not an indication of cleaning efficiency, the ability of a shampoo to provide a rich, copious lather is one of the first performance evaluations made by a consumer. A formulator should be aware that a consumer is likely to perceive a shampoo's lathering potential to reflect its efficacy. With this stated, there are multiple facts of the lathering attribute to consider, such as the speed with which lather is generated, the volume, the quality (i.e., loose or creamy), and the stability of the lather on the hair.

Several methods for generating foam have been reported including the use of kitchen blenders [20,164], shaking or rotation [165,166], by dropping from a height into a flask (Ross–Miles) [167], and by bubbling inert gas into the solution [168].

Hart and DeGeorge [21] employed a kitchen mixer to create lathers and measured the rate of foam drainage. They distinguished foam from lather, indicating that lather is a particular type of foam comprised of small, densely packed bubbles and is generated during shampooing and other processes. They also found that sebum significantly lowered lather quality and stability. This explains why a shampoo at the second application generally lathers better than during the first application.

Domingo Campos and Druguet Tantina [22] compared six methods of measuring foam and concluded that the Hart–DeGeorge method [21] and the Moldovanyi–Hungerbuhler method [168] correlate better with actual in-salon testing than the other four methods. They also concluded that the test should be carried out at a high concentration of about 60 g/l and preferably with the incorporation of a soil.

Robbins suggested that the rate of lather generation and the feel of the foam are two other important elements of shampoo lather. Methods for evaluating these two elements have not yet been well developed [3].

## C.  Ease of Rinsing

Following lather generation, it is important for a shampoo to be easily rinsed out of the hair. The ease of rinsing of a shampoo is affected by lather consistency, adsorption of surfactant to the hair, water conditions, such as temperature, hardness, and rinsing rate, and the quantity of hair [2]. A good shampoo formula should not contain ingredients that form precipitates in hard water, nor should it leave excessive residue on the hair surface.

## D.  Mildness

Mildness is an important concern for any type of shampoo and is especially crucial for baby shampoos due to the fact that the product could easily come into contact with skin, scalp, eyes, lips, and nose during the shampooing and rinsing processes. Surfactants and sensitizing agents used in personal care products are the main contributors of irritation.

A variety of test procedures exist for determining the relative mildness of personal cleaning products on human skin. The overall categories for the methods include patch testing, exaggerated use test, consumer use tests, and flex wash test [169–174].

## VII.   EVALUATING COSMETIC PROPERTIES OF SHAMPOOS AND CONDITIONERS

As described earlier, conditioning is a broad term for several desirable properties a conditioner can offer, including ease of wet and dry combing, shine, softness, manageability, and flyaway reduction. The assessment of these desirable conditioner properties is discussed in the following sections.

### A.   Ease of Combing

Ease of combing is, of course, one of the main benefits imparted to hair by a conditioner or conditioning shampoo. As such, the measured decrease in combing forces is frequently used to evaluate the effectiveness of a conditioning product.

Quantitative combing measurements can be performed using a Diastron or Instron tensile tester on both dry and wet fibers [175–178]. In this procedure, a comb is passed through a hair tress and combing force is recorded as a function of distance. In a typical combing curve, a sudden increase in combing force is generally observed at the fiber ends (end-peak force) as a result of entanglement of fiber tips. Kamath and Weigmann [179] have reported a double-comb method to prealign hair tresses that was claimed to eliminate the entanglement of fiber ends, and as a result produce a more reliable and consistent end-peak force.

A different technique, called spatially resolved combing analysis, was developed by Jachowicz and Helioff [180] to study the conditioning effects and substantivity of Polyquaternium-11 on different types of hairs. They found that bleaching results in a several-fold increase in combing forces compared to untreated hair. Subsequent application of a polymer solution was found to decrease friction against the hair surface and reduce combing forces.

### B.   Luster or Shine

The luster of hair is another attribute that is very important to consumers. Perhaps in part because of advertising messages, consumers tend to associate shine or luster with hair that is healthy and in good condition.

There are a number of factors that can affect the shine of hair. Shampoos and other hair care products that leave dulling deposits on the hair surface can reduce shine. Luster can also be lost as a result of abrasion, bleaching, grooming, or other stresses that damage and roughen the hair surface. Conditioning agents that reduce chipping and uplifting at scale edges or a shampoo that removes dulling deposits from hair will leave hair in a shinier state.

Hair luster has been evaluated subjectively and instrumentally. Subjective tests can be done on tresses or human volunteers. A number of instrumental methods for measuring hair shine have been reported [181–187]. Among these, light scattering has been used most extensively. Reich and Robbins [181] used a goniophotometer to measure hair shine quantitatively and found that it is a sensitive means of following changes to the hair surface including deposition (soiling and buildup), particle removal (cleaning), and even interactions on the fiber surface.

A quantitative measure of luster can be calculated from a light scattering curve (Figure 10.5) by the following relationship [181]:

$$L = S/DW_{(1/2)} \tag{3}$$

where $L$ = luster or shine, $D$ = integrated diffuse reflectance, $S$ = integrated specular reflectance, and $W_{(1/2)}$ = width of the specular peak at half its height.

The above equation was found to correlate well with panelists' subjective ranking of hair tresses. Note that the equation cannot be generally applied, since in cases where $D$ equals zero the expression goes to infinity. It is convenient to apply this equation to hair, however, since diffuse scattering from scale edges ensures that $D$ will not have a zero value.

## C. Body or Volume

From the consumer's standpoint, body is associated with fullness, volume, springiness, and bounce. Clarke *et al.* [188] stated that the visual impact of voluminous hair moving in a controlled manner is a universal description of hair with body. Another definition of hair body given by Hough *et al.* [189] is "a measure of a hair mass's resistance to and recovery from externally induced deformation."

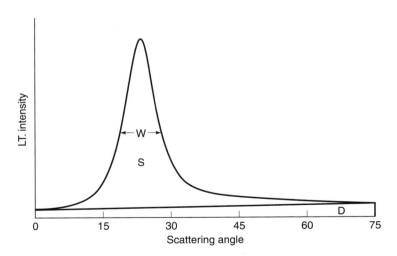

**FIG. 10.5** Typical light scattering curve for a virgin oriental hair.

The effects of shampoos and conditioners on hair body depend on the ingre-dients and the type of hair. A cleaning shampoo can provide body by removing soils or styling residues that weigh down the hair. Alternatively, shampoos and conditioners can deposit materials on the hair surface that can depress hair vol-ume. Straight fine hair is more sensitive to these effects than is curly coarse hair.

A number of instrumental methods have been reported for evaluating variables associated with hair body [189–193]. These methods measure changes in fiber friction, stiffness, curvature, diameter, weight, cohesion, and length. Treatments that increase the hair's curvature or diameter, increase the friction forces between fibers, or make the hair stiffer will increase body. Those that increase the cohesion between fibers or weigh them down will decrease hair body.

## D.  Surface Conditioning and Wettability

Wettability determinations provide important information on surface properties of different materials. A healthy hair fiber, for example, is covered by a hydropho-bic lipid layer that lowers the energy of the hair surface and prevents the fiber from being wetted by liquids with a high surface tension, such as water. The surface energy of a clean hair fiber, therefore, as reflected in its wettability, provides information on the degree to which the surface layer is intact and the hair is undamaged. Wettability measurements also provide information on the substantivity of conditioning agents, and the uniformity of deposited films.

The wettability of hair surfaces can be determined using the Wilhelmy tech-nique, in which the force exerted by the wetting liquid (usually water) on an individual fiber is scanned along the fiber length [194,195]. The wetting force ($F_w$) or contact angle ($\theta$) is then given by the Wihelmy equation:

$$\cos \theta = F_w / l\gamma \tag{4}$$

where $\theta$ = solid–liquid contact angle, $F_w$ = wetting force at any submersion position, $l$ = wetted circumference of the fiber, and $\gamma$ = surface tension of the liquid, usually water.

In general, the advancing contact angle for undamaged hair is greater than 90°, reflecting a hydrophobic surface [194] and a low value of $F_w$ in the above equa-tion. Weathering and chemical treatments typically make hair more hydrophilic, resulting in a decrease in the contact angle and an increase in wetting force.

## E.  Hair Strength

Chemical treatments, UV radiation, chlorine compounds from swimming pools, and oxidative pollutant compounds are all known to damage hair and reduce its strength [78–83].

The mechanical strength of a hair is determined by measuring its tensile properties using a Diastron or Instron tensile tester. The slope of the post-yield and the breaking force obtained from a typical stress–strain curve have been found to relate to the loss of mechanical strength of the hair and to the percent f reduction of disulfide bonds [79,82,196].

## F.  Flyaway Reduction

The phenomenon of flyaway is a result of charge repulsion between hair fibers, which makes hair hard to comb or to keep in place. This problem occurs when hair is combed or brushed, particularly at low humidity. The generation of static charges is due to an unequal transfer of charges across the interface between materials in contact.

Lunn and Evans [197] found that the density of charge is greatest near the tips of fibers, which corresponded to the highest combing force on the hair and that the forces acting between fibers and comb are directly related to the mutual area of contact and determine the magnitude of charging. This means that the magnitude of friction is directly involved in electrostatic charging. They concluded that quaternary ammonium compounds reduced static charge by reducing combing forces and that the half-life of charge mobility is dependent upon the concentration of quats. Jachowicz et al. [27] suggested that an increase of surface conductivity by long-chain alkyl quaternary ammonium salts is another mechanism by which these materials reduce static charge.

The measurement and control of electrostatic charge has been studied more extensively on textile materials than on hair fibers. Methods reported in the literature for static charge measurements include resistivity measurements, Faraday cage [197], dielectric losses, and TSC (thermally stimulated current) spectroscopy [198].

## G.  Manageability

Manageability, an important term in hair cosmetics, is a complicated attribute and is difficult to evaluate with a single parameter. The assessment of hair manageability is further complicated by other variables, such as the type of hair, humidity, or hairstyle.

Robbins et al. [199] defined manageability as the "ease of arranging hair in place and its temporary ability to stay in place" and suggested that hair manageability consists of three critical hair assembly properties: style arrangement manageability (combing or brushing of hair), style retention manageability (style retention during styling process), and flyway manageability (static flyaway).

## VIII.   DAMAGE TO HAIR FROM SHAMPOOS

Shampoos can damage the hair in different ways. Okumura [200] claimed that damage to the cuticle is a result of fibers rubbing against each other during the lathering stage. The removal of sebum, a natural lubricant for hair, from the fiber surface during shampooing also increases the susceptibility of hair to subsequent surface damage from combing and grooming. Kelly and Robinson [201] suggested that combing, brushing, and towel drying damages hair more than the lathering process. They also found greater cuticle loss from wet combing than from dry combing.

Damage to the interior of the hair fiber can also result from exposure to shampoos. Many studies have indicated that the hair structure is disrupted by surfactants that can slowly dissolve or remove structural proteinaceous materials. The number, percent, and mean area of voids or holes appearing in the endocuticle region after repeated washing and drying have been observed [202,203] using imaging analysis and TEM (transmission electron microscopy).

Sandhu and Robbins [204] found that hair fibers shaken with SLS and ALS solutions lost more protein than in water alone. Marshall and Ley [205] were able to extract protein from cuticle cells after a short 1 minute agitation in 1% SLS.

Recently, there has been increasing evidence that interior lipid material, much of which is structural, can be removed by shampooing [202,206]. The internal lipids removed over time by repeated shampooing are most likely from the inert ($\beta$) layers of the cell membrane complex which is the major pathway for entry of surfactants into the fibers.

## IX.   SAFETY CONCERNS

Very few adverse effects have been reported for shampoos and conditioners, especially when usage instructions are followed.

The potential health risks of personal care products that can occur include skin or eye irritation, ingestion, and inhalation. In general, surfactants used in shampoos do not demonstrate serious oral and ingestion toxicity.

Perfumes, preservatives, and emollients in cosmetic products are reported to be the major sources of adverse effects [207]. The problems that have been reported include temporary hair loss, contact dermatitis, scalp seborrhea, photosensitivity reaction, and mild acneform eruptions and folliculitis [208]. Bergfield attributes these problems either to preservatives or medicated ingredients rather than the active ingredients of hair products.

## X.   FORMULATION TECHNOLOGY
## A.   Formulation

The requirements for an effective commercial shampoo and conditioner have been touched on and discussed in some detail in the preceding sections of this chapter.

Essentially a good commercial shampoo formulation:

- Must clean effectively
- Must provide acceptable lather
- Must be stable with respect to phase separation, microorganisms, color, and fragrance
- Must be sufficiently easy to manufacture

Except for cleaning and lather, the above considerations also apply to a commercial conditioner. In the following sections, each of the above requirements is discussed.

## B. Formulation Requirements

### 1. Cleaning

As discussed earlier, shampoos have relatively high surfactant concentrations to ensure high lather and viscosity. For most types of soils, this also ensures adequate cleaning by shampoo detergents, including the two most common types of surfactant, ALS and ALES or SLES (see Section V).

In many systems, ALS delivers better lather and is easier to thicken than ALES or SLES, while the latter surfactants are milder and clean hydrophobic materials somewhat better, especially in hard water [152–154]. To take advantage of these differences, many formulations employ combinations of the two types of surfactant as their primary surfactant systems.

### 2. Lather

Generation of adequate lather is one of those product attributes that are crucial for a shampoo to be acceptable to consumers. Rich lathers can generally be attained through use of sufficiently high concentration of a primary, anionic surfactant together with a suitable secondary surfactant.

Common secondary surfactants employed to boost or modify lather include cocamidopropyl betaine, cocamide DEA, cocamide MEA, and cocamidopropylamine oxide. Anionics, such as disodium lauryl and laureth sulfosuccinates are also employed in some products as foam boosters.

Some ingredients in shampoos, notably dimethicone, can depress lather. The presence of these materials must be compensated for, by increasing surfactant concentration, changing the secondary surfactant, or adding additional foam boosters such as cetyl alcohol.

The particular materials chosen depend upon the formulation and its requirements. The effect of a particular foam booster on surfactant structure can vary with the formulation, especially if it is a complex emulsion containing many structure-modifying ingredients. In this case, several different lather modifiers should be investigated to optimize product lather.

## 3.   Viscosity and Rheological Characteristics

As stated earlier, both shampoos and conditioners must have sufficiently high viscosities to remain easily in the hand. In addition, however, the product must also have the correct rheological profile. Thus, it must have good stand-up in the hand, but must be sufficiently shear thinning that it is dispensed easily from the package.

In addition, shampoos and conditioners must spread easily over the hair, must not be stringy, and must not get too thick or too thin with changing temperature.

For simple cleaning shampoos, adequate rheological properties can generally be obtained by the same methods used for lather: sufficiently high primary surfactant plus the same secondary surfactants listed above as lather modifiers.

A particular viscosity level is then attained through addition of salt, usually sodium chloride (Section II.C.2). Salt increases viscosity up to a point, after which addition of more salt reduces viscosity. Generally, shampoo viscosity is adjusted on the rising portion of the salt curve.

For more complex shampoos, including emulsions, a variety of thickeners is available besides salt, including gums, associative thickeners, synthetic polymers, and long-chain alcohols (Section II.C.2). These materials are used to attain a desired viscosity, to stabilize a product, and to attain a desirable rheological profile. Since they affect product structure, they can also affect lather. Viscosity modifiers, therefore, should be chosen to give the best mix of lather and rheological properties.

## C.   Stability
## 1.   Types of Product Instability

Commercial products should have a shelf life of the order of two years. Over this time period viscosity, color, and fragrance should not change appreciably, and phases should not separate or materials precipitate out of the product.

The products most at risk are emulsions and colored formulations in clear packages. The latter are subject to exposure to radiation from sunlight resulting in possible color fading and also photodegradation of other light-sensitive components of the formula, especially in the fragrance. This type of instability can be handled by removing any nonessential light-sensitive materials and addition of antioxidants and UV blockers (Sections II.C.4 and II.C.5). The last mentioned can be added to the formulation or incorporated in the walls of the package.

Most conditioners and two-in-one shampoos are oil-in-water emulsions that are subject to phase separation of insoluble hydrophobic components. Standard emulsion technology must be employed when phase separation occurs. This can include changing orders of addition, adding energy in the form of increased temperature or higher stirring rates, different mixer configurations, increasing concentrations of stabilizing agent, adding a new or second stabilizing agent, etc.

## 2. Instability Testing

In general, a formulator cannot wait while a product ages for a year or two to determine its stability. Accelerated aging tests are therefore run, usually at elevated temperatures, to predict shampoo and conditioner stability. There are no standard accelerated aging tests in the hair care industry; each manufacturer has developed its own tests based on research and experience.

For evaluating physical stability, Robbins [3] has suggested running accelerated tests by placing formulations in ovens at 40 and 50°C for periods ranging from three to six months, along with aging at 25°C for one year. Products should also be subjected to one or more freeze–thaw cycles to evaluate stability at low temperatures. Note that elevated temperatures cannot be used to predict physical stability if a phase change occurs at the higher temperature, since one is no longer evaluating the same product. Under such conditions, it is even possible for a product to be unstable at room temperature and stable at higher temperatures.

For sunlight stability, products can be evaluated after exposure in their packages to direct sunlight for various periods of time. Accelerated aging can also be evaluated in instruments such as the Atlas Weatherometer in which products can be subjected to intense radiation at elevated temperatures.

## D. Manufacturing Ease

The different components in a formulation and their interactions and compatibility with each other all have an effect on the ease with which a product is manufactured.

Generally, simple cleaning shampoos with no insoluble ingredients are easiest to manufacture: they are simply mixed together in a vessel at room temperature. Even in this case, however, the order of addition may be important. Adjustment of pH should be carried out at the end of the process. If possible, adjustment of viscosity should be the last manufacturing step, since mixing solutions having high viscosity requires greater energy and can lead to excessive aeration.

Adding solid amides to modify lather and viscosity increases manufacturing complexity since heating above the melting point will generally be necessary. Even detergent-soluble solids may require heating in order to dissolve these materials in a reasonable amount of time. When possible, such materials should be predispersed or dissolved in the fragrance.

Conditioners and two-in-one shampoos, which are generally oil-in-water emulsions, are more difficult to manufacture than simple cleaning shampoos. Emulsions require energy to break up properly an oil phase and disperse it in water. This can be accomplished by energetic mixing, often aided by heating. Obtaining sufficiently energetic mixing may require different mixer types and configurations than would be required for a simple solution.

Order of addition is ordinarily important for obtaining the desired emulsion. With certain types of thickeners, this may mean that viscosity rises in the middle of the manufacturing process, making mixing and oil phase dispersion more difficult.

Many stabilizing agents, including some for dimethicone, are insoluble solids. Use of these requires heating above the melting point for both the water and oil phases of the formulation. The higher the melting point of the solid, of course, the higher the manufacturing temperature, and the more time and energy will be required to heat the formula and to cool it down. Choosing a predispersed or low-melting solid will improve manufacturing ease, but may increase formula cost.

Finally, many desirable silicone conditioning agents present handling difficulties. Dimethicone, while liquid, may have a viscosity so high that dispersal in the water phase along with cleaning after manufacture becomes difficult. Similar problems arise with solid silicones and gums. As a result of these difficulties, many manufactures offer pre-emulsified silicones having low viscosity. As with other predispersed materials, increased manufacturing ease must be balanced against increased cost.

## XI.   NEW PRODUCTS AND FUTURE TRENDS
## A.   Increased Use of Natural Materials

The market for natural personal care products has been growing at double-digit rates in the past few years and the trend is expected to continue for many years to come [209]. Natural and nonsynthetically derived ingredients are perceived by many consumers as being milder, safer, and more environmentally friendly.

Natural materials have been gaining popularity in shampoos and conditioners to support a wide variety of claims including conditioning, strengthening, luster, body, antidandruff, color retention, moisturizing, and photo protection.

Many natural materials provide real benefits and may be used either alone or in conjunction with their synthetic counterparts. The benefits from other materials may not be well documented or claims may be based on folklore or on benefits provided to other parts of the body, such as skin, rather than to the hair. These materials may be added to the product as marketing ingredients. As such, their purpose would be to differentiate the product and to capitalize on an ingredient's natural source or its association with health.

Table 10.18 lists some natural ingredients in the patent literature or in use in current hair care products along with their claimed benefits.

## B.   Increased Need for the Aging Population

The "baby boomers" born between 1943 and 1964 are moving into middle and old age. In 2003 they represented 27.7% of the U.S. population [227]. This group is

**TABLE 10.18** Natural Ingredients Found in the Patent Literature or Current Hair Care Products and their Claimed Benefits

| Natural ingredient | Claimed benefits |
|---|---|
| Keratin and derivatives, hydrolyzed protein, collagen protein, silk, and hydrolysate | Strengthening damaged and weathered hair fibers [210–213] |
| Essential oils and fragrances | Antistress, calming, and aromatherapeutic |
| Rosemary, sage, echinacea, camomile | Add sheen to hair |
| Hemp seed oil | Shine and conditioning effects [214] |
| Zingiber zerumbet extract (awapuhi) | Cleaning and conditioning [215] |
| Fruit concentrates containing fruit acids, vitamins B3 and B6, fructose, and glucose | Nourish and smooth the hair |
| Vitamins C and E, ginger root, green tea, rosemary, cranberry fruit, and grapeseed | Antioxidants and antiphotoaging [216] |
| Green tea extract, sage, Indian hemp, and rosemary | Anti-inflammatory, scalp itching or irritation control [217] |
| Grapefruit, lemongrass, and green tea | Lift sebum and flakes away |
| Basil, sage, mint extracts | Minimize oil production, add luster |
| Saw palmetto, lotus, and honey | Slow down hair loss, promote hair growth [218–222] |
| Chitosan | Improve hair strength and smoothness [223], reduce skin irritation [224], intensify coloration of hair dyes [225] |
| 18-Methyleicosanoic acid (18-MEA) | Restore shine, softness, and silkiness to long hair [226] |

more aware of the importance of a healthy life style than the preceding generation at the same age and is willing to pay more for premium products.

This generation is encountering a higher incidence of hair thinning, loss, and graying. Hair products that claim to provide antiaging benefits, including hair loss prevention [218–222,228–234], hair darkening [235,236], hair thickening [237], and graying reduction [219], are becoming more important to this generation.

Procter & Gamble and L'Oreal have been very active in investigating compounds and formulas for promoting hair growth or slowing down hair loss [228–234]. Currently, in this segment of the market leave-on products are dominant.

## C. Ethnic Hair Care Market

Hair care products have traditionally been formulated for Caucasian hair. In 2003 African-Americans made up 13% of the U.S. population and accounted for 30%

of hair care sales. The ethnic hair market requires very different styling, coloring, and conditioning products as a result of the very different hair textures and styling habits of African-Americans.

African-American hair is tightly twisted with a large degree of curl. This hair tends to be dry due to the structure of the hair and its curly nature. In addition, some African-Americans like to wear their hair relaxed, a process that causes excessive damage to the hair fiber. Products that provide moisturization, frizz control, breakage reduction, and shine are most desirable for African-Americans.

## D. Increased Demand for Specialty Products

Products for special, individual needs have been gaining dramatic acceptance among consumers with specific cosmetic or health concerns. The demand for specialty products is driven by race, age, gender, image, personality, lifestyle, health, well-being, fashion, etc. New specialty shampoos that have been developed and are emerging into the market include volume control [238–241], color protection [242], sun protection [84,243–246], revitalization or repair for damaged hair or split ends [247], frizz and flyaway reduction [248,249], styling control [250–260], etc. Currently, because they provide much greater actives delivery, leave-in products are more likely to provide real added benefits than rinse-off formulations.

## E. Nanotechnology

Nanotechnology is becoming a very promising technology in many areas. The use of nanoparticles of ingredients makes it easier to suspend water-insoluble actives in the medium, thus improving product stability and actives delivery. NanoSal, a controlled delivery system based on nanospheres, has been patented as an effective delivery vehicle for a broad range of ingredients and sensory markers onto skin, hair, and hair follicles. A prolonged release rate of active materials has been claimed [261]. This technology is especially valuable for rinse-off products, which have generally a very low amount of deposition compared to leave-on products. The systems can also provide heat-triggered release of active agents and yield a high-impact fragrance burst upon blow-drying of the hair.

L'Oreal has been granted several patents on nanoemulsions based on polymers, anionic polymers, and fatty esters of phosphoric or glycerol [262–268]. They have also patented nanocapsules that protect actives from premature release before reaching the targeted site for cosmetic, dermatological, and pharmaceutical applications [269–272]. Sunscreen formulations containing nanopigments based on metal oxides, like titanium oxide and zinc oxide, for improvement of sun protection factors (SPFs) of skin and hair products have also been extensively covered [273–278]. In addition, the use of ammonium oxide nanoparticles [279,280] in shampoos has been claimed to deliver very good hair retention and styling benefits as well as to harden and straighten hair fibers.

Procter & Gamble [281] has developed very recently a network of nanosized fibers containing chitosan and claimed that this composition would be useful in many areas, including hair care, skin care, oral care, water treatment, and drug delivery.

Up to now, nanotechnology has been applied more for skin care than hair care products. However, based on the similarities of the surface properties between skin and hair, the application of nanotechnology across a wide range of cosmetic products including hair products is expected to grow in the near future.

# REFERENCES

1. Barker, G., in *Surfactants in Cosmetics*, 1st ed., Rieger, M.M., Ed., Surfactant Science Series, Vol. 16, Marcel Dekker, New York, 1985, pp. 251–292.
2. Robbins, C.R., in *Liquid Detergents*, 1st ed., Lai, K.Y., Ed., Surfactant Science Series, Vol. 67, Marcel Dekker, New York, 1997, pp. 381–408.
3. Robbins, C.R., in *Chemical and Physical Behavior of Human Hair*, 4th ed., Robbins, C.R., Ed., Springer-Verlag, New York, 2002, pp. 193–310.
4. Schwuger, M.J., in *Structure/Performance Relationships in Surfactants*, Rosen, M., Ed., American Chemical Society, Washington D.C., 1984, pp. 1–26.
5. Kligman, A.M. and Wooding, W.A., *J. Invest. Dermatol.*, 49, 78–94, 1967.
6. Frosch, P.J. and Kligman, A.M., *Contact Dermatitis*, 2, 314–324, 1976.
7. Frosch, P.J. and Kligman, A.M., in *Cutaneous Toxicity*, Drill, V.A. and Lazar, P., Eds., Academic Press, New York, 1977, p. 127.
8. Lansdown, A.B. and Grasso, P., *Br. J. Dermatol.*, 86, 361–373, 1972.
9. Imokawa, G., Sumura, K., and Katsumi, M., *J. Am. Oil Chem. Soc.*, 52, 479–483, 1975.
10. Alpha olefin sulfonates for shampoos, Gulf Oil Chemicals Dept., New York, 1968.
11. Barker, G., Barabash, M., and Sosis, P., *Soap Cosmet. Chem. Specialties*, March, 38–42, 1978.
12. Schoenberg, T.G., *Soap Cosmet. Chem. Specialties*, May, 54–62, 1980.
13. Rieger, M., *Cosmet. Toiletries*, 103, 59–72, 1988.
14. Fort, T., Jr., Billica, H.R., and Grindstaff, T.H., *J. Am. Oil Chem. Soc.*, 45, 354–361, 1968.
15. McGuire, S.E. and Matson, T.P., *J. Am. Oil Chem. Soc.*, 52, 411–414, 1975.
16. Rosen, M.J., *Surfactants and Interfacial Phenomena*, 2nd ed., Wiley-Interscience, New York, 1989, pp. 276–303.
17. Goettem, E., *Tenside*, 5, 328, 1968.
18. Felletschin, G., *Tenside*, 7, 16, 1970.
19. U.S. Patent 3,086,943 to Procter & Gamble Co., 1963.
20. Gerstein, T., *Cosmet. Perfum.*, 90, 40, 1975.
21. Hart, J.R. and DeGeorge, M.T., *J. Soc. Cosmet. Chem.*, 31, 223–236, 1980.
22. Domingo Campos, F.J. and Druguet Tantina, R.M., *Cosmet. Toiletries*, 98, 121–130, 1983.
23. Reich, C. and Chupa, J., in *Harry's Cosmeticology*, Rieger, M.M., Ed., Chemical Publishing, New York, 2000, pp. 601–634.

24. Quack, J.M., *Cosmet. Toiletries*, 91, 35–52, 1976.
25. Gerstein, T., *Cosmet. Toiletries*, 94, 32–41, 1979.
26. Hunting, A.L.L., *Encyclopedia of Conditioning Rinse Ingredients*, Micelle Press, Cranford, NJ, 1987.
27. Foerster, T. and Schwuger, M.J., *Prog. Colloid Polym. Sci.*, 83, 104–109, 1990.
28. Jachowicz, J., Wis-Surel, G., and Garcia, M.L., *J. Soc. Cosmet. Chem.*, 36, 189–212, 1985.
29. Reich, C., in *Surfactants in Cosmetics*, 2nd ed., Rieger, M.M. and Rhein, L.D., Eds., Surfactant Science Series, Vol. 68, Marcel Dekker, New York, 1997, pp. 357–384.
30. Jurczyk, M.F., Berger, D.R., and Damaso, G.R., *Cosmet. Toiletries*, 106, 63–68, 1991.
31. Finkelstein, P. and Laden, KK., *Appl. Polym. Symp.*, 18, 673–680, 1971.
32. Jachowicz, J., *J. Soc. Cosmet. Chem.*, 46, 100–116, 1995.
33. Spiess, E., *Parfumerie Kosmetik*, 72, 370–374, 1991.
34. Scott, G.V., Robbins, C.R., and Barnhurst, J.D., *J. Soc. Cosmet. Chem.*, 20, 135–152, 1969.
35. Robbins, C.R., Reich, C., and Patel, A., *J. Soc. Cosmet. Chem.*, 45, 85–94, 1994.
36. Garcia, M.L. and Diaz, J., *J. Soc. Cosmet. Chem.*, 27, 379–398, 1976.
37. Ohbu, K., Tamura, T., Mizushima, V., and Fukuda, M., *Colloid Polym. Sci.*, 264, 798–802, 1986.
38. Stapleton, I.W., *J. Soc. Cosmet. Chem.*, 34, 285–300, 1983.
39. Polovsky, S.B., *Cosmet. Toiletries*, 106, 59–65, 1991.
40. Fukuchi, Y., Okoshi, M., and Murotani, I., *J. Soc. Cosmet. Chem.*, 40, 251–263, 1989.
41. Eccleston, G.M. and Florence, A.T., *Int. J. Cosmet. Chem.*, 7, 195–212, 1985.
42. Eccleston, G.M., *J. Colloid Interface Sci.*, 57, 66–74, 1976.
43. Barry, B.W. and Saunders, M.G., *J. Colloid Interface Sci.*, 41, 331–342, 1972.
44. Barry, B.W. and Saunders, M.G., *J. Colloid Interface Sci.*, 34, 300–315, 1970.
45. Barry, B.W. and Saunders, M.G., *J. Colloid Interface Sci.*, 36, 130–138, 1971.
46. Ucare Polymers: Cationic conditioners that revitalize hair and skin, Amercol, a subsidiary of the Dow Chemical Company, Piscataway, NJ. 2002.
47. Gafquat Quaternary Polymers, technical bulletin, GAF Corp., Wayne, NJ.
48. Hossel, P. and Pfrommer, E., in *Cosmetics Exhibition and Conference Proceedings*, Verlag fuer Chemische Industrie H. Ziokowsky, Augsburg, Germany, 1994, pp. 133–148.
49. Pfau, A., Hossel, P., Vogt, S., Sander, R., and Schrepp, W., *Macromol. Symp.*, 126, 241–252, 1997.
50. Sykes, A.R. and Hammes, P.A., *Drug Cosmet. Ind.*, 126, 35, 1980.
51. Hannah, R.B., Goddard, E.D., and Faucher, J.A., *Text. Res. J.*, 48, 57–58, 1978.
52. Luoma, A. and Kaea, R., Society of Cosmetic Chemists 1988 Spring Conference on Hair Care, London, April 21–23, 1998.
53. Berthiaume, M.D., Merrifield, J.H., and Riccio, D.A., *J. Soc. Cosmet. Chem.*, 46, 231–245, 1995.
54. Marchioretto, S., Dow Corning publication 22-1720-01, Brussels, 1998.
55. Thomason, B., Vincent, J., and Halloran, D., *Soap Cosmet. Chem. Specialties*, 68, 25–28, 1992.

56. U.S. Patent 4,898,595 to Dow Corning, February 6, 1990.
57. Wendel, S.R. and Disapio, A.J., *Cosmet. Toiletries*, 98, 103–106, 1983.
58. Yahagi, K., *J. Soc. Cosmet. Chem.*, 43, 275–284, 1992.
59. Nanavati, S. and Hami, A., *J. Soc. Cosmet. Chem.*, 43, 135–148, 1994.
60. Marchioretto, S. and Blakely, J., SOFW, October 2, 1997.
61. McCutcheon's *Emulsifiers and Detergents*, Vol. 1, North American ed., MC Publishing, Glen Rock, NJ, 1991.
62. Rieger, M.M., in *Harry's Cosmeticology*, Rieger, M.M., Ed., Chemical Publishing, New York, 2000, pp. 211–233.
63. Fox, C., *Cosmet. Toiletries*, 103, 25–58, 1988.
64. Lochhead, R.Y., *Cosmet. Toiletries*, 103, 99, 1988.
65. Glucamate DOE-120 technical bulletin, Amercol Corp., Edison, NJ.
66. Antil, technical bulletin, Degussa, Goldschmidt Personal Care.
67. Nagarajan, M.K. and Ambuter, H., in *Liquid Detergents*, 1st ed., Lai, K.Y., Ed., Surfactant Science Series, Vol. 67, Marcel Dekker, New York, 1997, pp. 129–177.
68. Lochhead, R.Y. and Warfield, D.S., *Soap Cosmet. Chem. Specialties*, 61, 46–54, 1985.
69. Formulating Shampoos with Carbopol, technical bulletin, BF Goodrich Co., Cleveland, OH.
70. Carbopol Aqual SF-1, technical data sheet, Noveon.
71. Rieger, M.M., in *Harry's Cosmeticology*, 8th ed., Rieger, M.M., Ed., Chemical Publishing, New York, 2000, pp. 247–259.
72. Weber, S.U., Sliou, C., Packer, L., and Lodge, J.K., in *Handbook of Cosmetic Science and Technology*, Barel, A.O., Paye, M., and Maibach, H.I., Eds., Marcel Dekker, New York, 2001, pp. 299–310.
73. Blac, H.S., *Nutr. Cancer*, 31, 212–217, 1998.
74. Levy, S.B., in *Handbook of Cosmetic Science and Technology*, Barel, A.O., Paye, M., and Maibach, H.I., Eds., Marcel Dekker, New York, 2001, pp. 451–462.
75. Asquith, R.S. and Brooke, K.E., *Ind. Chim. Belge*, 32, 437–440, 1967.
76. Korner, A., Schmidt, H., Merten, T., Peters, S., Thomas, H., and Hocker, H., 9th International Wool Textile Research Conference, 1995.
77. Klaic, B. and Friscic, V., *Tekstil*, 39, 717–720, 1990.
78. Ruetsch, S.B., Kamath, Y., and Weigmann, H.D., *J. Cosmet. Sci.*, 51, 103–125, 2000.
79. Nach, S., *Cosmet. Toiletries*, 105, 55–60, 1990.
80. Panda, C. and Jachowicz, J., *J. Soc. Cosmet. Chem.*, 44, 109, 1993.
81. Tolgysi, E., *Cosmet. Toiletries*, 98, 29–32, 1983.
82. Dubief, C., *Cosmet. Toiletries*, 107, 95–102, 1992.
83. Hoting, E., Zimmermann, M., and Hiterhaus-Bong, S., *J. Soc. Cosmet. Chem.*, 46, 85–99, 1995.
84. U.S. Patent 4,567,038 to Revlon, Inc., January 28, 1986.
85. U.S. Patent 4,061,730 to L'Oreal, December 6, 1977.
86. U.S. Patent 5,102,660 to L'Oreal, April 7, 1992.
87. U.S. Patent 5,616,331 to L'Oreal, April 1, 1997.
88. U.S. Patent 5,939,053 to L'Oreal, August 17, 1999.
89. U.S. Patent 6,159,456 to L'Oreal, December 12, 2000.

90. U.S. Patent 6,248,311 to L'Oreal, June 19, 2001.
91. U.S. Patent 6,221,343 to L'Oreal, April 24, 2001
92. U.S. Patent 6,703,002 to L'Oreal, March 9, 2004.
93. U.S. Patent 5,601,811 to Croda, Inc., February 11, 1997.
94. U.S. Patent 5,633,403 to Croda, Inc., May 27, 1997.
95. Crodasome UV-A/B: Sun Protectant for Hair, data sheet, Croda, Inc., Edison, NJ.
96. Jackowicz, J., *Spring 1998 Symposium on Haircare*, Society of Cosmetic Scientists, p. 71.
97. Zviak, C., in *The Science of Hair Care*, Zviak, C. and Vanlerberghe, G., Eds., Marcel Dekker, New York, 1986, pp. 49–86.
98. Orth, D., in *Harry's Cosmeticology*, 8th ed., Rieger, M.M., Ed., Chemical Publishing, New York, 2000, pp. 273–304.
99. McCarthy, T.J., in *Cosmetic and Drug Preservation Principles and Practices*, Kabara, J.J., Ed., Marcel Dekker, New York, 1984, pp. 359–387.
100. Busch, P., Hase, C., Hensen, H., and Lorenz, P., *Cosmet. Toiletries*, 102, 49–60, 1987.
101. Zviak, C. and Camp, M., in *The Science of Hair Care*, Zviak, C., and Vanlerberghe, G., Eds., Marcel Dekker, New York, 1986, pp. 309–344.
102. Wolfram, L.J. and Lindemann, M.K.O., *J. Soc. Cosmet. Chem.*, 22, 839–850, 1971.
103. Negri, A.P., Cornell, H.J., and Rivett, D.E., *Text. Res. J.*, 63, 109–115, 1993.
104. Ward, R.J., Willis, H.A., George, G.A., Guise, G.B., Denning, R.J., Evans, D.J., and Short, R.D., *Text. Res. J.*, 63, 362–368, 1993.
105. Kamath, Y.K., Dansizer, C.J., and Weigmann, H.D., *J. Soc. Cosmet. Chem.*, 28, 273–284, 1977.
106. Wertz, P.W. and Downing, D.T., *Lipids*, 23, 878–881, 1988.
107. Wertz, P.W. and Downing, D.T., *Comp. Biochem. Physiol.*, 92B, 759–761, 1989.
108. Logan, R.I., Johns, L.N., and Rivett, D.E., in *Proceedings of the 8th International Wool Textile Research Conference*, Crawshaw, G.H., Ed., Christchurch, 1990, p. 408.
109. Negri, A.P., Cornell, H.J., and Rivett, D.E., *Aust. J. Agric. Res.*, 42, 1285–1292, 1991.
110. Korner, A., Hocker, H., and Rivett, D.E., *Fresenius' J. Anal. Chem.*, 344, 501–509, 1992.
111. Wilkerson, V.J., *J. Biol. Chem.*, 112, 329, 1936.
112. Robbins, C.R. and Bahl, M., *J. Soc. Cosmet. Chem.*, 35, 379–390, 1984.
113. Goddard, E.D. and Harris, W.C., *J. Soc. Cosmet. Chem.*, 38, 233–246, 1987.
114. Powe, W.C., in *Detergency Theory and Test Methods*, Part I, Cutler, W.G. and Davis, R.C., Eds., Surfactant Science Series, Vol. 5, Marcel Dekker, New York, 1972, pp. 31–63.
115. Bruer, M., *J. Soc. Cosmet. Chem.*, 32, 437–458, 1981.
116. Krupp, H., *Adv. Colloid Interface Sci.*, 1, 122, 1967.
117. Kissa, E., *Text. Res. J.*, 43, 86–95, 1973.
118. Powe, W.C., *Text. Res. J.*, 29, 879–884, 1959.
119. Schott, H., in *Detergency Theory and Test Methods*, Part I, Cutler, W.G. and Davis, R.C., Eds., Surfactant Science Series, Vol. 5, Marcel Dekker, New York, 1972, pp. 153–235.

120. Schwartz, A.M., in *Surface and Colloid Science*, Vol, 5, Matijevic, E., Ed., Wiley-Interscience, New York, 1972, pp. 209–210.
121. Lange, H., in *Solvent Properties of Surfactant Solutions*, Shinoda, K., Ed., Marcel Dekker, New York, 1967, pp. 117–188.
122. Kissa, E., in *Detergency Theory and Technology*, Cutler, W.G. and Kissa, E., Eds., Surfactant Science Series, Vol. 20, Marcel Dekker, New York, 1987, pp. 1–89.
123. Adam, N.K., *J. Soc. Dyers Colour.*, 53, 121, 1937.
124. Kling, W., Lange, E., and Haussner, I., *Melliand Text.*, 25, 198, 1945.
125. Nagarajan, R., *Adv. Colloid Interface Sci.*, 26, 205, 1986.
126. McBain, M.E.L. and Hutchinson, E., *Solubilization and Related Phenomena*, Academic Press, New York, 1955.
127. Elworthy, P.H., Florence, A.T., and Macfarlane, C.B., *Solubilization by Surface Active Agents*, Chapman and Hall, London,1968.
128. Chan, A.F.C., Evans, D.F., and Cussler, E.L., *AIChE J.*, 22, 1006, 1976.
129. Shaeiwitz, J.A., Chan, A.F.C., Cussler, E.L., and Evans, D.F., *J. Colloid Interface Sci.*, 84, 47–56, 1981.
130. Ginn, M.E. and Harris, J.C., *J. Am. Oil Chem. Soc.*, 38, 605–609, 1961.
131. Mankowich, A.M., *J. Am. Oil Chem. Soc.*, 38, 589–594, 1961.
132. Mazer, N.A., Benedek, G.B., and Carey, M.C., *J. Phys. Chem.*, 80, 1075–1085, 1976.
133. Missel, P.J., Mazer, N.A., Benedek, B.G., Young, C.Y., and Carey, M.C., *J. Phys. Chem.*, 84, 1044–1057, 1980.
134. Phillips, J.N., *Trans. Faraday Soc.*, 51, 561–569, 1955.
135. Hayashi, S. and Ikeda, S., *J. Phys. Chem.*, 84, 744–751, 1980.
136. Tartar, H.V. and Lelong, A., *J. Phys. Chem.*, 59, 1185–1190, 1955.
137. Kuneida, H. and Shinoda, K., *J. Colloid Interface Sci.*, 70, 577–583, 1979.
138. Mino, J., in *Detergency Theory and Technology*, Cutler, W.G. and Kissa, E., Eds., Surfactant Science Series, Vol. 20, Marcel Dekker, New York, 1987, pp. 481–529.
139. Minegishi, Y., Takeuchi, T., Arai, H., and Yukagatsu, *J. Japan Oil Chem. Soc.*, 20, 60, 1971.
140. Lawrence, A.S.C., *Disc. Faraday Soc.*, 25, 51–58, 1958.
141. Lawrence, A.S.C., Bingham, A., Capper, C.B., and Hume, K., *J. Phys. Chem.*, 68, 3470–3476, 1964.
142. Lawrence, A.S.C., in *Surface Activity and Detergency*, Durham, K., Ed., Macmillan, London, 1961, chap 7.
143. Scott, B.A., *J. Appl. Chem.*, 13, 133, 1963.
144. Fort, T., Jr., Billica, H.R., and Grindstaff, T.H., *J. Am. Oil Chem. Soc.*, 45, 354–361, 1968.
145. Fort, T., Jr., Billica, H.R., and Grindstaff, T.H., *Text. Res. J.*, 36, 99, 1966.
146. Kligman, A.M. and Shelley, W.D., *J. Invest. Dermatol.*, 30, 99–125, 1958.
147. Cunliffe, W.L. and Shuster, S., *Br. J. Dermatol.*, 81, 697, 1969.
148. Eberhardt, H., *J. Soc. Cosmet. Chem.*, 27, 235–239, 1976.
149. Fort, T., Jr., Billica, H.R. and Sloan, C.K., *Text. Res. J.*, 36, 7, 1966.
150. Bouillon, C., *Clin. Dermatol.*, 6, 83–92, 1988.
151. Clarke, J., Robbins, C.R., and Schroff, B., *J. Soc. Cosmet. Chem.*, 40, 309, 1989.
152. Shaw, D.A., *Int. J. Cosmet. Sci.*, 1, 317–328, 1979.

153. Thompson, D., Lemaster, C., Allen, R., and Whittam, J., *J. Soc. Cosmet. Chem.*, 36, 271–286, 1985.
154. Clarke, J., Robbins, C.R., and Schroff, B., *J. Soc. Cosmet. Chem.*, 41, 335–345, 1990.
155. Rosen, M.J., *Surfactants and Interfacial Phenomena*, 2nd ed., Wiley-Interscience, New York, 1989, pp. 363–392.
156. Bala, F., Colgate-Palmolive Co., unpublished results.
157. Reich, C. and Robbins, C.R., *J. Soc. Cosmet. Chem.*, 44, 263–278, 1993.
158. Robbins, C.R., Reich, C., and Clarke, J., *J. Soc. Cosmet. Chem.*, 40, 205–214, 1989.
159. Goddard, E.D. and Hannah, R.B., *J. Colloid Interface Sci.*, 55, 73–79, 1976.
160. Goddard, E.D., *Cosmet. Toiletries*, 102, 71, 1987.
161. Goddard, E.D., Faucher, J.A., Scott, R.J., and Turney, M.E., *J. Soc. Cosmet. Chem.*, 26, 539–550, 1975.
162. Sendelback, G., Liefke, M., Schwan, A., and Lang, G., *Int. J. Cosmet. Sci.*, 15, 175–180, 1993.
163. Rushton, H., Gummer, C.L., and Flasch, H., *Skin Pharmacol.*, 7, 78–83, 1994.
164. Neu, G.E., *J. Soc. Cosmet. Chem.*, 11, 390–414, 1960.
165. Stiepel, C., *Seifensieder ztg.*, 41, 347, 1914.
166. Beh, H.H. and James, K.C., *Cosmet. Toiletries*, 92, 21, 1977.
167. Ross, J. and Miles, G.D., *Oil Soap*, 18, 99, 1941.
168. Moldovanyi, L., Hungerbuhler, W., and Lange, B., *Kosmetika*, 5, 37, 1975.
169. Frosch, P. and Kligman, A.M., *J. Am. Acad. Dermatol.*, 1, 35–41, 1979.
170. Cardillo, A. and Morganti, P., *J. Appl. Cosmetol.*, 12, 11–16, 1994.
171. Peck, S.H., Morse, J., Cornbleet, T., Mandel, E., and Kantor, I., *Skin*, 1, 261, 1962.
172. Justice, J.D., Travers, J.J., and Vinson, I.J., *Soap Chem. Specialties*, 37, 53–56, 1962.
173. Frosch, P.J., in *Principles of Cosmetics for the Dermatologist*, Frosch, P. and Horwitz, S.N., Eds, CV Mosby, St. Louis, 1982, pp. 5–12.
174. Strube, D.D., Koontz, S.W., Murahata, R.I., and Theiler, R.F., *J. Soc. Cosmet. Chem.*, 40, 297–306, 1989.
175. Newman, W., Cohen, G.L., and Hayes, C., *J. Soc. Cosmet. Chem.*, 24, 773–782, 1973.
176. Garcia, M.L. and Diaz, J., *J. Soc. Cosmet. Chem.*, 27, 379–398, 1976.
177. Robbins, C.R. and Reich, C., *J. Soc. Cosmet. Chem.*, 29, 783–792, 1978.
178. Jachowicz, J. and Bethiaume, M.D., *Cosmet. Toiletries*, 108, 65–72, 1993.
179. Kamath, Y.K. and Weigmann, H.D., *J. Soc. Cosmet. Chem.*, 37, 111–124, 1986.
180. Jachowicz, J. and Helioff, M., *J. Soc. Cosmet. Chem.*, 48, 93–105, 1997.
181. Reich, C. and Robbins, C.R., *J. Soc. Cosmet. Chem.*, 44, 221–234, 1993.
182. Bustard, H.K. and Smith, R.W., *Int. J. Cosmet. Sci.*, 12, 121–133, 1990.
183. Scanavez, C., Zoega, M., Barbosa, A., and Joekes, I., *J. Cosmet. Sci.*, 51, 289–302, 2000.
184. Czepluch, W., Hohm, G., and Tolkiehn, K., *J. Soc. Cosmet. Chem.*, 44, 299–318, 1993.
185. Tango, Y. and Shimmoto, K., *J. Cosmet. Sci.*, 52, 237–250, 2001.
186. Keis, K., Ramaprasad, K.R., and Kamath, Y.K., *J. Cosmet. Sci.*, 55, 49–63, 2004.

187. Ramaprasad, K.R., Note on Research No. 515, Textile Research Institute, Princeton, NJ, 1997.
188. Clarke, J., Robbins, C.R., and Reich, C., *J. Soc. Cosmet. Chem.*, 42, 341–350, 1991.
189. Hough, P.S., Huey, J.E., and Tolgyesi, W.S., *J. Soc. Cosmet. Chem.*, 27, 571–578, 1976.
190. Yin, N.E., Kissinger, R.H., Tolgyesi, W.S., and Cottington, E.M., *J. Soc. Cosmet. Chem.*, 28, 139–150, 1977.
191. Garcia, M.L. and Wolfram, L.J., 10th IFSCC Congress, Sydney, Australia, 1978.
192. Robbins, C.R. and Crawford, R.J., *J. Soc. Cosmet. Chem.*, 35, 369–377, 1984.
193. Clarke, J., Robbins, C.R., and Reich, C., *J. Soc. Cosmet. Chem.*, 42, 341–350, 1991.
194. Kamath, Y.K., Dansizer, C.J., and Weigmann, H.D., *J. Appl. Polym. Sci.*, 29, 1011–1026, 1984.
195. Rulison, C., Application Note no. 206, Kruss USA.
196. Cannel, D.W. and Carothers, L.E., *J. Soc. Cosmet. Chem.*, 29, 685–701, 1978.
197. Lunn, A.C. and Evens, R.E., *J. Soc. Cosmet. Chem.*, 28, 549–569, 1977.
198. Rajan, M. and Ellison, M.S., *Proceedings of the ESA 25th Annual Meeting*, Laplacian Press, Morgan Hill, CA, 1997, p. 62.
199. Robbins, C.R., Reich, C., and Clarke, J., *J. Soc. Cosmet. Chem.*, 37, 489–499, 1986.
200. Okumura, T., 4th International Hair Science Symposium, Syburg, Germany, Nov. 1984.
201. Kelly, S.E. and Robinson, V.N.E., *J. Soc. Cosmet. Chem.*, 33, 203–215, 1982.
202. Gould, J.G. and Sneath, R.L., *J. Soc. Cosmet. Chem.*, 36, 53–59, 1985.
203. Kaplin, I.J., Schwan, A., and Zahn, H., *Cosmet. Toiletries*, 97, 22–29, 1982.
204. Sandhu, S.S. and Robbins, C.R., *J. Soc. Cosmet. Chem.*, 44, 163–175, 1993.
205. Marshall, R.C. and Ley, K.F., *Text. Res. J.*, 56, 772–774, 1986.
206. Hilterhaus-Bong, S. and Zahn, H., *Int. J. Cosmet. Sci.*, 11, 167–174, 1989.
207. Smith, W.P., *Cosmet. Toiletries*, 108, 67–75, 1993.
208. Bergfeld, W.F., in *Hair Research*, Orfanos, C.E., Montagna, W., and Stuttgen, G., Eds., Springer-Verlag, Heidelberg, Berlin, 1981, pp. 507–512.
209. Global Report: Naturals, GCI, April 2004.
210. U.S. Patent 4,839,168 to Kao, June 13, 1989.
211. U.S. Patent 4,374,125 to Helene Curtis, February 15, 1983.
212. U.S. Patent 5,409,628 to Goldwell AG, April 25, 1995.
213. U.S. Patent 6,649,177 to Ingredients MGP, Inc., November 18, 2003.
214. U.S. Patent 6,063,369 to Alterna, Inc., May 16, 2000.
215. U.S. Patent 5,925,615 to Nu Skin, Inc., June 20, 2000.
216. U.S. Patent 6,068,848 to Color Access, Inc., May 30, 1999.
217. U.S. Patent 4,511,555, April 16, 1985.
218. U.S. Patent 6,596,266 to National Science, July 22, 2003.
219. U.S. Patent 6,333,057, December 25, 2001.
220. U.S. Patent 5,972,345, October 26, 1999.
221. U.S. Patent 5,217,711, June 8, 1993.
222. U.S. Patent 6,171,604 to MA Mousa, January 9, 2001.
223. Hui-Xian, C.S., Peixuan, P., and Ling, L.H., Singapore Science and Engineering Fair (SSEF), 2004.
224. U.S. Patent 6,719,961 to Cognis, April 13, 2004.

225. U.S. Patent 6,599,328 to L'Oreal, July 29, 2004.
226. U.S. Patent 6,562,328 to Croda, Inc., May 13, 2003.
227. Demographic Profile: The Baby Boomers in 2003, Mature Market Institute, Metlife.
228. U.S. Patent 6,653,317 to L'Oreal, November 25, 2003.
229. U.S. Patent 6,645,477 to L'Oreal, November 11, 2003.
230. U.S. Patent 6,541,507 to L'Oreal, April 1, 2003.
231. U.S. Patent 6,291,532 to L'Oreal, September 18, 2001.
232. U.S. Patent 6,667,028 to Wella, December 23, 2003.
233. U.S. Patent 6,307,049 to Procter & Gamble, October 23, 2001.
234. U.S. Patent 6,124,362 to Procter & Gamble, September 26, 2000.
235. U.S. Patent 6,696,417 to Chesebrough-Pond's, February 24, 2004.
236. U.S. Patent 5,422,031 to Kao, June 6, 1995.
237. U.S. Patent 5,270,035 to Lexin International Marketing Corp., December 14, 1993.
238. U.S. Patent 6,689,347 to Unilever, February 10, 2004.
239. U.S. Patent 6,432,393 to Helene Curtis, August 13, 2002.
240. U.S. Patent 6,348,439 to Bristol-Myers Squibb, November 21, 2002.
241. U.S. Patent 6,294,159 to Colgate-Palmolive, September 25, 2001.
242. U.S. Patent 5,045,307 to Colgate-Palmolive, September 3, 1991.
243. U.S. Patent 6,399,045 to Unilever, June 4, 2002.
244. U.S. Patent 6,362,146 to Unilever, March 26, 2002.
245. U.S. Patent 5,605,704 to Chesebrough-Pond's, February 25, 1997.
246. U.S. Patent 5,186,928 to Chesebrough-Pond's, February 16, 1993.
247. U.S. Patent 6,723,310, April 20, 2004.
248. U.S. Patent 6,709,648 to Procter & Gamble, March 23, 2004.
249. U.S. Patent 4,676,915 to Colgate-Palmolive, June 30, 1987.
250. U.S. Patent 6,524,563 to Unilever, February 25, 2003.
251. U.S. Patent 6,297,203 to Procter & Gamble, October 2, 2001.
252. U.S. Patent 6,268,431 to Procter & Gamble, July 31, 2001.
253. U.S. Patent 6,248,317 to Procter & Gamble, June 19, 2001.
254. U.S. Patent 6,218,346 to Stepan, April 17, 2001.
255. U.S. Patent 6,177,390 to Procter & Gamble, January 23, 2001.
256. U.S. Patent 6,113,890 to Procter & Gamble, September 5, 2000.
257. U.S. Patent 5,883,058 to Procter & Gamble, March 16, 1999.
258. U.S. Patent 5,391,368 to Revlon, February 21, 1995.
259. U.S. Patent 5,120,532 to Procter & Gamble, June 9, 1992.
260. U.S. Patent 5,776,444 to Chesebrough-Pond's, July 7, 1998.
261. U.S. Patent 6,491,902 to Salvona LLC, December 10, 2002.
262. U.S. Patent 6,689,371 to L'Oreal, February 10, 2004.
263. U.S. Patent 6,562,356 to L'Oreal, May 13, 2003.
264. U.S. Patent 6,541,018 to L'Oreal, April 1, 2003.
265. U.S. Patent 6,464,990 to L'Oreal, October 15, 2002.
266. U.S. Patent 6,375,960 to L'Oreal, April 23, 2002.
267. U.S. Patent 6,274,150 to L'Oreal, August 14, 2001.
268. U.S. Patent 6,120,775 to L'Oreal, September 19, 2000.
269. U.S. Patent 6,565,886 to L'Oreal, May 20, 2003.
270. U.S. Patent 6,379,683 to L'Oreal, April 30, 2002.

271. U.S. Patent 6,066,328 to L'Oreal, May 23, 2000.
272. U.S. Patent 5,919,487 to L'Oreal, July 6, 1999.
273. U.S. Patent 6,375,936 to L'Oreal, April 23, 2002.
274. U.S. Patent 6,261,542 to L'Oreal, July 17, 2001.
275. U.S. Patent 6,174,517 to L'Oreal, January 16, 2001.
276. U.S. Patent 6,060,041 to L'Oreal, May 9, 2000.
277. U.S. Patent 5,939,053 to L'Oreal, August 17, 1999.
278. U.S. Patent 5,553,630 to L'Oreal, September 10, 1996.
279. U.S. Patent 6,692,730 to L'Oreal, February 17, 2004.
280. U.S. Patent 6,617,292 to L'Oreal, September 9, 2003.
281. U.S. Patent 6,638,918 to Procter & Gamble, October 28, 2003.

# 11

# Liquid Hand Soap and Body Wash

**SUBHASH HARMALKER and KUO-YANN LAI** Global Technology, Colgate-Palmolive Company, Piscataway, New Jersey

## I. INTRODUCTION

Liquid soap was first developed in the 1940s and was primarily used in hospitals and institutions for washing hands. In the late 1970s with the launch of Softsoap brand liquid soap in the U.S. by Minnetonka, Inc., liquid soap has gained increasing popularity. Liquid soaps offer convenience and are considered to be more hygienic than bar soap especially in public places.

Body wash was first introduced in the U.S. in 1993 by Jergens followed by Dove (Unilever), Caress (Unilever), and Oil of Olay (Procter & Gamble) in 1994.

Sales of bar soaps in the U.S. and Europe have been declining while sales of liquid soap, shower gel, and liquid body wash have been increasing for the last few years [1].

In the last decade the functions of body washes have been extended from basic cleansing to include antibacterial action, skin moisturization, deodorant properties, exfoliation, aromatherapy, delivery of emollients to soften the skin, delivery of nourishing vitamins and other ingredients, and delivery of film-formers such as chitosan to create a protective barrier between the skin and the environment. There have even been attempts to make a body wash that provides UV protection.

Figure 11.1 and Figure 11.2 show some representative commercial liquid hand soap and liquid body wash/shower gel products from around the world.

This chapter attempts to give a thorough review of all aspects of liquid soaps and body washes, from typical composition and ingredients to the test methods and performance evaluations, formulation technology, and new products and future trends. The chapter that appeared in the first edition [2] has been rewritten, updated, and expanded to reflect the significant evolution and advances in these products.

**FIG. 11.1**   Commercial liquid hand soap products from around the world.

**FIG. 11.2**  Commercial shower gel/liquid body wash products from around the world.

## II.  TYPICAL COMPOSITION AND INGREDIENTS
### A.  Typical Composition

Liquid hand soap and body wash formulations are similar in composition. The essential ingredients are skin cleaning agents, skin conditioning agents, rheology modifiers, color, fragrance, preservatives, and other additives such as antibacterial agents, vitamins, and herbal extracts. A list of some of these ingredients and the suppliers can be found in the book by Flick [3]. The physicochemical properties, chemical structures, applications, and safety of these ingredients are well summarized by Hunting [4] and by Barel *et al.* [5]. A typical composition is shown in Table 11.1. Reever [2] provides a detailed review and summary of the compositions of some commercial liquid hand soaps.

### B.  Ingredients
#### 1.  Skin Cleaning Agents

Surfactants or mixtures of surfactants are the main ingredients for skin cleaning. There are two types of surfactants used for the formulation, soap-based surfactants and synthetic surfactants. The soap-based formulations provide voluminous,

**TABLE 11.1**  Typical Liquid Hand Soap and Body Wash Composition

| Ingredient | Amount (wt%) | Purpose |
|---|---|---|
| Surfactants | 10–40 | Cleaning, foaming |
| Emollients | 1–30 | Moisturizing, skin conditioning |
| Rheology modifiers | 1–5 | Viscosity control |
| Preservatives | <1.0 | Microbial stability |
| Fragrances | 0.3–1.5 | Aesthetics |
| Coloring agents | <0.1 | Aesthetics |
| Other additives | 0–3 | Antibacterial, exfoliating, antiaging, whitening |
| Water | Balance | Solubilizer/carrier vehicle |

creamy lather and a "squeaky clean" feel to the skin. However, they lead to drying and irritation of the skin in cold weather. Therefore, these surfactants are used mostly in products that are sold in tropical regions where the climate is hot and humid. The commonly used soap surfactants are based on potassium soaps of lauric, myristic, and palmitic acids. Generally, these surfactants are used in combination with each other to provide good lather and skin feel. Due to alkalinity of the soaps, the pH of these formulations is very high (9 to 10) and hence is not very mild to skin. Furthermore, this high pH limits the choice of skin care agents and fragrances that can be incorporated in these products due to hydrolytic instability.

Synthetic surfactants are used almost exclusively in the liquid products that are sold in Europe and North America. These surfactants are much milder to skin than the soap-based surfactants and can be formulated in products with skin neutral pH (5.5). Furthermore, the products formulated with these surfactants are much more compatible with a wide variety of skin care agents and fragrances.

Of the four types of synthetic surfactants, anionic, cationic, amphoteric, and nonionic, the anionic surfactants provide maximum lather and hence are used as major components in liquid products. The active ingredients used in the major brands of liquid soaps are described by Dyer and Hassapis [6].

A widely available anionic surfactant is sodium lauryl sulfate. This surfactant is fairly irritating to skin. Ethoxylation of this surfactant lowers its irritation potential but it also lowers its lather. The optimal mildness and lather is obtained with a degree of ethoxylation of 2 or 3. This surfactant is widely used with 2 moles of ethoxylation (laureth 2-sulfate) and is available as 28% active or 70% active. In some cases it is also used with 3 moles of ethoxylation (laureth 3-sulfate). The loss of lather due to ethoxylation is generally compensated for by using foam boosters such as lauramide diethanolamide, cocamide diethanolamide, and amine oxides. Alfa olefin sulfonate is another cost-effective surfactant emerging as a popular surfactant for liquid soap formulations [7].

**TABLE 11.2** Surfactants Commonly Used in Liquid Hand Soap and Body Wash for Skin Cleansing and Foaming

| Type | Surfactant |
|---|---|
| Anionic | Ammonium lauryl sulfate (ALS) |
| | Ammonium laureth sulfate |
| | Sodium lauryl sulfate (SLS) |
| | Sodium lauryl ether sulfate (SLES) |
| | Alpha olefin sulfonate (AOS) |
| | Sodium cocoyl isethionate |
| | Sodium isethionate |
| | Sodium alkylbenzene sulfonate |
| | Sodium lauryl sacorsinate |
| | Sodium lauryl lactate |
| | Sodium lauroamphoacetate |
| Nonionic | Alkylpolyglucoside |
| Amphoteric | Cocoamidopropyl betaine |

Table 11.2 lists the surfactants commonly used in liquid hand soap and body wash.

A typical liquid soap and body wash product is generally comprised of a mixture of these different types of surfactants to achieve the desired cleaning and foaming performance.

## 2. Skin Conditioning Agents

Skin feel is an important attribute to the users of liquid hand soap and body wash products aside from cleaning. Generally, it is desirable to have a feeling of smoothness but not sticky or feeling of residual materials on the skin after washing and rinsing of hands and body. Many different ingredients have been used to impart a good skin feel and often with moisturizing benefits. These include humectants, such as glycerin and protein, and skin refatting agents, such as PEG-7 glyceryl cocoate. Another widely used class of material is water-soluble cationic polymers such Polyquarternium-7, Polyquarternium-10, and guar hydroxypropyltrimonium chloride. One key benefit of this class of material appears to be the high degree of skin feel provided at a very low formula concentration used [8].

Table 11.3 lists a number of skin conditioning agents used in some major brands of liquid hand soaps or shower gels.

## 3. Rheology Modifiers

Liquid hand soap and body wash products are typically formulated in a thick liquid or gel form. Consumers look for the convenience in dispensing these products but they should not be "runny" or slip through the fingers. The viscosity of liquid hand

**TABLE 11.3**  Examples of Skin Conditioning Agents Used in Major Brands of Shower Gel

| Skin conditioning agent | Shower gel brand |
|---|---|
| Soybean oil | Olay |
| Sunflower seed oil | Dove, Sauve |
| Polyquaternium-7 | Palmolive, Softsoap, Fa |
| Polyquaternium-10 | Caress, Sauve, Dial, Olay |
| Glycerin | Dove, Caress, Sauve, Olay, Dial |
| Maleated soybean oil | Olay |
| Petrolatum | Dove, Olay |
| Hydrolyzed wheat protein | Caress |
| Seaweed extract, ceramides | Dove |
| Polyethylene (exfoliant) | Dove, Palmolive, Softsoap |
| Vitamin E acetate | Dove, Olay, Dial, Softsoap |
| PEG-6 caprylic/capric glycerides | Sauve, Olay |
| PEG-7 glyceryl cocoate | Dial, Softsoap, Nivea, Fa |
| Castor seed oil | Caress |
| Hydrogenated coco-glycerides | Dove |
| Retinyl palmitate | Olay |
| Niacinamide | Olay |
| PEG-200 hydrogenated glyceryl palmate | Nivea |
| Glyceryl oleate | Fa |
| Caffeine | Fa |

soap products is generally in the range 3,000 to 5,000 cP while the viscosity of shower gel products is typically in the range 5,000 to 20,000 cP. To achieve good physical and flow properties, rheology modifiers are typically used. These can be simple salts such as sodium chloride, potassium chloride, and ammonium sulfate or water-soluble polymers (see Chapter 5).

## 4.  Aesthetic Modifiers

The color and the overall appearance of the product are very important attributes in getting consumers interested in the product and also conveying the functionality of the product to some degree. For example, gold or amber color is associated with antibacterial benefit, a clear, colorless, or pearly product tends to signal mildness or ultra mildness, and a milky product signals moisturization. Therefore, the selection of product color will often depend on the positioning of the product and is typically based on consumer research.

The fragrance of a product is probably even more critical and often the deciding factor for consumers to purchase and use the product. How the product smells in the bottle, during use, and after use is very important. The fragrance should match the product concept or positioning and appeal to users. Some fragrances are capable of counteracting kitchen malodors [9,10]. More recently, fragrances have been used in products to provide experiential benefits such as aromatherapy to consumers (see Section V).

## 5. Preservatives

Typical liquid hand soap and body wash products contain significant amounts of water (60 to 85%) and many ingredients are also sensitive to microbial attack. To ensure the integrity of the product against microbiological contamination during manufacture and extended usage time, it is necessary to add antimicrobial preservatives. The level of these preservatives that is generally required is at a fraction of a percent. These preservatives are generally miscible in the surfactants that are used in the formulations. Microbiological preservation testing is done by injecting the formulation with certain microorganism commonly known to cause contamination and incubating them to make sure that no organism survives in the presence of these preservatives [11]. Preservatives commonly used include DMDM hydantoin, tetrasodium EDTA (preservation efficacy booster), sodium benzoate, sodium salicylate, methylchloroisothiazolinone, methylisothiozolinone, benzyl salicylate, butylphenyl methylpropional, hydroxylsohexyl 3-cyclohexene carboxyaldehyde, methylparaben, and propylparaben.

## 6. Other Additives

In addition to the basic ingredients for cleaning and skin conditioning, many other additives are added to impart other benefits such as antibacterial, exfoliating, anti-aging, and whitening (Section V). Examples of these additives are triclosan, sugar, salts, hydrated silica, nut shells, dried fruit particles, herbal extracts, vitamins, etc.

## III. TEST METHODS AND PERFORMANCE EVALUATION

There are a number of laboratory tests used by formulators to evaluate the various aspects of liquid hand soap or shower gel products. These include the evaluation of physical properties and various performance attributes. To validate the design of the product, consumer tests are usually necessary. Barel *et al.* [5] present an extensive and detailed discussion of various test methods and performance evaluations for cosmetic products. Most of these test methods also apply to liquid hand soap and shower gel.

The following provides a brief summary of some of the major test methods.

## A. Physical Properties

The key physical properties for liquid hand soap and shower gel products that need to be defined and characterized are rheology, lather, pH, and color.

## 1.  Rheology

Liquid hand soaps and shower gels are formulated so that they are thick in the bottle, become thinner during dispensing (for ease of dispensing), thicken back after dispensing (so that they are not runny during use), and can be spread easily during application. The rheological term for this property is "thixotropy" (see Chapter 4), i.e., a decrease of viscosity under shear stresses or shear rate, followed by gradual recovery when the shear stress or shear rate is removed. These properties are imparted to the formulations by components such as surfactants and rheology additives (see Chapter 5). The clear and opaque gels discussed in the preceding section fall under this category.

The viscosity is typically measured by a viscometer. The SI unit of viscosity is the pascal second (Pa s). However, the widely used unit is poise (P) or centipoise (cP) (1 cP = 1 mPa s). The routine laboratory testing of viscosity is typically done using Brookfield viscometers and reported in cP. For non-Newtonian fluids such as those discussed above, the viscosity is dependent on the spindle size and type (number) and speed of rotation (rotation per minute or r/min) of the viscometer.

Structured gel and emulsion formulations, designed to suspend particles or oils, are generally viscoelastic. They have both viscous and elastic properties. Such formulations are characterized by their elastic modulus ($G'$) and loss modulus ($G''$). The elastic modulus (elastic component) is a measure of energy storage and the loss modulus (viscous component) is a measure of energy dissipation. For viscoelastic fluids $G' > G''$ and for viscous fluids $G' < G''$. For a suspension or emulsion to be stable $G'$ should be greater than $G''$ over the range of temperature required for stability.

The viscoelastic parameters are generally measured by dynamic oscillatory measurements. Apparatus of three different configurations can be used: cone and plate, parallel plates, or concentric cylinders. In the case of cone and plate geometry, the test material is contained between a cone and a plate with the angle between cone and plate being small ($<4°$). The bottom member undergoes forced harmonic oscillations about its axis and this motion is transmitted through the test material to the top member, the motion of which is constrained by a torsion bar. The relevant measurements are the amplitude ratio of the motions of the two members and the associated phase lag. From this information it is relatively simple to determine $G'$ and $G''$.

There are a number of other rheological methods used for characterizing viscoelastic fluids. A detailed discussion can be found in Chapter 4 or a rheology text such as Barnes *et al*. [12].

## 2.  Lather

One of the most important attributes of shower gel formulations is lather. Shower gels are applied to the body either by hand or by body sponge (pouf) or wash cloth. In all cases the lather should be able to generate quickly and in sufficient amount, be creamy and dense so as to provide sensorial feel during rubbing, and be stable until it is rinsed off. Generally, a shower gel that provides creamy and voluminous lather is perceived to be efficacious by consumers. In the case of liquid hand soap, more easily rinsable lather is desirable.

In the laboratory, lather properties of liquid hand soap and shower gel are measured as flash foam (speed of foam generation), maximum foam (quantity and stability), and foam creaminess (drainage time).

The standard (ASTM) methods for measuring foams fall under two categories: static (pouring, shaking, beating, and stirring) and dynamic (air injection). These methods are described by Tomura and Masuda [13]. However, due to the complexity of foams, new methods are being developed to provide evaluation of foam characteristics that is more representative of foams generated during consumer usage.

A commonly used method for evaluation of foam in shower gels and liquid hand soap is a static method, "cylinder shake." In this method a certain amount of water with predetermined levels of water hardness, sebum content, and temperature is placed in a graduated cylinder. An appropriate amount of product is then added to this water so that the foam generated during the shaking will be contained in the cylinder. Generally the amount of water is between 100 and 200 ml and the amount of product is 10 to 20 g. The cylinder is then rotated in a vertical plane, 180° up and down for a certain period of time. The volume of foam is then measured. This is *flash foam*. The cylinder is then shaken for another period of time and the foam volume is measured again. This is *maximum foam*. The foam is then allowed to drain to the original level and the time taken to do so is noted. This is the *drainage time*.

## 3. pH

The pH of shower gel formulations is adjusted around skin pH 5.5, and the pH of liquid hand soap formulations is generally adjusted from 5.5 to 7. (The pH of hand soap formulations is sometimes higher than skin pH because of technical requirements for viscosity and the type of surfactant used.) A change in pH over time is often indicative of some sort of chemical interaction between the ingredients and hence is monitored during testing the product for stability to make sure that the pH stays in the specified range. The pH changes are generally controlled by using buffering agents such as citric acid/sodium citrate. The pH is measured using conventional pH meters.

## 4. Color

The dyes used in shower gel and liquid hand soap formulations are required by regulations in U.S. and Europe to be FD&C (food, drug, and cosmetic) or D&C (drug and cosmetic). Generally, two or three dyes are blended together to obtain a desired color. Initial color is then measured using conventional instruments such as Macbeth colorimeters and the results are reported in CIELAB color space [14]. The color is then monitored during stability testing to ascertain that change ($\Delta E$) in the chromaticity (intensity) and hue (shade) stays within the specified range. Although $\Delta E$ is still widely used, it is being replaced by more efficient parameters such as $\Delta E_{cmc}$ [15].

## B. Mildness

One of the key attributes of skin cleansing products such as shower gels and liquid hand soaps is that they are nonirritating and mild to skin. Testing for this starts from the early stages of formulations. The ingredients or combination of ingredients in the formula are evaluated for mildness and safety based on the information from the literature and available *in vitro* data. Some of the commonly used surfactants such as sodium laureth sulfate and cocoamidopropyl betaine show synergistic behavior toward mildness.

As a next step, the formulation is tested using *in vitro* skin irritation tests such as the Zion test [16], collagen swelling test [17], or pH rise test [18]. If the results of these tests show that the product is suitable for human clinical research, then it undergoes the following tests.

*Exaggerated arm wash test* [19]. This method is useful in evaluating relative irritation potential or mildness of personal cleansing products and is generally used for formulations that have undergone minor modifications. The protocol is based on consumer washing habits. In a standard arm wash test only two products are compared. However, in a split arm design of this test, four products can be tested simultaneously. The test involves 12 to 16 panelists. Two sites on the forearm are treated with the experimental formula, the third site is treated with water only, and the fourth site is the untreated control. The sites are washed twice a day for 6 days and evaluated for erythema and dryness by visual assessment and by bioengineering techniques such as TEWL (transepidermal water loss), corneometry (instrumental measurement of redness), and skin capacitance (skin surface hydration and dehydration). Generally, the instruments used for this purpose are Skicon 200, Corneometer CM820, and Nova DPM 9003 Dermaphase Meter, respectively.

*Modified soap chamber test* [20]. This is a more elaborate test and is generally used in the case of a major modification of the formulation or use of new ingredients. It involves 25 to 30 panelists. In this test the solutions of the products are applied to panelists under occlusion for two 24-hour periods and visual assessment of erythema and dryness is done by the same techniques as for the exaggerated arm wash test. The instrumental measurement of redness is done after 3 hours of each application and the TEWL measurement is done after 3 hours of first application. The measurement of skin capacitance is done 3 to 5 days after the second application [21].

*Cumulative irritation test.* This is another test that is commonly used to evaluate the absence of skin reactions to the formulation. It involves a minimum of 25 panelists. The solutions of test products as well as reference products with well-known skin tolerance are applied to the backs of panelists under occlusion for several consecutive days (generally 5 or 21 days) and the absence of skin reaction is then evaluated by trained professionals [22].

## C. Moisturization

Skin moisturization or hydration is measured by the water holding capacities of the stratum corneum and one of the techniques used to measure it is the water sorption–desorption test [23]. Instruments such as Skicon, Corneometer, and Dermaphase Meter are used in this case. The base line electrical measurement is taken with these instruments before the skin is hydrated (PHS, prehydration state). The skin surface is then hydrated with distilled water/sample for less than a minute, blotted dry, and electrical capacitance is recorded again which represents the hygroscopicity or the ability of the stratum corneum to take up water (sorption). Measurements are then taken at intervals ranging from 20 to 30 seconds for a period of 2 minutes as the electrical conductance rapidly falls. Water holding capacity or moisturization is then determined by plotting percentage sorption vs. time and calculating the area under the curve. The water holding capacity of moisturizers such as lotions is found to be higher than the control (cream O/W, 627DPM.min vs. water 273 DPM.min). Generally, water sorption is similar in both cases but the desorption is slower in the case of moisturizers.

## D. Antibacterial

Antibacterial agents are used in many liquid hand soap products but are not so common in shower gels. The commonly used antibacterial agents in liquid hand soap formulations are Triclosan, TCC (Trichlorcarban), and PCMX (*para*-chloro-*meta*-xylenol). These agents are incorporated in the formulation generally at a very low concentration (fraction of a percent). Triclosan provides broad-spectrum antibacterial activity whereas the other two are limited in their scope, generally effective in killing gram-positive bacteria. The methods outlined below are used to evaluate the antimicrobial activity of formulations. Further details of these methods can be found in Barel *et al.* [24].

*Minimal inhibitory concentration (MIC) test.* In this test, various concentrations of test product in a growth medium are inoculated with the test strain. After incubation, the lowest concentration that does not exhibit bacterial growth gives the MIC level.

*Zone of Inhibition test.* Antibacterial products are applied to a substrate, generally an agar plate, previously seeded with the test bacteria. During the incubation, the test product diffuses into the agar layer and creates a zone that inhibits microbial growth. The larger the inhibition zone, the higher the antibacterial efficacy.

*Time kill test.* In this test, a diluted antibacterial formulation is inoculated with specified bacteria and kept in contact for a certain period of time. The antibacterial agent is then inactivated by dilution into a neutralizing broth and the reduction in bacteria is counted on solid culture media. A 90% reduction in bacteria is reported as 1 log reduction, 99% reduction in bacteria is reported as 2 log reduction, and 99.9% reduction is reported as 3 log reduction.

## E.  Consumer Test

A consumer validation of the various benefits of shower gel and liquid hand soap formulations, evaluated in the various tests mentioned above, is critical to the success of the product in the marketplace. The consumer test not only provides this validation, but it also provides much more valuable information, such as consumer likes and dislikes about product attributes and aesthetics and purchase intent.

Consumer tests involve a large number of panelists, generally over 100. The panelists are selected from a target group to which the product is intended to be sold. The product is then tested with these panelists in a home use test. If a single product is to be tested, it is generally done in a monadic test. If a test product is to be compared to a reference product, the testing is done by either comparing monadic tests of the product and its reference, or sequential monadic test, i.e., one product is used for a period of time, followed by the other product. A number of other designs are also used depending on the desired information. At the completion of the test, the panelists are debriefed via a written questionnaire, interviews, or focus groups. If the results of this test show a strong consumer appeal and strong purchase intent, then the decision is made to commercialize the product. Also, as discussed earlier, the results of such tests provide a detailed evaluation of various product attributes and are used as a diagnostic tool to improve the weaknesses.

## IV.  FORMULATION TECHNOLOGY
## A.  Formulation Considerations

The basic requirements for a liquid hand soap or body wash/shower gel product are listed in Table 11.4. The formulation of these kinds of products starts with the consideration to satisfy these basic requirements.

In recent years more and more new benefits have been added to these products (see Section V) to appeal to consumers. To deliver these added special benefits, additional considerations will include:

1.  Capability to emulsify skin conditioning agents which are generally hydrophobic in nature.

**TABLE 11.4**   Basic Requirements for a Liquid Hand Soap or Body Wash/Shower Gel

Effective cleansing
Sufficient lather
Mild to skin
Appealing aesthetics, such as color and fragrance
Skin care benefits, such as moisturization
Preservation — stable against contamination by microorganisms
Adequate pH for formulation stability and skin mildness
Adequate viscosity for ease of dispensing and use
Product stability — stable under extreme environmental conditions such as heat and cold
At least two years of shelf life
Comply with government regulations if the product is regulated, such as antibacterial
   liquid hand soap
Competitive cost

2. Capability to suspend particulate matter for visual aesthetics or to deliver
   benefits such as exfoliation.
3. Capability to suspend particles encapsulating active ingredients to deliver
   various skin benefits.
4. Capability to deliver actives onto skin to provide benefits that are consumer
   perceptible and able to be clinically documented.

The last requirement is very challenging for rinse-off products such as liquid hand
soaps and shower gels.

The ingredients used to satisfy these requirements are discussed in Section II.
A good product is a result of the careful selection and balance of these ingredients.

## B.  Formulation

An extensive list of shower gel and liquid hand soap formulations can be found in
Flick [3].

The liquid hand soap and body wash/shower gel products sold by major
manufacturers can be broadly classified into four categories:

1. Clear gels. Typically for experiential products containing variety of colors and
   fragrances. These gels are of the lowest cost but do not have the capability to
   suspend particles.
2. Opaque/pearlized gels. Typically for delivering moisturizing aesthetics, with
   no particle-suspending properties.
3. Structured gels, clear or opaque. Typically with suspended particles for
   aesthetic and functional benefits.
4. Emulsions. Usually creamy, with skin conditioning oils, occlusive agents, or
   particulates emulsified or suspended.

## 1.  Clear and Opaque/Pearlized Gels

The main distinction between clear and opaque/pearlized gels lies in the presence of an opacifier or pearling agent in the latter. These liquid hand soap or shower gel formulations generally contain about 50 to 80% water. The surfactants are generally miscible with water and are incorporated in these formulations at a level of 5 to 15% by simply mixing with water at room or slightly elevated temperature. Other additives such as color, fragrance, extracts, skin conditioning agents, opacifying agents, preservatives, and pearlizing agents are then added to this formulation. Antibacterial agents such as triclosan are often added to the formulation via dissolution in fragrance. The pH is then adjusted to the desired value, generally between 5 and 7, except for soap-based formulations, in which case the pH is on the higher side (9 to 10). Finally the viscosity of the formulations is built either by addition of an appropriate salt or various thickening polymers. Sodium chloride, being very inexpensive, is widely used. The order of addition of the above ingredients is sometimes critical to obtain optimal mixing and to prevent undesired interaction. The typical viscosity range for such products is 5,000 to 15,000 cP.

Typical compositions of a clear gel and an opaque or pearlized gel prepared using these methods are summarized in Table 11.5 and Table 11.6.

## 2.  Structured Gels (Clear and Opaque)

The clear and opaque formulations discussed above do not have the capability to suspend particles. Although it may seem that a "thick" formula should suspend particles, high viscosity is not a sufficient condition for suspension. In order to suspend particles, a structured gel is created by incorporating structuring agents

**TABLE 11.5**  Typical Clear Gel Composition
Ingredients are generally listed as per regulations in descending order of predominance

| Ingredient | Function |
| --- | --- |
| Water | Vehicle |
| Sodium C12–C13 pareth sulfate, cocamidopropyl betaine, lauryl polyglucoside | Cleaning, foaming |
| Perfume | Fragrance |
| Sodium chloride | Viscosity builder |
| DMDM hydantoin | Preservation |
| Tetrasodium EDTA | Preservation efficacy booster |
| Polyquaternium-7 | Skin conditioner |
| Benzophenone-4 | UV absorber/color protector |
| Canaga odoranta, pogostemon cabin, lavandla angustifolia | Essential oils |
| CI 17200, CI 60730 (Dyes) | Coloring |

**TABLE 11.6**  Typical Opaque or Pearlized Gel Composition
Ingredients are generally listed as per regulations in descending order of predominance

| Ingredient | Function |
|---|---|
| Water | Vehicle |
| Sodium laureth sulfate, cocamidopropyl betaine | Cleaning, foaming |
| PEG-7 glyceryl cocoate | Skin conditioning agent |
| Disodium cocoyl, glutamate | Skin conditioning agent |
| Citric acid | pH adjuster |
| PEG-40 hydrogenated castor oil | Fragrance solubilizer |
| Sodium chloride | Rheology modifier |
| Sodium benzoate, sodium salicylate | Preservation |
| Styrene/acrylate copolymers | Opacifiers |
| Polyquaternium-10 | Skin conditioning agent |
| PEG-90 glyceryl isostearate | Skin conditioning agent |
| PEG-200 hydrogenated glyceryl palmate | Skin conditioning agent |
| Laureth-2 | Skin conditioning agent |
| Perfume | Fragrance |

such as xanthan gum, guar gum, or acrylate copolymers. These polymers provide a gel network with significant strength to have an adequate yield point to suspend the particles. The yield point is the minimum force required to make the gel flow. The polymers that only thicken but do not suspend particles do not have such a yield point. Rheologically, the structuring polymers are characterized by $G' > G''$ as described in Section III.A.1. These ingredients are generally solids, although acrylate copolymers are now available as dispersions. These ingredients are carefully dispersed at a level of 1 to 3% (solids) in water to obtain a homogenous gel. Surfactants are then added, followed by other ingredients as described above.

Acrylate copolymers are supplied in the acidic form and have to be neutralized in order to form the gel. Until neutralized, the formulation stays thin, thus facilitating the incorporation of minor additives. The neutralization is generally done after most of the ingredients are added to the formulation. These gels are fairly viscous and possess high yield point. The latter is critical to suspend particles.

The above mentioned structuring agents are anionic in nature and hence interact with cationic surfactants and are precipitated out. Therefore cationic surfactants should be avoided if high-clarity gels are desired.

The manufacturing complexity and cost of these formulations are generally greater than the clear and opaque gels and therefore they are generally used in applications where particle suspension is required. A typical structured gel formulation is shown in Table 11.7.

**TABLE 11.7**  Typical Structured Gel Composition

| Ingredient | Function |
|---|---|
| Water | Vehicle |
| Ammonium lauryl sulfate, ammonium laureth sulfate, cocamidopropyl betaine | Cleaning, foaming |
| Propylene glycol | Viscosity modifier |
| Acrylate copolymers | Structuring agents |
| Perfume | Fragrance |
| Glycerin, Polyquaternium-10 | Skin conditioning agents |
| Cocamide monoethanolamide | Foam stabilizer |
| Methylcellulose | Rheology modifier |
| Benzophenone-4 | UV absorber |
| Tetrasodium EDTA | Water softener, preservative |
| Carbomer | Structuring agent |
| Methylchloroisothiazolinone, methylisothizolinone, etidronic acid, guanine | Preservation |
| CI 75710, mica (CI 77019), red 33 (CI 17200), titanium dioxide (CI 77891) | Colorants, pearlizers |

## 3.  Emulsions

These formulations deliver very high levels of skin conditioning oils and hence are the most complex and expensive of all the formulations discussed above. Some of the major brands containing these types of formulations are Dove and Olay, with high levels of sunflower seed oil and soybean oil. Petrolatum is another water-insoluble ingredient used in this kind of formulation.

In order to suspend high levels of oils and hydrocarbons such as petrolatum in these formulations, a combination of emulsifiers and sometimes structuring agents are used. The oils are dispersed in the formulation in the presence of emulsifiers and then homogenized to break them down into fine particles. A proper balance of emulsifiers (determined by HLB, the hydrophile/lipophile balance) is required in order to obtain a stable emulsion. These emulsions are tested for stability at various temperatures for an extended period of time. Accelerated stability testing can be done using techniques such as centrifugation, coupled with measurement of particle size distribution as a function of time. It is a general principle of emulsions that those with smaller droplets are generally more stable, and agglomeration of the droplets into larger particles is a step toward destabilization of the system.

The theoretical background behind stabilizing emulsions has been discussed in the literature very extensively. Two such sources are Tadros [25] and Becher [26].

Table 11.8 shows an example of a typical emulsion formulation.

**TABLE 11.8** Typical Emulsion Composition

| Ingredient | Function |
|---|---|
| Water | Vehicle |
| Sodium laureth sulfate | Cleaning, foaming |
| Glycine soja (soybean oil) | Skin conditioning oil |
| Sodium lauroamphoacetate | Cleaning, emulsifying agent, foam booster |
| Glycerin | Humectant |
| Cocamide monoethanolamide | Foaming agent |
| Palm kernel acid | Emulsifying agent |
| PEG-6 caprylic/capric glycerides | Emulsifying agent |
| Citric acid | pH adjuster |
| Magnesium sulfate | Rheology modifier |
| Perfume | Fragrance |
| Maleated soybean oil | Skin conditioning agent |
| Tocopheryl acetate, niacinamide, retinyl palmitate | Vitamins |
| Polyquaternium-10 | Moisturizing agent |
| Sodium benzoate, DMDM hydantoin, disodium EDTA, benzyl salicylate, butylpheryl methylpropional, hydroxylsohexyl 3-cyclohexene carboxaldehyde | Preservation |

## V. NEW PRODUCTS AND FUTURE TRENDS

For liquid hand soap, new products introduced to the market in the last decade continue to focus on superior cleaning plus antibacterial and skin moisturization benefits. Some new benefits introduced to liquid body wash products, as discussed below, have also been extended to hand soaps such as Softsoap's Aromatherapy Hand Soap by Colgate-Palmolive. Table 11.9 lists some representative liquid hand soap products on the market around the world and the benefits these products claim.

For liquid body wash, there has been an explosion of new products in the marketplace. The rapid pace of innovation in the bath and shower market in the last decade has transformed the traditional bathing and showering practice from the necessity of basic cleaning and hygiene to pampering and caring for well being of body and mind. The high-end products that were being sold only in specialty stores are now coming to mass market, to deliver special skin care benefits. Relaxation of body and mind is being offered in the shower by the introduction of aromatherapy shower gels based on essential oils, traditionally known to soothe the nerves

**TABLE 11.9**   Commercial Liquid Hand Soaps

| Brand | Manufacturer | Ingredients | Claims |
|---|---|---|---|
| Foamy Liquid Gel Soap (Argentina) | Foamy | Vegetable proteins and essential oils | Deep cleans and leaves a pleasant sensation of softness and freshness |
| Shampoo for Hand and Body (Argentina) | Palacios | Irgasan DP 300 | Prevents contamination by germs |
| Campbell Bathroom Handwash (Australia) | Campbell | Vitamins E and B5 | Cleans and moisturizes hands |
| Softwash Liquid Handwash (Australia) | Colgate-Palmolive | | Kills germs; moisturizes |
| Country Life Hand Wash (Australia) | Faulding | | Cleans and moisturizes |
| Soft as Soap Handwash (Australia) | Reckitt Benckiser | Aloe vera with moisturizer | Gently cleanses and moisturizes leaving the skin feeling clean, soft, and refreshed |
| Softsoap Liquid Hand Soap (U.S. and Canada) | Colgate-Palmolive | Triclosan | Provides both thorough cleansing and light moisturizing. Its antibacterial action combined with an exclusive blend of light moisturizers leaves hands fresh, clean, and soft |
| Alpen Secrets Hand Sanitizer (Canada) | Delhar Group | | Kills bacteria and germs on contact where water and soap are not available |
| Pooh Extra Gentle Antibacterial Soap (Canada) | Funcare | Royal honey | Combats the spreading of germs transmitted through hand-to-hand contact |
| Jergens Antibacterial plus Cream Hand Soap (Canada) | Kao | | Kills bacteria; contains moisturizer |
| Ivory Skin Cleansing Liqui-gel (Germany) | Procter & Gamble | | Specially designed to wash away dirt and bacteria; so gentle to the skin that it can be used on the face |
| Palmolive Liquid Soap (Czech Republic) | Colgate-Palmolive | Keratin for hands and nails | Neutral pH, delicately washes and dries out skin. Strengthens nails and cares for hands |
| Carex Antibacterial Moisturizing Handwash (Czech Republic) | Cussons | Aloe vera and eucalyptus | Removes germs and stubborn odors |
| Palmolive Pouss Mousse Hyperallergenic Hand Soap (France) | Colgate-Palmolive | | Protects skin against dryness and helps to minimize risks of allergy. pH neutral for skin. Suitable for the most sensitive skin types |

*(continued)*

**TABLE 11.9** (Contd.)

| Brand | Manufacturer | Ingredients | Claims |
|---|---|---|---|
| Fa Spirit of Freshness Fluid Soap (Germany) | Henkel | Hydro-Balance system | Pleases the body, protects the skin. Leaves skin noticeably fresh and supple. With Hydro-Balance system to prevent dryness |
| CD Wash Lotion (Germany) | Unilever | Avocado extracts; calming, natural ingredients | For sensitive skin. pH neutral. Colorant free. Effective beauty care. Naturally mild and soap free |
| Lux Moisturing Cream Soap (Germany) | Unilever | Natural water lily extracts and vitamin E | Helps skin to maintain its natural moisture balance. Gently cleanses to leave skin feeling soft, smooth, and supple |
| Palmolive Fresh & Clean Liquid Soap (Italy) | Colgate-Palmolive | | Cleans hands in a nonaggressive way with a unique action against bad odor. The active ingredient neutralizes all odors of garlic, onion, fish, gas, bleach, etc. Ideal for the kitchen. pH neutral and helps to maintain the natural balance of skin. Moisturizers help prevent drying due to frequent washing |
| Palmolive Douss' Douss' (France) | Colgate-Palmolive | | Ideal for kitchen use. Unique formula, pH neutral, helps eliminate persistent odors from hands. Leaves a fresh lemon scent |
| Palmolive Vitamins Liquid Cleanser (Italy) | Colgate-Palmolive | Vitamin E + A complex | For healthy- and younger-looking hands. With a special complex with vitamin E which replenishes hands' supply of that vitamin, known for its antiaging properties |
| Badedas Super Soap (Italy) | Sara Lee | Proteins and vitamins E and F | Ideal for use in the kitchen. Eliminates from hands persistent odors like garlic, onion, and fish. Combines antiodor properties with emollient and moisturizing properties, leaving skin soft and gently fragrant |
| Pouss Mousse Liquid Body Cleanser (Japan) | Colgate-Palmolive | | Thoroughly cleans the skin. Refreshing and smooth finishing |
| Kazoku Seiketsu Hand Soap (Japan) | Kao | Tea essence | Medicated. Contains tea essence. Thoroughly sterilizes and does leave fragrance on hands |

*(continued)*

**TABLE 11.9**  (Contd.)

| Brand | Manufacturer | Ingredients | Claims |
|---|---|---|---|
| Naïve Hand Soap (Japan) | Kanebo | Aloe extract | Preserves moisture in the skin while cleansing hands thoroughly |
| Kao White Medicated Hand Soap (Japan) | Kao | Vitamin E | Maintains the skin smooth, preserving water and preventing chapping and drying. Contains vitamin E, which promotes blood circulation, and glycylrecine acid stearyl, which controls inflammation. Nonsticky type |
| Kirei Kirei Liquid Hand Soap (Japan) | Lion | Triclo acid | Clean hands with sterilizing ingredient. Rich lather quickly rinses off. 100% plant cleansing ingredients |
| Kitchen Lime SOAP Liquid (Japan) | Lion | Lime oil | Eliminates odor of fish, meat, and onions on hands. Especially made for kitchen use. It leaves not even the smell of soap on hands after rinsing |
| Dove Hand Care Wash (Japan) | Unilever | Contains 1/4 moisture milk | Moisture milk wash. Prevents chapping and drying |
| Dial Liquid Soap (Mexico) | Dial | Vitamin E | Kills bacteria and moisturizes at the same time |
| Natusan pH 5.5 Liquid Soap (Sweden) | Johnson & Johnson | Glycerin | Has a pH of 5.5, and therefore it does not interrupt the skin's natural protection. Remoisturizes the skin |
| Palmolive Fruit Essentials Liquid Hand Soap (Thailand) | Colgate-Palmolive | | Refreshes and leaves hands feeling soft |
| Protex Liquid Hand Soap (Thailand) | Colgate-Palmolive | Triclosan | Keeps hands clean and protected. It contains moisturizers to leave hands feeling soft and smooth |
| Dettol Liquid Hand Soap (Thailand) | Reckitt Benckiser | Extra moisturizer | Protects and cleans sensitive hands from unseen bacteria and dirt. With extra moisturizer that is pH balanced |
| Lifebuoy Plus Liquid Handwash (Thailand) | Unilever | Puralin Plus and moisturizer | Protection and care for healthy skin |
| Boots Antibacterial Handwash (U.K.) | Boots | Triclosan | Removes germs and leaves hands feeling refreshed |
| Palmolive Nourishing Liquid Handwash (U.K.) | Colgate-Palmolive | | Nourishes the skin after washing hands, leaving it supple and soft to the touch |

*(continued)*

**TABLE 11.9** (Contd.)

| Brand | Manufacturer | Ingredients | Claims |
|---|---|---|---|
| Johnson's pH 5.5 Handwash (U.K.) | Johnson & Johnson | | pH 5.5. Soap free. Extra gentle. Fragrance free |
| Carex Antibacterial Moisturizing Handwash (U.K.) | Cussons | Aloe vera and eucalyptus | Removes dirt, germs, and stubborn odors. Added moisturizers help protect against moisture loss and actively condition the skin |
| Dettol Fresh Moisturizing Handwash (U.K.) | Reckitt & Colman | | Antibacterial protection from germs and dryness. Kills germs, including *E. coli* and salmonella. Actively moisturizes the skin |
| Softsoap Liquid Hand Soap (U.S.) | Colgate-Palmolive | Triclosan | Antibacterial with light moisturizers. More than just cleans — cares for the skin |
| Softsoap Antibacterial Hand Soap (U.S.) | Colgate-Palmolive | Triclosan | Provides strong antibacterial protection. Contains light moisturizers to help leave hands clean and soft |
| Softsoap 2 in 1 Antibacterial Hand Soap (U.S.) | Colgate-Palmolive | Triclosan, Polyquarternium-7, and Polyquarternium-39 | Combines proven antibacterial formula with real moisturizing lotion; helps retain more of the skin's natural moisture |
| Softsoap Lavender & Chamomile Liquid Hand Soap (U.S.) | Colgate-Palmolive | | Helps one feel relaxed while leaving the skin feeling silky and smooth after washing the hands |
| Liquid Dial for Kids (U.S.) | Dial | | Kills germs. Fun to use |
| Softsoap FoamWorks Foaming Hand Soap (U.S.) | Colgate-Palmolive | | Makes hand washing fun and easy for children and the whole family |
| Lysol Antibacterial Hand Gel (U.S.) | Reckitt & Colman | | Helps reduce the risk of illness by killing 99.9% of the germs on hands without water. It contains emollients that moisturize hands |
| Suave Antibacterial Hand Sanitizer (U.S.) | Unilever | Vitamin E | Kills germs instantly without water |

and relax the muscles. Desire for youthful appearance and willingness to pay for products that would promise such benefit is leading to the development and introduction of a multitude of antiaging shower products such as those offering firming, exfoliation, etc.

## A.    Aromatherapy Products

Aromatherapy is becoming a very popular trend in cleansing products. The aromatic fragrances make the showering or bathing experience an indulgence rather than a chore. A number of products have been introduced in the market under brands such as Palmolive, Dove, and Ohm (Olay). A number of experiential benefits such as relaxing, soothing, and energizing are being delivered via various essential oils such as lavender, chamomile, and ginseng. Table 11.10 lists some aromatherapy products on the market, the essential oils contained in the products, and the benefits claimed.

Essential oils have been widely investigated in the medical field for their effect on brain and heart and the therapeutic benefits they deliver. They have been used for centuries by the Chinese, Egyptians, Greeks, and others and are well covered in the patent literature. A U.S. patent by Fletcher *et al.* [27] presents a good cross-reference for essential oils.

Essential oils are highly scented droplets found in minute quantities in the flowers, stems, leaves, roots, and barks of aromatic plants. It takes 440 lb of fresh lavender flowers to produce 2.5 lb of lavender essential oil.

These oils are complex mixtures of terpenes, alcohols, esters, aldehydes, ketones, and phenols. The chemical composition of the oils is strongly related to the season, month, and time of the day and therefore these oils need to be extracted at the right moment. These factors make the oils very expensive and scarce and therefore synthetic oils are made with the predominant constituents of the oils while still maintaining the aroma of the natural oil.

The effectiveness of a perfume containing valerian oil as an active ingredient in the reduction of stress is described in a European patent by Shoji and Sakai [28] and a measurement of such a reduction is stress is described in a U.S. patent application by El-Nokaly *et al.* [29]. The latter describes a method and apparatus to measure the stress level resulting from an application of stimuli such as fragrance, flavor, or product or while test subjects are performing an activity task. The method involves measuring both physiological and psychological responses of humans. The physiological measurements include electrocardiography and blood volume pulse and the psychological data are collected via questionnaires.

Besides reduction of stress, essential oils are known to provide a number of other benefits. Table 11.11 [30] gives examples of some of the essential oils and their corresponding potential benefits.

## B.    Exfoliating Body Wash

Incorporation of cosmetic benefits into body wash and liquid hand soap products is becoming a trend of the future. Exfoliation was used in the past as a beauty treatment, delivered as a facial scrubber. It is now being introduced for the whole body via body wash. These products have begun appearing in specialty stores,

| Product | Manufacturer | Essential oils/herbs | Claims |
|---|---|---|---|
| Palmolive Aromatherapy Body Wash — Antistress (France, U.K., Italy, Hungary) | Colgate-Palmolive | Lavender, ylang-ylang, and patchouli | Antistress. Helps to give radiance to skin as it softens it. Its calming fragrance envelops the user in an aura of peace and tranquility |
| Palmolive Aromatherapy Body Wash — Energy (France, U.K., Russia, India, Thailand) | Colgate-Palmolive | Mandarin, ginger, and green tea extract | Helps skin feel revitalized and softened while the awakening scent gives an energy boost to body and soul |
| Palmolive Aromatherapy Sensual Shower Gel (South Africa) | Colgate-Palmolive | *Rosa damascene* extract, *Jasminum officinale* oil, *Vanilla plantifolia* fruit extract | Creates an aura of sensuality and warmth |
| Herbaflor Aromatherapy Herbal Shower Gel (Canada) | Bellmira | Essential oil of rosemary | Feel refreshed after bathing |
| Switch On Aromatherapy Body Wash (U.K.) | Tesco | Grapefruit and mandarin | Helps revitalize and invigorate skin. The refreshing and uplifting effects of grapefruit and mandarin working together to help awaken the mind and body |
| NO-AD.Aroma Bath and Shower Therapy Soothing Body Wash (U.S.) | Solar Cosmetics | Lavender and chamomile | Stimulates the senses and cleanses the skin. Helps promote a sense of emotional well-being. Brings a sense of balance, comfort, and relaxation |
| Vitamin & Herbal Indulgence Energizing Ginseng Body Wash (U.S.) | Vogue | Ginseng | Revitalizes the mind and body. Stimulates the senses and restores health and glow to the skin. Revives and nourishes the skin |
| Vitamin & Herbal Indulgence sensual Sunflower Body Wash (U.S.) | Vogue | Sunflower | Provides a romantic escape from daily stress. Cleanses without drying |
| St. Ives Swiss Formula Body Wash (Australia) | Alberto Culver | Vanilla and vitamins E and A | Conditions and soothes skin |
| Beautiful Bath Bath Gel (U.S.) | Freeman | French vanilla and tangerine | Calms and soothes a tense body and sore muscles. Leaves skin soft, smooth, and radiant |
| Ohm Body Wash (U.S.) | Procter & Gamble | Jasmine and rose extracts | Cleanses skin gently and then calms it with a moisturizing recipe. For soothed and supple skin |
| Ohm Body Wash (U.S.) | Procter & Gamble | Sandalwood and chamomile extracts | Cleanses skin gently and then restores skin's serenity. For soft skin and harmonious spirit |
| Ohm Body Wash (U.S.) | Procter & Gamble | Citrus and ginger | Cleanses and helps restore the look and feel of younger, healthy skin. Reenergizes the mind and helps restore skin's youthful appearance |

**TABLE 11.11**   Examples of Essential Oils and Their Possible Benefits

| Essential oil | Traditional uses (possible benefits) |
|---|---|
| Angelica root | For relieving fatigue, migraines; to ease anxiety and nervous tension; to regulate menstrual cycles and relieve dysmenorrhoea; for coughs; to restore sense of smell; for releasing accumulated toxins in the body |
| Atlas cedar | To relax tense muscles, calm emotions, help breathing; for enhancing meditation, easing pain, repelling insects; for hair loss |
| Balsam fir | To relieve muscle aches and pains; for relieving anxiety and stress-related conditions; to fight colds, flu, infections; for relieving bronchitis and coughs; said to ground one mentally |
| Bay | As a stimulant for hair growth; for relieving muscle spasms and strains; to improve circulation; to relieve melancholy, nervous exhaustion; as an insect repellent |
| Bay laurel | As an immune system stimulant, to regulate the lymphatic system; for relieving melancholy and anxiety; to stimulate the mind; for healing bronchitis and sinus infection |
| Bergamot | Balancing nervous system; relieving anxiety and stress; lifting melancholy; for restful sleep; as an antiviral; treatment of cold sores, psoriasis, eczema; insect repellent |
| Black pepper | To energize, for increasing circulation, to warm and relieve muscle aches and stiffness, for fighting colds, flu, infections |
| Calendula | All skin complaints; varicose veins; for treating enlarged lymph nodes, cysts, skin lesions |
| Cardamon | Relieving mental fatigue, nervous strain, and heartburn; for healing coughs and bronchitis, anorexia; to uplift and warm; as an aphrodisiac |
| Carrot seed | For toning and rejuvenating mature skin, wrinkles, scars; for eczema and psoriasis; as a stimulant to immune and lymphatic systems; for relieving premenstrual syndrome and regulating monthly cycles; to ease anxiety and stress |
| Citronella | As a mosquito repellent; for colds, flu, neuralgia; to relieve pain of rheumatism and arthritis; to relieve melancholy. Use on sensitive or damaged skin should be avoided |
| Clary sage | Relieving stress and tension; lifting melancholy; easing pain; for restful sleep; as an aphrodisiac; contains estrogen-like hormone, for menopause and premenstrual syndrome; relieving nervous exhaustion |
| Clove bud | For toothache, colds, flu, fungal infections; as a mosquito repellent; to relieve fatigue and melancholy; as an aphrodisiac |
| Coriander | Relieving muscular aches and pains; increasing circulation; for colds, flu, rheumatism; for help with sleep and nervous exhaustion |

*(continued)*

**TABLE 11.11** (Contd.)

| Essential oil | Traditional uses (possible benefits) |
| --- | --- |
| Cypress | To increase circulation; relieve muscular cramps, bronchitis, whooping cough, painful periods; reduce nervous tension and other stress-related problems; as an immune stimulant |
| Eucalyptus | For colds, as a decongestant; to relieve asthma and fevers; for its bactericidal and antiviral actions; to ease aching joints |
| Frankincense | To calm, enhance meditation, elevate mind and spirit; to help breathing; for psychic cleansing; for care of mature skin and scars |
| Geranium | Reducing stress and tension; easing pain, balancing emotions and hormones, premenstrual syndrome; relieve fatigue and nervous exhaustion; to lift melancholy; reduce fluid retention; repel insects |
| German chamomile | To relieve muscular pain; to heal skin inflammations, acne, wounds; as a sedative; to ease anxiety and nervous tension; to help with sleeplessness |
| Ginger | Reducing muscular aches and pains; increasing circulation; relieving bronchitis and whooping cough; for nervous exhaustion; in healing colds, flu, fever; to stimulate appetite |
| Grapefruit | To lift melancholy; relieve muscle fatigue; as an astringent for oily skin, to refresh and energize the body; stimulate detoxification; as an airborne disinfectant |
| Helichrysum | To heal bruises (internal and external), wounds, scars; to detoxify the body, cleanse the blood, increase lymphatic drainage; for healing colds, flu, sinusitis, bronchitis; to relieve melancholy, migraines, stress, tension |
| Hyssop | To heal bruises; for healing respiratory complaints and bronchitis, low or high blood pressure, indigestion, stress, tension |
| Jasmine | To lift melancholy; for muscular spasm, painful periods, labor pains; to relieve anxiety and nervous exhaustion; an aphrodisiac |
| Juniper berry | To energize and relieve exhaustion; ease inflammation and spasms; for improving mental clarity and memory; purifying the body; to reduce fluid retention; for disinfecting |
| Lavender | Balancing emotions; relieving stress, tension, headache; to promote restful sleep; heal the skin; to lower high blood pressure; help breathing; for disinfecting |
| Lemon | To balance the nervous system; as a disinfectant; to refresh and uplift; for purifying the body |
| Lemongrass | As an insect repellent and deodorizer; for athlete's foot; as a tissue toner; to relieve muscular pain (from sports); increase circulation; for headaches; for nervous exhaustion and other stress-related problems |
| Lime | To purify the air; for alertness; to relieve coughs or congestion; for uplifting and cheering the spirit; to heal colds, flu, inflammations |

*(continued)*

**TABLE 11.11** (Contd.)

| Essential oil | Traditional uses (possible benefits) |
|---|---|
| Myrrh | To heal wounds and nurture mature skin; for bronchitis and colds; to relieve apathy and calm |
| Neroli | For healing thread veins and scars; nourishing mature skin; increasing circulation; for relieving anxiety, melancholy, nervous tension, bronchitis; as an aphrodisiac |
| Nutmeg | For warming muscles; easing muscle aches and pains; to invigorate or stimulate the mind; an aphrodisiac; to stimulate heart and circulation; for relieving nervous fatigue |
| Oregano | As a muscle relaxant and to ease muscle aches and pains; to heal colds, flu, bronchitis; as a stimulant, to energize the mind and body; for relieving headaches |
| Palmarosa | To stimulate cellular regeneration and moisturize skin; for nervous exhaustion and stress conditions; to calm and uplift |
| Patchouli | For athlete's foot; as an aphrodisiac; to relieve stress and nervous exhaustion. Emotional profile: to relieve indecision, lethargy, mood swings |
| Peppermint | For energy and brighter mood; reducing pain; to help breathing; improve mental clarity and memory |
| Petitgrain | For relieving respiratory infections; to ease nervous tension muscle spasms; for relieving joint inflammation; to balance the central nervous system; for stress relief and restful sleep |
| Pine | To ease breathing; as an immune system stimulant; to increase energy; for relieving muscle and joint aches; to repel lice and fleas |
| Roman chamomile | To relieve muscular pain; as a sedative; to ease anxiety and nervous tension; to help with sleeplessness |
| Rose maroc absolute | For brighter mood; menopause; to help reduce wrinkles; for calming and reducing nervous tension; to promote restful sleep; as an aphrodisiac |
| Rosemary | To energize; for muscle pains, cramps, sprains; brighten mood; for improving mental clarity and memory; easing pain; to relieve headaches; for disinfecting |
| Rosewood | To relieve stress and balance the central nervous system; for easing jet lag; to create a calm for meditation; for easing colds and coughs; to stimulate the immune system; as an aphrodisiac; in skin care |
| Sandalwood | To lift melancholy, enhance meditation, heal the skin; help breathing; for calming and reducing stress; restful sleep; for disinfecting; as an aphrodisiac |
| Spearmint | For relieving bronchitis and sinusitis; to ease nausea and headaches; for relieving colds or flu; to stimulate, energize, relieve fatigue |
| Spikenard | To relieve migraines, stress, tension; for rejuvenation of mature skin; to calm and promote restful sleep; for wounds; to inspire devotion |

*(continued)*

**TABLE 11.11** (Contd.)

| Essential oil | Traditional uses (possible benefits) |
|---|---|
| St. John's Wort | For fungal infections, oily hair, dandruff, sinusitis, sore muscles |
| Sweet basil | To brighten mood, strengthen nervous system, improve mental clarity and memory; for relieving headache and sinusitis |
| Sweet fennel | For neuromuscular spasms, rheumatism and arthritis, bronchitis, whooping cough; as a nerve tonic in relieving stress and nervous tension |
| Sweet marjoram | To relax tense muscles and relieve spasms; calm and promote restful sleep; ease migraine headache; for comforting the heart; lowering high blood pressure; to help breathing; for disinfecting |
| Sweet orange | To brighten mood, calm, reduce stress; as an environmental disinfectant |
| Tangerine | For relieving muscle spasms; to soothe and calm nerves; for stress relief and relaxation; to stimulate the liver and increase lymphatic drainage |
| Tea tree | As an immunostimulant particularly against bacteria, viruses, fungi; for relieving inflammation; as a disinfectant |
| Thyme | To heal colds and bronchitis; for relieving muscle aches and pains; to aid concentration and memory; for relieving fatigue; said to heal anthrax |
| Vetiver | For muscular aches; to increase circulation; to relieve melancholy and nervous tension; for restful sleep |
| Ylang ylang | Brightening mood; relieving anger and anxiety; relaxing tense muscles; to calm and promote restful sleep; lower high blood pressure; as an aphrodisiac |

*Source*: From Essential Oil Details, Ancient Healing Art, www.AromaMarket.com. With permission.

containing exfoliating agents such as sugars, salt, and rice, well known in ancient cultures. In the mass market, the Dove brand has taken a lead in this category by introducing an exfoliating body wash and Dial has introduced a liquid hand soap. Besides offering cleansing and moisturizing benefits, these products promise exfoliation of skin during the shower by abrasive particles incorporated in the formula. Table 11.12 lists some of the commercial body wash products that offer exfoliating benefits.

The practice of exfoliation to remove dead cells and oils from skin was well known in ancient cultures. Crushed nut seeds such as walnut were commonly used to provide abrasion as well as nutritional benefits. It is believed that removal of dead cells leads to improved skin elasticity and firmness by regeneration of epidermal tissues thus making the skin look smoother, supple, healthier, and younger.

**TABLE 11.12** Commercial Exfoliating Liquid Body Wash Products

| Brand | Manufacturer | Ingredients | Claims |
|---|---|---|---|
| Neutrogena Visibly Even Exfoliating Body Wash (U.S.) | Johnson & Johnson | Soybean seed extract, grapefruit extract, *Morus bombycis* root extract, *Scutellaria baicalensis* root extract | Provides toning, moisturizing, and exfoliating benefits to make skin look more radiant |
| John's pH 5.5 Daily Exfoliating Body Wash (U.K., South Africa) | Johnson & Johnson | Hydrogenated jojoba oil, sodium/styrene/acrylate copolymer, acrylates/C10–C30 alkylacrylate cross-polymer | The massaging action of the natural jojoba beads removes rough skin to reveal new, silky smooth skin after every shower |
| Herbal Essences Daily Body Smoother Exfoliating Body Wash (U.S.) | Clairol | Hydrogenated jojoba oil, coneflower (*Echinacea purpurea*) extract, *Magnolia acuminata* extract | Formulated to smooth and soften the skin, cleanse, and gently polish away dry rough skin |
| Sue Devitt Studio Lavender Exfoliating Body Wash (U.S.) | New Kingdom | Sea plant loofah and bamboo shoots | Naturally exfoliates skin. It is infused with hints of lavender and silver shimmers "to add subtle gleam" |
| Farmaervas Celulan Exfoliating Body Wash (Brazil) | Labolatorio Farmaervas | Jojoba microspheres | Removes dead cells and hydrates and cleanses skin. Helps to lighten skin |
| Aromamor Exfoliating Body Wash (Brazil) | VLD | Merguard, fennel extract, glycolic extract of fennel, fennel seed, vegetable luffa | Gently exfoliates with essential oils |
| St. Ives Apricot Exfoliating Body Wash (Australia) | St. Ives (Alberto Culver) | More than 50% naturally derived ingredients with jojoba beads, moisturizing apricot, and soothing Swiss botanicals | Softens, soothes, and refreshes skin as it cleanses and moisturizes the skin, improving the look and feel. It is proven to reveal healthier, brighter-looking skin |
| Lux 2-in-1 Skin Expert and Scrub Exfoliating Shower Gel (Netherlands) | Unilever | Exfoliant granules | Gently cleanses the skin as its exfoliant granules stimulate the skin, thus offering a refreshing sensation. Refreshes the skin while moisturizing it thoroughly |
| Ocean Potion Before Sunless Exfoliating Body Wash (U.S.) | Sun & Skin Care Research Inc. | Hydrogenated jojoba oil beads | An all natural, organically based product that removes impurities and dead skin cells paving the way for skin revitalization |
| Nivea Bath Care Exfoliating Body Wash (Canada) | Beiersdorf | Vitamin E beads, magnesium aluminum silicate | Aids in revealing radiant, healthy-looking skin |
| Simple Exfoliating Body Wash (U.K.) | Accantia Health & Beauty | Natural loofah and chamomile oil | Gently exfoliates and cleanses, leaving skin soft and smooth |

| Product | Manufacturer | Ingredients | Description |
|---|---|---|---|
| Les Actifs Marins Exfoliant Shower Gel (Belgium) | Sarbec | Microparticles, seaweed extracts, and antifree radicals | Gently eliminates impurities and dead cells which may asphyxiate and dull the skin. Gently cleanses the skin, returning to it luster, softness, and firmness |
| Obao Exfoliating Shower Gel (France) | L'Oreal | Microbeads, AHA | Softens the roughness of skin and eliminates dead cells. New skin effect contains AHA to help the renewal of the cells. Exfoliated skin sees its texture transformed. A skin like new, soft, velvety, toned |
| Veet Exfoliant Shower Gel (Germany) | Boyle | Finely ground granules | Speeds up the removal of dead skin cells and softens rough areas on the skin. Sheds off the old skin cells when rubbed over the skin. Makes skin smooth and soft |
| Neutro Roberts Shower Foam (Italy) | Manetti-Roberts | Microgranules derived from natural jojoba oil, vitamin E + B5 complex | Eliminates impurities, renews skin. Ideal for sensitive skin. Cleans the skin, giving it vitality and moisturization. Protects and reintegrates the moisturization of the skin surface, reinforcing the skin, leaving it soft and elastic |
| Profil Exfoliate Shower Gel (Germany) | Gemey | Microparticles | Eliminates dead skin and softens, revitalizes, and cleans |
| Citric Essentials Body Wash (U.K.) | Boots | Luffa cylindrica | Cleanses the skin to leave it toned and refreshed |
| Dove Exfoliating Body Wash (Australia) | Unilever | Ultrafine exfoliants (hydrated silica) | Gently smoothes away dull, lifeless skin to reveal beautifully fresh, new skin. Moisturizes skin, leaving it smooth to the touch |
| Suave Naturals Body Wash (U.S.) | Unilever | Vitamin E, aloe vera, and natural apricot exfoliants (apricot seed powder) | Gently cleanses and smoothes skin. Helps restore natural moisture while gently exfoliating to remove rough, dull surface cells and reveal the healthier skin below |

The formulation of skin cleansing products with exfoliating benefits has been subject of several patents. The materials claimed to provide exfoliation in these products include sugars, inorganic salts, calcite [31] and silica, clays, polymeric materials such as polyethylene powders [32], and crushed seed powders from walnut, apricot kernel, and almond.

## 1.  Salts and Sugars

A patent by Hramchenko and Sibley [33] describes an exfoliating composition containing salts and sugars. The composition is an anhydrous cream with uniformly dispersed fine particles of inorganic salts or sugars. The salts are so chosen that their solubility and particle size make them last long enough, preferably 15 seconds, in the presence of water to provide scrubbing action and eventually dissolve in water for a clean rinse. The preferred solubility is less than 30 wt% at 40°C and less than 10 wt% at 20°C. An average particle size of about 125 to 750 μm is preferred with a hardness of 1.5 to 4 on the Mohs scale. The preferred salts are sodium tetraborate decahydrate and potassium pentaborate octahydrate, sodium citrate, monobasic sodium phosphate, and sodium pyrophosphate. The preferred sugar is sucrose.

The other ingredients of these compositions are surfactants, foam boosters, benefit agents, and water. The water is added in amounts of 10 to 30 % so that the salts are not dissolved.

## 2.  Silica

Cordery et al. in a patent assigned to Unilever [34] describe a cosmetic composition containing silica as an exfoliating/massaging agent for use on scalp and body. The particle size and strength of silica is so chosen that it provides sufficient abrasion/exfoliation during normal use and thereafter breaks down into fine particles by shear and/or crush forces normally produced during use. The desired size and shape of silica is obtained by its structural modification. The preferred material is structurally modified silica derived from Sident 200 (Degussa) or Zeosyl 200 (Zeofinn), or Tixosil 333 (Rhodia). The preferred particle size of this material is in the range 0.1 to 1 mm, porosity in the region of 2 $cm^3g^{-1}$, surface area about 250 $m^2g^{-1}$, and crush strength of 10 to 34 MPa at 50% room humidity, breaking down after use to 40 μm. The composition contains silica, surfactants or surfactant mixtures, suspending agents such as clays, polyacrylates (e.g., Carbopol 910, 934, 940, and 941), heteropolysaccharide gums (e.g., xanthan gum, guar gum), and certain cellulose derivatives such as carboxymethyl cellulose, preservatives, and other aesthetic agents.

## 3.  Nut Shells/Kernels, Dried Fruit Particles

Japanese [35] and U.S. [36] patents describe the use of natural abrasives such as ground walnut shells, apricot shells, and olive kernels to provide gentle skin cleansing without scratching.

A U.S. patent application [37] describes the use of dried and fresh fruit particles in skin care preparations. This patent describes compositions containing powder or flakes of size 100 to 200 $\mu$m and density between 0.2 and 0.45 g/ml. These powders or fakes are derived from fruit peels or cores dried by techniques such as sun drying, vacuum drying, or freeze drying and other components such as suspending agents, surfactants, and emulsifiers. The various fruits described are peach, lemon, strawberry, pear, cherry, apricot, blackberry, papaya, mango, orange, apple, cranberry, mango, kiwi, banana, etc., supplied by sources such as International Botanical Specialty Products Inc. of Wisconsin and Freeman Industries, Tuckahoe, NY. Besides providing the scrubbing action during use, these fruits also add nutritive value since they generally contain mono-, oligo-, and polysaccharides that can provide moisturizing benefits, fibers, macro- and micronutrients, vitamins, and phenolic compounds such as flavonoids which can act as antioxidants.

## C. Toning and Skin Firming Products

A number of body wash products are making claims for toning skin and body and firming the skin. These products incorporate plant or seaweed extracts that are known to impart benefits to skin and body. Table 11.13 lists some examples of commercial products on the market.

## D. Antiaging Products

Cleansing products that promise a youthful and younger-looking appearance are beginning to appear on the market in the U.S. and Europe. Table 11.14 lists some of the commercial body wash products on the market that claim to offer antiaging benefits.

Generally, it is a challenge to deliver the antiaging agents on to skin via cleansing agents, since they tend to get washed away in the presence of the large amounts of surfactants in these products. However, the industry has begun devising various carriers to deliver these actives onto skin via cleansing products.

A youthful state of skin is generally measured by its elasticity and tautness. Collagen and elastin in the dermis contribute to these properties. However, with age and adverse environmental effects such as UV radiation, lack of moisture, etc., the elasticity and tautness of skin decreases and the skin wrinkles and shows sign of aging. Exposure to UV radiation and aging leads to a decrease in hyaluronic acid and polysaccharides and excessive production of an enzyme called elastase. This enzyme destroys elastin and leads to loss of skin elasticity.

A European patent by Inomata [38] describes the use of *Uncaria gambir roxburgh* extract in effectively suppressing the activity of elastase and restoring the tautness and elasticity of skin. Also, a world patent by Fransoni [39] describes the benefit of a complex of hyaluronic acid and carnitine and its derivatives

**TABLE 11.13**   Commercial Toning and Skin Firming Liquid Body Wash Products

| Brand | Manufacturer | Ingredients | Claims |
|---|---|---|---|
| Palmolive Naturals Toning Shower Gel (Australia) | Colgate-Palmolive | Natural extracts of kiwi and mango | Contains kiwi and mango extracts which are known to help preserve toned skin. Leaves skin clean, firm, and noticeably soft |
| Ricette dell'Erborista Elixir D'Aromes Shower Gel (Italy) | Conter | *Santalum album*, *Poppgostemon cablin* | Stimulates and tones the body. Cleanses the skin leaving it soft and fragranced |
| Prismalis Bain-Douche Bois Exotiques (France) | Prismalis | Mandarin orange, lavender, and mint | Stimulating shower gel that stimulates and protects the skin |
| Collistar Linea Uomo Gel Docci Tonificante Shower Gel (Italy) | Collistar | Ultra delicate skin purifying ingredient, panthenol, lime extract, and wheat germ protein | Skin toning. Alleviates muscle strain |
| Cosmence Douche — Soin Toning Shower Gel for a Stimulating Massage (U.S.) | LeClub des Createurs de Beaute | *Ruscus aculeatus* root extract, *Hedera helix* extract, *Arnica montana* flower extract | Helps stimulate circulation to areas of the body that have cellulite |
| Ushuaia Douche Tonifiante (Belgium) | Laboratoires Garnier | Atlas cedarwood | Tones skin and body |
| Dove Body Firming Shower Gel (U.K., Switzerland) | Unilever | Seaweed extract and ceramides | A skin firming shower gel with a new thalasso formula. For tauter skin |
| Jergens Skin Firming Body Wash (U.S. and Canada) | Hergens | Seaweed extract and essential moisturizers | Tightens and tones uneven skin |
| Bourjois Grains of Beauty Shower Gel and Draining Massage (France) | Bourjois | Holly and ivy extracts | Gently washes, reduces "cottage cheese" aspect, restructuring effect. With holly and ivy extracts known for their draining and firming properties |
| Venus Multi-active anticellulite shower gel (Italy) | Kelamata | Extracts of *Theobroma cacao*, *Aesculus hyppocastanum*, *Arnica Montana*, and *Centella asiatica* | With plant extracts that have been found to have dramatic effects on cellulite reduction, inhibit the body's production of cellulite tissue, and strengthen capillary walls near the surface of the skin and thus stimulate greater skin permeability. Also known to favor the removal of unnecessary water in cell tissue, reduce inflammation, swelling, and bloating, and produce an intense firming action on skin cell tissue |

**TABLE 11.14**  Commercial Antiaging Liquid Body Wash Products

| Brand | Manufacturer | Ingredients | Claims |
|---|---|---|---|
| Dove Nutrium Age-Defying Body Wash (U.S.) | Unilever | Green tea extracts, vitamins A and C | Dual formula that goes beyond cleansing and moisturizing to reduce the visible signs of aging. Smoothes fine lines and dry, rough areas to leave skin soft and younger looking |
| Les Actifs Marins Exfoliant Shower Gel (Belgium) | Sarbec | Seaweed extracts and antifree radicals | Gently eliminates impurities and dead cells which may asphyxiate and dull the skin. Deeply cleanses the skin, returning to it luster, softness, and firmness |
| Higiporo Liquid Soap (Brazil) | Davene | Sage, provitamin B5, and vitamin E | Adds softness to skin and regulates T-zone oiliness. Combats premature skin aging. Cleans and moisturizes |
| Health Basics Aloe Vera Body Wash (Australia) | Pharmaceutical Sales and Marketing | Aloe vera and vitamin E | Soothes and nourishes dry skin. Prevents premature aging. Protects against pollution |
| Rexona Shower Gel (Switzerland) | Unilever | Liposomes, vitamin E | Prevents premature epidermal signs of aging |
| Lux Shower & Gel (Hungary) | Unilever | Vitamin E | Slows down the skin aging process |

in antiaging, and restoring or maintaining activity on skin elasticity. A U.S. patent by Arraudeau and Aubert [40] describes the use of gentisic acid and 2,3-dihydroxybenzoic acid in combination with alpha- and beta-hydroxyl acids, keto acids, or retenoids to stimulate the process of epidermal cell renewal and beneficial effects to combat the main clinical signs of aging of the skin, i.e., formation of wrinkles or fine lines and blemishes of the skin.

A European patent by Ishikawa [41] describes a method for screening antiaging agents. As the cell ages its capacity to divide itself decreases and the cell becomes senescent. This patent describes ways to produce such cells rapidly (novel transformed cell) and then testing them with the antiaging agent to determine if the agent helps to regenerate the ability of these cells to perform cell division, as measured by an aging index.

A U.S. patent by Miller *et al.* [42] describes the use of legume products for topical applications for the good health of skin. Legumes such as soybeans contain high levels of proteins, lipids, and carbohydrates, and are considered very good nutrients for maintaining skin tone and texture.

## E.  Spa Products

Recently a number of new "spa" lines of body wash products have emerged on the market. Colgate-Palmolive, for example, introduced three products to the European market under the Palmolive Thermal Spa Shower Gel name — the Purifying variant with sea salt, the Massage variant with white clay, and the Hydrating variant with sea algae. The Art of Beauty introduced Qtica Smart Spa Shower Gel to the U.S. market claiming to sooth, repair, and condition skin. The product is said to contain fortified vitamins C and E. Pharmagis introduced Necca Spa Shower Gel to the Israeli market that contains 33% moisturizer, cetyl palmitate and beheneth-10 hydrogenated castor oil, and glyceryl stearate that claims to moisturize and refresh skin and with a fragrance lasting 24 hours.

## F.  Skin Whitening Products

Skin whitening products are becoming very popular in Asian countries. Skin darkening is believed to be due to deposition of high levels of melanin in the skin, which is generated by activation of melanocyte by UV radiation. Whitening agents such as L-ascorbic acid, hydroquinone derivatives, glutathione, and colloidal sulfur have been used to inhibit the production of melanin.

A European patent by Kuno and Matsumoto [43] describes the use of olive plant extract as an effective and stable whitening and antiaging agent. This extract is believed to have strong active oxygen elimination function such that it can eliminate superoxide and hydroxyl radicals and also effectively inhibit melanin production.

A world patent by Wakamatsu *et al.* [44] describes the use of ascorbic acid and its derivatives together with purine nucleic acid to provide antiaging benefits and improve skin pigmentation.

## G.  Products for Men

Traditionally, the shower gel market has been for women. In recent years men have become increasingly interested in their appearance and grooming. A number of shower products specifically formulated for men are making their way onto the mass market, such as Palmolive and Softsoap's Men's Active Body Wash by Colgate-Palmolive, Old Spice Body Wash by Procter & Gamble, and Suave Body Wash by Unilever. A number of skin care products for men are also being introduced by Nivea. These products generally differ from women's products in color, fragrance, and packaging. They have darker colors, and masculine fragrances and packaging.

## VI. CONCLUSION

As consumers continue to look for more and more experiential and therapeutic benefits from bath products, liquid soap and body wash products will continue to incorporate more and more of these benefits to satisfy such demands in the future. The future trend is to bring the benefits that are currently being delivered by high-end cosmetic products to bath and shower products, for the well being of body and mind.

## REFERENCES

1. Euromonitor, International (www.euromonitor.com).
2. Reever, R.E., in *Liquid Detergents*, Surfactant Science Series, Vol. 67, Lai, K.Y., Ed., Marcel Dekker, New York, 1996, chap. 11.
3. Flick, E., *Cosmetic and Toiletry Formulations*, 2nd ed., Noyes Publications, Park Ridge, NJ, 1989.
4. Hunting, A.L.L., *Encyclopedia of Shampoo Ingredients*, Micelle Press, Cranford, NJ, 1991.
5. Barel, A.O., Paye, M., and Maibach, H.I., *Handbook of Cosmetic Science and Technology*, Marcel Dekker, New York, 2001.
6. Dyer, D.W. and Hassapis, T., *Soap Cosmet. Chem. Specialties*, 59, 36–40, 1983.
7. Diez, R., Smith, R., and Levanduski, S., AOCS Meeting, October 1997.
8. Reever, R.E., in *Liquid Detergents*, 1st ed., Surfactant Science Series, Vol. 67, Lai, K.Y., Ed., Marcel Dekker, New York, 1997, pp. 413–418.
9. Schleppnik, A., U.S. Patent 4,187,251 to Bush, Boake and Allen, Inc., 1980.
10. Schleppnik, A., U.S. Patent 4,310,512 to Bush, Boake and Allen, Inc., 1982.
11. Brannan, D.K., Ed., *Cosmetic Microbiology*, CRC Press, New York, 1997.
12. Barnes, H.A., Hutton, J.F., and Walters, K., *An Introduction to Rheology*, Elsevier, Amsterdam, 1989.
13. Tamura, T. and Masuda, M., in *Handbook of Cosmetic Science and Technology*, Barel, A.O., Paye, M., and Maibach, H.I., Eds., Marcel Dekker, New York, 2001, pp. 417–450.
14. Hunter, R.S. and Hunter, R.W., *The Measurement of Appearance*, 2nd ed., Wiley-Interscience, 1987.
15. www.xrite.com
16. Gotte, E., *Aesthetische Medizin*, 10, 313–319, 1966.
17. Blake-Haskins, J.C., Scala, D., Rhein, L.D., and Robbins, C.R., *J. Soc. Cosmet. Chem.*, 37, 199–210, 1986.
18. Tavss, E.A., Eigen, E., and Kligman, A.M., *J. Soc. Cosmet. Chem.*, 39, 267–272, 1988.
19. Sharko, P.T., Murahata, R.I., Leyden, J.J., and Grove, G.L., *J. Dermatol. Clin. Evaluation Soc.*, 43, 187–192. 1992.

20. Simion, F.A., Rhein, L.D., Grove, G.L., Wojtkowski, J.M., Cagan, R.H., and Scala, D., *Contact Dermatitis*, 25, 242–249, 1991.

21. Paye, M., Van de Gaer, D., and Morrison, B.M., Jr., *Skin Res. Technol.*, 1, 123–127, 1995.

22. Lanman, B.M., Elvers, E.B., and Howard, C.J., *Joint Conference on Cosmetic Science. The Toilet Goods Association,* CTFA, Washington D.C., 1968, pp. 21–23.

23. Lee, C.M. and Maibach, H.I., *Exogenous Dermatol.*, 1, 269–275, 2002.

24. Barel, A.O., Paye, M., and Maibach, H.I., *Handbook of Cosmetic Science and Technology*, Marcel Dekker, New York, 2001, pp. 245–252.

25. Tadros, T.F., Ed., *Solid/Liquid Dispersions*, Academic Press, New York, 1987.

26. Becher, P., *Emulsions: Theory and Practice*, Oxford University Press, New York, 2000.

27. Fletcher, J., Hargreaves, R., and Michael, J., U.S. Patent 6,280,751, 2001.

28. Shoji, K. and Sakai, K., EP 1 293,554 A1 to Shiseido Co. Ltd, 2003.

29. El-Nokaly, M., Hilton, M.L., Doyle, K.L., Schaiper, D.R., Saud, A., and Prickel, D.L., U.S. Patent 2003/0236451 A1 to Procter & Gamble, 2003.

30. www.AromaMarket.com

31. Stanley, R. and Serridge, D., U.S. Patent 6,294,179 B1 to Lever Brothers Co., 2001.

32. Harrold, A.M. and Vargas, O., U.S. Patent 4,272,519 to Eli Lilly & Company, 1981.

33. Haramchenko, J. and Sibley, M.J., U.S. Patent 4,084,123 to Barnes-Hind Pharmaceuticals, 1977.

34. Cordery, C.S. and Dawson, P.L., EU 0670712B1 to Unilever PLC, 1996.

35. Kitamura, R., Kawamoto, R., and Takahashi, K., JP 11043428A2 (1982); Arita, J., JP 11043428A2 (1982).

36. Bouillon, G., Daniel, G., Denzer, H., Peppmoller, R., and Frazen, M., U.S. Patent 5,830,445 to Chemische Fabrik Stockhausen GmbH, 1998.

37. Cho, S., Zehntner, B., and Tuck, A., U.S. Patent 2004/0022818A1 to Fish & Richardson PC, 2004.

38. Inomata, S., EP 0919 223 A1 to Shiseido Company Ltd, Tokyo, 1999.

39. Fransoni, M., WO 00/29030 to Continental Projects Ltd, Ireland, 2000.

40. Arraudeau, J.P. and Aubert, L., U.S. Patent 5,766,613 to L'Oreal France, 1998.

41. Ishikawa, F., EP 1408109 A1, 2002.

42. Miller, J., *et al.*, U.S. Patent 6,555,143 B2 to Johnson and Johnson Consumer Products Inc., 2003.

43. Kuno, N. and Matsumoto, M., U.S. Patent 6,682,763 B2 to Nisshin Oil Mills Ltd Tokyo, 2004.

44. Wakamatsu, K., Harano, F., Koba, T., and Shinohara, S., WO 2004/016238 A1 to Otsuka Pharmaceutical Co. Ltd, Japan, 2004.

# 12

# Fabric Softeners

**ANDRÉ CRUTZEN** Advanced Technology Department, Colgate-Palmolive Research and Development, Inc., Milmort (Herstal), Belgium

# I.  INTRODUCTION

Fabric softness refers to a pleasant feel when using garments, which is maintained by regularly treating the laundry with appropriate products. Fabric softeners, however, deliver much more than a soft, fluffy, luxurious feel to most fabrics. They reduce the static cling and electric shock generated by static electricity buildup. They decrease fabric wrinkling and make ironing easier and drying time shorter. They reduce fiber damage. Moreover, due to their fragrance, they impart a pleasant smell to the washed fabric. They may also deliver various actives such as soil release agents, whitening agents, and antiwrinkle agents to the fabric.

After considering the justification for fabric softeners, the origins of the need and benefits delivered, this chapter reviews the technology of these products — chemistry and process — and the physical chemistry of fabric softening.

# II.  HISTORY
## A.  Origin of the Need for Fabric Softening

Textiles in contact with the skin must have a pleasant feel. Since natural fibers are harsh, textile manufacturers coat them with a finish. Before World War II, natural oils and fats were used as fiber finishes. Garments were essentially washed by hand and dried outdoors. Domestic washing was carried out with laundry soap, which, in hard water, forms insoluble lime soap that deposits on and softens fabrics.

The revolution in laundering started in the late 1940s. Because of the raw material shortage, many sulfonated oil substitutes were developed during World War II and the subsequent growth of the petrochemical industry made them available at a reasonable price. They were more compatible with acidity and with water hardness than soap; they were also more efficient in removing fatty soils. Hence, from the early 1950s soap was gradually replaced by the much more efficient but aggressive alkaline built detergents. These synthetic detergents were based on alkylbenzenesulfonates and builders such as phosphates, carbonates, or citrates to prevent the deposition of the insoluble alkaline earth salts of surfactants.

At the same time, fabrics were no longer hand-washed, but laundered in washing machines, undergoing hot washing and strong mechanical agitation. These new conditions were so efficient that they gradually washed out the finish and all natural lubricants from the surface of the fibers without leaving any beneficial residues.

Moreover, the strong mechanical stress degrades the individual fibers and makes them less flexible. Because of the higher washing temperatures, fabrics shrink and become more wrinkled.

The situation was not as bad in the U.S., where washers were much larger and wash cycles shorter. Loads, however, contained more and more synthetic fibers that must have the same feel as natural ones. Also, electrical tumble dryers were — and still are — more popular. They impart a perceivable softness to garments, but at the end of tumble drying the items cling together because of the static electricity generated in the dryer. Hence a need arose for static control.

Coating the fibers with a greasy material counteracts the damage generated by the more aggressive washing conditions. The coating may be applied to the garments during the rinse, during the drying, or now even during the wash. Best softness results are obtained by introducing the softener during the last rinse. Since the product then undergoes a huge dilution, the actives must exhibit a large affinity for the substrate (the affinity is defined as the partition coefficient between fabric and liquor). Therefore most fabric softeners were based on cationic surfactants, which exhibit an outstanding affinity for fabrics. Moreover they are extremely efficient in neutralizing static electricity.

## B.    Pioneering Companies and Products

Cationic surfactants appeared on the market in 1933. They were originally used as dye leveling agents in the textile industry, to improve the water fastness of direct and acid dyes on cellulose [1–3]. Some of the first cationic actives were synthesized by Ciba (Switzerland) and commercialized as Sapamines [1]. Very quickly, the soft feel delivered by long-chain derivatives was noticed and exploited to restore the fabric finish.

In the late 1940s cationic surfactants were widely used as finishing agents in commercial laundries. The multifarious benefits they delivered — improvement in feel, pleasant scent, and static control — attracted much interest for developing a new line of household products. The first cationic-based liquid rinse products for domestic use appeared in local markets in the U.S. in 1955, and were nationally launched in 1957 [4].

The first European product was launched in Germany in 1963 [2,4], which quickly became the largest market outside the U.S. [1].

From the beginning, the history of fabric softeners has been driven by inno-vation, by the producers' voluntary commitment to propose more efficient or convenient products delivering additional benefits, by technical changes in the production processes and appliance technologies, and also by legal constraints.

The first products were made of 4 to 6% active, a fragrance, and a viscosity modifier [5]. The dispersions of cationic actives indeed remain easily pourable as far as their concentration does not exceed 7% by weight; since the softening efficacy levels off at concentrations above 6%, this was not a concern. In the mid-1970s improved softening systems made of two actives appeared on the market. They were still based on the same quaternaries, but synergistically combined with other fatty materials called "co-softeners" to enhance their performance/cost ratio.

A new era started for brand relaunches, which had been until then limited to claim-ing new perfumes [6]. The development of products based on these double-active systems led to true product improvements and opened the door to the formulation of concentrated products. The first concentrates were introduced in the German market in 1979. Five years later, they were available in most other European countries and in North America. They contained about three times the usual level of softening actives and were usually positioned as extensions of the traditional brands.

The incentive to the launch of concentrates was threefold: convenience, cost, and subsequently environmental profile [7,8]. Being less bulky, they are indeed much more easily handled than the large bottles in which regular softeners are usually sold. They also enable a greater plant throughput with existing equipment and require less shelf space in each stage between production and retail outlet and in the washing area — a true advantage in the home. They lead to a reduction in shipping costs since less water is transported. Finally, they contribute to the reduction of plastic bottle waste in the environment. A serious drawback is that they are more difficult to dose correctly. Because of consumers' sensitivity to cost per unit volume of fabric softeners, they never became very popular in many countries such as those of southern Europe.

The 1980s were rich in innovation. As cationic actives precipitate in the pres-ence of anionic surfactants, thereby losing most of their efficacy, the anionic surfactant concentration in the liquor must be kept as low as possible. There-fore, the fabric softener had to be introduced in the last rinse of the wash cycle, when the detergent carryover is at a minimum. That represented a true constraint if the washer did not contain a dispenser for softener. The user had to stay near the washer to introduce the product at the beginning of the last rinse or had to run an extra rinse at the end of the laundering.

Many efforts have been devoted to overcome the technical difficulties, leading to several alternative systems: dryer cycle fabric softeners, which appeared on the shelves in the 1970s [9], wash cycle fabric softeners, and finally detergents con-taining the fabric softener or "softergents." Effective fabric softening in the wash cycle, however, supposes using alternative cleaning and/or softening actives. Each system presents advantages and shortcomings over the others; they are discussed below.

The next milestone in fabric softener history was the reconsideration of the use of di-hard tallow dimethylammonium chloride (DHTDMAC) as one of the most prominent softening ingredients. A dialogue took place between Dutch and German authorities with the industry. This dialogue focused on the existence of environmental data on DHTDMAC, covering the information available on aquatic toxicity and biodegradability. These contacts allowed identification of the much more cleavable esterquat that the industry selected to rebuild its softener compo-sitions. By 1993, DHTDMAC consumption in Europe fell by 70%. In the U.S., where its environmental profile had not been questioned, the reduction was only 20% [5,6,10].

$$R \diagdown \diagup CH_3$$
$$N^+ \quad Cl^-$$
$$R \diagup \diagdown CH_3$$

Di-hard tallow dimethylammonium chloride (DHTDMAC), where R is a hydrogenated tallow alkyl ($C_{16}/C_{18}$ chain).

The replacement of DHTDMAC by esterquat in fabric softening compositions represented a turning point. Before then, only two actives accounted for 95% of the cationic softeners in use. Since then, every manufacturer has had its own active ingredient [11,12].

Innovation in the fabric softener business not only focused on compositions, but also on packaging. As an alternative to plastic bottles, heat-sealed flexible polyethylene pouches were introduced to the market in the early 1980s [7], followed by several other containers: free-standing flexible pouches with a solid base, different "bag in a box" rigid units, and refill cartons coated with chemically resistant polymers. These novelties led to a 40% reduction in plastic bottle consumption, the use of more biodegradable, renewable, easily recycled material, and a decrease in the packaging and distribution cost. New bottles made of 100% "post consumer recycled" plastic sandwiched between two layers of virgin resin also became available [7].

## Short Historical Survey of Companies and Brands

In the late 1940s Hagge and Quaedvlieg patented DHTDMAC for imparting a soft feel and increased durability to cotton (DP902610 cited in [13]). The claim relied on fabric abrasion measurements. Ten years later, Harshaw Chemical, a raw material supplier, launched the first household fabric softener [1]; the company needed several years to invest in handling hot raw materials and succeed in getting satisfactory dispersions. Later, it sold the formulation to the Corn Products Company [1], which launched the product nationally under the trade name Nu-Soft in 1957 [4]. Afterwards, A.E. Staley introduced Sta-Puf and Procter & Gamble Downy. Eventually, the major detergent companies dominated the market. Within six years, domestic fabric conditioners represented a $30 million market, and $300 million market six years later [1].

In the 1960s brands such as Comfort (Lever), Soupline (Colgate-Palmolive), Lenor (Procter & Gamble), Silan (Henkel), and Orincil (Nobel Bozel) were launched in Europe, and Humming (Kao) in Japan [7]. All these systems were 3 to 8 wt% aqueous dispersions of cationic softener active. They mainly differed in the level of active and in their presentation (color, perfume, etc.). They delivered from 1.0 to 2.5 g of softener solids per kg of dry fabric [3].

In 1972 S.C. Johnson & Son, Inc. launched Rain Barrel in the U.S. [1]. This quaternary-based wash cycle fabric softener was intended for use with the consumer's choice of detergent. Because of their poor performance/cost ratio, this type of product never became very popular.

In the early 1980s two-in-one detergent softener compositions such as Bold 3, Axion 2, and Dynamo 2 were on sale on the European market. Fab Total, the latest generation, was still available in Latin America a few years ago.

The first dryer-added system appeared on the market in 1976 [14]. Brands such as Cling Free, Bounce, Snuggle, Sta-Puf, and Toss'nSoft sheets quickly represented 40% of the U.S. fabric softener market [7].

Among liquid products sold in polyethylene pouches, Add-Soft (Colgate-Palmolive) quickly gained 50% market share in Australia and Minidou (Lesieur-Cotelle) 65% in France [7].

## C.  Consumer and Producer Needs and Expectations

Defining consumer needs is not an easy task. Developing compositions that meet their expectations while fulfilling the safety and environmental requirements is even more challenging!

Fabric softeners are the most cosmetic of the household products. For consumers, their benefits are functional and emotional. Both types of attributes justify their use.

The softener performance perceived by consumers is the balance between the absolute efficacy determined in the laboratory and the product aesthetics. In other words, the consumer perception of the product performance is heavily influenced by aesthetic attributes such as fragrance and viscosity. Consumer tests indeed show that perfume, and more precisely perfume substantivity* on fabrics, is the main reason for preferring one product among several delivering the same softness. Consumers appreciate both the odor of the product itself, which generates the appeal and causes the purchase intent, and the smell of the laundered fabrics, which settles the repurchase intent.

Among the aesthetic attributes, viscosity also deserves special mention. The final viscosity of a rinse cycle fabric softener is indeed critical for the perception of product performance. Thick usually means rich in the consumer's mind. Thickness also affects the performance. Too thick a product is indeed not easily poured from the bottle and may disperse badly in the rinse, leaving residues in the dispenser of

---

*The substantivity may be defined as the tendency to adsorb onto the surface of various materials; it is measured by the proportion of product introduced in the rinse that is still present on the laundry after drying.

the washer; its efficacy will be altered. If, in contrast, the product does not exhibit enough consistency, it is also difficult to handle and to dose; it will be perceived as a poor performer and not economical to use. Cook [15] even considers the product appearance as the most attractive characteristics to consumers; afterward comes the feel and the touch, and eventually the absence of unpleasant odor.

Consumer expectations vary over time. In 1973 it was admitted that fabric softeners must fulfill the following functional requirements [15]:

1. They must keep white fabrics bright and should not cause dulling, yellowing, or graying.
2. They must not alter the shade of colored fabrics.
3. They should not impair the fabric affinity for water.
4. They should not induce corrosion of metal equipment.
5. They should not induce rash or dermatitis when in contact with human skin.

Nowadays, consumers ask for highly convenient products with the best price/performance ratio. Major functional benefits of fabric softeners are [10,16]:

1. To deliver a pleasant feel (softness) and smell.
2. To control the static electricity that impairs the comfort of handling and wearing clothes when ambient humidity is low.
3. To exhibit strong fabric care properties (fiber protection, looks new longer).
4. To make ironing easier.

Besides varying over time, the relative importance of the various functional benefits also varies with geographical location. They are not the same everywhere in the world, depending on washing habits and procedures. They are also linked to cultural, psychological, climatic, and lifestyle-related factors. For instance, laundering is fully automated in developed countries while washing by hand is still very popular in emerging markets; tumble drying is frequent in North America, while line drying is more common in the rest of the world.

It is admitted everywhere that the most important roles of fabric softeners are delivering softness and perfume. However, the next most important expectation varies from region to region: ease of ironing in Europe and Latin America, antistatic properties in North America and Asia.

The care aspects, fiber care and color care (and stain guard in Latin America), are much the same everywhere. Other benefits, closer to personal care attributes, are more specific to regions: long-lasting freshness and deodorization in Europe and North America, clean freshness in Asia and antibacterial activity in Latin America; skin mildness in Europe and Asia, luxury/comfort in Europe, and antimildew in Asia [16].

The major emotional benefits are pleasure, sense of task accomplishment, and caring for loved ones. These emotional attributes are generally reinforced through the product aesthetics, the package labels, and the advertising.

As expected, the relative importance of emotional benefits also varies from one region to another. Environmental considerations are essential in Europe (and increasingly in the rest of the world), and ease of use and price/performance ratio prevail in the U.S. The attitude is functional and technology-driven in Asia and emotional in Latin America. In all regions, using a fabric softener means to display personal commitment to laundering because of the importance of personal hygiene and the social image of clean, fresh-smelling clothes [16]. This is why the right choice of color, fragrance, and even texture to fit the product concept is so important and may markedly vary from region to region. The acceptability of a possible candidate must consequently be confirmed through different consumer tests in the countries of launch.

Fabric softener manufacturers also have specific needs:

1. Active molecules must be polyvalent because of the great variety of fibers and use conditions.
2. The fulfillment of the various requirements.
3. The availability of the technologies.

To be selected, a softening active must consequently fulfill the following conditions [5,17–19]:

1. It must be effective, delivering a pleasant touch to textiles without imparting a greasy feel or impairing their rewetting properties. It cannot alter the fabric color and must exhibit an antistatic efficacy. It should also deliver new consumer-perceivable benefits, if possible.
2. It must allow the formulation of stable, regular or concentrate finished products, with easy viscosity control. The finished products must be readily dispersible in water and deposit immediately when in contact with the washed fabrics, to get a uniform deposition within the short rinse cycle duration.
3. It must be chemically stable to avoid loss of performance on storage and any generation of undesirable odor or color.
4. It must be industrially available at the right quantity and quality, with an acceptable ecological and toxicological profile. In Europe, a valuable candidate must also satisfy the directives of the European Union (formerly European Community, EC).
5. It must exhibit a better cost/performance ratio than existing actives; manufacturers are indeed facing severe cost constraints.

No active offers all the characteristics of an ideal ingredient.

## III. BENEFITS AND DRAWBACKS

The importance of softeners in fiber treatment has long been recognized. They were routinely used in the textile industry for the lubricity and flexibility they impart to yarns, protecting them from damage during textile processing. They also impart a pleasant touch that enhances the consumer desire for a textile item.

From a consumer standpoint, fabric softeners deliver multifarious benefits. They may also present some side effects.

## A. Benefits

### 1. Basic Benefits

*(a) Softness.* Softness has been defined by Mallinson (quoted by Datyner [20]) as an alteration in feel making the item more pleasant to the hand. In other words, it is a pleasant feel perceived when the fabric is in contact with the skin. Fabric softeners prevent textile stiffening, usually observed after a wash with a detergent in a washer, and keep the garments in the state wanted by the user.

The improvement of the fabric feel and comfort is particularly noticeable on cotton items, but softeners' beneficial effects are also perceived on other fabrics such as wool, viscose, acetate, polyamide, and polyester.

*(b) Antistatic.* Cellulosic fibers such as cotton and viscose do not develop static charges under normal relative humidity. The situation is quite different for synthetic fibers at low ambient relative humidity such as that encountered in winter months or when an automatic tumble dryer has been used. The well-known static cling takes place upon tumble drying and an electric shock may even occur when removing the items from the dryer. In areas with a dry climate, friction can also generate electricity upon wearing garments, causing synthetic fibers to stick to the skin and to attract charged dirt present in the air.

These effects were very unpleasant to consumers, and the problem became more acute as synthetic fibers became more popular. Much more serious are the fire and explosion hazards created by static charge if clothes produce electric sparks in an atmosphere of a flammable solvent.

These inconveniences are overcome using a fabric softener.

*(c) Perfume.* It is commonly believed among softener manufacturers that many users purchase the product only for its fragrance. Whether this is true or not, the product scent is certainly one of its key characteristics since a pleasant fragrance is the first signal of the softener efficacy. It differentiates fabric softeners from one another and sustains the claims for new products (e.g., effectiveness, freshness, more softness, new and improved). Perfume suppliers even claim that the differentiation resulting from the incorporation of a higher quality — hence more expensive — fragrance is enough to provide larger market shares [21]!

Considerable time, effort, and money are devoted to the development of a softener fragrance. Fabric softeners must be nicely perfumed in the bottle, and impart a typical and pleasant smell to the laundry that is immediately and repeatedly perceived by the user at various stages of the laundering process. These include when pouring the softener into the washing machine, when removing wet laundry from the washing machine, when removing dry laundry from the line or from the dryer, when folding or ironing the clothes, and when using them. That is achieved by carefully designing the perfume composition.

## 2. Additional Benefits

*(a)  Smoothness and Easier Ironing.*   Hot ironing is usually necessary to remove the wrinkles from pure cotton garments. Fabric softeners improve the ease and efficiency of the ironing process. Their actives work as lubricants and favor fiber slipperiness. As a result, the garments are less wrinkled and the friction between the fabrics and the iron is reduced, thereby facilitating gliding of the iron. A 10 to 20% reduction of the time necessary for ironing may be achieved [13,22], which is especially meaningful in industrial laundries.

Benefits such as ease of ironing and wrinkle reduction are, however, less easily perceived by consumers.

Recently a new family of fabric care products appeared on the market. Called fabric conditioners, they are actually fabric softeners with enhanced antiwrinkle properties. By facilitating the ease of ironing, they address the basic consumer need of spending less time in one of the most tedious household chores. One of these products moreover exhibits completely different aesthetics from traditional fabric softeners, drawing attention to its specificity.

Compared to usual fabric softeners, the improvement delivered by fabric conditioners is not clearly consumer-perceivable. Consumers want more: no ironing at all. As a result, their market share remains low.

*(b)  Drying Time.*   Because of the hydrophobicity of their actives, softeners make fibers to bind less water; moreover, softened fabrics retain water less firmly. The extent of the effect varies. Bräuer *et al.* report about 10% less water linked to the fibers [23]; the spinning time is consequently reduced by 40% [22]. Lang and Berenbold report a 7 to 15% [24] or even a 15 to 20% [13] decrease of residual humidity after final spinning of the fabric. The drying time in tumble dryers is also decreased [11,18,24–26]. Barth *et al.* find that the effect remains marginal [4] while Berenbold reports an about 14% cut of the drying time [11], leading to a 12% reduction of the energy consumption [24].

Paradoxically, softener-treated cotton also exhibits an improved permeability to water vapor, leading to an improved comfort in wear.

*(c)  Fiber Protection.*   In the washer and dryer, and during wear, fabrics undergo severe mechanical and chemical constraints that can damage the fibers. Fabric softeners replace the finish removed by the detergent and lubricate the fibers, reducing the interfiber friction. This results in a reduction of the fiber damage [3,11]. Although the protection only takes place when the garments are dried and worn, not in the wash, fabric softeners increase their life span. Clothes look better and newer after repeated launderings [27]. Fiber damage reduction is illustrated in Figure 12.1.

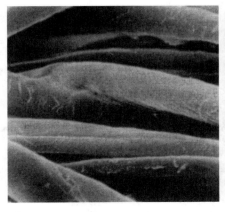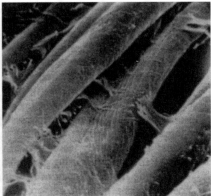

**FIG. 12.1** Fiber damage reduction by fabric softeners. These electron micrographs are of cotton bath towels after 12 cumulative wash cycles with fabric softener (left) and without fabric softener (right).

*(d) Antibacterial Activity.* Because of the trend for lower washing temperatures, the microbial threat becomes increasingly probable. Bacteria and fungi are detrimental since they degrade textiles, produce malodors, and generate skin irritation or infection [20].

Domagk [28] first reported the bacteriostatic activity of cationic surfactants in 1935. As most fabric softener compositions are based on cationic actives, it sounds logical to expect some biological protection from these products. Not all authors, however, agree on their exact efficacy. According to Martins *et al.* [29] and White [30], they are antibacterial agents. Laughlin speaks about a germicidal effect [26] and Milwidsky considers them as moderately good bacteriostats [18]. Chalmers stresses the fatty chain length effect: a bactericidal activity is observed for a $C_{12}/C_{14}$ chain length, longer chains exhibiting bacteriostatic properties only [1]. In contrast, Barth *et al.* report that cationic surfactants used as fabric softeners do not exhibit any antibacterial activity [4].

These differences are possibly due to variations in the experimental conditions of the various studies (e.g., ionic strength). Nevertheless, the antibacterial efficacy of softener actives under normal use conditions remains questionable.

## B.  Damage

Fabric softeners are safe for all washable fabrics. Some side effects are, however, possible in the case of misuse. For instance, pouring the softener directly onto garments may cause fabric staining while strongly overdosing the product may lead to a greasy feel and reduced affinity for water and/or to color alteration [15,31].

## 1.  Color Alteration

By color alteration is meant graying or yellowing of whites, and hue alteration or color fading of dyed items. These problems have their origin in various phenomena:

1.  Cationic actives interact with anionic fluorescent brighteners and reduce their whitening efficacy [2,26]. Whether this is visually perceptible or not varies among reports. Crutcher *et al.* [32] and Wilson [33] conclude that, after a larger number of cumulative launderings, the loss is visually perceptible. On the contrary, Baumert and Cox Crews report that the decrease of whiteness index is significant, but not visually perceptible [34].
2.  Cationic surfactants cause deposition of detergent residues loaded with soil, which are present in the liquor because of incomplete rinsing [1]. Some alteration of the fabric appearance may occur even in the absence of softener. Some soil redeposition on clothes may indeed take place in the wash, resulting in lightly soiled whites turning gray or yellowish and colors to become dull [35].
3.  Colored impurities such as iron, nickel, cobalt, or copper salts are present in the rinse liquor [1].

The whiteness index of softened fabrics depends on the fabric softener treatment, on the number of launderings, and on the fabric type. Dryer softeners significantly improve the whiteness index of cotton [34].

## 2.  Hydrophobicity

*(a)  Wettability.*  The intrinsic affinity of fibers for water depends on their chemical nature. The absorption capacity of a given amount of hydrophilic fibers such as cotton is much larger than that of the same amount of hydrophobic fibers such as polyester. Adsorbing chemicals such as surfactants may modify that characteristic. For instance, softeners make hydrophilic fibers more hydrophobic. Hence, it is not surprising that they reduce the wettability upon use. The effect is stronger with mixed fabrics than with plain cotton [24]. This characteristic is important, as softened textiles must absorb the humidity of the skin with which they are in contact. That is a true drawback in the case of terry towels that may exhibit less wiping efficacy. In fact, it is the water absorption rate that is impaired by the presence of softener at the fiber surface, with the absorption capacity remaining unaffected [1,36].

This problem can be avoided by limiting the amount of softener in the rinse. Because of the softener buildup with time, an increase of the cationic concentration at the fiber surface is nevertheless unavoidable. At the levels used under practical conditions, no difference exists between treated and nontreated garments.

As already mentioned, softener-treated cotton items exhibit enhanced permeability to water vapor. That paradoxical effect is due to the reduction of the water content inside the fibers. Since they swell less, the fabric is more permeable to vapor [13]. This effect decreases as the amount of softener on the fibers increases.

*(b)   Greasy Feel.*   Since fabric softeners are made of greasy material, it is not surprising that fabrics treated with an excess of fabric softener exhibit a greasy touch.

## 3.   Compatibility with Anionic Surfactants and Dyes

Cationic surfactants are incompatible with anionic surfactants. They are precipitated by various ingredients that can be present in a wash or rinse liquor, such as bentonite, $TiO_2$, starch, and phosphates.

## 4.   Flameproof Treatment

It is often said that cationic softeners should not to be used on baby clothes. This is because of a possible negative effect on the flameproof treatment.

## IV.   FABRIC SOFTENER MARKET

The most important markets for fabric softeners are Europe, the U.S., and Japan. In all three the in-home penetration is high (>60%) and also the consumption (17 l/household/year) [16]. In 1997 the worldwide fabric softener market accounted for $3.5 billion ($1.1 billion in the U.S.).

Many authors (e.g., [14]) explain the growth of the softener market by the evolution of the type of fabrics in laundering. Natural fibers must be softened and static electricity must be countered on synthetic fibers. On top of these criteria, the evolution of consumer needs and expectations discussed above has had major consequences for the evolution of the softener market. Effective cleaning is a must, fabric softening a pleasure.

The factors influencing the appearance of new products have changed over time [7] and with region:

1.   Prior to the 1980s, the prevalent parameters were the identification of the need, of a population with a discretionary income, and of a cost-effective distribution and sale.
2.   From 1984, the development driver has been improved consumer convenience. It led to the appearance of concentrates.
3.   From 1988, environmental considerations have taken precedence, causing the reformulation of European fabric softeners.
4.   Today, building profitable market shares faces low penetration/low net income in high-growth markets and severe competition in developed markets. The major trend is a growth in low-cost packaging, with low-cost bottles and refills in developed markets and low-cost bottles and unit dose sachets in high-growth markets [16].

Several examples illustrate the key role of some parameters on the success of products exhibiting objective advantages. In the early 1980s Germany was the leading fabric softener market in Europe. Consumption reached a maximum in 1983. From 1985, the public debate on the relative benefit delivered by rinse cycle fabric softeners relative to the water pollution that they allegedly caused led to a steady fall in consumption. From 1988, stagnation was also observed in other "green" European countries such as The Netherlands, Denmark, Austria, and Switzerland [11].

Concentrated products lead to a reduction of plastic waste and distribution costs. They enable space savings on the shelves at the point of sale and at home. Their success, however, varies strongly from country to country: introduced in Germany in 1980 [14], they represented 95% of the German rinse cycle fabric softener market in 1994 and only 10% of the Spanish and Italian ones [21]. In France and the U.K. they are at parity with regular products [5]. Likewise, they appeared on the Japanese market in 1988 and represented 50% of sales six years after [5]. On the contrary, in emerging countries, regular products still remain the most popular, as consumers feel they get more for their money.

Of course, differences of penetration are also linked to objective parameters. For instance, electrical tumble dryers have always been more common in North America than anywhere else in the world. In 1983, 65% of U.S. households owned a dryer, the proportion being only 10% in Europe [14]. In 1994 the figures were 75% in the U.S. and 20% in Germany [11]. Today, the proportions are 70 to 80% in the U.S., 21% in Western Europe, only about 15% in Japan, and still less in other parts of the world [3].

Because of the static electricity imparted to synthetic fabrics by tumble drying, and the increasing proportion of synthetic fabrics in the U.S., the sales of dryer-sheet fabric softener rose much more in the U.S. than in the rest of the world. By 1983, these products accounted for 40% of all household fabric softener sales. Today, tumble dryer sheets and liquid softener sales are of the same order of magnitude [5,6,16]. In contrast, rinse-added softeners had a much larger impact in Europe, where 100% cotton items have always been dominant and line drying is still standard practice; hence, the softener effect is more noticeable to consumers.

## V. COMPOSITIONS

Fabric softener compositions have been regularly modified, as a result of variations in the performance needs, in the expected secondary benefits, and in regulations. They have always been fascinating and challenging products, as they must be stable in the bottle and destabilize upon dilution to deposit onto fabrics during the rinse. This is achieved by carefully choosing the ingredients.

## A.   Softening Ingredients

Fabric softener actives can be classified into three groups: organic, inorganic, and silicones. To the organic class belong cationic surfactants, lime soaps, and oils. Cationic surfactants are by far the most frequently used. They are found in wash, rinse, and dryer softeners. Montmorillonite clay is the main inorganic softener and is essentially used in softergents. Silicones are much less common and are generally used as a minor component in combination with organic softeners.

To be worth considering, a fabric softener candidate must fulfill several requirements, which have been reviewed above.

## 1.   Organic Actives

*(a)   Cationic Surfactants.*   For 40 years most rinse cycle fabric softeners have been built from cationic surfactants, because of their high degree of substantivity and high exhaustion rates from dilute solution. Thousands of patents exist, covering hundreds of different molecules and of mixture compositions. Only a few of them have been of practical importance.

The characteristics of each active result from the details of the molecule structure. This includes the number and length of the alkyl chain(s), degree of saturation, and presence of oxygen atoms. All products have their own strengths and weaknesses; none of the existing actives meets all criteria of quick biodegradability, low toxicity, acceptable cost, good stability, good softening, antistatic efficacy, etc. The most efficient softening molecules are the ones bearing two long alkyl chains. Straight chains are preferred to branched ones and saturated chains to unsaturated ones.

From a softening standpoint, the most efficient alkyl chain length is $C_{18}$. Industrial raw materials are consequently prepared from natural tallow, in which $C_{16}$ and $C_{18}$ chain lengths predominate. Tallow contains 5% $C_{14}$, 35% $C_{16}$, and 60% $C_{18}$ (stearic and oleic). The exact proportion of fatty chain length varies from delivery to delivery, according to the origin (beef, mutton, palm oil), on the season, and on the amount of rain [18]. These molecules are extremely substantive, impart to fabrics outstanding draping properties, and deliver excellent abrasion resistance and static control to synthetic fibers. Once on the fabrics, however, their water absorbency is less than other cationic softeners. They are also more difficult to formulate since they exhibit higher melting points, requiring higher temperatures to handle and disperse, and leading to more viscous aqueous dispersions than shorter-chain derivatives. With unsaturated fatty chains, the softening effect is somewhat reduced but the fabric is still left with a dry and very supple, flexible feel.

The usual counterions are chloride (in Europe and the U.S.) and methylsulfate (in the U.S.).

For 25 years almost all softeners were made of ammonium ion derivatives bearing two straight fatty chains. The ammonium ion makes the molecule water-dispersible, while the alkyl chains account for the tendency of molecules to deposit onto the fabrics and for the softening efficacy. Developed in the 1940s, DHTD-MAC was the earliest commercial active, and the most popular one. It is commonly referred to simply as "quat."

Many authors give details of the synthetic procedures for the manufacture of DHTDMAC [3,37,38]. It is prepared by a rather complicated process, detailed in Figure 12.2.

The resulting solid is crystalline. The chains are fully hydrogenated, and they do not bear any functional groups to hinder crystal formation. Pure dioctade-cyldimethylammonium chloride (DODMAC) melts and decomposes at 147°C [26]. The melting temperature depends on the level of residual isopropyl alcohol in the raw material. The higher the content, the lower the melting point. This is why the raw material always contains some alcohol, usually isopropanol, some-times ethanol in the U.S. The amount is critical since the melting point remains too high when the alcohol level is too low, and the active ingredient disperses poorly in water. Organic solvents must be maintained at a low level since, in addition to their unpleasant smell perceptible in the finished product, they interact with the hydrophobic layer of the softener particles, causing the membrane structure to disrupt and the particles to stick together [39]. Hence the active ingredient is generally diluted with 25 wt% of an alcohol–water mixture, corresponding to 15% isopropyl alcohol. As a result, DHTDMAC is commercially available as a waxy solid at room temperature, which becomes fluid at 50°C and can be easily pumped and handled at 60°C.

Although DHTDMAC fully meets all the needs and expectations criteria listed above, its use has dramatically decreased because of changes in European reg-ulations, which led to its replacement by esterquats in the early 1990s to avoid adverse labeling. Moreover, the formulation of DHTDMAC-based concentrated softeners was not possible without the help of cosofteners. With other actives the technical constraints do not exist or are less stringent.

The chemical structure of esterquats is very versatile. The structure may vary by the alkyl chain length and saturation extent, by the mono-, di-, and triester ratio, and by the quaternization degree. It is similar to the DHTDMAC structure in that they essentially bear two hard tallow chains and an ammonium ion. How-ever, in esterquats at least one of the fatty chains is linked to the cationic nitrogen through ester bonds. This linkage is a point of weakness making biological degra-dation easier and faster [40–44]. Microorganisms in sewage treatments readily cleave the ester bonds, depriving the molecule of its substantivity. As a result, the biodegradability profile is dramatically improved.

The most common esterquats in fabric softener formulation are quater-nized di-tallow esters of methyltriethanolamine, dimethyldiethanolamine, or

(a)

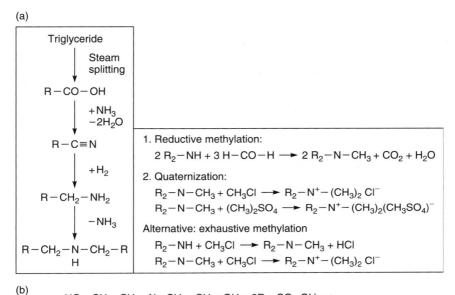

(b)

$$HO-CH_2-CH_2-N-CH_2-CH_2-OH + 2R-CO-OH \longrightarrow$$
$$\qquad\qquad\qquad | $$
$$\qquad\qquad CH_2-CH_2-OH$$

$$R-CO-O-CH_2-CH_2-N-CH_2-O-CO-R$$
$$\qquad\qquad\qquad\qquad |$$
$$\qquad\qquad\qquad CH_2-CH_2-OH$$
$$+R-CO-O-CH_2-CH_2-N-CH_2-CH_2-OH$$
$$\qquad\qquad\qquad\qquad\quad |$$
$$\qquad\qquad\qquad\quad CH_2-CH_2-OH$$
$$+R-CO-O-CH_2-CH_2-N-CH_2-CH_2-O-CO-R$$
$$\qquad\qquad\qquad\qquad\quad |$$
$$\qquad\qquad\qquad\quad CH_2-CH_2-O-CO-R$$

$$\qquad\qquad\qquad\qquad\qquad CH_3$$
$$\qquad\qquad\qquad\qquad\qquad |$$
$$R-CO-O-CH_2-CH_2-N^+-CH_2-CH_2-O-CO-R\ (CH_3SO_4)^-$$
$$\qquad\qquad\qquad\qquad\qquad |$$
$$\qquad\qquad\qquad\qquad\quad CH_2-CH_2-OH$$

$$\qquad\qquad\qquad\qquad\qquad CH_3$$
$$\qquad\qquad\qquad\qquad\qquad |$$
$$+(CH_3)_2SO_4 \longrightarrow +R-CO-O-CH_2-CH_2-N^+-CH_2-CH_2-OH\ (CH_3SO_4)^-$$
$$\qquad\qquad\qquad\qquad\qquad\qquad |$$
$$\qquad\qquad\qquad\qquad\qquad\quad CH_2-CH_2-OH$$

$$\qquad\qquad\qquad\qquad\qquad CH_3$$
$$\qquad\qquad\qquad\qquad\qquad |$$
$$+R-CO-O-CH_2-CH_2-N^+-CH_2-CH_2-O-CO-R\ (CH_3SO_4)^-$$
$$\qquad\qquad\qquad\qquad\qquad |$$
$$\qquad\qquad\qquad\qquad\quad CH_2-CH_2-O-CO-R$$

**FIG. 12.2** Synthesis of softening active molecules: (a) DHTDMAC; (b) esterquat.

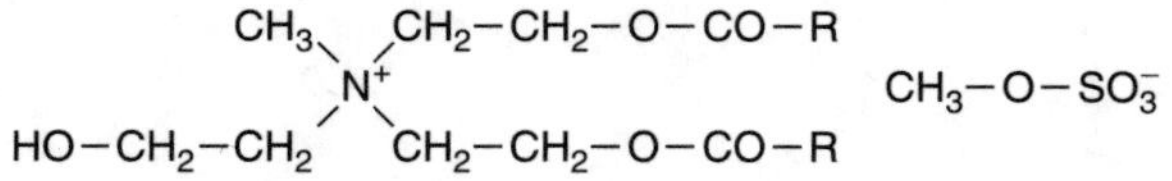

$$CH_3\diagdown \quad CH_2-CH_2-O-CO-R$$
$$N^+$$
$$HO-CH_2-CH_2 \diagup \quad CH_2-CH_2-O-CO-R \qquad CH_3-O-SO_3^-$$

Ditallow ester of methyltriethanolammonium methylsulfate

$$CH_3\diagdown \quad CH_2-CH_2-O-CO-R$$
$$N^+ \qquad Cl^-$$
$$CH_3 \diagup \quad CH_2-CH_2-O-CO-R$$

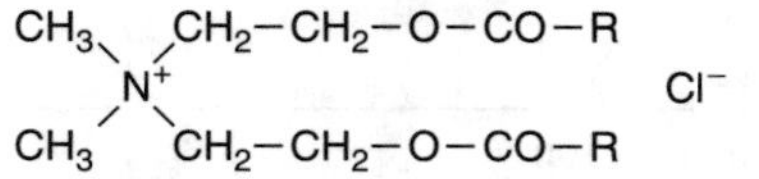

Ditallow ester of dimethyldiethanolammonium chloride

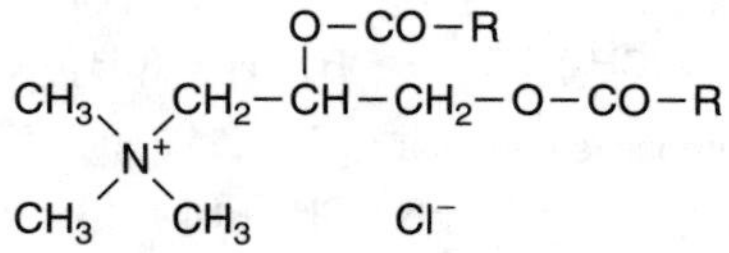

$$O-CO-R$$
$$CH_3\diagdown \quad CH_2-CH-CH_2-O-CO-R$$
$$N^+$$
$$CH_3 \diagup \quad CH_3 \qquad Cl^-$$

Ditallow ester of trimethyldihydroxypropylammonium chloride

**FIG. 12.3** Structures of the most common esterquats used in fabric softeners.

trimethyldihydroxypropylamine (Figure 12.3). Ester amidoamines are also used, mainly in Japan, and di-tallow imidazoline ester worldwide but to a lesser degree [5,16].

Esterquats molecules are not new. Alkanolamine-based molecules exhibiting a better cost performance were patented much before the environmental controversy. Patents covering esters of methyldiethanolamine were issued around 1970 by BASF and Hoechst while Stepan Company commercialized diethyloxyester dimethylammonium methylsulfate in 1974. The latter molecule exhibits an excellent softening performance but does not readily form stable dispersions. Three years later a patent disclosing the esterquat built on *N,N*-dimethyl-3-aminopropane-1,2-diol [45] was issued. This molecule appeared on the European market because of its excellent biodegradability and aquatic toxicity profiles with no compromise in softening performance [3]. The patent covering the use of triethanolamine-based esterquats as fabric softeners was also issued as early as 1975 [46]. These compounds are prepared by esterifying triethanolamine with fatty acids and quaternizing the resulting esteramines with methyl chloride or dimethylsulfate. The raw material is consequently a mixture of quaternized and nonquaternized mono-, di-, and triesteramines, whose statistical distribution is thermodynamically controlled [3]. The exact composition of the raw material in terms of esterification and quaternization degrees requires the isolation of the individual constituents by solid phase extraction and the elucidation of

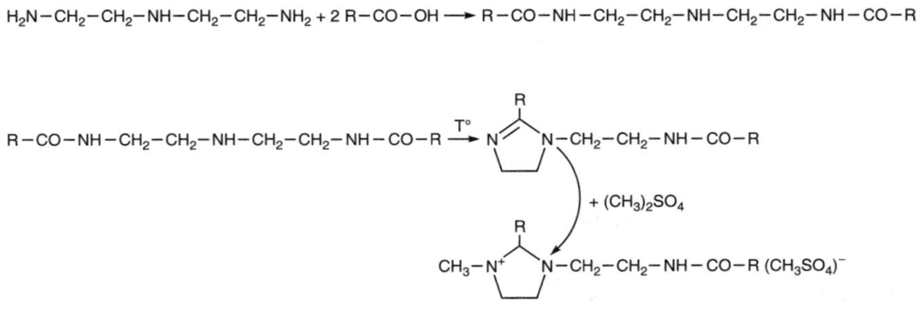

$$H_2N-CH_2-CH_2-NH-CH_2-CH_2-NH_2 + 2\,R-CO-OH \longrightarrow R-CO-NH-CH_2-CH_2-NH-CH_2-CH_2-NH-CO-R$$

**FIG. 12.4** Synthesis of DHTIMS.

their structure by [1]H-NMR [47]. Typical examples of these new molecules are N-methyl-N,N-di(2-($C_{16}$/$C_{18}$-acyloxy)-ethyl)-N-(2-hydroxyethyl)ammonium methylsulfate [48], 2,3-di($C_{16}$/$C_{18}$-acyloxy)propyltrimethylammonium chloride [49], and 2-($C_{16}$/$C_{18}$-alkyl)-3-($C_{16}$/$C_{18}$-acyloxy)ethylimidazoline [50].

Among the many other molecules used for fabric softening, only two have been a commercial success: imidazolinium salts and diamidoquaternary ammonium salts [26]. Imidazolinium methylsulfate (DHTIMS) has been a useful alternative to DHTDMAC in the U.S., and in Europe to a lesser extent. Both actives indeed deliver similar benefits. The counterion here is always methylsulfate.

Imidazolinium salts are also derived from diethylenetriamine. The diamidoamine that results from the esterification of this molecule is dehydrated into an imidazoline, then quaternized with dimethylsulfate, as shown in Figure 12.4 [51]. The resulting molecule is no longer susceptible to hydrolysis. Despite the low price of the material from which they derive, imidazolinium salts are expensive because of their manufacturing cost. This is due to the costly conditions necessary to convert the diamidoamine into imidazoline [3]. Details of the synthesis may be found in Egan [37] or Billenstein and Blaschke [38].

Saturated derivatives are almost as efficient as DHTDMAC for softening, but exhibit the same limitations for formulating concentrated softeners. They are mainly found on dryer-added softening sheets. If the imidazoline is dispersed in water containing enough acid to neutralize the amines, no phase separation occurs after a month of storage at room temperature. Once on the fabric, the ring is hydrolyzed during tumble drying, forming diamidoamine, which delivers more softening [3].

Fatty amides — (R–CO–NH–CH$_2$–CH$_2$)$_2$–N$^+$ (CH$_3$)(CH$_2$–CH$_2$–O)$_n$H — also require low pH (3.5 to 5.0) to be dispersed since they do not bear a permanent charge. They are the third most popular active in North America. Because of the European restrictions on using ethylene oxide derivatives, they are much less used in Europe [7]. They are easily formulated in concentrated products.

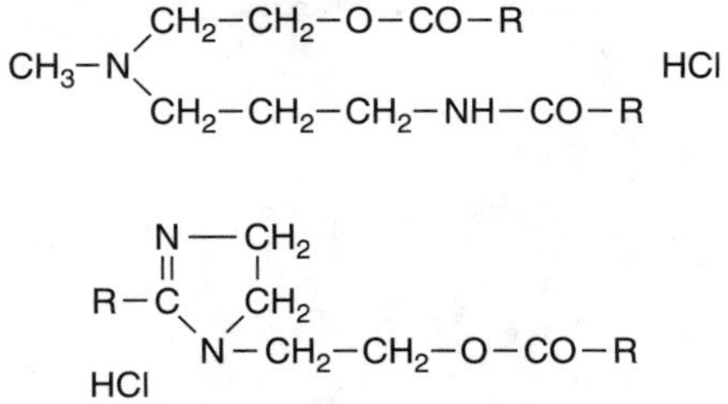

$$CH_3-N\begin{matrix} CH_2-CH_2-O-CO-R \\ \\ CH_2-CH_2-CH_2-NH-CO-R \end{matrix} \qquad HCl$$

$$R-C\begin{matrix} N=CH_2 \\ \\ N-CH_2-CH_2-O-CO-R \end{matrix} \\ HCl$$

**FIG. 12.5**   Acid salts of diesteramines, used in fabric softeners.

Other actives (V, VI, VII, and VIII in McConnell [7]) meet European environmental safety regulations (Figure 12.5). They are expected to deliver the same softness as esterquats, but they are more expensive [7]. They are used by Unilever and by Kao [7]. Puchta also mentions the amide as a valuable alternative to esterquats [8].

Amphoterics — substituted aminoacids, sulphobetaines, amine oxides — are less substantive and more expensive than cationic surfactants [52]. They deliver more softness and a better static control than nonionic surfactants; they also better withstand the subsequent launderings.

Softening actives cannot be compared on the basis of their softening performance only. To draw meaningful conclusions, several other parameters must also be considered, such as their ease of handling and processability, their ease of formulation as concentrates, their stability upon storage and under use conditions, and their cost [16].

They should not be too sensitive to salting-out by electrolytes or by high temperatures, causing the formation of scum or precipitates, and they should not contain colored impurities that may alter the shade or the light fastness of the dyes or the brightness of whites. When the active bears unsaturated chains, the softener may cause a malodor to develop because of the rancidity, and iron and other heavy metals present in hard water may interact with the double bonds causing the appearance of yellow spots on the items [5]. However, unsaturated chains usually make the fibers less hydrophobic than saturated ones. More generally, compatibility problems may also arise from the presence of the positive charge of the softener in the anionic environment of washing; they may interact with anionic surfactants and dyes and precipitate. The softening effect is then lost or at least strongly impaired.

Alkyl quaternaries are the most efficient fabric softeners. On a weight basis, they deliver more softness than any other system. They exhibit a strong affinity to almost all fibers, and usually impart a durable effect. The water absorbency of the fabrics is usually only slightly impaired by the presence of the softener but the effect gets worse upon cumulative treatments. Their antistatic efficacy is usually less than that of other cationic surfactants [52].

Chloride derivatives of the quaternaries have the drawback of causing corrosion of storage tanks, manufacturing vessels, and tumble dryers when dryer cycle fabric softeners are used (see below). Moreover, these chemicals are waxy pastes and must be melted before being dispersed in water. These problems are overcome by using methylsulfate derivatives, which do not corrode stainless steel and are usually liquid at room temperature [1].

The main limitation linked to DHTDMAC is that formulating concentrates with a solid content exceeding 15% is generally not possible without using co-softeners. To exceed this concentration, amidoamine quats or imidazolines must be introduced in the system [6].

Esterquats cause fewer problems for formulating ultras and concentrates [7]. Carefully selecting structural details of the active such as the mono-, di-, and triester ratio, the presence of unsaturation on the alkyl groups and their *cis–trans* configuration, the pH value, and the particle size of the dispersion enables one to get stable, low-viscosity softeners containing 20 to 25% active [6]. Another advantage of esterquats is that they do not stain fabrics [7].

The presence of the ester function in the molecular structure facilitates rapid biodegradation in sewage but also threatens its chemical stability on storage in the bottle and in the rinse, where it has to work at a slightly alkaline pH. At pH 6, for instance, the molecule is completely degraded after four weeks of storage at 50°C [6]. Paradoxically the hydrolytic stability is much enhanced by keeping the pH below 3.5.

Since esterquats were accepted in Europe in the early 1990s because they fulfill European environmental regulations, it was on the European market that the first ultra concentrated products appeared, packaged in plastic pouches or sachets. Subsequently they spread around the world [3].

Imidazolinium salts are said not to impair the wettability of cotton when over-dosed in the rinse and to control efficiently static electricity. They are more easily processed than other cationic surfactants, especially in the formulation of concentrates, but they are more expensive, less efficient from a softening standpoint, and possibly cause more yellowing [7,8,22,52,53].

If the amine is used instead of the ammonium derivative, and if the pH is kept low enough to maintain a positive charge on the nitrogen atom, the imidazolinium active becomes more efficient than DHTDMAC for softening, there is less yellowing, and the ability to be concentrated is excellent [7]. Likewise, the oleyl-substituted imidazolinium salt enables an easy formulation of concentrated products, and makes fibers less hydrophobic than its saturated counterpart [8].

Amidoamines exhibit a softening performance close to that of the unsaturated tallow-imidazolinium salts, but are less difficult to formulate in concentrated products exceeding 20% solids without using special additives [3]. They are very mild to the skin, are more biodegradable than DHTDMAC, and they cause little or no corrosion [7]. The molecule efficacy can be fine-tuned by minor structure modifications. For instance, EO/PO variations affect the ease of formulation, the rate of deposition, the feel of the softened fabric, the static control efficacy, the affinity of treated fibers for water, and the durability of the effects. Some derivatives are expensive.

Various authors have ranked the softening actives [19,34,36,54]. They proposed the following sequence of decreasing efficacy: dialkyldimethylammonium > imidazolinium > diamido alkoxylated ammonium.

*(b)* *Anionic and Nonionic Surfactants.* The oldest anionic softeners are soap derivatives. $R–SO_3$ and $R–O–SO_3$, sulfosuccinate, and soap have been reported to exhibit some softening efficacy [52]. Their affinity to fabrics remains limited, as their hydrophobicity is moderate and fibers are negatively charged in water. Consequently, they are usually applied by padding [20,52].

Mineral oils, paraffin and other waxes, polyethylene, polyethylene glycol, ethoxylated glycerides, ethoxylated fatty amines, and esters of fatty alcohols and acids have also been used in fabric softening [52], and also nonionic actives such as glycols, sorbitol, and urea, but in combination with a charged active.

Because of their good affinity for water, all these compounds are efficient antistatic agents (especially ethoxylated nonionic actives). They also exhibit a good stability to heat [52]. Polyglycol fatty esters deliver good softness and static control without any drawback [15,55].

Nonionic and even anionic surfactants have also been added in small amounts to DHTDMAC to boost a product's softening efficacy. For example, it was shown in the late 1970s that the performance of a 6% DHTDMAC composition is matched by a mixture of 4.4% DHTDMAC and 0.6% anionic [56]: 1.6% DHTDMAC could then be replaced by 0.6% anionic, which is less expensive. That was quite unexpected, as it was generally accepted that fabric softeners must be introduced in the last rinse of the laundering process to avoid their neutralization by the anionic detergent residues on the fabric, which causes the formation of insoluble species.

Likewise, a dispersion of 3.6% DHTDMAC–0.9% tallow alcohol blend delivers as much softness as a 6% DHTDMAC dispersion [57]. From a performance standpoint, 0.9% fatty alcohol in the mixture is equivalent to 2.4% DHTDMAC, while it contains as many fatty chains as 0.8% DHTDMAC (tallow alcohol: mol wt = 259, 1 chain; DHTDMAC: mol wt = 569, 2 chains). Fatty alcohols alone are not better than DHTDMAC in fabric softening. Due to the insolubility of fatty alcohols in water, this equivalence has been evidenced by spraying alcoholic solutions of actives onto textiles. Actually, the DHTDMAC–cosoftener synergy results from a modification of the DHTDMAC dispersion structure. A dispersion of straight DHTDMAC is made of large multilamellar vesicles while only small unilamellar vesicles are formed in the presence of cosoftener. It has been shown that the formation of these smaller vesicles is made possible by the insertion of the cosoftener molecules between the ammonium ions at the external surface of the vesicles. The formation of smaller vesicles causes a more even fiber coating, hence a greater softening efficacy.

This synergy has been systematically investigated and exploited to enhance the cost efficiency of fabric softeners, delivering either better performance at equal cost or the same performance at improved cost. It has also opened the door to the formulation of DHTDMAC-based concentrates, which without cosofteners is not possible for viscosity reasons.

The main nonionic cosofteners are fatty alcohols, fatty acid esters, ethoxylated fatty amines, or lanolin derivatives. To the cationic–anionic systems belong ether sulfates, alkyl sulfonates, or fatty acids [5,6,10]. In all the blends, the weight ratio of DHTDMAC to the cosoftener is always greater than unity [10].

## 2. Inorganic Actives

*(a)   Silicones.*   Silicones were first used by the textile industry as lubricants [34]. They also improve permanent press finish durability and garment wear life [5,26]. In softeners, they were considered solely for their unique softening properties. The benefit they deliver is, however, much broader, as suggested by the numerous patents that have been filed. They are claimed to reduce fabric wrinkling in the washer, to facilitate ironing by improving the glide of the iron, and to enhance the fabric water absorbency. They also strengthen color protection, shield fibers from staining, and help maintain the shape of garments [27].

Several chemistries have been developed: polydimethyl siloxane polymers (PDMS), amine- or amide-functional polydimethyl siloxanes, and silicone gum-in-cyclic blends. They are usually supplied as emulsions, offering a large choice of candidates to achieve the desired performance and physicochemical properties such as viscosity.

Polydimethyl siloxanes deliver a particular, very well appreciated feel referred to as "silicone-touch." This is due to the strong reduction of the cotton friction coefficient, which also facilitates gliding of the iron during pressing. These effects are probably due to the flexibility of the siloxane backbone and to the free orientation of the methyl groups at the polymer surface [27].

In esterquat-based and more so in DHTDMAC-based softening compositions, PDMS improves the wettability of the treated cotton at low PDMS/quaternary ratio. This property is surprising, considering the strong hydrophobicity of the material. Since PDMS is not very substantive, it is easily removed in the subsequent wash.

Amine- and amide-functional silicones resist launderings better, as they react with cotton hydroxyl groups through the amine moiety. They are more efficient than conventional nonreactive silicones in boosting the softening efficacy [58,59], the ease of ironing, and the resistance to wrinkling [60]. They also deliver antistatic benefits [60]. Although they have never been intensively used because of their high cost, amine silicones do bring a consumer-perceptible new dimension to rinse cycle fabric softeners.

Silicone gum-in-cyclic blends are dispersions of very high-molecular-weight silicone polymers in volatile silicone. Cyclomethicone helps the polymer to spread on fibers.

Emulsion characteristics such as the type of emulsion and its particle size are important to determine the nature of the benefit the product will deliver [27]:

- Macroemulsions remain on the external surface of fabrics. They achieve an excellent lubrication through the decrease of the dynamic coefficient of friction (see below). An excellent softness results.
- Microemulsions (<150 nm) can penetrate into the yarn and deposit onto the fibers. They deliver a dry lubrication and feel. They probably reduce the static coefficient of friction.
- Polymer emulsions (150 to 250 nm) deposit on the external surface of fabrics and on fibers. They improve the softness and the ease of ironing since they reduce both static and dynamic coefficients of friction.

The surfactant system of the emulsion can influence silicone deposition. This exceeds 80% when the surfactant is cationic and falls in the 60 to 80% range when the surfactant is nonionic [27]. Anionic emulsifiers are incompatible with cationic fabric softeners.

The silicone level must be adjusted for the final benefit required. More material is needed for ease of ironing than for improving rewettability. The silicone concentration in a regular composition is typically between 0.5 and 1.5%. Silicones usually disperse very well in the composition when introduced at the end of the formulation.

Silicones are expensive materials. Their cost and their very high stability, hence their rather poor biodegradability, have often restricted their use as basic raw materials of fabric softeners. Changing market forces have resulted in silicone removal from most household fabric softener compositions.

*(b)   Clays.*   Although some work has been done to develop clay-based rinse cycle fabric softeners, the application of clays as fabric softeners is essentially limited to softergents. These products combine a standard heavy-duty built anionic detergent composition with clay softeners. Commercial products such as the Australian Fab and U.S. Bold were based on this technology.

Clays involved in fabric softening are most often of the montmorillonite type, particularly sodium and calcium montmorillonite. These clays, also referred to as bentonite, are unique in that their particles swell in water, readily forming colloids whose size is between a few hundredths of a micrometer and several micrometers.

Clay particles are actually made of stacks of three to four platelets, each consisting of a sheet of hydrated alumina sandwiched between two sheets of silica. Isomorphous replacement of Si by Al and of Al by Mg imparts negative charges on the surface of the platelets.

Platelets are held together by cations. They impart a positive charge to the edge of the particles. These interlayer cations play a key role in the physicochemical properties of bentonite and in the stability of aqueous dispersions. Normally calcium is predominant and the clay swells to a moderate extent when dispersed in water. When Ca ions are replaced by Na, e.g., by reacting with $Na_2CO_3$, the bentonite is said to be "activated." This activation makes the clay much more swellable.

The swelling of clays is a two-step process. First, hydration of the platelet surface occurs, leading to a slight volume increase. Second, repulsion takes place between the electric double layers, leading to the complete separation of the platelets; this is so-called osmotic swelling.

Sodium montmorillonite quickly and irreversibly deposits onto cellulose at extremely low clay concentrations [61–63]. It exhibits some fiber lubrication properties.

The multilayered swellable clay particles are overall negatively charged, as shown by electrokinetic studies [64,65]. Hence bentonites are quite compatible with anionic-based detergents.

Cheap clays are generally colored by impurities to light brown or gray. This has never been reported to alter fabric color, but mixing powdered clay with the detergent results in an unaesthetic product. For this reason, clay powder is agglomerated into detergent-sized aggregates, which are afterwards added to the detergent without affecting its aesthetics.

An important fiber–fiber friction takes place in the washing machine and, to a lesser extent, afterwards when the clothing is used. Hence, to get the best prevention of damage caused by the wash, the protection must start from the beginning of the laundering process. In that respect, traditional rinse cycle fabric softeners come too late in the process: they protect the laundry only when dried and worn. The situation is completely different with clay-based fabric softeners, which protect the laundry throughout the entire wash, hence preventing the damage caused by the wash (Figure 12.6).

Fiber damage reduction has been demonstrated in the laboratory, according to the yarn-to-yarn abrasion test carried out in the washing liquor. Details of the test procedure may be found in Azoz [66]. Evidence of fabric protection by clay/PDT has also been collected in consumer tests.

In the first washes, the softness obtained with clay, although being perceivable, remains far from that delivered by a rinse cycle fabric softener, despite the high clay levels involved – up to 20% of the product and more. These levels are possible because of the relatively low price of bentonites.

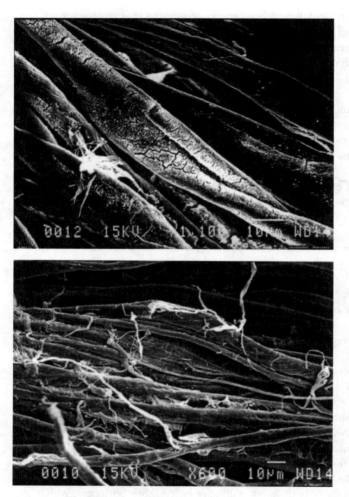

**FIG. 12.6** Cotton fiber protection by a clay-based wash cycle fabric softener after 40 washes at 40°C in hard water (European washer). Top: washed with a clay-based softergent; bottom: washed with a regular detergent.

Another limitation with clay softeners is that they do not exhibit any antistatic activity. This is a serious drawback in North America, where tumble dryers are very common. Several approaches have been followed to address the problem, such as introducing a neutral amine softener in the composition or encapsulating a cationic softener to avoid its interaction with clay and detergent components [10].

## 3. Enzymes

The use of cellulases is an alternative way to deliver fabric softness in the wash cycle. They catalyze the hydrolysis of the β-1,4-glycosidic linkages of cellulose. Detergent cellulases are mixtures of endocellulase, which degrades cellulose randomly in the chains, and of exocellulase, which attacks the chains at their ends, releasing glucose and cellobiose, a disaccharide; the latter inhibits the exocellulase activity. Therefore β-glucosidase, which transforms cellobiose into glucose, is also present [67].

Upon use and during successive washes of fabrics, microfibrils, called "pills" when they gather into visible balls, are generated on cotton fibers, which gradually lose their smooth structure. As microfibrils and pills scatter light, whites become grayish and bright colors dull and fuzzy. Microfibrils are also sites where soils are entrapped and $CaCO_3$ crystallizes from hard water, giving rise to encrustations. By removing the pills and the microfibrils from the surface of cotton fibers, cellulases improve the touch and the appearance of cotton [68,69].

The enzyme performance depends on its concentration. Increasing the enzyme concentration increases the effect, which eventually levels off. It also strongly depends on temperature, which affects the enzyme dissolution rate, activity, and stability in the washing bath. These effects are cumulative and increase considerably with the number of wash cycles carried out [70].

## B. Other Ingredients

The aesthetic characteristics of fabric softeners have always attracted much attention from formulators. Each product is personalized by some specific ingredients, which determine its appearance.

An appropriate perfume is usually added to fabric softeners. Neutralizing the base odor and imparting to the product a pleasant smell, which illustrates the benefits delivered by the product, is not the only challenge faced by perfume houses when they develop a fragrance for fabric softeners. The perfume has also to fulfill other important requirements, such as remaining stable in the product — the stability window of many perfume ingredients is rather narrow — and keeping the product rheology and storage stability unchanged. No phase separation, modification of the viscosity, or chemical degradation should occur. Once the fine-tuning is completed and the above conditions fulfilled in a particular composition, the perfume is optimized to receive a positive consumer reaction.

Many fabric softeners are also dyed to cover the yellowish aspect of the product. Traditionally, they were blue, because of the success of Comfort, one of Lever's earliest brands [1]. Nowadays, various dyes are used to adjust the shade and strength of the color desired and any color may be found on supermarket shelves. Suitable dyes, however, must be safe for the composition, for fabrics, and for the environment. The dyes must be water soluble and biodegradable, and must exhibit nonstaining properties.

Most products are stabilized against biological degradation by preservatives. Despite the antibacterial activity of cationic surfactants, there is some risk of bacterial contamination of fabric softeners.

Most generally, deionized water should be used to formulate fabric softeners. In this way a possible cause of electrolyte variation due to water quality is eliminated. Moreover, it is a way of avoiding the presence of ferric ions, which can cause fabric yellowing. Appropriate treatment of water helps eliminate the initial contamination of the product.

If nondeionized water is used, product viscosity remains low and a thickener must be incorporated. Excellent thickening effects have been obtained with nonionic cellulose derivatives, cationic guar gum, or crosslinked cationic polyacrylates.

Fabric softeners may also contain antistatic agents (anionic or ethoxylated nonionic surfactants) and/or humectants to increase the moisture level at the fabric surface [52].

As stated by Levinson, rinse-added softeners also provide a way of introducing valuable ingredients in the laundering process that may not be compatible with laundry detergents [3]. Several patents have been filed covering the incorporation of polyethylene terephthalate soil release agents, nitrogen-containing polymers or polyvinylpyrrolidone (PVP)-type polymers for dye transfer inhibition and soil release [16], biocidal quaternaries to suppress mildew and odor formation on wet fabrics, and many others [3].

Some ingredients have also been used in some fabric softeners to deliver particular benefits. Among them are the following:

*Bluing agents* to counteract the yellowing tendency of cationic actives.

*Cosolvents* and *hydrotropes* as formulating aids and to make the product easily dosed and dispersed in the rinse liquor without affecting the storage stability [52].

*Exhaustion aids* to ensure that the best conditions are achieved in the wash bath to get full deposition of the softener. Among these systems are pH controllers [52]. Nonionic surfactants have the reverse effect, as they tend to retain the softening agent in the rinse liquor.

*Antifreezing agents.* When it thaws after having been frozen, a rinse cycle fabric softener usually undergoes a huge increase of viscosity that remains even after prolonged storage at room temperature. This problem is avoided by introducing an antifreezing agent. The addition of 4 to 7% methanol, ethanol, (poly)ethylene glycol, or glycol ethers protects the product from freezing down to −7°C; without protection, the product starts freezing at around −2°C. For safety reasons, the amount of antifreezing agent should be kept as low as possible. Methanol, which is cheaper than ethylene glycol, is registered as a poison in many countries. Imidazolinium softeners are less subject to freeze–thaw instability than DHTDMAC.

*Optical brightener.* Because of the screening effect of the active, softened fibers may reflect less light and be somewhat less bright. To overcome the problem, manufacturers add an optical brightener to some products. A benefit can indeed be displayed in the laboratory by fluorescence and pair comparison tests; users, who do not perform pair comparisons, hardly detect the difference. The benefit on colored items is more obvious and the consumer usually perceives the color revival. Selecting a candidate requires a preliminary assessment of possible hue changes.

The overall characteristics of the finished product are the result of the presence of the additives (nonionic surfactants, electrolytes, cationic polymers), of the nature and amount of the perfume, and of the manufacturing procedure (temperature, stirring) [5].

## C.  Types of Products

Cationic actives are usually used at a low level (0.5 to 5.0 g active/kg textile). They may be applied on the fabrics by padding, which is usually reserved for industrial applications. In domestic use, they are rather deposited by exhaustion from liquor or by transfer from a substrate [10].

Household fabric softeners may be applied to fabrics at three different steps of the laundering process: in the main wash, together with the detergent; in the last rinse after the wash; or in the tumble dryer, together with the wet fabrics before drying. The corresponding physicochemical environment of the softener varies greatly and the compositions must be adapted accordingly. DHTDMAC and DHTIMS have been used in formulating all three types of products. The best softness is obtained with DHTDMAC-based rinse cycle fabric softeners.

### 1.  Rinse Cycle Fabric Softeners

The first fabric softeners were used by introducing them in the last rinse of the wash cycle, when most soil and detergent residues have been eliminated. Otherwise part of the active was wasted in forming water-insoluble complexes with anionic materials. These rinse cycle fabric softeners were 225 ml whole-cup "economy" or 115 ml half-cup "premium" products. Today, rinse cycle fabric softeners are still the most popular forms used by consumers. Indeed, they are the fabric softener form that delivers the greatest level of perceived softening efficacy. Markets are moving toward using smaller amounts of more efficient products. Modern washing machines in North America are equipped with fabric softener dispensers holding 115 ml or less. Using 30 to 45 ml of ultra products (see below) is now very well accepted by consumers in many industrialized countries [3].

(a)  *Regulars.*   These products are intended for direct use, without preliminary dilution. For many years the preferred actives for regular rinse cycle fabric softeners have been DHTDMAC, usually at a level of about 6%, or the tallow derivatives of diamidoalkoxylated ammonium or of imidazolinium [9]. Regular products were usually sold in 4 l bottles. The optimal softener concentration in the rinse depends on several factors such as the nature of the clothes to be treated, the drying conditions, and the level of softness expected by consumers.

The best trade-off between softening and affinity of the treated fabric for water is obtained at 1 g DHTDMAC/kg dry load [13,18,71].

Today, the average content of softening actives in European and in U.S. products varies from 3.3 to 5% solid, on a weight basis. Recommended dosages are usually 1/3 to 1/2 cup (80 to 120 ml) per wash depending on load size (U.S. conditions) [35] and 110 g per wash under European conditions. This represents 1.3 g/kg load under U.S. conditions (80 g product, 3.5% active, 2.2 kg load) and 1.1 g/kg load under European conditions (110 g product, 3.3% active, 3.3 kg load).

The various products present on the market differ from one another by the active ingredient system (structure of the quaternary and level in the composition, type of cosoftener, when present) and by the additives and perfume that fine-tune the performance and aesthetics.

(b)  *Concentrates, Ultras, Compacts.*   Concentrates and ultra products are up to 10 times concentrated. Usually they correspond to aqueous dispersions of 15 to 30% DHTDMAC, i.e., are 3 to 5 times more efficient than regular products. As already stated, the performance of these formulations is made possible by the synergy developed between softeners and cosofteners, which deliver more softness at the same level of active. In this way the viscosity problems encountered when the DHTDMAC concentration increases are bypassed. Alternatively, the concentration of actives may be raised when saturated chains are partly or totally replaced by unsaturated ones; however, some loss of softening occurs. In this approach, a trade-off must consequently be defined between advantages of concentration and the best softening.

Concentrates are dilute-before-use products. They must be first dispersed in three or four times the amount of tap water and shaken vigorously to regenerate the 5 to 8% concentrate dispersion and avoid clogging the fabric softener dispenser [3,35]. This is not necessary with ultras, which can be added directly in the rinse without preliminary dilution.

The stability of these products is governed not only by their composition but also by mechanical factors. The finished product is shear sensitive. From a formulation standpoint, usual actives do not enable the production of storage-stable concentrated products without incorporating emulsifiers such as fatty esters, ethoxylated fatty amines, or amides, and the viscosity must be further adjusted using inorganic salts. Moreover, the formulation of concentrates requires alcohol-reduced grades of quaternaries [39], 15% organic solvent in the raw material being then already too much.

*(c) Solid Softeners.* These products are made of a spray-dried powder of DHTDMAC. They may also be made of DHTDMAC sprayed on urea, since they easily form inclusion compounds [8,18]. Usually, some nonionic is incorporated to help dispersion in water.

They are not very popular, since they disperse slowly in water and are not very suitable for washers, which are designed for liquid products.

## 2. Wash Cycle Fabric Softeners, Softergents

Introducing the fabric softener in the last rinse cycle is a true constraint for users of vertical-axis washers that are not equipped with a fabric softener dispenser. Indeed, the consumer must then either stay in the vicinity of the washer during the wash, to pour in the product at the beginning of the final rinse, or run an extra rinse cycle after laundering completion. This constraint does not exist with horizontal-axis washers, since a compartment for automatically releasing the fabric softener at the last rinse cycle is incorporated in their dispenser.

From the beginning, softener manufacturers sought to eliminate this constraint by introducing the fabric softener in the wash with the detergent rather than in the rinse. The requirements of rinse cycle and wash cycle fabric softening are completely different. In the wash liquor, the cationic softener is neutralized by the anionic surfactants present in most detergents, causing its precipitation. As a result, both the softener efficacy and the detergent cleaning performance are impaired and more active is required to get satisfactory results from a wash cycle fabric softener [26,37]. Hughes *et al.* report that the softening effect of the cationic active neutralized by anionic surfactants is not necessarily completely lost, as the electrically neutral fatty complex may deposit and lubricate the fibers [72]. Milwidsky adds that the corresponding softness is, however, not as good as that obtained with a pure cationic [18]. In contrast, Bräuer *et al.* [73] claim that if the anionic surfactant concentration in the wash exactly corresponds to the DHTDMAC concentration, the deposition of the latter is then more important than in a rinse. This proposal is difficult to reconcile with the common observation that the usual softener deposition is practically complete in the rinse under real use conditions.

Another way to achieve wash cycle fabric softening is to replace cationic actives by amphoteric or zwitterionic actives.

Practically, to avoid excessive cost, the cationic concentration was limited to 9 to 15% [8,9]. At these levels the delivered benefits did not match user expectations. Consumers did not accept the compromise between convenience and efficacy.

Effort was maintained, and a few years later detergents with incorporated softener (softergents) appeared on the market. The two-in-one-product approach is much more acceptable since the formulator has control of both the detergent and the softener systems, which can and must be adapted to coexist in the same composition.

A first attempt consisted of fully replacing the detergent anionic surfactants by nonionic ones [8]. DHTDMAC or the hard tallow derivatives of imidazolinium or of diamidoalkoxylated ammonium were used in powder softergents, and the corresponding soft tallow or oleyl derivatives in clear liquids. The introduction of high levels of nonionic in a powder remains difficult and this approach has been limited to liquid detergents [18]. Moreover, the amount of DHTDMAC adsorbed on fabrics decreases as the nonionic concentration increases, because of the competition that takes place between these two species to deposit [23]. Other detergent ingredients, except carboxymethylcellulose [14], exhibit only limited interaction with the softener [72]. Hence the softener level had to be increased two or three times to match the performance of a rinse cycle fabric softener [9,14].

A valuable alternative to introducing a softener excess is to prevent or at least to reduce the contact between oppositely charged surfactants by exploiting their characteristic difference of solubility [8]. The effect is maximized by dry-blending the softener with very water-soluble spray-dried particles containing the anionic surfactants [14]. In the washing liquor, the anionic surfactants disperse first and have enough time to remove the soils before the cationic active deposits onto the fabric surfaces [14]. Afterwards, the detergency declines as the anionic and cationic surfactants interact and deposit onto the fabrics.

Alternatively physical separation may be realized by encapsulating the softener in a high-melting matrix such as paraffin wax, high-molecular-weight polyethylene glycol, or fatty acid triglycerides. The softener is released in the liquor as the temperature rises at the end of the wash, when the soil has already been removed.

Another approach is to replace quaternaries by the corresponding tertiary amines. Amines are electrically neutral at alkaline pH; as such, they are entrapped in the textile structure during the wash. In the rinse liquor, the pH is neutral and they bear a positive charge because of a proton binding to the nitrogen atom ($pK_a \approx 9.0$); these compounds withstand the rinse because of their water insolubility [8]. The first product of this type was made by incorporating alkylamide in a detergent composition (U.S. Patent 3,795,611 to Colgate-Palmolive, referenced in Chalmers [1]).

Other systems exist using montmorillonite-type clays as softeners [5]. Clay alone is much less efficient in fabric softening than the usual cationic actives. In their latest version, these softergents match the softening performance of fabric softeners, due to the clay being coated with a large fatty molecule such as a fatty alcohol or pentaerythritol di-tallow [74–76]. In this system the softening is achieved by the organic molecule and the clay works as a carrier.

Clay-based softergents offer the unique advantage of protecting fibers against the mechanical and chemical aggressions from the beginning of the washing cycle at no cleaning penalty. Ethoxylated alcohol surfactants must, however, be avoided in these compositions because of their great efficacy in removing bentonite from cellulose and in preventing its deposition [77].

The first softergents appeared on the market in the early 1970s. They were quite popular in the U.S. between 1981 and 1989 or so, but their popularity declined in the 1990s as a result of the increased softening efficacy of rinse-added softeners and the stronger cleaning efficacy of detergents [10].

Procter & Gamble developed a completely different approach to remedy the lack of fabric softener in washers. They developed a sphere dispenser for ultra fabric softeners designed to be incorporated in the laundry before the wash and to deliver 28 ml of product in the heart of the load at the beginning of the rinse. The Downy Ball has a cap that opens during the spin cycle because of centrifugal force, enabling the softener to flow out [78].

## 3.  Dryer Cycle Fabric Softeners

Electrical tumble dryers are much more popular in the U.S. than in Europe (see Section IV). Consequently the market penetration of dryer cycle fabric softeners is much more important in North America than anywhere else. They are usually based on the same types of actives as the rinse cycle fabric softeners [35]. Two methods of delivery exist. Nonwoven sheets are impregnated with cationic softener that are incorporated in the wet load at the start of tumble drying or a fabric conditioner is sprayed from an aerosol onto the empty dryer drum before the drying process [16].

*(a)  Sheets.*  Dryer sheets are nonwoven sheets of synthetic fabrics coated with the softening ingredients. They are added to the damp laundry in the tumble dryer before drying. The nonwoven material simply works as a carrier. The active must have a melting point slightly below the dryer temperature. When the temperature rises and the impregnated sheet is rubbed against the humid load items, the softener is slowly transferred from the sheet to the hot and moist clothes. Afterwards, the softener spreads on the various parts of the fabrics for the same reason. The softener does not diffuse into the fibers but remains on their outer surface. This is the best way to deliver the correct dose of softener [14] — and for a time it was the only successful dryer softener form on the market [8].

Hard tallow ammonium compounds, especially DHTDMAMS, are usually preferred for this application [9]. Methylsulfate derivatives, which do not attack the epoxy resin usually found in U.S. dryers [7], are preferred to chlorides, which are said to corrode the metal parts of dryers during drying [26].

Too quick a release may cause fabric staining [14]. The active is therefore usually mixed with softening and melting point modifiers, polyethylene glycol esters or fatty amine soap, to optimize its release [7]. Incorporating a small amount of nonionic surfactant improves the spreading of the active onto fabrics; it cannot be too hydrophilic, to remain soluble enough in the base and disperse efficiently. The resulting softening and antistatic effects are stronger, and there is less dusting and fabric staining.

Dryer sheets work by replacing the finish, which is otherwise gradually lost with repeated washing. They are convenient to use and deliver a better static control than rinse cycle products (95% instead of 50 to 80% [14]) with a lower level of active, usually 22 to 45% of the amount delivered by a rinse cycle softener [3]. This is probably linked to the fact that the softener does not penetrate inside the fibers. They are also superior to rinse cycle softeners as regards the soil release effect [34].

From a softening standpoint, dryer sheets are less efficient than rinse-added products but slightly more efficient than wash cycle fabric softeners. Besides the lower amount of solids delivered, the deposition in the dryer is much less uniform than in the rinse [9,34,37].

The dryer sheet market is enhanced by the change from natural to synthetic fibers.

*(b)   Sprays.*   In 1976 aerosol products became available to be sprayed directly onto the wall of the dryer drum before introducing the wet textiles. The foam substrate allowed the active ingredient to melt and be transferred by mechanical rubbing onto the fabrics [3]. They were efficient and addressed the drawbacks of sheet fabric softeners (staining, dusting, and tackiness) but caused corrosion of the dryer walls. They also carried moisture into the sensing devices of the dryer, causing malfunctions [14].

An alternative is an aerosol foam to be sprayed onto one laundry item in the dryer. This approach is close to that of sheet softeners described above.

## VI.   REGULATORY ISSUES AND SAFETY
## A.   Worker Safety

Quaternaries and esterquats contain some alcohol. Their flash point is low (below 37°C, 100°F). The tendency of a liquid to ignite is commonly linked to the existence of a flash point, defined as the lowest temperature at which a liquid generates enough vapor to form an ignitable mixture with air. Here, however, the vapor mixture exhibits a low heat of combustion and releases heat slowly. Hence, fire hazard is limited to the headspace of heated storage tanks and to spillage upon transfer to the formulation tanks. In case of fire, quaternaries tend to self extinguish, as the amount of flammable component is limited. What can also happen is a fire caused by these vapors igniting neighboring flammable material. Naked flames and high temperatures should therefore be avoided in their vicinity; vapors should rather be exhausted to a safe location by an efficient ventilation system [79].

When formulating, workers may be exposed to flammable vapors. They should wear face shields, respiratory protection, and impervious gloves.

In the case of spillage, absorbent materials are generally not necessary because of the high viscosity of quaternaries at room temperature. Applicable regulations regarding chemicals disposal must be followed.

## B. User Safety

Cationic fabric softeners are practically nontoxic by oral or dermal administration [3]. Consumers with sensitive skin may suspect softener-treated fabrics to possibly cause adverse effects. Textiles in contact with the skin can have dermatological effects such as allergic reactions and irritations. Skin irritation, however, depends on many parameters such as textile properties, skin sensitivity, and conditions of exposure (e.g., duration, perspiration, and environmental conditions such as climate) [80].

No problems could be evidenced from visual, instrumental, and self evaluations of intact and damaged skin exposed to treated and untreated fabrics [80,81]. Hence, no adverse effect on human health is associated with the use of DHTDMAC, amidoamine, imidazoline, or ester-based quaternary softeners [3]. In contrast, fabrics repeatedly washed with detergent in hot water become harsh and may cause irritation. Fabric softeners eliminate this problem [1]. Significant differences of stratum corneum, skin barrier integrity, and water holding capacity have been observed after rubbing the skin with softened or nonsoftened fabrics. The differences were always in favor of softened fabrics [81]. Likewise, Tronnier has shown that sensitive baby skin is less irritated when diapers have been treated with fabric softeners that are formulated so as not to sensitize or irritate the skin [80]. This is not surprising, considering the reduction of the skin–fiber friction delivered by the softener.

## C. Environmental Safety

Like all other detergent ingredients, fabric softener actives are discharged into sewers and possibly end up in the environment. Their effect on aquatic or terrestrial organisms depends on their concentration and toxicity [6].

Quats are usually less easily biodegraded than anionic or nonionic surfactants, or esterquats, as they are less prone to be attacked by bacteria. Nevertheless softeners are not detrimental to the environment, since, besides their elimination by biological degradation, their concentration in effluents remains very low [15].

The amounts found in the environment depend on several factors, among which the use level, the population density, and the waste treatment process are the most important [3]. Three mechanisms account for their removal from sewage [8]:

1. They are neutralized by anionic surfactants, whose concentration is much larger.
2. They concentrate at the bottom of surface waters as colloids because of their interaction with minerals.
3. They are eliminated by adsorption onto particulate activated sludge. The adsorbed portion is subsequently further broken down by bacteria if the sludge is incorporated into agricultural land.

A two-step waste treatment process has been shown to reduce softener actives by more than 90% [3]. The rest is diluted in surface waters.

For these reasons surfactants such as DHTDMAC have been safely used world-wide for decades. For instance, despite an annual DHTDMAC consumption of 27,000 tons in 1980 in Germany, there has been no clear evidence of any negative impact on the environment [10].

Nevertheless, as already mentioned, esterquats replaced DHTDMAC in the early 1990s. This move, a voluntary initiative from the industry, allowed the use of materials not classified as dangerous for the environment instead of materials classified as very toxic to aquatic organisms and potentially having long-term effects on the environment.

Microorganisms in sewage treatment readily break down the ester bonds of esterquats, releasing fatty acids and quaternized di- or triethanolamine. The fatty acids ultimately generate carbon dioxide and water [5] while the smaller cationic molecules are not further degraded but are not toxic [5,10]. An example of an environmental study of esterquat biodegradation may be found in Giolando et al. [82].

Replacing DHTDMAC by esterquats required a full reformulation of fabric softeners to keep the softening performance and the aesthetic attributes of products. For instance, esterquat hydrolysis may occur upon storage of a product. The reaction is hindered by maintaining the product at slightly acidic pH values (2 to 3.5). Since 1996, the rest of the world has also started to remove DHTDMAC from softening compositions [16]. As a result, DHTDMAC annual consumption has dropped by over 70%.

Packaging reduction has been achieved by simultaneously following several approaches. These include the use of recyclable plastics to reduce the use of virgin plastic and create a market for recycled plastic, introduction of lightweight bottles and refills, which use less plastic than conventional containers, and development of concentrated products. Recycled paper has also been used for cartons made from paper, to reduce the amount of wood fiber used and provide a market for recycled paper [35]. This enabled the launch of concentrated fabric softeners, which were successfully formulated due to product and packaging improvements (e.g., introduction of self-draining caps to reduce bottle messiness) [6]. These trends have been repeated in the U.S. and Japan [6].

The most striking success in the field of packaging reduction has probably been the sachet developed by Cotelle in France. In 1957 it launched concentrated hypochlorite bleach in a sachet. Twenty-five year later, it used the technology for a fabric softener under the Minidou trade name. The weight of a 250 ml sachet is 10 g, 3.5 times less than a 1 l plastic bottle. Its cost is less than half the price of a rigid polyethylene bottle. From the trade's standpoint, the sachet packaging reduced distribution and warehousing costs.

The product had to be prediluted four times by the consumer, preferably in a 1 l fabric softener bottle. The recommended procedure was to pour the concentrate into a bottle half filled with warm water, then to fill up with warm water and shake vigorously. The bundle was described as a "clever" product that works as well as a bulky and expensive bottle but is more convenient: It is efficient, economical, and easy to carry, store at home, and use. The product was extremely successful and, within a year, it reached the second position on the French market with 22% market share [83].

## D. Regulatory

In Europe the environmental safety of detergents is assessed by the PEC–PNEC system. PEC is the predicted environmental concentration of each ingredient of the detergent. PNEC, the predicted no-effect concentration, is the highest concentration at which an ingredient does not affect an organisms exposed to it in relevant environmental situations [5].

PEC and PNEC values are determined experimentally and/or by model calculations. For any ingredient, PEC should not exceed PNEC. Data generated by the European industry under simulated field conditions showed that the concentrations of cationic surfactants found in the environment were significantly below harmful threshold for even the most sensitive organisms [6,11].

Fabric softeners and their ingredients are governed by several directives:

- Existing substances regulation (793/93). Distearyldimethylammonium has been put on first priority list and a risk assessment for humans has been carried out (CAS 107-64-2).
- The dangerous substances directive (67/548/EEC) and the dangerous preparations directive (1999/45/EC). Their purpose is to classify and label raw materials ("substances") and formulated products ("preparations") on the basis of their intrinsic properties: physical chemical hazards, hazards for humans, and hazards for the environment. Labeling consists of a symbol of danger (black icon on orange background), risk phrases, and safety phrases. Esterquat raw material is not classified as dangerous for the aquatic environment. As a consequence, it does not contribute to the classification

of fabric softeners as potential dangerous preparations from the aquatic environmental perspective.
* The recommendation for labeling detergents (89/542/EEC). The presence of cationic surfactants must be put on the label.

A new detergent regulation is now in preparation. It will cover (and replace some previous directives):

* The labeling (composition).
* The biodegradation of surfactants (surfactants will have to be readily biodegradable).

## VII.  PRODUCT FORMULATION AND MANUFACTURE

Liquid fabric softeners are formulated by dispersing the melted raw material in well-stirred hot water. Although DHTDMAC aqueous dispersions are not emulsions in the strict sense, chemical and mechanical principles of emulsification apply to control the viscosity and phase stability [10].

In concentrated products, the stability and viscosity depend on other variables such as the type and concentration of solvent in the raw materials, the perfume composition and concentration, and the salt concentration. The order of ingredient addition also influences the product characteristics [3,26].

The product physical and chemical properties consequently depend on many parameters.

### A.  Principles
### 1.  Chemical Factors
The choice of a softening system relies on several factors [9]:

* Chemical composition of the softening raw material. Since they are synthesized from natural feedstock, which are mixtures of molecules bearing fatty chains of different lengths, the softening raw materials are not pure compounds. Moreover industrial synthesis leads to the formation of mono-, di-, and sometimes tri-chain compounds. Fabric softeners made from these raw materials are more easily formulated than those based on pure double-chain derivatives. Variations in the nature and concentration of the byproducts formed in the synthesis modify the characteristics of the finished product.
* Handling characteristics. These essentially depend on the fatty chain composition. Oleyl derivatives are liquid at room temperature, tallow derivatives are opaque liquids that become clear at 38°C (100°F), and hard tallow derivatives are opaque pastes that become pourable at 50°C (120°F). To get the maximum stability for the dispersion, oleyl derivatives can be

dispersed in water at 21 to 27°C (70 to 80°F) [14], tallow derivatives must be heated at 32 to 49°C (90 to 120°F), and hard tallow derivatives at 49 to 60°C (120 to 140°F) [9].

- Formulation parameters. Formulations and processes depend on the level of actives. As a basic rule, formulations are split in two main categories: regular and concentrated. To each category corresponds a well-defined process. The active concentration must exceed 4%, except tallow derivatives, for which 3% is enough [9]. The viscosity of the dispersion increases with the chain length and saturation extent. It is fine-tuned by adding salts such as sodium chloride, sulfate, or acetate.
- Performance properties. In the presence of solvents, oleyl derivatives give clear stable dispersions with good freeze–thaw stability [14]. Aqueous dispersions of hard tallow derivatives are milky.
- Price. Oleyl derivatives are more expensive than their saturated counterparts.

## 2. Mechanical Factors

The manufacturing conditions cover both the procedures and the equipment.

A correct dispersion of the quaternary in water requires water heated at 60 to 70°C. The lower the water temperature, the more viscous the product, and below 35°C the dispersion is no longer homogeneous. Above 70°C the final viscosity also increases. The product temperature must be maintained at a minimum to reduce energy usage necessary for heating and to reduce the cooling time at the end of the process.

The particle size distribution of the dispersion depends on the magnitude of the shear and of the flow applied at each step of the formulation. The wrong conditions may induce two kinds of product instability: thickening or clearing upon storage.

Besides their detrimental effect on viscosity, long shears also cause air to be incorporated in the product. The resulting foaming of the product may create problems in the filling step. Foaming can be avoided by reducing the speed or modifying the design of the device. Otherwise, addition of 30 ppm of a silicone antifoam emulsion will solve the problem.

Air incorporation also eventually causes phase separation in the product upon storage. The risk completely disappears when the density of the finished product is close to 0.99. Clearing during aging is considered below, in the discussion on physical stability.

The mixing conditions must consequently be carefully optimized when defining the manufacturing procedure by scale-up experimentation.

## B. Process

Fabric softeners are prepared either in a batch or in a continuous process.

## 1.  Batch formulation

Water-insoluble chemicals are premixed in a tank, then pumped into the main tank, where they are dispersed in hot water. The hot product is then cooled.

Since a structured liquid must be obtained, the formulation procedure is stringent. It is a four-step process:

1.  Dispersion of water-soluble ingredients (dye, nonionic surfactant, etc.) in hot water (60°C).
2.  Dispersion of the hot premix of DHTDMAC or esterquat, cosoftener, fragrance, and other water-insoluble ingredients, if any, in the well-stirred aqueous phase. Overheating should be avoided, as it is detrimental to the product's stability. The stirring is maintained until a homogeneous dispersion is obtained.
3.  Cooling step. The final cooling step is the bottleneck of the process. It may be achieved by circulating water at 16°C in the double jacket of the mixers. The cooling speed is slow. During the 3 to 4 hours needed to bring the product back to room temperature, it undergoes a shear that may influence its viscosity profile and may cause the aeration of the product. Using a heat plate exchanger to cool the hot product to room temperature drastically reduces the cooling time (3 to 4 times), hence the shear applied to the product.
4.  Addition of minors such as buffer or preservative. Optionally a structuring polymer or some electrolyte can be introduced at the end to adjust the viscosity.

The product is then transferred to the storage tank or to the filling lines.

Another way to enhance the cooling speed is to use the low-energy emulsification (LEE) procedure [84]. Only 25 to 50% of the total amount of water is heated to 60°C. The remaining part is kept at 16°C and slowly added afterwards, when the dispersion of the oil phase is completed. Besides reducing the shear undergone by the product upon cooling, this procedure also leads to a large reduction of energy consumption.

For concentrates enhancing the solid level in the composition only requires an adjustment of the active addition speed to avoid lump formation and the adaptation of the shear applied to the system. Good stability of the product upon aging is observed when the particle size remains below 10 $\mu$m. Here, the addition of electrolytes is necessary to reduce the viscosity of the finished product.

Numerous improvements have been gradually incorporated in the basic batch equipment, which have led to fully computer-controlled processes. Large equipment is necessary to reduce the formulation time. The equipment is most often at rest as it operates intermittently. These periods permit changeover and cleaning of the installation between two different compositions.

## 2. Continuous Formulation

In the continuous process, most mixing operations are achieved with static mixers; only highly viscous or difficult materials are mixed with dynamic mixers.

Raw materials are all stored in separate tanks. Water-insoluble and water-soluble ingredients are mixed in two separate mixers and then mixed together in a third one. The product is then cooled to 25°C in a heat exchanger. The minors and the thickener and/or salt are then incorporated into the product, which is pumped to a storage tank or filling lines.

The amounts are adjusted either by volumetric metering or mass-flow metering [85]. Solid ingredients must first be solubilized, to be metered with a pump (in the batch process, they are directly added into the mixer).

In the volumetric metering approach, the quantities are determined through the pumps used to send them to the mixing lines. Densities of all raw materials must be known under the particular conditions of temperature and formulation, and the accuracy of the pumping rate regularly checked to get the right mass flow rate from the volumetric flow rates.

In the mass-flow metering approach, the control unit adjusts the pump speeds to deliver the right amount of ingredient. These systems are highly accurate and reliable as they measure true mass flows.

In continuous formulation, the temperature constraints are of course the same as in the batch process. The major advantage over the batch process is that the required equipment can be smaller as it operates continuously (mixing pipelines are purged with the subsequent product). Moreover, no aeration can occur in the system, which is closed and pressurized.

## C.  Aging Studies

The long-term stability of finished products is assessed under accelerated conditions. Stability test protocols vary among laboratories but prototypes are generally aged at 4°C, room temperature, 35°C, and 40 to 43°C for up to 3 months. Freeze–thaw stability is assessed by submitting the samples to three (24 hours frozen, 24 hours thawing) cycles. The physical characteristics that are most usually followed during aging are the product appearance, viscosity, pH, odor, and dispensability. Frequently, the dye and perfume stability to light is also assessed in the sun test.

These methods only give a rough idea of product stability under real conditions.

## D.  Analytical Evaluations

Several methods exist to determine the level of active matter in raw materials and finished products. Quaternaries can be assayed by standard two-phase titration.

Auerbach's method of quat–methylene blue complex extraction by chloroform is also very popular [86,87].

## VIII.  PHYSICAL CHEMISTRY OF FABRIC CONDITIONING

Because of their very high affinity for fabrics and relatively low cost, cationic surfactants are the workhorses of fabric softening. Moreover, they are easily formulated. From a softening standpoint, the best results are obtained when the ammonium ion bears two saturated $C_{18}$ alkyl chains. Consequently, fabric softener actives are usually made of hydrogenated ("hardened") tallow acid derivatives.

## A.  Structure of Liquid Fabric Softeners

Since the chain length of tallow components is essentially between $C_{14}$ and $C_{18}$, the total number of different structures in the softener active exceeds 15 [26]. The raw material also contains mono- and usually tri-tallow derivatives. Most of those molecules are not water soluble and do not associate into micelles, but form stable colloidal dispersions in water. Maltese crosses can be observed using an optical microscope under polarized light (Figure 12.7), revealing the presence of strongly birefringent particles, typical of liquid crystalline phases [88,89].

The particles are usually composed of lamellar or gel phases [3,90,91]. Electron micrographs (TEM or freeze fracture) show that most DHTDMAC particles exhibit a liposome-like structure called a vesicle (Figure 12.8). In this structure, the surfactant fatty chains form one or several highly organized hydrophobic areas

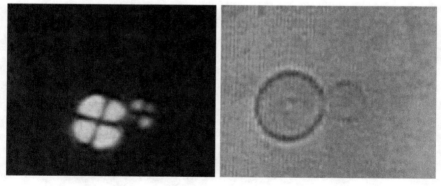

**FIG. 12.7**  DHTDMAC dispersion viewed with an optical microscope (magnification ×800). Left: under polarized light, Maltese crosses show the presence of multilamellar vesicles. Right: the same under nonpolarized light.

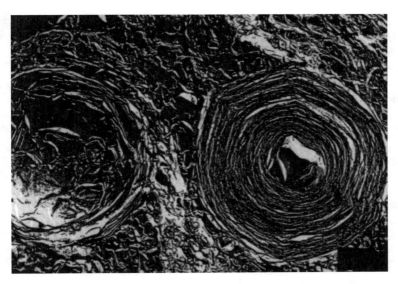

**FIG. 12.8** Electron micrograph of a DHTDMAC aqueous dispersion. An intact vesicle is seen on the left, a cross section on the right, in which the lamellar structure is apparent (freeze fracture, magnification ×2500).

called bilayers, entrapping a large amount of water inside. The ammonium ions are at the internal and at the external surfaces of the bilayer, in contact with water [92].

Vesicles can be unilamellar or multilamellar. The exact structure of the aggregates formed in solution by a type of surfactant results from geometric constraints. It is determined by the relative volumes of the hydrophilic head and of the hydrophobic group of the molecule (e.g., see Israelachvili [93]). That is why di- and tri-tallow derivatives adopt a vesicle structure, while mono-tallow ammonium chloride forms micelles [26]. The structure of the dispersions based on other actives such as esterquats is more complicated.

The liquid crystalline phase makes DHTMAC extremely viscous when the concentration exceeds 5%. Because of this inherently high viscosity and the high volume fraction of colloidal particles, the formulation of concentrated products is difficult. In practice, DHTDMAC is never used at concentrations above 15%. In contrast, esterquats or other quaternaries bearing unsaturated fatty chains are more suitable for producing concentrated dispersions. As a result of their more disordered fatty chain packing, the particles they form are cubic or isotropic but not lamellar, and the resulting phase viscosity is lower and more stable [3].

In aqueous dispersion DHTDMAC spontaneously adopts the vesicle structure. This is uncommon, as vesicle formation not only requires molecular characteristics (double fatty chain molecules, no strong repulsion) but usually also a strong mechanical energy supply (sonication). No such energy supply is required for DHTDMAC to form vesicles [26].

DHTDMAC vesicles have been characterized by Okumura *et al.* [92]. The width of the bimolecular layer is 50 Å, and the interlamellar spacing is between 100 and 400 Å. Each DHTDMAC molecule is hydrated with 7 water molecules. The particle size distribution, measured by dynamic light scattering and optical microscopy, is very broad, ranging from 0.1 to 10 μm. This may be assigned to the presence of both unilamellar and multilamellar vesicles in the dispersion.

Although a fabric softener is not an emulsion but a suspension of charged particles, it exhibits the same instabilities as an emulsion. The physical stability is twofold: phase separation and viscosity.

## 1. Phase Stability

Under some conditions, concentrated softeners exhibit a kind of phase separation called clearing. In a product that undergoes clearing, two layers appear on aging, the bottom layer usually being more translucent and clearer than the top one. The viscosity of each layer remains quite stable and low. This phenomenon disappears as soon as the sample is gently moved.

This instability is due to the presence of a bimodal population. The small, sub-micrometer particles remain evenly distributed throughout the dispersion whereas the larger particles move up, because of the difference in density between the dispersed and the continuous phases. If the interactions between the particles are weak, the viscosity remains low and clearing occurs according to Stokes' law [10].

Another mechanism leading to phase separation is coacervation. The symptoms are the same as above but here the destabilization is sensitive to the ionic strength. Coacervation was extensively studied by Bungenberg de Jong and Kruyt and co-workers, among others, in the 1930s (e.g., see [94]).

The phase stability of the DHTDMAC or esterquat dispersions relies on electrostatic repulsion between the particles. The DLVO theory predicts that when the ionic strength increases, the electrical double layer of each particle is compressed and the electrostatic field at any given distance of the particles is reduced. As a result, the electrostatic repulsion between the particles decreases; they may come into closer contact before undergoing the electrostatic repulsion from their neighbors. At the same time, the charged head–counterion interaction is strengthened; as a result, the solubility of the particles decreases. This effect has been quantified by following the equivalent conductivity of DHTDMAC as a function of its concentration. When the particle concentration is enhanced, the increase of ionic strength results from the larger counterion concentration. In DHTDMAC dispersions, particles are too large and the individual quaternary ion concentration is much too low to contribute significantly to the current transportation. Hence the conductivity of the quat dispersion essentially depends on the free counterion concentration, i.e., on the extent of the vesicle ionization. The ionization extent may be quantified by considering the mobility of chloride ions.

Coming from a fairly high concentration to infinite dilution enhances the proportion of ionized quat from 8 to 27% in a plain quat dispersion, which contains unilamellar and multilamellar vesicles, and from 13 to 47% in quat–fatty alcohol dispersions where only unilamellar vesicles are present. The difference between the two systems results from the larger proportion of quat molecules in contact with the continuous phase.

The particles concentrate at the top or at the bottom of the solution, forming two phases. One of them is rich in colloid, corresponding to a cluster of loosely stacked particles weakly repelling each other and the other is a dilute suspension. Since the repulsion between the particles is much reduced but still present, they do not interact as in a flocculate. Consequently, they are very easily homogenized by gentle shaking, even after long storage periods.

Two phenomena may reinforce the destabilization by coacervation:

1. Ions in solution may affect the particle hydrophilicity according to their position in the lyotropic (Hofmeister) series. As a result, particles are salted-out and aggregate.
2. The decrease of surfactant solubility in solution may also modify the phase equilibria in the system [26].

To improve the uniformity of the dispersions, which must remain constant from batch to batch, an emulsifier such as an ethoxylated fatty alcohol can be incorporated. Emulsifiers form an interfacial film around the dispersed particles and prevent them from interacting with each other, as long as the emulsifier is well located at the interface. Medium-length hydrocarbon chains and long ethoxylated chains usually give the best results. The selection of the right emulsifier can be achieved with Griffin's HLB (hydrophilic/lipophilic balance) system [95]. Jacques and Schramm give an illustration of the method [10]. Once the emulsifier is identified, its concentration, usually in the 0.5 to 1.0% range, must also be fine-tuned to get stable dispersions.

Additional stability improvements can also be obtained by limiting the solvent content in the raw materials and by using hydrophilic polymers, such as polyethylene glycol, which prevent coalescence by steric stabilization.

## 2. Viscosity

A key attribute of rinse cycle fabric softeners is their viscosity. The control of the finished product viscosity is very delicate, as its value at the end of the formulation depends on several parameters linked to the ingredient composition and the manufacturing procedure. Also, viscosity may vary upon storage.

The effect is particularly critical in concentrated fabric softeners since, according to the particle size distribution, a given composition may lead to a thin liquid or an unpourable gel [10].

Since a large amount of water is trapped in the vesicles, the volume fraction of the dispersed phase is much larger than the volume of lipid and the viscosity strongly depends on the conditions adopted to disperse the active (temperature and stirring speed), i.e., the shear undergone by the product. It also depends on the type of colloidal structure, on the electrolyte content, and on the nature and concentration of other ingredients [91].

Introducing small amounts of salts such as sodium or calcium chloride reduces the viscosity. Bilayers are impermeable to inorganic ions and work as semipermeable membranes, which let only water pass freely. Because of the salt addition, the electrolyte concentration in the inner core of the vesicles is smaller than outside, and water migrates from the interior to the continuous phase to restore the osmotic equilibrium. As a result, the vesicles shrink and the volume of the continuous phase increases. Since more solvent is available to disperse the now smaller particles, the distance between neighboring particles increases and the interparticle interactions (electrostatic repulsion and van der Waals attraction) decrease [39]. These effects reduce the viscosity of the dispersion and hinder phase separation. The strongest effect is obtained by adding the salt at the end of the formulation, so that the electrolyte remains in the continuous phase. Above a critical temperature corresponding to the "melting point" of the fatty chains in the bilayer (30 to 40°C for DHTDMAC), water and electrolyte can pass freely through the hydrophobic layer of the vesicles.

The salt concentration must, however, be maintained at the low side, as levels above 100 ppm induce coacervation in the product. Moreover the particle size reduction effect is limited by the electrostatic repulsion between the head groups in adjacent layers, which increases as the space between the vesicle bilayers decreases [91]. Consequently, the formulator has to identify the electrolyte concentration that decreases the size of the particles as much as possible without affecting the physical stability. Some electrolytes are introduced through water and raw materials, especially the quaternary itself, and their amounts vary from one delivery to the other. Using deionized water eliminates one source of variation.

The temperature effect is considered in Section VII.A.2.

Product viscosity is strongly affected by the shear undergone during the formulation and in the subsequent handling steps (pumping, filling). Because of the larger ionic strength that results from a particle size decrease, smaller particles can come in a closer proximity and interact more strongly with one another than large particles. As a result, a network floc gradually forms and the product viscosity rises.

According to Okumura et al., dispersions whose particle size falls in the micrometer range are less subject to phase separation or viscosity change [92].

They assign the stabilizing effect of micronizing to the formation of the network structure in the dispersion, which counteracts the difference of specific gravity between the dispersed and continuous phases.

This mechanism is contested by Laughlin [26], who stresses that DHTDMAC particles remain independent in solution and do not form a network, as the product structure corresponds to a sol not to a gel [72].

Surfactants also modify product viscosity. At a low level, such as 0.15%, the viscosity increases, becoming hard to stabilize. At higher levels the product is stable and its viscosity decreases. Any excess of surfactant is not detrimental but useless and expensive.

Product consistency is usually adjusted and maintained in a well-defined range of viscosity by post-adding a thickening agent (often a highly charged cationic polymer). When the thickener is dispersed in water, electrostatic repulsion takes place between the charges on the chains. Linear chains unfold and occupy the volume between the particles, imparting a rheological structure to the continuous phase. The best results are obtained with crosslinked chains.

As the fragrance is essentially made of organic compounds, it also interacts with the hydrophobic layer of the softener particles, causing them to stick together [39]. One way of avoiding the fragrance interaction with the particles is to introduce an emulsifying agent in the composition. An alternative, which enables one to avoid the increased cost linked to the introduction of an additional ingredient, is to disperse the fragrance in the melted active prior to its dispersion in water [39].

## B.  Deposition and Desorption

### 1.  Deposition

The success of cationic surfactants results from their strong efficacy at low concentration, which at least partly results from their huge substantivity on textiles. Once dispersed in the rinse liquor, the softener exists as a very diluted dispersion (250 ppm under European conditions, 100ppm under U.S. conditions; Okumura *et al.* even worked on 33 ppm solutions [92]). The cationic active nevertheless deposits almost quantitatively and coats garments more or less evenly within a few minutes. Some 97% of the DHTDMAC introduced in the liquor is found on cotton at the end of the rinse [96]. This value was confirmed over a wide range of pH and temperature by colorimetric and radiometric evaluations [97]. The homogeneity of the quaternary deposition may be visualized by immersing dry softener-treated fabrics in a 0.01% bromphenol blue solution, followed by rinsing and drying. Softener covered areas appear as blue spots, the dye excess, which is not linked to the cationic active, being released in the rinse.

Substantivity is of course not limited to cationic surfactants. Other surfactants also adsorb onto surfaces — otherwise the detergency would not be possible — but to a lesser extent and their desorption is practically complete upon the following wash [25,98].

*(a)   Mechanism of Fabric Softener Deposition in the Rinse Liquor: Electrostatic Model.*   Since the early 1960s the deposition of fabric softeners has been extensively studied to understand the huge substantivity of cationic actives. Among other methods, it has been shown by electrokinetic potential measurements that a cationic surfactant such as dodecylpyridine bromide (DPB) deposits mainly because of electrostatic attraction [99]. Since the softener counterions do not adsorb onto fabrics, Hughes *et al.* conclude that the softener deposition results essentially from an ion exchange mechanism [100]. According to Sexsmith and White, the cation binding proceeds by ion exchange and by adsorption of ion pairs [101]. The binding of soluble cationic surfactants such as cetyltrimethylammonium bromide (CTAB) to cotton proceeds by ion exchange first, at pH values below p$K$, and by physical adsorption afterwards.

Most work has been done with monoalkyl quaternaries, and the conclusions extended to the dialkyl cationic surfactants. This process is incorrect since the monoalkyl derivatives are much more water soluble than the corresponding dialkyl derivatives and they behave differently. Only a few studies have been carried out on DHTDMAC. They are discussed by Crutzen [102]. A very different approach was that of Kunieda and Shinoda [88] and Laughlin and co-workers [89,103,104] who studied the phase diagram of dioctadecyldimethylammonium chloride as a model for DHTDMAC deposition.

The general picture arising these studies is that the deposition of cationic actives onto cotton is essentially due to the electrostatic attraction of the positively charged vesicles by the negative charges borne by cotton in water. As a result, the ammonium ions form ion pairs with the carboxyl groups of cotton [25].

The pattern of the curves of quaternary deposition corresponds to a high-affinity adsorption. For many authors, who link the surface affinity for softeners to the ion exchange capacity, such a pattern reinforces the electrostatic attraction mechanism [23,24,30].

*(b)   Mechanism of Fabric Softener Deposition in the Rinse Liquor: Hydrophobic Model.*   Several experimental evidences refute the electrostatic model:

1.   The deposition of a cationic softener onto cotton is always practically quantitative under real conditions, irrespective of the number of carboxylic groups [105].
2.   The affinity of cations for the negative charges borne by cellulose increases according to the following sequence: $Na^+ < CTA^+ < Ca^{2+}$ [106]. The affinity of softener actives for fabrics in solution consequently markedly exceeds that arising from pure electrostatic interactions [30].

3. The deposition also occurs onto more neutral fibers [73]. Therefore Laughlin suggested an ion exchange reaction between the counterions of cotton carboxyl groups and the cationic salts, coupled with physical adsorption [26].
4. More recently, it has been shown that electric charges on cellulose are not even necessary for the deposition of DHTDMAC [102]. This was achieved by following the electrokinetic potential of cellulose in the presence of increasing amounts of DHTDMAC, and by evaluating the effect of an organic solvent on the stability of the DHTDMAC–cotton adduct.
5. The deposition of DHTDMAC is favored by water hardness despite the screening of electric charges that reduces the electrostatic interactions.

The reality is consequently more complex than the simple electrostatic model. The adsorption of surfactants onto surfaces is the result of various factors: characteristics of the surfactant and of the surface, lateral interaction between the fatty chains of the adsorbed surfactant molecules, solvation of the surfactant and of the surface, etc. It is not the type of active ingredient–surface interaction that accounts for the deposition onto fabrics.

DHTDMAC is dispersed in water as vesicles, in which fatty chains are sheltered from the solvent. Upon dilution, a modification of the vesicle structure causes the deposition of the DHTDMAC molecules onto the available solid surfaces. Coating cotton by quaternaries leads to the release in the bulk of many water molecules initially interacting with cellulose. As a result, the system entropy increases. This phenomenon is known as hydrophobic interaction. Consequently, the driver of the softener deposition is the hydrophobic ejection out of the aqueous phase and the resulting huge increase of the system entropy.

Once on the fibers, DHTDMAC has little tendency to go back into solution because of its insolubility. Moreover, it interacts with the fabrics through dispersion forces, and electrostatic interactions when charges are present. Among the mechanisms reviewed by Rosen [107], ion exchange and ion pairing (charge neutralization) are more specific of the interactions with cotton, while interactions with synthetics rather involve dispersion forces and hydrophobic bonding, and the polarization of $\pi$-electrons to a lesser extent.

Since the deposition of DHTDMAC results from its insolubility in water, it is not surprising that no stoichiometric relationship exists between the charge quantity and the deposition extent. The amount of DHTDMAC on the fabrics depends on the specific surface area of the fibers, and on the softener concentration in the liquor to a lesser extent. As the deposition continues, repulsion gradually takes place between the ammonium ions on the fibers and the oncoming ions. This causes a decrease of affinity and its leveling off.

The neutralization of some ammonium ions by the negative charges enables additional softener molecules to deposit at the same place. This effect accounts for the larger amounts of softener present on charged fabrics (compared to uncharged

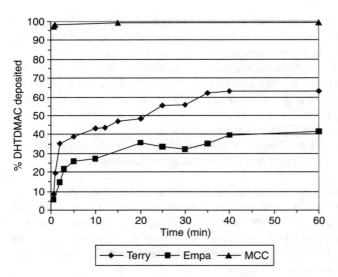

**FIG. 12.9** Kinetics of DHTDMAC deposition onto various celluloses as measured by adsorption isotherms. Amounts deposited on the fabrics are expressed as the percentage of DHTDMAC introduced in the liquor (17.3 mg DHTDMAC/g fabric). Terry, bath towels; Empa, short-napped cotton; MCC, microcrystalline cellulose.

fabrics). It also explains why DHTDMAC cannot be completely removed from softened cotton, even by organic solvents such as isopropyl alcohol.

Comparing the adsorption kinetics of DHTDMAC–fatty alcohol on microcrystalline cellulose, terry towel, and short-napped cotton showed evidence that the high substantivity on cotton is due to its very large specific surface area, not to the presence of negative charges. The larger the surface area, the more quickly and completely the DHTDMAC deposits (Figure 12.9). Microcrystalline cellulose is at once saturated, while terry towels adsorb more quickly than short-napped cotton.

The deposition of DHTDMAC onto a charge-free surface with a very large specific area is consequently much larger and quicker than on the same amount of negatively charged cotton with smaller specific area! Previously Sexsmith and White [101] had also found that DHTDMAC adsorbs much more on microcrystalline cellulose than on cotton.

These conclusions are not limited to laboratory experiments. When terry towels are laundered in a washing machine under real conditions (wash with a detergent, rinse with a commercial fabric softener), up to 95% of the DHTDMAC present on the fibers is linked to cotton by hydrophobic interactions [102].

*(c)  Parameters Governing Fabric Softener Deposition.*   From a number of studies [36,72,92,96,98,108,109], it appears that several parameters influence the rate, extent, and evenness of deposition.

*Structure of the active.* The most important parameter in the adsorption–desorption characteristics of cationic surfactants is the number of fatty chains linked to the cationic hydrophilic group and the number of carbon atoms of which they are made [25]. It is this bulky part that makes the molecule very hydrophobic and boosts its deposition onto fabrics, independently of the negative charge on the target fibers [16,35].

No consensus exists on the relative speed of deposition of the active. For instance, Linfield *et al.* reports that imidazolinium salts deposit much less quickly than ammonium salts [36], while, according to Hughes *et al.*, they deposit more rapidly at low concentration and at the same rate at higher concentration [54].

The type of counterion, chloride or methylsulfate, has little influence [36]. A more or less extensive ion exchange probably takes place in the rinse liquor with the anionic surfactants coming from the detergent carryover and the anions of water hardness. As a result, it is very likely that no strong difference exists between the two actives once on fabrics [26]. This is not in line with other results showing that anionic counterions of the surfactant are present on the fibers, which suggest an ion pair type adsorption [30]. For instance, iodide is much more substantive than chloride [98].

It should be noted that relative affinities of the different actives for cotton hold in water only. DHTDMAC affinity for cellulose is small in organic solvents such as isopropyl alcohol or chlorinated solvents, which readily dissolve it; higher amounts of active must then be involved to observe a significant deposition [15].

*Structure of the dispersion.* A good example of the dispersion structure effect may be found in the softener–cosoftener synergy. It is most probably by allowing the formation of small particles, hence improving the uniformity of DHTDMAC deposition, that cosofteners enhance the softener performance (see above).

*Type of fibers.* The chemical structure of the fabric strongly affects the softener affinity. The following sequence of increasing affinity can be found in the literature [2,4,23,73]: polyacrylonitrile < polyamide < cotton < viscose < wool. Acetate [2,4] and polyester cotton [23,73] are placed between polyamide and cotton. The ranking of the affinity for polyester varies according to the study: less than polyacrylonitrile [23,73], or just larger [2, 4], or even larger than polyamide [99]. These discrepancies may be due to variations of the specific surface area of the materials used in the various studies, as illustrated by the evaluations of the adsorption kinetics of DHTDMAC–fatty alcohol on microcrystalline cellulose, terry towel, and short-napped cotton (Figure 12.9). The experimental conditions, especially the pH of the liquor, should also influence the sequence.

Not surprisingly in the light of these results, softener deposition has been reported to vary with the nature of the fabric finish [34,105]. This aspect has already been discussed [102].

Some authors have also quantified the deposition of DHTDMAC onto various substrates (Table 12.1). Evans [2] reported the deposition leveling off at 1 mg DHTDMAC/g cotton.

**TABLE 12.1** Quantification of the Deposition of DHTDMAC onto Various Substrates

| Substrate | Fabric saturation (mg quat/g load) [23] | Deposition from a liquor (conc. = 1.5 g quat/kg load) [11] |
|---|---|---|
| Polyester | 0.6 | 60% |
| Polyacrylonitrile | 1.1 | 47% |
| Polyamide | 1.8 | |
| Polyester cotton | 4.0 | 73% |
| Cotton | 7.5 | 77% |
| Wool | 7.7 | 100% |

*pH of the liquor.* The deposition extent moderately increases when the pH rises from 2 to 9 [8,26,52,96], and is a maximum at pH 8 [18,97,98]. This may be assigned to the increase of the overall negative charge of the fibers that neutralizes a larger number of ammonium ions on the fabric. More DHTDMAC will consequently adsorb before the electrostatic repulsion prevents any further deposition.

The evenness of common softener deposition (DTDMA and imidazolinium chlorides and methylsulfates) is improved when the exhaustion slows down [1,24,36], i.e., when the pH decreases, and is best at neutral pH [1,8,18].

*Temperature.* The higher the temperature, the more and quicker the deposition [1,25,36,52,96,98]. The optimal temperature to get almost complete deposition of the softener is 25 to 32°C [97]. The effect has been assigned to a solubilization phenomenon [96]. This proposal is paradoxical, as deposition and solubility are at first glance antagonistic. In contrast, CTAB adsorption onto cotton increases only slightly when the temperature rises [106]. An alternative explanation is that DHTDMAC deposition involves entropy-driven phenomena, such as hydrophobic ejection and hydrophobic interaction.

*Duration.* The deposition onto the various fibers is almost complete in less than 5 minutes [1,92,97] and reaches a maximum after 10 to 15 minutes [96]. It has also been reported that the deposition of a monolayer over the whole surface of the fiber is usually achieved in 2 minutes, although reaching the equilibrium can take up to an hour [1]. Some sorption also probably takes place inside the fibers [73], which requires time for the fiber to swell.

*Water hardness/ionic strength of the liquor.* Water hardness favors softener deposition onto fabrics. Increasing the ionic strength can double DHTDMAC deposition onto cotton or polyester [23]. A plateau is reached at a hardness of 300 ppm [100]. The ionic strength of the liquor also influences the rate of deposition [52].

All the effects discussed above are quite consistent with the hydrophobic model. The various observations are explained by the electrical double layer compression.

In the hydrophobic model, quaternaries deposit onto textiles until their concentration on the surface is large enough to develop an electrostatic repulsion that prevents any further softener deposition. The introduction of neutral salts in the medium strengthens the interaction of the ammonium ions with their counterions, decreasing the softener solubility in water and increasing its tendency to deposit. At the same time, the electrical double layers of the particles are compressed, causing the reduction of electric field at any given distance of the particles and decreasing the repulsion by the molecules on the fabric. Consequently, a larger number of softener molecules can deposit per unit area.

Some authors also assign the enhanced deposition at high ionic strength to an increase of the adsorption by ion pair uptake [72,101] at the expense of the ion exchange mechanism [101]. Alternative explanations are a salting-out of the softener [96], or the neutralization of anionic surfactants by bivalent cations, leaving more softener available to deposit.

Finally, an osmotic shock due to the high electrolyte content of hard water has also been suggested as causing a decrease of the softener particle size, thereby enhancing the softening efficacy.

Softener deposition is drastically affected by anionic surfactants such as linear alkylbenzene sulfonate (LAS) or sodium lauryl sulfate (SLS) because of the immediate formation of water-insoluble complexes. For instance, if increasing amounts of SLS are added to a softener dispersion, DHTDMAC deposition is nearly quantitative until both concentrations are equal. The amount of softener on the fabric always corresponds to the concentrations of free softener and of catanionic complex. It seems likely that the complex is held on cotton by mechanical entrapment in the cotton fibers [72]. Once the SLS concentration exceeds that of the softener, the complex is increasingly solubilized by the surfactant excess and deposits less and less [72,92,108].

The extent of the anionic surfactant effect depends on the procedure adopted. If the anionic surfactant is added to the rinse liquor after quaternaries, the amount of DHTDMAC present on cotton is reduced to 19% of its initial value. It is reduced to 6% if the anionic and cationic surfactants are mixed in equimolar amounts before their introduction in the rinse liquor and no quaternary deposits if there is an excess of anionic surfactant [100].

Nonionic surfactants such as ethoxylated fatty alcohols also impair softener deposition, presumably because they improve the solubility of the softener particles and compete to deposit onto fabrics [37,73,110].

Surfactants are usually more detrimental in U.S. than in European washing machines. The amount of anionic detergent residues is much lower in European washers, which achieve several rinse cycles before introducing the fabric softener in the liquor.

As regards wash cycle fabric softening, the speed of softening in the wash is not the same as in the rinse. The differences are due to the competition for deposition that takes place between the various substances present in the liquor. In the beginning, the softening of hydrophilic fibers, cotton and wool, is somewhat quicker in the wash than in the rinse and slower afterwards. In contrast, the softening of synthetic fibers is at least as quick in the rinse as in the wash.

Soils present in the washing liquor decrease the amount of DHTDMAC that deposits onto fibers, since DHTDMAC also adsorbs onto the soil particles, which are negatively charged. As a result, the softening efficacy is impaired. Furthermore, because of quaternary adsorption, the negative charges on the soil particles and on the fabrics disappear, favoring soil redeposition.

These observations have also found application in softergents. The harder the water, the more the cationic softener deposition from an anionic detergent [9], since the compression of electrical double layers leads to a reduction of the anionic–cationic interaction. The deposition from a nonionic detergent may drop from 100% to 50% [9] as the softener, becoming less soluble, tends to interact more strongly with the hardness-insensitive surfactant.

The best softening is observed at a washing temperature of 40 to 45°C [14].

## 2. Desorption

Once on the fabric, the quaternaries withstand several rinses in plain water [1,97,98]. They are partly removed by anionic surfactants in subsequent washes [6,31], but quat removal is never complete, even after two washes with a detergent [26,96]. The situation is probably different with esterquats because of the hydrolysis that takes place under the alkaline conditions of the wash. In fact, only 25% of the DHTDMAC present on the fibers is removed during a wash with SLS, 75% remaining as a catanionic complex [72]. Molecules with the longest alkyl chain derivatives best resist cumulative rinses [20].

Quaternaries are also not totally extracted from cotton with chloroform [98] or with isopropyl alcohol. The residual amount most probably corresponds to the molecules that form ion pairs with the carboxyl groups at the fiber surface.

Under real conditions, the overall amount of quaternaries on fabrics slowly increases upon cumulative launderings [26,98,100].

## C. Softening
### 1. Mechanism

Fabric softness is the result of the skin–fabric interaction and of the fabric mechanical properties. Fabrics are made of yarns, yarns being assemblies of twisted fibers. Fabric mechanical properties depend on the fabric geometry, on the yarn–yarn friction, and on the yarn mechanical properties. The mechanical properties of the yarn are determined by the fiber rigidity, by the fiber–fiber friction, and by the yarn geometry [26].

During successive launderings and wear, the fabric finish is removed and the fibers are degraded by chemical and mechanical attack. Cotton fibers gradually unravel and break into microfibrils. In hard water areas, insoluble salts deposit onto the microfibrils during the wash and rinse. The lime scale buildup gives the fibers a "sandpaper-like" structure, making the textiles harsh to the skin. This phenomenon is known as fiber encrustation.

Upon drying, the microfibrils remain up, away from the fiber bundle. Fibers bristled with these microfibrils no longer slip easily over one another and, when garments are dried indoors, sheltered from the wind, no force counteracts the capillarity which brings neighboring fibers into contact. As a result, they interact much more strongly together, loosing flexibility and slipperiness [26].

Fabric softeners remedy these degradations and improve the feel by coating the fibers with a protective film of fatty material that reconstitutes the finish, maintains the microfibrils close to the fibers [23], masks fiber alterations, and covers the mineral deposits. Hence, the main mechanism for fabric softening is fiber lubrication.

According to Chalmers [1], the softener forms a film over the entire surface of the fabric, with the fatty chains pointing away from the surface. There is some evidence for the correlation between the soft touch and the reduction of friction, friction between fabric and skin and between the fibers themselves [20]. For instance, Röder and colleagues report an excellent correlation between softening efficacy and the interfiber friction coefficient. The latter was assessed by fixing a piece of treated textile to a metal block and measuring the force applied to move the metal along a fixed piece of the same textile [31]. Once the fiber–fiber friction coefficient decreases, the fibers move more independently of one another [3,27]. The overall result is a 20% decrease of fiber abrasion [13] and fabrics are perceived as more flexible [3]. The friction coefficient, however, increases when the relative humidity rises above 90% [20].

The ability of a film to reduce the friction coefficient depends on two factors: the energy of adhesion of the polar group to the surface and the energy of intermolecular cohesion in the fatty layer [20]. The former must be large to ensure a good film adhesion to the surface; the latter must remain small so that the films are not torn from the surface upon rubbing. To separate adequately the two surfaces, the fatty chains must have at least 16 carbon atoms [20]. A good softener, however, does not always give the lowest friction coefficient. Mooney mentions that unsaturated derivatives better lubricate than their saturated counterparts, as estimated from the friction coefficient. The contradiction is probably due to the lower affinity of unsaturated fatty molecules for fabrics, leading to lower concentrations on fibers [52].

At this point, a distinction must be made between static and dynamic friction coefficients. The force required to overcome the resistance to starting a movement is not the same as the one necessary to maintain an existing movement. The former corresponds to the static friction coefficient, the latter to the dynamic friction coefficient [52]. For instance, silk exhibits high static and low dynamic friction coefficients. A good softener must reduce both coefficients, but especially the dynamic one [20].

According to Berenbold [13], the presence of softener at the fiber surface also reduces the encrustations due to repeated washes by 35%. The weight increase of terry towels after 25 cumulative washes decreases from 7% (w/w) without softener to 3% (w/w) in the presence of softener [13].

The importance of frictional properties largely exceeds the area of domestic fabric softeners. They are also key in the textile industry, as they condition the slipping of fibers over each other in all mechanical processes [20]. Friction causes breakage of threads and generates static electricity. More generally, softened fabrics are pressed and sewed more easily, as there is less resistance to the metal movement. They also relax more quickly at the dry state since fibers slide over one another more easily. However, it is more difficult to cut fabrics with scissors as the blade slips over the fibers and fewer fibers are torn during cutting [20].

Reducing the friction is not the only way to improve fabric softness. Plasticizing the fibers is another approach. Plasticizing agents can restore the fiber flexibility. To be effective, the molecules of the plasticizing agent must be small enough to penetrate the fibers. The most common plasticizer is water; it is a good softening agent as moist fibers are less harsh than dried ones. Humectants such as urea, potassium acetate, glycerin, and other polyhydric alcohols, which moisten the fibers, can be regarded as possible softening actives [52]. However, because of their very large solubility in water, they are not prone to deposit from the rinse onto the fabrics. This prevents formulators from considering them as ingredients for rinse cycle fabric softening.

## 2. Evaluation

Some quantitative, instrumental methods exist for assessing fabric softness, but they are not sensitive enough to assess differences between softeners. The evaluation is usually carried out by sensory perception. The feeling of a textile is very subjective as, besides the basic physical properties of the fabrics on which it depends, it is also influenced by many evaluator-linked unquantifiable parameters such as mood and tactile sensitivity [52].

Most commonly, experienced judges are asked to compare the softness of fabrics treated with an experimental softener to that of a reference. The latter may be an untreated sample of the same fabric or a sample treated with the benchmark under the same conditions. The evaluation is conducted in blind pair comparison and in random order.

Pieces of terry towels have been found to be particularly suitable to evidence slight differences. The difference of softness is rated (e.g., 1 = weak, 2 = medium, 3 = strong difference); it is actually an overall estimation of the surface slipperiness, fluffiness, and texture. A statistical treatment gives a significance to the difference and, to some extent, quantifies the softening efficacy of the prototype.

The assessment is sometimes carried out by ranking groups of four to six towels but the evaluation is more difficult and the results less precise.

An interesting but more time consuming alternative is to quantify the softening efficacy of a system by determining the concentration at which it is equivalent to a reference. In another alternative, mentioned by Levinson, towels treated with softeners A and B are compared and ranked against a high standard and an untreated control [3]. A is judged superior to B at the 95% confidence interval if it is preferred by at least 15 out of 20 panelists. Quantification results must be interpreted with care, since the plot of softness magnitude against softener concentration is not linear but an S-shaped curve.

Bücking *et al.* [96] have quantified the effect of alkyl chain length on softening efficacy. They propose the following values (expressed as percentages of the $C_{18}$ softening efficacy): $C_{12} = 0\%$, $C_{14} = 40\%$, $C_{16} = 80\%$, $C_{18} = 100\%$, $C_{20} = 110\%$, $C_{22} = $ less. Williams evaluated the effect of the softener active structure [9]. He proposes the following equivalence: $0.10\%$ DHTDMAC $= 0.135\%$ tallow imidazolinium chloride $= 0.18\%$ tallow diamidoalkoxylated quat. It is important to note that these results were obtained in the laboratory, on clean loads in the absence of detergent. Differences under actual use conditions should be less because of the presence of soil and detergent residues in the liquor.

Several methods have been developed to quantify the softening efficacy of a system. Most are intended to quantify the various parameters that relate to softness. None is fully satisfactory. For instance, Laughlin mentions the determination of the fiber lubricity, and the determination of the compressibility and of the resistance to folding [26].

Good results are obtained with the Kawabata approach, which measures several different mechanical properties of the fabric at the same time. The Kawabata evaluation system for fabrics uses four devices measuring the tensile and shearing, pure bending, and compressional properties, and surface characteristics of fabrics. Key parameters are the applied force, the rate of deformation, and the tension on the sample. By comparison with subjectively evaluated standards, statistical correlations can be drawn, leading to the objective quantification of fabric softness. The method, however, is too complex for routine work [26]. Some other methods are reviewed by Mooney [52]:

- The Flesher pin method measures the force applied to push several pins into a fabric.
- The Taylor sound meter measures the sound generated by a textile when it is drawn over glass pins.
- The Kakiage method measures the thickness of a fabric as a function of the applied pressure. The pressure of human grip is, according to Mooney, 16 g/cm$^2$.

## D. Static Control

Static electricity may be generated in different ways [20,111]:

1. When separating two surfaces in contact. The quantity of charge, which is transferred from one surface to the other, depends on their relative affinity for the charge. The phenomenon is very common as different materials always have different affinities.
2. When rubbing surfaces of the same material if their microstructure is different. The large temperature gradient that appears at the interface between rough and smooth surfaces causes particles to move from the hot to the cold side, leading to a transfer of charge if they are charged.
3. When rubbing two identical surfaces of different size. Repeatedly rubbing the surfaces enhances the charge transfer.

The static charge is dissipated in the environment after the separation of the surfaces. The kinetic depends on the surface characteristics. It takes less than one second if the material is a conductor, minutes or even hours if it is an insulator [111]. Static charge occurrence also depends on the ambient relative humidity. The charge on a textile can decrease markedly when the relative humidity rises from 10 to 90%. It decreases by a factor of 7 for every 10% of humidity increase [111].

Dry cotton and wool do not conduct electricity very well, but, at ambient relative humidity above 60%, they adsorb enough moisture to dissipate static charge [20]. The situation is quite different with low-polarity fibers such as synthetic fibers, e.g., polyester (Dacron), polyamide (nylon), polyacrylonitrile (acrylics), and vinyl [4, 10]. Even at 60% relative humidity, they remain poor conductors.

The problems linked to static electricity occur when synthetic fibers are subject to friction under conditions of low relative humidity such as using the garments in a dry climate or drying in an automatic tumble dryer. The static problem is not solved by preventing the formation of charges but by dissipating them as quickly as possible, before the problem appears [111]. In other words, the buildup of static electricity is avoided by creating a layer that conducts electricity on the fibers and enables the charges to leak away. This is achieved by enhancing the surface conductivity and/or the humidity at the fiber surface.

To enhance the surface conductivity, a chemical is deposited that transports the current. Adsorbing a hygroscopic material increases the humidity at the fiber surface [31].

An efficient antistatic agent must consequently exhibit the following properties [23]:

1. High substantivity and homogeneous deposition onto fabrics.
2. Good moisture uptake capacity from ambient air.

3.  Ions released on the surface that can move freely in an electric field — polyions that interact more or less strongly with the fabric are less efficient in enhancing the conductivity of the surface [111].

All surfactants develop an antistatic activity. Their efficacy relies both on their ionic character and their capacity to bind water [23]. However, the most efficient antistatic agents are cationic softeners, because of their high affinity for fabrics [20]. Moreover they also reduce the generation of static electricity by lubricating fibers and reducing interfiber friction [10].

Softener antistatic activity is easily quantified by measuring the electrostatic charge or the surface resistance of items directly after the drying cycle. It depends on several parameters:

1.  The chemical structure of the cation, which determines the ionic and hygroscopic characteristics. Enhancing the hydrophilicity, e.g., by introducing double bond(s), ethylene oxide chain(s), or hydroxyl group(s) in the molecular structure, can increase the antistatic efficacy of an active [20].
2.  The type of counterion. Methylsulfate derivatives are somewhat more efficient than chloride ones [1].
3.  The amount present on fabrics. The higher the content, the better the static control.
4.  The particle size of the softener dispersion, since the coating is more homogeneous when the dispersion is micronized [92].
5.  The type of fiber [36].

Softening actives exhibit the following sequence of increasing antistatic activity in rinse cycle products: tallow imidazolinium < tallow ammonium chloride < tallow diamidoalkoxylated quat. Triethanolammonium oleate has also been reported to be a very efficient antistatic agent [20].

The type of softener is also important. Williams [9] proposed the following sequence of decreasing efficacy: dryer softener ≥ (liquid nonionic) wash cycle softener > rinse cycle softener.

The difficulties encountered by the textile industry are even greater. The friction of hot air moving along fibers produces static electricity. Besides electric shocks, the problems caused by static electricity are twofold: it may cause malfunctions in the operation of electronic equipment and it generates sparks that may be hazardous in the presence of flammable vapor [20]. Here also, softeners are extremely efficient in fighting static charges on fibers.

## E.  Others

### 1.  Ironing

Because of their long fatty chains that form a film at the fiber surface, softening actives work as lubricants and decrease the frictions between fibers. As a result, they facilitate the flexibility of the fibers and their sliding over each other, thereby improving fabric smoothness. Many [2,26,112] but not all [34] authors found the effect perceivable. The discrepancy is probably due to differences in experimental conditions, such as a different number of cumulative launderings. Even more efficient than the usual softener actives in improving fabric smoothness are polydimethylsiloxane and partially oxidized polyethylene [17].

The antiwrinkle effect of softeners is usually quantified by the wrinkle recovery angle, the residual angle exhibited by a fabric after having been creased and compressed. Better results are usually observed for softener-treated fabrics [34].

Softeners also reduce the dynamic friction coefficient between the fabrics and the iron, thereby making gliding easier [27]. The resistance to an iron is easily quantified by measuring the height of an inclined plane at which the iron starts to move: it decreases as the number of rinses with a fabric softener increases; the opposite is observed in the absence of softener [4].

## 2. Wettability of Softened Textiles

The intrinsic affinity of fibers for water is easily quantified by measuring the sinking time of a piece of fabric carefully deposited on water. A better alternative is the dye wicking method in which a fabric strip is suspended with the extremity dipping in a 0.05% aqueous solution of methylene blue. The rise of the liquid is recorded as a function of time and compared to that of a control without softener.

Water absorption by porous solids depends on the water–solid contact angle and on the liquid surface tension. The spreading of water in fabrics is the result of wetting the fiber surface, penetration into the fibers, and capillary pressure [20]. The wetting of yarns depends on their surface energy and the interfiber space.

The elaboration of a model to describe cotton wicking is very complicated, although the effect of quaternaries on the wetting of fibers is easily seen: the softener enhances the interfacial tension strongly. Since the fiber surface energy and the surface tension of water are not affected, the spreading coefficient is decreased, and so is the wetting of the fiber surface.

The spreading coefficient is given by:

$$S_{L/S} = -\left(\frac{\Delta G_{\text{spreading}}}{S}\right) = \gamma_{SA} - (\gamma_{SL} + \gamma_{LA})$$

where $S$ = surface area of the fiber, $\gamma_{SA}$ = surface energy of the fiber, $\gamma_{SL}$ = fiber–water interfacial tension, and $\gamma_{LA}$ = water surface tension. When an angle is formed between the solid and the liquid, the equation is:

$$\gamma_{LA} \cos \theta = \gamma_{SA} - \gamma_{SL} \text{ or } S_{L/S} = \gamma_{LA}(\cos \theta - 1).$$

Because of the presence of pores in cotton, no real capillary exists in or between the yarns and no true meniscus is formed. Consequently, Jurin's law does not apply. This explains why the elaboration of a model to describe cotton wicking is so complicated.

In real life, the decrease of water absorbency by fabrics is significant only when an excess of fabric softener is present on the fibers. Modifying the structure of the softening active to make it less hydrophobic restores the wettability. For example, one can use actives bearing shorter or unsaturated alkyl chains; the former approach is much more efficient than the latter. Introducing ether groups in the structure also helps in restoring the fiber affinity for water. All these modifications, however, induce a loss of softening efficacy! As a general rule, any structure modification that enhances the affinity of the softened fabric for water impairs its softness. The only exception is silicones, which, at low levels, increase both softness and water absorption.

An alternative way of restoring the affinity of cotton for water is to introduce small amounts of nonionic surfactant in the softener composition.

## 3. Perfume

Perfumes are a key ingredient of fabric softeners. They are complex mixtures of water-insoluble organic molecules, which remedy the intrinsic malodor of the product by integrating the base odor in the scent. They do not cover the base odor, they eliminate it [21].

When dispersed in water together with the melted cationic active, the perfume is solubilized in the hydrophobic area of the softener particles, which transport it onto fabrics.

Upon storage of the laundered garments, the perfume is slowly released into the air. As the release of the volatile components from the fatty layer at the fiber surface is slower than when the perfume is simply sprayed on the fabrics, the pleasant smell lasts longer.

Escher and Oliveros systematically studied the effect of various parameters on fragrance adsorption onto fabrics [113]. They found that the affinity of fragrance for fabrics is mainly determined by the type of fiber (cotton > polyacrylonitrile) and, to a lesser extent, by the type of single-chain surfactant (cationic > anionic and nonionic). These factors are interdependent (the effect of the type of surfactant on the affinity for polyacrylonitrile is weak). The effect of temperature and of surfactant concentration is less.

The tenacity of an ingredient, defined as the proportion of product present on wet laundry that remains after drying, essentially depends on the type of fabric and is much larger on cotton. This is tentatively assigned to the swelling of cotton fibers, leading to a better penetration of the fragrance into the fibers.

It would have been extremely interesting to include DHTDMAC in the study, as it is expected to outperform CTAB in transporting fragrance onto fabrics.

## 4. Antibacterial Activity

Cationic surfactants are extremely harmful to enzymes. The most efficient germicidal agents are single alkyl chain derivatives of ammonium or pyridinium [18]. They are known to interact immediately with the negatively charged amino acid residues of proteins. As a result, the electrostatic repulsion between the charged groups at the enzyme surface is lost and its structure collapses. This and the neutralization of the ionized groups in the enzyme active center cause the complete loss of the catalytic activity [69].

This is the mechanism proposed by Datyner to account for CTAC antibacterial, and antifungal to a lesser extent, activity he reports [20]. Martins *et al.* invoke the ability of the surfactant to disrupt the cell membrane and form mixed micelles with its lipids to account for CTAB and SLS antibacterial activity [29]. In contrast, the bacteria-killing efficacy of double-chain surfactants such as DHTDMAC is assigned to an alteration of the membrane protein function resulting from the adsorption of vesicles onto the bacterial membrane. This study was carried out under laboratory conditions (very low ionic strength), and the antibacterial efficacy of fabric softeners under realistic use conditions remains highly questionable.

## 5. Color Protection

Fabric softeners exhibiting color protection properties are also found on the market. Some contain ingredients that prevent dye bleeding or dye transfer from colored items onto whites. Others keep dissolved minerals in solution, thereby preventing their deposition onto the fabric and the resulting dull look [35].

## IX. FUTURE TRENDS

Trends for future fabric softeners may, to some extent, be found in the patent literature [10]. In a worldwide survey of 280 patents related to fabric softeners, 151 (54%) cover new softening molecules (Figure 12.10a). The main claims are color protection (color fading, dye transfer inhibition, and ultraviolet protection) and better fragrance perception (new perfume or improved delivery or longer lasting). Miscellaneous benefits cover improved aesthetics, increased convenience (ease of ironing and reduced wrinkling), soil release, and enhanced wettability. Reduction of malodor and disinfection are also claimed; they are particularly critical in equatorial areas (Figure 12.10b) [16].

Many of the new molecules are claimed to deliver more softness, leading to more cost effective or more easily concentrated products. To be considered for incorporation in a softening composition, the candidates must be available to the manufacturer on an industrial scale and cannot be hindered by patents. They must fulfill the human safety requirements and in Europe the environmental regulations.

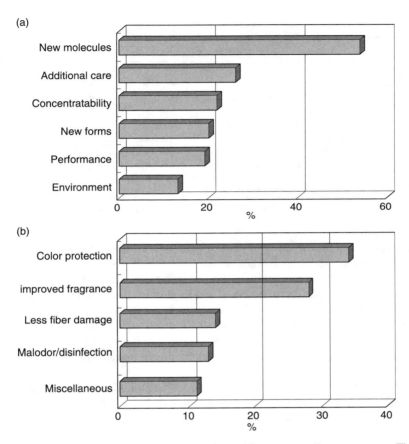

**FIG. 12.10** Trends for the future, as estimated from a patent literature survey. The survey involved 280 patents; percentages are the proportion of patents claiming the corresponding benefit. (a) Softener characteristics; (b) Softener attributes.

The cost of testing safety and environmental risk is high, usually over U.S.$200,000 or €155,000 [3].

Another important driver of the softener market is the globalization of formulations to reduce costs. Besides scale reduction, globalization enables the centralization of product development. Habits, preferences, and devices, however, vary from country to country. A global technology has to be adjusted to local habits as well as the different washing equipment.

Another source of growth may be found in concentrates, which enable a reduction of the packaging material. They also deliver more convenience and exhibit a

**TABLE 12.2**  Levinson's Vision of the Future for Fabric Softeners [3]

|  | Europe | North America | Latin America | Asia |
|---|---|---|---|---|
| Expected growth rate (%/year) | 2 | 3 | >5 | 1–2 |
| Primary market driver | New softening molecules | Cost/performance ratio | Performance; strong fragrance | Biodegradation[a] |
| Remarks | Esterquat fulfills environmental profile requirements | Actives are removed by waste treatments and considered as biodegradable | Line drying predominates; refill to reduce packaging waste | Japan accounts for 50% consumption |

[a] Biodegradation is a concern because of the high individual consumption and population density.

better environmental profile, but they may be perceived as offering less value for money (small bottle, thin product when diluted).

Major volume and value growths in the future are expected to take place in high-growth markets: Asia, Latin America, and Eastern Europe. The present consumption for these markets is low. The fabric softener in-home penetration is less than 30% and the consumption is only 7 l/user household/year. For fabric softener use to increase requires a certain economic well being for consumers to spend their income on products that go beyond their basic needs, which is simply cleaning [16].

In mature markets, added-value benefits may become a must for further business growth [16]. A complementary vision of the future of fabric softeners has been proposed by Levinson [3]. This is summarized in Table 12.2.

## REFERENCES

1. Chalmers, L., *Chemical Specialties, Domestic and Industrial*, 2nd ed., revised by Bathe, P., George Godwin, London, 1978.
2. Evans, W.P., Chem. Ind., 27, 893, 1969.
3. Levinson, M.I., *J. Surf. Det.*, 2, 223, 1999.
4. Barth, H., Heinz, W.D., and Mugele, H., *Fette Seifen Anstrichmittel*, 68, 48, 1966.
5. Whalley, G.R., *HAPPI*, 55, 1995.
6. Sebold, U., in *Proceedings of the 3rd World Conference on Detergents: Global Perspectives*, Cahn, A., Ed., AOCS Press, Champaign, IL, 1993, p. 88.
7. McConnell, R.B., *INFORM*, 5, 76, 1994.
8. Puchta, R., *J. Am. Oil Chem. Soc.*, 61, 367, 1984.
9. Williams, J.A., *Soap Cosmet. Chem. Specialties*, 59, 28, 1982.
10. Jacques, A., and Schramm, C.J., in *Liquid Detergents*, Lai, K.Y., Ed., Surfactant Science Series, Vol. 67, Marcel Dekker, New York, 1997, p. 433.
11. Berenbold, H., *INFORM*, 5, 82, 1994.

12. Foley, J., *Soap Cosmet. Chem. Specialties*, 55, 25, 1978.
13. Berenbold, H., *Comm. Jorn. Com. Esp. Deterg, Barcelona*, 18, 49, 1987.
14. Sherex-Rewo, *Manufacturing Chemist*, 30, 1983.
15. Cook, A.A., *American Dyestuff Reporter*, 24, 1973.
16. Schröder, U., in *Proceedings of the 4th World Conference on Detergents: Strategies for the 21st Century*, Cahn, A., Ed., October 4–8, 1998, Montreux, Switzerland, p. 142.
17. Middelhauve, B. and Penninger, J., *SÖFW J.*, 127, 3, 2001.
18. Milwidsky, B., *HAPPI*, 9, 40, 1987; 10, 40, 1987; 11, 96, 1987.
19. Ginn, M.E., Schenach, T.A., and Jungermann, E., *J. Am. Oil Chem. Soc.*, 42, 1084, 1965.
20. Datyner, A., *Surfactants in Textile Processing*, Surfactant Science Series, Vol. 14, Marcel Dekker, New York, 1983, p. 121.
21. Vogel, M., *HAPPI*, 29, 1987.
22. Recherche Blanchisserie CTTN, *Entretien des Textiles Ecully*, 43, 34, 1982.
23. Braüer, K., Nüßlein, H., and Puchta, R., *Seifen Öle Fette Wachse*, 111, 337, 1985.
24. Lang, F. and Berenbold, H., *Seifen Öle Fette Wachse*, 117, 690, 1991.
25. Müller, H. and Krempl, E., *Fette Seifen Anstrichmittel*, 65, 532, 1963.
26. Laughlin, R.G., in *Cationic Surfactants: Physical Chemistry*, Rubingh, D.N. and Holland, P.M., Eds., Surfactant Science Series, Vol. 37, Marcel Dekker, New York, 1991, p. 449.
27. Henault, B. and Elms, R., *Soap Cosmet.*, 34, 2001.
28. Domagk, G., *Dtsch. Med. Wochenschr.*, 21, 829, 1935; quoted by Puchta, R., *J. Am. Oil Chem. Soc.*, 61, 367, 1984; or 61, 829, 1935; quoted by Martins, L.M.S., Mamizuka, E.M., and Carmona-Ribeiro, A.M., *Langmuir*, 13, 5583, 1997; and 61, 250, 1935; according to http://www.dutly.ch/domagk/dom.html and http://www.asmusa.org/mbrsrc/archive/significant.htm.
29. Martins, L.M.S., Mamizuka, E.M., and Carmona-Ribeiro, A.M., *Langmuir*, 13, 5583, 1997.
30. White, H.J., *Cationic Surfactants*, Surfactant Science Series, Vol. 4, Marcel Dekker, New York, 1970, p. 311.
31. Schwartz, A.M., Perry, J.W., and Berch, J., *Surface Active Agents and Detergents*, Vol. II, Interscience, New York, 1958, p. 262.
32. Crutcher, T., Smith, K.R., Borland, J.E., Sauer, J.D., and Perine, J.W., *J. Am. Oil Chem. Soc.*, 69, 682 and 688, 1992.
33. Wilson, P.A., AATCC Book of Papers, International Conference and Exhibition, 1987.
34. Baumert, K.J. and Cox Crews, P., AATCC Book of Papers, International Conference and Exhibition, 1994, p. 140.
35. Soap and Detergent Association internet site (http://www.sdahq.org/laundry/fabriccare).
36. Linfield, W.M., Sherill, J.C., Davis, G.A., and Raschke, R.M., *J. Am. Oil Chem. Soc.*, 35, 590, 1958.
37. Egan, R.R., *J. Am. Oil Chem. Soc.*, 55, 118, 1978.
38. Billenstein, S. and Blaschke, G., *J. Am. Oil Chem. Soc.*, 61, 353, 1984.
39. Ramsbotham, J., *Soap Perfum. Cosmet.*, 158, 1986.

40.  Jacques, A. and Grandmaire, J.P., U.S. Patent 4,844,823 to Colgate-Palmolive, 1988.
41.  Waters, J., irral, B., Kleiser, H.H., How, M.J., Barratt, M.D., Birch, R.R., Fletcher, R.J., Haigh, S.D., Hales, S.G., Marshall, S.J., and Pestell, T.C., *Tenside Surfactants Detergents*, 28, 460, 1991.
42.  Puchta, R., Krings, P., and Sandkuhler, P., *Tenside Surfactants Detergents*, 30, 186, 1993.
43.  Singh, A.P., Turner, G.A., and Willis, E., European Patent 0,0409,504 to Unilever, 1989.
44.  Demeyre, H.J.M., U.S. Patent 4,933,096 to Procter & Gamble, 1988.
45.  Naik, A.R., Todt, K.H., and Wells, M.A., U.S. Patent 4,137,180 to Lever Brothers, 1979.
46.  Kang, H.H., Peters, R.G., and Knaggs, E.A., U.S. Patent 3,915,867 to Stepan, 1975.
47.  Wilkes, A.J., Jacobs, C., Walraven, G., and Talbot, J.M., in *Proceedings of the 4th World Surfactants Congress*, Barcelona, 1996, p. 389.
48.  Marsan, M.S., U.S. Patent 3,989,631 to Procter & Gamble, 1976.
49.  Oldshue, J.Y., *Fluid Mixing Technology*, McGraw Hill, New York, 1983.
50.  Grandmaire, J.P. and Jacques, A., GB Patent 2,139,658 to Colgate-Palmolive, 1984.
51.  Earl, G.W., in *Cationic Surfactants, Organic Chemistry*, Richmond, J.M., Ed., Surfactant Science Series, Vol. 34, Marcel Dekker, New York, 1990, p. 101.
52.  Mooney, W., *Textile Month*, 32, 1980.
53.  Ramsbotham, J., *Fabric Conditioners*, International Flavor and Fragrance, Hilversum, 1989, p. 2.
54.  Hughes, L., Leiby, J.M., and Deviney, M.L., *Soap Cosmet. Chem. Specialties*, 56, 1975.
55.  Crutzen, A. and Wouters, F., U.S. Patent 5,599,473 to Colgate-Palmolive, 1994; Crutzen, A., U.S. Patent 5,419,842 to Colgate-Palmolive, 1994.
56.  Jacques, A., GB Patent 2,133,043 to Colgate-Palmolive, 1982.
57.  Grandmaire, J.P., U.S. Patent 4,772,403 to Colgate-Palmolive, 1985.
58.  Turner, J.D., *Text. Chem. Color*, 20, 36, 1988.
59.  Union Carbide Silicones, technical literature SUI-554, 1985.
60.  Yang, K.O. and Yeh, K., *Text. Res. J.*, 63, 557, 1993.
61.  Schott, H., *Text. Res. J.*, 35, 612, 1965.
62.  Schott, H., *J. Am. Oil Chem. Soc.*, 45, 414, 1968.
63.  Schott, H., *Text. Res. J.*, 40, 924, 1970.
64.  Schott, H., *Text. Res. J.*, 35, 1120, 1965.
65.  Delgado, A., Gonzales-Caballero, F., and Bruque, J.M., *J. Colloid Interface Sci.*, 113, 203, 1986.
66.  Azoz, N.E.N., European Patent 0,585,039 or European Patent 0,585,040 to Unilever, 1994.
67.  Boyce, C.O.L., *Handbook of Practical Biotechnology*, 2nd ed., Novo Nordisk, Bagsvaerd, Denmark,1986.
68.  Christensen, P.N., Thomsen, K., and Branner, S., in *Proceedings of the 2nd World Conference on Detergents: Looking Towards the 90s*, Montreux, Switzerland, 1986.
69.  Crutzen, A. and Douglass, M.L., *Handbook of Detergents, Part A: Properties*, Broze, G., Ed., Surfactant Science Series, Vol. 82, Marcel Dekker, New York, 1999, p. 639.

70. Whalley, G.R., *HAPPI*, 10, 55, 1994.
71. Linfield, W.M., in *Cationic Surfactants*, Jungermann, E., Ed., Surfactant Science Series, Vol. 4, Marcel Dekker, New York, 1970, p. 9.
72. Hughes, L., Hsing, L.H., Simmons, B.L., Leiby, J.M., and Deviney, M.L., *Text. Chem. Color*, 10, 88, 1978.
73. Brauer, K., Fehr, H., and Puchta, R., *Tenside Surfactants Detergents*, 17, 281, 1980.
74. Puentes Bravo, E., Hermosilla, A., Grandmaire, J.P., and Tack, V., U.S. Patent 5,126,060 to Colgate-Palmolive, 1992.
75. Doms, J.R.P., Gillis, M.J.E.G., Lambert, P.M., Heckles, P.A., Puentes Bravo, E.E., Hermosilla, A.M., Grandmaire, J.P.M.H.F., and Tack, V.E.A., European Patent 0,570,237 to Colgate-Palmolive, 1993.
76. Crutzen, A. and Wouters, F., U.S. Patent 6,194,374 to Colgate-Palmolive, 2000.
77. Schott, H., *J. Colloid Interface Sci.*, 26, 133, 1968.
78. McKibben, G.E., U.S. Patent 5,768,918 to Procter & Gamble, 1998.
79. Brodt, M., internal communication.
80. Tronnier, H., *SÖFW J.*, 128, 8, 2002.
81. Piérard, G.E., Arrese, J.E., Rodriguez, C., and Daskaleros, P., *Contact Dermatitis*, 30, 286, 1994.
82. Giolando, S.T., Rapaport, R.A., Larson, R.J., Federle, T.W., Stalmans, M., and Masscheleyn, P., *Chemosphere*, 30, 1067, 1995.
83. Roux, E., internal communication.
84. Lin, T.J., in *Surfactants in Cosmetics*, Rieger, M.M., Ed., Surfactant Science Series, Vol. 16, Marcel Dekker, New York, 1985, p. 87.
85. Cascio, T., internal communication.
86. Auerbach, M.E., *Ind. Eng. Chem. Anal. Ed.*, 15, 49, 1943.
87. Auerbach, M.E., *Ind. Eng. Chem. Anal. Ed.*, 16, 739, 1944.
88. Kunieda, H. and Shinoda, K., *J. Phys. Chem.*, 82, 1710, 1978.
89. Laughlin, R.G., Munyon, R.L., Fu, Y.C., and Fehl, A.J., *J. Phys. Chem.*, 94, 2546, 1990.
90. Tanford, C., *The Hydrophobic Effect: Formation of Micelles and Biological Membranes*, John Wiley, New York, 1980, pp. 2, 24, 40.
91. James, A.D. and Ogden, P.H., *SD&C*, 50, 542, 1979.
92. Okumura, O., Ohbu, K., Yokoi, K., Yamada, K., and Saika, D., *J. Am. Oil Chem. Soc.*, 60, 1699, 1983.
93. Israelachvili, J., *Intermolecular and Surface Forces*, 2nd ed., Academic Press, London, 1997, p. 375.
94. Bungenberg de Jong, H.G., in *Crystallization — Coacervation — Flocculation in Colloid Science*, Vol. II, Kruyt, H.R., Ed., Elsevier, New York, 1949, p. 232.
95. Griffin, W.C., *J. Soc. Cosmet. Chem.*, 1, 311, 1949.
96. Bücking, H.W., Lötzsch, K., and Taüber, G., *Tenside Surfactants Detergents*, 16, 2, 1979.
97. Hughes, G.K. and Koch, S.D., *Soap Chem. Specialties*, 109, 1965.
98. Hughes, G.K. and Koch, S.D., 52nd Annual Meeting of the Chemical Specialties Manufacturers Association, Washington D.C., December 1965.
99. Suzawa, T. and Yuzawa, M., *Yukagaku*, 15, 20, 1966; *Chem. Abstr.*, 64, 12864b, 1966.

100. Hughes, L., Leiby, J.M., and Deviney, M.L., *Soap Cosmet. Chem. Specialties*, 44, 1976.
101. Sexsmith, F.H. and White, H.J., *J. Colloid Sci.*, 14, 598, 1959.
102. Crutzen, A., *J. Am. Oil Chem. Soc.*, 72, 137, 1995.
103. Laughlin, R.G., Munyon, R.L., Fu, Y.C., and Emge, T.J., *J. Phys. Chem.*, 95, 3852, 1991.
104. Laughlin, R.G., Munyon, R., Burns, J.L., Coffindaffer, T.W., and Talmon, Y., *J. Phys. Chem.*, 96, 374, 1992.
105. Beal, C.M., Olson, L.A., and Wentz, M., *J. Am. Oil Chem. Soc.*, 67, 689, 1990.
106. van Senden, K.G. and Koning, J., *Fette Seifen Anstrichmittel*, 70, 36, 1968.
107. Rosen, M.J., *Surfactants and Interfacial Phenomena*, 2nd ed., John Wiley, New York, 1989.
108. Smith, D.L., Cox, M.F., Russell, G.L., Earl, G.W., Lacke, P.M., Mays, R.L., and Fink, J.M., *J. Am. Oil Chem. Soc.*, 66, 718, 1989.
109. Hughes, L., Leiby, J.M., and Leviney, M., *Soap Cosmet. Chem. Specialties*, 51, 44, 1974.
110. Yanaba, S., Kanao, H., and Okumura, O., *J. Am. Chem. Soc.*, 65, 1977, 1988.
111. Hadert, H., *Seifen Öle Fette Wachse*, 90, 261, 1964.
112. Obendorf, S.K., *Text. Chem. Colorist*, 20, 11, 1988.
113. Escher, S.D. and Oliveros, E., *J. Am. Oil Chem. Soc.*, 71, 31, 1994.

# 13

# Specialty Liquid Household Surface Cleaners

**KAREN WISNIEWSKI** Global Technology Research and Development, Colgate-Palmolive Company, Piscataway, New Jersey

## I. INTRODUCTION

Household hard surface cleaners are defined in this discussion as formulations, powder or liquid, used to clean hard surfaces in the home, excluding dishes. Therefore, cleaners used on "soft" surfaces in the home — upholstery and carpet cleaners, fabric stain cleaners, etc. — are not discussed here. Also excluded from this discussion are household products that are used primarily as treatments rather than cleaners *per se* — polishes, floor waxes, tarnish removers, and drain cleaners (decloggers). Also not included are air fresheners, which are not cleaners but are often included in market analysis as part of the household cleaning market. Metal cleaners, surface descalers, and other such industrial liquid cleaners will not be discussed. The formulation of liquid bleach products is an art in itself so bleach will be discussed here briefly as an adjunct to all-purpose cleaning.

Household surface cleaners are now moving to added benefits beyond simple cleaning of the hard surface. Added benefits for cleaners are well established in other areas (e.g., conditioning with shampooing, softening with laundering, tartar control with tooth brushing) but such advantages are arriving late for hard surface cleaning.

This list of benefits can be described as changes made to the surface beyond cleaning, where something is left behind on the surface to give it desired properties beyond being clean. Cleaning only removes soil, therefore returning the surface to its native state, including whatever wear-and-tear the surface has incurred. Only very recently has treatment of the surface beyond cleaning been incorporated into household cleaners. This development is mostly taking place in bathroom/toilet cleaners and not so much in other areas of the house. (Polishes are, of course, the main surface modifiers in other areas of the household, but their main purpose is the surface modification and not cleaning. As noted before, they are therefore not part of this discussion.)

The other recent development in cleaners is the use of ingredients with "name recognition." In the personal care area, certain ingredients such as alphahydroxy acids or aloe are recognized by consumers as having specific benefits for the skin. In the case of aloe, this is the result of long years of folk tradition and word of mouth. In the case of alphahydroxy acids, this is the result of intensive advertising and education on the part of cosmetic companies. The aura around an ingredient can be achieved either way, or by a combination of ways; the point is that consumers recognize the ingredient and infer their own ideas of how that ingredient improves the performance of the product. Examples of such ingredients in the hard surface cleaning area are orange oil (for cleaning) and Teflon® (for surface improvement). These will be described in more detail in the discussion of product areas where they play the largest role.

The starting point for this discussion is the history of the development of household cleaners [1]. Powder cleaners will be covered briefly as part of the evolution of this field of cleaners. In general, powder cleaners tend to have large market share in developing countries, while liquid all-purpose cleaners and cream cleansers dominate in Western Europe, and liquid cleaners, especially those dispensed through trigger sprayers, enjoy popularity in North America and Australia and New Zealand. Therefore, this discussion, especially with respect to recent developments, focuses on developments in North America and Western Europe [2]. The area of household cleaning may be seen as one of the most challenging for the formulator, as the household cleaning regime can be said to have the most varied chemistry of any cleaning field. This is in response to the variety of cleaning tasks in the home, and the demands of the consumer.

To illustrate the problem for the product developer, one only has to enumerate the soils and surfaces. The soils can vary from simple dust and hair to dirt, hard water spots, and fingerprints to hardened grease, soap scum, and excrement. Although the usual household cleaning tasks are concentrated in only two rooms of the house, kitchen and bathroom, the number of different surfaces encountered are many. In the U.S., for example, there may be Formica®, ceramic tiles, grout, lacquered wood, vinyl flooring, painted surfaces, brass, stainless steel, enamel, porcelain, aluminum, chrome, glass, marble, methyl methacrylate, and other types of plastics. All of these materials may occur within only two rooms of the same home!

From a scientific point of view, one can see that these surfaces run the gamut from high-energy (ceramic and metal) to low-energy (plastic) surfaces. Soils, also, can vary, from very nonpolar (motor oil) to very polar (lime scale), and combinations of everything in between. The tenacity of the soil adherence will therefore vary according to the combination of soil and surface. In general, the better the soil wets the surface, the better the adherence. It is a well accepted principle of adhesion that two substances in intimate contact with each other tend to adhere very well, this being a necessary (but not sufficient) condition. High-energy surfaces tend to be easy to wet, making them generally easy to soil. An example is the relative tenacity of soap scum on ceramic as opposed to plastic surfaces (so called "fiberglass" bath enclosures that are made of methyl methacrylate or other acrylates).

Once wetting has occurred, the soil can then "bond" to the surface. What is often forgotten in adhesion is that van der Waals forces can be strong enough to account for the adhesion of soils to surfaces. Simple dispersion forces are about 5 kcal/mol, and hydrogen bonds between 4 and 40 kcal/mol, whereas covalent bonds can be as weak as 15 kcal/mol [3]. It can be seen from these numbers that if good molecular contact is made between the soil and surface, a bond can be made. This is especially easy if the soil is liquid or deposited from a liquid medium. Of course, the contributions of other mechanisms such as electrostatic attraction tend to strengthen the bond between soil and surface, if they are present. (Also, if the surface tends to be rough, then there also exists the possibility of purely mechanical adhesion, with the soil physically located in nooks and crannies of the surface.) If an attempt is made to break an adhesive boundary, a likely course is that one or the other of the materials tends to break within itself. Therefore, in cleaning, the soil can be broken down into successively thin layers removed from the surface, unless the fundamental bond between the soil and the surface can be compromised. Very thin layers can be even more difficult to remove than the original thicker layer [4]. For household cleaning this would imply that the most tenaciously held soil is that most intimately in contact with the surface and this should be the target of truly efficacious cleaning. Upper layers of the soil are relatively easy, by this analysis, to remove compared to the fundamental layer. In the beginning of cleaning history, the soil was simply abraded off, which inevitably damaged the unsoiled areas of the surface. In recent times, the discovery of more chemical, rather than mechanical, means of removing soil has greatly improved this situation.

A well-formulated modern cleaner avoids abrasion as a primary mechanism of cleaning and depends on more chemical rather than mechanical means. This has obvious advantages in terms of wear and tear on the surface. However, care must also be taken to make sure that chemical compatibility with the surface is also observed. For instance, acid cleaners generally have advantages in cleaning soap scum residues, but these would not be good cleaning formulations to use on a marble bathroom sink. Even though the acid will not greatly damage the structural integrity of the sink, it would surely remove the polished shine of the stone surface, minutely dissolving and therefore pitting and roughening the calcium carbonate. Therefore, a good knowledge of the chemical susceptibility of various household surfaces is necessary to the successful formulator.

Implements should also be considered in the components of the cleaning task. Much of the abrasiveness of the early cleaning process came not from the cleaner but from the implement, often a heavy scrub brush. Consumers in developed markets have a wide variety of implements to use in the cleaning process including cellulose sponges, brushes, cleaning cloths, paper towels, and plastic and metal scrubbing pads. The first three of these are also used in developing markets. These implements supply different amounts of abrasion to the soiled surface during the cleaning process and can blur the differences between cleaning formulations if the implement is highly abrasive.

Cleaning is generally accomplished in three steps: wetting, penetration, and removal. In some ways, water may be looked upon as the primary cleaning element. Given enough time, enough volume, and the right temperature, water is capable of cleaning almost any soil/surface combination. However, the times involved can be of the order of days, or the temperatures required close to boiling. The volumes of water also needed to rid the surface of the soil are also increasingly a concern nowadays.

Cleaning solutions can be viewed as water with ingredients added to speed the cleaning action, decrease the water volume, or lower the temperature at which effective water-based cleaning takes place. Organic solvents are, of course, very effective cleaners, particularly for nonpolar (greasy) soils, but the most effective solvents can have toxicity concerns [5]. (Various chlorinated solvents were long used by the electronics industry to degrease circuit boards, but due to increasing health and environmental concerns there is now a large patent literature on safer cleaning formulations.) Solvents are largely used in household cleaners as important, but not predominant, ingredients, except in glass cleaners and pine cleaners.

In pine cleaners, the pine oil component can be in the 15 to 30% range, and in glass cleaners, the solvent far exceeds the amount of surfactant in the formulation. However, even in these types, water is still the predominant ingredient, as it is in the rest of the household cleaners.

Surfactants, of course, lower the surface tension of water, thereby increasing the wetting of the soil by the cleaning solution. This is especially important for hydrophobic soils like grease. Solvents help the cleaning solution to penetrate into the soil, softening it, sometimes even partially dissolving the soil. (Surfactants also help to incorporate hydrophobic ingredients like solvents into water-based formulations by solubilizing them.) Other active ingredients, acids, bleaches, alkaline compounds, etc., can then more effectively react with the soil to change its composition to make it more liquid, less polymerized, more tenacious, etc., and easier for the cleaning solution to remove. The surfactant then helps to emulsify or otherwise lift the soil from the surface into the cleaning solution. The mechanical action of wiping or scrubbing also aids the wetting, penetration, and lifting of the soil by spreading, roughening, and breaking up the soil. The cleaning solution is then removed, leaving a clean surface. The best cleaners are effective at performing these tasks on even the most fundamental layers of soil, restoring the surface to its original state.

A further challenge for the formulator is the incorporation of surface-modifying ingredients into the cleaner. The ingredient must be stabilized into the formulation, but must destabilize to be deposited on the surface. Also, the deposition of the surface-protective agent cannot interfere with the cleaning process.

A cleaning task for the consumer will usually be one or two soils on any one of the surfaces. In addition, besides removing the soil, one must consider the safety of the chemical strategy used to remove the soil on the underlying surface. This has grown to become a more desired benefit in recent time [6]. Germ killing is also considered part of the household surface cleaning task by many consumers, especially on bathroom or food preparation surfaces. To meet some of these target concerns, the chemistry of the cleaner may be focused, but this can also limit the useful scope of the product.

As will be seen in this discussion, the evolution of cleaners developed from simple soap powders to liquid formulations to products that are more specialized. "All-purpose cleaners" are the backbone of this development. Along with special-ized formulas for specific cleaning problems, in some cases these products are augmented by specialized packaging. Generally, the packaging contributes more to the convenience of the product than the efficacy.

This is especially true of the new product form that has taken a large share of the market in North American and to an even larger extent in Europe. This product form is the wipe. Wipes have existed for a number of years in the personal care area (particularly as baby wipes) but have only recently become a preferred form in household cleaning. The structure and the formulation around this form is covered in the appropriate sections of this chapter.

The trend in the U.S. and Europe has continued over the years to more "niche" products as toilet bowl cleaners and dedicated bathroom cleaners. These types of products tend to show the greatest growth, and have maintained this high growth position for about ten years [7,8]. Abrasive-type cleaners have been more or less flat for ten years. Liquid all-purpose cleaners continue to sell well. Spray cleaners, bathroom cleaners, and toilet bowl cleaners all increased dramatically in 1999 and then more or less had a fairly constant level for three years. It may be that the sales were then also similar the following three years (2002 to 2004), but Wal-Mart data were no longer available. The largest part of the decrease in the sales numbers is due probably to the subtraction of Wal-Mart's contribution to the sales volume. The jump in bathroom cleaner sales in 1999 may be due in part to the launch of shower rinsing products. Toilet bowl cleaners have also seen a large increase in novelty and types in the same period. Glass cleaners probably include not only glass cleaners and ammonia cleaners (a very minor part of the market) but also the glass and surface sprays. (All-purpose sprays are probably included in the all-purpose cleaners category.) These spray versions can sometimes be considered "niche" products themselves, and can account for as much as 33% of the all-purpose cleaner segment. In Europe also special-purpose cleaners, such as bathroom cleaners, were the largest growth segment [9] until the launches of household cleaning wipe products in the late 1990s, growing from $578 million in 1997 to $1.8 billion in 2002 [10]. Glass cleaner wipes sales (in dollars) grew 405% in the U.S. from 2000 to 2001, and 344% from 2001 to 2002, and all-purpose cleaner wipes were up 85% and 14% in the same periods [11]. U.S. sales trends for various cleaning products are shown in Figure 13.1.

All-purpose cleaners have been launched in "ultras" — formulas that give double or triple the cleaning strength of the formulas already common in the marketplace. This has already happened in Europe and North America. These products use less packaging, occupy less storage space, and give consumers more flexibility in the dilution of the product. However, they also require more careful measuring on the part of the consumer if they are to reap the full value of the product. As will be seen in the section on all-purpose cleaners, the range of concentration of active ingredients in recent all-purpose cleaner patents is sufficient for these products to be formulated either as normal strength products or as ultras. These concentrated formulas have largely been rejected by North American consumers and the "regular" concentration is what is seen on store shelves.

Nevertheless, all-purpose cleaners are generally the beginning points for entry and for specialization in a given market. The niche products are the fastest growing part of household cleaning in mature markets, and yet they are starting to appear in developing markets as well. Strangely, all-purpose cleaners (or APCs) can be

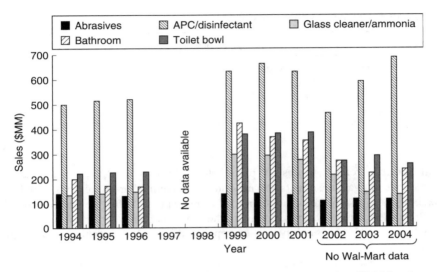

**FIG. 13.1**   Sales amounts for household cleaners in North America. Wal-Mart data are not included after 2001, which accounts for the apparent decrease in volume in 2002. (Source data all from *Household and Personal Products Industry*, November 1994–1996; December 1999–2001; April 2002–2004.)

considered specialty products themselves, growing out of the real all-purpose cleaner of the past — a heavy-duty cleaning powder. This can still be true in less mature cleaning markets.

## II.   ALL-PURPOSE CLEANERS
## A.   Historical Background
### 1.   Powder Cleaners

The evolution of household cleaners begins with all-purpose cleaners. Specialization to handle the multiple problems of household cleaning has arisen relatively recently. Before the 1930s [12], consumers had only soap powders with which to do all their household cleaning, which included not only kitchen and bath surfaces but also laundry and dishes. Glass windows were nearly the only surface that could not be effectively cleaned with this product. This multiple use of a basic cleaning product continues in many developing regions of the world.

In the 1920s powdered products began to appear in the U.S. market that were formulated especially for general household cleaning. These were generally highly built, very alkaline formulations designed to be dissolved in warm or hot water for tasks such as floor mopping, grease removal, and bathroom cleaning.

**TABLE 13.1**   Powdered All-Purpose Cleaner Formulas

| Ingredient | Examples | Amount (wt%) |
| --- | --- | --- |
| Anionic surfactant | Soap, alkylbenzene sulfonate (usually one only) | 1–10 |
| Builder | Phosphates, carbonates, silicates (usually a mixture) | 50–60 |
| Sodium sulfate | Processing and delivery aid | 30–50 |
| Perfume, color, etc. | | 0.5–1.0 |

Built formulas refer to the use of phosphates, silicates, carbonates, etc. These ingredients mitigate the effects of hard water by chelating hard water ions, supply alkalinity, and also buffer the system to high pH values. At first powdered products still used soap but later began to use the modern anionic surfactants (such as alkylbenzene sulfonates). A typical example of this kind of formula is given in Table 13.1.

These cleaners were more effective than their predecessors, but they also required a large amount of rinsing. The builders, which boosted cleaning efficacy, also increased the amount of residue left behind when the cleaning solution dried on a surface. Depending on water temperature and hardness, these cleaners could also be difficult to dissolve completely in a bucket of water. Being somewhat hygroscopic powders, they also tended to cake and solidify once their container, usually a cardboard box, was opened.

## 2. Cleansers

As far back as the 1880s a product was sold as a pressed cake of soap with abrasive [13]. However, modern powder cleansers also started to appear at roughly the same time as powder all-purpose cleaners (approximately 1930 to 1935). The addition of abrasives to the basic cleaning product helped use mechanical as well as chemical energy to do cleaning, but obviously made these products unsuitable for general use. In this sense, cleansers can be thought of as some of the first "specialty" cleaners because the presence of abrasive made them appropriate for very tough cleaning jobs, but also limited their usefulness because they could scratch softer surfaces. (Typical cleanser formula amounts are shown in Table 13.2.)

This tends to be the continuing theme of specialty cleaners — the formulation delivers more directed power at a particular cleaning problem, and then disqualifies itself from other tasks due to this adaptation. Consumers often comment that they want a truly "all-purpose" cleaner, but tend to buy targeted products [14].

Both powder all-purpose cleaner and powder cleansers are still in the U.S. market and maintain large shares of the market in developing areas of the world. Powder cleansers are still a major part of the abrasive cleaning subcategory, but

**TABLE 13.2** Powdered Cleanser Formulas

| Ingredient | Examples | Amount (wt%) |
|---|---|---|
| Anionic surfactant | Soap, alkylbenzene sulfonate (usually one only) | 1–5 |
| Builder | Phosphates, carbonates, silicates (usually a mixture) | 5–30 |
| Abrasive | Silica, feldspar, calcite | 60–90 |
| Disinfectant | Usually hypochlorite-generating compound (e.g., trichlorocyanuric acid) | 0–2 |
| Perfume, color, etc. | | 0.25–0.5 |

powder all-purpose cleaners in the U.S. were represented recently only by one major brand (Spic and Span®).

The formulas have had to react to modern pressures, the largest of which were the limitations and bans of phosphates and on branched alkyl aryl sulfonates. Phosphate builders are usually not used, or used in very small quantities, in most household cleaners outside of automatic dishwashing detergents. Usually carbonates, bicarbonates, and silicates are used along with more modern ingredients such as EDTA (ethylenediamine tetraacetic acid) and its various derivatives, NTA (nitrilotriacetate), citrates, and polyacrylates. (Many of these compounds were under toxicity and environmental investigations.) [15]. Some small amount of nonionic surfactants may also be used, although alkylbenzene sulfonates now dominate as the surfactant of choice. Hydrotropes such as sodium cumene sulfonate and sodium xylene sulfonate may also be added to help aid the dissolution and cleaning of the main surfactant. In these choices, household surface cleaner formulation has mirrored the developments in fabric care.

Cleansers have also undergone some formulation changes that are specific to their group. One major change was the addition of bleach to major cleanser brands in the 1950s. Cleansers had been used in the past to "sand out" stains, with some degree of surface damage. With the introduction of ingredients such as the isocyanurates (like trichlorocyanuric acid, TCCA), usually in the range of 0.5% available chlorine, stains could be removed due to hypochlorite bleaching rather than muscle. This addition also opens the possibility of disinfectant claims for the product. However, this addition also brings with it more demands on the formulator, as the formula is harder to stabilize. Compounds such as TCCA decompose with water to form hypochlorite bleach. This is good for the usual use conditions but confers water sensitivity on the formula. Usually small amounts ($<5\%$) of water-adsorbing compounds are added to prevent premature activation of the bleach. However, two factors work against the formulator. The normal packaging of cleansers is cardboard cylinders that are not able to be resealed tightly, and the

**TABLE 13.3**  Mohs Hardness of Some Abrasives and Household Surfaces

| Surface | Mohs hardness |
|---------|---------------|
| Corundum (alumina) | 9 |
| Quartz (sand or fused silica) | 7 |
| Glass | 5–7 |
| Feldspar | 6 |
| Steel | 5–6 |
| Aluminum | 3 |
| Calcite (marble, limestone) | 3 |
| Plastics | 2–5 |

product is usually stored under a sink. Under these conditions prolonged storage of the product usually results in loss of bleaching ability.

The largest recent change in cleansers has come in abrasives. These have generally become softer as time goes on, going from sand originally to calcium carbonate (calcite) now in the major brands. (Some smaller brands continue to use feldspar or similar compounds as abrasives, which rate on the Mohs scale around 5.) This constitutes a change on the Mohs hardness scale from about 7 to 3 [16]. (Table 13.3 lists the Mohs scale hardness of some abrasives and household surfaces. Mohs hardness is not always diagnostic of surfaces with elasticity where impact hardness tests are usually used. However, with respect to scratching, the Mohs scale can give some indications.) This is also a reaction to changing times — many household surfaces are now plastics of various kinds and are easily scratched by silica. For instance, where glazed ceramics in the bathroom could be scrubbed for years without seeing signs of wear, today's methyl methacrylate shower enclosures would show damage after a single vigorous use of a silica cleanser. The major exception to this trend is the increasing use of solid polymer countertop, such as Corian®, which encourages the use of strong abrasives to eliminate nicks and stains in the surface.

Mention should also be made here of the low market share products that are acidic cleansers. Some make a claim for rust stain removal. This is based on the inclusion in the formula of oxalic acid that is particularly good at chelating iron ions [17]. However, this also brings a particular problem: oxalic acid is moderately toxic (evidently due to upset of the ion balance in the body) which might mean warnings on the label. Naturally, the low pH is supplied by the acid content of the product. It would be expected, given the low surfactant concentration (similar to other cleansers) and the low pH, that grease cleaning by this type of cleanser would be less than that of the alkaline cleansers due to the effect of alkalinity on grease cleaning.

## B. Cream Cleansers

Both cleansers and all-purpose cleaners are now also available in liquid forms, which were the next stage of their evolution. The liquid form has two main advantages. Liquids can be formulated in a concentrated form that can be diluted by the consumer before use to the desired strength. This dilution operation is easier for the consumer because the liquid form mixes easier and dissolves better than the powder form that preceded it. The liquids can also be used straight from the container on heavily soiled areas; the powder cleaners had to be made into a paste before they could be applied.

From the formulator's point of view, there is a mixture of advantages and disadvantages. Water is an even less expensive diluent to use than the sodium sulfate or abrasives that were used in powder forms. However, because all the ingredients are dissolved or dispersed, this makes possible more interactions between the ingredients than in a powder. Fortunately the liquid form broadens the choice of surfactant available because they do not have to be available in powder form.

Liquid cleaners with suspended abrasives — cream cleansers — first started to appear in the U.S. and Europe in the 1980s. This chapter does not go into detail on the formulation of these products because although they are fluid household surface cleaners, they should be properly considered as suspensions and not as true liquid formulas. Their arrival on the market so recently is due to the difficulty of producing stable suspensions of abrasive particles; the advancement of polymer science and clay technology during the last 30 years has played a large role in the successful formulation of these products.

Unlike their parent product, powder cleansers, the cream cleansers usually use the gentler calcite abrasive. This, combined with their liquid form, helps to convey the image of less harsh cleaning to the consumer. This is especially important to consumers who have softer, plastic surfaces in their bathrooms such as the fiberglass (polymethyl methacrylate) shower enclosures which are much more easily marred than the traditional vitreous materials [18].

The patent art for this kind of cleaner begins in the mid-1960s [19–21], but there continues to be abundant patents written up to the present time [22]. The main point of most art is the stable suspension of the abrasive particles and this is achieved largely by raising the viscosity of the system. Imanura [23] uses crosslinked polyacrylic acid as an aid to suspending the abrasive. Two patents give examples of two polymers mixed together, one of which is a polysaccharide [24,25]. Brierly and Scott's patent [26] gives an example of the use of clays to thicken the formula, their aim being the stabilization of the formula during high shear processing steps. Another patent gives the alternative of either clay or fumed silica as the thickening agent [27], as well as combinations of polymer with clay [28]. Alternatively, the same inventors also give a method for stabilizing the cleanser at high shear using monoalkylolamides [29]. Other recent art uses colloidal alumina associated with a surfactant to suspend the abrasive [30–32]. Another approach is to use stearate soap to thicken a formula in which the aluminum oxide is used as the abrasive [33]. A somewhat novel approach by two groups of inventors [34,35] uses neither of the usual approaches, but instead generates a liquid crystal phase to suspend the abrasive. There are also examples of nonthickener viscous systems, a mixture of surfactant, nonpolar solvent, and electrolyte [36] or a mixture of surfactants [37,38]. In all cases the viscosity of the product is quite high, of the order of 500 to 5000 cP.

The most challenging formulations also contain hypochlorite bleach as well as trying to suspend an abrasive. As many suspending or thickening agents are sensitive to oxidation, it becomes difficult to put together a product lacking syneresis. For cream cleansers, the most successful approach for bleach-containing formulas seems to be alumina/surfactant thickening systems [39,40]. The alumina particle size is very small and calcium carbonate is used as the abrasive. However, in recent years the art of using polymer thickeners such as polyacrylates with bleach has improved with more patent art appearing on this [41,42]. One of the approaches is to limit the ionic strength of the formulation [43].

There is also an example using oxalic acid and suspended abrasive to give iron stain removal, more aimed at bathroom soils [44].

Sometimes the abrasive is given as an optional ingredient in an otherwise completely liquid cleaning formula. In one example of this type, the abrasive is combined into a terpene/limonene solvent-based cleaner while remaining stable [45]. In another example, the surfactant instead of the abrasive is largely eliminated [46].

A generalized formula for liquid cleansers is given in Table 13.4. Most of the recent patent art concerns the thickening system, while another significant part deals with "soluble abrasive" [47–52], including one that uses a soluble form of borax [53,54]. The solubility of the abrasive is probably intended to defeat the largest consumer complaint about cleansers, either powder or liquid: the difficulty of completely rinsing away the abrasive after use. However, one unusual example uses plastic particles as the abrasive, suitable for scrubbing plastic surfaces (like methacrylate shower enclosures) but are still bleach stable [55].

Cream cleansers can be even more difficult to rinse away because of the agents used to keep the abrasive suspended, such as clays. One unusual example uses silicone compounds to make the residue less visible and make the surface more glossy and to feel smoother [56].

Another example of silicon compound use is in the very specialized area of glass-topped stove cleaner/conditioners [57]. These can be thought of as a special kind of cream cleanser with an added benefit. The strength of an abrasive cleaner is needed to remove the baked-on soil encountered on a stove top, and as long as the abrasive is less than Mohs hardness 6 the glass top will not be scratched.

**TABLE 13.4**   Cream Cleanser Formulas

| Ingredient | Examples | Amount (wt%) |
|---|---|---|
| Anionic surfactant | Soap (stearate), alkylbenzene sulfonate | 1–15 |
| Nonionic surfactant | Ethoxylated alcohols, amine oxides | 0–10 |
| Builder | Phosphates, carbonates, silicates (usually mixture) | 1–20 |
| Solvent | Alcohol, glycol ether, limonene | 0–7 |
| Thickening agent | Polyacrylate polymer, clay, alumina | 0–10 |
| Abrasive (soluble compounds used at levels beyond their solubility) | Usually calcite ($CaCO_3$), but also sodium carbonate or bicarbonate, potassium sulfate, sodium citrate, borax | 10–90 |
| Bleach | usually hypochlorite but can be peroxide | 0–1 |
| Hydrotrope | sodium cumene sulfonate, sodium xylene sulfonate | 0–5 |
| Perfume, color, etc. | | 0–1 |
| Water | | 40–80 |

However, polysiloxanes will react with the glass surface to give a treatment that reduces the tenacity of new soils ("conditioning").

## C.   Gel Cleaners

A natural development of thickening systems is to develop transparent or translucent thickened systems. Such a form is generally called a "gel" although it may or may not conform to the technical rheological definition of a gel. Because such products are thickened, they lend themselves to the addition of abrasive because the structure of the gel can suspend the solid. However, some cleaners are formulated as gels to achieve benefits like cling on a surface. One of the ways of achieving thickening in the gel is through the surfactant system [58], especially a liquid crystal system [59]. There are largely aesthetic advantages to this type of approach. In ordinary cream cleansers, the appearance is white and opaque — the abrasive is largely invisible. In a gel cleaner, the abrasive could be made to stand out from the background because the medium is clear. In one case, the abrasive is made a different color than the medium [60].

## D.   Liquid All-Purpose Cleaners
### 1.   Historical Background

Simple all-purpose cleaners were introduced in liquid form starting in the 1930s. Liquid all-purpose cleaners were for many years differentiated mainly by the specific active ingredient that they contained. The simplest liquid cleaner was ammonia with some added soap which has been used for nearly a hundred years [61]. This is an effective grease cleaner, but is very unpleasant and harsh to use. Simple household bleach solutions date back to the early 1900s [62]. These are effective stain removers, and have some effect on proteinaceous soils, but are not particularly good for tough oily soils. Liquid disinfectant solutions date back to the 1870s [63], but these were more targeted to the elimination of germs than to soil removal. The closest products to modern formulations begin with pine oil formulas, dating back to 1929 [64]. This would have given good grease cleaning with a more pleasant (and safer) odor. However, there are consumers who dislike pine odor and would prefer a different cleaner. Modern formulas without pine oil were introduced between 1955 and 1965 in the U.S.

Liquid all-purpose cleaners at that time still incorporated many of the charac-teristics of their dry predecessors. They still used the popular anionic surfactants such as alkylbenzene sulfonate and builders with high alkalinity to achieve their goals. However, they had three important differences. First, liquids dissolve (or disperse) more quickly than powders. This means the cleaning solution can be prepared by the consumer quickly at ambient water temperatures. Second, they are not as limited by their solubility. Higher amounts can be added to the clean-ing solution without the precipitation that could be encountered with the older powder cleaners. Third, they are neater to dispense and store. The development of plastic bottles has been a huge boon for this product form. Liquid household cleaners originally came in glass bottles, which are heavier and easier to break than plastic bottles. The cardboard boxes of powder cleaners would have been an advantage for that product form until the commercialization of plastic bottles. The availability of higher density polyethylene combined with blow molding tech-nology (developed for glass) resulted in the widespread use of plastic bottles in the late 1950s [65]. This accelerated in the mid-1960s when household bleach led the conversion in the household area by converting from glass to high-density polyethylene (HDPE) [66]. The development of polyethylene terephthalate (PET) in the 1980s for carbonated beverages expanded the choice for rigid plastic bot-tles [67]. With these developments the advantages for liquid formulations now dominate: plastic bottles are light, durable, and easily reclosable.

However, despite these advantages there are some drawbacks as well. Although they might have had a weakness on greasy soil, powder all-purpose cleaners were not particularly deficient in cleaning; they were deficient in convenience. In gen-eral, this tends to be the biggest component in the evolution of household cleaners. A major change in form or formulation is not motivated by claims of superior cleaning; these claims tend to be made among cleaners of the same form. The major motivation seems to be increasing the convenience for consumers while maintaining the cleaning efficacy. This may be achieved by more convenient dis-pensing, or shortening the number of steps in the cleaning process. It may be the transfer from muscular effort on the part of the consumer to chemical energy supplied by the cleaner. These trends will be seen in all the various types of cleaners discussed here.

Part of the high efficacy of powder detergents was their high concentration of builder salts. To formulate the same level of builders into liquid detergents required even higher levels of surfactant. This usually results in higher foaming. Although high foam may be preferred in applications like shampoos and dishwashing detergents, such sustained foam is undesirable in household cleaning. Many surfaces that are washed with all-purpose cleaners are large and horizontal, e.g., floors and countertops. If a cleaner has a slowly collapsing foam, the consumer must laboriously rinse the surface. Even in areas where rinsing is relatively easy, such as a bath or shower enclosure, the extra effort and time spent rinsing a high-foaming product is undesirable to a consumer. The immediate "flash" foam produced when the cleaner is first used serves as a signal of its efficacy. Once this message is communicated, the foam should collapse to give easier rinsing and some formulations are made to optimize this, usually involving the use of soap/fatty acids [68,69].

Continuing to use the high levels of builders used in powders would also have meant continuing another rinsing problem: residue if the cleaning solution is left to dry on surfaces. A consumer will not consider a surface clean unless it shines. Residue from the cleaner left on the surface, even if all the soil has been removed, will diminish a consumer's evaluation of the cleanliness of the surface. Residues from crystalline compounds like builder salts tend to dull the native shine of a smooth surface. The degree of shine left on the surface is an important indicator to the consumer of the degree of cleanliness of the surface. In an effort to counteract the foaming and residue effects, formulators began decreasing builders, using solvents, and putting more effort into finding surfactant synergies.

## 2. Solvents

Solvents became useful when the product form changed to liquids. Their main role is to penetrate and soften grease to facilitate its removal by the surfactant. Their fluid form made them attractive to use in liquid formulations, although the solubility of good grease cutting solvents in water is very low. Some examples of these are pine oil and D-limonene (see Figure 13.2). These are still very popular today as grease cutting solvents, not only for their efficacy, but also more recently due to their "natural" origins. This reflects on the current marketing ploy of playing on some consumers' opinion that vegetable-extracted chemicals are intrinsically safer or more ecologically sound than petroleum-based ones. Although the first pine cleaners appeared in the early part of the twentieth century, formulations built around pine oil still occur in the patent art, most recently as an alternative to more volatile organic solvents [70]. Pine oil content in cleaners has dropped significantly, from 70 to 90% originally to 10 to 30% currently [71].

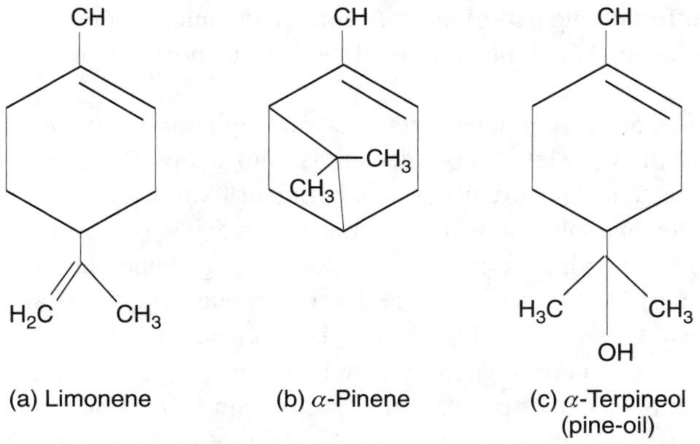

(a) Limonene      (b) α-Pinene      (c) α-Terpineol
(pine-oil)

**FIG. 13.2**  Structure of naturally derived grease cutting solvents (terpenes).

A recent trend has developed around "orange oil" cleaners [72]. In this case, most sellers of these cleaners mean D-limonene, as limonene is indeed obtained by extraction from the rinds of oranges (and to a lesser extent other citrus fruits) used in the juice industry. This is a very effective grease cutting solvent, similar to α-pinene of pine oil. "Natural" cleaner companies have spent some time and energy educating consumers to the efficacy of this ingredient as an alternative to other cleaning additives. The solvent power of limonene is similar to pinene, and to some consumers the orangey or lemony odor of the orange oil is preferable to the smell of pine oil. Used in sufficient amount the orange oil can indeed contribute to the cleaning efficacy of a formulation. However, in many cases the amount of actual orange oil present in the cleaner is below 0.5% and it contributes mainly to the fragrance of the product rather than its efficacy. One can find a number of examples in the patent art where limonene is highlighted as a significant ingredient [73–75]. The orange oil trend has been most strong in all-purpose cleaning products in several forms (dilutable cleaner, spray cleaner, wipes) as well as some polishing products that wish to imply more cleaning power.

Much more common and more easily formulated because of their higher water solubility are the glycol ether solvents. This approach dates back as far as the early 1970s [76,77]. Earlier formulations made use of the simpler ethylene glycol monoalkyl ethers (Cellosolves®), but this use has been largely discontinued because of health hazards [78]. The diethylene glycol monobutyl ether is most favored, although the use of the propylene glycols is increasing. As the chain length increases the health hazards decrease [79], but the solubility in water also decreases. This increases the difficulty of formulating stable products.

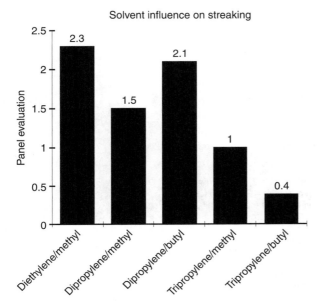

**FIG. 13.3**  Streaking/filming as a function of glycol ether. (Source: Michael, D.W., U.S. Patent 5290472, 1994.)

However, stable products are made, as evidenced by recent patent art which mentions the use of tripropylene glycols and their ethers for concentrated floor cleaning products [80,81]. In this case, the glycol ether is claimed not only as part of the cleaning system, but also as a way of enhancing the shine on a vinyl flooring surface. In Figure 13.3, a comparison is made of the streaking/filming characteristics based on solvent changes alone. There is also mention in other art of longer chain lengths in the ether portion of the glycol ethers also being used [82].

Other solvents can also be found in the literature. In particular, when the surface to be cleaned is painted or vanished (and therefore susceptible to solvent damage) the solvent must be chosen with care [83].

It can be seen from the structure of the glycol ethers (Figure 13.4) that if the alkyl chain is extended sufficiently then it begins to approximate the structure of simple nonionic surfactants, those of the ethoxylated alcohol family. This has led some chemical producers in recent times to introduce compounds meant to be hybrid chemicals with the properties of both solvent and surfactant. They are not at present used to any great extent in household cleaners, but remain a possibility for future formulations.

Monoglycol monoether:

$$HO-\underset{\underset{R}{|}}{C}H-CH_2-O-[CH_2]_n-CH_3$$

Ethylene glycol monobutyl ether: R = H, $n = 3$
Propylene glycol monobutyl ether: R = $CH_3$, $n = 3$
Propylene glycol monoethyl ether: R = $CH_3$, $n = 1$

Diglycol monoether:

$$H[-O-\underset{\underset{R}{|}}{C}H-CH_2]_m-O-[CH_2]_n-CH_3$$

Diethylene glycol monobutyl ether: $m = 2$, R = H, $n = 3$
Dipropylene glycol monobutyl ether: $m = 2$, R = $CH_3$, $n = 3$
Tripropylene glycol monobutyl ether: $m = 3$, R = $CH_3$, $n = 3$

Monoglycol diether:

$$CH_3-[CH_2]_m-O-CH_2-CH_2-O-[CH_2]_n-CH_3$$

Ethylene glycol dimethyl ether: $m = 0$, $n = 0$
Ethylene glycol dipropyl ether: $m = 2$, $n = 2$

**FIG. 13.4**　Generalized structures of glycol ether solvents.

Alcohols are sometimes used in dilutable all-purpose cleaners, but they are usually tertiary or benzyl alcohols [84,85]. The grease cutting ability of the lower alcohols is inferior to these and to the glycol ethers. Lower carbon alcohols (especially ethanol and isopropanol) find their main use in glass and light-duty cleaners. In these products the greasy soil load is lighter and the volatility of the short-chain alcohols has advantages.

One other restriction on the formulator with regard to solvents is the passage of environmental regulations in some localities on restrictions on amounts of volatile organic compounds (VOCs) in household products. The limitations vary from government to government, as well as the definition of what is volatile. In the U.S. several individual states as well as the federal government have restrictions on VOCs. Formulators intending their product for pan-North American sale would do well to abide by the most restrictive set of rules in their formulating. Similar considerations apply for pan-European sales and rules in individual European countries.

The aim in modern formulation is to minimize builder concentration due to the problems in rinsing and residue cited above, as well as the environmental restrictions on phosphates. Part of this is done with the choice of solvent and part with the surfactant synergies.

## 3.   Surfactant Innovations

One important area is actually that of surfactants working in negative synergy so as to give quick-breaking foams even with high total surfactant concentrations. This is most commonly done by including a small amount of soap in the formulation. Interest has also been shown in the counterion, a potassium soap combined with paraffin solvent claimed to give a much reduced tendency for stable foam [86].

This brings up the important point that the choice of counterion is also key for efficacy of the main anionic surfactants used in the cleaning formula. It has been known for some time that divalent metal salts of alkylbenzene sulfonate, paraffin sulfonates, and the like are better grease cleaners than the analogous sodium salts [87,88]. These are more difficult to use due to their lower water stability, but they can be formulated with some of the more effective grease cutting solvents [89–90]. It has also been claimed that if ammonium salts of the anionic surfactants are used less residue is left on the surface [91].

Many formulations continue to depend mainly on anionic surfactants for detergency. The major change in the last 50 years has been the change from branched alkyl sulfonates to linear ones, based on concerns about biodegradability [92]. (This was essentially complete in the U.S., Europe, and Japan by the late 1960s, but has taken place more slowly in the developing world.) However, there is a growing trend toward the use of other surfactants beyond the workhorse anionics that have served for so long [93]. There are, for example, all-purpose cleaners that depend more on nonionic surfactants [94,95]. Nonionics are less sensitive to water hardness, can be synthesized to target hydrophilic/lipophilic balances (HLBs) [96], and lead to less foaming. Much of the pioneering work in this area was and is done by the primary suppliers of the materials, such as the work of Shell on primary alcohol ethoxylates [97] and the more recent work by Henkel and Kao on alkylpolyglycosides [98–100]. The sugar-derived polyglucosides [101] show good surfactant qualities and are becoming favored because of the environmental claims that can be made for vegetable rather than petroleum feedstock material. They also show mildness to the skin. In contrast, ethoxylated alkyl phenols are falling out of favor due to their low biodegradability and resulting health concerns [102,103].

Some formulators claim that it is possible to achieve effective cleaning with nonionics while completely excluding anionic materials [104–108]. It is now also claimed that some short-chain nonionics can give superior cleaning [109]. The claim for all nonionic formulas is that they are less sensitive to water hardness. There are also cases where longer chain, block copolymer nonionics are used. In one particular case [110] the nonionic is used not only for its cleaning, but also for its residual spotting/filming characteristics. The general practice, however, is to use a combination of both anionic and nonionic components [111]. In part, this may be because nonionic surfactants have a lower tendency to foam, and foam is expected by consumers as a sign that the "active" is working. Therefore, including a certain amount of anionic surfactant helps to signal the consumer that the detergent is present [112].

It can also be argued that anionic surfactants give the best grease cutting efficacy/price ratio [113]; linear alkylbenzene sulfonate is a commodity chemical, for all intents and purposes, being the highest tonnage surfactant (estimated 3 million metric tons) in the world, due in large part to its use in laundry detergents as well as in household cleaners. Nonionic surfactants are more expensive and of lower tonnage than anionics but are growing in popularity. The alkoxylated alcohols are generally lower foaming than the anionics which can be either an advantage or a disadvantage. Nonionics are usually significantly milder to skin than most anionics, but this is not usually high on the list of preferred benefits for household cleaners. Fortunately, a mixture of nonionic surfactant with anionic surfactant increases the mildness of the formulation over the anionic alone. In general, due to the synergy between surfactants, particularly those of different charge groups, it is advantageous to use mixtures of types rather than all one kind.

As noted above, grease cutting solvents are included in nearly every all-purpose cleaning formula. These solvents are generally kept to low levels because of volatility and general safety considerations. It is interesting to note, therefore, that recent claims are being made for surfactants to also have solvent-like cleaning functions [114,115].

A new subclass of formulations in this area is the microemulsions. These special phases were used in areas such as oil recovery as early as the 1970s, and have been exploited more recently in secondary oil recovery, organic synthesis, and analytical extractions [116]. However, microemulsions have come much more slowly to the household cleaning area, although there are anecdotes of Australians making what is essentially a microemulsion cleaner in the early 1900s to wash wool. (This was reported to be a mixture of white spirit, soap (surfactant), and eucalyptus oil (oil) in water [117].) Much has been written recently on microemulsions [118,119], their advantages over regular emulsions being increased stability, spontaneous formation, and oil solubilizing potential. They are also transparent (unlike emulsions which are generally milky), therefore giving the aesthetics of a true solution. There are examples in the literature that are formulas intended for industrial cleaning of computer components or metal parts [120,121]. In these formulas, the highly hydrophobic solvent (usually a chlorinated hydrocarbon) is delivered to the surface in an aqueous media. Microemulsions, when used neat, also exhibit ultralow interfacial tensions; this increased contact should give them an advantage in the first step of cleaning. The clear advantage of the microemulsion in most cases is the ability to stabilize the solvent in a mixture with other hydrophilic components. The cleaner is used neat only in certain cases, whereas it should be used dilute more often where the interfacial advantages would be lost on dilution but the solvent would contribute more strongly to the cleaning.

The usual approach in consumer products (as opposed to industrial cleaners) is to deliver high amounts of a hydrophobic, but more consumer-friendly, grease cutting solvent such as pine oil or D-limonene. It should be expected that almost any water-based cleaning formulation that contains over 15% hydrophobic solvent and yet comes as a clear transparent liquid is a microemulsion. (This can easily be determined with quasielastic light scattering experiments or other such techniques to determine if droplets of the proper size distribution exist in the liquid [122].) Examples may be found that use these insoluble grease removal solvents as the essential oily material in microemulsions, sometimes with glycol ether solvents added as the cosurfactant [123,124], although more water-insoluble solvents have also been used [125]. All microemulsions formulated for household cleaning so far have far more hydrophilic components than hydrophobic, opposite to the industrial examples. Surfactant types vary, with one of these examples using an anionic/nonionic mixture while the other is all nonionic. Still another example uses a anionic/nonionic surfactant mix with glycol ether cosurfactant but uses the perfume as the oily material essential for forming the microemulsion [126–128]. Another uses a nonionic surfactant as the oil component of the microemulsion [129,130].

## 4. Added Benefits and Disinfectancy

The addition of any added benefits tends to narrow the scope of a cleaner, and so there are a few general types included in all-purpose cleaners. The most popular is the presence of disinfecting or antimicrobial action. An excellent overview of disinfectants in general and a more detailed treatment is given in Block [131,132]. Biguanides, alcohols, aldehydes, and phenols as well as quaternary ammonium surfactants ("quats") and oxidizing agents (bleaches, largely hypochlorite bleach) are used [133]. Phenols, the basis of original household disinfecting cleaners (e.g., the original Lysol disinfectant, a mixture of cresols with soap to solubilize them in a water solution), have fallen out of common use in household cleaners as disinfecting agents due to toxicity concerns [134–136], although the original Lysol formula may still be sold in Europe. (This Lysol disinfectant should not be confused with the plethora of other Lysol brand products, many of which are disinfectant *and* cleaning products.) Phenol derivatives such as *o*-benzyl-*p*-chlorophenol or *o*-phenylphenol are still somewhat used in household cleaners, but not very widely [137]. They are more widely used in very low concentrations as preservatives. Similarly, aldyhydes (such as gluteraldehyde and formaldehyde) are not generally used because of the difficulty of making such reactive organic compounds stable for long periods in water-based cleaners. However, compounds that break down to produce formaldehyde are used in low levels as preservatives in liquid cleaners.

In view of the recent popularity of "natural" products it is unsurprising that there is also patent art on using essential oils with antimicrobial properties [138,139], of which pine oil is a very old example.

The most commonly used of the disinfectants are quats and alkali metal hypochlorites (rather than oxygen bleaches which are weaker disinfectants). Sometimes they can be combined, as in the case of quats and alcohols [140,141] or pine oil and acid [142]. In fact, pine oil, which was long used as a disinfecting agent, requires high concentrations and is a fairly narrow-spectrum compound. (Spectrum denotes the variety of organisms a disinfectant can kill.) In practice nowadays pine oil is almost never used by itself; the most common combination is pine oil and quats where they compensate for each other's weakness. Acids are also gaining popularity as disinfecting agents [143,144]. Biguanides are used more often in personal care products and are seldom used in household cleaners because of their relatively lower efficacy, but patent art does exist [145]. There is even an example based completely on solvents [146]. Ethanol and isopropanol have long been used as disinfecting agents but have to be formulated at relatively high concentrations [147]. Pine oil gives the most freedom of formulation, bleach the least.

The effective concentration of the disinfecting agent varies widely with the type, and depends on the anticipated use conditions — contact time, soil load, and the amount the consumer is likely to use. The conditions for testing the effectiveness of the disinfectant dictate the label use instructions. If the test was run for 10 minutes, then the consumer is directed on the label to contact the surface to be disinfected for 10 minutes before wiping or rinsing. Table 13.5 gives a brief summary of some of the characteristics of common disinfecting compounds.

The main restriction on using quats is that association of the quats in solution with most anionic surfactants inactivates the disinfecting action of this cationic compound. The action of quats [148] is well known to be boosted by the addition of common chelating agents [149,150], with the sodium salts of EDTA most commonly used [151]. It is also claimed that their effectiveness can be increased through the choice of cleaning surfactant used with them [152,153]. Ethoxylated quats are also starting to appear in the patent art [154–156]. Quats are usually formulated at high pH values, although they seem to have activity down to pH 3 [157]. However, acid pH formulations do exist [158]. Most of the quat-containing disinfectant cleaners on the market tend to be formulated in the pH range 8 to 9. Pine oil can have a narrow spectrum of antimicrobial effectiveness which can be broadened by adjunct compounds such as quats or some organic acids [159].

The broadest spectrum disinfection, at a low price, is delivered by bleach, which is often a compensating factor for the difficulty of formulation. Formulation with bleach usually prevents the use of most common nonionics, paraffin sulfonates, alkylbenzene sulfonates, betaines, long-chain unsaturated quats, etc. The surfactants most often used with hypochlorite bleach are amine oxides, soaps,

**TABLE 13.5** Characteristics of Commonly Used Disinfecting Agents

| Compound (example) | Spectrum of disinfection (types of organisms killed) | pH range for effective kill | Typical concentration needed | Typical formulation concentration (%) | Chemical incompatibilities/ comments |
|---|---|---|---|---|---|
| Alcohol (ethanol, isopropanol) | Many types of bacteria and fungi; not spores | | 50–80% | 60–80 | Not used neat as evaporates too quickly to kill effectively |
| Hypochlorite bleach (sodium hypochlorite) | Bacteria, fungi, viruses, protozoa, and spores | 6–8 | 50–500 ppm | 1–5 | Catalyzed decomposition by heavy metals like iron and copper; unstable at effective pH; active form is hypochlorous acid (HOCl) |
| Peroxide (hydrogen peroxide, peracetic acid) | Bacteria, fungi, viruses, and spores | 3–7.5 | 50–500 ppm | 0.5–5.5 | Catalyzed decomposition by heavy metals like iron |
| Pine oil (α-terpeneol) | Gram-negative bacteria | >7 | ~5000 ppm[a] | 30–80 | Usually not used alone; most often combined with quats or alcohol |
| Quats (benzalkonium chloride) | Bacteria (less effective on gram-negative bacteria), yeasts, and mold | 3–10.5 | 200–1000 ppm | 0.2–3.0 | Inactivated by anionics and hard water salts |
| Phenolics (2-phenylphenol) | Bacteria, fungi, viruses, and protozoa | 3–9 | 400–1500 ppm | 3–20 | Not compatible with quats; inactivated by nonionics |
| Organic acids (citric, lactic, acetic) | Bacteria (both gram negative and gram positive) | 2–3 | 70–1500 ppm | 1–5 | Often need mineral acid to lower pH; active agent is nonionized form |

[a] Lee, W.H. and Rieman, H., The inhibition and destruction of enterobacteriaceae of pathogenic and public health significance, in *Inhibition and Destruction of the Microbial Cell*, Hugo, W.B., Ed., Academic Press, New York, p. 411.

*Source:* Data taken from Block, S.S., Disinfectants, in *Kirk-Othmer Encyclopedia of Chemical Technology*, 4th ed., Vol. 8, Kroschwitz, J.I. and Howe-Grant, M., Eds., Wiley, New York, 1993, p. 237; Richter, F.L. and Cords, B.R., Formulation of sanitizers and disinfectants, in *Disinfection, Sterilization and Preservation*, Block, S.S., Ed., Lippincott Williams & Wilkins, New York, 2001, pp. 477, 480.

and sodium alkyl sulfates, although there is one example using a nonionic [160]. It has been known for some time that amine oxides interact with anionic surfactants at certain ionic strengths to generate liquids with dynamic viscosities from 10 to about 5000 cP [161–164], or, in one case, the interaction of an amine oxide with a polycarboxylate polymer [165]. In this way, bleach-containing formulas can be thickened by the use of the surfactants alone [166]. Therefore, bleach all-purpose cleaners can have three benefits in addition to the primary cleaning: removal of oxidizable stains, disinfection, and increased cling or residence on vertical surfaces.

Most bleach cleaners, however, are simple, water-thin solutions. The most common formulations are a simple combination of hypochlorite bleach, sodium hydroxide (to achieve a pH of 10 to 12), amine oxide surfactant, and a low quantity of perfume. However, despite their simplicity, these types of products are very effective stain removers and disinfectants.

There is an interest in nonhypochlorite disinfection, as evidenced by patent art for institutional use. Examples can be found that use peroxyacids to achieve antimicrobial effects [167–169]. The advantage of these formulas would be lack of odor and corrosion that can be encountered in hypochlorite formulas.

As disinfectant cleaners are being used throughout the home (as all-purpose cleaners), the inclusion of characteristics such as low streaking (important for cleaning shiny surfaces) is also being claimed in antimicrobial formulations [170,171].

Although disinfectant cleaners are still very popular in North America, there are questions about the need for these types of cleaners in Europe where there is more awareness of "good" versus "bad" bacteria, and therefore more question about the wisdom of wholesale killing of bacteria. Disinfectant cleaners were first promoted and sold in times when a household might actually have a member infected with a serious infectious disease. This serious need for disinfection in the household seldom occurs in the developed world now. This is not true, of course, in some developing nations where serious epidemics can justify a need for broad-spectrum disinfection. However, several events of the past two decades have fostered an ever increasing interest among consumers in disinfection or germ-killing cleaning:

- The AIDS epidemic of the 1980s.
- Information campaigns about salmonella and *E. coli* food-borne infections.
- Information about the "germiness" of sponges and kitchen cleaning implements.
- Outbreaks/epidemics of contagious diseases (severe acute respiratory syndrome (SARS), Ebola virus, hantavirus).
- The anthrax scares of 2001 (mistakenly). (Household disinfectants are generally ineffective on spores.)

However, disinfectant household cleaners are also more likely now to be mentioned in newspaper and magazine articles as contributing to the problems of antibiotic-resistant bacteria; these tend to be reports in the popular press where all antimicrobial products tend to be mentioned, despite no demonstrated scientific link between the household cleaner disinfectants and the problem [172–174]. As noted, disinfectant cleaners continue to have popularity, but given some of the trends in microbial control issues, they may be either more heavily regulated or limited in the future.

Bleach can be added to an all-purpose cleaner simply for the stain removal properties as well, with no disinfectant claims made. (In fact, bleach is one of those ingredients, like aloe, that carry its own strong associations; many consumers would assume a bleach cleaner was a disinfectant whether it was claimed on the label or not.) Consumers often make "witches' brews" of household chemicals to make their own cleaning solutions. Most consumers are well aware of the "don't mix ammonia with bleach" rule, although they are generally unaware of other chemical interactions that can take place. Certainly if a regular household cleaner is mixed with household hypochlorite bleach, the bleach will start degrading most surfactants. As noted before, only a limited subset of surfactants are compatible with bleach. A commercially formulated product gives the stain-removing property that consumers are looking for in their homemade versions with considerably more safety and effectiveness.

This is another recent trend in "ingredient stories" — the rise of oxygen bleach cleaners. This is another fad in ingredients that is too recent to know if it has staying power or not. Oxygen bleaches are demonstrably less effective on destaining, but they are also safer to surfaces. It is this last trait that is most mentioned — destaining with surface safety. This started because of the sale of tubs of sodium sulfate/hydrotropes with sodium percarbonate [175]. This was meant to be used as a laundry additive and as a household stain remover when dissolved in a recommended amount of water. Consumers use this powder product largely as a laundry additive but it evidently ignited an appetite for oxygen bleach products. Most of the products have come in spray cleaner form rather than dilutable, and so the particulars of their formulation are detailed in the spray cleaner section.

Besides disinfection and stain removal, there are few other added benefits of major market importance. The next most important is the special class in which a polymer film is left behind on the surface. The most common example of this is not all-purpose cleaners, but are the "mop and shine" products for floor cleaning. This subset is intended solely for floor cleaning and leave behind a film intended to mimic the shine and soil resistance of waxing a floor. Unlike waxes, however, these polymer films do not need to be buffed to make them glossy [176]. The only drawback to this kind of formula is the possible buildup of polymer on the surface. These polymer films tend to be slightly colored, and so repeated layers can yellow the surface. The aim of inventions in the field is therefore to give polymers that deposit on the surface readily in formulations that avoid buildup [177].

Another benefit that also seems to have the benefit aimed at floor cleaning (because it may be about the only cleaning job still done with a bucket in developed markets) is the idea of coagulating the soil in the bottom of the bucket. This should mean that the wash water remains cleaner (if the mop is not put all the way to the bottom of the bucket) and soil removed from the floor is not redeposited. Although the patent art goes back to the late 1980s [178], only recently has a cleaner been marketed in the U.S. making exactly this claim.

This concept of polymer adsorption for greater shine and soil resistance has since been extended to other household surfaces, going from specialized floor products back to all-purpose cleaners. These generally use nonionic polymers such as polyethylene glycol, polyvinyl pyrrolidone, or other film formers [179,180]. One very interesting invention states that by the use of a polyalkoxylated alkylolalkane and vegetable oil surfactant a formula may be made that shines shiny surfaces but which imparts no gloss to matte surfaces [181] and a similar one that also uses a vegetable oil surfactant [182]. The concept of preventing soiling can also be extended to preventing subsequent microbial growth and so combine antisoiling with germ prevention claims [183,184].

Another recent interesting invention is the formulation of an insect repellent into an all-purpose cleaner [185–187]. The key here is that the active ingredient is not an insecticide but simply a repellent which makes it possible to leave it behind routinely in the cleaning process without concerns for repeated human contact.

There are also examples of formulations that include a dye so that soil can be visualized so as to signal to the consumer when "complete" cleaning has taken place by the absence of the color [188]. Admittedly, some soils, like soap scum and grease, are apparent on light-colored household surfaces mainly by their dulling of the natural shine of a surface. In this case, the dye is sensitive to the presence of protein, so if the soil was a pure grease soil (like many kitchen soils) it would fail to react, whereas it would probably be highly indicative of bathroom soils.

## 5. Formulation Technology

Table 13.6 gives ranges for common ingredients for all-purpose cleaner formulation. Although it is not explicit from the previous discussion, "regular" all-purpose cleaners are intended to be dilutable. The consumer may use them full strength from the bottle for cleaning a very heavily soiled small area, or may dilute them in the range of 1:32 to 1:128, with a dilution rate of 1 part cleaner to 64 parts solution being most common. There are all-purpose cleaners launched in North America and Europe that operate at the top of the dilution range; these are the "ultra" or concentrated products, similar in concept to the ultra laundry detergents. These have largely disappeared from the North American market. There are also some patents appearing for household surface cleaners that are mentioning

**TABLE 13.6** Liquid Dilutable All-Purpose Cleaner Formulas

| Ingredient | Examples | Amount (wt%) |
|---|---|---|
| Anionic surfactant | Alkylbenzene sulfonate, paraffin sulfonate, ethoxylated alcohol sulfate, soap | 0–35 |
| Nonionic surfactant | Ethoxylated alcohol, amine oxide, alkanolamide fatty acid | 1–35 |
| Hydrotropes | Sodium cumene sulfonate, sodium xylene sulfonate | 0–10 |
| Builder | Carbonates (more rarely phosphates), silicates, citrates, EDTA salts, polyacrylates | 0–10 |
| pH adjuster | Ammonia, sodium hydroxide, magnesium hydroxide, alkanolamines | 0–10 |
| Solvent | alcohol (pine oil, benzyl alcohol, or lower carbon number alcohols), glycol ether (Carbitol®, Dowanol®, etc.), D-limonene | 0.5–50 |
| Disinfectant | Hypochlorite bleach, pine oil, other low carbon number alcohols, quaternary ammonium compounds | 0–15 |
| "Shine" polymers and other benefits | Polyacrylate, polyethylene glycol, polyvinyl pyrrolidone, organosilanes | 0–25 |
| Perfume, color, etc. | | 0.1–3 |
| Water | | QS |

the concentration of the formula [189–191]. The largest difficulty in formulating such products is usually keeping high concentrations of surfactant from creaming or separating. The use of hydrotropes and solvents, and the minimization of electrolytes, is especially important in achieving this goal. If such concentrated formulas can be made stable, the manufacturer gains the advantages of less packaging, smaller shipping weights, and less storage space. The consumer gains advantages in more easily stored containers and less packaging to recycle while the absolute cleaning potential in its delivered form increases. The difficulty for consumers is in changing their habit of dosing so as not to waste the concentration of the new formulations. These types of concentrated all-purpose cleaners should be considered as existing toward the top of the surfactant concentrations listed in Table 13.6.

With higher concentrations of surfactant, the formulas also tend to become thicker. In most cases, the increase in viscosity will be minor, but in some cases it can become significant. If the product is meant to be dilutable by the consumer, then the perception of pouring and dispersability of the product cannot be adversely affected by the viscosity. The ways to decrease the viscosity in these cases is to decrease the surfactant association that is responsible. Many of the commonly used solvents can affect the solubility of the surfactant and therefore redistribute the relative amount of monomer versus associated forms. Changes due to ionic strength can have the same effect and so the addition of simple salts to the formula can also remedy the problem. Of course, if it is desirable to thicken the product, this can be done most easily through the addition of various polymers such as polyacrylates, polyethylene oxides, cellulose gums, polyglycols, etc., depending on which is most compatible with the surfactant system.

The pH of most all-purpose cleaners is between 8 and 12. Generally this is the best range for grease cleaning in that the alkalinity can (to a small degree) saponify some portions of a grease, thereby assisting the surfactant/solvent system in removing soil. However, high pH can also damage some sensitive surfaces, such as aluminum, as well as being irritating to skin. In the interest of giving consumers more advantages, formulators strive to work at pH values as close to neutral as possible to reduce these negative effects. This means that more care has to go into optimizing the surfactant system and sometimes more reliance on the grease cutting solvents. For instance, pine cleaners tend to be acidic, but the pine oil, more than the pH, contributes to the grease cleaning. This approach also has its dangers, as some surfaces such as paint and wall coverings can be sensitive to solvents.

As can be seen from the preceding discussion, all-purpose cleaners started with the oldest surfactant, soap, and have progressed to more powerful surfactants and then further developed sophisticated surfactant synergies. As these developments are made, there is less and less reliance on the old inorganic builders and more interest in solvents, particularly those with grease cutting ability. Only concerns about human toxicity and environmental regulations limit the choice of solvent.

## 6.   Aesthetic Ingredients

Fragrance is a very important part of household cleaners, often overlooked in the technological drive for formula performance. However, fragrance can often be the driving attribute in a consumer's evaluation of a product. Fragrance can sometimes be difficult to incorporate stably into a product, due to its low solubility (oily nature) and chemical reactivity (presence of aldehydes, esters, ketones, etc.). One of the ways of easing the addition of a perfume oil into a formula is to premix the fragrance with either surfactant (a lower HLB component) or with the solvent. Higher concentration formulas, rich in surfactant, often have enough solubilization power to make the addition of the fragrance less difficult than in more dilute products like glass cleaners.

Colors can also sometimes be difficult to stabilize in a cleaner. Obviously, this is particularly a problem in bleach-containing cleaners. The strategy in coloring bleach products, if they are thickened, is to color them with pigments that can then be stably suspended by the thickening system of the product.

(Powder cleansers may also use pigments, but this is significantly less of a problem in nonliquid systems.) Most liquid cleaners use dyes as their coloring system, and, as with fragrances, the primary chemical compatibilities must be considered when picking a color system.

## 7. Minor Ingredients

In the course of formulating all-purpose cleaners there may be a need for other minor ingredients in addition to the main cleaners (the surfactant, solvent, or builder). These ingredients include hydrotropes, hard water control, and buffers. These ingredients do not include any that are added for special benefits such as shine enhancement, disinfection, soil release, etc.

When builders are used in a formula, they also fill the function of hard water controls and buffering agents. However, the high electrolyte concentration imparted by the use of these builders may make it necessary to use a hydrotrope (see Chapter 2). In general, hydrotropes are organic compounds that enhance the solubility of other species. In cleaning formulations they facilitate the dissolution and continued solubility of the main detergent surfactant in a liquid formula. Many times the solubility of the surfactant is limited by high salt content or other factors. Examples of hydrotropes are given in Table 13.6. As the use of inorganic builders decreases over time, the use of hydrotropes is also decreasing. However, hypochlorite bleaches also increase the electrolyte concentration, as can the inorganic thickeners in cream cleansers. In these cases, the use of a hydrotrope may also be required.

When, as in many modern formulas, builders are excluded or limited to decrease visible cleaner residue, other means are necessary to control water hardness and buffer the formula. In some areas where the water hardness is very high (above 250 ppm as $CaCO_3$), even modern anionic surfactants can be partially precipitated. Soap is very easily precipitated. The most common remedy is to add one of the salts of EDTA or NTA to chelate the water hardness ions and therefore maintain the anionic surfactant efficacy. (Although these compounds may be classed as builders because they control water hardness ions, they usually do not supply significant alkalinity to the formula.) These components can still contribute to residue and need to be limited. In recent times these compounds have come under scrutiny for toxicity concerns, with some governments considering legal limitations. These environmental and toxicological concerns are anticipated in the literature [192]. Of course, this problem of divalent cation precipitation does not occur with nonionic or cationic surfactants, which is an advantage when formulating.

The other problem when builders are eliminated is stabilizing the pH of the formula. Many anionic surfactants can impart a slightly acid pH to the solution when dissolved. If the aim of the formula is largely grease cleaning, this is most efficiently done at basic pH. Therefore, the pH can be adjusted using common bases such as ammonia or sodium hydroxide. Ammonia is useful in that it is volatile and therefore leaves no residue, but it also imparts a distinctive odor that is not pleasing to some consumers. However, neither of these choices is a good buffering agent. If the pH of the formula is to be maintained in a lower range (pH 8 to 9) or if the formula pH tends to drift over time, then alkaline buffering agents, usually one of the alkanolamines, are used.

## E.  Floor Cleaners

All-purpose dilutable cleaners are often used for floor cleaning. This results in a laborious task: mixing of the solution in a bucket, washing of the floor with a mop (which must then be cleaned), rinsing of the cleaned surface (difficult on a large horizontal surface), and then cleaning the solution bucket in which the removed soil resides.

In an effect to shorten this task, one of the recent developments was the "ready to use" floor cleaners. These are even more dilute than spray cleaners, formulated at the high dilution that conventional cleaners are used for floor cleaning (Table 13.6, diluted 1:32 or 1:64). They are usually packaged in plastic bottles with push/pull or flip-top caps. This means that the solution making and bucket cleaning steps are eliminated. In addition, it is contended that these low-dilution cleaners do not need rinsing, also eliminating that step. (The most recent development, that of the Swiffer® type floor cleaning systems, is discussed in the section on wipes.)

Wood floor cleaning also seems to be a special case for some consumers with cleaners formulated for this particular use [193]. Certainly, wood floors have a higher sensitivity to water-based cleaners than almost any other type of flooring. This is one of the few types of dilutable cleaner in which soap is seen as the main surfactant [194].

## F.  Test Methods

There are several key performance areas to test for all-purpose cleaners. As may be deduced from the previous discussion, these are cleaning, ability to foam, and residue/shine. Unfortunately, very few published standardized methods exist in this area, especially residue/shine. Although foam height and soil removal are easily quantified, the impact of various amounts of residue and its distribution on a surface are not. Most residue/shine tests are based on the evaluation on scales from 1 to 10 by panels of observers of prepared samples. Usually the method used is described in the corresponding patent. A usual general method is to apply the cleaning solution, either by wiping or by pipetting, onto a clean glossy surface, usually of a dark color. Black ceramic wall tiles are convenient for this purpose. The solution is left to dry on the surface, and a panel of observers rates their impression of residue on the resulting pattern.

## 1. Cleaning Tests

For cleaning tests, the methods for applying soil, simulating the cleaning process, and judging the result are well established [195,196]. What is not well established is the precise identity of the soil used. For all-purpose cleaning, the target soil usually investigated is grease. This is meant to replicate the soil left on kitchen surfaces due to normal household cooking practices. Of course, the type of oil or fat used in cooking can vary widely around the world, from vegetable oils and margarine to beef fat, lard, and butter. Even among the vegetable oils, there can be differences between olive and corn oils, due to the distribution of chain lengths and unsaturation. These factors affect how the grease is changed by heating and exposure to the air in thin films [197].

In many cases, a pure grease soil is not used. Sometimes the soil is colored with carbon black to make it more visible. However, this also has the effect of introducing a solid particulate into the soil mix [198,199]. Other particulates have also been introduced such as humus, clay, ferrous oxide, soot, and filtered vacuum cleaner dust [200,201]. However, the grease component of the soil usually predominates. If the grease soil is liquid (such as vegetable oil), then it can be sprayed on the surface neat. Mixtures of particulates with oils are also spread on surfaces using paint rollers. However, liquid soils usually require longer aging periods before they can be used. Soils can also be prepared by dissolving solid greases (tallow) in various solvents (naphthenic hydrocarbons, chloroform, etc.) and then spraying the solution or dispersion. The solvent flashes off, leaving a solid grease layer (usually without particulate). These soils need shorter aging times because their solid form makes them more difficult to clean than oily soils. However, there are two concerns with this type of procedure: (1) the proper protection of laboratory workers from these hazardous solvents and (2) the contamination of the greasy soil with any residual solvent that might influence the cleaning process.

The soil is applied to typical kitchen surfaces: vinyl floor tile, sections of Formica, ceramic tile, pieces of enamel, aluminum, stainless steel, painted wall sections, etc. The local point of sale is the determiner of the choice of surface, so knowledge of local materials of construction is necessary. It is sometimes necessary to alter the surface in order to make the soil tenacious. This is sometimes done with a light sanding of the surface to roughen it, or with chemical etching such as strong acid or strong base treatments of susceptible surfaces. It is preferred to alter the composition or aging of the soil to increase tenacity only where necessary, although surface roughening is sometimes considered accelerated "aging" of a surface. It is, indeed, the daily wear and tear of living with the surfaces that results in their alteration, and any changes in the surface should be done with a view to mimicking the natural aging changes that take place in the surface.

Once the soil has been aged, either at ambient temperature or by heating and drying in an oven, cleaning experiments can be carried out. The usual apparatus for this testing is a Gardener abrader (Figure 13.5). This consists of a testing platform

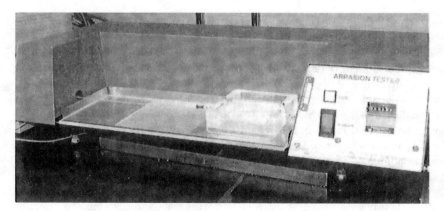

**FIG. 13.5**   Gardener straight line abrader machine.

and a carrier for holding cleaning utensils that is driven by a motor back and forth along the platform. Cleaning utensils fitted in the machine are also chosen according to local habits and practices. They can be sponges, mohair, folded cloth, folded paper towel, or scrub brushes. Small pieces of sponge wrapped around solid blocks are the usual choice. The utensil holder is usually made to hold two separate sponges so that the cleaners can be tested side by side. Often, variations in the individual quality of a surface and its applied soil make it necessary to do direct comparisons of cleaners on the identical item. Also, a standard amount of weight is applied to the utensil holder to simulate the force a person employs in the wiping process. This force is usually in the range of 200 to 500 g per sponge including the weight of the carrier and the loaded sponges.

   In some cases, such as spray cleaners, the test product can be applied directly to the surface and a wetted utensil used for cleaning, but for most general or more concentrated all-purpose cleaners the solution is loaded onto the utensil. The utensil is usually wetted so that it is wet but not dripping water. The cleaning solution can be used neat or dilute depending on the intention of use. If floor cleaning is simulated, the cleaner is diluted according to label instructions usually in the range of 1:64. If tough soil spot cleaning is simulated, then the cleaner is used neat. It can be applied to the utensil in two ways. Either a specified amount of cleaner is poured or pipetted onto the sponge, or the utensil mounted in its holder is soaked in a shallow pool of the cleaner for a specified time. The utensil is then fixed in the holder on the abrader.

   The abrader is then set in motion, making reciprocal sweeps back and forth over the soiled surface which is fixed on the testing platform. Cleaning can be done in two ways: a fixed number of strokes can be used, or the process continued until one or both sides of the testing surface is completely clean. The first method can be used to compare cleaners at different number of strokes to generate data on the "kinetics" of cleaning. The second method mimics the practice of the consumer, which is to wipe until the surface is clean. In this case, the number of stokes needed to give 100% cleaning is tallied for at least one cleaner. The slower cleaner can then be continued until it is completely clean and its higher number of strokes recorded, or its amount of lesser soil removal tallied when the superior side reaches 100%.

Soil removal can be judged by eye, but the more common methods use a reflectometer. First, the "new" surface is measured before soiling. Most conveniently this is done on a white surface. The surface is soiled, usually with a colored soil. As mentioned before, the soil may be colored with carbon black or other particulates giving a gray or brown appearance. It may also be colored with oil soluble dyes. After soiling, another reflectometer reading is made. The cleaning is done, and the surface is then usually rinsed to eliminate cleaner residues or loosely held soil. Then the final "cleaned" measurement is made. The percentage soil removal is calculated as:

$$\% \text{ Soil removal} = \frac{R_c - R_s}{R_n - R_s} \times 100$$

where $R_c$ is the reflectance of the cleaned surface, $R_s$ is the reflectance of the soiled surface, and $R_n$ is the reflectance of the surface before soiling.

Another way of doing grease testing was shown by a consumer organization [202]. Although the choice of ingredients for the grease soil in this test is not the most consumer relevant, depending on mineral oil and petroleum jelly rather than household kitchen grease, the method by which the grease was removed is interesting. (This soil also included a fair proportion of particulates in the soil.) In this test, the greasy soil was applied in a narrow strip perpendicular to the travel of the sponges in the abrader apparatus. There were clean areas on both sides of the soiled strip. In this way, these experimenters measured not only the soil removal from the greasy area, but also the tendency of the cleaner to smear the soil onto previously clean areas. So a cleaner that performs poorly on soil redeposition or one that adsorbs the soil poorly into the cleaning implement will be judged inferior by this test.

Other, special, soils may also be tested. A "sticky" kitchen soil has been cited in the patent literature [203], consisting of vegetable shortening and all-purpose flour. This soil was baked on. Heating grease tends to oxidize and, to some extent, polymerize it into a resin-like coating. This can also happen over long periods at room temperature, resulting in a very tenacious soil. (Even dust soils can become more tenacious with time where it can no longer be simply vacuumed but must be removed by wiping.) A variety of soils was used by a consumer organization to test sprays and dilutable cleaners, meant to show a variety of cleaning problems (grape juice, ketchup, vegetable grease, and baked lard) [204]. In the case of cleaners containing bleach it is desirable to test the stain removal ability of the cleaner. Relevant household stains should be researched and chosen as the test soil. Only oxidizable stains, of course, will react to the bleach. One common widely used test uses tea stain on an enamel surface. Plates made of enamel on steel are boiled in a concentrated tea solution. They are air dried, rinsed with deionized water, and this is then repeated until the uniformity and degree of staining desired is reached [205]. This soil is cleaned and evaluated in the same manner as the grease soil.

Static soaking tests can also be done which eliminate the contribution of the mechanical action of the abrader test. The surface has a volume of the cleaning solution trapped within a ring (like a rubber washer), and covered to stop evaporation. After a set time the cleaning solution is poured off, the surface rinsed, and the area evaluated (either by eye or by reflectometer) to determine degree of stain removal. This can also be used as a test for damage to the surface by the cleaner if done on an unsoiled surface.

Although grease is the main soil target, followed by particulate soil, for all-purpose cleaners they may also be tested against other nonkitchen problems such as soap scum. Soap scum testing is described in the section on bathroom cleaners.

## 2.  Foam Level Testing

Foam tests of all-purpose cleaners are done similarly to other fields. One of the most common tests is the cylinder test. The cleaner may be placed, neat or diluted, in a glass graduated cylinder. The cylinder is then inverted a specified number of times, and the resulting foam height noted. This immediate reading is referred to as "flash foam." The cylinder then may sit undisturbed for various lengths of time, and the gradual collapse of the foam recorded in decreasing foam heights. Another test is the Ross Miles foam test [206]. In this method the solution is dropped over a specified distance into a receiver. The foam produced by this fall is measured immediately and after 5 minutes. Different foam aesthetics are preferred around the world, although generally it is preferable that the foam does collapse, as this is perceived to decrease the effort of rinsing.

## 3.  Surface Safety

Another area of investigation is surface safety. Households contain many different surfaces that may be soiled. If a consumer uses these products as true all-purpose cleaners, they will be carried from room to room, encountering many of these surfaces. It is wise to test the effect of a cleaning formulation on various items, depending on local materials. This is done most simply by immersing a solid block of material in the cleaner, or by letting a pool of the cleaner contact a representative surface. The length of time for the test is left to the experimenter's discretion. The compatibility of a cleaner with a variety of surfaces is part of the designation of an all-purpose cleaner. Evaluating surface safety becomes even more important when aggressive substances such as strong solvents or bleach are included in the formula. Inclusion of these types of ingredients tends to limit the formulation's use, relegating it to the category of a specialty cleaner in some consumers' minds. This is especially true of bleach cleaners, whose aggressiveness to many colored surfaces is well known. Cleaners containing high concentrations or more efficacious solvents may be aggressive to coated, plastic, or paper surfaces such as paint, shellac (used on wicker and some metal surfaces), or various types of wallpaper. It is also important to test low or high pH cleaners on metal surfaces such as aluminum.

## 4.  Disinfectancy

Disinfectancy tests are usually regulated by local government. For example, in the U.S. the rules and procedures for disinfectancy are set out by the Environmental Protection Agency, and in France they are given by AFNOR (Association Française de Normalisation). (However, since the formation of the European Committee for Standardization (CEN), CEN-TC216, the Technical Committee for Disinfectants and Antiseptics, has issued methods for all of the European Union.) The usual test for disinfectancy in the U.S., for both disinfectant compounds and cleaning formulations, is the use dilution test [207,208]. These tests usually consist of challenging a use dilution of the cleaner with specified microorganisms, followed by incubation. The disinfectancy of the formula is determined by the number of surviving cultures at the end of the incubation period. For the use dilution test, for example, the usual score for successful disinfection is lack of subsequent growth in 59 out of 60 tubes. U.S. regulations specify three different levels of disinfection claim, depending on the organism(s) used in the test [209]. "Sanitization" (in the U.S.) is usually defined as a lower level of kill than disinfection. Similar tests are laid out in the European tests [210]. As disinfection is very dependent on how the solution is applied and the contact time, these are usually carefully designated in the test method. The method has to be modified for spray cleaners as they are delivered ready to use (needing no dilution), and are sprayed on the surface in a thin coating of droplets [211]. Regulatory agencies usually require preview and negotiation of the test method if any changes are made to their standard procedure.

## 5.  Miscellaneous Testing

If the product is not a disinfectant formula, it may be advisable to conduct adequacy of preservation tests. In these tests, the formula, as made for sale, is challenged with various microorganisms. This test sample is incubated for a time, and the amount of growth measured. Aging tests for the shelf life of the formula are also advisable, usually run for up to 12 to 18 weeks at both room temperature and elevated temperatures.

Other tests may also be required by various governments. Eye irritancy warnings on the product, for example, may be required depending on the outcome of standard irritancy tests. There are also many regulations or labeling procedures regarding the biodegradability of cleaning formulas and their ingredients. In these ecologically aware days, many primary suppliers of surfactants and other active ingredients do their own biodegradation tests. An excellent overview of the field of biodegradability and various testing methods is given by Swisher [212]. It is wise for the formulator to inquire about these tests, especially if the product is intended for use in Western Europe or other ecologically conscious areas. There may also be regulations relating to the shipping and handling of large amounts of the cleaner and so some tests may have to be done regarding, for instance, the flammability of the cleaner. The closed or open cup flash point test is a standard for this [213].

These are standard tests, used for standard attributes of the cleaner. If special new added benefits, such as shine enhancement, are invented for a cleaner, then new testing methods must also be invented to measure them.

## III.  SPRAY ALL-PURPOSE CLEANERS
## A.  Historical Background

Spray all-purpose cleaners began to emerge on the U.S. markets in the 1950s. In the beginning they were pump sprayers, but the late 1970s saw the development of the more ergonomic trigger sprayers. These are now coupled with shaped bottles to give convenient gripping (Figure 13.6). These formulations extend the convenience of the liquid form by marrying it to a very convenient dispensing container. The trigger sprayers deliver more product to the surface than the older pump forms and with reduced hand fatigue [214]. The formula in the spray bottle is generally in ready-to-use concentration as opposed to the formula in the regular bottle which is meant to be diluted by the consumer. These spray cleaners are used predominately for spot cleaning and special needs rather than for larger area cleaning (e.g., floors). They are also generally used for lighter soil loads (finger prints, thin films of oil) than for tougher soils (thick layers of aged grease on range hoods) which are reserved for more concentrated products, the regular dilutable cleaners. This marriage of cleaning formula to specific package form, tailoring the action of the cleaner to the way it is dispensed, is an important trend in household surface cleaning.

All-purpose cleaners, as powders, were dispensed from boxes or bags. This could be a messy operation, and spills are difficult to clean up. Liquid all-purpose cleaners are dispensed from bottles, usually equipped with screw off or flip-top caps. Therefore, these were not only less messy to dispense, but they were easier to close tightly to eliminate spills. The caps could also be used to measure the product

(a)

(b)

**FIG. 13.6** Modern spray trigger packaging: (a) glass and surface cleaners, (b) mold and mildew and bathroom cleaners.

so that the appropriate amount of cleaner could be specified by the formulator without the need of a measuring device separate from the container. As noted before, plastic bottles, of light weight, inexpensive, and nearly unbreakable, were another packaging innovation that made use and dispensing of the cleaner easier and more convenient. The more convenient formula is accompanied by the more convenient dispensing system.

Spray cleaners are also of this type. The formula is already at the concentration appropriate for use, and the dispenser easily spreads a small amount of cleaner.

Aerosol packaging has played a smaller role in this area than in others like furniture polishes, air fresheners, and hair care. It has survived most often when a thick foam meant to stick to vertical surfaces is needed as in the case of oven cleaning or bathtub enclosures. In future it can be expected that as packaging innovation continues, specialized formulas will be matched to them to create more convenient and targeted cleaning systems for consumers to use.

In theory, almost any dilutable cleaning formula, including pine, disinfectant, grease cleaning, bleach, etc., can be "watered down" to give an effective spray cleaner. The trigger sprayers used with these formulas tend to exaggerate the foaming qualities of the cleaner, so the surfactant levels are generally at the lower end of the ranges given in Table 13.6. (Even at these concentrations, a spray cleaner is much more concentrated than the solutions generally used for floor cleaning where the cleaner has been diluted 30 to 60 times.) Also, these convenience cleaners are generally used without rinsing, so minimum ingredients have to be used to minimize residue. Any crystalline ingredients also have to be minimized to reduce buildup on the trigger itself when residual cleaner dries on the nozzle.

These are general restrictions on spray cleaners. About the same time that all-purpose cleaners were developing into sprays, formulas were also becoming less "all purpose." One of the first specialties to appear was glass cleaners. At the present time there are also formulas adapted for bathroom cleaning, stain removal, carpet cleaning, oven cleaning, etc. These types of specialization also impose their own special restrictions. One of the largest areas of specialization is grease cleaning, which in very developed markets can be subdivided into three different soil loads.

## B.  All-Purpose Spray Cleaners

Most closely related to the dilutable cleaners packaged in bottles are the all-purpose spray cleaners. These are used for the heaviest soil in spot cleaning — small greasy areas like stove tops, small spills, sticky spots like drops of jelly on countertops, etc. The small, quick nature of the job does not justify getting out a bottle and a bucket; the difficulty of the soil load calls for something close to the concentration of the dilutable all-purpose cleaner. As a general trend, these spray formulations are richer in solvent and poorer in surfactant than their dilutable counterparts. Also in common with the dilutable cleaners, the main trends in formulation are a greater emphasis on safer solvents, increasing use of nonionic surfactants, and decreasing use of builders and other salts. Typical formula ranges for this type of cleaner are given in Table 13.7. One unusual example provides for formulation of the cleaner at either acid or alkaline pH [215]. It would be useful therefore, as either a typical all-purpose spray cleaner or as a "vinegar" glass cleaner, depending on the pH. The compositions of these formulations fall between those of Table 13.7 and Table 13.8 (due to the possible inclusion of acetic acid, not included in Table 13.7).

**TABLE 13.7**   Spray All-Purpose Cleaner Formulas

| Ingredient | Examples | Amount (wt%) |
|---|---|---|
| Anionic surfactant | Alkylbenzene sulfonate, paraffin sulfonate, ethoxylated alcohol sulfate | 0–10 |
| Nonionic surfactant | Ethoxylated alcohol, alkanolamide fatty acid, amine oxide | 1–10 |
| Builder | Carbonates (more rarely phosphates), silicates, citrates, EDTA salts, NTA | 0–10 |
| pH adjuster | Ammonia, sodium hydroxide, magnesium hydroxide, alkanolamines or citric acid | 0.1–10 |
| Solvent | Alcohol (pine oil, benzyl alcohol, or lower carbon number alcohols), glycol ether (Carbitol®, Dowanol®, etc.), D-limonene | 0.5–0 |
| Disinfectant | Pine oil, C2–C3 alcohol, quaternary ammonium compounds | 0–5 |
| Bleach | Hydrogen peroxide | 0–10 |
| Antistreak polymers | Polystyrene/maleic anhydride, polyethylene glycol, etc. | 0–5 |
| Perfume, color, etc. | | 0.1–2 |
| Water | | QS |

**TABLE 13.8**   Spray Glass Cleaner Formulas

| Ingredient | Examples | Amount (wt%) |
|---|---|---|
| Anionic surfactant | Alkylbenzene sulfonate, paraffin sulfonate, ethoxylated alcohol sulfate | 0–1 |
| Nonionic surfactant | Ethoxylated alcohol, alkanolamide fatty acid, carbamates, amine oxide | 0.01–3 |
| Amphoteric surfactant | Betaines, sulfobetaines | 0–10 |
| Builder | Carbonates, silicates, citrates | 0–2 |
| pH adjuster | Ammonia, sodium hydroxide, alkanolamines or acetic acid | 0–5 |
| Solvent | Lower carbon number alcohols, glycol ether (Carbitol®, Dowanol®, etc.) | 0.5–40 |
| Antistreak, antifog polymers | Silanes, ethoxylated silicones, polyethylene glycol, polyvinyl alcohols | 0–1 |
| Perfume, color, etc. | | 0.001–0.5 |
| Water | | QS |

A survey of the literature reveals a preoccupation with streak-free cleaning when the all-purpose cleaners are in spray form [216–219]. The last citation is unusual for its inclusion of soap as a component that, in combination with a non-ionic surfactant, gives better streaking properties. Usually soap is formulated into the cleaner, if at all, as a foam breaker [220]. One of the drawbacks of these spray cleaners is that even a low level of builders, chelators, buffers, etc., still leaves a residue which is perceptible on very shiny or transparent surfaces. The traditional answer to this problem was to then go back and clean these types of surfaces with a glass or window cleaner to remove the all-purpose cleaner residue. Work on these formulations concentrates on optimizing surfactants [221–223] especially if it helps to minimize the solvents [224,225]. Minimizing all ingredients helps to minimize residue and streaking, but minimizing the solvent also has the attributes of avoiding VOCs and reducing the odor of the cleaner. Odor can be more of an issue in cleaners that are sprayed because of the aerosolization of the cleaner.

One unusual example of a hard surface cleaner of this type claims a residual effect on the surface that reducing the cleaning effort needed on subsequent cleanings. The surprising part of this invention is that the claim is based, in part, on the presence of lipase in the formula [226]. This is a rare example of the use of an enzyme in a consumer-intended hard surface cleaning formula. The use of enzymes in hand dishwashing, automatic dishwashing, and laundry is quite common.

Disinfectant cleaning had largely been limited to bathroom cleaners and dilutable all-purpose cleaners, but made a breakthrough with the introduction of an "antibacterial" spray cleaner for kitchens in 1994 [227]. At this time disinfectant was the usual term used in the product name of bathroom cleaners and dilutable all-purpose cleaners making germ-killing claims. The ones that have "antibacterial" in the name of the product usually have the term "disinfectant" somewhere else on the label. These are largely marketing distinctions in the U.S.; some localities may have laws about the exact wording that can be used. In practice, household cleaners with both names (at least in the U.S.) use the same ingredients and make the same label claims. Consumers seem to have slightly different views of the two terms and so it is a marketing, and not a formulating, choice [228].

## C.  Glass Cleaners

Glass cleaners are made to have the least possible residue. (Formulas of this kind go back to the late 1960s [229].) However, this low residue is usually accomplished by ultralow levels of ingredients which results in very light-duty cleaning. Glass cleaners have sufficient ingredients to remove common window soils such as fingerprints, dust, water spots, etc. They are not intended for heavy-duty soil loads like kitchen grease or sticky food spots. The main consideration for glass cleaners is that they deliver the minimum cleaning while leaving no streaks or residues that would be readily apparent on transparent surfaces. For this reason, volatile ingredients are desired in glass cleaners. This will be limited due to VOC legislation.

The trend previously noted for grease or all-purpose sprays continues for glass cleaners. Glass cleaners depend even more on solvent content and less on surfactants than the all-purpose sprays, which depend more on solvent and less on surfactant than the dilutable all-purpose cleaners. This is readily apparent if Table 13.8 is compared to Table 13.6 and Table 13.7. Two other differences are also apparent. One is that the more powerful grease cutting solvents (pine oil, D-limonene, etc.) tend not to be used in glass cleaners because of their lower volatility and oily character. Volatility is an important consideration in glass cleaners. The solvents that are used in glass cleaners, especially the low carbon number alcohols, may not be the best grease cutting solvents, but they give very quick drying of the cleaner. Also, given the ultralow concentration of surfactant, it would be very difficult to solubilize more nonpolar solvents. Ethanol or isopropanol used in glass cleaners are water miscible as well as being very volatile which makes them ideal glass cleaning solvents. However, much of the recent patent art devotes itself to the use of other solvents to improve cleaning without contributing to streaking [230,231]. If the cleaner does not dry quickly, the cleaner film may not be evenly rubbed out by the user, resulting in streaking. The streaks will not be apparent until the cleaner completely dries, and so it is desirable that this happen while the consumer is engaged in the cleaning task, not later.

The other difference noticeable from comparison of the tables is the use of different groups of surfactants: amphoterics [232–235] and amido nonionics, although the use of the more mundane anionics is more common. There seems to be advantages to these types of surfactants for less streaking, which is of special concern in glass cleaning. In the first example cited, it is claimed that because of the way the formula is constructed, it does not lose its performance (cleaning and lack of streaking) when perfume is added. It should be commented that perfume can be a source of streaking/residue problems in glass cleaners, and in most glass cleaners the perfume is kept to a minimum.

A signal in hard surface cleaning that is often highlighted in commercials (especially those for glass cleaners) is the "squeak." Greasy or dirty surfaces will not squeak when the surface is rubbed due to the lubrication of the surface; consumers assume that a surface that squeaks is very clean. In one patent, it is claimed that the surfactant and buffer can be chosen to foster this effect in the cleaned surface [236,237].

The ultralow amount of surfactant and the weak degreasing character of the solvents used generally results in significantly less cleaning power than in the other spray all-purpose cleaners. Of course, most of the soil levels encountered in glass cleaning are low, so this does not usually constitute a problem. However, another ingredient can help the cleaning. Many glass cleaners are also alkaline, to aid in cleaning the most common window soil — greasy fingerprints. The alkalinity is usually produced by "fugitive" compounds such as ammonia, which minimizes residue, obviously an important consideration in glass cleaning. However, the formula sometimes needs better buffering than ammonia can provide so the use of alkanolamines and carbamates is also well established. The key is to avoid crystalline compounds (traditional builders) to avoid noticeable residue on the transparent glass surface. The only acid glass cleaners are those that contain acetic acid. These depend on the reputation for good window cleaning developed by the home remedy of vinegar and newspaper for cleaning windows. (This is still a widely popular way to clean glass, although there is speculation as to whether or not the modern vegetable-based newsprint inks are more or less effective than the old petroleum-based inks.)

Streaking is caused by the drying of the residual product on glass in droplets larger than 0.25 $\mu$m, which can scatter visible light [238]. Only if there is no residue, or the residue breaks up into droplets smaller than this size, can streaking be avoided. Lubricity is also a factor in window cleaning. Unlike other cleaners, window cleaners are formulated to evaporate quite rapidly. This can cause some difficulty in wiping the cleaner in its final stages. Ammonia salts of surfactants and builders tend to be favored in window cleaners. Not only is ammonia a volatile compound that can conveniently be used to adjust the pH, it also seems that these ammonia salts increase the lubricity of the formula during wiping [239,240].

These cleaners are also trying to deliver added benefits. They usually fall in the category of antifogging, of which there is voluminous patent art [241,242]. Antifogging consists in preventing the formation of water droplets that scatter visible light and result in the "fog" on the glass. There are two mutually opposite approaches: either to make the surface so hydrophobic that all the water is repelled and drains off the surface or to make the surface very hydrophilic so that the water wets the surfaces very well in a continuous film and avoids the formation of droplets. There are examples of both approaches in the patent literature. Some of these claim that in addition to the antifogging effects, usually achieved through the deposition of a polymer film, there are also antisoiling benefits. Antisoiling benefits alone can be achieved with silanes [243] which can react with siliceous surfaces. Another patent claims the use of polyglycols as both adjuncts to the surfactant system as well as giving the antifogging/ antisoiling benefits [244,245] or amine oxide polymers [246]. Antifogging can also evidently be achieved by synergistic mixtures of surfactants [247]. Another benefit that is claimed is the uniform draining of water from the glass surface [248] which tends to decrease water spotting. This was also achieved with a film using polyvinyl alcohol and/or cationic polymers. There is also an example using polycarboxylate polymer to impart a lasting sheeting action to glass [249]. This is a major benefit when the cleaner is also intended to be used on automobile windshields or to clean outside windows.

Another glass cleaning variation that was launched as a product is a no-drip application. Glass cleaners, of their nature, have the viscosity and flow properties of water. This gives them a tendency to drip if the spray is concentrated in one area. Glass surfaces are most often vertical, and so thickening the cleaner to give it more cling might be attractive to some consumers. Most of the formulations use polymers to thicken the glass cleaner [250,251]. The problem cited in these patents is that the presence of thickening polymers usually makes the cleaner harder to rub out, which could cause streaks, undesirable in glass cleaning.

These cleaners, in common with the all-purpose sprays, are also delivered in trigger spray packages. They were at one time packaged as aerosol sprays, but protest against aerosol packaging has largely eliminated them. The major recent change in window cleaners has been the development of wipe products, which is discussed in its own section.

## D.   Glass and Surface Cleaners

Another relatively recent development in the marketplace is the introduction of glass and surface cleaners. These are cleaners that are presented as being able to clean greasy soil, and yet leave no residue. This gives the advantage to the consumer of having to only buy one product to do both the jobs of glass cleaner and all-purpose cleaner. In practice, they cut grease less than all-purpose cleaners and streak more than window cleaners. However, they also streak less than the all-purpose sprays and cut grease better than do window cleaners. In general their formulations are between the all-purpose cleaners and window cleaners. They share with the window cleaners high solvent levels and minimal builder concentrations. However, they also have surfactant levels closer to those of the all-purpose cleaners. There are some advantages claimed for betaines in the literature of these cleaners [252–254]. These inventions are usually synergistic mixtures of betaines and other surfactants that are claimed to give good grease cleaning while minimizing streaking or residue. Modified sulfobetaines have also been claimed in glass/general cleaning formulations [255] where it is pointed out that the solvent/buffering system of the product also has a role with the surfactant in keeping filming and streaking to a minimum. More exotic surfactants have also been used, for the benefits of good cleaning with less streaking, such as amido-substituted soaps [256]. Like the glass cleaners, glass and surface cleaners also may contain polymeric ingredients to decrease streaking [257]. They are also delivered in bottles equipped with trigger sprayers. These have also been recently formulated as wipe products as with glass cleaners and all-purpose cleaners. These are discussed in their own section.

## E.   Test Methods
### 1.   Cleaning Methods

The test methods for these spray cleaners are similar to those described for dilutable all-purpose cleaners. Grease cleaning is tested the same way with surfaces.

Amounts of soil are sometimes adjusted to lower levels, especially for glass or glass and surface cleaners. Another adjustment that can be made is that the product may be sprayed directly on the soiled test surface rather than applied to the sponge. The cleaning tests may also use paper towels or mohair cloth (to stand in for cleaning cloths) instead of sponges, depending on the local consumer habit.

For window or glass and surface cleaners, the test substrate is often glass. This allows not only for testing soil removal on the surface of interest, but then the same surface may be evaluated for residue [258]. The soil tested may also be changed for glass cleaners. In this case, the main task is not cooking grease, but the grease of fingerprints. Therefore, the soil is changed from animal/vegetable fat to synthetic sebum [259], sometimes mixed with dust [260]. For glass and surface cleaners, either soil may be used because it is used for both glass and general-purpose cleaning.

## 2.  Streak/Residue Testing

The major difference in this area, especially for any cleaner promoted for use on glass, is the emphasis on residue/streak testing. The most challenging surface is that of glass mirrors because of their reflectivity. Streaking or residue is readily apparent. A given amount of cleaner is wiped on the surface in a specified number of strokes, sometimes as few as one. If more than one stroke is used, all the strokes are done in the same direction — no perpendicular or circular wiping is done. The usual applicator is a soft, lintless cloth or paper towel [261]. The area is left to dry. A panel of observers rates the prepared surfaces, taking care to view the surface from several angles. Streaks or residue are not always apparent from a single lighting condition, and the surface should be tipped several ways for proper observation. Many of the methods outlined in the patent literature also specify the humidity under which the evaluation takes place as this and the room temperature are said to influence the evaluation. Alternatively, streaking can also be evaluated using the product on black ceramic tiles and using a glossmeter to measure the residue on the surface [262].

## 3.  Other Testing

Foam level becomes a question of the interaction of the formulation with the spray trigger. The inherent foaming character of the formulation can be tested by the methods used for other household cleaners. However, this foam profile may be changed by the trigger used. The degree of foaming should also be evaluated by spraying the product out of the trigger.

There are other characteristics that are also due entirely to being a sprayed product. Although some of these attributes are part of the testing of the actual packaging itself, they should also be done with the formulation. It is desirable to test the area covered by a single spray, and the volume of product delivered.

One could also measure the amount of time it takes the product to run down the surface or its cling.

Surface safety is evaluated in the same manner as for the dilutable cleaners, by letting the product sit on a surface for a predetermined amount of time. Safety profiles for spray cleaners can be quite different from dilutable cleaners due to the different proportions of solvent/surfactant generally used.

The antifogging qualities of glass or glass and surface cleaners can be tested by exposing them to steam and noting whether a fog forms on the surface.

Some of the usual product testing becomes even more important with these dilute spray cleaners. For instance, higher concentrations of solvents could change the flammability of the product, and therefore the shipping of the product. The abundance of water and the low concentration of surfactant might make the product more susceptible to microbial degradation and therefore adequacy of preservation is more important.

## F.  Household Cleaning Wipes

The biggest change around the turn of the millennium in household cleaners (largely in the developed markets of Europe and North America) was the rise of wipes as a product form. These take the theme of convenience even further, presenting the cleaner at its use concentration (like spray cleaners) but already impregnated in the cleaning implement. Wipes constitute yet another delivery system for liquid cleaners.

The major uses for nonwovens traditionally are areas where a significantly less expensive fabric is needed, such as linings for footwear, linings on upholstered furniture, and barrier layers in road building. They have also been very useful in the filtration industry and in surgical drapes/apparel where disposability is important. Nonwovens in consumer products have a longer history of use in personal care products. Baby wipes have been a significant consumer product in developed markets for over two decades. (Nonwovens also figure prominently in the fabrication of feminine hygiene and incontinence products.) More recently, the innovation in personal care use of nonwovens has been as facial cleanser wipes.

By contrast the use of nonwovens in household cleaning products is much more recent (compared to baby wipes) although there were a few introductions of this product form as long as 15 years ago [263]. The literature gives examples of formulation going back 20 years [264]. These product entries, as glass cleaning wipes, silver polishing cloths, and toilet wipes were largely unsuccessful at the time. However, wipes as a household cleaning form have experienced unprecedented growth and success recently, particularly in Europe. As the field of wipes is relatively new, a short discussion of the nonwoven substrate is given here. As might be expected, dry wipes, used predominately for dusting, are not discussed here as they lack a significant liquid cleaning component in use, although they might have been treated with surfactants, polymers, etc., during manufacture.

## 1. Nonwoven Substrates

A nonwoven is exactly what the name implies: a fabric or substrate made by bonding or interlocking fibers (by mechanical, chemical, or solvent means, or combinations of these [265]) rather than by weaving them. Additionally it is usually made from individual cut fibers or continuous filaments rather than from a continuous yarn [266]. Depending on one's definitions, nonwovens could be considered ancient as both paper and felt can be considered nonwovens. (For more complete coverage of the topic of nonwovens and their uses, the reader is referred to the publications of INDA, the association of the nonwovens industry [267].) There is extensive art and science in the construction of nonwoven materials that will not be discussed here. Key considerations in the design of a nonwoven are the length, denier (diameter), surface roughness, and cross-section of fibers, and their chemical composition. Suffice to say that the common synthetics used in nonwovens, polyethylene terepthalate and polypropylene, are well known to household cleaner formulators as packaging materials. (Polyethylene is usually not used because of its comparative brittleness.) The challenge can begin when the wipes contain a certain level of cellulose fibers, which are often added for either absorbency or biodegradability. The chemical interactions between the cleaner formulation and cellulose can be significant especially as the cleaner (being largely water) will be absorbed into the cellulose fibers.

Nonwovens, due to their high surface area, cellulose content, and some of the manufacturing processes (such as hydroentangling), can carry a bio-burden that could tax the preservative in the cleaning solution of the ensuing wipe product. If the product needs to be preserved, attention should be paid to the nonwoven substrate as well as the solution.

## 2. Cleaning Solution

The substrate is one half of the "formulation" of a wiping product. The other half is the cleaning solution on the nonwoven. In general these are very dilute systems, but some can contain large amounts of solvents. Wipes are usually fully saturated with the cleaning solution, and the packaging usually tries to maintain this. The coating level is usually of the order of 150 to 500% of the weight of the nonwoven substrate being coated, depending on its absorbency. Therefore wipes are usually packaged in plastic tubs, canisters, or laminated film flow wrap to maintain the high level of liquid on the wipe.

As the cleaning solutions are largely water it is reasonably easy to impregnate the nonwoven substrate with the solution. This can be done in two different ways. In the first method, the nonwoven is unreeled from numerous rolls, each separate length of substrate being wetted with the solution. The separate lengths are brought together, one on top of the other in the number needed for final product. This continuous stream of stacked, wetted substrates is then cut to the length desired to yield the individual stacks to be inserted into packages. In the other method, the nonwoven is unreeled, and collated as dry fabric. Then, either before or after it is cut into individual stacks it is soaked with the cleaning solution. This can happen on the manufacturing line or in the package.

## 3. All-Purpose and Glass Cleaning Wipes

Table 13.9 gives typical ingredients for the cleaning solution of wipes in general. These are supplied at use dilution, and this concentration is similar to that supplied

**TABLE 13.9**  Generalized Formula for Impregnating a Wipe

| Ingredient | Examples | Amount for APC, glass, disinfecting wipes (wt%) | Amount for floor cleaner wipes (wt%) |
|---|---|---|---|
| Anionic surfactant | Alkyl sulfate, alkylbenzene sulfonate, ethoxylated alkyl sulfate, soap | 0–10 | 0–0.3 |
| Nonionic surfactant | APG, ethoxylated alcohol, amine oxide | 0–14 | 0–0.07 |
| Amphoteric surfactant | Betaine, sultaine | 0–10 | 0–0.01 |
| pH adjuster | Citric acid, triethanolamine, morpholine, ammonia, sodium hydroxide | 0–2 | QS |
| Hard water chelator | EDTA | 0 | 0–0.4 |
| Solvent | Ethanol, isopropanol, limonene, glycol ether | 0–30 | 0–4 |
| Disinfectant | Quat, biguanides, organic acid | 0–5 | 0–0.03 |
| Biocide (quat) release agent | Potassium citrate, magnesium sulfate, ammonium chloride | 0–5 (disinfecting) | 0 |
| Suds control | Soap, silicone | 0–0.1 | 0–0.5 |
| Specialty polymers | Polyacrylic acid, polyethylene glycol | 0–2 | 0–0.04 |
| Perfume, color, etc. | | 0.001–0.5 | 0.001–0.5 |
| Distilled or deionized water | | QS | QS |

in a spray bottle product [268–271]. The typical array of surfactants and solvents is used in these types of products as in their bottled predecessors. (The trend noted before continues here: the innovation is not so much in chemical formulation as in the delivery/packaging of the product.) Much of the patent literature is concerned with streak-free cleaning. In general, wipes are not used for heavy-duty cleaning but more for routine cleaning or touch ups where either the soil level is low or the soil is easily moved (like dust). It is assumed in the use of wipes that the surface will not have to be rinsed (as with dilutable cleaners) nor will it be wiped to dryness with a separate implement (as with spray cleaners used with sponges/paper towels). Therefore the expectation on the wipe to do streak-free cleaning that dries quickly is high. Volatile solvents are an easy way to do effective cleaning with no residue, but they contribute significantly to the odor of the product and can be limited by VOC considerations. Therefore, some inventions are concerned with lower levels of solvent [272].

The all-purpose wipe products cover the gamut of general household cleaning, being positioned as all-purpose cleaner, glass and surface cleaner, and glass cleaner despite little change in the formulations. The biggest change tends to be that the glass cleaners contain significantly higher solvent concentrations and lower surfactant levels, as in the bottled products. There is a distinct advantage with these types of products, since the formulator supplies the implement, to give the consumer better residue profile because the implement can be essentially lintless. Therefore, wipes are often aimed at shiny surfaces, glass being the ultimate example of the shiny surface [273–275].

One recent entry in the wipes category leverages the packaging of the wipe to contribute to its efficacy. A wipe is impregnated with a typical dilute cleaning solution, but it is individually packed in a thin flow wrap bag. In use, the single packaged wipe is put into a microwave oven and heated according to the directions. The vaporizing solution inflates the bag and eventually pops it, releasing the hot water and solvent vapors into the microwave. This is intended to soften and loosen any baked on soils in the microwave. The user can then take the heated wipe and clean the microwave surfaces. The idea of microwave-heated cleaners does appear in the patent literature [276].

## 4. Floor Cleaning Systems

The area in which these types of nonwoven products have made the biggest impact is, surprisingly, floor cleaning. The main advantage to these systems is that they represent essentially "bucketless" floor cleaning, which started almost 10 years ago in the literature [277]. There are two different styles of product: wet and dry wipes. Both are used in conjunction with a "mop:" a resilient slightly spongy pad on the end of a long handle that supplies reach for floor cleaning.

In the wet system, wipes are supplied saturated with the cleaning solution. The wet wipe is secured to the bottom of the pad to clean the floor [278,279].

In the "dry" system, dry nonwoven pads are supplied separately from the cleaning solution, which is bottled. The dry nonwoven is attached to the bottom of the pad on the mop, and the cleaning solution is fixed in some way to the mop, either in a holder for the bottle or in a reservoir. In the cleaning process, the consumer sprays the cleaning solution by activating a device on the mop and then wipes the mop over the wetted area [280]. In either case, the consumer uses the wipe until no more soil can be picked up from the floor.

This type of system has undoubtedly been one of the largest changes in consumer cleaning habit and practice in the last ten years. First, the system makes floor cleaning immediately available, cutting out the setup phase of getting out a bucket, cleaner, and mop and then making a solution. Second, it also eliminates the cleanup of the mop and bucket. Third, because minimal solution is used on the floor and the wipe is highly absorbent, the claim is that the cleaned floor does not need rinsing, so that time consuming and laborious step also is eliminated. For many consumers this has completely changed the way they clean floors.

The formulations of both the liquid impregnated on the wipes and the liquid supplied in a bottle are similar. They are more similar to glass cleaning formulas in their surfactant concentrations, although they tend to have lower solvent content than do glass cleaners. They are likely to contain suds suppressors (such as silicones) because excess foam would leave the impression that the surface needed to be rinsed.

True to their use as floor cleaners, similar trends are seen in the formulation of floor cleaning wipes. There are wipes where soap is the main cleaning surfactant [281], and ones where there are formula ingredients to entrain the particulate soils [282].

## 5. Disinfectant Wipes

Another popular class of wipes is one used simultaneously to both clean and disinfect surfaces. The advantage is that the solution in the wipe is applied to the surface without rinsing or wiping to dryness. The disinfectant solution is therefore applied at its proper strength directly to the surface and left there. A variety of antibacterial agents and solvents are used [283–285], especially the typical quats used in other household disinfectant cleaners. They are used to both spot clean and disinfect small household areas such as countertops and tables. These also make spot disinfection very convenient, available instantly as the wipe is pulled from its container. Again, a disposable implement is supplied with the cleaner, which means that the "germy" soil that has been cleaned up can be thrown away on the implement. This is not much of an advantage for a consumer that uses paper towels, but for those that use sponges or woven cloths that would have to be cleaned after use the wipes are a significant increase in convenience.

The same features that make a wipe a good delivery system for disinfection would also make it appropriate for delivering other treatments for surfaces such as decreasing soiling or decreasing dust deposition [286]. The advantage once again is that the treatment is applied to the surface at its intended concentration and is left to dry without wiping to dryness. This antisoiling benefit is now present in toilet wipes that have been commercialized.

## 6. Test Methods

Wipe cleaners can be tested in one of two ways: either just the cleaning solution can be tested (using the methods outlined above for spray or dilutable cleaners) or the final wipe itself with cleaning solution on the nonwoven substrate can be tested. The testing of the wipe for cleaning performance would have to be, because of the form, abrader testing. In this case, however, there would be no question of how to apply the cleaner, or how much, if wet wipes are used.

In a similar way, the wipe can be used on a glossy surface (such as black ceramic tiles) or on a mirror surface to test for streaking. The testing would be done in the manner previously outlined.

## G.  Bleach Spray Cleaners

Bleach spray cleaners for general household use have emerged with the bleach containing all-purpose dilutable cleaners. These constitute a yet even more specialized niche than the glass or glass/surface cleaners due to the sensitivity of many household surfaces to bleach. The majority of such sprays on the market are hypochlorite bleach based and are close or identical in formula to the dilutable bleach all-purpose cleaners. That is, they contain about 1 to 2% hypochlorite bleach, a low level of bleach-compatible surfactant (usually amine oxide), an alkalinity agent like sodium hydroxide, and possibly some builder salt such as phosphate or silicate. These cleaners combine a medium level of grease cleaning with the obvious stain-removing properties of the bleach. The trigger sprayer used with the product can deliver either the usual aerosol or, more usually, a loose foam. Bearing in mind that the ingredients have to be hypochlorite stable, the formulations would have most in common with those in Table 13.7.

The other group, of recent entry, is the "oxygen" spray cleaners, as noted in the dilutable cleaner section. These are very similar in formulation to glass and surface spray cleaners, but with the addition of a quantity of a peroxide-producing species ($<7\%$) and a lower pH to stabilize the peroxide [287]. (The surfactant used, of course, has to be oxygen bleach stable.) The cleaning solutions are actually colorless themselves probably owing to the difficulty of stabilizing a low level of dye in a bleach-containing solution, and the impossibility of suspending a pigment in such low-viscosity formula. However, they are sold in colored bottles to give consumers the familiar look of a colored cleaning solution. The performance capabilities of the formulas should be similar to those of the all-purpose cleaners with the addition of destaining ability owing to the peroxide. Although they do not have the stain removal potential of the hypochlorite formulas, they do have significantly greater surface safety and so are better for general household cleaning. As noted above, these products are formulated at low pH in order to stabilize the peroxide as the more traditional bleach cleaners are formulated at high pH in order to stabilize the hypochlorite.

These cleaners are tested in a similar manner to other spray all-purpose cleaners with the addition of tests for destaining ability.

## IV. BATHROOM CLEANERS

Bathroom cleaners, along with bleach cleaners, are the largest category of specialty liquid cleaners. These products are formulated and packaged with the specific soil and cleaning problems associated with modern bathrooms. All-purpose cleaners are also used to clean bathroom surfaces, but they are not targeted at soils such as soap scum and so suffer some deficiency when compared with the specialty products. As already mentioned, this specialization is a two-edged sword. Although these cleaners are very efficient on targeted soils, the ingredients to accomplish this often limit their use in other circumstances. Most often, as will be seen, the concern is with surface safety. Also, there are psychological barriers. There is no reason why an acid toilet bowl cleaner could not be used to clean hard water stains from a kitchen sink, but very few consumers would be willing to do this.

Three categories of bathroom cleaner are discussed here: general bathroom cleaners, mildew removers (with some cross-over to bleach cleaners), and toilet bowl cleaners. "Automatic" toilet bowl cleaners are not discussed due to the dominance of solid, and not liquid, forms in this group.

## A. General Bathroom Cleaners

There seems to be a worldwide consensus that the main problem in bathroom cleaning is soap scum, followed by the related problem of hard water deposits [288]. All-purpose cleaners have some effect on soap scum, but tend to have difficulty with hard water spots. General bathroom cleaners tend to target these problems and tailor their chemistry accordingly. These types of cleaners are usually moderately alkaline or strongly acidic. Some make disinfectant claims and others do not. Although bathroom cleaners are commonly packaged similarly to dilutable all-purpose cleaners in squared or handled bottles, in North America (and increasingly in Europe) this particular subset of specialty cleaners is dominated by trigger spray packaging.

Bathroom cleaners are the predominant area where soil prevention treatments are important. There are very few kitchen or general-purpose cleaners that make claims to make general household surfaces easier to clean, although there is patent literature to that effect. However, this is a growing and increasingly important benefit in bathroom cleaning. Are bathroom surfaces that much harder to clean than kitchen surfaces? Are bathroom soils more difficult? Are consumers more accepting of surface treatments in the bathroom than in other rooms of the house? Whatever the answers are, there seems to be a perception that bathroom cleaning is difficult and laborious, and any tool that can decrease this factor is welcome.

It should be noted that bathroom cleaners, while predominately trigger spray cleaners in the U.S., and either trigger sprays or in pour out bottles in Europe, still appear as aerosol sprays in some parts of the U.S. These generally use the typical aerosol propellants [289] appropriate to the types of cleaner (alkaline, acid, or neutral).

## 1. Alkaline Bathroom Cleaners

Alkaline bathroom cleaners are direct descendants of the all-purpose cleaners. They tend to have somewhat higher builder levels than modern all-purpose cleaners, presumably to try to chelate some of the hard water ions that contribute so significantly to tough bathroom soils. One example attributes the cleaning to the form of EDTA used in the cleaner [290,291]. Many of these types of cleaners also have disinfectant claims. Bleach bathroom cleaners fit this general description, but due to their destaining ability they are discussed in a later section. This category includes general bathroom cleaners that use alkyl dimethylbenzylammonium chlorides as their disinfecting agent. Use of this quat to achieve disinfectant places the constraint of nonanionic surfactants on the formulation. Generally nonionic surfactants are used. One cleaner formulation claims the use of polymers to retain the disinfectant on the surface and so prolong the action [292]. These cleaners, like their disinfectant and nondisinfectant all-purpose cleaning forerunners, have moderate soap scum removal abilities and poor water stain cleaning. Still, there are claims in the literature for effective soap scum cleaning with alkaline systems [293]; in a thickened system such as this, the added cling time on the vertical surface would be an advantage. However, they do find application in areas where many surfaces are acid sensitive [294], such as countries where many bathroom surfaces are marble. Still, the most activity in bathroom cleaners has been in acidic bathroom cleaners [295].

## 2. Bathroom Shower Treatments

An exception will be made here to discuss a product that is largely a surface treatment rather than a cleaner *per se* as it illuminates the surface protection trend evident in bathroom cleaning. These were first introduced in the mid-1990s.

Shower treatment products are liquids intended to be used immediately after showering to *prevent* soils from occurring and "setting" on the shower surfaces. Therefore these products are intended to be used while the surfaces are still wet from showering. Before they are used the first time, the consumer is often directed to first clean the surface, and then apply the shower rinse. Although the patent art for these products says that the product can also be used to clean the shower surfaces, this shows that they do not have the power to clean a significantly soiled surface, and are intended more as daily applications to work as preventatives.

The formulations are clear, essentially colorless, water-thin liquids supplied in trigger spray bottles. The formulas are reasonably similar, being based on chelation, soil softening, and water film formation on the surface [296,297]. The mechanism seems to be to chelate the hard water ions in the water left on the surfaces to prevent soap scum formation. The surfactant used is specified to be nonionic (both of the typical ethoxylated alcohols and of the sugar-based surfactants [298,299]) so that it can interact (in a solubilizing not precipitating way) with both anionic and cationic surfactant residues. Also, the small amount of surfactant and/or solvent helps to make whatever water is left on surfaces spread and wet the surfaces, thereby forming a film that conducts the potentially soiling components down the wall and then to the drain as well as preventing water droplets. Water droplets result in more concentrated areas of residue after they dry where a uniform film of water will leave behind a thinner, more uniform, and less noticeable coating, even if the soiling components were not moved to the drain. The surface can therefore appear cleaner when in fact the same amount of residue can be present on the surface. (This is similar to the approach used in automatic dishwasher products to prevent water spotting.)

An interesting specification in the original patent art is the use of distilled or deionized water, where the water makes up over 90% of the formulation. In practice the hardness of formulation water must usually be controlled, but that would be much more critical in a product of this type since the main aim is to rinse hard water residues off the shower surface. Table 13.10 gives some examples of typical ingredients and ranges.

The formula is also usually formulated to be slightly acid/neutral (pH 6 to 7). Depending on the acidity of the other ingredients added (such as the chelant) the agents used to adjust the pH are volatile (fugitive) compounds and, unlike

**TABLE 13.10** Shower Treatments

| Ingredient | Examples | Amount (wt%) |
|---|---|---|
| Nonionic surfactant | Ethoxylated alcohol, APG, amine oxide | 0.5–3 |
| Additional surfactant | Betaine, alkyl sulfate, octyl pyrrolidone | 0–10 |
| pH adjuster | Ammonium hydroxide, morpholine, citric acid | QS |
| Hard water chelator | EDTA, NTA | 0.1–3 |
| Solvent | Lower carbon number alcohols, glycol ether (Carbitol®, Dowanol®, etc.) | 1–8 |
| Disinfectant | Quaternary ammonium | 0–0.5 |
| Perfume, color, etc. | | 0.0005–0.001 |
| Distilled or deionized water | | QS |

sodium hydroxide, would not tend to contribute to residues on the surface. In some ways these formulations owe as much to glass cleaner as to bathroom cleaner formulations.

Another interesting aspect is that shower enclosures are usually either ceramic (grouted tiles) or plastic (methacrylate, or Plexiglas®). One patent emphasizes the choice of surfactant and its effect on plastic. Nonionic surfactants have been said to cause "crazing" in plastic surfaces [300], and one shower rinse formulation claims the use of amphoteric surfactants for superior soil removal and greater safety to plastic surfaces [301].

## 3.  Acidic Bathroom Cleaners

Acidic bathroom cleaners have some distinct advantages on common bathroom soils. First, the main matrix for the soil referred to as soap scum is soap that has been precipitated by water hardness ions. Imbedded in this matrix may be skin flakes, lint, dirt, etc. (see Figure 13.7), but the waxy precipitated soap serves to hold the mass together and make it adhere to surfaces. Acids can work to reverse this chemical reaction, turning some part of the soap fatty acids into liquid components (notably oleic acid). This serves to soften the soil overall and thereby make it more easily removed. Second, if there were any ion bridging of the soil to a receptive

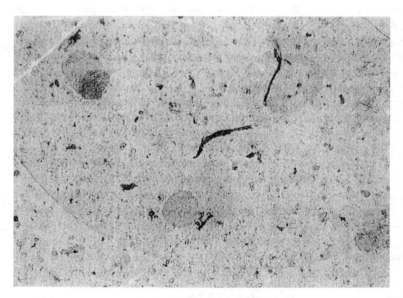

**FIG. 13.7** Photomicrograph of home-generated soap scum. Visible are water spots (large circles), skin flakes (dark speckling), and fabric fibers within the soap scum matrix (magnification ×50).

surface (ceramic), similar to the bridging effects found in ore flotation, then strong acid could be used to disrupt this bonding, freeing the soil from the surface. Acid is also effective at dissolving hard water spots, stains, and encrustations, these being mostly $CaCO_3$, $MgCO_3$, and similar salts.

There are also some minor advantages to acid cleaners. Although an acid cleaner can sting any open wound, in general moderate acid pH is kinder to skin than alkaline pH. Acids are generally more effective at removing rust or copper stains than alkaline products. The most effective of these acids, such as oxalic acid, have chelating effects that also aid in the cleaning action similar to EDTA salts at high pH. Disinfectant quaternary ammonium surfactants are also compatible with acids, although the same restriction as to choice of cleaning surfactant holds at low pH as well as at high [302]. It was long thought that the most effective pH for quaternary disinfectant action was high pH, hence its common use in alkaline bathroom and all-purpose cleaners. However, disinfection has been documented at a variety of pH values depending on the organism and the formula [303].

More examples arise that give the acid as the disinfecting agent itself, therefore getting double duty out of the acid: low pH for soap scum and hard water stain cleaning, and disinfection [304].

However, acid also has one main disadvantage: many bathroom surfaces may be acid sensitive. Certainly marble fixtures top the list; the beautiful shine of a well-polished marble surface is easily destroyed in even one application of an acid cleaner. The cement grout between wall tiles is another sensitive surface. The modern addition of latex to the grout mixture helps resist acid attack, but cannot stop it completely. In Europe many ceramic tiles and enamel surfaces are also acid sensitive, prone to accelerated wear if an acid cleaner is used. As seen in the examples cited below, the use of more moderate pH (in the range 2.5 to 5) will slow the damage, but it will not stop it completely.

There is also a minor disadvantage to acids if used in trigger spray products. In these cases, the respirable mist produced by the sprayer may irritate the nose and throat of the consumer. This can be mediated to some extent by the delivery of the product, discussed below.

The beginnings of this field are, predictably, centered around the kind of acid used. Very strong mineral acids are generally avoided in favor of better buffering organic acids. Indeed, several patents give claims that the performance of a higher $pK_a$ acid is superior to that of a strong acid at comparable pH levels [305]. A synergy between different acid mixtures [306], or the advantages of a particular acid are usually claimed [307–309]. The patent literature also gives examples where this fundamental weakness of acids, their attack on certain surfaces, is claimed to be circumvented. Usually this includes the use of phosphoric acid or derivatives [310–313]. Some patents claim the use of microemulsions. One recent invention cites the use of an esterase to generate acid under mild pH conditions to enhance cleaning performance [314].

Interestingly, many of these bathroom cleaning formulas were first developed using zwitterionic, amido nonionic, and ethoxylated alkyl sulfates just about the time that such surfactants were becoming popular in spray all-purpose/glass and surface cleaners [315–317]. They also use similar solvents to the all-purpose sprays, leaning heavily on the glycol ethers. Table 13.11 gives a comparison of alkaline and acid general bathroom cleaners. Figure 13.8 shows a general comparison of the soap scum cleaning abilities of the two types of formulas.

**TABLE 13.11**   Spray Bathroom Cleaner Formulas

| Ingredient | Examples | Acid cleaner amount (wt%) | Alkaline cleaner amount (wt%) |
|---|---|---|---|
| Anionic surfactant | Alkylbenzene sulfonate, paraffin sulfonate, alkyl sulfate, ethoxylated alcohol sulfate | 0–6 | 0 |
| Nonionic surfactant | Ethoxylated alcohol, alkanolamide fatty acid, carbamates, amine oxide | 0–3 | 1–5 |
| Amphoteric surfactant | Betaines, sulfobetaines | 0–2 | 0–2 |
| Builder | Carbonates, citrates | 0 | 0–2 |
| Chelator | EDTA | 0 | 0–15 |
| Alkalinity | Sodium hydroxide, alkanolamines, sodium carbonate | 0 | 0.25–5 |
| Acid | Phosphoric, dicarboxylic (like glutaric), citric, sulfamic, acetic | 0.5–10 | 0 |
| Solvent | Lower carbon number alcohols, glycol ether (Carbitol®, Dowanol®, etc.) | 0–10 | 0–10 |
| Disinfectant | Quaternary ammonium surfactants | 0.1–3 | 0.1–3 |
| Bleach (may also disinfect) | Acid: peroxide; alkaline: hypochlorite bleach | 0–3 | 0–3 |
| Polymers for thickening, water sheeting, etc. | Xanthan gum, polyacrylate, polyvinylpyrrolidone | 0–0.1 | 0–0.1 |
| Perfume, color, etc. | | 0.05–1 | 0.05–1 |
| Water | | QS | QS |

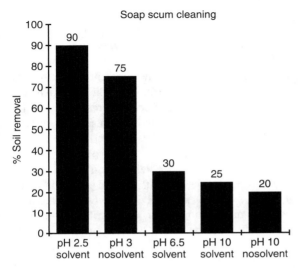

**FIG. 13.8** Relative soap scum cleaning; experiments done with formulas at different pH with and without glycol ether solvents.

Both kinds of bathroom cleaner also usually use solvents as part of the formulation. This also helps to soften and loosen the soil, particularly in the case of the alkaline cleaners that do not have the acid conversion of the calcium salts to the acid form. One unusual example in the patent literature uses a silicone surfactant and solvent to lift soap scum; there are essentially no other examples of this type of formulation for this use [318].

One of the other problems of bathroom cleaning is that many of the surfaces with tenacious soil are vertical. There should be an advantage to increasing the residence time of the cleaner on such surfaces. One way to do this is to produce a thick foam with the cleaner by combining the formulation with the right delivery system, usually a trigger spray or aerosol. The cleaner is formulated to stabilize the foam formed [319]. However, the foam cannot be too persistent or it becomes a rinsing problem [320]. Another way to increase the residence time is to thicken the product, usually with surfactants [321–323], although the use of polymers is not unknown [324]. Once again, care must be taken to make sure that the thickened product rinses easily. One very interesting alternative approach gives a thixotropic gel that forms a water-impermeable skin in use, preventing the cleaner from drying out and thereby increasing the time of action on the soil [325].

There are also cleaners that are formulated with polymer thickeners to reduce misting [326]. In this case, the object is to increase the particle size of the droplets formed upon spraying so as to decrease the number of droplets that will continue to float in the air. The longer a particle floats, the more likely that a consumer will breathe it. Considering the aggressive chemistry used in many bathroom cleaners, this can be irritating and unpleasant. Thickening the product increases the resistance of the liquid to moving through the spray head therefore increasing the energy needed to divide the liquid into droplets and so producing fewer, larger droplets. Any cling given to the product on the surface then becomes a side benefit to reducing the irritation. However, thickening can also slow the diffusion of active ingredients to the soil or stain, thus counteracting the effect of increasing residence time. Therefore, the benefits of reduced misting and increased residence must be balanced with ease of rinsing and slowed diffusion. These properties can also be important for bleach sprays, as discussed below.

Added benefits are also arriving in this field paralleling their arrival in all-purpose cleaning. Disinfectancy, the oldest added benefit, is found less in the acidic formulations than in the alkaline ones. However, as previously noted, quaternary ammonium surfactants are active as disinfecting agents at low pH as well as at alkaline pH [327,328]. The newest of added benefits, as in all-purpose cleaning, is the inhibition of soiling [329–331]. One entry into the North American market in the early 1990s claims to waterproof surfaces "to keep dirt from sticking and building up" [332]. This would be a desirable trait in bathroom cleaning: many soils are tenacious, the surface area to be cleaned is large, and the opportunity to resoil intrinsic to the use of the room. This trend was noted above with the introduction and growth of a whole class of shower enclosure treatments. The trend continues with the recent entry of a bathroom cleaner that claims the inclusion of Teflon® to make the surface more resistant to soiling. This combines two trends in one: the antisoiling treatment and the use of a name recognition ingredient. The same brand in the U.S. has also launched a toilet bowl cleaner making similar claims. However, as noted in a *Chemical and Engineering News* article of January 2005, the technology used in this product is not fluoropolymer, despite the use of the Teflon® name which is strongly associated with fluoropolymers. The patent art contains numerous examples using a variety of surface treatments including siloxanes [333], anionic polymers [334], fluorosurfactants [335], and zwitterionics [336] in which the antisoiling effect is presumably the change in the surface due to deposition of active ingredient. There is one example where it is claimed that even an anionic hydrotrope can change the surface energy and prevent soiling [337]. Another claims that the polymer in the formula chelates the hard water ions to prevent subsequent soiling [338]. There is even one example intended to be used at alkaline pH, remarking on the good cleaning achieved, since it is noted that the acid cleaners generally give better cleaning [339]. Some examples use polymers, but only claim better wetting and lack of streaking without claiming antisoiling [340].

## B.  Mildew Removers

Bleach-free cleaners generally show little effectiveness against the black stains caused by mold/mildew. Bleach cleaners are effective at removing this stain, and usually have the added benefit that they tend to kill the offending organism at the same time. Oxygen bleach products are not apparent in this category, as they often lack the speed and effectiveness of hypochlorite bleach that the consumer is accustomed to using. Mildew removers are actually a subset of the general bleach cleaners described before, although they usually predate the bleach cleaners intended for general household use. The advantages of bleach cleaning specifically in the bathroom are that most of the major surfaces tend to be bleach resistant, the disinfectancy supplied by the bleach is highly desirable, and there is a prevalent, highly colored stain to be removed.

Mildew removers are very closely related to the spray bleach cleaners discussed above. The main distinction between general household bleach cleaners and mildew removers is the concentration of bleach. While in the household cleaners the bleach level rarely exceeds 2% available chlorine, in mildew cleaners the level may reach as high as 3%. This is testament to the tenacity of the melanin stain that molds and mildews are able to produce, particularly in porous substrates like grout. Beyond this difference, the types and amounts of surfactants tend to be similar, as are the choice of alkalinity agent and the presence of any builders.

These cleaners should not only be tested for their stain-removing ability, but also for their soap scum cleaning. Although such alkaline products generally show poor soap scum cleaning compared to the acid bathroom cleaners, many consumers use them for general bathroom/tile cleaning.

One of the main problems with these types of cleaners has been the mist produced by the trigger sprayer. The situation is similar to the irritation described above for general bathroom cleaners. Although this mode of delivery contributes much to the convenience of the product, it also makes the product very unpleasant to use. The respirable particles produced with their high alkalinity and bleach can be very irritating to the consumer. One way to combat this is to thicken the product, and there is literature to show that this can be done with polymers [341] or with surfactants [342–346], or a combination of both [347]. The important aspects to balance are reducing the mist produced by the product while still achieving a consumer-acceptable spray pattern. If the product becomes very thick, the spray pattern often collapses to a narrow stream. For some uses, this kind of pinpoint application may be preferable, but for other consumers, a broader spray pattern is expected.

Newer products are also claiming to not only remove mildew stain or kill mildew, but also to keep it from recurring. Once again, this is another incidence of the trend in household cleaners to give added benefits. Many products are so efficient at removing the mildew stain that a further step to ease the cleaning problem must be taken to differentiate new products. A commercial entry made using this claim also uses a packaging innovation relatively new to household cleaning: a dual bottle with a single trigger sprayer (Figure 13.6b). Evidently the chemistries used in this product are not compatible on storage. One bottle contains the solution used to remove the mildew stain (bleach) and the other contains the ingredients used to keep the mildew from recurring [348]. Both solutions are delivered simultaneously when the single trigger, with dual dip tubes, is pulled. Once again, packaging works with chemistry to give a new product benefit. This approach to bathroom cleaning evidently proved too bulky or not effective enough to justify the cost of this packaging.

Another place where dual packaging also comes into play is the use of peroxide for mold cleaning. Peroxide has a higher oxidation potential than hypochlorite ($-1.36$ eV for sodium hypochlorite and $-1.8$ eV for hydrogen peroxide), but paradoxically does not work with the speed of hypochlorite on bathroom mildew. However, peroxide has essentially no odor, in contrast to hypochlorite that has a distinctive and unpleasant odor. Without activation, peroxide bleach would take longer to do the job that hypochlorite does more quickly. One patent, therefore, incorporates an activator into a peroxide mold cleaner, but puts the formulation into a dual-chamber package so as to ease the job of stabilizing the formula on the shelf [349].

Another recent change is a product that claims to penetrate and remove mildew "from the root." It is claimed on the package that foaming action is what allows the product to penetrate.

It is likely that new claims are on the horizon for bleach mold cleaner as a study has found that (hypochlorite) bleach solutions neutralize indoor mold allergens [350]. Hypochlorite bleach would also inactivate many types of protein residues (like those from dust mites) by denaturing the protein. Indoor allergen cleaners were launched and on the market briefly in 1999 to 2000, which were nonbleach formulas (depending instead on benzyl benzoate, an acaricide). These products were recalled and discontinued due to consumer complaints.

The "holy grail" of bathroom cleaning would be to be able to clean effectively soap scum and to remove mildew stains. As stated above, generally the best soap scum/hard water stain cleaning is found at low pH where hypochlorite bleach, the most effective mildew decolorizer, is unstable. Formulators continue to try to combine the two and there is an example of a formulation with bleach for destaining that claims to also give good soap scum removal by ion exchange [351]. There is another example where the acidic and hypochlorite portions of the cleaner are kept in separate parts of a dual-chamber package until dispensed onto the surface [352].

There are not a large number of examples of bathroom surface cleaning wipes, although an example in the patent literature stresses mold and mildew inhibition [353]. This is a natural extension of wipe usage as it is meant to be used without rinsing.

## C. Test Methods
### 1. Soap Scum and Hard Water Cleaning Methods
The test methods for evaluating bathroom cleaners are very similar to those for evaluating all-purpose cleaners, and the ASTM published a soap scum cleaning

method first in 1993, revised in 1997 [354]. The Gardener straight line abrader is still used, with sponges or other appropriate utensils for cleaning tests. The substrate usually used for evaluation is a ceramic tile, this surface being generally representative of ceramic, porcelain, and enamel. The same equation is used to measure the percentage soil removed after measurements with either a reflectometer or with a glossmeter. In this case, as in the case of all-purpose cleaning, the main discrepancy among cleaning methods is the choice of soil composition and application. The soil in many methods is applied by being dissolved or dispersed in a solvent, isopropanol, chloroform, etc., and then sprayed on the surface. However, the method in the new ASTM procedure is for the soil to be "painted" on the surface while hot and melted. The soap scum is usually calcium stearate, calcium oleate, or calcium palmitate, or a mixture of all three. It may or may not be mixed with other soil components. The most common is carbon black or charcoal that also colors the soil so that it can be seen on a white tile [355]. The ASTM soil is a very complex mix using synthetic sebum, carbon black, potting soil, and two separate mixtures of stearate. One of the stearate mixtures uses all the common hardness ions found in household water, the usual calcium and magnesium plus the less usual inclusion of iron. Another test uses soil made entirely of calcium stearate, artificial body soil (including sebum), and carbon lamp black [356]. Conversely the pure white calcium fatty acid can be used on glossy black tiles. This last choice has the added advantage that glossy black tiles are often used for residue/shine tests.

Hard water spotting is usually a much easier soil to clean than soap scum or mildew. It is related to soap scum in that it has the same root cause: hard water. Water spots may be produced by applying hard water (150 ppm as $CaCO_3$, either all calcium salts or with a set Ca to Mg ratio) to a glossy tile surface. This should be allowed to air dry. Multiple applications of the hard water may be necessary to build a tenacious, visible soil. The ease of removal of the spots gives the strength of the cleaner. For more difficult hard water problems, such as lime buildup around water faucets, a quick test can be done. If a cleaner can dissolve a piece of chalk, it is a good indication that it can remove these kinds of scale. Therefore one way of estimating the cleaning performance on hard water deposits is to measure the weight loss of cubic marble chips soaked in the cleaner. (In the reverse sense this can also be a test of the cleaner's safety when used on an acid-sensitive bathroom surface.)

Ease of rinsing is another test that can be performed. In one case, this is described as scrubbing a sink with a set amount of product to generate foam. This is then rinsed with moderately hard water, collecting the rinse water under the sink to quantify the amount needed until no foam is visible in the sink [357].

## 2. Mold and Mildew Cleaning

An important bathroom soil is mold/mildew stain. A distinction has to be made with regard to this soil. Cleaning tests are a measure of the cleaner's ability to remove the melanin stain produced by the organism. This is not necessarily a measure of the cleaner's ability to kill or retard the organism causing the stain. To do that, formal disinfection or mildicide tests have to be performed. The usual place for mold growth is the grout between wall tiles, so it is important that grout, or some other such porous surface, as well as tiles are included as the cleaning surface. A nutrient medium is applied to the test surface and this is inoculated with the organisms of interest (usually *Aspergillus niger*). The culture is incubated for several weeks, and then the surfaces are inspected for mildew growth. Cleaning tests may then be performed, with special attention going to the cleaning on the grout surfaces [358]. Alternatively, cleaners can also be tested for mold inhibition. In this case, the cleaners are applied to the surface before the introduction of the medium and microorganisms. The tiles are allowed to incubate, with periodic checks on the growth (or nongrowth) of the mold.

For claims pertaining to killing mildew, controlling its growth, or general disinfecting action, local government regulations should be consulted. In the U.S. this means following the tests set out by the Environmental Protection Agency (EPA) [359]. The tests usually require production samples of cleaner to be used against test organisms grown according to specified methods. The application of the cleaner and the time of contact are usually also described.

## 3. Other Tests

Residue/shine, foam, disinfectancy, and surface safety tests are done as described in the section under all-purpose cleaning. Grease cleaning tests may also be performed with bathroom cleaners, as bathroom soils such as lipstick and bath oils have an oily component. The other nonperformance tests (eye irritancy, biodegradability, etc.) are also outlined in the section on all-purpose cleaners.

A test of resoiling tendency should also be done for products claiming this. The simplest approach would be to clean a surface with the formulation and then try to generate the usual bathroom soil on the surface afterwards. This can also be done as a combined cleaning/resoiling test if the test is done as a conventional cleaning test followed by the new application of the soil.

## D. Toilet Bowl Cleaners

Toilet bowl cleaners are also a product where great specialization has taken place in the form of the dispenser. The most modern of the toilet bowl packages feature shaped necks that allow the product to be squirted directly under the rim of the toilet (Figure 13.9). This allows users to keep themselves, and their hands, out of the toilet bowl. With more traditional packaging, the user would have to reach down into the toilet in order to squirt the product under the rim and try to maneuver the bottle inside the confines of the bowl space.

Toilet bowl cleaners also include products placed in the toilet tank to be released on flushing into the toilet bowl. Many of these products are solid "pucks" or

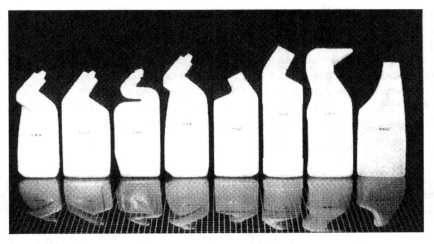

**FIG. 13.9**  Modern toilet bowl cleaner packaging.

blocks and are beyond the scope of this review. Others are liquid formulas with very elaborate dispensing devices. The device can be built to release the cleaning liquid at the beginning or end of the flushing cycle. If delivered at the end of the flush, it usually is held with the water in the tank, thereby imparting its benefits to the tank as well as the bowl. The cleaning liquids contained in the devices are generally very simple low-viscosity formulas. Their compositions are given in patents mainly devoted to the intricacies of the mechanical devices [360,361]. Typically they contain traditional anionic or nonionic surfactants (0.5 to 10%), perfume (0.01 to 1%), and large amounts of dye (1 to 10%). The concentrations are higher than might be expected because they will be significantly diluted in use, either in the bowl or in the tank. The color of the water is used as a signal to the consumer that the cleaner is present (and is aesthetically pleasing to some) and so the level of coloring agent is much higher than generally found in household cleaners. Sometimes disinfectant quaternary ammonium surfactants (0 to 5%) are used. Often, these kinds of products will also make extended benefit claims (by virtue of their residence in the bowl) for inhibiting bacteria growth [362] or inhibiting stains [363].

Originally, toilet bowl cleaners, like all-purpose cleaners, were powders based largely on sodium bisulfate [364]. They were packaged in dispensers very much like powder abrasive cleansers. In fact, many products that have been mentioned in this review are used to clean the toilet. General bathroom cleaners, liquid and powder abrasive scourers, all-purpose cleaners, and even simple household bleach are used by consumers for this task. Modern cleaners specialized for toilet bowl cleaning, however, have one factor in common that these other formulas generally lack: high viscosity. The viscosity of toilet bowl cleaners is generally between 300 and 1000 cP. Only cream cleansers have higher viscosities. When applied to the rim of the bowl they gradually flow down the sides. The high viscosity also increases the contact time of the cleaner with the surface aiding in soil removal. The consumer does not have to spread the cleaner over the surface and scrubbing should be decreased. Other than this common denominator, toilet bowl cleaners, like general bathroom cleaners, are either acid or alkaline, although there is an unusual example formulated at neutral pH [365].

## 1. Acidic Toilet Bowl Cleaners

Acidic toilet bowl cleaners are by far the largest group. The greatest cleaning problem, outside of the obvious organic soils, is hard water buildup. Toilets function most of the time as water storage tanks and therefore suffer the same evaporation/scale problem as other tanks. Acids are especially efficacious against this type of problem. The usual array of acids are used in this field (similar to those used in general bathroom cleaners) although lower pH values are targeted. The use of oxalic acid, which is particularly good at removing rust, can be especially beneficial in this type of cleaner [366] if rust stains are a local problem, although other acids can also be recommended [367]. More attention has been given recently to the choice of acid used in these cleaners, citing environmental, or surface safety issues [368,369]. One set of researchers are adding enzymes to their acidic cleaners, to fight biofilm formation [370,371]. (This is one of the few examples of enzymes in household cleaners, only the second cited in this review.) An outline of the formulas used for acidic toilet bowl cleaners is given in Table 13.12.

Newer toilet bowl cleaner formulations are moving to the newer classes of surfactants with the rest of household cleaning [372]. More nonionic surfactants than anionic surfactants are used. These acid formulas are generally self-thickening,

**TABLE 13.12**  Acid Toilet Bowl Cleaners

| Ingredient | Examples | Amount (wt%) |
|---|---|---|
| Anionic surfactant | Alkylbenzene sulfonate, paraffin sulfonate, ethoxylated alcohol sulfate | 0–3 |
| Nonionic surfactant | Ethoxylated alcohol | 2–20 |
| Cationic surfactant | Quaternary ammonium | 0–2 |
| pH adjuster | Phosphoric, hydrochloric, oxalic, citric acids | 0.5–20 |
| Electrolyte | Nitrate, chloride | 0–10 |
| Bleach | Persulfate salts, hydrogen peroxide | 0–10 |
| Thickening agents | Polyoxyethylene, cellulose gums | 0–1 |
| Abrasive | Calcite, silica | 0–15 |
| Perfume, color, etc. | | 0.01–0.5 |
| Water | | QS |

using a high concentration of electrolytes combined with the surfactant to produce the high viscosity. It is also important that toilet bowl cleaners give lower foam than do all-purpose cleaners. It is desirable to have all the foam disappear quickly. Non-ionic surfactants are preferred over anionics for both these purposes. Thickeners, such as polyacrylates, are sometimes used to supplement the natural viscosity of the surfactant system. As noted at the beginning of this section, cling to the vertical surfaces of the bowl is important and is usually achieved through thickening of the formula, although foaming can foster cling. In one application, the clinging foam is produced through the reaction of acid and base, but the invention is also careful to have the foam collapse as well [373]. This is also one area in which peroxide compounds are widely used. Persulfate salts have long been used in powder toilet bowl cleaners so that bleaching and acidic cleaning were combined in a single formula. Persulfate salts are now also making appearances in liquid formulas. In one formula, it is claimed that the dilution on adding to the toilet bowl helps the cleaning because the pH rises rapidly destabilizing the bleach [374].

Disinfection is also a benefit desired by consumers in toilet bowl cleaning. This can be achieved with quaternary ammonium surfactants, as in bathroom and all-purpose cleaners. Quats are effective bactericides at both low (1 to 4) and high (8 to 12) pH and so are compatible with very acidic toilet bowl cleaners. One of the problems with disinfection is knowing whether the product has been used at the proper dilution, and one toilet bowl cleaner formula gives the signal via a pH-dependent dye [375].

Similar to the technology used to produce cream cleansers, there are also formulas that can produce liquid toilet bowl cleaners with suspended abrasives [376,377]. The main difference between the cream cleansers and this type of product is that the suspending system should be acid stable instead of alkaline and/or bleach stable. Suspended particles are appearing in more and more of the toilet bowl cleaners. This has been commercialized in a gel form, which shows the suspended particles.

As noted in the bathroom cleaning section, antisoiling claims are also coming to toilet bowl cleaning. The commercialized formula is labeled to contain Teflon® (but does not use fluoropolymer as mentioned above), and a patent uses a fluorosurfactant [378].

In a similar vein (although not precisely soil prevention) is a toilet bowl cleaner meant to be used daily and allowed to sit for as long as overnight [379]. In some senses this is similar to the shower rinse products, meant to be used daily to prevent the buildup of soil. The examples in patents are given both as acid and alkaline.

## 2. Bleach Toilet Bowl Cleaners

The other category of popular toilet bowl cleaners is that containing bleach. The chemistry to produce these formulas is very similar to that for thickened bleach all-purpose cleaners. The most common example uses amine oxide/anionic surfactant combinations to thicken the formula, sometimes supplemented by a polymer [380]. The formula can also be self-thickened, as in the case of acid systems, by the interaction of surfactant and electrolyte [381]. The same surfactants are used in these formulas as with other bleach formulas, and similar alkalinity agents. One formula claims the use of the chelator to help remove the hard water salts that are a problem in toilet bowl cleaning [382]; the high pH necessary to stabilize hypochlorite would be a drawback in the cleaning of hard water scale. The bleach level tends to be in the range 0.5 to 5% available chlorine.

New formulas have appeared in the patent art using oxygen bleaches instead of hypochlorite. These formulas would have the advantage of low pH, and therefore be more useful for the typical toilet bowl problems such as minerals, as well as having bleaching ability that consumers find attractive. One of these formulas uses persulfate salts, long known as powerful oxidizing agents, but difficult to stabilize as liquid systems [383,384]. Another approach is to use peroxyacids, such as peracetic acid, where the bleaching agent and the limescale remover are combined in the same compound, thereby combating the two most common toilet cleaning problems [385].

As mentioned in the section on bathroom cleaning, there has been a recent entry in toilet bowl cleaners that claims the inclusion of Teflon® to prevent soil formation on the cleaned surfaces. This is the first cleaning formulation to claim this, having only been claimed before in the slow-release products (that do minimal cleaning).

## 3. Toilet Cleaning Wipes

Toilet cleaning wipes, as opposed to the toilet bowl cleaners, seem to be intended more for the toilet surfaces outside of the bowl, as using a thin wipe with a dilute solution would not seem to be the best product for under the water line inside the bowl. In the case of toilet wipes, the substrate can vary from that of the usual, largely synthetic fiber nonwoven used in other household wipes. It is desirable to be able to flush the toilet cleaning wipe, which means it should break up (so as not to clog pipes) and be biodegradable (so as to not damage septic systems). This is often accomplished through the extensive use of cellulosic material [386]. If the point is only to be flushable (and not necessarily biodegradable) then fibers other than cellulose can be used [387]. Most recent entries in this area have claimed that they are flushable, but this is something of a hot topic in the nonwovens industry [388]. This eliminates the toilet-soiled wipe from the bathroom. Another recent wipe, a companion to the liquid in-bowl cleaner and bathroom cleaner, is a wipe that contains Teflon® to make surfaces soilless.

## 4. Test Methods

Not much has been published on the subject of testing the efficacy of toilet bowl cleaners. References to "toilet soil" are made in some patent literature, but the details of the ingredients of the soil are not made clear. The efficacy testing is done by soil solubilization efficiency, and if bleach is present, soil discoloration. Tests can be made against hard water stains, as described in the bathroom cleaning section. As with other household cleaners, the normal disinfectancy tests for a locality can be used.

One test that has more relevance for toilet bowls, especially in high mineral areas, is a test for iron or manganese staining. Either ferric chloride or manganese(II) solution is spread on a light etched ceramic tile. The toilet cleaners were tested as static soaking tests with no mechanical action [389].

For in-tank automatic cleaners, the most important test is usually lifetime of the product. In the most brute force method, the product is installed in a real toilet. The toilet is then repeatedly flushed to determine how long the product will last, measuring either the persistence of some ingredient such as color or bleach in the bowl water.

## V.  CONCLUSION AND FUTURE TRENDS

The area of household hard surface cleaning has advanced at a rapid rate in 70 years. From its beginnings in simple soap powders, it has branched out into many types of specialized cleaners (see Figure 13.10). It is at present a dynamic field, full of both formulation and packaging innovation. These advances tend to work in synergism, as in the development of trigger sprayers and spray all-purpose cleaners. The following summarizes some of the highlights and future trends.

(1) In well-developed markets, the drive is to supply the consumer with greater convenience through added benefits such as two-in-one products and the recent boom in wipes. Included in this trend is the combination of bleach with various all-purpose cleaners and cleansers that has now evolved into an interest in peroxygen compounds. Bleach cleaners give the consumer more effective stain removal combined with traditional cleaning (solid soil removal). Although cleaning is always the main requirement, the household surface cleaners of the future will do more: prevent tenacious soil adhesion, reduce fogging, give more shine, prolong disinfection, etc. These kinds of added benefits are just making their appearance in household hard surface cleaning but are gaining more acceptance in the market as shown by the shower treatment sprays.

(2) Giving more benefits is at times at odds with the concomitant trend to give easier cleaning through greater chemical targeting of soils (grease or mildew) or special needs, e.g., streak-free window cleaners. The chemistry needed to target these problems often limits the scope of their use in the household. An example is the inadequacy of window cleaners for overall grease cleaning in the household. Another example is the compromise of bleach cleaners: greater stain removal, but limited to use on bleach-resistant surfaces.

(3) The overall direction is, however, to let the chemistry do more work and relieve consumers of their mechanical contribution to the cleaning process. This is

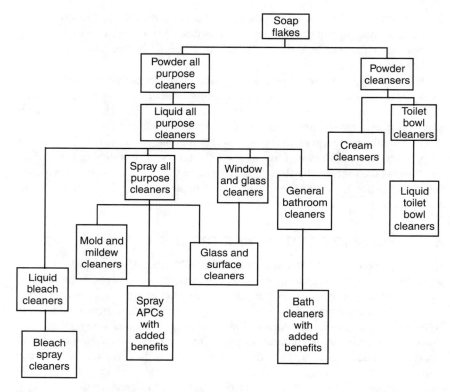

**FIG. 13.10**  Lineage of specialty household cleaners.

being achieved through continuing invention of chelants and solvents, studies of ingredient synergies, and the increasing use of polymeric ingredients for various added benefits. This is coupled with the invention of new household surfaces that are increasingly more difficult to soil and/or easier to clean. This would account for the decrease in consumers' use of abrasive cleaners. Liquid cleaners are now sufficient for nearly every cleaning job, leaving only special "tough" jobs for cleansers such as very worn surfaces or baked-on soils. Abrasive cleaners will probably never completely disappear but represent an increasingly smaller portion of the market.

(4) For a while dilutable all-purpose cleaners were tending to become more concentrated, giving the consumer use and storage advantages. However, the consumer seems to have rejected these attempts at concentration, perhaps not seeing the value of these formulas through a failure to adapt their use methods to take advantage of them. At the same time there has been a proliferation of very

dilute cleaners that exist as convenience spray products, largely as targeted cleaners mentioned above. This goes to the extreme with the ready-to-use bottled floor cleaners and the liquids in wipes where the concentration of the active ingredients (excluding solvents) is less than 2%.

(5) Powder products for household cleaners are now in the minority of products, and are decreasing every year. The household cleaning category in developed countries is dominated by the ease and convenience of liquid products. It is expected that the same trend will eventually emerge in developing countries.

(6) Packaging continues to evolve in this area in synergy with changing chemistries. Redesigns of toilet bowl cleaner bottles and the invention of the trigger sprayer have already greatly changed the array of products in the hard surface category. New dual-chamber packages offer the most exciting possibilities. Mixtures that have limited shelf life can be stored separately and dispensed simultaneously. These are prime examples of how new containers can augment the convenience, safety, or delivery of a cleaner to its target. In some ways nonwoven substrates used in wipes can be considered as "packages" to deliver the cleaner, continuing the innovation in packaging trends. The importance of packaging innovations to the field of household cleaners cannot be underestimated and seems to continue primarily in the direction of delivering convenience.

(7) Packaging continues to be a more innovative area rather than the actual ingredients in cleaning formulations. To a large extent, the surfactants used 20 years ago are still used, supplemented by the recent addition of betaines and amine oxides. It remains to be seen if surfactants like alpha olefin sulfonates (AOS) will eventually be used in this area in preference to the benzene sulfonates used today. Gemini surfactants, so prevalent in the academic literature, are almost entirely absent from the commercial literature (with the exception of Dowfaxes®). Enzymes are rarely used as ingredients in hard surface cleaners, although they appear extensively in laundry detergents and to some degree in dishwashing products. They tend to appear more often in powder products, thus the higher use in laundry detergents and automatic dishwashing products. Enzymes are relatively expensive and unstable (in liquids) whose real value to the efficacy of a formula must be carefully evaluated. "Natural" ingredients are likely to become more important in formulations, particularly as ingredients flagged on labeling to consumers. As more is learned about them and the production techniques (consistent supply, harvesting, analytical qualification) become more advanced it is anticipated that targeted uses will be developed for these ingredients.

(8) One of the biggest current questions is whether the recent boom in wipe products will continue in the future. Wipe products are very expensive per use compared to even the combination of a spray cleaner combined with a paper towel, never mind the comparison to a dilutable bottled cleaner and a wash cloth (which can be washed and reused). However, products with higher per use price but much improved convenience, such as trigger spray cleaners, have survived

and thrived. As long as consumers are willing to pay the mark-up for convenience at least some of the wipes will survive in household cleaning, as baby wipes have succeeded in personal care.

(9) More cleaners will be targeted in the future at controlling environmental health situations like allergens. This is a natural development of the ever increasing disinfection cleaner trend. It is also extremely topical given the subliminal concern over bio-warfare threats such as the anthrax scare of 2001 to 2002 in the U.S. and the global outbreak of SARS in 2003.

(10) More attention is being paid to the safety of household surface cleaners. There are at least three aspects of safety: less irritating to the consumer, less aggressive to household surfaces, and more environmentally acceptable.

(11) One of the biggest changes in surfaces in the home in developing nations over the past 15 years has been the increasing presence of electronic gadgetry: sound equipment, more elaborate televisions, faxes, and computers and their associated printers and monitors. Currently there is a very limited selection of products targeted at cleaning these electronics, but their maintenance will probably become more important in the future.

(12) Other than this, most surface types in the home — finished woods, stone, concrete, plastics like acrylates and terepthalates, steel, and ceramics (porcelain and glass) — seem to have stabilized in the developed world, although the relative amounts of these surfaces vary according to fashion trends. Some of the newer surfaces still are not well distributed throughout the developing world but will become increasingly so.

The sheer volume and variety of chemistry in this field is challenging, and only promises to continue. Likewise, the increase in the number of products and their overlapping claims should also continue.

## REFERENCES

1. See also Heitland, H. and H. Marsen, in *Surfactants in Consumer Products*, Falbe, J., Ed., Springer-Verlag, New York, 1987, p. 333.
2. It should also be noted that the references given for various formulations are intended as exemplars and not as an exhaustive list of all instances of this type of formulation.
3. Wu, S., *Polymer Interface and Adhesion*, Marcel Dekker, New York, 1982, p. 343.
4. Wu, S., *Polymer Interface and Adhesion*, Marcel Dekker, New York, 1982, p. 347.
5. Cavender, F., Alicyclic hydrocarbons, in *Patty's Industrial Hygiene and Toxicology*, Vol. II, Part B, Clayton, G.D. and Clayton, F.E., Eds., Wiley, 1994, p. 1267.
6. Somers, A., in *Proceedings of the 3rd World Conference on Detergents: Global Perspectives*, Cahn, A., Ed., AOCS Press, Champaign, IL, 1994, p. 101.
7. *Soap Cosmetic Chemical Specialties*, Sept., 43, 1985.
8. Branna, T., *Household and Personal Products Industry*, Nov., 71, 1994.
9. Somers, A., in *Proceedings of the 3rd World Conference on Detergents: Global Perspectives*, Cahn, A., Ed., AOCS Press, Champaign, IL, 1994, p. 105.

10. Bitz, K., Consumer wipes continue to boom, *Nonwovens Industry,* 35, 36, 2004.
11. Branna, T., Plenty of opportunities in household cleaning, *Household and Personal Products Industry (HAPPI),* 38, 56, 2001; What's up with wipes?, *Household and Personal Products Industry,* 39, 64, 2002.
12. Verbeek, H., in *Surfactants in Consumer Products,* J. Falbe, Ed., Springer-Verlag, New York, 1987, p. 1.
13. Jorgensen, J., Ed., *Encyclopedia of Consumer Brands,* Vol. 2, St. James Press, Detroit, 1994, p. 66.
14. Smith, R., Household cleaning made easy, *Soap/Cosmetics/Chemical Specialties,* March, 46, 1996.
15. Johnson, M.T. and B. Marcus, Competitive Implications of Environmental Regulation in the Laundry Detergent Industry, prepared for the U.S. Environmental Protection Agency, 1996, p. 7.
16. Weast, R.C., Ed., *CRC Handbook of Chemistry and Physics,* CRC Press, West Palm Beach, FL, 1977, p. B215.
17. Cavanagh, J.W. and R.P. Manzo, U.S. Patent 5460742, 1995.
18. Odioso, R.C. in *Detergents in Depth, 1980,* Soap and Detergent Association, New York, 1980, p. 71.
19. Zmoda, B.J., U.S. Patent 3149078, 1964.
20. Gangwisch, W.J., U.S. Patent 3210285, 1965.
21. Gangwisch, W.J., U.S. Patent 3214380, 1965.
22. Choy, C.K., B.P. Argo, K.J. Brodbeck, and L.M. Hearn, U.S. Patent 5470499, 1995.
23. Imanura, T., U.S. Patent 4284533, 1981.
24. Allan, A. and A.J. Fry, WO 9405757, 1994.
25. Rennie, G.K. and N.P. Randle, U.S. Patent 5286405, 1994.
26. Brierly, J.M. and M. Scott, U.S. Patent 4397755, 1983.
27. Rayner, G.G., U.S. Patent 3966432, 1976.
28. Luciani, A., B. Antelme, S. Fauvet, V. Manget, and J.-F. Lhoste, U.S. Patent 6268325, 2001.
29. Brierley, J.M. and M. Scott, U.S. Patent 4530775, 1985.
30. Argo, B.P. and C. K. Choy, U.S. Patent 5279755, 1994.
31. Choy, C.K., F. I. Keen, and A. Garabedian, U.S. Patent 5298181, 1994; U.S. Patent 5376297, 1994.
32. Choy, C.K., F.I. Keen, and A. Garabedian, EP 336651, 1994.
33. Chapman, F.E., U.S. Patent 4240919, 1980.
34. Straw, A., E. Cropper, and A. Dillarstone, U.S. Patent 4302347, 1981.
35. Denis, J.-P., N. Andries, and P. Fonsny, U.S. Patent 4869842, 1989.
36. Braganza, L.F., CA 1332908, 1994.
37. Machin, D. and J.F. Helliwell, U.S. Patent 4614606, 1986.
38. Allan, A., EP 977826A1, 2000.
39. Argo, B.P. and C.K. Choy, U.S. Patent 5279755, 1994.
40. Argo, B.P. and C.K. Choy, U.S. Patent 5346641, 1994; U.S. Patent 5346641, 1994.
41. Brodbeck, K.J., A. Garabedian, B.P. Argo, A.M. Penticoff, and C. K. Choy, U.S. Patent 5827810, 1998.
42. Luciani, A., B. Antelme, S. Fauvet, V. Manget, and J.-F. L'hoste, EP 862610B1, 2003.

43.  Garabedian, A. and C.K. Choy, U.S. Patent 5688756, 1997.
44.  Cavanagh, J.W. and R.P. Manzo, U.S. Patent 5460742, 1995.
45.  Faber, R.D., U.S. Patent 5281354, 1994.
46.  Curtis, R.J., J. Duffy, P. Graham, G. Newbold, C.F. Stanley, and S.I. Gianoli, EP 1196535A1, 2002.
47.  Lancz, A.J., U.S. Patent 4784788, 1988.
48.  Instone, T., D.P. Jones, K.L. Rabone, and M. Shana'a, EP 524762, 1993.
49.  Reed, D.A., EP 334566 A2, 1989.
50.  Instone, T., D.P. Jones, K.L. Rabone, and M. Shana'a, EP 524762A3, 1993.
51.  Haylett, N., EP 958340B1, 2003.
52.  Ormerod, I.V. and C. Raleigh, EP 821721B1, 1999.
53.  Blum, R.L., D.A. Garner, and C.M. Kling, U.S. Patent 5962393, 1999.
54.  Garner, D.A., J.R. Latham, D.K. Swatling, K.A. Henderson, and D.G. Cohen, U.S. Patent 6037316, 2000.
55.  Weibel, A.T., U.S. Patent 5821214, 1998.
56.  John, V.B., U.S. Patent 5316692, 1994.
57.  Bindl, J., F. Wimmer, and R. Kaufmann, U.S. Patent 5080824, 1992.
58.  Argo, B.P., C.K. Choy, and A. Garabedian, U.S. Patent 6100228, 2000.
59.  Yianakopoulos, G., G. Blandiaux, and M. Mondin, U.S. Patent 5741770, 1998.
60.  Baggi, P., H. Burgess, C. Fontana, and T. Inamura, EP 1337616A1, 2003.
61.  Milwidsky, B., Hard surface cleaners, *Household and Personal Products Industry (HAPPI)*, 25, 78 and 97, 1988.
62.  Jorgensen, J., Ed., *Encyclopedia of Consumer Brands*, Vol. 2, St. James Press, Detroit, 1994, p. 131.
63.  Jorgensen, J., Ed., *Encyclopedia of Consumer Brands*, Vol. 2, St. James Press, Detroit, 1994, p. 348.
64.  Jorgensen, J., Ed., *Encyclopedia of Consumer Brands*, Vol. 2, St. James Press, Detroit, 1994, p. 425.
65.  Mandel, A.S., Bottle design, plastic, in *Encyclopedia of Packaging Technology*, Brody, A.L. and Marsh, K.S., Eds., Wiley, New York, 1997, p. 93.
66.  Brody, A., Packaging materials, in *Encyclopedia of Polymer Science and Engineering*, Vol. 10, Klingsberg, A., Piccininni, R.M., Salvatore, A., and Mannarino, E., Eds., Wiley, New York, 1987, p. 696.
67.  Irwin, C., Blow molding, in *Encyclopedia of Polymer Science and Engineering*, Vol. 2, Klingsberg, A., Piccininni, R.M., Salvatore, A., and Mannarino, E., Eds. Wiley, New York, 1985, p. 447.
68.  Garrett, P.R., T. Instone, F.M. Puerari, D. Roscoe, and P.J. Sams, EP 559472B1, 1996.
69.  Frusi, A.B., A.J. Fry, and D.P. Jones, EP 670883B2, 2000.
70.  Lazarowitz, V.L. and A.D. Urfer, U.S. Patent 5308531, 1994.
71.  Bledsoe, J.O., Terpenoids, in *Kirk-Othmer Encyclopedia of Chemical Technology*, 4th ed., Vol. 23, Howe-Grant, M., Ed., Wiley, New York, 1997, p. 853.
72.  Brana, T., Why is orange glo so appealing, *Household and Personal Products Industry (HAPPI)*, 39, 46, 2002.
73.  Lu, R.Z. and A.A. Kloeppel, U.S. Patent 6110295, 2000.
74.  Kennedy, M.S., U.S. Patent 5856289, 1999.

75. Faber, R.D., U.S. Patent 0137658A1, 2002.
76. Clayton, E.T., R.E. Johnston, and J.M. Rector, U.S. Patent 3882038, 1975.
77. Hirano, S., J. Tsumura, I. Imaseki, and Y. Kawasaki, U.S. Patent 3935130, 1976.
78. Rowe, V.K., *Industrial Hygiene and Toxicology*, Vol. 2, Patty, F.A., Ed., Interscience, New York, 1962, p. 1543.
79. Gingell, R., R.J. Boatman, J.S. Bus, T.J. Cawley, J.B. Knaak, W.J. Krasavage, N.P. Skoulis, C.R. Stack, and T.R. Tyler, Glycol ethers and other selected glycol derivatives, *Patty's Industrial Hygiene and Toxicology,* Vol. II, Part D, Clayton, G.D. and Clayton, F.E., Eds., Wiley, New York, 1994, p. 2763.
80. Michael, D.W., U.S. Patent 5290472, 1994.
81. Michael, D.W., U.S. Patent 5538664, 1996.
82. Kacher, M.L., U.S. Patent 4749509, 1988.
83. Callaghan, I.C., B. Adat, and J.A. Freeman, U.S. Patent 5958149, 1999.
84. Siklosi, M.P., U.S. Patent 4287080, 1981.
85. Goffinet, P.C.E., U.S. Patent 4414128, 1983.
86. Garrett, P.R., T. Instone, F.M. Puerari, D. Roscoe, and P.J. Sams, WO 9404639, 1994.
87. Demessemaaekers, E.M., G. Bognoto, and G. L. Spadini, U.S. Patent 4017409, 1977.
88. Instone, T., D.P. Jones, D. Roscoe, P.J. Sams, and M. Sharples, WO 94/04644, 1994.
89. Herbots, I., J.P. Johnston, and J. R. Walker, EP 160762, 1989.
90. Tsukuda, K., M. Toda, M. Saito, and M. Tsumadori, U.S. Patent 5013485, 1991.
91. Ellis, R.D., Y. Demangeon, and A. Jacques, U.S. Patent 4486329, 1984.
92. Keifer, D.M., The tide turns for soap, *Today's Chemist at Work,* Oct., 70, 1996.
93. Ainsworth, S.J., *Chemical and Engineering News*, Jan. 23, 1995, p. 36.
94. Malihi, F.B. and N.J. Sparacio, U.S. Patent 4921629, 1990.
95. Branna, T., Surfactant market update, *Household and Personal Products Industry (HAPPI)*, 40, 67, 2003.
96. Becher, P., in *Emulsification*, Surfactant Science Series, Vol. 1, Schick, M.J., Ed., Marcel Dekker, New York, 1967, p. 607.
97. Lutz, E.F., D.L. Wood, and H.E. Kubitschek, U.S. Patent 4474678, 1984.
98. JP 94031402 B2, 1994.
99. Hees, U., R. Jeschke, and M. Weuthen, WO 9422998, 1994.
100. Soldanski, H.-D., M. Kalibe, and J. Noglich, WO 20595, 1994.
101. Brancq, B., in *Proceedings of the 3rd World Conference on Detergents: Global Perspectives*, Cahn, A., Ed., AOCS Press, Champaign, IL, 1994, p. 147.
102. Rosen, M.J., *Surfactants and Interfacial Phenomena*, 2nd ed., Wiley, New York, 1989, p. 21.
103. http://pubs.acs.org/hotartcl/est/97/jul/euro.html, for example.
104. Wegneer, J., EP 17149, 1980.
105. Cardola, S., S. Scialla, P.R.J. Geboes, L.G. Scott, D. Rapisarda, and M. Trani, EP 598973, 1994.
106. Scialla, S. and S. Cardola, EP 598973, 1994.
107. Geboesp, R.J. and L.G. Scott, EP 600847, 1994.
108. de Waele, J.K.E. and L.G. Scott, EP 561103B1, 2000.
109. Carrie, M.J., W.A. Cilley, P.R.J. Geboes, M. Morini, L.G. Scott, E. Vos, and R.A. Woo, EP 616028, 1994.

110.  Michael, D.W. and M.S. Maile, U.S. Patent 5382376, 1995.
111.  Scott, L.G. and P.R.J. Geboes, EP 600847, 1994.
112.  Maile, M.S. and D.W. Michael, U.S. Patent 5350541, 1994.
113.  Morse, P.M., Soap and detergents, *Chemical and Engineering News*, 77, 35.
114.  Cilley, W.A. and D.R. Brown, WO 9410272, 1994.
115.  Cilley, W.A. and D.R. Brown, U.S. Patent 6180583, 2001.
116.  Ree, G.D. and B.H. Robinson, *Advanced Materials*, 5, 608, 1993.
117.  Langevin, D., *Accounts of Chemical Research*, 21, 255, 1988.
118.  Friberg, S.E. and P. Bothorel, Eds., *Microemulsions: Structure and Dynamics*, CRC Press, Boca Raton, FL, 1987.
119.  Overbeek, J.T.G., P.L. de Bruyn, and F. Verhoeckx, in *Surfactants*, Tadros, Th.F., Ed., Academic Press, New York, 1984, p. 111.
120.  Gautier, J.C. and J. Kamornicki, U.S. Patent 4540448, 1985.
121.  Klier, J., G.M. Strandburg, and C.J. Tucker, WO 9423012, 1994.
122.  Smith, D.H., Microemulsions, in *Kirk-Othmer Encyclopedia of Chemical Technology*, 4th ed., Suppl. Vol., Howe-Grant, M., Ed., Wiley, New York, 1998, p. 310.
123.  Herbots, I., J.P. Johnston, and J. R. Walker, EP 160762, 1989.
124.  Faber, R.D., U.S. Patent 5281354, 1994.
125.  Farnworth, D.M. and A. Martin, EP 753050B1, 1999.
126.  Blanvalet, C., M. Loth, and B. Valange, U.S. Patent 5075026, 1991.
127.  Adamy, S.T. and B.J. Thomas, EP 620271, 1994.
128.  Masters, R.A., M.S. Maile, and T.C. Roetker, WO 95/13345, 1995.
129.  Blum, R.L., M.H. Robbins, L.M. Hearn, and S.L. Nelson, U.S. Patent 5854187, 1998.
130.  Robbins, M.H., L.M. Hearn, R.L. Blum, and A.B. Edsinger, U.S. Patent 6147047, 2000.
131.  Block, S.S., Disinfectants, in *Kirk-Othmer Encyclopedia of Chemical Technology*, 4th ed., Vol. 8, Howe-Grant, M., Ed., Wiley, New York, 1993, p. 237.
132.  Block, S.S., Ed., *Disinfection, Sterilization and Preservation*, Lippincott Williams & Wilkins, New York, 2001.
133.  Heitland, H. and H. Marsen, *Surfactants in Consumer Products*, Falbe, J., Ed., Springer-Verlag, New York, 1987, p. 340.
134.  Allan, R.E., Phenols and phenolic compounds, in *Patty's Industrial Hygiene and Toxicology*, Vol. II, Part B, Clayton, G.D. and Clayton, F.E., Eds., Wiley, New York, 1994, pp. 1568, 1575.
135.  Goddard, P.A. and K.A. McCue, Phenolic compounds, in *Disinfection, Sterilization and Preservation*, Block, S.S., Ed., Lippincott Williams & Wilkins, New York, 2001, p. 274.
136.  Richter, F.L. and B.R. Cords, Formulation of sanitizers and disinfectants, in *Disinfection, Sterilization and Preservation*, Block, S.S., Ed., Lippincott Williams & Wilkins, New York, 2001, p. 481.
137.  Block, S.S., Disinfectants, in *Kirk-Othmer Encyclopedia of Chemical Technology*, 4th ed., Vol. 8, Howe-Grant, M., Ed., Wiley, New York, 1993, p. 249.
138.  Romano, N., M. Trani, and K.H. Baker, WO 9802044, 1998.
139.  Das, J.R. and K.L. Rabone, EP 912678B1, 2003.

140. Jones, M.V., CA 1315196, 1993.
141. Colodney, D. and D.B. Kenkare, U.S. Patent 4597887, 1986.
142. Spaulding, L., A. Rebarber, and E. Wiese, U.S. Patent 4867898, 1989.
143. Alexander, A., I.M. George, and K.L. Rabone, U.S. Patent 5631218, 1997.
144. Allan, A., K.L. Rabone, M. Sharples, EP 796315A1, 1997.
145. Colclough, V.L., U.S. Patent 6303557, 2001.
146. Conway, M.J., U.S. Patent 6565804, 2003.
147. Ali, Y., M.J. Polan, E.J. Fendler, and E. L. Larson, Alcohols, in *Disinfection, Sterilization and Preservation,* Block, S.S., Ed., Lippincott Williams & Wilkins, New York, 2001, p. 235.
148. Merianor, J.J., Surface active agents, in *Disinfection, Sterilization and Preservation,* Block, S.S., Ed., Lippincott Williams & Wilkins, New York, 2001, p. 315.
149. Lancz, A.J., U.S. Patent 3965026, 1976.
150. Muia, R.A. and N.S. Sherwood, AU 48983, 1993.
151. Cremieux, A., J. Freney, and A. Davin-Regli, Methods of testing disinfectants, in *Disinfection, Sterilization and Preservation,* Block, S.S., Ed., Lippincott Williams & Wilkins, New York, 2001, p. 1309.
152. van Buskirk, G., K-H. Disch, C. Friese, E. Kiewert, and B. Middelhauve, U.S. Patent 5856290, 1999.
153. Burt, D.J., K.A. Harrison, A.M. Lynch, and J.M. Weller, U.S. Patent 6090771, 2000.
154. Fong, R.A., S.B. Kong, and D. Peterson, U.S. Patent 6605584, 2003.
155. Cheung, T.W. and D.T. Smialowicz, U.S. Patent 6693070, 2004.
156. Rabone, K.L., EP 874887B1, 2000.
157. Richter, F.L. and B.R. Cords, Formulation of sanitizers and disinfectants, in *Disinfection, Sterilization and Preservation,* Block, S.S., Ed., Lippincott Williams & Wilkins, New York, 2001, p. 477.
158. Barger, B. and T.J. Wierenga, U.S. Patent 6255270, 2001.
159. Spaulding, L., A. Rebarber, and E. Wiese, CA 1329103, 1994.
160. Fontana, C. and L. Novita, U.S. Patent 6511953, 2003.
161. Carlton, P. and D. Davison, EP 137871, 1985.
162. Schilp, U., U.S. Patent 4337163, 1982.
163. Stoddart, B.S., U.S. Patent 4783283, 1988.
164. Hynam, B.N., J.L. Wilby, and J.R. Young, U.S. Patent 3684722, 1971.
165. Finley, L.M. and S. H. Iding, EP 373864 A2, 1990.
166. Citrone, A.M. and S.B. Pontin, U.S. Patent 4282109, 1981.
167. Boufford, T.G. and T.R. Oakes, WO 9423575, 1994.
168. Monticello, M.V. and G.R. Mayerhauser, U.S. Patent 6106774, 2000.
169. Thompson, K.M. and D.W. Thornthwaite, EP 1288283A1, 2003.
170. Zhou, B. and A.G. Stanislowski, U.S. Patent 6013615, 2000.
171. Rennie, G.K. and P. Bernardi, EP 342997B2, 1997.
172. Morse, P.M., Antibacterial products grow despite controversy, *Chemical and Engineering News,* Feb. 1, 1999, p. 40.
173. Storck, W., Antibiotic resistance still hot topic in personal care, *Chemical and Engineering News,* Jan. 21, 2002, p. 22.
174. Ochs, D., Biocidal resistance, *Household and Personal Products Industry (HAPPI),* 36, 103, 1999.

175. Branna, T., A big boost for Oxiclean, *Household and Personal Products Industry (HAPPI)*, 40, 50, 2003.
176. Heitland, H. and H. Marsen, *Surfactants in Consumer Products*, Falbe, J., Ed., Springer-Verlag, New York, 1987, p. 346.
177. Connolly, A.D., J.M. Eshleman, and E.W. Knaub, U.S. Patent 4230605, 1980.
178. Wile, R.G. and I.F. Middien, U.S. Patent 4820450, 1989.
179. Baker, H.R., E.S. Thrower, and D.C. Simpson, U.S. Patent 4690779, 1987.
180. Maile, M.S. and D.W. Michael, CA 2107203, 1994.
181. Becker, B.D., U.S. Patent 4822514, 1988.
182. Blanvalet, C., M.C. Brauchli, J. Dautas, and C. Marchese, U.S. Patent 5380452, 1995.
183. Norman, C.L., U.S. Patent 6087319, 2000.
184. Norman, C.L., U.S. Patent 6451755, 2002.
185. Colodney, D., T.C. Hendrickson, J.H. Puckhaber, and R.J. Steltenkamp, EP 525892, 1993.
186. Steltenkamp, R.J., R.L. Hamilton, R.A. Cooper, and C. Schal, *Journal of Medical Entomology*, 29, 141, 1992.
187. Connors, T.F., MNDA: achieving repellency benefits in a cleaning product, *5th European Congress of Entomology*, York, U.K., Aug. 29–Sept. 2, 1994.
188. Rabone, K.L. and Z. Haq, EP 631610B1, 1997.
189. Evers, M.F.T., P.R.J. Geboes, D.W. Michael, M. Morini, N.J. Policicchio, V. Reniers, and L.G. Scott, EP 616027, 1994.
190. Evers, M.F.T., P.R.J. Geboes, M. Morini, V. Reniers, and L.G. Scott, EP 616026, 1994.
191. Haley, K.S. and J.J. Fischer, EP 673992, 1995.
192. Culshaw, S. and E. Vos, CA 1332217, 1994.
193. Fusiak, F. and K. Narayanan, WO 95/00611, 1995.
194. Lu, R.Z., U.S. Patent 5700768, 1997.
195. ASTM D 4488-89, Guide for testing cleaning performance of products intended for use on resilient flooring and washable walls, *Annual Book of ASTM Standards*, Vol. 15.04, American Society for Testing and Materials, Philadelphia, PA, 2000, p. 474.
196. ASTM D 4828-92, Test method for practical washability of organic coatings, *Annual Book of ASTM Standards*, Vol. 06.01, American Society for Testing and Materials, Philadelphia, PA, 2000, p. 376.
197. Malihi, F.B. and N.J. Sparacio, U.S. Patent 4921629, 1990.
198. Lutz, E.F., D.L. Wood, and H.E. Kubitschek, U.S. Patent 4474678, 1984.
199. Ellis, R.D., Y. Demangeon, and A. Jacques, U.S. Patent 4486329, 1984.
200. Hartman, W.L., U.S. Patent 4005027, 1977.
201. Michael, D.W., U.S. Patent 5342549, 1994.
202. All Purpose Cleaners, Consumer Reports, September 1993, p. 589.
203. Rubin, F.K., D.V. Blarcom, D.J. Fox, U.S. Patent 4396525, 1983.
204. Household Cleaners, Consumer Reports, July 2000.
205. McClain, C.P., in *Detergency: Theory and Test Methods*, Part III, Cutler, W.G. and Davis, R.G., Eds., Surfactant Science Series, Vol. 5, Marcel Dekker, New York, 1975, p. 542.

206. ASTM D 1173-53, Ross Miles foam test, *Annual Book of ASTM Standards*, Vol. 15.04, American Society for Testing and Materials, Philadelphia, PA, 1992.

207. Fredell, D.L., in *Cationic Surfactants: Analytical and Biological Evaluation*, Cross, J. and Singer, E.J., Eds., Surfactant Science Series, Vol. 53, Marcel Dekker, New York, 1994, p. 3638.

208. Engler, R., Disinfectants, in *Official Methods of Analysis of the Association of Official Analytical Chemists*, Williams, S., Ed., Association of Official Analytical Chemists, Arlington, VA, 1984, p. 67.

209. Code of Federal Regulations, Data Requirements for the Registration of Pesticides, 40CFR, Parts 152, 153, 156, 158, 162 and 163, U.S. Government Printing Office.

210. Cremieux, A., J. Freney, and A. Davin-Regli, Methods of testing disinfectants, *Disinfection, Sterilization and Preservation,* Block, S.S., Ed., Lippincott Williams & Wilkins, New York, 2001, p. 1318.

211. Engler, R., Disinfectants, in *Official Methods of Analysis of the Association of Official Analytical Chemists*, Williams, S., Ed., Association of Official Analytical Chemists, Arlington, VA, 1984, p. 71.

212. Swisher, R.D., *Surfactant Biodegradation*, 2nd ed., Surfactant Science Series, Vol. 18, Marcel Dekker, New York, 1987.

213. ASTM 1310-86, Vol. 6.01, ASTM D93-99b, Vol. 5.01, D 6450-99, Vol. 5.04, *Annual Book of ASTM Standards*, American Society for Testing and Materials, West Conshohocken, PA, 2000.

214. Odioso, R.C., in *Detergents in Depth, 1980*, Soap and Detergent Association, New York, 1980, p. 70.

215. Anthony, R.A., EP 623669 A2, 1994.

216. Clarke, D.E., U.S. Patent 4508635, 1985.

217. Baker, H.R., E.S. Thrower, and D.C. Simpson, U.S. Patent 4690779, 1987.

218. Rennie, G.K. and P. Bernardi, EP 342997 A2 1989.

219. Michael, D.W. and T.A. Borcher, U.S. Patent 5604192, 1997.

220. Michael, D.W., U.S. Patent 5376298, 1994.

221. Garabedian, A., S.C. Mills, W.P. Sibert, and C.K. Clement, U.S. Patent 5817615, 1998.

222. Choy, C.K., A. Garabedian, J.C. Julian, and G.L. Robinson, U.S. Patent 5851981, 1998.

223. Cheung, T.W., D.T. Smialowicz, and M.H. Mehta, U.S. Patent 6136770, 2000.

224. Wagers, K.J., U.S. Patent 6384010, 2002.

225. Cheung, T.W., D.T. Smialowicz, and M.H. Mehta, U.S. Patent 6440916, 2002.

226. Martin, A. and D. Roscoe, WO 94/12607, 1994.

227. J. Toscano, Household cleaning: a convenience revolution, *Soap & Cosmetics*, Nov. 1999, p. 48.

228. Kintish, L., Disinfectants market: putting a new face on germs, *Soap/Cosmetics/ Chemical Specialties,* Jan. 36, 1998.

229. Stonebraker, M.E. and S.P.Wise, U.S. Patent 3463735, 1969.

230. Cummings, G., U.S. Patent 5750482, 1998.

231. Neumiller, P.J. and S.M. Ziemelis, U.S. Patent 5849681, 1998.

232. Garabedian, A., C.S. Mills, and W.P. Sibert, U.S. Patent 5252245, 1993.

233. Maile, M.S. and R.A. Masters, CA 2109188 A, 1994.

234. Michael, D.W., D.C. Underwood, G.E. Dostie, and P. Stiros, WO 93/15173, 1993; U.S. Patent 5454983, 1995.
235. Garabedian, A., C.S. Mills, and W.P. Sibert, U.S. Patent 5437807, 1995.
236. Cable, E.A. and A. Garabedian, U.S. Patent 6399553, 2002.
237. Cable, E.A., U.S. Patent 6432897, 2002.
238. Barby, D., D.E. Clarke, J.L. Lloyd, and Z. Haq, U.S. Patent 4448704, 1984.
239. Church, P.K., U.S. Patent 4213873, 1980.
240. Church, P.K., U.S. Patent 4315828, 1982.
241. Kiewert, E., K. Disch, and J. Wegner, U.S. Patent 4343725, 1982.
242. Barone, P., M.T. Endres, and S.B. Patel, U.S. Patent 5254284, 1993.
243. Heckert, D.C. and D. M. Watt, U.S. Patent 4005028, 1977.
244. Church, P.K., U.S. Patent 4213873, 1980.
245. Church, P.K., U.S. Patent 4315828, 1982.
246. Wiley, A.D., R.A. Masters, and M.S. Maile, WO 97/33963, 1997.
247. Neumiller, P.J., WO 95/33812, 1995.
248. Alvarez, V.E. and D.L. Conkey, U.S. Patent 4539145, 1985.
249. Masters, R.A. and M.S. Maile, WO 9604358, 1996.
250. Svoboda, G.J., U.S. Patent 5798324, 1998.
251. Aszman, H. and A. Kugler, U.S. Patent 6649580, 2003.
252. Michael, D.W., U.S. Patent 5108660, 1992.
253. Masters, R.A. and M.S. Maile, CA 2109188, 1994.
254. Michael, D.W., U.S. Patent 5342549, 1994.
255. Michael, D.W. and D.R. Bacon, WO 9111505, 1991.
256. Maile, M.S. and R.A. Masters, EP 595383, 1994.
257. Burke, J.J. and R.R. Roelofs, U.S. Patent 5126068, 1992.
258. Michael, D.W., U.S. Patent 5342549, 1994.
259. Barone, P., M.T. Endres, and S.B. Patel, U.S. Patent 5254284, 1993.
260. Cahn, A. and G.C. Feighner, Performance evaluation of household detergents, in *Soaps and Detergents, A Theoretical and Practical Review*, Spitz, L., Ed., AOCS Press, Champaign, IL, 1996, p. 400.
261. Master, R.A. and M.S. Maile, CA 2109188, 1994.
262. Baker, H.R., E.S. Thrower, and D.C. Simpson, U.S. Patent 4690779, 1987.
263. Branna, T., The nonwovens phenomena, *Household and Personal Products Industry (HAPPI)*, 40, 79, 2003.
264. Clark, W.A. and D. Pregozen, U.S. Patent 4666621, 1987.
265. ASTM D 1117-99, Standard guide for evaluating nonwoven fabrics, *Annual Book of ASTM Standards*, Vol. 7.01, American Society for Testing and Materials, West Conshohocken, PA, 2000, p. 264.
266. Vaughn, E.A., Nonwoven Fabrics Sampler and Technology Reference, INDA, Cary, NC, 1998, p. i.
267. INDA, PO Box 1288, Cary, NC, www.inda.org.
268. Leonard, I., D. Dormal, and J. Julemont, U.S. Patent 6495508, 2002.
269. Leonard, I., D. Dormal, and J. Julemont, U.S. Patent 6429183, 2002.
270. Takano, R., U.S. Patent 6624135, 2003.
271. Deleo, M.A., R.L. Blum, M.G. Ochomogo, P.A. Pappalardo, and E.N. Swayne, U.S. Patent 6340663, 2002.

272. Skrobala, C.J., K.J. Edgett, and P.A. Siracusa, EP 604996B1, 2001.
273. Evers, M. and L. Galvagno, EP 1063284A1, 2000.
274. Julemont, J., U.S. Patent 6680264, 2004.
275. Clark, W.A. and D. Pregozen, U.S. Patent 4666621, 1987.
276. Cherry, J.-P.F., U.S. Patent 6656288, 2003.
277. Silvenis, S.A. and D.C. Wilson, U.S. Patent 5071489, 1991.
278. For example Holt, S.A., R.A. Masters, and V.S. Ping, U.S. Patent 5960508, 1999.
279. Julemont, J., U.S. Patent 6384003, 2002.
280. For example Policicchio, N.J., P.J. Rhamy, M.W. Dusing, K.W. Willman, and R.J. Jackson, U.S. Patent 23133740A1, 2003; U.S. Patent 23126709, 2003.
281. Aszman, H. and J. Puckhaber, U.S. Patent 6495499, 2002.
282. Gofroid, R.A., K.W. Willman, and C.J. Binski, U.S. Patent 6653274, 2003.
283. Coruzzi, M., T. Inamura, A.S. Jamieson, F.T. Van de Scheur, I. Trombetta, and V. Vijayakrishnan, EP 1348016A1, 2003.
284. Mahieu, M., G. Zocchi, and Y. Cartiaux, U.S. Patent 6596681, 2003.
285. Mitra, S., R.E. Simon, W.B. Scott, K.L. Vieira, G.A. Shaffer, and A. Kilkenny, U.S. Patent 20030216273A1, 2003.
286. Julemont, J., U.S. Patent 6376443, 2002.
287. Monticello, M.V. and G.R. Mayerhauser, U.S. Patent 5891392, 1999.
288. Somers, A., in *Proceedings of the 3rd World Conference on Detergents: Global Perspectives*, Cahn, A., Ed., AOCS Press, Champaign, IL, 1994, p. 99.
289. Chang, J., M.G. Ochomogo, W.B. Scott, and M.H. Robbins, U.S. Patent 5948742, 1999.
290. Mills, S.C. and J.C. Julian, U.S. Patent 5814591, 1998.
291. Ochomogo, M., T. Brandtjen, S.C. Mills, J.C. Julian, and M.H.Robbins, U.S. Patent 5948741, 1999.
292. Avery, R.W., S.L. Bakich, R.A. Wick, and H.E. Bryant, U.S. Patent 23073600A1, 2003.
293. Strandberg, G.M., D.H. Haight, J.M. Gardner, K.J. Wagers, and E.D. O'Driscoll, U.S. Patent 6200941, 2001.
294. Blandiaux, G., M. Loth, and J. Massaux, U.S. Patent 5254290, 1993.
295. Sherry, A.L., N.J. Policicchio, and J.M. Knight, U.S. Patent 6627590, 2003.
296. Black, R.H., U.S. Patent 5536452, 1996.
297. Black, R.H., U.S. Patent 5837664, 1998.
298. Robbins, M.H., D. Peterson, C.K. Choy, J.W. Chu, and J.R. Latham, U.S. Patent 6242402, 2001.
299. Robbins, M.H., D. Peterson, C.K. Choy, J.W. Chu, and J.R. Latham, U.S. Patent 6159916, 2000.
300. Leach, M.J., G. Newbold, M. Sharples, and J.E. Turner, EP 854909A1, 1998.
301. Dahanayake, M.S., WO 0012662A1, 2000.
302. Sherry, A.E., J.L. Flora, and J.M. Knight, U.S. Patent 5962388, 1999.
303. Fredell, D.L., in *Surfactant Science Series*, Vol 53, Cross, J. and Singer, E.J., Eds., Marcel Dekker, New York, 1994, p. 50.
304. Crisanti, M.G., R.Z. Lu, R.P Manzo, and G.P. Kline, U.S. Patent 6221823, 2001.
305. Blum, R.L. and S.B. Kong, EP 606712, 1994.
306. Casey, S.K., EP 0130786, 1984.

307. Vos, E., EP 0496188, 1992.
308. Blandiaux, G., M. Thomas, B. Valange, and T. Michel, U.S. Patent 5294364, 1994.
309. Riehm, T.L., U.S. Patent 4699728, 1987.
310. Aszman, H.W., C.E. Buck, and C.H. Everhart, U.S. Patent 4501680, 1985.
311. Petersen, A.W., A. Cimiluca, and L. Hirschberger, U.S. Patent 4028261, 1977.
312. Blandiaux, G., M. Thomas, and B. Valange, U.S. Patent 5039441, 1991.
313. Thomas, M., G. Blandiaux, and B. Valange, U.S. Patent 5192460, 1993.
314. Kaiserman, H.B. and M.T. Tallman, U.S. Patent 5308529, 1994.
315. Cilley, W.A. and C.G. Linares, U.S. Patent 5061393, 1991.
316. Carrie, M.J., W.A. Cilley, R.A. Masters, D.W. Michael, and E. Vos, WO 9421772, 1994.
317. Woo, R.A., M.J. Carrie, W.A. Cilley, R.A. Masters, D.W. Michael, and E. Vos, U.S. Patent 5384063, 1995.
318. Wisniewski, K.L. and M.L. Knudson, U.S. Patent 4960533, 1990.
319. Schramm, C.H., U.S. Patent 4692276, 1987.
320. Frusi, A.B., A.J. Fry, and D.P. Jones, WO 9409108, 1994.
321. Carrie, M.J., A. Koenig, and E. Vos, EP 0601990, 1994.
322. Blum, R.L. and S.B. Kong, EP 0606712, 1994.
323. Rorig, H. and R. Stephan, Cationic surfactants in organic acid-based hard surface cleaners, *Commun. Sorn. Com. Esp. Deterg.*, 21, 191, 1990.
324. Ormerod, R.C., A.G. Benecke, and D.R. Meese, U.S. Patent 6211124, 2001.
325. Bolan, J.A., U.S. Patent 4207215, 1980.
326. Bona, G.T., C.A. Keller, S.E. Lentsch, and V.F. Man, U.S. Patent 5364551, 1994.
327. Cook, W.J., K.L. Wisniewski, N.S. Dixit, and N.S. Rao, U.S. Patent 5008030, 1991.
328. Carrie, M.J. and A. Koenig, WO 94/13769, 1994.
329. Lohr, R.H., and L.W. Morgan, U.S. Patent 4347151, 1982.
330. Wisniewski, K.L., N.S. Dixit, and N.S. Rao, EP 467472 A2, 1992.
331. Sharples, M., WO 94/26858, 1994.
332. Carson, H.C., *Household and Personal Products Industry*, Nov. 1993, p. 42.
333. Paszek, L.E., U.S. Patent 5439609, 1995.
334. Sharples, M., EP 699226B2, 2001.
335. Leach, M.J. and Niwata, Y., EP 866115A3, 1999.
336. DeLeo, M., D.R. Scheuing, A. Garabedian, S. Morales, and P. Pappalardo, U.S. Patent 20030216281A1, 2003.
337. Cartoletti, M.M.L., G.V. Bolzoni, E. Ferro, M. Galli, and R.M. Morris, U.S. Patent 1023423B1, 2004.
338. Blanvalet, C. and I. Capron, U.S. Patent 6034046, 2000.
339. Das, J.R., K.L. Rabone, and M. Sharples, EP 971997B1, 2003.
340. Sherry, A.E., J.L. Flora, and J.M. Knight, EP 1047763A1, 2000.
341. Choy, C.K. and A. Garabedian, EP 0606707, 1994.
342. Choy, C.K. and P.F. Reboa, CA 2104817, 1994.
343. Humfress, B.G., J.M. Jones, H. Martin, and C. McGrath, EP 694061B1, 1997.
344. Choy, C.K. and P.F. Reboa, U.S. Patent 5462689, 1995.
345. Choy, C.K. and B.P. Argo, U.S. Patent 5728665, 1998.
346. Choy, C.K. and P.F. Reboa, U.S. Patent 5916859, 1999.
347. Orlandini, L., M. Petri, and G. Sirianni, U.S. Patent 6066614, 2000.

348. Leathers, A.Z. and S. Greer, U.S. Patent 4806263, 1987.
349. Smith, G. and R.R. Smith, EP 733097B1, 1998.
350. Branna, T., New product driven, *Household and Personal Product Industry (HAPPI)*, 41, 66, 2004.
351. Strandburg, G.M., D.H. Haigh, J.M. Gardner, K.J. Wagers, and E.D. O'Driscoll, U.S. Patent 6200941, 2001.
352. Coyle-Rees, M., U.S. Patent 6200941, 1999.
353. Roughi, M., Shower cleaners, *Chemical and Engineering News*, Dec. 3, 2001, p. 39.
354. ASTM D 5343-97, Guide for evaluating cleaning performance of ceramic tile cleaners, *Annual Book of ASTM Standards*, Vol. 15.04, American Society for Testing and Materials, West Conshohocken, PA, 2000, p. 554.
355. Siklosi, M.P., EP 179574, 1986.
356. Ormerod, R.C., U.S. Patent 5763384, 1998.
357. Sherry, A.E., N.J. Policicchio, and J.M. Knight, U.S. Patent 6627590, 2003.
358. *Consumer Reports*, 56, 603, 1991; also Household cleaners, *Consumer Reports*, July 2000.
359. Data requirements for the registration of pesticides, Code of Federal Regulations, 40CFR, Parts 152, 153, 156, 158, 162, 163, U.S. Government Printing Office.
360. Gatarz, G.M., U.S. Patent 4302580, 1989.
361. Leardi, A., U.S. Patent 4296503, 1981.
362. Carmello, R., B.A. Salka, and G.G. Corey, U.S. Patent 3897357, 1975.
363. Kaplan, R.I., U.S. Patent 4861511, 1987.
364. Odioso, R.C., in *Detergents in Depth, 1980*, Soap and Detergent Association, New York, 1980, p. 72.
365. Aszman, H. and A. Kugler, U.S. Patent 6667287, 2003.
366. Rahfield, S. and B. Newman, U.S. Patent 4828743, 1989.
367. Neumiller, P.J. and W.M. Rees, U.S. Patent 5895781, 1999.
368. Hahn, H., U.S. Patent 5877135, 1999.
369. Callaghan, I.C., G. Abamba, R. Kinsley, T.I. Moodycliffe, and R.P. Woodbury, U.S. Patent 6310021, 2001.
370. Callaghan, I.C., G. Abamba, R. Kinsley, T.I. Moodycliffe, and R.P. Woodbury, U.S. Patent 5877132, 1999.
371. Callaghan, I.C., G. Abamba, R. Kinsley, T.I. Moodycliffe, and R.P. Woodbury, U.S. Patent 6420329, 2002.
372. Malik, A.H. and A.D. Urfer, U.S. Patent 4683074, 1987.
373. Klinkhammer, M.E., U.S. Patent 6583103, 2003.
374. Bianchetti, G.O., S. Campestrini, F. DiFuria, and S. Scialla, EP 598694, 1994.
375. Stirling, T., CA 1329751 C, 1994.
376. Choy, C.K. and B.A. Sudbury, U.S. Patent 4561993, 1985.
377. Sirine, G.F., I.A.J. Day, and S.J. Kahn, U.S. Patent 3997460, 1976.
378. Nayar, B.C., R.A. Carroll, and K.J. Ward, U.S. Patent 6255267, 2001.
379. Klinkhammer, M.E., T.I. Moodycliffe, and V.M. Hempel, U.S. Patent 6425406, 2002.
380. Finley, L.M. and S.H. Iding, U.S. Patent 5348682, 1994.
381. Choy, C.K., U.S. Patent 5279758, 1994.
382. Callaghan, I.C., T.I. Moodycliffe, and R.W. Avery, U.S. Patent 6291411, 2001.

383.  Bianchetti, G.O., S. Campestrini, F. DiFuria, and S. Scialla, EP 598694, 1994.
384.  Overton, C. and P.E. Figdore, EP 199385, 1986.
385.  Cooper, N.F., GB 2273105, 1993.
386.  For example, Manning, J.H., J.H. Miller, and T.E. Quantrille, U.S. Patent 4755421, 1988.
387.  For example, Yeo, R.S., U.S. Patent 5509913, 1996.
388.  Bitz, K., Consumer wipes continue to boom, *Nonwovens Industry,* 35, 40, 2004.
389.  Neumiller, P.J. and W.M. Rees, U.S. Patent 5895781, 1999.

# 14

# The Manufacture of Liquid Detergents

**R.S. ROUNDS**   Fluid Dynamics, Inc., Flemington, New Jersey

## I. INTRODUCTION

Commercial liquid detergents are available to consumers as low-, moderate-, and high-viscosity Newtonian and non-Newtonian solutions, free flowing or thick, opaque dispersions, gels, and pastes. Despite the differences in composition and consistency of these diverse delivery systems, the manufacturing processes typically involve the same fundamental unit operations. Viscous non-Newtonian dental creams and low-viscosity Newtonian hard surface cleaners, for example, both require dispersive and distributive mixing, dissolution of various components, heat transfer for heating and cooling, solids and liquids conveying, pipeline transport, filtration, and filling. The primary differentiation in the processing of these various products lies in the industrial equipment that is required for each unit operation and the difficulty of each operation.

Many, if not all, transport functions and corresponding unit operations in the processing of liquid detergents are linked to rheology. This is most apparent from mathematical simulations and dimensional analyses used to describe these phenomena in manufacturing. Depending on the delivery system of a liquid detergent selected for a specific consumer application, mass transfer, heat exchange, and fluid flow or mixing characteristics can be cumbersome and, generally, manufacturing conditions are selected to minimize any obstacles created by adverse fluid dynamics.

This chapter reviews basic process requirements for both structured and unstructured liquid detergents. In addition to an overview of the process patent literature, a general review is provided for the manufacture and handling of both Newtonian and non-Newtonian fluid compositions. Also included are a limited number of practical aspects of the manufacturing process and many references that should be consulted when a production system is to be designed.

## II. STRUCTURED AND UNSTRUCTURED LIQUID DETERGENT DELIVERY SYSTEMS

The physicochemical state of a liquid detergent frequently determines manufacturing requirements and, for the purposes of this review, we will adopt the nomenclature of van de Pas [1], and broadly partition liquid detergents into two general material categories: structured and unstructured liquid detergents. The unstructured fluid category includes both Newtonian and marginally non-Newtonian single- and multiple-phase detergents, where the dispersed phases are not highly interactive and the volume fraction of the total dispersed phase is relatively low. These products may show minor deviation from Newtonian behavior but display neither significant elasticity nor time-dependent shear effects. Fluids of this type can generally be processed as Newtonian fluids. This broad liquid detergent classification includes many, but certainly not all, personal and household care liquid detergents, including certain shampoos, conditioners, light-duty liquid laundry detergents, hard surface cleaners, and hand dishwashing detergents. The second category, structured detergents, refers to highly non-Newtonian, viscoelastic dispersions, including physically or chemically crosslinked gels, which is an increasingly popular form of both personal and household care products. These complex fluids may exhibit yield stresses and shear effects, such as thixotropy, rheopexy, pseudoplasticity, and dilatancy, and generally will be viscoelastic products with appreciable elasticity. Dispersions and emulsions are common within this product group. For example, dental creams exemplify the "structured detergent" category, in addition to phosphate and certain nonphosphate built heavy-duty detergents, fabric softeners, and select shampoos, conditioning shampoos, conditioners, automatic dishwashing liquids, etc.

Experience has shown structured fluids to be more difficult to manufacture, due to the complexity of their rheological profiles. In addition to elasticity, dilatancy, and rheopexy, certain structured fluid compositions may exhibit solid-like properties in the quiescent state and other flow anomalies under specific flow conditions. For emulsions and solid particulate dispersions, near the maximum packing volume fraction of the dispersed phase, for example, yield stresses may be excessive, severely limiting or prohibiting flow under gravity, demanding special consideration in nearly all unit operations. Such fluids pose problems in pumping, mixing, filling, filtration, and in storage or holding vessels, with potential for negative cumulative effects on both heat and mass transfer. In addition, impaired drainage characteristics can contribute to material loss during production, increasing operating costs substantially.

It is understood that manufacturing of liquid detergents that are unstructured in their commercial form may involve intermediate streams which are, in fact, structured fluids, such as surfactant solutions at high active concentrations, within anisotropic mesophase boundaries, or concentrated polymeric solutions and gels. Whether the source is raw material, premix, or final product, manufacturing operations for each of these classifications are discussed with a focus on any specific requirements or limitations due to the physicochemical form.

## III. LIQUID DETERGENT PROCESS PATENT TECHNOLOGY

Patent activity is very aggressive in the personal and household care detergent industry, based on the total number of worldwide patents issued annually. A review of the current patent literature highlights the complexity of liquid detergent compositions and their manufacturing requirements. In process technology, the influence of process variables on product efficacy, stability, and viscosity control is common patent subject matter, disclosed for both structured and unstructured systems.

Liquid detergent process patents frequently define both compositional and process requirements, such as raw material concentrations and specifications, order of addition of critical components, thermal history, premix or adjuvant preparation methods, product/process stabilizers, distributive and dispersive mixing requirements, and process instrumentation. These patents apply to the production of primary raw material constituents, such as surfactants, builders, conditioning agents, rheology regulators, hydrotropes, disinfectants, bleach additives, etc., in addition to the specification of fully formulated detergent systems.

One patent for the manufacture of a liquid detergent composition, containing surfactant and water insolubles, describes air injection for increased dispersion stability [2]. The preparation of admixtures is disclosed, in addition to the process for air incorporation. Also issued is a process patent for the production of a pearlescent aqueous dispersion, containing fatty acid glycol ester and a wetting agent, for use in shampoos, hair rinses, cosmetics, and other detergents [3]. The primary advantage of the process described is pearlescence achieved in the absence of crystallization. In a further example, a patent has been granted for the production of an opalescent, stable dispersion obtained through multistage emulsion polymerization of $n$-vinyl-pyrrolidone and styrene, using both anionic and nonionic emulsifiers, for use in bath foams, shampoos, and various cosmetic preparations [4].

Process requirements maximizing product stability are often disclosed in the liquid detergent patent literature. In one example, Neutrogena Corporation has been assigned a patent for a coal tar shampoo prepared with a novel, reproducible, and specific process whereby detergent clarity, color, and viscosity are maintained for extended periods of time [5]. A patent describing the process for the production of a stable liquid detergent containing surfactant, aluminosilicate, a water-soluble detergent builder, and a stabilizing agent discloses the partial gelatinization of an aqueous zeolite mixture to promote dispersion stability [6]. Dispersion stability is also the subject of a patent issued to the Colgate-Palmolive Company for stable fabric softening heavy-duty liquid detergents, including the process for their manufacture [7,8]. Further, a patent has been issued to Lever Brothers for the process of making a colorfast heavy-duty liquid detergent, whereby the sequence of addition of the builder is specified [9]. It is noted that the builder is required to be added to the batch process vessel prior to the neutralization of the anionic detergent, from acid to salt, by an alkali metal hydroxide. Advantages include rapid neutralization, with a potential for reduced batch cycle time. A process for the preparation of an aqueous liquid detergent composition formulated with clay as a fabric softener is described in a patent issued to Conoco Inc., yielding a stable finished product with no undesirable increase in viscosity following clay incorporation [10].

In structured fluid detergent delivery systems, considerable effort is directed at maintaining or enhancing product shelf life and phase stability and the patent literature contains various methods intended to increase the physical stability of surfactant-based compositions. One technology presented in the patent literature imparts an internal physicochemical microstructure within the detergent system, for the retardation of phase separation. For these detergent systems, processing requirements are frequently vital to the formation of the required internal ordering of product components. In this notable example, the use of aqueous dispersions of a multilayered lamellar liquid crystal phase to stabilize structured liquid detergent systems has been proposed and several patents issued [11–15]. In the proposed examples, which are anionic/nonionic surfactant paired compositions, both rheopexy and thixotropy are found to occur. High shear is required during detergent manufacture to obtain the appropriate lamellar liquid crystal particle size distribution and it is suggested that a high-shear device in a recirculation loop can be applied, if the shear rate is greater than $1,000 \text{ sec}^{-1}$, preferably within the range of 4,000 to $15,000 \text{ sec}^{-1}$. As mentioned in these patents, mixing is the strategic engineering element for the successful production of this stable, internally structured liquid detergent. Several patents have also been assigned to the Colgate-Palmolive Company for a linear viscoelastic aqueous liquid automatic dishwasher detergent composition with exceptionally good physical stability [16,17] and a process for producing the linear viscoelastic detergent [18]. This patent discloses the dispersal of a crosslinked polyacrylic acid in water, neutralization and gelation with alkali metal hydroxide, addition of builder and silicate, and emulsification of fatty acid and detergent in an aqueous solvent. Further, air incorporation to the gel is disclosed to further promote dispersion stability.

The importance of the order of addition of detergent components throughout the mixing process as a critical process variable is demonstrated in a patent issued to Unilever, for the incorporation of perfumes in liquid detergents in laundry and personal care products via a structured emulsified liquid crystalline or vesicle delivery vehicle [19]. The selective premixing of various components is further cited as a prerequisite for the process of manufacturing structured lamellar concentrated heavy-duty liquid detergent compositions, in an additional patent issued to Lever Brothers [20]. By strict adherence to component order of addition and premix compositions, viscosity can be reduced and draining characteristics improved. Steady shear viscosity data at 21 $sec^{-1}$ is included in the patent, defining the criteria for stability.

Lever Brothers has also been assigned a patent for the process of producing an anticorrosion aqueous liquid detergent composition containing particulate alkali metal silicate. This detergent contains 5 to 25% of a soap and/or synthetic detergent and 5 to 40% of a detergent builder [21]. A process patent has also been issued for the making of a silicone-containing shampoo, detailing the thermal requirements of various adjuvants and the final mixing process [22].

Numerous patents exist for the manufacture of specific surfactants and other raw materials used in the formulation of liquid detergents. For example, a patent has been granted for the design of a reactor and process of saponification [23] claimed to be applicable to the preparation of various liquid detergent cleaning agents. In this patent, saponification is described for batch processing on a semicontinuous basis. Another example is a patent describing the efficient manufacture of an amphoteric surfactant for use in shampoos by reacting amino-containing compounds with acid halide alkali salt [24]. Further, a patent has been issued for the production of a fatty acid monoglyceride monosulfate salt surfactant describing the complete sulfation of glycerol, reaction with fatty acid, hydrolysis, and neutralization [25]. Advantages of this process include reduced concentration of sulfating agent and high active concentration with low inorganic sulfate concentration. Procter & Gamble Company also holds a patent for the production of alkyl ethoxycarboxylate surfactant compounds through the reaction of an ethoxylated fatty alcohol with a hindered base and anhydrous chloroacetic acid or its salt [26]. Use of this surfactant is found in cleaning compositions such as shampoos, laundry detergents, and liquid dishwashing products. An additional Procter & Gamble patent has been assigned for the co-sulfation of ethoxylated and unsaturated fatty alcohols producing acid sulfate compounds which, upon neutralization, form mixed surfactant systems for use in heavy- and light-duty liquid detergents, shampoos, and other cleaning compositions [27].

Liquid detergents contain many product components, including surfactants, salts, soluble and insoluble builders, polymers, viscosity modifiers, fragrances,

colorants, stabilizers, hydrotropes, and other ingredients, which are often interactive and capable of affecting product efficacy and synergistically influencing
rheological attributes. Throughout the process patent literature, the manufacture
of liquid detergents appears to require a regimented order of addition of ingredients, with the appropriate shear and thermal history, to obtain the appropriate
consistency, appearance, stability, and performance, and minimize product aging
following manufacture.

## IV.  CONTINUOUS VERSUS BATCH PROCESSES

Structured and unstructured liquid detergents can be processed in batches or continuously, depending on the specific production/volume requirements. Unstructured
liquid detergents, especially those lacking significant solid components, are well
suited for continuous processing. Examples of such detergents may include certain shampoos and light-duty liquids, including hand dishwashing detergents. With
the development of high-precision mass flow meters, proportional metering systems, and in-line multiple stage dispersers, both dynamic and static, continuous
processing is frequently the optimum process selection. If well designed, the process can be adequately controlled to ensure adherence to specifications, meeting
high-volume production demands with favorable manufacturing costs.

There are various minor components in most liquid detergents, such as colorants, pH adjusters, and fragrance, and metering of these low-concentration, yet
critical, components can be achieved with good accuracy in continuous operations.
For pH adjustment, which frequently controls product viscosity, adequate sensors
and product controls are required to ensure consistent product quality.

Continuous processes depend on various types of mixing devices, both static
and dynamic, to disperse and/or blend formulation components. Several commercial examples of in-line static mixers, for both turbulent and lamellar flow, are
shown in Figure 14.1, exposing the internal flow elements, and an example of a
commercial flow configuration containing three in-line static mixers is provided
in Figure 14.2. The primary advantage of such mixers, namely minimum space
requirements, is clearly seen. Continuous processes are not restricted to unstructured detergents, and applications do also exist for structured systems. A dynamic
in-line mixer, applicable to the blending of high-viscosity fluids, containing a helical ribbon impeller is shown in Figure 14.3. In this figure, the dynamic mixing
element has been removed from the in-line flanged assembly and positioned on
supports in a horizontal position to show the mixer impeller. In operation, the
mixer would be rotated 90° and installed in the pipeline.

Liquid–liquid dispersion in a continuous system or recycling in a batch system
can be achieved by flow through an orifice. When very high energies are needed,
high-pressure homogenizers are used. For lower energies, a very interesting orifice
is a check valve or several valves in series. The characteristic of the check valve

(a)

(b)

**FIG. 14.1**   (a) Turbulent flow configuration in-line static mixer. (Courtesy of LIGHT-
NIN, a unit of General Signal Corporation.) (b) In-line static mixer with mixing elements.
(Courtesy of Chemineer, Inc.)

is that it maintains a constant pressure drop independent of the liquid flow rate.
It can be easily shown that the energy applied per unit of liquid volume is equal
to the pressure drop. Therefore, if the pressure drop is constant, so is the energy
per unit of liquid volume, or mass. This energy is what determines the size of the
dispersed phase droplet, known as the Kolmogoroff theory [28].

**FIG. 14.2**   Commercial application of three in-line static mixers. (Courtesy of LIGHT-NIN, a unit of General Signal Corporation.)

In industrial practice, commercial liquid detergent manufacturing processes may occur on a semicontinuous basis, through a combination of batch and in-line static and/or dynamic mixers. This may be the result of special process requirements in the preparation of product intermediates, such as:

1.   Inorganic solids dispersal and hydration
2.   Polymer hydration/swelling
3.   Surfactant neutralization
4.   Thermal gelation of select components
5.   Liquefaction of a component(s)
6.   Emulsification
7.   Thermal/temporal equilibration
8.   Ion exchange

**FIG. 14.3** In-line helical ribbon blender. (Courtesy of LIGHTNIN, a unit of General Signal Corporation.)

Powder addition has to be done in an atmospheric tank. For a continuous system, powder is predominantly predispersed in an agitated tank and then the suspension is injected into the continuous system. Occasionally it is added into a continuously stirred agitated tank which is in-line in the process. Liquid flow in and out of the tank has to be rigorously controlled.

An example of a complete continuous process with a multiple head metering pump suitable for liquid detergent manufacture is provided in Figure 14.4a. This flexible and continuous manufacturing process can produce multiple product variants. Viscosity measurements, pH adjustment, level control, feedback process control to the metering pumps from the level controller at the buffer tank feeding the filling line, and an in-place water flushing cleaning system demonstrate the advantages of such production systems, in addition to the limited space requirements. These metering systems have found successful application in the production of fabric softeners, shampoos, dishwashing detergents, and other liquid detergent products [29]. Figure 14.4b shows a typical flow diagram for a continuous unit with rotary feed pumps and Coriolis mass flow meters.

## V. UNIT OPERATIONS IN LIQUID DETERGENT MANUFACTURE

The manufacture of liquid detergents involves many of the basic engineering unit operations common throughout the chemical process industries. Depending on the specific detergent formulation, each unit operation can contribute significantly to

(a)

**FIG. 14.4** (a) Complete continuous manufacturing process. (Courtesy of Bran+Luebbe, Inc.) (b) Typical continuous unit with mass flow meters.

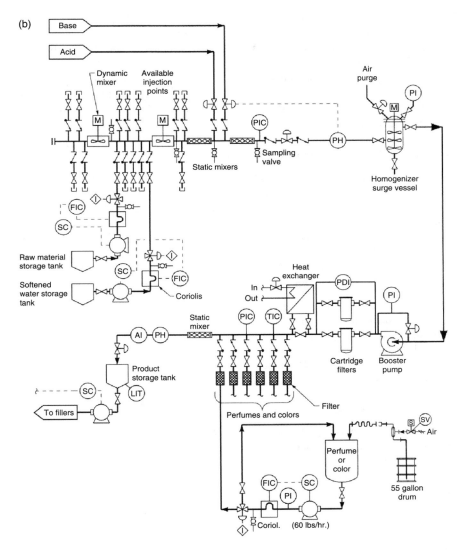

**FIG. 14.4**   (Contd.)

the physical, mechanical, and visual properties of the finished product. For this reason, quality control and manufacturing process controls are important, particularly for structured detergents. For structured liquid detergents containing multiple dispersed phases, factors such as particle size distribution, of both solid and immiscible liquid components, particle geometry, hydration kinetics and extent of hydration, interactions between components, kinetics governing each association under diffusion-controlled static conditions and shear environments, anisotropic surfactant phases, etc., can determine efficacy and many consumer-perceived product attributes.

Several critical unit operations are briefly reviewed with an emphasis on the process requirements of each liquid detergent classification, both structured and unstructured systems, emphasizing momentum, material, and energy transfer operations.

## A.  Transport Phenomena in Agitated Vessels

Mathematical simulation of heat, mass, or momentum transfer in agitated vessels is often untenable, due to the three-dimensional components of the material and energy balances, and the large number of material and process variables. In such cases, dimensional analyses are the preferred method of correlation. Numerous references are available for a review of dimensional analysis in engineering applications [30–32].

Dimensionless groups provide an excellent overview of the critical parameters influencing heat, mass, and momentum transfer, and several are defined below:

Reynolds number ($N_R$): $DV\rho/\mu$
Froude number ($N_{Fr}$): $V^2/gD$
Brinkman number ($N_{Br}$): $\mu V^2/k\Delta T$
Nusselt number (heat): $hD/k$
Nusselt number (mass): $k_x D/c\mathcal{D}_{AB}$
Prandtl number ($N_{Pr}$): $C_p\mu/k$
Peclet number ($N_{Pe}$): $N_R N_{Pr}$
Schmidt number ($N_{Sc}$): $\mu/\rho\mathcal{D}_{AB}$

Using such dimensionless groups, one can easily deduce the importance of fluid properties, namely, resistance to flow, on many of the production steps in the manufacture of a liquid detergent. For example, the power needed to provide agitation in a mixing vessel, known as the power number, $N_p$, can be expressed as [33]:

$$N_p = f(N_R, N_{Fr}, S_i)$$

where $S_i$ are factors relating to the design of the agitation system, e.g., agitation number, placement, and design. In mixing, fluid viscosity is clearly a significant material variable, influencing the power drawn during mixing.

Similarly, under nonisothermal conditions, as might be experienced in heat exchange by forced and free convection in an agitated vessel, the equations of change for the energy function can be expressed as:

$$DT^*/Dt^* = (N_R N_{Pr})^{-1}\Delta^{*2}T^* + N_{Br}(N_R N_{Pr})^{-1}\varphi_v^*$$

The time rate of temperature will be a function of the dimensionless groups which include resistance to flow, or viscosity, $\mu$, as one of the physical properties governing the heat exchange process.

In mass transfer, the primary variables to be considered include all physical properties, including density of relevant phases, viscosity, and diffusivity. Where liquid–solid mass transfer in agitated vessels is the interest, factors related to particle geometry, such as shape and size, need to be considered, as well as process design, including vessel geometry, agitator configurations, and speed [34]. Since fluid viscosity function is the distinguishing feature between structured and unstructured fluids, it is clear that rheology is a major factor in the processing of liquid detergents.

## 1. Momentum Transfer

Momentum transfer involves all unit operations where fluid motion occurs. The most common examples of this operation include pipeline transport, mixing, and the filling operation. In liquid detergent manufacture, mixing is undoubtedly the most important momentum transfer unit operation, occurring in-line or in agitated batch vessels [35]. While mixing appears to be a very simple and straightforward procedure, it can be extremely complex [36–40] and perhaps the most difficult of the unit operations used in liquid detergent manufacture.

Each mixer configuration imposes a strain distribution to the fluid being processed, which may influence the overall characteristics of the final product. As such, mixing can be the critical production step determining the physical and mechanical characteristics of the finished product. The physical stability of the product, immediate and long-term aging effects, efficacy, texture, appearance, and rheology are some of the important product characteristics that can be significantly altered by the total shear or strain history a product experiences during mixing. Scale-up from laboratory to pilot plant through to production volumes becomes a significant challenge, as it is difficult to reproduce exactly the fluid velocity profiles and residence times, or total strain, that a fluid experiences during its process history.

Equally important is the influence of fluid properties on the efficiency and power requirements of the mixing operation. This is especially true for structured liquid detergents with appreciable elasticity shear sensitivity. For these structured systems, there is a strong interdependence between rheology and mixing efficiencies. A basic understanding of the flow characteristics of the finished liquid detergent product and all intermediate streams or product components is a key to the selection and optimization of the mixing process.

Research devoted to the processing and mixing of complex non-Newtonian fluids has not been extensive. This is unfortunate since many commercial fluids, including structured liquid detergents, can be non-Newtonian fluids, with appreciable normal stresses, and guidelines or process design criteria for viscoelastic fluids are generally unavailable. Most dispersions at high solids content and gels fall within this category, creating significant challenges in the design of an efficient batch, continuous, or semicontinuous manufacturing process.

*(a)   Mixing of Structured Versus Unstructured Liquid Detergents.*   Several common radial and axial flow open impellers used in batch mixing of low- to medium-viscosity, unstructured or weakly structured liquid detergents are shown in Figure 14.5. In addition to the type of impeller, impeller diameter, vessel height and diameter, impeller locations, and baffles are design variables to be specified for a particular application. Placement of baffles to minimize vortexing and facilitate mixing, and the type and location of impellers will depend on the specific

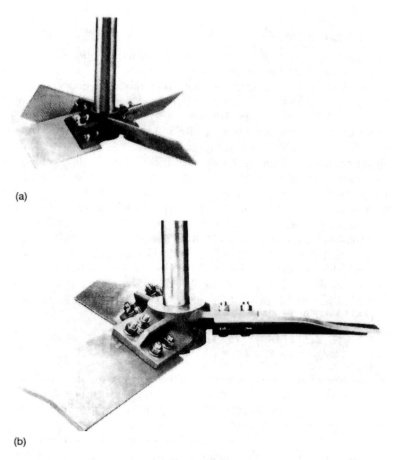

(a)

(b)

**FIG. 14.5**   (a) Pitch blade impeller, (b) high-efficiency impeller, (c) straight blade turbine impeller with stabilizer, (d) welded disc impeller. (Courtesy of Chemineer, Inc.)

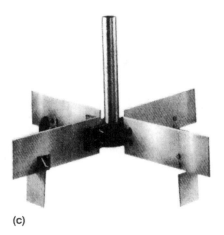

(c)

(d)

**FIG. 14.5** (Contd.)

mixing needs. Due to the size and scale of most industrial mixing vessels, multiple impellers are generally needed to obtain an adequate degree of mixing. A simple schematic of a batch mixer is provided in Figure 14.6, showing all relevant engineering dimensions.

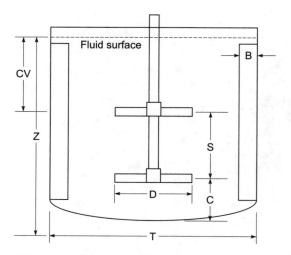

**FIG. 14.6**    Schematic of batch mixing vessel with baffles and dual impellers with indicated engineering dimensions.

For high-viscosity structured and unstructured surfactant systems, one example of an alternative mixer design, consisting of three top-entering coaxial agitators, surface-sweeping anchor, counter-rotational blades, and rotor/stator homogenizer, is provided in Figure 14.7. Details of this turboemulsifier are shown in Figure 14.8, showing the presence of a powder inductor through the vessel bottom near the high-speed homogenizer turbine, variable-speed drive, jacketing for heating and cooling, and provisions for vacuum or pressurized operation. For high-viscosity systems, common marine propellers and turbines are generally unsuitable, providing limited bulk flow to the process fluid under normal operating conditions. The efficiency of the mixing operation and the effectiveness of the flow field that is generated in the bulk of the fluid will have a significant impact on blending times and the kinetics of these operations that are governed by effective mass, heat, and momentum transfer. Figure 14.9 broadly summarizes the viscosity limitations of most commonly used industrial agitators [41,42].

Large, jacketed 316 stainless steel construction mixing vessels with variable drive agitators are expensive. Due to the high capital cost, space requirements, and high operating costs associated with such vessels in batch operations, they are typically required to be multifunctional and capable of performing many of the manufacturing elements of a liquid detergent. At large production batch volumes of 10 to 20 metric tons or larger, this puts a great demand on the impeller/drive selections and placement of these impellers and baffles within the mixing vessel.

For liquids with viscosities less than 200 cP, jet mixing is a very economical option [43].

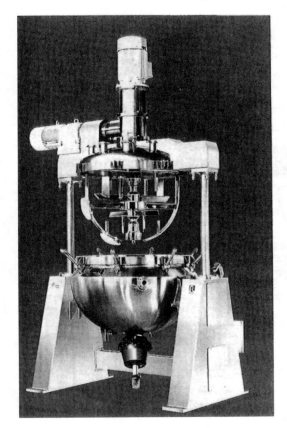

**FIG. 14.7**   Ross turboemulsifier. (Photo courtesy of Charles Ross and Son Company, Hauppauge, New York.)

Most industrial batch operations do not utilize single vessels for the manufacture of a liquid detergent; multiple vessels of various sizes and specifications are used to perform discrete functions. Vessels with and without temperature control may be selected to prepare polymeric solutions, or surfactant premixes, or dissolve various solid components uniformly prior to the final blending operation. Further, intermediate product components may be prepared in sufficient bulk to support multiple production batches. This is generally desirable when production of intermediates is time consuming, excessively increasing individual batch cycle times. In continuous operations, this approach may also be applied to various product intermediate streams metered through a final multistage in-line disperser. This may be achieved with static in-line mixers or dynamic multiple-stage dispersers, depending on the specific product requirements. Use of multiple supporting vessels, frequently of

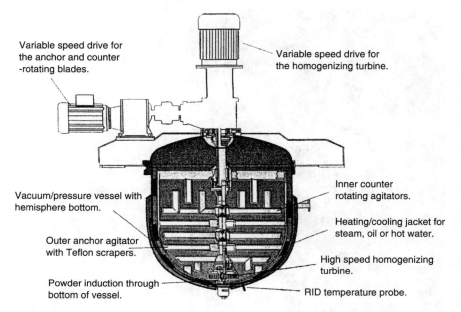

**FIG. 14.8** Detailed schematic of turboemulsifier. (Photo courtesy of Charles Ross and Son Company, Haupopage, New York.)

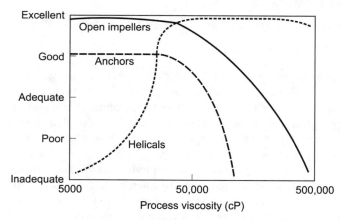

**FIG. 14.9** Viscosity limitations of various impeller configurations. (From Klinkenberg, A. and Mooy, H.H., *Chem. Eng. Proc.*, 44, 17, 1948.)

smaller size, can reduce cost by application of special requirements such as heat transfer or high shear only where needed.

In batch mixers, impellers impose flow, or momentum transfer, to the fluid mass contained within the vessel. The power consumption during mixing for a given fluid viscosity and density is proportional to the rotational speed of the impeller and the impeller diameter and flow is achieved by the momentum transferred to the fluid by the motion of the impellers. Placement of the agitator will be determined by the mixing requirements when the vessel is fully charged and also by requirements during batch filling and discharge. Agitator placement is coordinated with ingredient order of addition to limit, for example, excessive foaming in the presence of surfactants. The order of addition of formulation components can also be balanced to limit undesirable rheological properties and promote the formation of the desired microstructural state.

For unstructured liquid detergents, standard methods can generally be used to size and specify mixing equipment. Pseudoplastic behavior with low order of magnitude normal stresses would most likely not present serious mixing problems. Structured detergents, however, and intermediates, may require special consideration. Significant deviation from Newtonian behavior cannot be ignored in the specification of a production agitation system, as this can produce significant errors in the estimation of power requirements for a particular application. This is especially true for fluids capable of developing solid-like mechanical properties in the quiescent state.

The influence of elasticity on the mixing unit operation is well illustrated by Prud'homme and Shaqfeh [44]. A correlation of dimensionless mixing torque vs. Reynolds number is provided for three Newtonian fluids, two of which exhibit significant elasticity, as determined by the magnitude of the primary normal stress differences. For Rushton turbines, the results indicate a fourfold increase in torque during mixing, for the fluids exhibiting high normal stress differences, indicating the fluid rheology must be considered in the assessment of torque and power requirements for various agitation systems.

Power consumption is an important mixing design parameter, dependent upon impeller diameter ($D$), rotational speed ($N$), and fluid properties including viscosity ($\mu a$) and density ($\rho$), and power consumption of impellers is usually provided as correlations of power number, $N_p$, to Reynolds number, $N_r$. For fluids exhibiting time-independent power law viscosity functions, $\tau = K\gamma^n$, the generalized Reynolds number in agitation can be expressed as:

$$N_r = [\{D^2 N \rho (8N)^{1-n}\}/K]\{4n/(3n+1)\}^n \tag{1}$$

Power characteristics for the mixing of non-Newtonian fluids have been determined for various impellers and other critical mixer design variables, using pseudoplastic, dilatant, and Bingham slurries, and polymeric solutions frequently encountered in the manufacture of liquid detergents, such as clay dispersions and cellulosic and carbomer solutions [45,46]. This research has also provided correlations of the mean fluid shear rate and impeller speeds, for various impeller geometries and fluid viscosity functions. Typical agitators used in these investigations include anchors, paddles, fan paddles, and turbines in agitated vessels with and without baffling. Results clearly indicate that power requirements for mixing of non-Newtonian fluids can be much greater than for Newtonian systems.

Power consumption and blend times in the mixing and agitation of Newtonian and non-Newtonian fluids are not equivalent and, in fact, blend times can be much longer for non-Newtonian fluids when comparing fluids with comparable apparent viscosity values. Through dimensional analysis, the dimensionless blend or mix time, $\theta_m$, is expressed as:

$$\theta_m = f(N_R, N_{Fr}, S_i) \tag{2}$$

Su and Holland report power input per unit volume and mixing time are substantially higher for pseudoplastic fluids than for their Newtonian counterparts [47]. Godleski and Smith [48] report blend times nearly 10 to 50 times longer for non-Newtonian aqueous dispersions of hydroxyethyl cellulose when compared to equivalent viscosity Newtonian fluids, using flat-blade turbine agitators. Blending times for the pseudoplastic, time-independent cellulosic fluid are also noted to increase even further in baffled vessels. This study suggests a strong dependence of mixing efficiencies on fluid rheology.

There have been contradictory results reported in the literature, however, regarding the influence of fluid elasticity on the mixing unit operation [49–52]. Further research is apparently required to define adequately the influence of viscoelasticity on mixing in agitated vessels, for a broad range of fluid properties and mixer configurations.

Carreau et al. [53] have investigated the behavior of Newtonian, inelastic, and elastic non-Newtonian high-viscosity paste-like fluids in helical ribbon agitators, showing that the efficiency of mixing both psuedoplastic and viscoelastic fluids is lower than for Newtonian fluids, decreasing significantly with increasing fluid elasticity. Ranking the three fluid systems, the efficiency rating is such that viscoelastic < pseudoplastic < Newtonian.

As with open impellers, elastic fluids are apparently more difficult to process in helical ribbon mixers. To quantify this effect, the mixing efficiency of a highly elastic Separan solution is only 20 to 40% that of glycerol, which is Newtonian.

The effect of fluid rheology on the power consumption of helical ribbon agitators has also been evaluated [54] and power consumption as a function of generalized Reynolds number for shear thinning but inelastic fluids defined. When shear thinning effects are small, and elasticity is negligible, deviations from the Newtonian power curve are slight. At low Reynolds numbers, this is also true for viscoelastic fluids. At higher fluid velocities, fluid elasticity begins to dominate the power curve. Further, it appears that shear thinning delays the effect of elasticity on mixing efficiency for structured fluids. The transient and extensional nature of the flow field is stressed as a key factor in the increased energy required to obtain a required degree of mixing. Fluids that exhibit both shear thinning and elasticity show deviations from the Newtonian power curves at higher Reynolds numbers than viscoelastic fluids with minimal shear effects.

Power needed to maintain mixing may not be the same as the power needed at the inception of flow when processing structured liquid detergents. An interruption in the process sequence can introduce a transient power requirement quite different from the steady-state design criteria. The rheology of structured liquid detergents systems is quite complex and can introduce many variables not applicable to unstructured systems. When yield stresses are significant, and delays are expected at any part of the mixing process, slip drive couplings to the motor drive may be required. This may prevent damage to agitator motors and shaft assemblies when agitation is restarted. For pumping, mixing, and any fluid transport operation, fluids with yield stresses require shear stresses in excess of the yield stress to initiate and maintain flow. Depending on the magnitude of the yield stress, this can be problematic.

In a batch mixer, a sufficient shear stress exceeding the yield stress may occur only near the impeller, producing flow in the immediate vicinity of the impeller, but with stagnation zones throughout the remaining fluid bulk. This has an overwhelming impact on the efficiency of the mixing operation and can be further complicated in the presence of baffles. One solution is to utilize large surface sweeping agitators such as gate, anchor, or pattern mixers to minimize regions experiencing stagnation. Discharging vessels containing these types of fluids can also be difficult, resulting in residual material on the vessel surfaces which cannot be fully evacuated. For example, in discharging a vessel under gravity, a film thickness, $\delta(z, t)$, remaining on the vessel wall can be calculated as a function of draining time ($t$) and distance from the initial fluid height ($z$), as shown in Figure 14.10. For Newtonian fluids, from the unsteady state mass balance, the film thickness can be expressed as [55,56]:

$$\delta(z, t) = \{(\mu/\rho g)(z/t)\}^{1/2}$$

where $\mu$ = viscosity, $\rho$ = density, and $g$ = gravitational constant. As we would expect, film thickness is directly proportional to fluid viscosity, and inversely related to density.

For a comprehensive overview of the mixing unit operation, for both structured and unstructured fluids, various references are available describing the specific requirements for the design and specification of complete mixing systems [57–79].

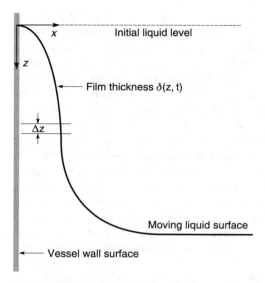

**FIG. 14.10**   Film thickness on vessel wall during drainage.

*(b)  Dispersive Mixing.*   There are liquid detergents that require specific particle size distributions of certain components to maximize efficacy and substantivity, for both solid–liquid and liquid–liquid dispersions. Examples of such products include hair conditioners, shampoos, conditioning shampoos, body cleansing bath gels, and fabric softeners. In conditioning shampoos containing silicone oils, substantivity and effective deposition on hair is a function of particle size, ionic charge of the particles, and silicone oil viscosity [80]. Formulation and processing of these systems can be extremely demanding.

Emulsions are difficult to process batchwise if a strict control on particle size distribution is required. Processing may also be hindered by the complex rheological properties these emulsions can exhibit. Very strict mixing controls are therefore not unusual to ensure that product during manufacture, at the filling line, and on the market shelf is within specification.

Fundamental mixing studies on simple two-component systems have provided insight into the effect of mixing parameters on critical emulsion properties such as particle size distribution. For example, Nagata [81] has shown the distribution of sizes of the dispersed liquid phase as a function of agitator speeds. As we might expect, a normal distribution occurs at higher speeds. In a similar study, the effect of surface tension was determined for several liquid dispersed phases from benzene to paraffin oil [82].

Due to the very broad distribution of shear rates that fluids experience in batch mixers, control of particle size distributions may not be possible. There are, however, alternative agitation systems that can be used in tandem to achieve a more controlled distribution of emulsion droplets or particulate solids. These include colloid mills, in-line dynamic dispersers in recirculating lines, and other high shear flow-through devices. These devices can be very successful in tailoring emulsion and dispersion characteristics. An example of a colloid mill with optional rotor/stator options is provided in Figure 14.11, capable of producing stable emulsions to the submicrometer range. This rotor/stator design provides four-stage shearing action for effective dispersion and de-agglomeration. Figure 14.12 details a two-stage tandem shear pipeline in-line mixer with two turbines and mating stators on a single shaft to provide greater high shear dispersal. Use of external in-line mixers positioned in batch mixer recirculation loops is an effective process method for achieving a high degree of dispersive mixing [83].

*(c) Pumping of Newtonian and Non-Newtonian Fluids.* For liquid detergent products known to be shear sensitive or containing particulates, pump selection is an important process variable. Whether driven by centrifugal force, volumetric displacement, mechanical impulse, or electromagnetic force, an understanding of fluid exposure to high shear in close clearances is required. An internal schematic of a rotary gear pump is shown in Figure 14.13, showing the close clearances between impeller surfaces and pump casing, with two alternative rotary screw pumps illustrated in Figure 14.14 and Figure 14.15.

## 2. Aeration Avoidance

Aeration and foaming are serious problems in liquid detergent manufacturing. Bottles cannot be filled with aerated liquid to a specified volume or mass. Most often aeration and foaming are produced by the process, and not by air contained in raw materials or water.

In agitated vessels, air is incorporated during improper liquid feed, splashing, through the vortex, or during powder addition. Agitation further disperses the air, making it more difficult to separate, producing foam as the air rises to the liquid–air interface. Avoiding incorporation of air rather than separation once it is incorporated should be pursued. To avoid air entrapment while feeding liquid, several process configurations are possible, including:

1. Deep tube feeding.
2. Bottom tank feeding.
3. Feed entering from the top of the tank, discharging against the tank wall tangentially and downwards in such a manner that the liquid spirals along the tank wall.

In the first two cases, no air slug should be in the pipe, lest it be dispersed into bubbles by the agitation. If air slugs are present, there should be no agitation at

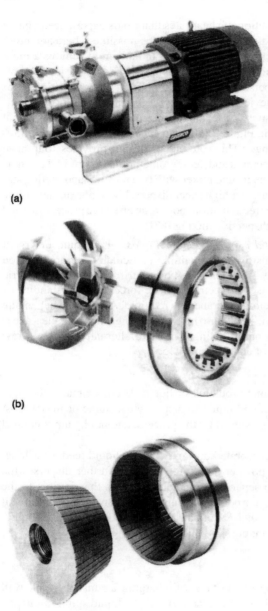

(a)

(b)

(c)

**FIG. 14.11**   (a) Greerco colloid mill. (b) Standard colloid mill rotor/stator combination with plain, smooth milling surfaces for most emulsions and dispersions. (c) Specialty rotor/stator that has grooved milling surfaces for viscous emulsions. (Courtesy of Greerco Corp., Hudson, NH.)

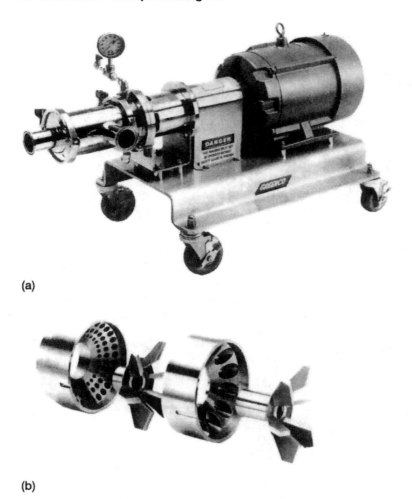

(a)

(b)

**FIG. 14.12**    (a) Greerco 4″ two-stage sanitary tandem shear pipeline mixer. (b) Exploded view of tandem shear turbine/stator assembly. (Courtesy of Greerco Corp., Hudson, NH.)

the time of addition since the slug will become a large bubble that rises quickly to the surface.

If aeration has occurred, the common way to de-aerate is to stop agitation and wait until the air rises and separates. More efficient than this is to design the tanks for de-aeration. This is done by laminar agitation at a Reynolds number of 45 using an eccentric shaft at one third the tank diameter and one half the turbine diameter from the bottom. This accelerates de-aeration while still mixing [84].

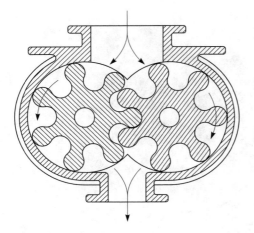

**FIG. 14.13**   Schematic of gear pump. (From Perry, R.H., Chilton, C.H., and Kirkpatrick, S.D., *Chemical Engineers' Handbook*, 4th ed., McGraw-Hill, New York, 1963.)

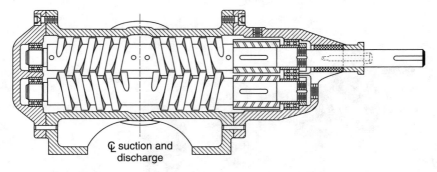

**℄ suction and discharge**

**FIG. 14.14**   Two-rotor screw pump. (From Perry, R.H., Chilton, C.H., and Kirkpatrick, S.D., *Chemical Engineers' Handbook*, 4th ed., McGraw-Hill, New York, 1963.)

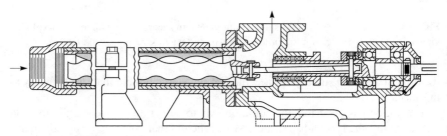

**FIG. 14.15**   Single-rotor screw pump with elastomeric lining. (From Perry, R.H., Chilton, C.H., and Kirkpatrick, S.D., *Chemical Engineers' Handbook*, 4th ed., McGraw-Hill, New York, 1963.)

Producing low pressures in the tank also accelerates de-aeration but the foam formed on the interface may be quite stable. Usually large tanks used for mixing liquids are not built to withstand vacuum. There are commercially available continuous de-aerators based on the formation of a film on a vessel wall and subjecting it to vacuum and/or centrifugal force. They tend to fail when the foam produced is well stabilized.

High-density foam floating on the interface can be destroyed mechanically or with hot air, or by spraying water (ethanol or other solvent, if possible). High-density foams are more difficult to break and frequently impossible. Again, de-aeration is much more difficult than aeration avoidance.

Continuous systems are mostly pressurized and normally no aeration occurs. A well-designed system will not incorporate air in the suction of the centrifugal pumps and will have no accidental venturi effect.

## 3. Heat Transfer

Forced-convection heat transfer is a common unit operation in the production of liquid detergents. Whether experienced in jacketed process vessels, agitated vessels with immersion coils, or other forms of heat exchange, there are multiple causes for thermal regulation during detergent manufacture. Temperature may be controlled to increase the dissolution rates of various components, facilitate mixing, accelerate hydration, moderate phase behavior of the product intermediates, regulate viscosity, reduce yield stresses, etc.

Many liquid detergent products contain components that serve as product viscosity modifiers, added to achieve the desired consistency of the commercial product. Cellulosic polymers, for instance, are an excellent example of such an additive and various polysaccharides are capable of gelation under specific thermal conditions. In such cases, heat transfer during manufacture may be required to complete hydration and effect the necessary conformational change in the select polymer system [85], in the appropriate aqueous environment. Products requiring controlled heat transfer processes may include various dental creams, shampoos, built liquid detergents, and hard surface cleaners.

Heat transfer may also be required to maintain isothermal or adiabatic conditions in the presence of endothermic and/or exothermic reactions, as the result of mixing product components, surfactant neutralization, and other chemical reactions. In these cases, heat transfer requirements may be severe to minimize exposure of the bulk fluid to high temperatures for extended time periods, resulting in irreversible thermal degradation.

## 4. Mass Transfer

Liquid–solid and liquid–liquid mass transfer is highly dependent upon surface area, or particle size. Mass transfer is involved in simple wetting, dissolution, hydration, swelling of product components, ion exchange, electric double layer formation, dispersion stabilization through adsorption or absorption, and surfactant phase equilibration, among others. Both momentum and heat transfer are frequently concurrent in the effectiveness and efficiency of most mass transfer operations.

*(a)   Solids Hydration: Builders and Polymers.*   Solid–liquid suspensions are frequently encountered in the production of liquid detergents. For example, the chelating agent sodium tripolyphosphate, in anhydrous form, is a common builder in both laundry and automatic dishwasher detergents, forming a hexahydrate when exposed to an aqueous environment. It is known that material and process variables can influence phosphate hydration kinetics [86–89]. The shear exerted on the slurry during hydration, rate of phosphate addition, order of addition of various components, electrolytic solution environment, temperature, tripolyphosphate characteristics including Phase I/Phase II crystalline form, particle size distribution, and pH, for example, can influence the rate and extent of sodium tripolyphosphate dissolution. Formation of the hexahydrate may result in an increase in consistency of the phosphate slurry, limiting the solids concentration during processing. The rheology of phosphate liquid detergents is critically dependent upon the characteristics of the anhydrous phosphate, Phase I/II ratio, which influence degree of hydration. Modification of the hydrating characteristics of Form II phosphate to minimize undesirable processing effects is the subject of a patent issued to Lever Brothers [90].

The agitation rate and solids suspension should be sufficient to maximize available surface area, especially where mass transfer is occurring. The dependence of the mass transfer coefficients on relative power is shown in Figure 14.16 [91]. The mass transfer coefficient is much higher when complete off-bottom solids

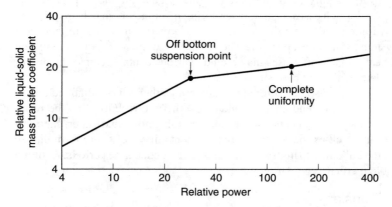

**FIG. 14.16**   Dependence of mass transfer coefficients on solids suspension. (Adapted from Oldshue, J.Y., *Fluid Mixing Technology*, McGraw-Hill, New York, 1983, p. 234. With permission.)

suspension is achieved. Even in cases where suspension of solids is the process objective, as in phosphate hydration, for example, it is necessary to determine if complete solids suspension and uniformity throughout the continuous phase is required or if off-bottom suspension with a solids gradient throughout the vessel is adequate. A general review of mass transfer in mechanically agitated vessels involving particulate suspensions is provided by Upadhyay and Kumar [92].

Phosphate hydration in the primary detergent mixing vessel represents a good example of the challenges facing a process development or manufacturing engineer in the specification of a multifunctional mixing vessel. If we assume that the phosphate hydration or other solids dispersal will occur early in the manufacturing process, the immediate requirements of the mixing vessel is that an agitator is adequately positioned to keep all solids suspended during the hydration process. If this occurs as a highly concentrated dispersion with a minimum of solvent, solids suspension may be difficult to achieve and the extent of hydration limited, placing constraints on the impeller selection and location within the mixing vessel.

A major difficulty associated with solids dispersion and hydration is settling in the event of a process interruption. Depending on the duration of the interruption, redispersal of solids may be difficult to achieve. An excellent example of this is provided in Figure 14.17, showing the agitation times required to redisperse solids after settling [93]. Depending on the nature of the solid being processing, particle

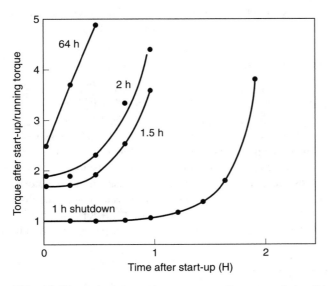

**FIG. 14.17** Agitation requirements to redisperse settled solids. (From Oldshue, J.Y., *Fluid Mixing Technology*, McGraw-Hill, New York, 1983, p. 234.)

size and shape, degree of compaction, and impeller location, redispersal may not be possible. Adequate characterization of the slurry during process development is needed to anticipate such difficulties. Depending on the impeller location relative to the compacted solids region, precautions may be necessary to protect the agitator motor drive. In certain instances, compaction may occur when agitator motion is reintroduced due to the forces exerted on the settled solids by the surrounding flow field.

Polymer processing during the manufacture of a liquid detergent represents another difficult processing step since certain polymers may undergo rapid hydration when introduced to an aqueous solvent. This complicates the mixing operation and may require a predispersion of the polymer with a second inert powdered ingredient, predispersion of the polymer solid in a nonaqueous solvent, or high-shear dispersion. This can also be achieved in a dynamic in-line mixer through a transfer or recirculating line attached to a mixing vessel, as previously discussed, or a powder inductor, as shown in Figure 14.8. High-shear dispersion reactors, or mechanical dry powder dispersers, are known to reduce favorably blending times, while increasing the concentration of polymer that can be dispersed, even for certain hydrophilic carbomer polymers [94].

When mixing hydrating species, the order of addition of components can again be significant. Polymer hydration, for example, can be significantly hindered in the presence of specific salts. In general, with liquid detergents of limited shelf life, the order of addition of ingredients can be critical in finished product attributes. This is especially true for structured high solids dispersions. The order of addition can influence phase stability, rheology, and many other product properties.

## B.  Microbial Contamination

As more restrictions on product preservatives have been set in the last 10 years, more instances of microbial contamination have appeared and liquid detergent process equipment and operations have approached those used in the food and pharmaceutical industries. Process equipment is being installed to a more sanitary level, which means easier to clean and disinfect. Predominantly the equipment is designed for "cleaning in place" (CIP) without the need to disassemble. This chiefly means that surfaces are polished, circulation dead spaces are avoided, and drainage is virtually perfect. Usually the equipment is washed with alkaline and acid solutions, and then with a disinfectant solution. Additional equipment to handle and recirculate disinfectant solutions becomes part of the system design.

The predominant material of construction is 316 and 304 stainless steel or variations of both, but plastics such as polyolefins and fluorinated hydrocarbons are also used. Extensive literature in this area is available even if directed more toward foods and pharmaceuticals processing [95–97].

## C.  Product Shelf Life

Liquid detergents can be dynamic systems in a metastable thermodynamic state during and following manufacture. Depending on the complexity of the formulation and concentration of surfactants, polymers, and other additives, changes in consistency, texture, appearance, color, etc., can be experienced following manufacture and this is not unusual for many detergent systems. Controlling this effect, however, is necessary to ensure consistent product throughout the shelf life and represents the underlying necessity of effective heat, mass, and momentum transfer during product manufacture.

For example, a product may exhibit a Brookfield viscosity of several thousand centipoises at the filling line, yet viscosity may continue to increase or decrease following production, for a period of time. Most undesirable is progressive thinning or thickening for extended time periods. Depending on the mechanism responsible for the thickening or thinning behavior, this may be accelerated through effective heat and mass transfer at the relevant manufacturing step. Possible causes of such phenomena include continuing hydration of polymeric and solid species, occlusion of solvent in porous solids, surfactant phase equilibration, etc. Heat transfer may well be used to increase the rate of each of these effects during manufacture, thereby moderating any changes in consistency following production, driving the product to a more stable pseudoequilibrium or steady state.

Heat transfer can occur in either batch or continuous configurations. Both types of processes require fluid motion to obtain an effective heat transfer to the bulk of the fluid. In batch processing using jacketed vessels, helical coils, or coils in a baffle configuration, for example, sufficient agitation is required for heat transfer through the medium while continuous systems rely on flow rate to achieve effective heat transfer to satisfy process requirements. Effective heat transfer in batch operations for structured liquid detergents may require scrapers or anchor-type impellers to increase heat transfer coefficients in jacketed vessels.

Effective mass transfer is as important since product stability can be seriously compromised in colloids and suspensions, both liquid–solid and liquid–liquid phases, if the morphology of the interface is not properly formed, and interactions sufficiently developed. Phase separation is the major consideration in such complex systems, and is easily affected by poor process history.

## VI.  SUMMARY

Liquid detergents are seldom in equilibrium during processing or throughout their shelf life. Few of the reactions are driven to completion during manufacture; they continue throughout product shelf life. Ion exchange, crystallization, phase equilibration, adsorption, absorption, diffusion, etc., may continue to occur from the point of manufacture to the point of use. If these proceed without significant change in physical properties, there may be little reason for concern, unless efficacy is impaired. Unfortunately, however, these can result in viscosity increases with product age, induce physical phase separation, shift particle size distributions, alter color and fragrance, etc., leading to product changes that are consumer perceptible. With a rigorous statistical experimental design during process development, many of these characteristics can be understood and successfully controlled.

Both structured and unstructured liquid detergents have process requirements and limitations. The order of addition of ingredients and the shear history experienced during processing can determine the physical state of the detergent and ultimate stability. Any additional unit operations such as pumping, pipeline transfer, and filling need to be defined in a manner that does not irreversibly alter the structured phase. The objective throughout manufacturing is to deliver a consistent product to the consumer with minimal variability encountered during production.

Specification of raw materials used in the manufacture of liquid detergents is critical to controlling process effects. Although some detergents are relatively insensitive to broad fluctuations in raw material characteristics, others are extremely sensitive to minor variability. Apparent in the patent literature, surfactant chemistry can be a critical component. For example, when manufacturing with a surfactant/solvent composition near a phase boundary, a minor change in electrolyte concentration, surfactant composition, and concentration can significantly alter product characteristics.

Mixing is an important unit operation in the production of a liquid detergent. Effective mixing of liquid detergents requires a basic understanding of the rheology of the system being manufactured. A preliminary investigation of the fluid properties such as viscosity behavior, normal stress differences, time dependence or shear effects, yield stresses, and structural kinetics including deformation and recovery are relevant and necessary in anticipation of specific agitation requirements. As we have seen, the elastic effects are particularly important to identify and monitor during processing of liquid detergents, depending on the physical form of the product, and can be a significant engineering challenge.

## ACKNOWLEDGMENTS

The author wishes to thank Mr. Roberto Chorny for his contributions to this chapter.

## REFERENCES

1. van de Pas, J.C., *Tenside Surf. Det.*, 28, 158, 1991.
2. Japanese Patent 60,076,598 to the Lion Corp., 1985.
3. Yamamoto, H., Kinosita, M., and Konisi, S., Japanese Patent 61,268,797 to the Lion Corp., 1986.
4. Auweter, H., Nuber, A., Tschang, C.J., and Sanner, A., DE 3,818,868 to BASF AG, 1989.

5. Fong, J., Noukarikia, V., and Jungermann, E., WO 8,603,679 to the Neutrogena Corporation 1986.
6. Kukzel, M. and Leikhim, J.W., U.S. Patent 4,405,483 to the Procter & Gamble Co., 1982.
7. Grand, P.A. and Ramachandran, P.N., U.S. Patent 4,619,774 to the Colgate-Palmolive Company, 1984.
8. Grand, P.A. and Ramachandran, P.N., U.S. Patent 4,469,605 to the Colgate-Palmolive Company, 1982.
9. Mazzola, L.R., U.S. Patent 4,800,037 to Lever Brothers Co., 1987.
10. Green, R.J. and van de Pas, J.C., U.S. Patent 5,002,688 to Conoco Inc., 1990.
11. Dawson, P.L., Upton, C., and van de Pas, J.C., WO 91/09108 to Unilever PLC, 1991.
12. Hales, S.G., Khoshdel, E., Montague, P.G., van de Pas, J.C., and Visser, A., WO 91/09109 to Unilever PLC, 1991.
13. Buytenhek, C.J., Mohammadi, M.S., van de Pas, J.C., and De Vries, C.L., WO 91/09107 to Unilever PLC, 1991.
14. Van de Pas, J.C., *Tenside Surf. Det.*, 28, 3, 1991.
15. Jurgens, A., *Tenside Surf. Det.*, 26, 3, 1989.
16. Delsignore, A., Dixit, N.S., Rounds, R., and Makarand, S., U.S. Patent 5,252,241 to the Colgate-Palmolive Company, 1991.
17. Dixit, N.S., Farooq, A., Rounds, R.S., and Makarand, S., U.S. Patent 5,232,621 to the Colgate-Palmolive Company, 1993.
18. Delsignore, A., Dixit, N., Rounds, R., and Makarand, S., U.S. Patent 5,202,046 to the Colgate-Palmolive Company, 1993.
19. Behan, J.M., Nejss, J.N., and Perring, K.D., U.S. Patent 5,288,423 to Unilever, 1994.
20. Potocki, J., U.S. Patent 5,397,493 to Lever Brothers Co., 1995.
21. Boskamp, J.V., U.S. Patent 4,539,133 to Lever Brothers Co., 1984.
22. Fieler, G.M. and Stacy, L.V., U.S. Patent 4,728,457 to the Procter & Gamble Co., 1986.
23. Bereiter, B.A., U.S. Patent 4,671,892 to the Henkel Corp., 1986.
24. Matsumoto, Y.S., Japanese Patent 63,012,333, 1988.
25. Ahmed, F.U., U.S. Patent 4,832,876 to the Colgate-Palmolive Company, 1989.
26. Cripe, T.A., European Patent 399,751 to the Procter & Gamble Co., 1990.
27. Matthews, R.S. and Ward, J.F., WO 9,113,057 to the Procter & Gamble Co., 1991.
28. Friedlander, T., *Turbulence Classical Papers on Statistical Theory*, London, 1961.
29. Gray, J., *Soap Cosmet. Chem. Specialties*, April 1986.
30. Bird, R.B., Stewart, W.E., and Lightfoot, E.N., *Transport Phenomena*, Wiley, New York, 1960.
31. Massey, B.S., *Units, Dimensional Analysis and Physical Similarity*, Van Nostrand Reinhold, New York, 1971.
32. Klinkenberg, A. and Mooy, H.H., *Chem. Eng. Proc.*, 44, 17, 1948.
33. Coulson, A.M. and Richardson, J.F., *Chemical Engineering*, 4th ed., Vol. 1, Pergamon Press, New York, 1990.
34. Treybal, R.E., *Mass Transfer Operations*, 3rd ed., McGraw-Hill, New York, 1980.
35. Fallbe, J., Ed., *Surfactants in Consumer Products: Theory, Technology and Application*, Springer-Verlag, New York, 1986.
36. Nagata, S., *Mixing: Principles and Applications*, Wiley, New York, 1975.

37. Oldshue, J.Y., *Fluid Mixing Technology*, McGraw-Hill, New York, 1983.
38. Uhl, V.W. and Gray, J.B., *Mixing: Theory and Practice*, Academic Press, New York, 1966.
39. Skelland, A.J.P., *Non-Newtonian Flow and Heat Transfer*, Wiley, New York, 1961.
40. Bourne, H.J.R. and Butler, H., *Mixing-Theory Related to Practice, 89*, AIChE–IChemE Symposium Series No. 10, 1965.
41. Oldshue, J.Y., *Fluid Mixing Technology*, McGraw-Hill, New York, 1983, p. 325.
42. Harnby, N., Edwards, M.F., and Nienow, A.W., *Mixing in the Process Industries*, Butterworths, London, 1985, chaps. 7, 8.
43. Harnby, N., Edwards, M.F., and Nienow, A.W., *Mixing in the Process Industries*, Butterworths, London, 1985, chap. 9.
44. Prud'homme, R.K. and Shaqfeh, E., *AIChE J.* 30, 485, 1984.
45. Metzner, A.B. and Otto, R.E., *AIChE J.*, 3, 3, 1957.
46. Metzner, A.B., Feehs, R.H., Ramos, H.L., Otto, R.E., and Tuthill, J.D., *AIChE J.*, 7, 3, 1961.
47. Su, Y.S. and Holland, F.A., *Chem. Proc. Eng.*, September 1968.
48. Godleski, E.S. and Smith, J.C., *AIChE J.*, 8, 617, 1962.
49. Chavan, V.V., Arumugan, M., and Ulbrecht, J., *AIChE J.*, 21, 613, 1975.
50. Hall, K.R. and Godfrey, J.C., *Trans. Inst. Chem. Eng.*, 46, 205, 1968.
51. Skelland, A.H.P., *Handbook of Fluids in Motion*, Cheremisinoff, N.P. and Gulpta, R., Eds., Ann Arbor Science, Ann Arbor, MI, 1983.
52. Chavan, V.V. and Mashelkar, R.A., *Adv. Transport Proc.*, 1, 210, 1980.
53. Yap, C.Y., Patterson, W.I., and Carreau, P.J., *AIChE J.*, 25, 516, 1979.
54. Carreau, P.J., Chhabra, R.P., and Cheng, J., *AIChE J.*, 39, 1421, 1993.
55. Bird, R.B., Stewart, W.E., and Lightfoot, E.N., *Transport Phenomena*, Wiley, New York, 1960.
56. van Rossum, J.J., *Appl. Sci. Res.*, A7, 121, 1958.
57. Calderbank, P.H. and Moo-Young, M.B., *Trans. Inst. Chem. Eng.*, 39, 22, 1961.
58. Patterson, W.I., Carreau, P.J., and Yap, C.Y., *AIChE J.*, 25, 508, 1979.
59. Meyer, W.S. and Kime, D.L., *Chem. Eng.*, Sept., 27, 1976.
60. Gates, L.E., Morton, J.R., and Fondy, P.L., *Chem. Eng.*, May, 24, 1976.
61. Oldshue, J.Y., *Chem. Eng.*, June, 13, 1983.
62. Rautzen, R.R., Corpstein, R.R., and Dickey, D.S., *Chem. Eng.*, Oct., 25, 1976.
63. Perry, R.H., Chilton, C.H., and Kirkpatrick, S.D., *Chemical Engineers' Handbook*, 4th ed., McGraw-Hill, New York, 1963.
64. Bates, R.L., Fonday, P.L., and Corpstein, R.R., *I&EC Process Design Develop.*, 2, 311, 1963.
65. Fasano, J.B. and Penney, W.R., *Chem. Eng. Prog.*, October 1991.
66. Metzner, A.B. and Otto, R.E., *AIChE J.*, 3, 1, 1957.
67. Metzner, A.B., and Taylor, J.S., *AIChE J.*, 6, 1, 1960.
68. Metzner, A.B., Feehs, R.H., Ramos, H.L., Otto, R.E., and Tuthill, J.D., *AIChE J.*, 7, 14, 1961.
69. Chemisinoff, N.P., Ed., *Encyclopedia of Fluid Mechanics*, Gulf Publishing, Houston, TX, 1986.
70. Holland, F.A. and Chapman, F.S., *Liquid Mixing and Processing in Stirred Tanks*, Reinhold, New York, 1966.

71. Skelland, A.H.P., *Non-Newtonian Flow and Heat Transfer*, Wiley, New York, p. 1067.
72. Ulbrecht, J.J., *Chem. Eng.*, 286, 347, 1974.
73. Astarita, G., *J. Non-Newtonian Fluid Mech.*, 4, 285, 1979.
74. Beckner, J.M., *Trans. Inst. Chem. Eng.*, 44, 224, 1966.
75. Edwards, M.F., Godfrey, J.C., and Kashani, M.M., *J. Non-Newtonian Fluid Mech.*, 1, 309, 1976.
76. Kale, D.D., Mashelkar, R.A., and Ulbrecht, J., *J. Chem. Eng. Technol.*, 46, 69, 1974.
77. Gates, L.W. and Henley, T.L., *Chem. Eng.*, Dec., 9, 1975.
78. Coulson, J.M. and Richardson, J.F., *Chemical Engineering*, 4th ed., Vol. 1, Pergamon Press, New York, 1990, chap. 7.
79. van den Tempel, M., *Chem. Eng.*, February 1977.
80. Lochhead, R.Y., *Soap Cosmet. Chem. Specialties*, October 1992.
81. Nagata, S., *Mixing: Principles and Applications*, Wiley, New York, 1975, p. 126.
82. Nagata, S., *Mixing: Principles and Applications*, Wiley, New York, 1975, p. 127.
83. Baker, M.R., *Chem. Eng. Sci.*, 48, 3829, 1993.
84. Muzzio, F.J. and Alvarez, M.M., Rutgers University Communication, New Jersey, 2000.
85. Davidson, R., Ed., *Handbook of Water-Soluble Gums and Resins*, McGraw-Hill, New York, 1980.
86. Guerrero, A.F., Rodriguez Patino, J.M., Flores, V., and Gallegos, C., *Chem. Biochem. Eng. Q.*, 5, 3, 1991.
87. Guerrero, A.F., Rodriguez Patino, J.M., Albea, L., Flores, V., and Gallegos, C., *JAOCS*, 66, 261, 1989.
88. Shen, C.Y. and Metcalf, J.S., *Ind. Eng. Chem. Product Res. Dev.* 4, 107, 1977.
89. Nielen, H.D. and Landgraber, H., *Tenside Deterg.*, 14, 205, 1977.
90. Ryer, F.V., U.S. Patent 3,056,652 to Lever Brothers, 1962.
91. Oldshue, J.Y., *Fluid Mixing Technology*, McGraw-Hill, New York, 1983, p. 234.
92. Upadhyay, B.N. and Kumar, V., *Rev. Chem. Eng.*, 10, 1, 1994.
93. Oldshue, J.Y., *Fluid Mixing Technology*, McGraw-Hill, New York, 1983, p. 113.
94. Hodel, A.E., *Chemical Processing*, July 1991.
95. 3-A Sanitary Standards, Ad Hoc Committee, Milwaukee, WI, 2002.
96. Clean in Place (CIP) VR Carlson, Cedar Rapids, IA, 1995.
97. Mouks, D.M., Pharmaceutical Eng. An Integrated Approach to CIP/SIP Design, March 1999.

# Index

**673**